Air Pollution Modeling and Its Application X

NATO · Challenges of Modern Society

A series of edited volumes comprising mulitfaceted studies of contemporary problems facing our society, assembled in cooperation with NATO Committee on the Challenges of Modern Society.

Volume 1   AIR POLLUTION MODELING AND ITS APPLICATION I
           Edited by C. De Wispelaere

Volume 2   AIR POLLUTION: Assessment Methodology and Modeling
           Edited by Erich Weber

Volume 3   AIR POLLUTION MODELING AND ITS APPLICATION II
           Edited by C. De Wispelaere

Volume 4   HAZARDOUS WASTE DISPOSAL
           Edited by John P. Lehman

Volume 5   AIR POLLUTION MODELING AND ITS APPLICATION III
           Edited by C. De Wispelaere

Volume 6   REMOTE SENSING FOR THE CONTROL OF MARINE POLLUTION
           Edited by Jean-Marie Massin

Volume 7   AIR POLLUTION MODELING AND ITS APPLICATION IV
           Edited by C. De Wispelaere

Volume 8   CONTAMINATED LAND: Reclamation and Treatment
           Edited by Michael A. Smith

Volume 9   INTERREGIONAL AIR POLLUTION MODELING: The State of the Art
           Edited by S. Zwerver and J. van Ham

Volume 10 AIR POLLUTION MODELING AND ITS APPLICATION V
           Edited by C. De Wispelaere, Francis A. Schiermeier, and Noor V. Gillani

Volume 11 AIR POLLUTION MODELING AND ITS APPLICATION VI
           Edited by Han van Dop

Volume 12 RISK MANAGEMENT OF CHEMICALS IN THE ENVIRONMENT
           Edited by Hans M. Seip and Anders B. Heiberg

Volume 13 AIR POLLUTION MODELING AND ITS APPLICATION VII
           Edited by Han van Dop

Volume 14 HEALTH AND MEDICAL ASPECTS OF DISASTER PREPAREDNESS
           Edited by John C. Duffy

Volume 15 AIR POLLUTION MODELING AND ITS APPLICATION VIII
           Edited by Han van Dop and Douw G. Steyn

Volume 16 DIOXIN PERSPECTIVES: A Pilot Study on International Information Exchange on Dioxins and Related Compounds
           Edited by Erich W. Bretthauer, Heinrich W. Kraus, and Alessandro di Domenico

Volume 17 AIR POLLUTION MODELING AND ITS APPLICATION IX
           Edited by Han van Dop and George Kallos

Volume 18 AIR POLLUTION MODELING AND ITS APPLICATION X
           Edited by Sven-Erik Gryning and Millán M. Millán

Air Pollution Modeling and Its Application X

Edited by

Sven-Erik Gryning

Risø National Laboratory
Roskilde, Denmark

and

Millán M. Millán

Centre for Environmental Studies of the Mediterranean (CEAM)
Valencia, Spain

SPRINGER SCIENCE+BUSINESS MEDIA, LLC

Library of Congress Cataloging-in-Publication Data

Air pollution modeling and its application X / edited by Sven-Erik
Gryning and Millán M. Millán.
    p.   cm. -- (NATO challenges of modern society ; v. 18)
   "Published in cooperation with NATO Committee on the Challenges of
Modern Society."
   "International Technical Meeting on Air Pollution Modeling and Its
Application was held in Valencia, Spain during late 1993"--Pref.
   Includes bibliographical references and index.
   ISBN 978-1-4613-5734-6   ISBN 978-1-4615-1817-4 (eBook)
   DOI 10.1007/978-1-4615-1817-4
   1. Atmospheric diffusion--Mathematical models--Congresses.
2. Air--Pollution--Meteorological aspects--Mathematical models-
-Congresses.   I. Gryning, Sven-Erik. II. Millán, Millán M.
III. North Atlantic Treaty Organization. Committee on the
Challenges of Modern Society. IV. NATO/CCMS International Technical
Meeting on Air Pollution Modeling and Its Application (20th : 1993 :
Valencia, Spain) V. Series.
QC880.4.D44A372 1995
628.5'3'015118--dc20   94-47051
                                                        CIP

Proceedings of the Twentieth NATO/CCMS International Technical Meeting
on Air Pollution Modeling and Its Application,
held November 29–December 3, 1993, in Valencia, Spain

ISBN 978-1-4613-5734-6

© 1994 Springer Science+Business Media New York
Originally published by Plenum Press, New York in 1994
Softcover reprint of the hardcover 1st edition 1994

PREFACE

The 20th International Technical Meeting on Air Pollution Modelling and Its Application was held in Valencia, Spain, during late 1993. At this conference, a new record of abstracts was submitted, a new record of scientists participated, and a new record of countries was represented. This clearly indicates society's continuous and growing interest in, as well as importance of, the complexities associated with the modelling of air pollution.

The conference addressed the following main subjects: integrated regional modelling, global and long-range transport, new modelling developments, accidental releases, and model assessment and verification. In addition, two project-oriented workshops were organized as part of the conference.

The many contributing authors and scientists taking active part in the discussions following the papers, have made this proceeding a record of the current status in the field of air pollution modelling. We want to express our gratitude to their efforts. We also wish to extend our gratitude to the sponsors that made this conference possible. In addition to financial support from NATO/CCMS the conference received contributions from CEAM, the European Association for the Science of Air Pollution, Danish Center for Air Research, and Risø National Laboratory. A special grant was given by NATO/CCMS to facilitate attendance of scientists from Central and Eastern Europe.

We also wish to express our gratitude to Rosa Salvador and Pilar Zamora of CEAM, who laboriously organized the conference pre-proceedings, and to Anne Nørregaard and Ulla Riis Christiansen of Risø National Laboratory, who seved as conference secretariat.

Sven-Erik Gryning
Millan M. Millan

March 1994

The Scientific Committee of the International Technical Meeting on Air Pollution Modelling and Its Application:

G. Schayes, Belgium
D.G. Steyn, Canada
J.L. Walmsley, Canada
R. Berkowicz, Denmark
S.E.Gryning (chairman), Denmark
W. Klug, Germany
H. Meinl, Germany
N. Chaumerliac, France
G.B. Kallos, Greece

D. Anfossi, Italy
R.M. van Aalst, The Netherlands
H. van Dop, The Netherlands
T. Iversen, Norway
M.M. Millan, Spain
S.E. Ulug, Turkey
M.L. Williams, United Kingdom
F.A. Schiermeier, U.S.A

CONTENTS

INTEGRATED REGIONAL MODELLING

Experimental and Numerical Study of High Ozone Events in the Alpine-Region 3
J. Graf, H. Schlager, and M. Krautstrunk

MATCH: A Mesoscale Atmospheric Dispersion Model and Its Application
    to Air Pollution Assessments in Sweden 9
C. Persson, J. Langner, and L. Robertson

Recent Applications of the RAMS Meteorological and the HYPACT Dispersion
    Models 19
W. A. Lyons, R. A. Pielke, W. R. Cotton, C. J. Tremback, R. L. Walko,
    M. Uliasz and J.I. Ibarra

Application of the Mesoscale Dispersion Modeling System to Investigation of
    Air Pollution Transport in Southern Poland 27
M.Uliasz, M. Bartochowska, A. Madany, H. Piwkowski, J.Parfiniewicz,
    and M. Rozkrut

Effects of the Selected Domain in Mesoscale Atmospheric Simulations and Dispersion
    Calculations 35
G. Kallos and P. Kassomenos

Summer Episodes of Pollution Dispersion Over the Coastal Area of Israel
    -A Numercal Study 45
J. Goldstein, Y. Tokar, Y. Balmor, E. Glaser, and P. Alpert

Numerical Simulation of Meso-Meteorological Circulations in the Lisbon Region 53
M. Coutinho, A. Rocha, and C. Borrego

Problems of Modelling the Long-Range Transport of Reduced Nitrogen 63
H. M. ApSimon, B. M. Barker, and S. Kayin

Coupling the Photochemical, Eulerian Transport and 'Big-leaf' Deposition
    Modelling in a Three Dimensional Mesoscale Context 73
R. San José, L. Rodríguez, M. Palacios, and J. Moreno

GEM: A Lagrangian Particle Model for the Dispersion of Primary Pollutants
    in Urban Canyons. Sensitivity Analysis and First Validation Trials 81
G. Lanzani and M. Tamponi

Verification of Urban Scale Time-dependent Dispersion Model with Subgrid Elements,
    in Oslo, Norway 91
    S. Larssen, K. E. Grønskei, F. Gram, L.O. Hagen, and S.-E. Walker

Observation and Simulation of Urban-Topography Barrier Effects on Boundary Layer
    Structure Using the Three-Dimensional TVM/URBMET Model 101
    R. Bornstein, P. Thunis, and G. Schayes

Wind Flow and Photochemical Smog in Thessaloniki: Model Results Compared
    with Observations 109
    N. Moussiopoulos, A. Proyou, and P. Sahm

GLOBAL AND LONG-RANGE TRANSPORT

A Three Dimensional Hemispheric Air Pollution Model Used for the Arctic 119
    J. Christensen

Model Simulation of the Atmospheric Input of Trace Metals into the North, Baltic,
    Mediterranean and Black Seas 129
    J. Alcamo, L. Bozó and, J. Bartnicki

Model of Long-range Pollutants Transport and Acidity of Precipitation for
    Baltic Region 137
    D. Syrakov, M. Kolarova, D. Perkauskas, K. Senuta, and A. Mikelinskene

Modelling the Long-range Transport and Deposition of Persistent Organic Pollutants
    over Europe and its Surrounding Marine Areas 143
    J. A. van Jaarsveld, W. A. J. van Pul, and F. A. A. M. de Leeuw

A New Model for Routine Calculation of Long-Range Transport of Air Pollutants 157
    T. Iversen and E. Berge

Atmospheric Mercury Species over Europe. Model Calculations and Comparison
    with Observations from the Nordic Air and Precipitation Network for 1987
    and 1988 167
    G. Petersen, Å. Iverfeldt, and J. Munthe

A Boundary Layer Parameterization for Global Dispersion Models 177
    C. J. Nappo and H. van Dop

Analysis of Mid-Tropospheric Carbon Monoxide Data Using a Three-Dimensional
    Global Atmospheric Chemistry Numerical Model 185
    R. C. Easter, R.D. Saylor, and E. G. Chapman

The Influence of Aqueous Phase Chemistry on Long Term Ozone Concentrations over
    Europe 195
    J. Matthijsen, P. J. H. Builtjes and M. G. M. Roemer

A Parameterization of the Effect of Clouds on Photodissociation Rates; Comparison
    with Observations 203
    M. van Weele, J. Vilà-Guerau de Arellano, and P. G. Duynkerke

Evaluation of Radiative Flux and Tropospheric Chemistry under Global Climate
    Change Scenarios 213
    K. C. Crist, G. Carmichael, and K. John

NEW DEVELOPMENTS

Similarity and Scaling for Convective Boundary Layers (Extended Summary) 223
    S. Zilitinkevich

A Fast Lagrangian Particle Model for use with Three-dimensional Mesoscale Models 235
    W. Physick and P. Hurley

A Random Walk Model for Atmospheric Dispersion in the Daytime Boundary Layer 243
    C. Tassone, S.-E. Gryning, and M. Rotach

Applied Model of the Height of the Daytime Mixed Layer Including the Capping
    Entrainment Zone 253
    E. Batchvarova and S. E. Gryning

Applications of the Mixed Spectral Finite-Difference (MSFD) Model and Its Nonlinear
    Extension (NLMSFD) to Wind Flow Over Blashaval Hill 263
    J. L. Walmsley, W. Weng, S. R. Karpik, D. Xu, and P. A. Taylor

Impact of a Fully Spectral Microphysical Scheme Upon Gas Scavenging in a Mesoscale
    Meteorological Model 273
    N. Huret, N. Chaumerliac, and S. Cautenet

A Numerical Study of DMS-Oxidation in the Marine Boundary Layer 279
    K. Suhre and R. Rosset

A Comparison of Fast Chemical Kinetic Solvers in a Simple Vertical Diffusion Model ...287
    O. Knoth and R. Wolke

Modelling Flux-Gradient Relationships for Chemically Reactive Species in the
    Atmospheric Surface Layer 295
    J. Vila-Guerau de Arellano, P. G. Duynkerke, and K. F. Zeller

Neural Network Techniques for SO_2 Episode Prediction 305
    S. J. Perantonis, N. Vassilas, G. T. Amanatidis, S. J. Varoufakis, and J. G. Bartzis

Source Footprint Analysis for Scalar Fluxes Measured in Flows over an Inhomogeneous
    Surface 315
    A. K. Luhar and K. S. Rao

Multiple Master Length Scales Derived From a Statistical Diffusion Theory 325
    G. A. Degrazia, A. P. de Oliveira, and O. L. L. Moraes

Development of a Lagrangian Stochastic Model for Dispersion in
    Complex Terrain 329
    G. Brusasca, G. Tinarelli, D. Anfossi, E. Ferrero, F. Tampieri
    and F. Trombetti

ACCIDENTAL RELEASE

Recurrence of Extreme Concentrations 341
    L. Kristensen

Major Industrial Hazards: The SEVEX Project-Source Terms and Dispersion
    Calculation in Complex Terrain 357
    C. Delvosalle, J-M. Levert and F. Benjelloun, G. Schayes, B. Moyaux, F. Ronday,
      E. Everberg, T. Bourouag, and J.P. Dzisiak

A Mesoscale Boundary Layer Meteorological Model for Inhomogeneous Terrain 367
    S. M. Daggupaty, R. S. Tangirala, and H. Sahota

Use of DMI-HIRLAM for Operational Dispersion Calculations 373
    J. H. Sørensen, L. Laursen, and A. Rasmussen

Experimental Evaluation of a PC-Based Real-Time Dispersion Modeling System for
    Accidental Releases in Complex Terrain 383
    S. Thykier-Nielsen, T. Mikkelsen, and J. M. Santabàrbara

Radioactive Dispersion Modelling and Emergency Response System at the German
    Weather Service 395
    B. Fay, H. Glaab, I. Jacobsen, and R. Schrodin

The Embedding of the Lagrangian Dispersion Model LASAT into a Monitoring
    System for Nuclear Power Plants 405
    L. Janicke

A Transport and Dispersion Model Performance Evaluation using the Results of
    a Tracer Experiment in Complex Terrain 413
    R. Lamprecht and D. Berlowitz

An Emergency Response and Local Weather Forecasting Software System 423
    C. J. Tremback, W. A. Lyons, W. P. Thorson, and R. L. Walko

Comparison of Models for Aerosol Vaporisation in the Dispersion of Heavy Clouds 431
    J. Kukkonen, M. Kulmala, J. Nikmo, T. Vesala, D. M. Webber and T. Wren

MDGP: A New Numerical Model for Dense Gas Dispersion. Sensitivity Analysis
    and First Validation Trials. 439
    R. Bellasio and M. Tamponi

Evaluation of the Atmospheric Release Advisory Capability Emergency Response
    Model for Explosive Sources 447
    R. L. Baskett, R. P. Freis, and J. S. Nasstrom

MODEL ASSESSMENT AND VERIFICATION

Improving the Science of Regulatory Dispersion Models for Short-Range Applications ...457
    J. C. Weil

European Coordinating Activities Concerning Local-Scale Regulatory Models 481
    H. R. Olesen

UK Atmospheric Dispersion Modelling System. Validation Studies 491
    D. J. Carruthers, C. A. McHugh, A. G. Robins, D. J. Thomson, B. Davies, and
    M. Montgomery

The Use of Simultaneous Confidence Intervals to Evaluate Carbon Monoxide (CO)
    Intersection Models 503
    D. C. DiCristofaro, D. G. Strimaitis T. N. Braverman, and W. M. Cox

A Study of the Dispersion of Air Pollutants Released from Major Elevated Sources
    Located near Athens, Greece 513
    P. Kassomenos, G. Kallos, M. Varinou, and A. Papadopoulos

Evaluation of Atmospheric Dispersion Models During the Guardo Experiment 523
    J. I. Ibarra

Dispersion Modelling and Observations from Elevated Sources in Coastal Terrain 533
    J. A. Noonan, W. L. Physick, J. N. Carras, and D. J. Williams

Sensitivity Analysis of the Urban Airshed Model to Wind Fields Derived from
    the Regional Oxidant Model, Diagnostic Wind Model, and the
    URBMET/TVM Mesoscale Model 541
    G. Sistla, S. T. Rao, R. Bornstein, F. Freedman, and P. Thunis

Assessment and Verification of Different Types of Dispersion Models in Complex
    Terrain 549
    V. R. D. Herrnberger and P. Doria

Examination of the Efficacy of VOC and NOx Emissions Reductions on Ozone
    Improvement in the New York Metropolitan Area 559
    K. John, S. T. Rao, G. Sistla, N. Zhou, W. Hao, K. Schere, S. Roselle,
    N. Possiel, and R. Scheffe

Deposition of Gases and Particles in the PBL: Evaluation of the Influence of a
    Vertical Resolution in Atmospheric Transport Models 569
    O. Hertel, J. Christensen, E. Runge, R. Berkowicz, W. A.H. Asman, K. Granby,
    M. F. Hovmand, and Ø. Hov

VIDEO SESSION

Designing an Air Quality Network for Brisbane Using a Mesoscale Model 583
    W. Physick, P. Best, K. Lunney, and G. Johnson

An Operational System for Emergency Response to Large Scale Releases of Pollutants
    in the Atmosphere 587
    R. D'Amours, M. Jean, J.-P. Toviessi, and S. Trudel

Real-Time Application Software "Tracer Imager Package" (TRIP) 589
    W. Weiß and E. Reimer

POSTER SESSION

Vertical Diffusion Parameter in the Atmospheric Boundary Layer 593
    L. E. Venegas

Trajectory Analysis of High-Alpine Air Pollution Data 595
    P. Seibert, H. Kromp-Kolb, U. Baltensperger, D. T. Jost, and M. Schwikowski

An Eddy Diffusivity Model from a Theoretical Spectral Model for the Stable
    Boundary Layer 597
    O. L. Moraes, G. A. Degrazia, A. P. Oliveira

On the Relationship Between Synoptic Scale Parameters and Pasquill Stability Classes
    for the Purposes of Air Pollution Modeling 601
    D. Yordanov, D. Syrakov, and M. Kolarova

A Lagrangian Model of Long-Range Transport of Sulphur 603
    Z. Klaić

Local Background Air Pollution in Response to Coastal Circulation 605
    E. Lončar and N. Šinik

Lagrangian Stochastic Dispersion Modelling for Varying Boundary Layer Stabilities 607
    M. Rotach, S.-E. Gryning, and C. Tassone

Wake Flows over Mountainous Areas 609
    G. Adrian and F. Fiedler

Fluid Particle Motion in Inhomogeneous and Non-Gaussian Turbulence. 611
    S. Heinz

Mesoscale Modelling of the Atmospheric Input into Coastal Waters 613
    K. H. Schlünzen, K. Bigalke, and U. Niemeier

Regional Scale Transport Model of Atmospheric Acid Compounds, and Its
    Application for Hungary 615
    K. E. Fekete

Regional Scale Modeling Case Studies for Aerosol Transport over Hungary 617
    E. Mészáros, L. Bozó, and A. Molnár

Fluxes over Complex Terrain, Analytical Evaluation 619
    G. A. Dalu, M. Baldi, R. A. Pielke, and G. Kallos

Lagrangian Model Simulation of 3-D Concentration Distribution in Complex Terrain 621
    G. Tinarelli, D. Anfossi, G. Brusasca, E. Ferrero, U. Giostra, M. G. Morselli,
        F. Tampieri, and F. Trombetti

Treatment of Transport in Moguntia 623
    M. C. Krol

Methodology for Mapping Local Scale Deposition of Acidifying Components Over
    Europe 625
    W. A. J. van Pul, J. W. Erisman, J. A. van Jaarsveld, and F. A. A. M. de Leeuw

Ground Level Concentrations of Ozone, Oxidant, PAN and Precursors in The
    Netherlands During the Last Two Decades and the Relation with the 629
    LOTOS-Model
    P. Esser, M. G. M Roemer, P. J. H. Builtjes, R. G. Guicherit, and Th. Thijsse

Meteorological Aspect of Chemical Composition of Precipitation/Deposition-Long
Range Pollution Transport in a Mountain Region 635
G. Kmieć , A. Zwoździak, K. Kacperczyk, and J. Zwoździak

Sea Breeze in Summer, along the West Coast of Portugal 637
R. A. C. Carvalho and V. Prior

Researches on Dispersion of the Pollutants Emitted into the Atmosphere from a
Nuclear Power Plant under Rugged Conditions in View of Validating
Varied Mathematical Models 639
T. Pop and L.-M. Pop

Long-Term Average Air Pollution over Cities: Operative Calculation Technique for
Elevated Sources 641
I. A. Krotova and L. G. Melikhova

Influence of the Traffic Conditions on the Air Quality of Barcelona during
the Olympic Games'92 643
J. M. Baldasano, M. Costa, L. Cremades, Th. Flassak, and M. Wortmann-
Vierthaler

Evaluation of the Three-Dimensional Distribution of Dense Gas Concentration
Estimated by Numerical Models 645
F. Martín, I. Palomino, and B. Aceña

Influence of the Topography on the Long-Term Average Concentration Computed
by Dispersion Models 647
F. Martín, I. Palomino, and R. Salvador

Coupling the Photochemical, Eulerian Transport and 'Big-leaf' Deposition
Modelling in a Three Dimensional Mesoscale Context 649
R. San José, L. Rodríguez, M. Palacios, and J. Moreno

Test of Mesoscale Numerical Model on Swiss Midlands 651
D. Schneiter and C. Thurre

Application of the Abatement Strategies Assesment Model, ASAM, to Abatement
of SO_2 Emissions in Europe 653
H. M. ApSimon and R. F. Warren

Review and Evaluation of the RATCHET Model Used for the Hanford Dose
Reconstruction Project 655
C. J. Nappo, W. R. Pendergrass, and R. M. Eckman

Linear Advection Scheme for Air Pollution Transport Modelling from Individual
Sources 657
M. Pekar

Neural Networks Predict Pollution 659
P. Mlakar, M. Božnar and, M. Lesjak

SPECIAL SESSION: ATHENIAN PHOTOCHEMICAL SMOG INTERCOMPARISON OF SIMULATIONS (APSIS)

Results of Nested Wind Flow Simulations for the Athens Basin
    Using the Non-Hydrostatic Model Memo. 663
    R. Kunz, and N. Moussiopoulos

Influence of the Sea Breeze on the Air Pollution over the Attica Peninsula 665
    K. Nester

Numerical Simulation of the Flow Regime in Athens Area 667
    D. Melas, I. Ziomas, and C. Zerefos

Prediction of Wind Flow and Ozone Formation in Athens for the APSIS B_2 Exercise
    Using the EUMAC Zooming Model 669
    N. Moussiopoulos and P. Sahm

Intercomparison on the Flow Field Over the Attic Peninsula with Two
    Models 671
    G. Schayes, H. Gallée, G. Graziani, and P. Thunis

Studies with the Three-dimensional Eulerian Photochemical Dispersion Model MARS
    for the APSIS B Case: Viability of the Inclusion of Horizontal Diffusion
    into the Implicit Solver of the Model 673
    D. Berlowitz and N. Moussiopoulos

SPECIAL SESSION: AIR POLLUTION TRANSPORT AND DIFFUSION OVER COASTAL URBAN AREAS (ATHENS)

A Numerical Study of Air Flow in Thessaloniki Area 677
    D. Melas, I. Ziomas, and C. Zerefos

PARTICIPANTS 679

AUTHOR INDEX 699

SUBJECT INDEX 703

INTEGRATED REGIONAL MODELLING

chairmen: S. Zilitinkevich
                        J.G. Kretzschmar

rapporteurs: H.R. Olesen
                        D. Melas

EXPERIMENTAL AND NUMERICAL STUDY OF HIGH OZONE EVENTS IN THE ALPINE-REGION

Jutta Graf, Hans Schlager, Monika Krautstrunk

Institut für Physik der Atmosphäre
Deutsche Forschungsanstalt für Luft- und Raumfahrt (DLR)
Oberpfaffenhofen, Germany

INTRODUCTION

In the alpine region as well as in other European areas the ozone concentration appears to be increasing (Warmbt, 1977). During this century the ozone levels in the atmospheric boundary layer have been almost doubled (Volz and Kley, 1988). After Puxbaum et al. (1991) the ozone concentration in the Zillertal (a small valley in the Austrian Alps) has been increased by a factor of 3 since 1950. Beside this long term trend high ozone values of 100 ppb and more are increasingly observed during pollution episodes. These episodes are associated with high pressure systems, when the meteorological condition, in particular warm temperatures, clear skies and low wind speeds, encourage the photochemical formation of ozone. Calms and inversions causing reduced ventilation occur more frequently in mountain valleys than over flat terrain (Dreiseitl and Weber, 1991). Therefore the meteorological conditions in the Alps conduce the ozone production.

Ozone is formed by photochemical oxidation of CO, CH_4 and higher hydrocarbons (HC) in the presence of NO_x. HC and NO_x are called ozone precursors and their concentration levels determine the ozone production. High levels of ozone are harmful to human health and to vegetation, they are assumed to contribute to the decline of forest in Europe. This fact is very important for the Alpine region because the ecosystem in this area is highly sensitive.

In this study we examine the origin and distribution of ozone and ozone precursors in the Alps by means of aircraft measurements and model simulations in the framework of ME-MOSA (" Messung und Modellierung des Schadstoffverhaltens im Alpenbereich "). The main object was to investigate whether ozone and its precursors are locally produced as an effect of the local emissions of NO_x and HC or transported into the Alpine region.

Air Pollution Modeling and Its Application X, Edited by S-V. Gryning
and M. M. Millán, Plenum Press, New York, 1994

3

AIRCRAFT MEASUREMENTS

The experiment MEMOSA has been performed during 1990 and 1991. Using two aircrafts the concentration of NO, NO_2, HC, O_3, SO_2 and meteorological parameters (temperature, humidity, wind) have been measured along diverse flight tracks within three intensive measering periods during spring and summer. To obtain the horizontal and vertical distribution of pollutants one airplane was measuring at a constant height (100 - 300 m above ground), while at the same time the second aircraft was flying vertical profiles.

One flight track pattern leads across the Alps from north to south along the Inn Valley, over the Brenner Pass, Bozen, Garda lake to Bergamo and vice versa. These flights have been realized during high pressure episodes with less advection of pollutants to study the local ozone production. The other flight tracks were north and south parallel to the Alps to investigate the transport of pollutants into the northern and southern alpine valleys under advective weather conditions. Additional measurements took place in valleys with different traffic in order to study the effect of vehicle emissions on the ozone production.

In general higher values of ozone and ozone precursors were found in the southern part of the Alps. The mean ozone maximum during the experiment was about 80 - 90 ppb in the northern Alps compared to 100 - 120 ppb in the southern Alps. A positive correlation between the local ozone and HC peak values was observed south of the main Alpine ridge. No correlation was found in the north. Everywhere we noticed significant differences between valleys with high and low traffic load. For instance the NO_x - concentration in the Eisack Valley which include the Brenner highway was by a factor of 3 higher than in the more or less rural Sarn Valley. From the composition of the analysed HC samples we could also conclude, that the emissions from motor vehicles are the main source of ozone precursors in the Alps.

In the southern part of the Alps a well defined mountain valley circulation has been observed which is influencing the pollutant dispersion. During the night a very stable inversion layer could develop as a consequence of cold drainage flow. The airmass above the inversion layer is decoupled from the surface layer and ozone can not be destroyed by NO - emissions or deposition processes. These so-called ozone reservoir layers has been observed very often during this experiment. For example in figure 1 one can see the ozone distribution at noon on the 7th of August 1991 as a vertical cross section showing the well established layer of high ozone values over the Adige Valley. Under convective conditions the airmass is mixed down and leads to an additional increase of the ozone concentration at the surface during a smog episode.

NUMERICAL SIMULATIONS

To complement the experimental work and for data interpretation we have performed simulations with numerical models. One set of simulations have been carried out to investigate the possible transport of pollutants into the Alps. Other simulations serve to classify the chemical compositon with respect to ozone reduction strategy.

Dynamic simulation

To investigate the transport into the Alpine region we used the three dimensional hydrostatic mesoscale model REWIH3D which is desribed in detail in Heimann (1990). For this study the model domain covers an area of approximately 1200 x 800 km² represented by 40 x 40 meshes, which implies a horizontal gridsize Δx of about 28 km and a Δy of about 20 km. The model domain covers the entire Alps as well as the surrounding industrial areas like the Po Valley in the south and the industrial areas of Bavaria in the north.

The dispersion of emissions from industrial areas located in the mudflat of the Alps was calculated for eight different winddirections. For all simulations we assumed a geostrophic

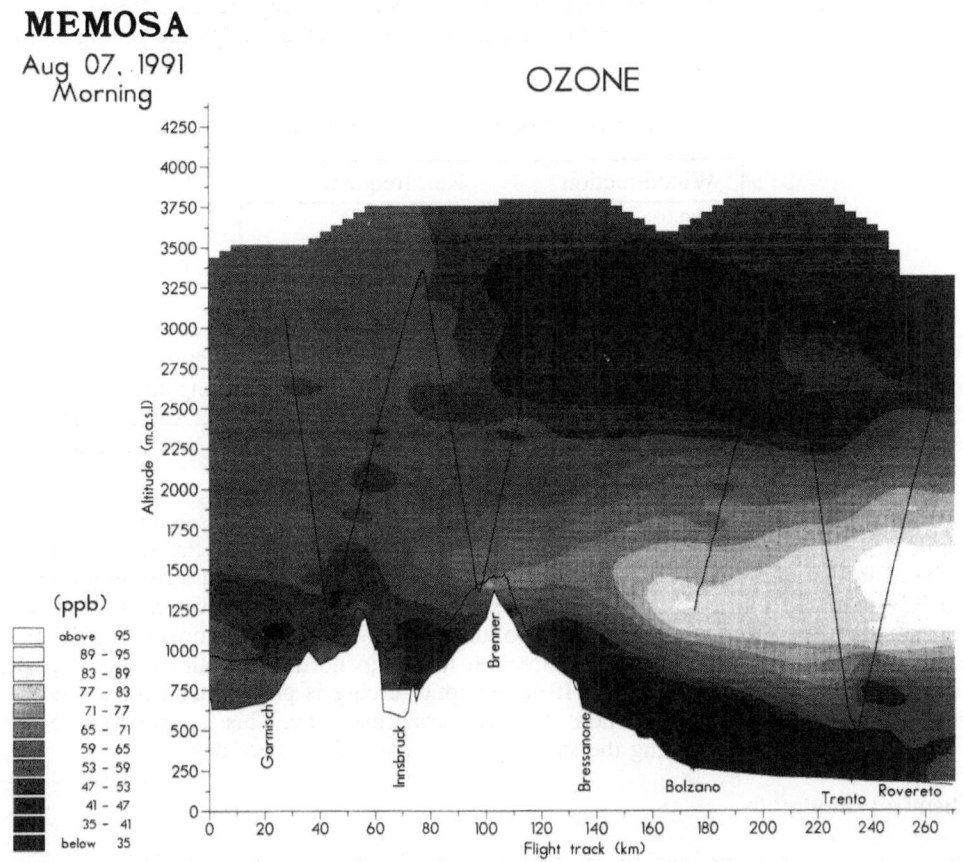

Figure 1. Vertical distribution of ozone along the north-south flight track at noon the 7th of August 1991.

forcing of 10 m/s and an initial adiabatic temperature profile. The nonstationarity of the runs was achieved by prescribing a daily dependent surface temperature. Each run was simulated a whole day, starting at midnight. This integration period is long enough for a possible transport of tracers over the Alpine chain. The pollutants are presumed to emerge with unit emission strength at a constant height of 125 m. In most cases no transport of pollutants into the Alps could be found. When the geostrophic wind has a north-easterly component pollutants are transported along the Inn Valley to the Brenner Pass. For south and south-westerly flows transport from the industrial areas in the Po Valley into the Alps is possible. In all other cases the pollutants are forced to go around the Alps due to chanelling effects of the mountains.

To investigate the transport path from north to south in more detail we performed a two day simulation with emission sources being located only north of the Alpine chain. This simulation shows clearly two predominent transport routes through the eastern Alps: the Brenner Pass and the Triest gap. This result is in agreement with Camuffo et al (1991).

In order to check the significance of the cases with transboundary transport in the cause of a year a statistic of the geostrophic flow was considered (Frey Buness, 1993). As obvious from table 1 the winddirections causing transport into the Alps are of minor importance in the annual average.

Table 1. Relative frequency of the winddirection

Winddirection	Rel. frequency
45°	5 %
180°	3 %
225°	10 %

Thermally induced circulations, for example mountain - valley circulation influence the dispersion processes. The measured data collected during a flight along the south track show, that pollutants emitted in the Po Valley were transported along the Adige Valley into the Alps. On this particular day the synoptic situation - clear sky and low wind speeds - was suitable for the development of a authochthonous circulation.

To simulate this real situation the model was initialized with an observed temperature profile and no synoptic forcing. Different heating of mountains and flat terrain can be achieved in the model using a radiation parameterization. The polluted airmass in the Po Valley was realized in form of a line source. In the course of the day when the flow was from south due to differential heating the pollutants were transported almost to the Brenner Pass. In the evening, when the flow direction turned to north the polluted airmass was transported back into the Po Valley. If no synoptic forcing is predominent the airmass is wobbling back and forth getting more polluted during that time. This fact was proofed by a three day simulation assuming the same synoptical situation every day.

Chemical simulation

The production of ozone depends on the concentration level of NO_x and HC (Silman et al., 1990). In unpolluted areas the ozone formation is mainly controlled by NO_x and almost insensitive to HC. Ozone increases as NO_x increases. When the concentration of NO_x exceeds a certain threshold ozone formation is controlled by both HC and NO_x. A one sided reduction of NO_x- emission leads to an increase of ozone.

The aim of the chemical simulations was a classification of the collected chemical composition with regard to the relation NO_x / HC and to identify the two NO_x - regimes for the Alpine region. For this purpose the photochemical formation of ozone due to anthropogenic emission of NO_x and HC was calculated using a one-dimensional chemical model (Graf, Moussiopoulos, 1991). The diurnal behaviour of ozone and NO_x, with maximum ozone levels in the late afternoon is produced by the model. We run the model for three days with various constant emissions of HC and NO_x covering a wide range of HC - and NO_x concentrations. Taking the concentrations obtained at the third day at 4 p.m. ozone isoplethes as a function of NO_x and HC are drawn for the southern and northern part of the Alps. Figure 2 shows a comparison of calculated isoplethes for the southern Alps with measurements. For the threshold between the two photochemical regimes NO_x volume mixing ratios of 2 to 6 ppb are obtained. The experimental NO_x data for the southern Alps are significant higher (8 to 14 ppb) which implies that the photochemical ozone production is HC-controlled. An analogous calculation for the north Alps with adapted photolysis rates produced similar results, but the measured NO_x data were closer to the threshold.

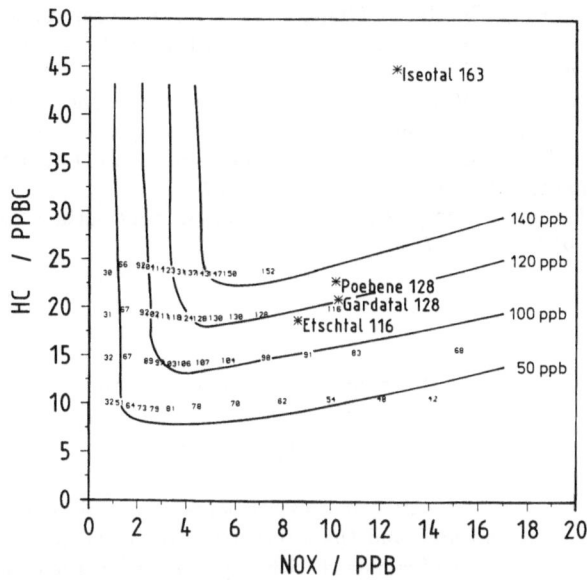

Figure 2. Ozone peak values as a function of NO_x and HC concentrations at 4 p.m. for the southern Alpine region. Crosses indicate the measured data.

CONCLUSIONS

Both experimental and numerical studies indicate that on the annual average transboundary transport is of minor importance with respect to the airpollution in the Alps. Especially in valleys with high traffic load we found high concentrations of ozone and ozone precursors and large differences to valleys with low traffic load. From a comparison of the analysed HC samples with a typical hydrocarbon composition of motor vehicle exhaust it can be concluded that the pollution in the Alps is to a great extent locally produced by traffic.

Sensitivity studies concerning the photochemical ozone formation suggest that ozone peak concentrations depends more on the HC concentrations in the soutern Alps while it depends more on the NO_x concentration in the northern Alps. Since the experimental data are close to the transition point between the two photochemical regimes an effective reduction strategy should include both NO_x and HC reductions.

LITERATURE

Camuffo, D., A. Bernardi, P. Bacci, 1991: Transboundary transport of atmospheric pollutants through the eastern Alps. *Atmos. Environm. Vol. 25A, No. 12, 2863-2871.*

Dreiseitl E., A. Weber, 1991: Auswirkungen des Straßenverkehrs suf dei Umwelt, Teilbericht: Immissions-klimatologische Besonderheiten des Inntals. *Bericht an den Tiroler Landtag.*

Frey-Buness A.,1993: Ein statistisch-dynamisches Verfahren zur Regionalisierung globaler Klimasimulationen. *Dissertation in press (DLR Forschungsbericht)*

Graf J., N. Moussiopoulos , 1991: Intercomparison of two models for the dispersion of chemically reacting pollutants. *Beitr. Phys. Atmosph., Vol. 64, No.1, 13-25.*

Heimann, D., 1990: Three-dimensional modelling of synthetic cold fronts approaching the Alps. *Meteorl.Atmos.Phys. 42, 197-219.*

Puxbaum H., K. Gabler , S. Smidt , F. Glattes , 1991: A one-year record of ozone profiles in an alpine valley (Zillertal/Tyrol, Austria, 600 - 2000 m a.s.l.). *Atm. Environment 25A, 1759.*

Sillman S., J.A. Logan, S.C. Wofsy, 1990: The sensitivity of ozone to nitrogen oxides and hydrocarbons in regional ozone episodes. *J. Geophys. Res., 95, 1837-1851.*

Volz A., D. Kley, 1988: Evaluation of the Montsouris series of ozone measurements made in the nineteenth century. *Nature 332, 240.*

Warmbt W., 1979: Results of long term measurements of near surface ozone in the GDR, *Z. Meteorol., 29,24-31.*

DISCUSSION

S.T. RAO: Is there a reason why NO_x and HC concentrations of 4pm are plotted against maximum ozone concentrations in the ozone isoplethes diagram.

J. GRAF: Aircraft messurements were taken during the late morning and the afternoon. To ensure an optimum agreement of simulatd with measured data, the ozone isoplethes for the afternoon situation were plotted. In addition the HC/NO_x-ratio obtained in the morning is an insufficient information for an effective reduction strategy. There exists no significant statistical relationships between the morning HC/NO_x and the maximum ozone level in the afternoon (Wolff, Korsog, 1992).

S.T. RAO: Wouldn't a plot of maximum NO_x and HC contrations, which usually occurs in the morning when the emissions is high, against maximum ozone, which usually occurs in the afternoon, be more meaningsful.

J. GRAF: We did'nt used a real emission inventory for the calculations, therefore the HC/NO_x-ratio for the emissions is not available. To cover a wide range of precursor concentrations various constant emissions were prescribed.

S.T. RAO: The HC/NO_x-ratio is very important to the formation of O_3 and for the development of control options. How do you reconcile the differences between observed and predicted HC/NO_x ratios? What is your HC/NO_x-ratio from the emissions inventory?

J. GRAF: The aim of the chemical simulations was to classify the measured chemical composition with regard to the HC/NO_x-ratio and the maximum ozone level. A simulation of a real episode was not subject of the project.

MATCH: A MESOSCALE ATMOSPHERIC DISPERSION MODEL AND ITS APPLICATION TO AIR POLLUTION ASSESSMENTS IN SWEDEN

Christer Persson, Joakim Langner, and Lennart Robertson

Swedish Meteorological and Hydrological Institute (SMHI)
S-601 76 Norrköping
Sweden

ABSTRACT

The MATCH (Mesoscale Atmospheric Transport and CHemistry) model has been developed as a tool for air pollution assessment studies on different geographical scales. It has been used as a basis for decision making concerning environmental protection within Sweden or subregions of Sweden.

MATCH is an Eulerian atmospheric dispersion model, including physical and chemical processes governing sources, atmospheric transport and sinks of oxidized sulfur and oxidized and reduced nitrogen. In its standard configuration the model has three levels in the vertical, where the second layer is variable in height and follows the mixing height. The mixing height is derived from standard meteorological fields and varies in both space and time.

Using the MATCH system, air pollution contributions from different source types like traffic, industry, shipping, farming etc can be obtained. Using a combination of air and precipitation chemistry measurements and the MATCH model, the contribution of air pollution and deposition from long range transport can be quantified in the model region. A first tentative comparison between the EMEP and MATCH calculations indicate that the relations between local/regional contributions and long range transport are similar for deposition of oxidized and reduced nitrogen, but for oxidized sulfur the MATCH system gives a somewhat smaller influence from local/regional sources compared to what is obtained from EMEP.

INTRODUCTION

The MATCH model is a 3-dimensional Eulerian atmospheric dispersion model. The model is designed with options for operational applications on:

Air Pollution Modeling and Its Application X, Edited by S-V. Gryning
and M. M. Millán, Plenum Press, New York, 1994

- meso-α scale (Europe, grid 55 x 55 km)
- meso-β scale (Sweden, grid 20 x 20 km)
- meso-γ scale (different subregions within Sweden, grid 5 x 5 km).

The handling of the vertical mixing, the horizontal grid size and parts of the chemistry essentially constitutes the difference between the various scales. In this paper a description is given of the model version, covering Sweden or subregions of Sweden, applied to air pollution assessment studies in Sweden. In Figure 1 the model areas and grid size resolutions used in the studies over Sweden are presented.

METEOROLOGICAL ANALYSES

Detailed objective meteorological analyses are performed on the relevant scale for the dispersion model. The parameter derived includes wind fields, temperature, precipitation, friction velocity, sensible heat flux, Monin-Obukovs length and mixing height. The analyses are performed at three hour intervals. A high resolution data base for topography and land use (fraction of forest, field, water, urban) is used to calculate ground roughness. The land use information is also applied in the dry deposition calculations. Figure 2 gives an illustration of the use of different meteorological observations for the meteorological analysis system.

DISPERSION MODEL

Horizontal and Vertical Structure

The horizontal advection is calculated using a fourth order flux correction scheme (Bott 1989a, 1989b). The scheme utilizes polynomial fitting between neighbouring grid points of the concentration field in order to simulate the advective fluxes through the boundaries of adjacent grid boxes. It is a positive definite mass conserving scheme with low numerical diffusion.

The basic feature of the vertical structure of the model is a space and time variable vertical resolution. In it´s standard configuration the model has three layers in the vertical. The first layer has a fixed height of 75 m. The two following layers are variable in thickness. The top of the second layer is taken to be the same as the mixing height, and the top of the third layer is fixed at a certain level (~1.5 km in winter, ~2.5 km in summer). It is possible to include additional layers if possible.

Vertical advection is calculated using a first order (upstream) scheme. Vertical diffusion between layers 1 and 2 is for the convective case described from a determination of the turn-over time for the boundary layer based on similarity theory, and for the neutral and stable case from a parametrization based on a 40 layers 1-D eddy diffusivity model. Vertical transport is also induced by the spatial and temporal variations of the mixing height.

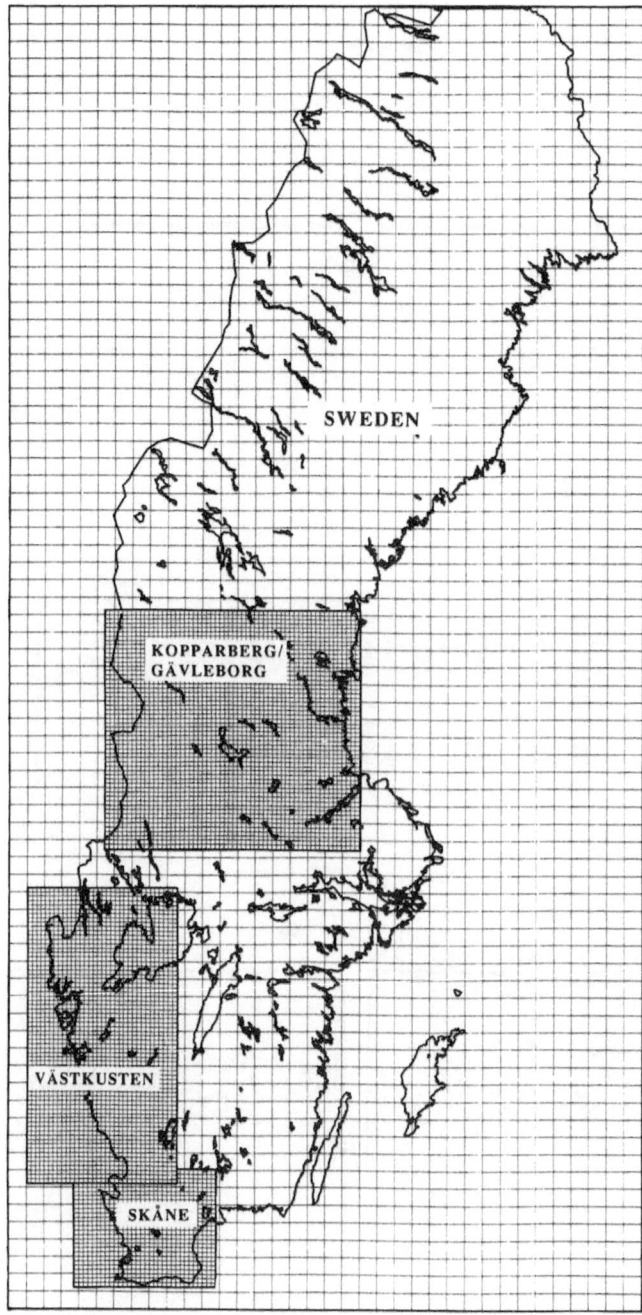

Figure 1. MATCH modelling areas used in air pollution assessment studies in Sweden. The model version covering the whole of Sweden has a grid size of 20 x 20 km, for subregions a grid size of 5 x 5 km is used.

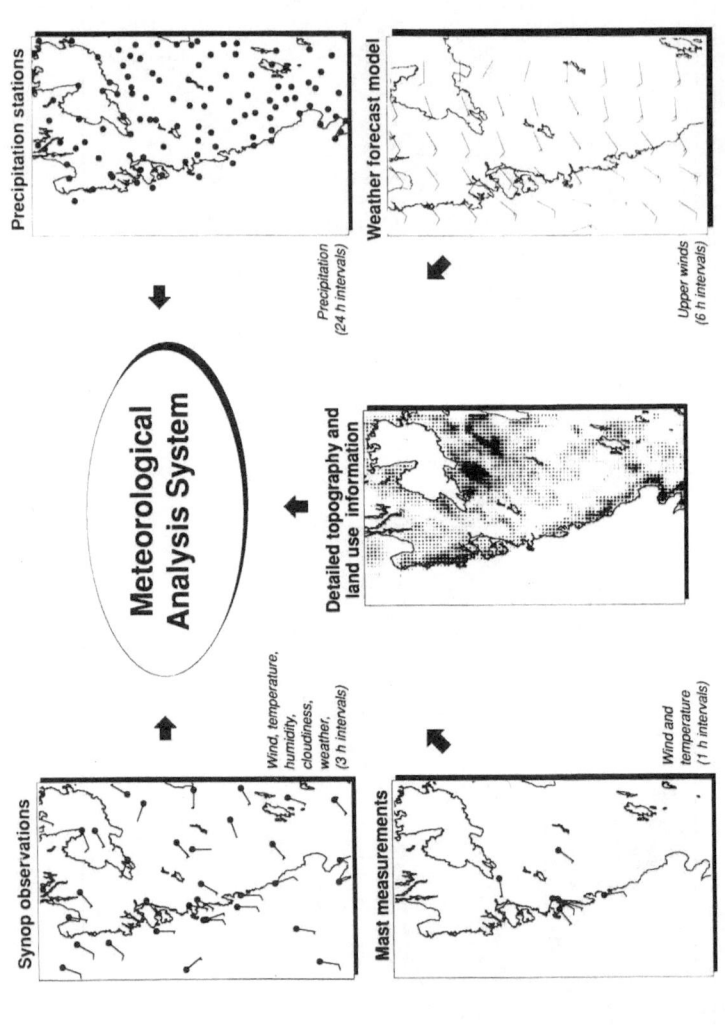

Figure 2. Illustration of data input to the MATCH meteorological analysis system, here exemplified with the MATCH-Västkusten model area.

12

Emissions

Emissions can be specified both as area and point sources and with a high resolution in time. Surface area sources are introduced into the lowest layer of the model. The initial dispersion from point sources is described with a Gaussian puff model including plume rise calculations. A new puff is introduced every hour for each source. The puffs are then advected until they have reached the size of the horizontal grid when they are merged into the large scale concentration field.

In air pollution applications the number of point sources are often large (several hundred). To reduce the amount of computations only the 30 largest sources are treated as individual point sources. The remaining point sources are classified in three classes according to stack height. Emission weighted source characteristics (stack diameter, exhaust gas flux and temperature) are calculated for each class. The weighted source characteristics are then used to calculate the plume rise and partition the emissions from each class in the three levels of the Eulerian model.

In the air pollution assessment studies emission data are separated into a large number of different source types like: car traffic, buses and trucks, industry, energy production, farming, shipping etc.

Chemistry and Deposition

The chemistry in the model deals with sulfur oxides and oxidized and reduced nitrogen. The following compounds are included: sulfur dioxide (SO_2) ammonium sulfate ((NH_4)$_2SO_4$ and NH_4HSO_4), other sulfate particles (SO_4^{2-}), nitric oxide (NO), nitrogen dioxide (NO_2), ammonium nitrate (NH_4NO_3), other nitrate particles (NO_3^-), nitric acid (HNO_3) and ammonia (NH_3). A local adjustment of the ozone (O_3) concentration with regard to local NO- and NO_2-concentration and solar radiation is also done.

For the numerical solution of the combined horizontal and vertical transport and chemistry an operator split time integration scheme is used.

Wet scavenging of the different species is proportional to the precipitation rate and a species specific scavenging coefficient. Dry deposition is proportional to the concentration and a species specific dry deposition velocity at 1 m height. Since the lowest model layer has a height of 75 m, the dry deposition flux calculation is transformed to the middle of that layer using standard similarity theory for the atmospheric surface layer. Dry deposition velocities are specified as a function of the surface characteristics (fraction forest, field etc).

RESULTS

In the air pollution assessment studies for the year 1991 results have been obtained for daily, monthly and annual values of concentration, dry-, wet- and total deposition of all included chemical compounds. These results refer to contributions from sources within the model area. Combining measured daily data from 6 - 8 background air and precipitation chemistry stations with model results from the MATCH area, also estimates of the long range transport contribution to the total air pollution situation can be obtained. In these calculations careful comparisons have been made with available independent air chemistry measurements in different parts of the model area.

Some examples of results obtained for the MATCH-Västkusten (Swedish West-Coast) study are presented in Figures 3 - 6. Figure 3 illustrates the annual mean concentrations of NO$_2$ from the region, compared to the long range transport. As can be seen, the emissions from the Swedish west-coast region accounts for a substantial fraction of the total NO$_2$-concentration in the region. In Figures 4 and 5 the same comparisons are made for wet deposition of total nitrogen and wet deposition of sulfur, respectively. For wet deposition the regional contributions are smaller - less than 20 % for nitrogen and less than 10 % for sulfur. In Figure 6 a comparison between some different source types for NO$_2$ within the model area is made.

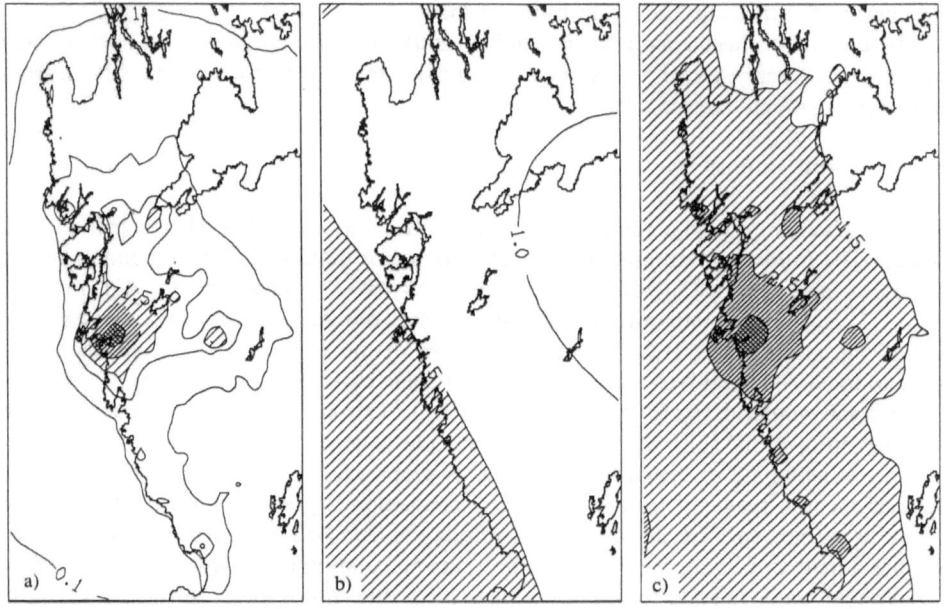

Figure 3. Calculated annual average (year 1991) NO$_2$-concentration: a) contribution from the Västkusten region, b) contribution from long range transport, c) total concentration. Light shading 1.5 - 2.5, medium shading 2.5 - 5.0 and heavy shading above 5.0 (μg N/m^3).

Using the MATCH system, we estimate the long range transport in an independent way compared to EMEP (Sandnes, 1993). A first tentative comparison between the EMEP and MATCH calculations indicate that the relations between regional contributions and long range transport are similar for deposition of oxidized and reduced nitrogen, but of course with a much more detailed resolution within the MATCH model. For oxidized sulfur the MATCH system gives a somewhat smaller influence from local/regional sources compared to what is obtained from EMEP.

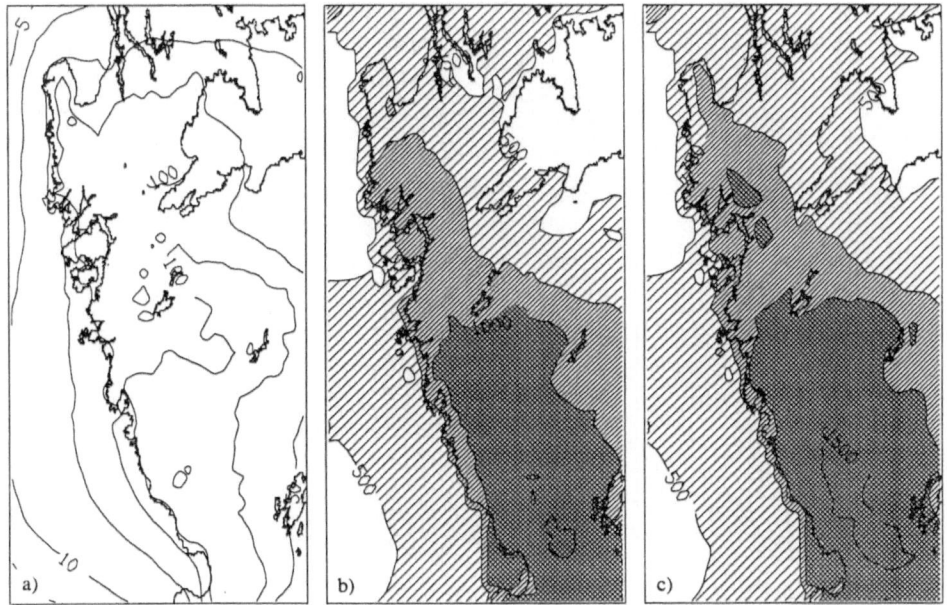

Figure 4. Calculated annual (year 1991) wet deposition of total nitrogen, oxidized plus reduced nitrogen: a) contribution from the Västkusten region, b) contribution from long range transport, c) total wet deposition. Light shading 500 - 750, medium shading 750 - 1000 and heavy shading above 1000 (mg N/m^2).

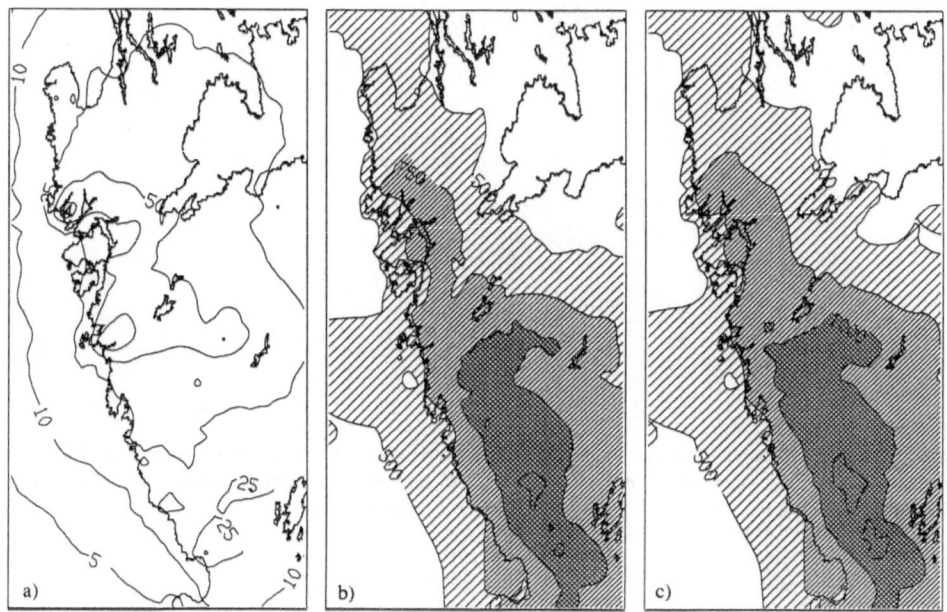

Figure 5. Calculated annual (year 1991) wet deposition of oxidized sulfur: a) contribution from the Västkusten region, b) contribution from long range transport, c) total wet deposition. Light shading 500 - 750, medium shading 750 - 1000 and heavy shading above 1000 (mg S/m^2).

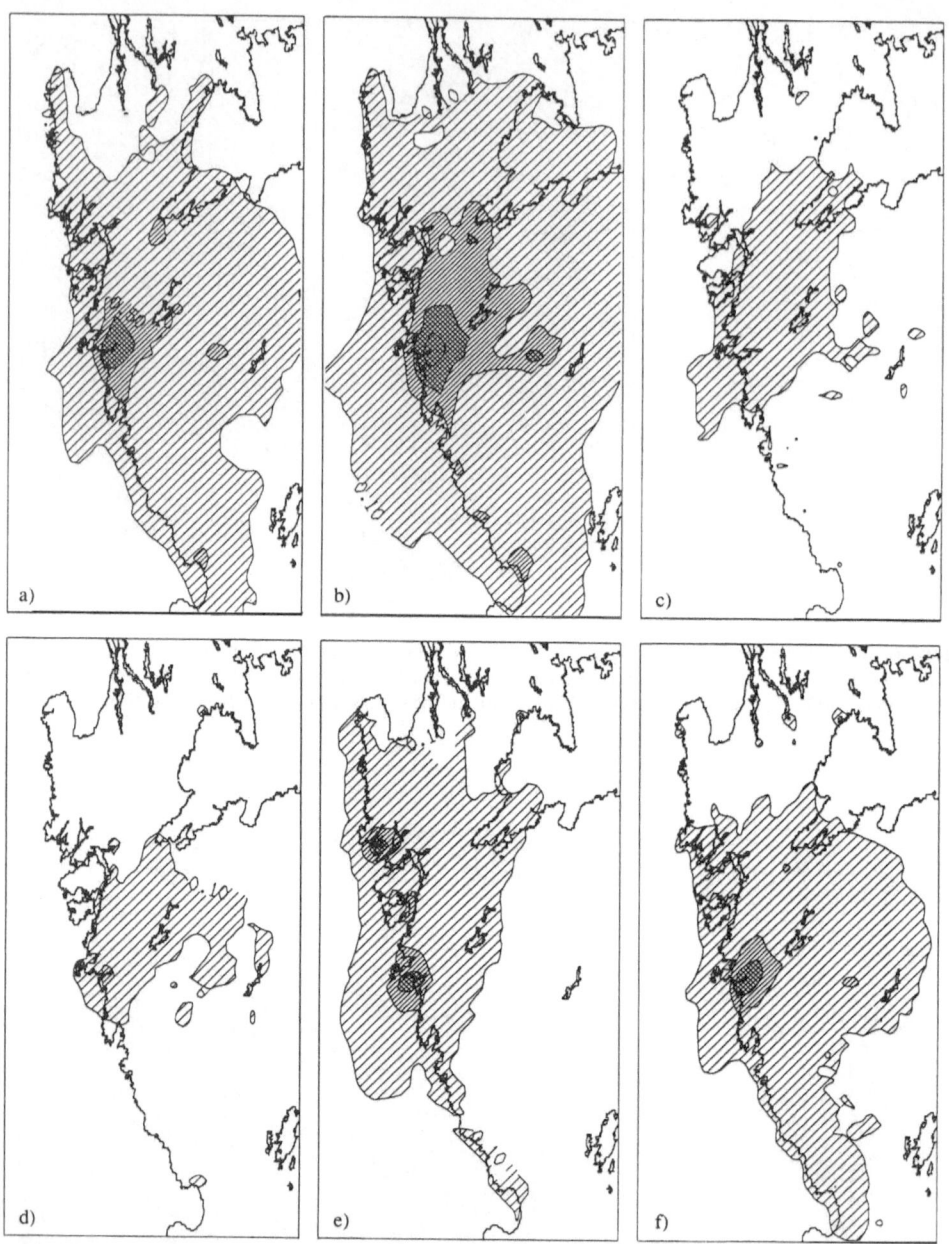

Figure 6. Calculated annual average (year 1991) contributions of NO$_2$-concentrations from a) trucks and busses, b) car traffic, c) industry, d) energy production, e) shipping, f) labour vehicles. Light shading 0.1 - 0.5, medium shading 0.5 - 1.0 and heavy shading above 1.0 (µg N/m^3).

REFERENCES

Bott, A., 1989a, A positive definite advection scheme obtained by nonlinear renormalization of the advective fluxes, *Mon. Wea. Rew.* 117, 1006 - 1015.

Bott, A., 1989b, Reply. Notes and correspondence, *Mon. Wea. Rew.* 117, 2633 - 2626.

Sandnes, H., 1993, Calculated budgets for airborne acidifying components in Europe, 1985 - 1992, EMEP/MSC-W Report 1/93, DNMI, Oslo, Norway.

DISCUSSION

H. KAMBEZIDIS:	Can you identify the remote air pollution sources contributing to the LRT?
J. LANGNER:	No, with our technique using observed air- and precipitation chemistry data it is not possible to identify e.g. from which country the pollution derives.
G. KALLOS:	How are you initializing your meteorological model over the different grids you showed to us?
J. LANGNER:	Our model is an "off line" model i.e it reads meteorological data from external archives, in this case data produced with our mesoscale meteorological analysis scheme. Therefore no initialization is necessary.
T. IVERSEN:	How are you treating subgrid-scale dispersion from emissions (point-sources) in your model?
J. LANGNER:	Emissions from the 30 largest point-sources are initialized using a puffmodel until the puff reaches the size of the horizontal grid. The remaining point-sources are classified according to stack height and are introduced as area sources but at different heights depending on the classification.
J. WALMSLEY:	Your results show low concentrations over Norway. Is that because there were no Norwegian emission sources in the model?
J. LANGNER:	Yes.
S.T. RAO:	What is the advantage of treating emissions from large elevated sources as puffs instead of as a continuous plume?
J. LANGNER:	I don't think that there is a difference, it is just a question of how you organize the calculations.
S. ZILITINKEVICH:	How do you calculate the mixed-layer height?
J. LANGNER:	For neutral and stable conditions we use a diagnostic formula from Bert Holtslag. For unstable conditions we use an energy balance method.
K. JUDA-REZLER:	I would like to ask about the vertical resolution of your model?
J. LANGNER	For applications over Sweden we use three layers in the vertical: A surface layer with a constant height of 75 m; a second layer with a height following the mixing height;

and a third layer with a fixed top and the lower boundary following the mixing height.

R. BORNSTEIN   Is the biosphere in the region sensitive to short term acid deposition amounts, and if so can you show some daily results?

J. LANGNER   I'm not sure about the sensitivity of the biosphere to short term acid deposition amounts, but we do have daily results on wet deposition both the regional contributions and from long range transport.

RECENT APPLICATIONS OF THE RAMS METEOROLOGICAL AND THE HYPACT DISPERSION MODELS

Walter A. Lyons, Roger A. Pielke, William R. Cotton
Craig J. Tremback, Robert L. Walko , M. Uliasz

ASTeR, Inc.
P.O. Box 466
Ft. Collins, CO 80522

Jose Ignacio Ibarra

IBERDROLA, S.A.
Madrid 28001, Spain

INTRODUCTION

The Regional Atmospheric Modeling System (RAMS) developed at Colorado State University is being widely applied in regulatory, operational forecasting and emergency response programs (Pielke et al., 1992). Its two-way nesting capability facilitates use of the model over relatively small regions employing a very fine innermost grid while at the same time treating the influences of inhomogeneous and time variable synoptic fields within which the mesoscale perturbations develop. RAMS has been applied using horizontal mesh sizes (Δx) from as large as 100 km to only 1 meter. The model predicts the basic state variables (U,V,W wind components, pressure, temperature and mixing ratio) as well as a wide variety of derived products, including planetary boundary layer (PBL) depth. RAMS also drives an advanced dispersion code called HYPACT (Hybrid Particle and Concentration Transport). HYPACT disperses emissions from a variety of source types using both Lagrangian particle and Eulerian methodologies (Lyons et al., 1993). Two recent RAMS applications will be summarized.

THE LAKE MICHIGAN OZONE STUDY

In spite of stringent emission controls, numerous exceedances of the U.S. ozone air quality standard have continued in the Lake Michigan region, especially during the very hot summers of 1987 and 1988. Analyses revealed that exceedances of the 120 PPB hourly standard were 400% more likely at monitors located within 20 km of the lake shore. While the role of Lake Michigan in exacerbating regional air quality problems has been investigated for almost 20 years (Lyons and Cole, 1976) the relative impacts of various phenomena upon regional photochemical air quality have yet to be quantified. In order to design a defensible regional emission control policy, LMOS sponsored the development of a comprehensive regional photochemical modeling system (Koerber et al., 1991). This system is comprised of an emission model, an advanced regional photochemical model, and a prognostic meteorological model.

The prognostic meteorological model component for LMOS is called CAL•RAMS, the Chicago and Lake Michigan configuration of RAMS (Pielke et al. 1992). CAL•RAMS is a non-

Air Pollution Modeling and Its Application X, Edited by S-V. Gryning
and M. M. Millán, Plenum Press, New York, 1994

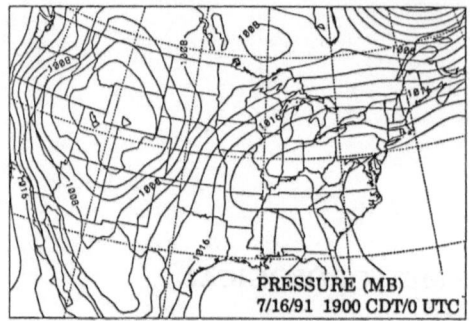

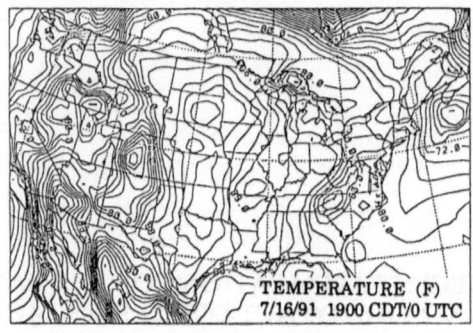

Figure 1. (upper left) CAL•RAMS-predicted sea level pressure field, 0000 UTC 17 July 1991, 2 mb isobars. This is the domain for the outermost grid which used a 32 km Δx.

Figure 2. (upper right) CAL•RAMS-predicted surface layer temperatures, 2°F isotherms, at 0000 UTC 17 July 1991.

```
SETTING DATE   19910716      180000
---------------------------------------------------
---------------------------------------------------
NUMBER OF POSSIBLE OBSERVATIONS:          923
NUMBER OF OBSERVATIONS:             685

AVERAGE OF OBSERVATIONS:       4.38
        STD DEV OF OBS:       2.02
AVERAGE OF PREDICTIONS:       4.05
        STD DEV OF PRED:      1.66
MAXIMUM OBSERVATION VALUE     14.90
MAXIMUM PREDICTED VALUE:       9.03
MINIMUM OBSERVATION VALUE      1.00
MINIMUM PREDICTED VALUE:        .03
MEAN ABS DIFFERENCE (P-O)      1.39
        STD DEV OF ABS DIFF:   1.36
===================================================
THE REMAINING STATISTICS ARE COMPUTED THROWING
OUT VALUES WHEN THE ABSOLUTE DIFFERENCE OF
PREDICTED MINUS OBSERVATIONS IS GREATER THAN
TWO STANDARD OF DEVIATIONS FROM THE MEAN
ABSOLUTE PREDICTED MINUS OBSERVATION DIFFERENCE
===================================================
        AVERAGE ERROR:       -.24
           RMS ERROR:       1.28
         AVERAGE BIAS:      2.56%
                NMSE:       .0037
  AVERAGE GROSS ERROR:        .31
       FRACTIONAL BIAS:      -.06
    INDEX OF AGREEMENT:       .79
HISTOGRAM (PREDICTED-OBSERVATION)
-14    0
-13    0
-12    0
-11    0
-10    0
 -9    1
 -8    1
 -7    1
 -6    6*
 -5    4*
 -4   20*****
 -3   49*************
 -2   98************************
 -1  145***************************************
  0  180*************************************************
  1  100***************************
  2   49*************
  3   28********
  4    2
  5    0
  6    1
  7    0
  8    0
  9    0
 10    0
```

Figure 3. Tool for evaluating the CAL•RAMS 32 km grid output by plotting the predicted minus observed differences at hourly surface weather reporting stations.

Figure 4. Statistical summaries of model performance evaluated at routine hourly aviation weather reporting stations in the eastern United States.

hydrostatic, 3-D, primitive equation code using multiple two-way nested grids. It allows a variety of options for treatments of surface energy budgets, radiation, cloud processes and initialization scenarios. CAL•RAMS simulated meteorological regimes during four major 1991 LMOS episodes. RAMS was configured for multi-day simulations (ten or more) of the synoptic-scale patterns over the entire U.S. Three nested grids were used, with mesh sizes (Δx) of 80 km, 16 km and 4 km, the finest being centered over Lake Michigan and its surrounding shore areas. In order to limit error growth, four-dimensional data assimilation (4DDA) using routine National Weather Service surface and upper air observations was applied at twelve hour intervals. Model output files served as input for two regional photochemical grid models, the Regional Oxidant Model (ROM) developed by the U.S. Environmental Protection Agency and an advanced version of the SAI Urban Airshed Model (UAM-V). All simulations were conducted on an IBM RS/6000-550 workstation.

The ROM model domain was enlarged to cover essentially all of the eastern United States. Its purpose was to provide boundary conditions of ozone and its precursors for the smaller domain of the UAM-V. In its large domain configuration, RAMS resembled a standard hemispheric prognostic model such as those run at NMC or ECMWF. The large scale runs were evaluated in part by comparing charts of such fields as surface pressure, wind vectors, temperature and dewpoint to synoptic scale analyses. Samples of such output are shown in Figures 1 and 2. More quantitative measures included comparison of surface layer values of model-generated wind speed and direction, temperature, dewpoint and pressure reduced to mean sea level to reports from 900+ operational weather reporting stations. Difference fields were plotted at six hourly intervals (Figure 3). Tables of summary statistics were also prepared, an example being shown in Figure 4. The simple device of plotting a histogram of predicted–observed differences proved useful in detecting model performance problems. For instance, if soil moisture were specified too high, this would be reflected by too cool daytime PBL temperatures, and the mode would shift from ±0° to -2° or -3°C.

The major emphasis on model evaluation was within the 16 km and 4 km domains surrounding Lake Michigan. Within this region, the LMOS project established several dozen additional surface reporting stations. Thus we have the opportunity of comparing the model output to a suite of independent surface weather reports not previously assimilated into the model. A wide variety of statistical and graphical measures of operational model performance were developed using the techniques proposed by Tesche (1991). Domain-wide statistics for surface vector wind speed, wind direction and temperature are shown plotted over a 48 hour period in Figure 5. Similar measures were applied in sub-regions and for individual stations. Additional measurements aloft included special rawinsondes, a Doppler sodar, 915 Mhz boundary layer wind profilers and as many as six aircraft on intensive measurement days. Techniques were developed which allowed a numerical "aircraft" to fly through the model output mimicking the flight path of an actual plane. Figure 6 shows such a comparison of model and airborne observations, in this case of temperature.

The lake breezes of 26 June and 16 July 1991, associated with some of the highest ozone values recorded during LMOS, were found to be extremely shallow - on the order of 200 m. Their fronts pushed inland no more than a few kilometers along the Wisconsin-Illinois border (though much further in northernWisconsin). CAL•RAMS was found to simulate these key flow field characteristics. On these intensive days, hourly plots of observations and model output were made (Figure 7). Outputs included surface wind streamlines, wind vectors, temperature, dewpoint, maximum upward model and subsidence in the PBL, mixing depth, surface layer Pasquill-Gifford stability class, etc. Figure 7 shows a typical plot of the CAL•RAMS observed and predicted surface layer winds. Such qualitative displays are helpful in assessing how well the model replicated key morphological features such as inland lake breeze penetration. The Doppler sodar at Zion, on the Wisconsin-Illinois shoreline between Chicago and Milwaukee, was ideally located to monitor the characteristics of the west shore lake breeze. Figure 8 shows the time series of lower PBL winds at Zion during the 16 July 1991 episode. The lake breeze wind shift to onshore (easterly) wind components lasted only 6.5 hours, with the inflow layer extending only 200-225 meters above the surface. The CAL•RAMS output was also examined in various east-west vertical planes, such as that passing through the Doppler sodar site. Figure 9 shows the UW streamlines at 1400 UTC in the plane passing over the sodar revealing an inflow depth on the order of 230 meters. The observed and predicted lake breeze inflow depth is shown in Figure 10, along with the observed and modeled inland penetration of the lake breeze front in this plane (Figure 11). The additional rawinsonde launches from a half dozen sites also permitted assessing the model's ability to predict mixing depths (Figure 12). On this day the model appeared to slightly underpredict some of the higher values. These were largely from a single station that had a mixing depth uncharacteristically higher than other sites. Table 1 provides a summary of key observations and RAMS model predictions for the 16 July 1991 lake breeze. Such parameters as the presence of a lake breeze at Chicago, or its onset and breakdown times, inflow depth and penetration at the Zion site are noted.

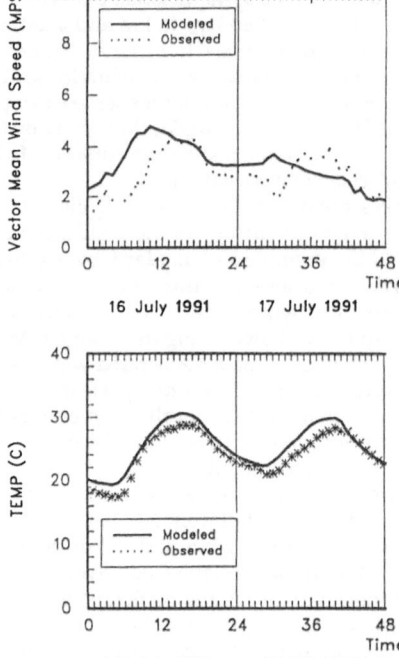

Figure 5. Domain-averaged observed versus model-predicted values of vector mean wind speed, wind direction and temperature (all in the surface layer) for 16 - 17 July 1991. The observations were acquired by special LMOS monitors not previously assimilated into the CAL•RAMS run. The domain is the 4 km grid. Prepared by T. Tesche (Alpine Geophysics).

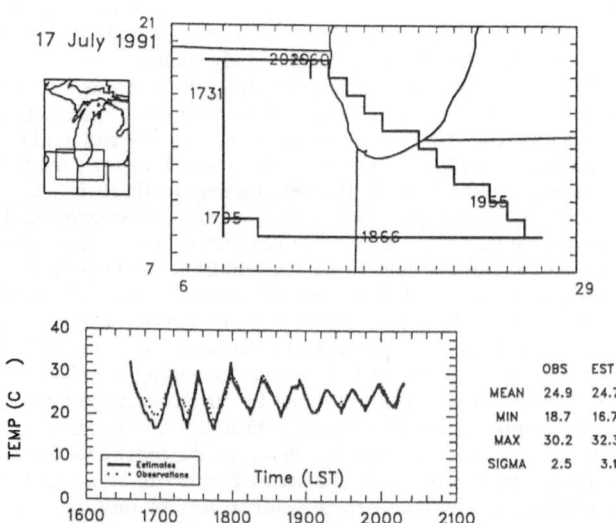

Figure 6. Evaluation of the CAL•RAMS model using data from instrumented aircraft. This shows the flight of one of the LMOS aircraft on an intensive measurement day (17 July) with the model-predicted and observed temperature versus time. Prepared by T. Tesche (Alpine Geophysics).

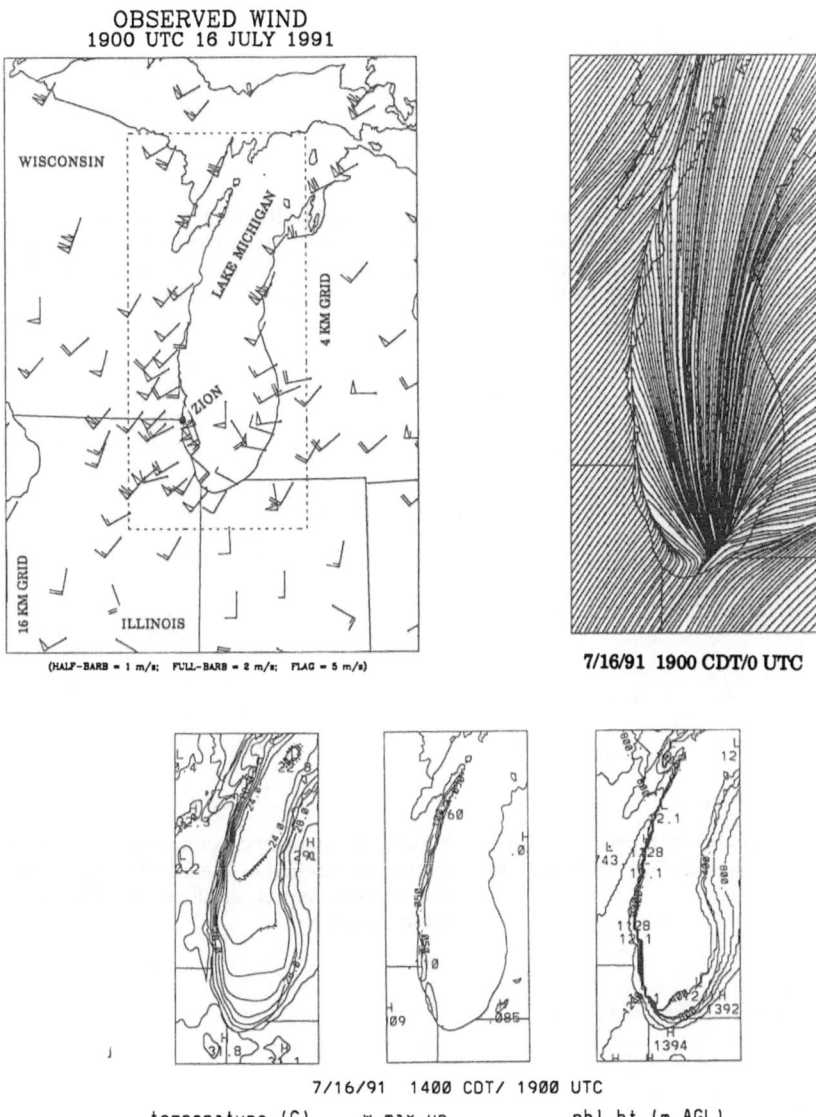

Figure 7. Plotted observed surface winds (upper left) at 1900 UTC 16 July 1991 showing the area coverage of the 16 km and 4 km grids (dashed line). CAL•RAMS output for the 4 km grid includes the surface layer wind streamlines, temperature (1°C isopleths), maximum PBL upward vertical motion (5 cm/sec isopleths) and the computed mixing depths (200 m isopleths).

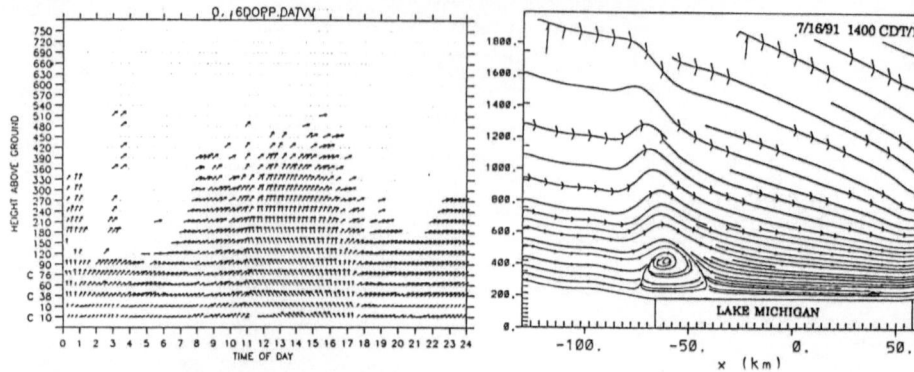

Figure 8. Doppler sodar winds measured at the Zion site on the western shore of Lake Michigan on 16 July 1991. An easterly component to the wind indicates the onshore flow of the lake breeze.

Figure 9. CAL•RAMS simulation of the UW streamlines in an east-west plane through the Zion sodar site at 1400 LT showing a 225 m deep lake breeze inflow at Zion.

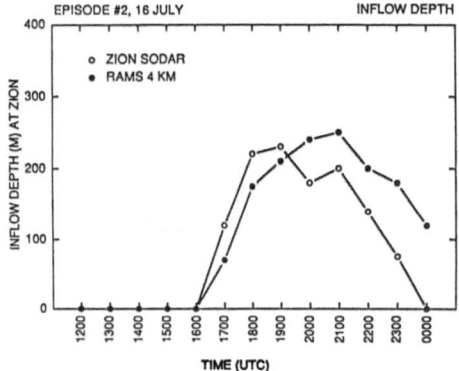

Figure 10. Measured versus predicted lake breeze inflow depth at the Zion shoreline site on 16 July 1991.

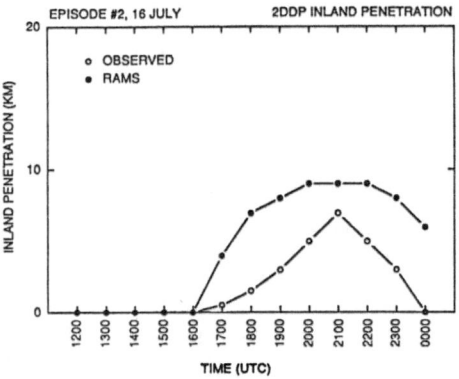

Figure 11. Measured versus model-predicted inland penetration of the lake breeze front on the western shoreline along the Wisconsin-Illinois border.

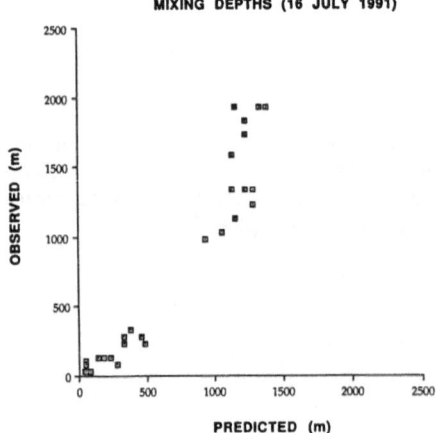

Figure 12. Measured versus predicted mixed layer depths in the LMOS domain during the day on 16 July 1991. The model performed well, except for an apparent underprediction of several of the highest values. These, however, are from a site demonstrating an uncharacteristically high mixed layer depth compared to other sites in the region.

Table 1. Summary of Observed Characteristics for the Lake Breeze of 16 July 1991 Compared to the RAMS Simulations from the Production Run and a Sensitivity Test Run in which the Water Temperature was Uniformly Decreased by 5°C

FEATURE	OBSERVED	RAMS	RAMS(-5°C)
LAKE BREEZE AT CHICAGO	NO	NO	NO
ZION ONSET TIME (Z)	1700	1630	1600
DEPTH INFLOW DEPTH (M)	230	250	250
ZION BREAKDOWN (Z)	2330	0300	0300
ZION 10 M MAX SPEED (M/S)	5.2	4.8	5.4
2DDP PENETRATION (KM)	7	9	10
OFFSHORE PENETRATION (KM)	75+	60	62
MAX VERTICAL MOTION (CM/S)	????	23	22
MAX SUBSIDENCE (CM/S)	????	11	15
MAX INLAND TEMP (°C)	34	34	34
MIN LAKE TEMP (°C)	20	20	18
MAX INLAND Zi (M AGL)	1900	1727	1727
MAX LAKE Zi (M AGL)	0	12	12
OVER LAKE MAX P-G CLASS	G	G	G

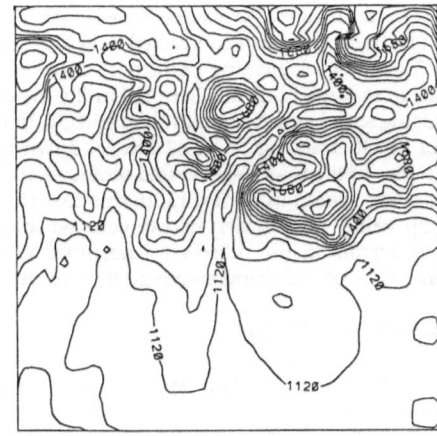
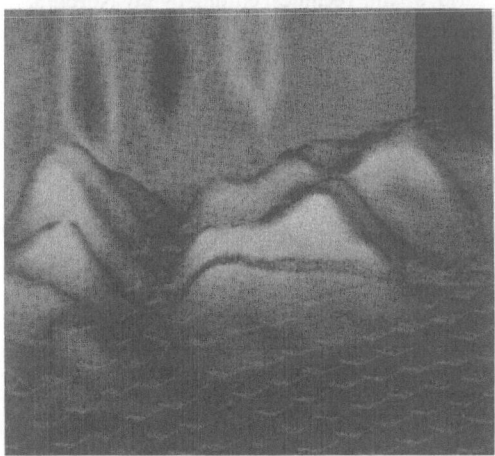
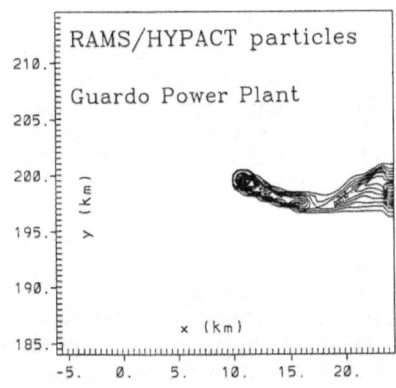

Figure 13. Application of RAMS/HYPACT to simulating flow in a 30x30 km mountain-valley region of northern Spain (upper right). The RAMS-predicted surface layer streamlines during early afternoon show an up valley wind component in the Guardo River valley. A visualization shows the predicted surface layer winds, the vertical motion field (on the backplain) and the dispersing plume from the tall stack. The surface layer concentration pattern is also shown. (Original in color).

In addition, a series of model sensitivity tests were conducted to assess the impact of uncertainties in model input. As an example, if the lake water temperatures were everywhere 5°C colder than used in the production run, it is found that there is virtually no effect upon the resulting lake breeze. Additional evaluations of model performance for the 16 July 1991 case compared HYPACT simulations to airborne measurements of a SF_6 tracer plume released from a shoreline source into the lake breeze front (Eastman et al., 1993). It was determined that a 4 km Δx grid on the western shoreline simulated the main features of the plume dispersion with considerable fidelity. A finer mesh size would have been desired if some of the details of the local flows needed to be included.

SOME OTHER APPLICATIONS

RAMS/HYPACT are being utilized in a prototype operational sea breeze forecasting system for emergency response at the Kennedy Space Center (Lyons et al., 1993). A dedicated workstation produces twice daily 24-hour forecasts of winds using a 3 km Δx in the vicinity of KSC. With a 3-D mesoscale forecast always available, in the event of an accidental release of air pollutants, the available observations will be assimilated using an objective combination technique producing an optimized wind field. Dispersion scenarios include evaporation from "cold spills" of toxic chemicals to Space Shuttle or missile launch explosions.

Model evaluation remains an ongoing issue. While displaying many obvious strengths, they undoubtedly also have some weaknesses. Efforts such as the LMOS program have shown the model to be reasonably robust in simulating sea breeze-type mesoscale regimes. Applications in complex terrain are also needed. One such is the Guardo-90 campaign, a meteorological monitoring and SF_6 tracer program conducted 9-30 November 1990 in the vicinity of the Guardo Power Plant. The 350 Mw coal burning plant with a 185 m stack is located at the mouth of the Carrion River valley in mountainous northern Spain (Lyons and Ibarra, 1993). A number of the 14 tracer releases are being simulated with RAMS/HYPACT. In this case, five nests are employed which encompass all of the Iberian region at 32 km Δx down to a 500 m Δx in the vicinity of the tracer release. Examples of the RAMS-generated surface wind fields and tracer concentrations from a stack-level release are shown in Figure 13. In this case the elevated plume does not become entrained within the up valley flow which had developed during the afternoon but rather advected around the higher terrain in the more west-northwesterly gradient flow aloft. Detailed evaluations of the meteorological and dispersion model performance are being conducted.

ACKNOWLEDGMENTS

This research was supported by the Lake Michigan Air Directors Consortium. We acknowledge the ongoing assistance of the LADCO staff (Steven Gerritson and Michael Koerber) as well as numerous LMOS investigators, particularly Thomas Tesche (Alpine Geophysics). Portions of this work were funded by IBERDROLA, S.A. Madrid, Spain. Technical editing by Liv Nordem.

REFERENCES

Eastman, J.L., R.A. Pielke and W.A. Lyons, 1993: "Numerical Study of the Effects of Grid Spacing on Dispersion and Tracer Evaluation," Paper 93-TP26B.02, 86th Annual Meeting, A&WMA.

Koerber, M., R. Kaleel, L. Pocalujka and L. Bruss, 1991: "An Overview of the Lake Michigan Ozone Study," Preprints, 7th Joint Conf. on Applications of Air Pollution Meteorology, AMS/AWMA, New Orleans, pp. 260-263.

Lyons, W.A. and H.S. Cole, 1976: "Photochemical Oxidant Transport: Mesoscale Lake Breeze and Synoptic Scale Aspects," J. Appl. Meteor., 15, 733-743.

Lyons, W.A., R.A. Pielke, W.R. Cotton, M. Uliasz, C.J. Tremback, R.L. Walko and J.L. Eastman, 1993: "The Applications of New Technologies to Modeling Mesoscale Dispersion in Coastal Zones and Complex Terrain," in, Air Pollution, P. Zannetti et al., Eds., Elsevier Applied Science, London, pp. 35-85.

Lyons, W.A. and J.I. Ibarra, 1993: "Evaluation of Complex Terrain Dispersion Predictions using a Fine Mesh Prognostic Mesoscale Model," Paper 93-A328, 86th Annual Meeting, A&WMA.

Pielke, R.A., W.R. Cotton, R.L. Walko, C.J. Tremback, W.A. Lyons, L. Grasso, M.E. Nicholls, M.D. Moran, D.A. Wesley, T.J., Lee, and J.H. Copeland, 1992: "A Comprehensive Meteorological Modeling System - RAMS," Meteorol. and Atmospheric Physics, 49, 69-91.

Tesche, T. W., 1991: "Evaluating procedures for using numerical meteorological models as input to photochemical models," Preprints, 7th Joint Conf. on Applications of Air Pollution Meteorology, AMS/AWMA, New Orleans, 4 pp.

APPLICATION OF THE MESOSCALE DISPERSION MODELING SYSTEM TO INVESTIGATION OF AIR POLLUTION TRANSPORT IN SOUTHERN POLAND

Marek Uliasz[1,2], Magda Bartochowska[2], Anna Madany[2]
Henryk Piwkowski[2], Jan Parfiniewicz[2] and Maciej Rozkrut[2]

[1]Department of Atmospheric Science, Colorado State University
Fort Collins, CO 80523, USA
[2]Institute of Environmental Engineering Systems, Warsaw University of Technology, Nowowiejska 24, 00-653 Warsaw, Poland

INTRODUCTION

Recent advances in computer technology have opened the door for a broad application of sophisticated numerical models for air pollution dispersion in complex terrain. Very promising opportunities for intensive air quality studies and a real time modeling are provided by Lagrangian particle dispersion models linked to 3-D mesoscale meteorological models (Pielke et al., 1991; Lyons et al., 1993; Uliasz, 1993). A real revolution in mesoscale dispersion applications has been introduced by powerful and affordable workstations which can be dedicated to specific tasks. It allows one not only design and perform short case studies but to use these models for extended periods of time as well. An example of such a computationally intensive application on modern workstations is a project MOHAVE (Uliasz et al., 1993) where daily meteorological and dispersion simulations for the southwestern United States are being performed for a year long study. The modeling methodology developed for the project MOHAVE is being applied to regions in southern Poland with very serious air pollution problems.

This paper presents preliminary simulations performed for a region where Poland, Germany and the Czech Republic meet (Figure 1). This region, called the "Black Triangle" is an area with extensive mining of hard coal and lignite, mostly by opencast methods coupled with numerous coal-fired power plants and heavy industry. There are also some national parks and large forest areas in this region. However, they have been already seriously damaged by air pollution. The largest forest destruction is observed in western part of the Sudety mountains, in the Izerskie Mountains. A chemical analysis of air pollution measurements indicates that coal combustion and metallurgy industry are the main contributors to these damages (Lisowski et al., 1989; Zwoździak and Zwoździak, 1990; Zwoździak and Zwoździak, 1991). Winds from SW to NW directions occur most frequently and associated with high concentrations of sulphur components.

High concentrations of sulphur components have been also observed during winds from N-E sector, however, such winds occur much less frequently than western ones.

The simulations discussed in this paper were performed for May and June of 1993 with the aid of a Mesoscale Dispersion Modeling System (MDMS). The presented results demonstrate the unique feature of the MDMS: a receptor-oriented dispersion modeling.

MESOSCALE DISPERSION MODELING SYSTEM

The MDMS was originally developed on PC in Warsaw University of Technology, Poland (Uliasz, 1990a; Uliasz, 1990b; Uliasz, 1993) and then used in different applications on Unix workstations at Colorado State University (CSU). A detailed description and discussion of MDMS equations can be found in Uliasz (1990a) and Uliasz and Pielke (1990, 1993). This system includes 3-D hydrostatic meteorological mesoscale model (MESO), Lagrangian Particle Dispersion (LPD) model and Eulerian Grid Dispersion (EGD) model. Special emphasis in the MESO model has been put on the parameterization of land surface processes including subgrid-scale terrain features (Uliasz and Pielke, 1992b). Turbulence parameterization is based on a simplified second-order closure, so called level 2.5 scheme, with a prognostic equation for turbulent kinetic energy proposed by Mellor and Yamada (1982) and modified for the case of growing turbulence (Helfand and Labraga, 1988). Several versions of the LPD model with different treatment of particle advection have been implemented. The simplest versions based on a fully random walking scheme may be recommended for mesoscale studies due to their computational efficiency (Uliasz and Pielke, 1993). The LPD model is used not only with the MESO model but also with other meteorological models including the Colorado State University Regional Atmospheric Modeling System (RAMS) (Pielke et al., 1992).

The MESO and LPD models have been evaluated against meteorological and tracer data from the Øresund 1984 field experiment performed over a land-water-land area (Uliasz, 1990b). The obtained results compare favorably with results obtained from another models used to simulate the Øresund data (Andrén, 1990). Several numerical simulations of local atmospheric circulation and tracer dispersion have been also carried out for a hilly coastal zone of the Baltic Sea in northern Poland. Recently, the MDMS was applied to investigate the influence of land surface characteristics on atmospheric dispersion (Uliasz and Pielke, 1992a; Pielke and Uliasz, 1993). The LPD model in both the source-oriented and the receptor-oriented mode was used to study impact of local emission sources on air pollution in Shenandoah National Park in the eastern United States during typical summertime stagnant conditions (Uliasz, 1993). Current applications of the MDMS include simulations of regional transport in the southwestern United States within the project MOHAVE using the LPD model linked to the RAMS.

SOURCE- AND RECEPTOR-ORIENTED DISPERSION MODELING

The source-oriented and receptor-oriented dispersion modeling techniques can be used as complementary tools in air quality studies (Uliasz and Pielke, 1990; Uliasz and Pielke, 1991). These two alternative techniques can be defined as follows:

$$\Phi[C] = \int \int R\,C\,dt\,d\mathbf{x} = \int \int C^*Q\,dt\,d\mathbf{x} \qquad (1)$$

where the source- and receptor-oriented approaches are represented by the first and second integral respectively. It is assumed that a final goal of dispersion modeling is to calculate a certain characteristic of air pollution at a given receptor $\Phi[C]$, which can be defined, in general, as the integral of concentration, $C(\mathbf{x},t)$ over time and space modeling domain. The receptor function, R, determines the geometry (point, area or volume) and location of the receptor and the sampling time of concentration at the receptor. Therefore, this integral expresses averaging of the concentration field over receptor and sampling time with the weight function R.

The traditional source-oriented approach consists in solving model equations forward in time for given emission sources of pollutant $Q(\mathbf{x},t)$ to obtain a time- and space-distributed concentration field $C(\mathbf{x},t)$. It allows us to calculate various air pollution characteristics, Φ, for any number of receptors located within the modeling domain. However, for any new emission scenario, the model solution must be repeated in order to calculate Φ.

In many practical applications when air pollution at the receptor is of primary interest, the alternate receptor-oriented modeling may be considered as a more effective approach. In this case, an influence function, $C^*(\mathbf{x},t)$ is calculated backward in time for a given receptor. The influence function, C^*, is defined by the second integral in equation (1). It depends on meteorology, deposition, and transformations of pollutant in the atmosphere but is independent of emission sources. The air pollution at the receptor, $\Phi[C]$, may now be calculated with the aid of the influence function directly from the emission field, $Q(\mathbf{x},t)$. It should be pointed out that these calculations can be repeated for any emission field or emission scenario, Q, without additional solving of model equations. However, a new influence function must be determined for each receptor.

In the particular case when the emission field, $Q = \sum_i e_i\delta(x-x_i)\delta(y-y_i)\delta(z-z_i)$, consists of multiple point sources with coordinates x_i, y_i, z_i and constant emission rates e_i, the average concentration at the receptor may be rewritten in a simple form:

$$\Phi[C] = \sum_i e_i \int C^*(x_i, y_i, z_i)dt \qquad (2)$$

This expression indicates that the influence function integrated over the simulation period may be used to characterize dispersion conditions for the receptor if emission sources are constant during this time. If the model initial and boundary conditions are taken into account, the influence function allows us to express $\Phi[C]$ as a sum of contributions from: (1) local sources within the modeling domain; (2) distant sources outside the modeling domain in terms of pollution flux across the model boundaries; and (3) initial pollution in the modeling domain. For a sufficiently long period of simulation the contribution from the initial concentration field is negligible and the time integrated influence function may be used to characterize the dispersion conditions in the atmosphere for the receptor.

The influence function is calculated from backward trajectories of particles in the LPD model where particles are released from the receptor during the assumed sampling

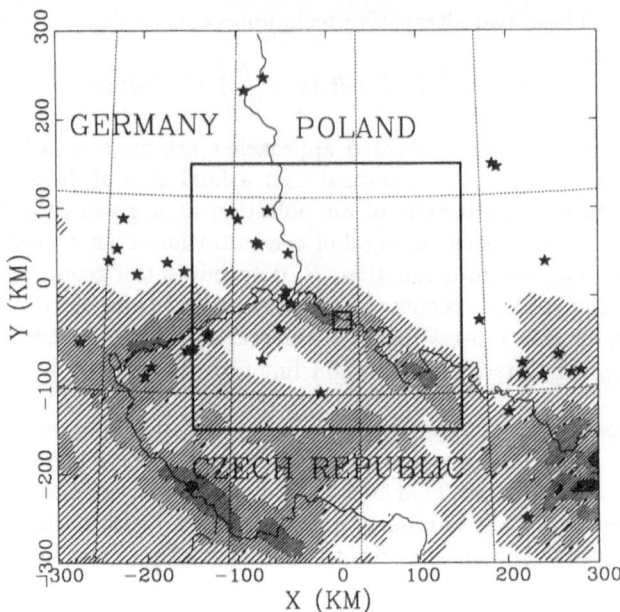

Figure 1. Black Triangle region: stars - major emission sources of SO_2, rectangulars - modeling domain and receptor in the Karkonosze Mountain National Park, terrain elevation contours - $250, 500, 750, \ldots$ m

time. In the case of the EGD model the influence function is obtained as a solution of the adjoint equations with the receptor function, R, as a source term. Applicability of the receptor-oriented option is limited to linear dispersion models.

BLACK TRIANGLE SIMULATIONS

The meteorological mesoscale model MESO together with the LPD model is being used for simulations of air pollution dispersion in the Black Triangle region. In the preliminary simulations reported here, the modeling domain $300 \times 300 \times 6$ km with $31 \times 31 \times 25$ gridpoints is centered in Sudety Mountains. One soil type (loam) is assumed in the whole modeling domain but four land-use categories are distinguished (water, urban, forest and agriculture areas). A series of 3-D meteorological simulations for May and June 1993 was performed assuming that synoptic fields vary in time but are horizontally homogeneous within the modeling domain. For each day of the study period, 36 hours of simulation was performed with the first 12 hours being used to initialize the mesoscale circulation. The next 24 hour period is then used to represent the diurnal cycle of a given day. Additionally, 1-D simulations are run for a flat and homogeneous terrain in order to study the sensitivity of pollution transport patterns in respect to representation of terrain and surface processes parameterization in the meteorological model. The initial soil moisture was assumed to be the same at the beginning of each simulation. One 3-D meteorological simulation needs 2 hours of computer time on an IBM RISC-6000/550 workstation.

The simplified random walk version of the LPD model was used in a receptor-oriented mode for a 20 × 20 × 0.2 km receptor at the Karkonosze Mountain National Park with the highest peak in the region (Śnieżka, 1602 m). Twelve hundred particles were released during each 12 hour sampling period and traced backward in time up to 72 hours. All simulations were performed for a passive tracer. The influence functions are calculated initially for 12 hour sampling periods: 0600-1800 and 1800-0600 GMT and then can be combined for any longer time period. The traditional source-oriented dispersion simulations would be difficult to perform at this time because of problems with availability of reliable emission data for this region.

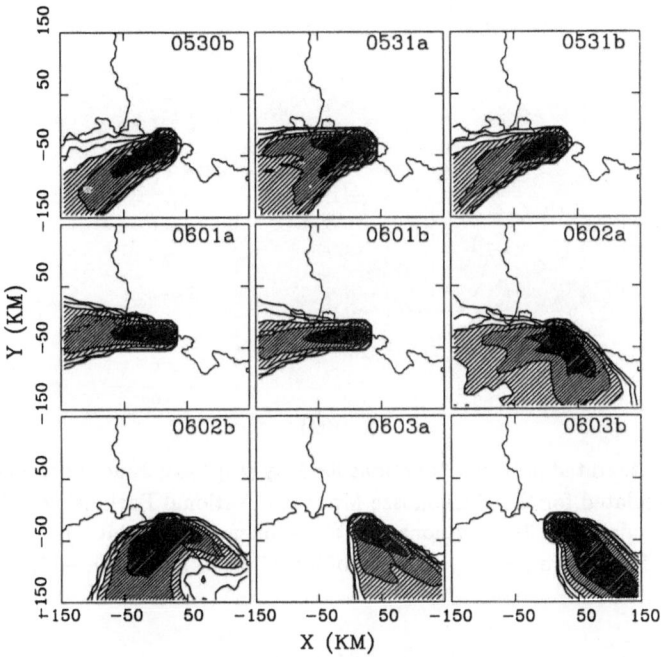

Figure 2. Example of time integrated influence functions for 12-hour sampling periods (a - 0600-1800 GMT, b - 1800-0600 GMT) in the layer 0-500 m calculated for the Karkonosze Mountain National Park (contours of $\log(C^*) = -12, -11.5, -11, \dots sm^{-3}$)

The value of the influence function at a given point multiplied by the emission rate from the source at this location provides the contribution of this point source to the concentration averaged at the receptor over the considered sampling period. Figure 2 shows example of distributions of the time integrated influence functions averaged in the layer from 0 to 500 m above the ground surface. These distributions characterize potential contributions of any emission sources located within this layer to the 12-hour average concentration at the receptor.

The influence functions determined for two monthly periods (May and June of 1993) using series of 1-D and 3-D meteorological simulations are shown in Figure 3. An additional series of 1-D meteorological simulations for May was performed using the increased initial soil moisture (60% of saturation state instead of 30% as in the rest of simulations). This picture clearly demonstrates the importance of mesoscale

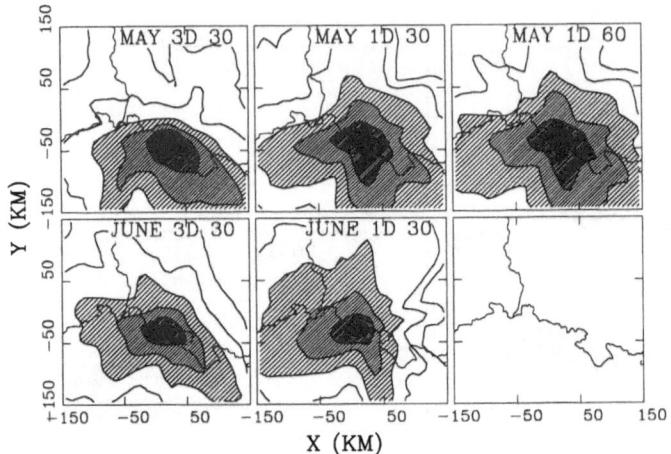

Figure 3. Time integrated influence functions for May (top) and June (bottom) of 1993 in the layer 0-500 m calculated for the Karkonosze Mountain National Park using different series of meteorological simulations: 3-dimensional (left), 1-dimensional (middle), and 1-dimensional with increased soil moisture (right); contours of $\log(C^*) = -12, -11.5, -11, ... \ sm^{-3}$

circulations in this mountainous region for dispersion modeling although terrain features are rather poorly resolved with the horizontal grid spacing of 10 km used in the current meteorological simulations.

It will be useful for applications to investigate pollution dispersion patterns as represented by the influence functions with the aid of a synoptic classification (e.g, Yu and Pielke, 1986). Our synoptic analysis includes four elements: type of pressure field, geostrophic wind velocity, synoptic scale flow direction on surface chart and type of air mass. During May and June 1993 the prevailing synoptic situations for the Sudety Mountains were represented by the 4th class of pressure type (the polar anticyclone or the ridge of polar anticyclone - 45 %), 4 - 8 m/s geostrophic wind (82 %) and the west flow (34 %).

CONCLUSIONS AND FURTHER RESEARCH

This paper demonstrates the value of using state-of-the-art meteorological and Lagrangian particle dispersion models on high performance workstations in order to assess air pollution impacts over mesoscale. The source- and receptor-oriented approaches provide a different insight into regional air pollution transport. The influence functions can be very useful in application to emission control problems, especially, in assessing the emission reduction scenarios in the eastern Europe.

In further research, the modeling domain will be extended in W-E direction to cover Lower and Upper Silesia in Poland, the region around Ostrava and northern Bohemia in Czech Republic, and the surrounding of Cottbus and Leipzig in east Germany. This larger domain may be called a "Sulphur Triangle" due to high emissions of SO_2. The simulations will be performed for a year long period. These simulations will require a better representation of 3-D synoptic fields and soil moisture.

REFERENCES

Andrén, A., 1990: A meso-scale plume dispersion model. Preliminary evaluation in a heterogeneous area. *Atmos. Environ.*, **24A**, 883–896.

Helfand, H. M. and J. C. Labraga, 1988: Design of a nonsingular level 2.5 second-order closure model for the prediction of atmospheric turbulence. *J. Atmos. Sci.*, **45**, 113–132.

Lisowski, A., A. B. Zwoździak, J. W. Zwoździak, and R. Jagieło, 1989: Pollution of the Karkonosze Mountains with the industrail gases and products of their transformations in the atmosphere. *Air Protection*, 4, 107–107. (in Polish).

Lyons, W. A., R. A. Pielke, W. R. Cotton, M. Uliasz, C. J. Tremback, R. L. Walko, and J. L. Eastman, 1993: The applications of new technologies to modeling mesoscale dispersion in coastal zones and complex terrain. In *Air pollution*, P. Zanetti, C. A. Brebia, J. E. G. and G. A. Milian, Editors, Computational Mechanics Publications, Southampton, 33-85.

Mellor, G. L. and T. Yamada, 1982: Development of a turbulence closure model for geophysical fluid problems. *Rev. Geophys. Space Phys.*, **20**, 851–875.

Pielke, R. A., W. R. Cotton, R. L. Walko, C. J. Tremback, M. E. Nicholls, M. D. Moran, D. A. Wesley, T. J. Lee, and J. H. Copeland, 1992: A comprehensive meteorological modeling system - RAMS. *Meteor. Atmos. Phys.*, **49**, 69–91.

Pielke, R. A., W. A. Lyons, R. T. McNider, M. D. Moran, D. A. Moon, R. A. Stocker, R. L. Walko, and M. Uliasz, 1991: Regional and mesoscale meteorological modeling as applied to air quality studies. In *Air Pollution Modeling and Its Application VIII*, van Dop, H. and D. G. Steyn, Editors, Plenum Press, New York, 259-290.

Pielke, R. A. and M. Uliasz, 1993: Influence of landscape variability on atmospheric dispersion. *J. Air & Waste Management* 989–994.

Uliasz, M., 1990a: Development of the mesoscale dispersion modeling system using personal computers. Part I: Models and computer implementation. *Zeitschrift für Meteorologie*, **40**, 104–114.

Uliasz, M., 1990b: Development of the mesoscale dispersion modeling system using personal computers. Part II: Numerical simulations. *Zeitschrift für Meteorologie*, 40, 285–298.

Uliasz, M., 1993: The atmospheric mesoscale dispersion modeling system. *J. Appl. Meteor*, **32**, 139–149.

Uliasz, M. and R. Pielke, 1990: Receptor-oriented Lagrangian-Eulerian model of mesoscale air pollution dispersion. In *Computer Techniques in Environmental Studies III*, Zanetti, P., Editor, Computational Mechanics Publications, Southampton and Springer-Verlag, Berlin, 57-68.

Uliasz, M. and R. A. Pielke, 1991: Application of the receptor oriented approach in mesoscale dispersion modeling. In *Air Pollution Modeling and Its Application VIII*, van Dop, H. and D. G. Steyn, Editors, Plenum Press, New York, 399-408.

Uliasz, M. and R. A. Pielke, 1992a: Effect of land surface representation on simulated mesoscale pollution dispersion. In *Air Pollution Modeling and Its Application IX*, van Dop, H. and G. Kallos, Editors, Plenum Press, New York, 163-170.

Uliasz, M. and R. A. Pielke, 1992b: Source- and receptor-oriented approaches as complementary tools in dispersion modeling. In *10th Symposium on Turbulence and Diffusion*, American Meteorological Society, Portland, OR, Sept. 29 - Oct. 4, 1992, 218-221.

Uliasz, M. and R. A. Pielke, 1993: Implementation of Lagrangian particle dispersion model for mesoscale and regional air quality studies. In *Air pollution*, P. Zanetti, C. A. Brebia, J. E. G. and G. A. Milian, Editors, Computational Mechanics Publications, Southampton, 157-164.

Uliasz, M., R. A. Stocker, and R. A. Pielke, 1993: Numerical modeling of atmospheric dispersion during the MOHAVE field study. In *Air pollution*, P. Zanetti, C. A. Brebia, J. E. G. and G. A. Milian, Editors, Computational Mechanics Publications, Southampton, 208-216.

Yu, C.-H. and R. A. Pielke, 1986: Mesoscale air quality under stagnant synoptic cold season conditions in the lake powell area. *Atmos. Environ.*, **20**, 1751–1762.

Zwoździak, J. W. and A. B. Zwoździak, 1990: Atmospheric sulphate formation and air pollution episodes in the upper parts of the Karkonosze Mountains, Poland. *Environment Protection Engineering*, **16**, 89–98.

Zwoździak, J. W. and A. B. Zwoździak, 1991: Identification of the emission sources categories which influence on pollution concentration in Sudeten Mountains. *Air Protection*, **5**, 127–129. (in Polish).

EFFECTS OF THE SELECTED DOMAIN IN MESOSCALE ATMOSPHERIC SIMULATIONS AND DISPERSION CALCULATIONS

George Kallos and Pavlos Kassomenos

University of Athens
Department of Applied Physics
Ippocratous 33, Athens 10680, Greece

INTRODUCTION

The role of the various-scale circulations (synoptic, regional, sub-regional, mesoscale, microscale) in the dispersion of air pollutants, released from different type of sources, is considered as very important because the 3-D meteorological and turbulence fields show significant temporal variations (e.g. Pielke et al., 1983; Glendening et al., 1986; Segal et al., 1988; Ulrickson and Mass, 1990; Kallos, 1990 among others). This is especially true in areas with significant physiographic variations. Today, allot of mesoscale models are able to provide meteorological and turbulence fields adequate for use in dispersion and/or transformation models (Pielke, 1988; Seigneur, 1989). The combined use of such models is considered as superior from the existing regulatory models because they should provide a more realistic description of the dispersion characteristics and the spatial and temporal variations of the concentration fields of air pollutants on more accurate manner. Despite this fact, the use of such models may lead in erroneous calculations for several reasons. The most important reasons are the mishandling from the models the physical processes taking place under certain circumstances and of course the improper use of them. The improper use may be due to the violations of physical rules which are not necessarily associated with the algorithms and the chosen initial and boundary conditions.

As it is known, a limited area atmospheric model is able to accurately describe atmospheric disturbances of a limited portion of the entire spectrum. The highest and the lowest frequencies described on an accurate way by such a model are strictly bounded by the domain dimensions, grid increments and filters used. The horizontal topographic, land-water and land-use variations always generate atmospheric circulations at the meso-alfa, beta or gamma scale of the spectrum. It is quite interesting the way the atmospheric disturbances of this portion of the spectrum interact with the larger-scale (synoptic or regional) ones.

During the last three decades, several mesoscale models have been developed and used successfully in order to provide detailed atmospheric and dispersion conditions in several areas. Regularly, in such kind of models the so-called primitive equations of motion, continuity, state and energy are solved with different numerical schemes.

For most of the simulated cases, especially for these related to air quality studies in urban areas, a model domain with horizontal dimensions less than 100x100 kmxkm and horizontal grid increment of 1 to 5 km was used. The initialization of such a model usually is done on a simplistic way, assuming that the larger scale of motions are not changing spatially and slightly changing (or not at all) during the simulation. The selection of such a domain and initialization imposes some limitations.

These limitations do not permit:

Air Pollution Modeling and Its Application X, Edited by S-V. Gryning
and M. M. Millán, Plenum Press, New York, 1994

35

i. The adequate representation of a large spectrum of the disturbances.
ii. The extension of the run for long period of time.
iii. The use of such models for simulations of a large variety of atmospheric phenomena.

The first category of these limitations is of great importance in the accurate representation of the flow fields and stability of the atmosphere. The wave to wave interaction is a highly non-linear process and therefore the effects of such omissions cannot be determined easily.

The upper limit of the frequency spectrum is determined from the grid increment. These high frequency disturbances are not always wiped out from the lower ones (e.g. frequent horizontal variations in elevation, land water distribution or land-use may create disturbances which might be important) (Dalu et al., 1992, 1993).

When a propagating atmospheric circulation reaches the model boundaries it is advised not to continue the simulation in order to avoid phenomena such is reflection etc. Therefore, a severe limitation in model simulations is imposed by a relatively small for the case model domain.

Because of the way these models are initialized, they cannot be used to simulate a large number of atmospheric phenomena. The most appropriate use of such models is the simulation of classical thermally driven atmospheric circulations (e.g. sea-breezes). Even such simulations cannot be successful when there are strong pressure gradients or when there are significant diurnal variations on larger scale phenomena (regional scale phenomena) or fast moving synoptic scale circulations.

Today, some of these problems could be surpassed with the use of models with capabilities such as the two-way interactive nesting. In this presentation an atmospheric modeling system with such capabilities was used in order to show some effects of the above mentioned problems and the way they should be avoided.

MODEL SIMULATIONS

Various-type atmospheric and dispersion simulations were performed over a large area of SE Europe for different types of nesting and grid increments. The main purposes of these simulations were to identify the interactions between the various-scale circulations in the large area around Athens and to examine possible influence in the results from the selected model domains and set-ups. Because of the air quality problems in the Greater Athens Area (GAA) it is essential to know first of all the main characteristics of the local atmospheric circulations in the region and their interaction with the larger scale ones (sub-regional, synoptic).

Models used: The atmospheric model used for these simulations is the Colorado State University Regional Atmospheric Modeling System (CSU-RAMS) (Walko and Tremback, 1991; Pielke et al., 1992). The CSU-RAMS is an advanced modeling system with several capabilities which make it unique for very complex atmospheric processes. Some of its capabilities, extremely useful for complex flow interactions, are the two-way interactive nesting, the various-complexity cloud microphysics, turbulence, radiative transfer and surface parameterization. It also has the capability to choose different kind of initialization, coordinate systems, upper and lateral boundary conditions.

The other model used in combination with RAMS is a Lagrangian Particle Dispersion Model (LPDM), (McNider, 1981; Kallos 1989; Moran, 1992). It is a Lagrangian-type dispersion model which uses as input the predicted fields from RAMS and some emission parameters.

Description of the area: In this study, the main area of interest is the Athens basin where the cities of Athens and Piraeus are located. This basin is approximately 25 km long and 17 km wide, at the SE part of mainland Greece (Fig. 1). It is surrounded by mountains from three sides (West, North and East) and open to the sea from the fourth (Saronic Gulf). The ventilation of the basin is mainly from the opening toward the sea and from the three gaps between the mountains. At the larger area around the Attic Peninsula are some topographic features which play important role in the final formation of the flow over the

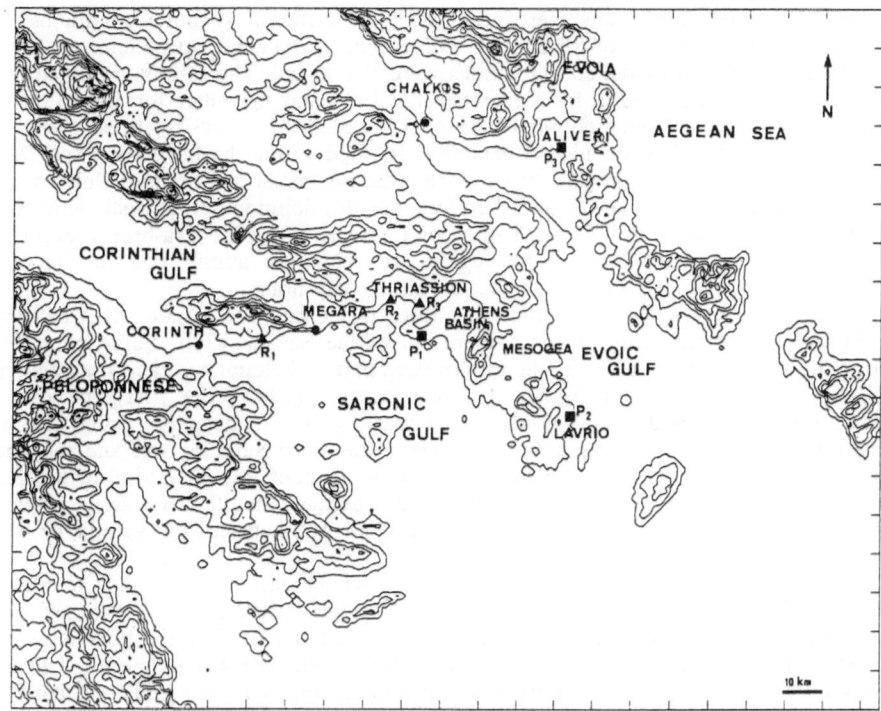

Fig.1 The topography of SE Greece. Contours are every 200 m.

area of interest. These include the southern part of the island of Evoia, the northeastern part of Peloponnese with the Isthmus of Corinth and the islands of Aegina and Salamis at the Saronic Gulf. At the sub-regional scale, the Greek and Asia Minor Peninsulas are forming a channel across the Aegean Sea which is closed to the N from the mountains of the Balkan area. The main opening to the NE is the area of Vosporus. To the W of the Greek Peninsula is the Ionian Sea which is the extension of the Adriatic Sea. Between the Mainland Greece and Peloponnese is a narrow strip of sea the Corinthian Gulf with orientation WNW-ESE. This land-water distribution, the major and minor topographic features and the variations in landscape are responsible for the formation of a series of local circulations (mechanical or thermal) at all scales (Kallos and Kassomenos, 1992). Some of these circulations are well known even from the ancient era (e.g. the various sea-breeze cells over the Attic Peninsula, the diurnal variation of the flow across the Aegean during the summer months which is known as Etesians). More information about these circulations someone could find in Kallos et al. (1993) and the references therein. As it was shown in Dalu et al. (1992) the local circulations of a Rossby radius smaller than a critical one should be wiped-out from the larger disturbances. The way and under which circumstances this occurs is considered as very important not only from the dynamic point of view. It is considered of great importance for air quality applications for local or regional and long-range.

Model Results: As it was found in Kallos and Kassomenos (1992) and Kallos et al. (1993) there is a strong interaction between the local circulations developed in the GAA and the sub-regional scale ones in the area of SE Europe and NE Mediterranean. This is true during all seasons but it is more pronounced during the warm period of the year. This occurs because the strong insolation and the complicated distribution of the various   physiographic characteristics in the area. These thermal circulations are responsible for the transport of air pollutants to distant and sometimes unexpected areas. Similar transportation of air pollutants is described in Millan et al. (1992) for the Iberian Peninsula which has several common climatological characteristics with the SE part of Europe.

Several model simulations were performed, for different synoptic conditions, over different domains in the area of SE Europe with emphasis over the Attic Peninsula and the area around it. Different model configurations were also tested. The results presented below are from two simulations: One for December 13, 1989 and the other for June 15, 1990. During the first case, the synoptic circulation over the SE Europe was relatively weak from WNW and the atmosphere was almost neutrally stratified within the lowest 1.5 km. During this winter-day the thermal circulations over the Attic Peninsula and Saronic Gulf are weak because of the limited insolation and therefore the development of a weak temperature difference between the land and sea. The second case is a typical sea-breeze event. The thermal circulations developed in the region are relatively strong due to the strong differential heating between the land and sea during the day-hours.

Most of the mesoscale simulations performed in the past for the area of Athens were covering only the Attic peninsula with a portion of the Saronic Gulf. Some others were covering the southern part of Evoia, the northeastern part of Peloponnese and the Isthmus of Corinth. As it was shown in Kallos and Kassomenos (1992), the first type of these simulations should provide wind fields which may be in a satisfactory agreement with surface observations in the Athens basin but erroneous over Saronic Gulf and may be at some other critical areas such as the gaps between the mountains. Such kind of mesoscale simulations cannot be used for dispersion calculations because they fail to describe important for such cases phenomena, like the recirculation of air pollutants or the ventilation of the Basin.

A simulation with RAMS was made for the winter case over a small area around Athens with horizontal grid increment of 2 km. The wind field at the SW corner of the domain is from WSW, relatively weak, which directly influenced from the boundaries (it is assumed to be sea indefinitely while in the reality is the NE Peloponnese). The same simulation was repeated, with the same initial conditions but with two grids: An inner grid which was the same as in the previous single-grid simulation and an outer grid which was covering a larger area around it (including the southern Evoia, NE Peloponnese and the Isthmus of Corinth) with horizontal grid increment 4 km. The results shown in Fig. 2. are from the outer grid at 11:00 LT. The horizontal wind field is quite different from the previous simulation because the covered domain is larger. The differences are more pronounced over the Saronic Gulf where the horizontal wind field is from WSW with

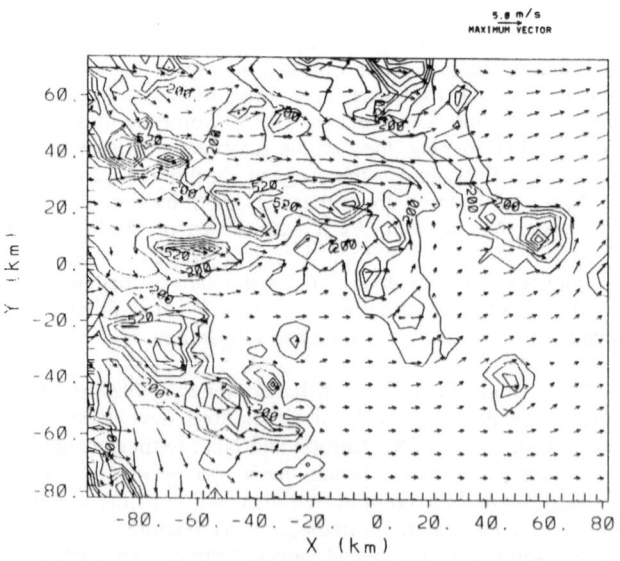

Fig. 2 The horizontal wind field 125 m above the ground, at 11:00 LT of December 13, 1989 for the outer grid from the two-grid simulation (arrows every second grid).

smaller speed. No significant differences were found over the Athens Basin. Significant differences were found at some critical locations (e.g. at the gaps between the mountains, over Mesogea plain). A third simulation was performed with three nested grids: The coarser one which covers the region from the Ionian sea to the western part of Asia Minor and from South of Crete to the northern part of the Balkan area with horizontal grid increment of 16 km. The other two grids are the same as in the previous run. As it is seen in Fig. 3 the wind field over the area of interest is completely different from the previous simulations. The results shown in this figure are for the intermediate grid, at the same hour as before just for comparison. Light winds are observed over the Saronic Gulf while in the previous two simulations a westerly flow was observed. The relatively strong westerly winds observed over the western Saronic Gulf at the second of the above simulations are due to the channelling of the flow at the Corinthian Gulf and the Isthmus of Corinth. In this simulation, the Corinthian Gulf is acting as a channel between the Peloponnese and Mainland Greece which is at the western model-boundaries. This has as an effect the artificial spin-up of the flow over the Isthmus of Corinth which cannot be opposed from the weak thermal circulations developed over the Saronic Gulf. The flow fields obtained from the third simulation with the three nested grids compare well with the surface and upper-air observations and in general are considered as more realistic. This goal has been achieved because the flow is balanced over the larger area and therefore the local spin-ups have been avoided. It is worth to mention that with simulations like the first or even the second from these presented above someone should claim that sea-breeze circulations are evident in the area even during winter. This of course is not true because the temperature gradients developed between the land and the sea were negligible during this simulation.

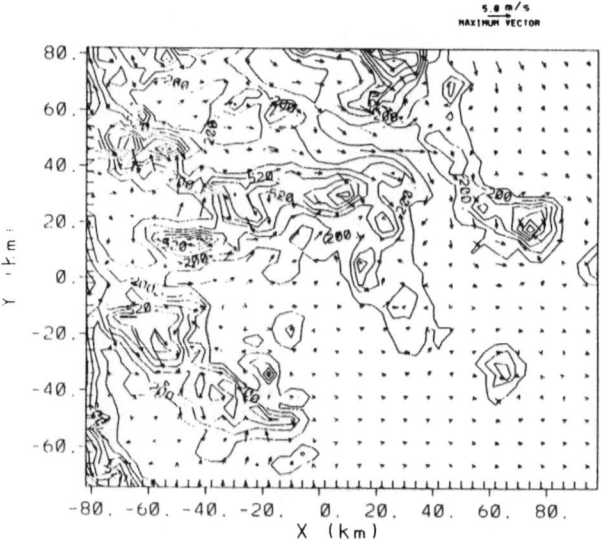

Fig. 3 The same as in Fig. 2 but for the intermediate grid from the three-grid simulation.

The same type of simulations were performed also for the second case. As it is seen in Figs 4 and 5 there are some significant differences between the wind fields estimated from the two and three nested grid simulations respectively. These differences are considered as important even form the morning hours, despite the fact that these cannot be considered as important over the Athens Basin. The most significant differences are shown over the Saronic Gulf where at the two-grid simulation (with horizontal grid increments 4 and 2 km) the flow is mainly from W, with speeds 4-5 m/s (see Fig. 4 for 10:00 LT), because of the channelling over the Isthmus of Corinth. At the three-grid simulation (with horizontal grid increments 16, 4 and 2 km) the flow over the Saronic Gulf is very weak, mainly from W (see Fig. 5) during the same hour of this day. Another characteristic of this simulation is the flow field at the coarsest grid which covers mainly the Greek Peninsula and the Aegean Sea. The flow is relatively weak from NW to NNW directions initially which starts veering toward the land, during the morning hours because of the development of the thermal circulations. It becomes from W at the southern Peloponnese, and from S to CE over the sea E of Peloponnese. This flow is directed toward the Saronic Gulf at around noon. This flow pattern was initially described in Kallos and Kassomenos (1992) for a case with very stable conditions in the lower troposphere. It was also found in a number of cases during all seasons. This characteristic flow pattern puts under revision the described in the past recirculation mechanism of air pollutants over the Athens Basin and Saronic Gulf.

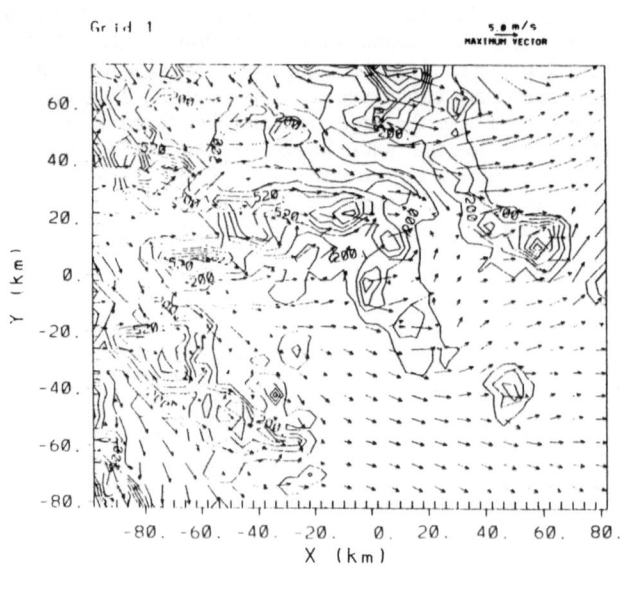

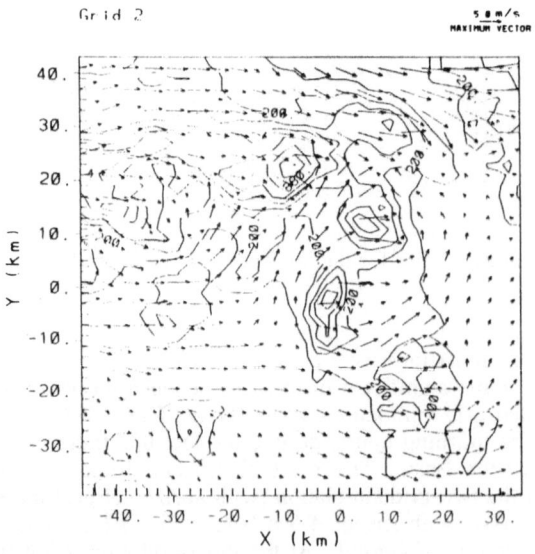

Fig. 4 Wind fields at the two model grids 125 m above the ground at 10:00 LT of June 15, 1990 (two-grid simulation).

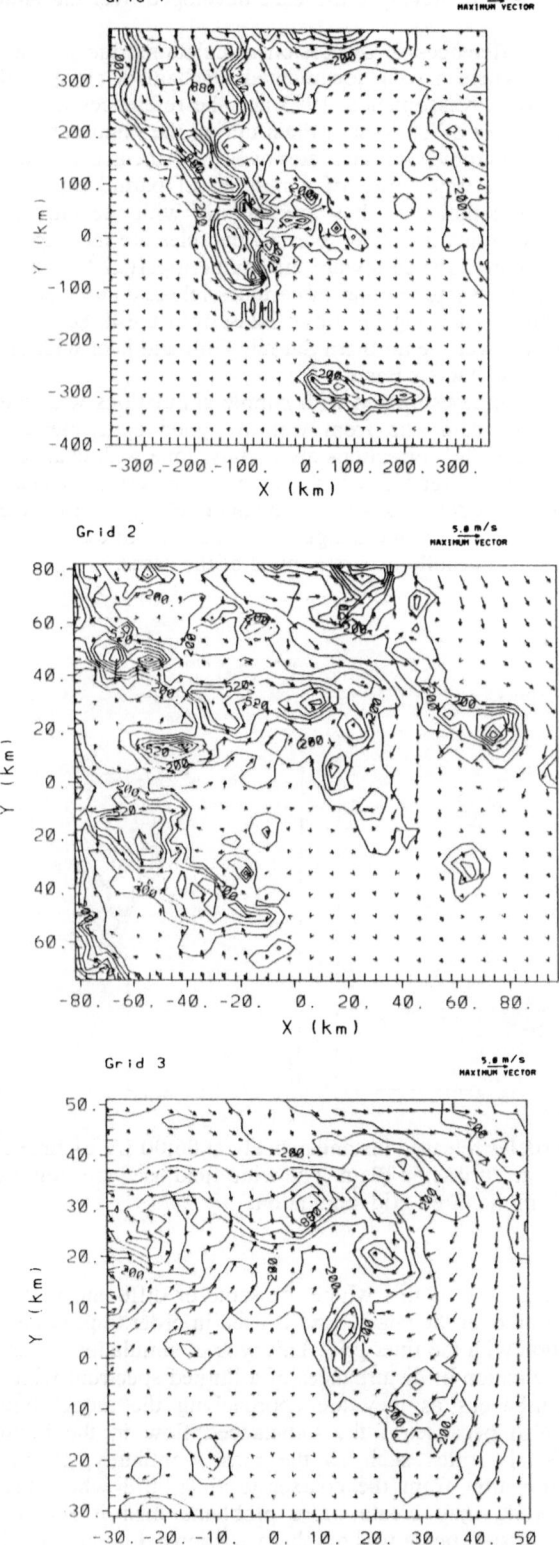

Fig. 5 The same as in Fig. 4 but for the three model grids (three-grid simulation)

Under revision is also the idea of having sea-breeze developed over the Athens Basin even during the winter months.

An example of the differences in dispersion calculations due to the selected model domain is given in Fig. 6 where a well documented mechanism is described. As it was found in previous studies (e.g. Kassomenos, 1993 and the references therein) air pollutants released from the industrial area of Thriassion Plain during the night and morning hours are initially transported over the Saronic Gulf and later over the Athens basin, with the aid of the sea-breeze. Fig. 6a shows the trajectories of a number of particles released early in the morning from the Thriassion plain, as they are moving with the aid of the wind field estimated from the two-grid simulation. The released particles are moving over the northern part of the Saronic Gulf and are directed toward the southern part of the Attic Peninsula. Fig. 6b shows the trajectories of the same number of particles released at the same place and time as in the previous simulation but using the flow fields from the three-grid simulation. The particles are initially directed over the northern Saronic Gulf and later with the aid of the sea-breeze are transported over the Athens Basin.

Such differences cannot be estimated with simple simulations with mesoscale models except in cases where large data sets from well designed experimental campaigns are available. Even for these cases the corrections are mainly done with changes in the assumed synoptic-scale wind speed and direction which are more or less provided on an arbitrary way. But this is a fit of the model results to the available observations rather than a model simulation of physical processes. It is too dangerous to be used in cases with little (or not at all) available observations as it usually occurs in most of the cases.

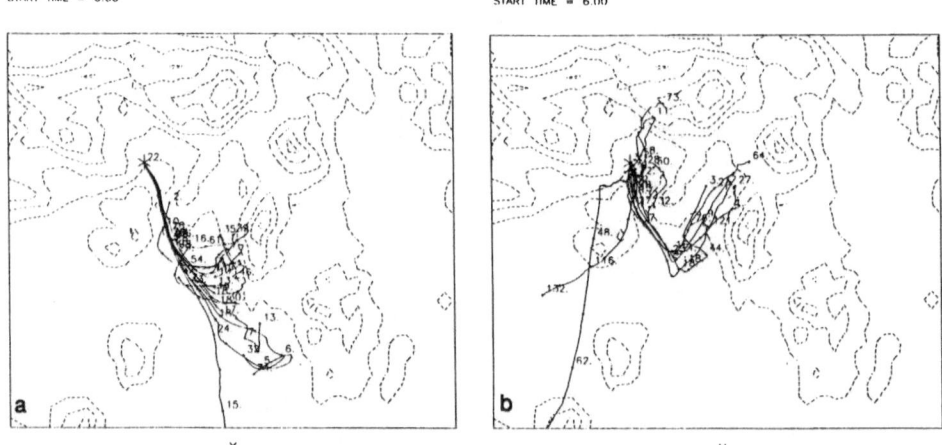

Fig 6. Trajectories of particles released instantaneously at 06:00 LT of June 15, 1990, 40 m above the ground, at Thriassion Plain. The wind field used are from (a) the two-grid simulation, (b) from the three-grid simulation.

Another type of simulation is this where the same or different models are used over nested grids. The simulation over the coarse grid is used in order to provide lateral boundary conditions to the simulation with the finer grid. This type of simulations does not solve either the problem of the representation of disturbances of a limited spectrum which was discussed above or the local disturbances as they are approaching the model boundaries. Major problems should arise also because of the unbalanced flow in the boundaries and the violation of basic physical rules such as the mass continuity. It may also lead to overdiffusion. The interpolation from the coarser to finer grid where the elevations are usually completely different, may create more problems than it should solve. This is especially true in areas with significant topographic variations at the boundaries of the model domains.

CONCLUSIONS

In this presentation, an attempt was made to highlight some problems which should arise when mesoscale atmospheric models are used in air quality studies. Some of these problems are concentrated to the way the model domain (or domains) is defined. The grid increments and the sizes of the selected domain must be defined for each case after a careful examination of the role they might play the different physiographic characteristics and their distribution in a large area around. The interaction between the different scales of atmospheric disturbances is considered as very important in air quality studies. This is considered as essential in areas like the southern Europe. Useful tools for such studies are the atmospheric models with the two-way interactive nesting capabilities.

ACKNOWLEDGEMENTS

This research has been partially supported by IBM-Hellas, the General Secretariat of Science and Technology of Greece and the Commission of the European Communities DG XII, contracts EV5VCT910050 (SECAP project) and AVI-CT92-0005 (AVICENNE).

REFERENCES

Dalu, G.A., R.A. Pielke, R. Avissar, G. Kallos, M. Baldi, and A. Guerrini, 1992: Linear impact of thermal inhomogeneites on mesoscale atmospheric flow with zerosynoptic wind. Ann. Geophysicae, 9, 641-647.

Dalu, G.A., M. Baldi, R.A. Pielke, and G. Kallos, 1993: Fluxes over complex terrain: Analytical evaluation. Preproceedings of the 20th ITM on Air Pollution Modeling and its Application.

Gledening, J.W., B.L. Ulrickson, and J.A. Businger, 1986: Mesoscale variability of boundary layer properties in the Los Angeles Basin. Mon. Wea. Rev., 114, 2537-2549.

Kallos, G., 1989: The use of mesoscale models on air pollution studies in industrial installations. Environmental Software, 4, 117-122.

Kallos, G., 1990: Mesoscale flow fields and dispersion in the coastal terrain of Eastern Corinthian Gulf in Greece during summer. Proceedings of the 18th ITM of NATO/CCMS on Air Pollution Modeling and its Application. Editors: H.v Dop and D.G. Steyn, Plenum Press, VIII, 285-292.

Kallos, G., and P. Kassomenos, 1992: Weather conditions during air pollution episodes in Athens, Greece: An overview of the problem. Proceedings of the 19th ITM of NATO/CCMS on Air Pollution Modeling and its Application. Editors: H.v Dop and G. Kallos, Plenum Press, IX, 77-102.

Kallos, G., P. Kassomenos, and R.A. Pielke, 1993: Synoptic and mesoscale weatherconditions during air pollution episodes in Athens, Greece. Boundary-Layer Meteorol., 62, 163-184.

Kassomenos, P., 1993: Analysis of the weather conditions during air pollution episodes in the Greater Athens Area. Ph.D. Thesis. University of Athens, Dept. of Applied Physics, Athens, Greece. pp 360 (in Greek).

McNider, R.T., 1981: Investigation of the impact of topographic circulations on the transport and dispersion of air pollutants. PhD Dissertation, University of Virginia, Charlottesville, Virginia, pp 210.

Millan, M.M., B. Artinano, L. Alonso, M. Castro, R. Fernandez-Petier, and J. Goberna, 1992: Mesometeorological Cycles of Air Pollution in the Iberian Peninsula. CEAM, Air Pollution Research Report 44, CEC-DG/XII/E-1, Contract EV4V-0097-E. pp 219.

Moran, M, 1992: Numerical modeling of mesoscale atmospheric dispersion. Ph.D. Dissertation, Colorado State University, Dept. of Atmospheric Science, Fort Collins, Colorado, pp 768.

Pielke, R.A., 1988: Status of mesoscale and subregional models. Volume 2: The availability of mesoscale meteorological models to realistically simulate atmospheric conditions in Northern California. RP2434-6. Pacific Gas and Electric Company, Research and Development Program. 139 pp.

Pielke, R.A., W.R. Cotton, R.L. Walko, C.J. Tremback, W.A. Lyons, L.D. Grasso, M.E. Nicholls, M.D. Moran, D.A. Wesley, T.J. Lee, and J.H. Copeland, 1992: A comprehensive meteorological modeling system - RAMS. Meteorol. Atmos. Phys., 49, 69-91.

Segal, M., C-H. Yu, and R.A. Pielke, 1988: Model evaluation of the impact of thermally induced valley circulation in the Lake Powel area on long-range transport. J. Air Poll. Control Ass., 38 (2), 163-170.

Seigneur, C., 1989: Status of subregional and mesoscale models. Volume 1: Air quality models. RP2434-6. Pacific Gas and Electric Company, Research and Development Program, 55 pp.

Ulrickson, B.L., and C.F. Mass, 1990: Numerical investigation of mesoscale circulations over the Los Angeles Basin, Part I: A verification study. Mon. Wea. Rev., 118, 2138-2161.

Walko, R.L., and C.J. Tremback, 1991: RAMS-The Regional Atmospheric Modeling System, Version 2c: User's Guide. Published by ASTeR Inc., P.O. Box 466, Fort Collins, Colorado, pp 86.

DISCUSSION

S. ZILITINKEVICH: How can this preliminary information about typical synoptic situations in the region considered be used in the numerical model.

G. KALLOS: Identification of the main characteristics of each synoptic class, estimation of the frequency of occurrence of each category and selection of representative case(s). Then, performing simulations. This methodology is very useful in cases where the available data are not sufficient for the examined area or are no data at all.

SUMMER EPISODES OF POLLUTION DISPERSION OVER THE COASTAL AREA OF ISRAEL – A NUMERICAL STUDY

Joan Goldstein[1], Yakov Tokar[1], Yakov Balmor[2],
Ed Glaser[3] and Pinhas Alpert[1]

[1] Department of Geophysics and Planetary Sciences
Tel-Aviv University
Ramat-Aviv 69978
[2] The Israel Electric Corporation, Ltd.
P.O.Box 25
Tel-Aviv
[3] The Association of Towns for Environmental Protection
P.O.Box 3041
Hadera
Israel

INTRODUCTION

The rising need to control industrial emissions over urban areas in Israel and to monitor the dispersion of pollutants (Graber, 1981) has stimulated extensive experimental and simulation studies in this field (Graber et al., 1984). The dependence of the dispersion of pollutants on mesoscale meteorological conditions was investigated with the aid of a recently developed model of air pollution transport over Israel (Tokar et al., 1993) and a mesometeorological model based on the PSU/NCAR mesoscale numerical model known as MM4 (Anthes et al., 1987). This model is suitable for forecasting flows with characteristic wavelengths of 10 to 2500 km under a variety of meteorological conditions. The models were implemented over a 324×270 km region with the Hadera power plant –the largest in Israel – at its center. Topography was included in the model. The MM4 model with 15 height levels and 6 km horizontal resolution was adapted to fit the Israeli conditions. The pollution transport model was applied for a smaller domain of 108×90 km with 2 km of horizontal grid spacing. The transport model uses the MM4 calculations as its input data.

MODEL DESCRIPTION

The present version of the MM4 model is described by Anthes et al. (1987). The vertical coordinate of the model is the terrain-following sigma coordinate . The set of equations contains: momentum equations; continuity; surface pressure tendency; thermodynamic energy and the hydrostatic equation. The closures on the sub-grid are reported as first order, and detailed boundary parameterization is according to Blackadar. A grid scaled diffusivity is also included for stability. Typical horizontal grid spacing for this model is 10 to 80 km resolution, but some runs have been made with resolution down to 5 km. The vertical grid spacing is from 30 meters at the ground to 1 km at the higher levels. The standard model has sixteen levels. The horizontal grid structure represent the so-called "Arakawa B" staggered grid (Arakawa and Lamb, 1977). In the present study the MM4 model was implemented for a 55×46 cell grid with a 6 km horizontal interval.

A full description of the Pollution Dispersion and Transport Model (PDTM) may be found in Tokar et al. (1993). The MM4 model is used to provide meteorological fields and thermodynamic parameters such as friction velocity, convective velocity scale, Monin-Obukhov length L, and the stability state. These data are needed to obtain the vertical eddy diffusivity, which was calculated according to the Deardorff (1970) scheme for unstable conditions. For neutral conditions, when $|L| \geq 10^5$, Lamb et al. (1975) model was applied. For stable conditions, i.e $10^5 > L > 0$, Businger and Arya (1974) derivations were used. The PDTM model was implemented for the nested 55×46 grid with the horizontal grid interval of 2 km, which is three times smaller than the grid interval used in the meteorological model. The pollution model area is placed at the center of the meteorological model domain in order to reduce the boundary effects. Initialization for the model runs was based on radiosonde data from two sites: one near Hadera power plant and the other, at the Israel Meteorological Service, in Beit-Dagan near Tel-Aviv, approximately 50 km to the south from Hadera.

Table 1 presents the typical parameters of four power plants stacks included in the model. The sites are located (Fig.1) in Hadera (two stacks), Haifa (two stacks), and Reading near Tel-Aviv (two stacks). These power plants are the main sources of SO_2 in the central part of Israel. The effective stack heights were calculated using the Briggs (1969) formula. This formula is quite applicable for the stacks under consideration. For example, in case of Hadera stack, assuming typical air temperature of 300K, the magnitude of stack parameter F is 552 $m^4 s^{-3}$. According to (Seinfeld, 1986), in case of F>55, the plume becomes passive at a distance of $119 F^{2/5} \approx 1500$ m, i.e within one grid cell. The calculations show that in the model conditions the plume rise may change considerably from 200 to 500 meters depending on the wind velocity.

Table 1. Nominal stacks parameters.

Site	Stack height (m)	Diameter (m)	Gas velocity (m/s)	SO_2 emission (kg/s)	Exit gas temperature(°C)
Hadera	250	7.0	24.0	2.00	130
	250	7.0	24.0	2.00	130
Haifa	80	3.9	9.8	0.52	165
	80	3.0	25.4	0.89	160
Reading	150	3.3	25.8	1.43	165
	150	3.7	21.1	0.42	142

Table 2. Vertical grid altitudes in the MM4 model.

Level #	1	2	3	4	5	6	7	8	9	10	11	12	13	14	15
Height, m	15	100	250	400	500	600	700	800	900	2000	5000	7000	9000	11000	15000

The initial data for the MM4 model are the domain topography and meteorological parameters. Fig.1 shows the topography maps used in the model calculations for: a) the MM4 model domain, b) the pollutions transport model domain. According to the MM4 boundary conditions treatment, 5 first grid layers neighboring the border serve as a sponge layer. To avoid the influence of the near–border inconsistencies of the MM4 model data on the pollution calculations, the MM4 model domain was taken larger than that of pollution model, both domain centers coincide. The vertical grid

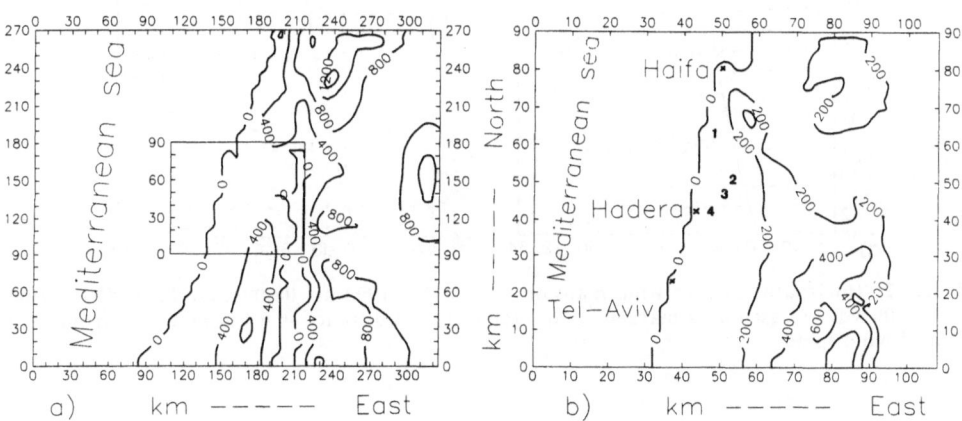

Figure 1. Topography maps of the MM4 model (a) and PDTM model (b) domains. The box on (a) indicates the location of the inner domain. The numbers on (b) indicate locations of air pollution monitoring stations: Zichron–Yakov (1), Givat–Ada (2), Pardes–Hana (3) and Hadera (4).

of the MM4 model was taken as presented in Table 2. Vertical grid of the pollution dispersion model consists of 25 levels with 80 m of grid spacing.

METEOROLOGICAL ENVIRONMENT

The above model was tested by simulating for four days on which radiosondes were released near Hadera power plant. The days were chosen in the hope of obtaining temperature profiles during episodes of relatively high SO_2 pollution concentrations.

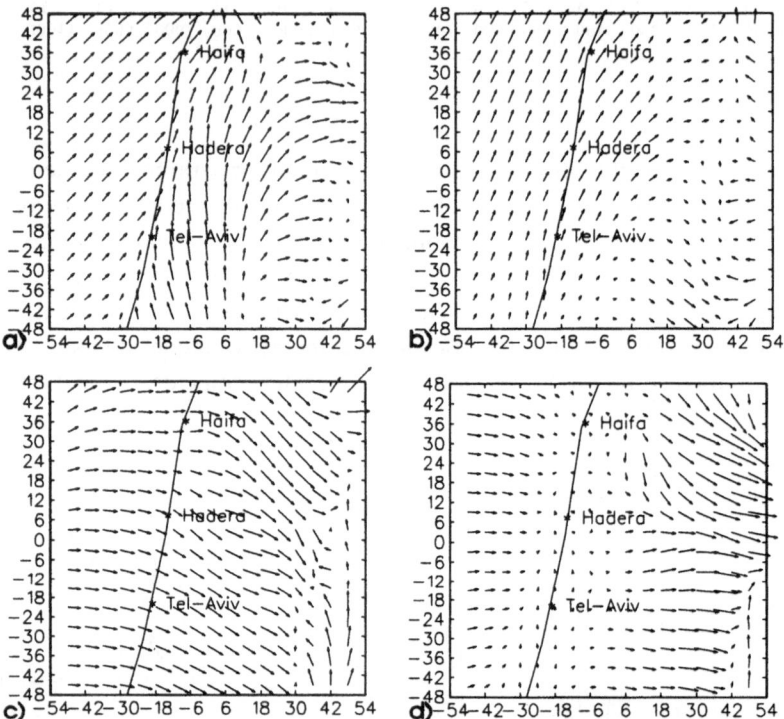

Figure. 2. Horizontal surface wind maps on August 10, 1992 at 08:00 (a), 11:00 (b), 17:00 (c), 20:00 (d). Coastline is marked solid. Vertical grid distance is 6 km and corresponds to 10 ms^{-1} wind speed.

The upper air synoptic system governing the Eastern Mediterranean during summer is the subtropical anticyclone(Alpert et al., 1992; Barkan and Felix, 1991). The lower troposphere over northern region is governed by a warm trough that is an extension of the Indian monsoon low. The local mesoscale system is controlled by the sea–land breeze. Fig. 2 presents diurnal wind field evolution obtained using the MM4 model for a typical summer day – August 10, 1992. It can be seen that the nocturnal wind (Fig. 2a) is typically southerly and the inland breeze starts at about 12 a.m. and reaches its maximum at about 3 p.m. At 9 p.m. the breeze weakens and the wind becomes northerly. By midnight the breeze is easterly and then turns to be southerly once again, thus completing the cycle. Cyclic wind behavior is confirmed also by other investigators (Alpert et al., 1984). Such a wind system stimulates temperature inversions which are indeed formed on a diurnal basis (Lieman and Alpert, 1993). The temperature inversion base is dependent on the breeze speed. According to calculations (Fig.3) the sea breeze development causes the decrease of the inversion base height and the rise of its temperature. It was found that early in the morning and during the first half of the day the height of the inversion base reaches its maximum at about 300-700 m. In the afternoon and evening it lowers down to 100m.

POLLUTION MODEL APPLICATION RESULTS

As mentioned above, four summer episodes were simulated. The dates of the episodes were chosen to be able to compare the calculations with the measurements. The monitoring stations which provided the SO_2 concentrations are located at Zichron-Yakov, Givat-Ada, Pardes-Hana and Hadera (see Fig.1). The main goal of the model runs was to investigate the role of the Hadera site as the most powerful SO_2 source in the region, in the pollution of the coastal area. Fig. 4 presents the diurnal time series of the ground SO_2 concentrations calculated on August 10, 1992 for four monitoring stations. The simulations are shown in comparison with the measurements. Peak concentration may be seen at about 15 LT both in calculations and measurements and corresponds to maximum wind sea breeze. The observed maximum SO_2 concentration of 44 ppb is somewhat higher than the calculated one (27 ppb). One of the reasons for this is that the model did not consider in the present run the alternative pollutant sources and the background concentrations. The observations, Fig.4, show the peak only at two of the four monitoring stations: Givat-Ada and Pardes-Hana (for locations see Fig. 1). The two stations are positioned exactly in the direction of the plume of the Hadera power plant which corresponds to the sea breeze direction. Though the model predictions are generally in good agreement, both qualitative and quantitative, with measurements, there are significant discrepancies for Hadera and Zichron-Yakov stations. They are supposed to originate partly from some internal artificial diffusion (though relatively small) of the adopted computational scheme of advection (Smolarkiewicz, 1992). Another experiment was performed in which the Hadera site was excluded (No-Hadera run). Fig. 5 demonstrates the contours of the ground SO_2 concentration calculated in the full (including Hadera) run for 15:00 and 21:00 local time. It is seen that the Hadera stacks produce in the coastal area rather moderate pollution concentrations of 10 to 20 ppb. Comparison of the two runs (not shown here) demonstrates that the main contribution of the Hadera plant is over the eastern part of the domain where the Haifa station produces 10 to 40 ppb whereas Hadera produces 40 to 70 ppb. Relatively high SO_2 concentrations of about 100 ppb are predicted also in the vicinities of the Haifa and Reading (Tel-Aviv) power plants.

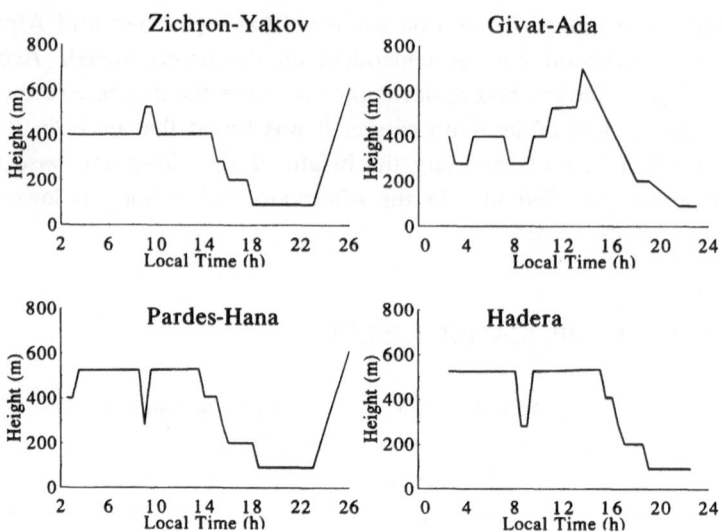

Figure 3. The height of the inversion base as a function of time on August 10, 1992.

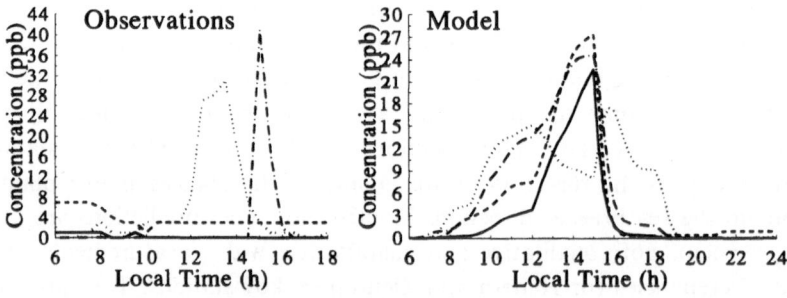

Figure 4. Ground SO_2 concentration as a function of time on August 10, 1992 at four measuring stations of Zichron-Yakov (solid), Givat-Ada (dots), Pardes-Hana (dashed-dots) and Hadera (dashed).

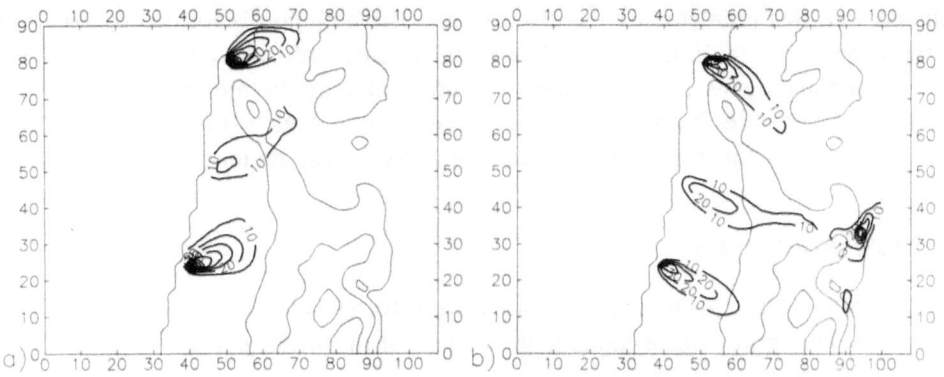

Figure 5. Ground SO₂ concentration maps at 12:00 (a), 18:00 (b) on August 10, 1992. The axes are measured in kilometers. The concentration is given in parts per billion (ppb).

CONCLUSIONS

The comparative analysis of the four model runs makes it possible to formulate the main features of the summer meteorological situation in the coastal area of Israel and the mechanism of forming of the pollution fields. The meteorological situation for all four episodes was similar. The wind is governed by the sea breeze circulation and may be typified as southerly in the morning and turning clockwise around noon. The temperature inversions were observed in the computational experiments all over the coastal area. The inversion base reaches its maximum height late at night and in the morning then decreases later in the day. The inversions, however, are strongly smoothed in the model runs probably due to the lack of the large-scale subsidence mechanism over the mesoscale domain. Future runs will incorporate this effect through lateral-boundary divergence.

The model shows relatively low SO₂ concentrations with maximum of about 30 to 40 ppb which is reached at the hottest hours of the day. These peak concentrations are not caused by variations in the operating conditions of the power stations, so the convective mechanism and the changing wind direction must be responsible for the temporary concentration rise. This conclusion is based on the model finding that the inversion thickness has in the middle of the day the same order of magnitude as the effective stack height. The SO₂ calculations for 10 August are in general agreement with measurements at two of four monitoring stations located nearby. The simulation indicates that Hadera power plant contributes mainly to the pollution of the eastern part of Israel – Judean mountains and Jordan valley, where a relatively high polluted area was predicted by the calculations. The pollution is formed by the joint effect of both the Haifa and the Hadera power plants. Unfortunately, because of the lack of measurements in this region, it is impossible at the moment to verify this prediction. There are, however, some measurements of high – up to 50 ppb – SO_2 concentrations in the south-eastern part of Israel-Gush Etzion (Peleg and Luria, 1993).

ACKNOWLEDGEMENT

The authors are grateful to the Israeli Ministry of Science and Technology and Ministry of Immigrant Absorption for the support of this investigation.

REFERENCES

Alpert, P., Abramsky, R., and Neeman, B.U. (1992). The prevailing summer synoptic system in Israel – Subtropical High not Persian Trough. *Israel J. of Earth Sci.*, **39**: 93-102.

Alpert, P., Kusuda, M., and Abe, N. (1984). Anti-clockwise rotation, eccentricity and tilt angle of the wind hodograph, Part II: Observational study. *J. Atmos. Sci.*, **41**: 3558-3573.

Anthes, R.A., Hsie, E.-Y., and Kuo Y.-H. (1987). "Description of the Penn State/NCAR Mesoscale Model Version 4 (MM4)". NCAR Technical Note, NCAR/TN-282 + STR.

Arakawa,A.,and Lamb,V.R. (1977). Computational design of the basic dynamical process of the UCLA general circulation model. *Methods in Computational Physics.*, Academic Press, **17**: 173-265.

Barkan, F., Felix, Y. (1993). Observations of the diurnal oscillation of the inversion over the Israeli coast. *Boundary-Layer Meteorol.*, **62**: 393-409.

Briggs, G.A. (1969). "Plume Rise". U.S.Atomic Energy Commission Critical Review Series T/D 25075.

Businger,J.A., Arya,S.P.S. (1974). Height of the mixed layer in the stably stratified planetary boundary layer. *Adv. Geophys.*, **18A**: 73-92.

Deardorff,J.W. (1970). A three-dimensional numerical investigation of the idealized planetary boundary layer. *Geophys. Fluid Dyn.*, **1**: 377-410.

Graber, M. (1981). Air quality in Israel in the years 1979 and 1980, in: " Developments in Arid Zone Ecology and Environmental Quality", H.Shuval, ed., Balaban International Science Service, Philadelphia, PA.

Graber, M., Dayan, U., and Laznow, J. (1984). Development of a dispersion model for power plant siting applications in coastal Israel, in: "Proc. 4th Joint Conf. on Applications of Air Pollution Meteorology", 16-19 Oct., 1984, Amer. Meteorol. Soc., Boston, MA, pp. 268-269.

Lamb,R.G., Chen,W.H., Seinfeld,J.H. (1975). Numerico-empirical analyses of atmospheric diffusion theories. *J.Atmos.Sci.*, **32**: 1794-1807.

Lieman, R., Alpert, P. (1993). Investigation of the planetary boundary layer height variation over complex terrain. *Bound. Layer Meteor.*, **62**: 129-142.

Peleg, M., and Luria, M. (1993). Transportation of air pollutants from the coastal area to the Judean mountains. Proc. of 24th meeting of Israel Soc. for Ecology and Environmental Quality Sciences, June 21-22, 1993, Tel-Aviv.

Seinfeld, J. (1986). "Atmospheric Chemistry and Physics of Air Pollution". John Wiley & Sons, 738pp.

Smolarkiewicz, P.K., Clark, T.L. (1986). The multidimensional positive definite advection transport algorithm: further development and applications. *J.Comp.Phys.*, **67**: 396-438.

Tokar, Y., Goldstein, J., Levin,Z. and Alpert,P. (1993). The use of a meso-gamma scale model for evaluation of pollution concentration over an industrial region in Israel (Hadera). *Boundary-Layer Meteorol.*, **62**: 185-193.

NUMERICAL SIMULATION OF MESO-METEOROLOGICAL

CIRCULATIONS IN THE LISBON REGION

Miguel Coutinho[1], Alfredo Rocha[2] and Carlos Borrego[1]

[1] Department of Environment and Planning
[2] Department of Physics
University of Aveiro
3800 AVEIRO, Portugal

INTRODUCTION

During the last 5 years, the University of Aveiro has directed significant research towards the analysis of regional air quality and problems related to the transport of airborne pollutants over Portugal (Coutinho et al., 1989; Borrego et al., 1991). One of the main conclusions of these studies was the necessity to consider mesoscale atmospheric circulations in studying atmospheric dispersion patterns in Portugal. Mesoscale meteorological models seem to be the most adequate tool to represent those circulations through the mathematical simulation of atmospheric physical processes.

This paper presents typical simulated 3-dimensional meso-meteorological flow-fields for the summer, over the region of Lisbon. Results were obtained using the non-hydrostatic mesoscale model MEMO, developed at the University of Karlsruhe. The selection of meteorological scenarios is framed over an analysis of summer synoptic conditions over the Iberian Peninsula. A special attention is drawn to the effect of the Iberian thermal low, on the atmospheric dispersion patterns that occur over Portugal.

MODEL DESCRIPTION

Within MEMO (Flassak, 1990), the conservation equations in the atmosphere for momentum, mass and scalar quantities such as potential temperature, turbulent kinetic energy and specific humidity are solved numerically. The result shown in this paper were obtained with the model version using the Boussinesq approximation. For the calculation of the turbulent diffusion, K-theory is applied. The exchange coefficients for momentum and scalars are computed with an one-equation turbulence model. At roughness height z_0 similariry theory is applied where u^* and θ^* are calculated from the Businger equations.

Air Pollution Modeling and Its Application X, Edited by S-V. Gryning
and M. M. Millán, Plenum Press, New York, 1994

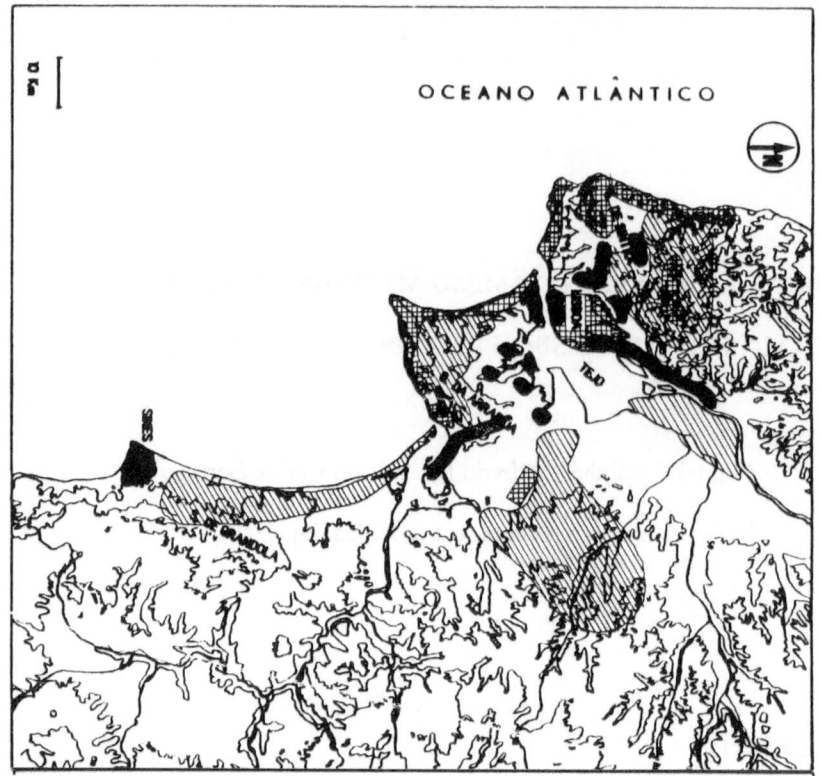

Figure 1. Topography of the Lisbon region. The domain is 150x150 km and altitude isopleths are contoured at 50, 100, 200, 300 and 400 m. Residentials areas are squared, industrial areas are solid and forest and agricultural areas are marked with diagonals.

The governing equations are solved in terrain-following coordinates. The model MEMO includes an efficient numerical scheme for the calculation of the atmospheric radiative heating/cooling rates and of the radiative fluxes at ground level for both clear and polluted or cloudy atmosphere (Moussiopoulos, 1987). The land surface temperature is computed from the surface heat budget equation. For the calculation of the soil temperature, an one-dimensional heat conduction equation for the soil is solved. Water temperature is kept constant during a simulation.

DESIGN OF EXPERIMENTS

The southern region of Portugal (Figure 1), including the Greater Lisbon Area, was choosen for this study taking into account the geographical distribution of the main industrial sources and of the most important urban centers, with a population of 3.5 million inhabitants.

Lisbon is built in a very complex topographic region, dominated by a large estuary and multiple hills, surrounded by small mountain ranges reaching heights over 400 m above sea level. The need to understand the behaviour of this airshed is most felt as a substantial increase air pollutants emissions in the region is foreseen for the near future mostly caused by the increase of roadway traffic.

A modeling domain of 150 km x 150 km x 6 km was selected and the horizontal grid spacing was fixed at 5 km. In the vertical direction the grid consists of 20 layers non-equidistant with a minimum grid spacing of 20 m near the ground.

Summer climatology of the region

The most frequent summer synoptic circulation type over the Iberian Peninsula is characterized by slightly above average mean sea level pressure and almost non-existent surface pressure gradients over the domain (Nadal, 1980). This circulation type called "Pantano barométrico" (barometric swamp), which is formed in about 70% of the summer days, is generally associated with weak winds in the lower troposphere, cloudless skies, high maximum temperatures and weak precipitation rates. Local thermal depressions develop in the afternoon but are extinguished at night due to radiative cooling.

A detailed report focussing on the analysis of mesometeorological cycles ocurring in the Iberian Peninsula under these conditions was recently published by M. Millan and co-authors (1992). During this report a special concern was directed to the atmospheric circulations over the Mediterranean coast.

Nevertheless geographic differences between the Eastern-Mediterranean coast of the Iberian Peninsula and the Portuguese coast might have an important effect over their climatology. The Portuguese coast is very affected by the vicinity of the Azores high pressure system. Moreover the Atlantic sea surface temperature is low (16-18 °C), compared to the Western Mediterranean where the water temperature reaches values between 22 and 26°C. Under these circunstances the pressure gradient between the centre of the Peninsula and the Atlantic coast is usually higher than the pressure gradient observed on the eastern coast of the Peninsula. These conditions cause a strong northerly quasi-geostrophic wind is developed over the Portuguese coast (Ferreira, 1984). Meanwhile the weaker gradient at the Eastern coast of the Iberian Peninsula allows the penetration onshore of Mediterranean and even Saharian air.

Episode selection

A detailed analysis of daily pressure synoptic charts was carried out for August of 1991 and 1992. The lower tropospheric pressure field for the 4th of August 1992 is a typical example of a "pantano barométrico" situation. An extension of the Azores antyciclone extended over the northern part of the Iberian peninsula and a low pressure system was located to the west of the British Islands.

The vertical profile of the atmosphere over Lisbon showed a quite marked thermal inversion of about 4°C at about 300 m and winds were generally weak throughout the lower troposphere (~3 m.s^{-1}) blowing from N/NE.

RESULTS AND DISCUSSION

Mesoscale airflow patterns

Prognostic air flow patterns generated by the MEMO model at 10 m Above Ground Gevel (AGL) are presented in Figure 2-5. Wind fields for 8.00, 14.00, 19.00 and 22.00 Local Standard Time (LST) were selected.

The airflow within the modeling region during the morning of 4 August 1992 (Fig. 2) was characterized by offshore-directed drainage and downslope flow along the Sintra, Arrabida and Grandola mountain ridges, a northeasterly flow over the Tejo valley and a northerly flow over the Atlantic Ocean.

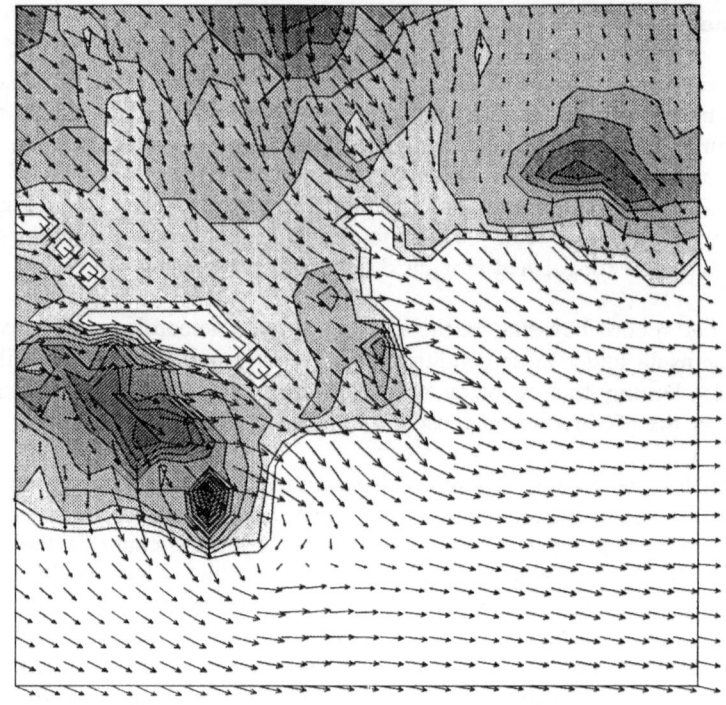

Figure 2. Wind field for 8.00 LST 4 August 1991 at 10 m AGL. Wind velocity is scaled at 4 m.s^{-1} per grid spacing.

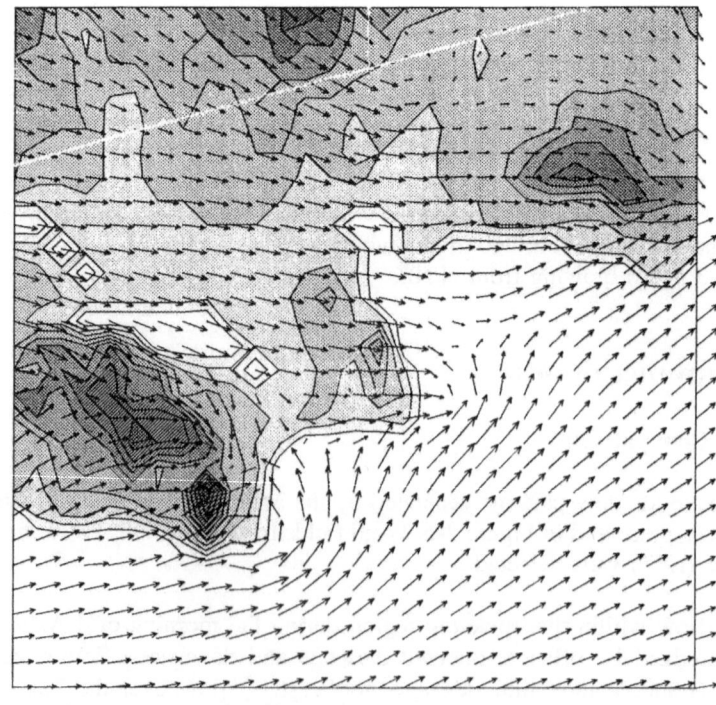

Figure 3. As in Fig.2, but for 12.00 LST

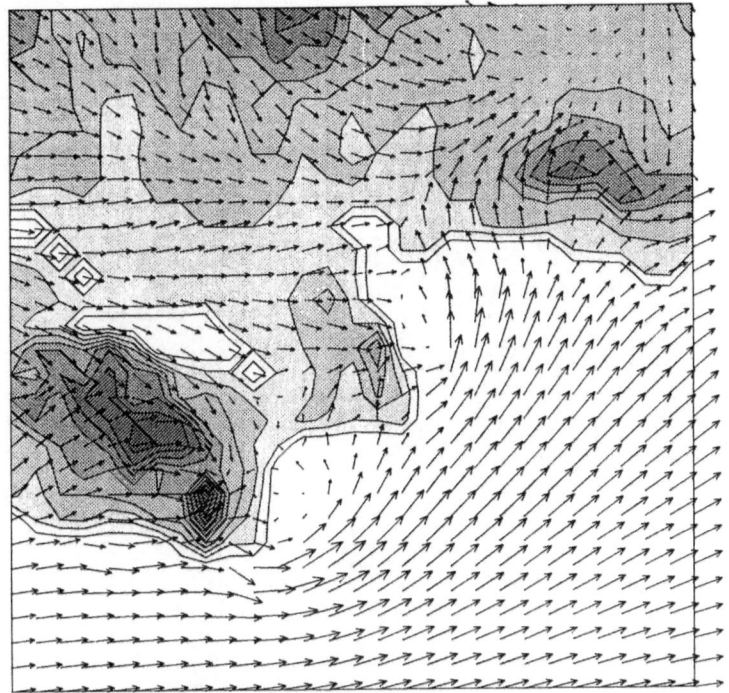

Figure 4. As in Fig.2, but for 19.00 LST

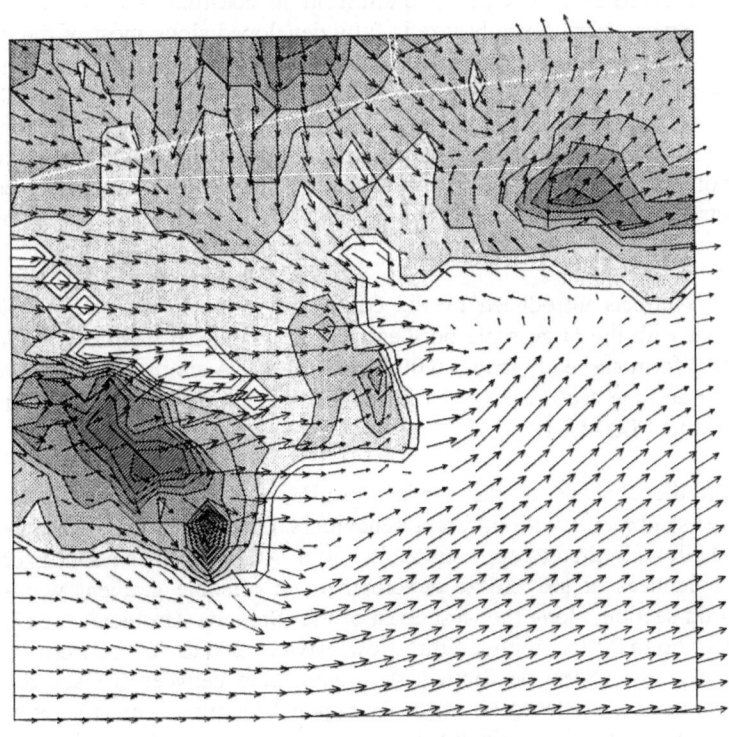

Figure 5. As in Fig.2, but for 22.00 LST

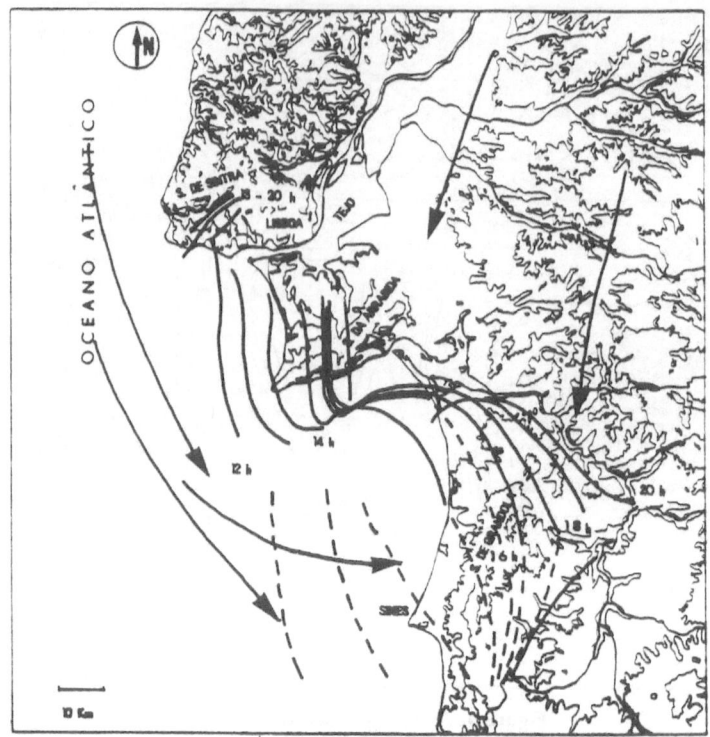

Figure 6. Evolution of the sea breeze front from 12.00 LST to 20.00 LST.

By 12.00 LST the wind over the Atlantic starts veering west and an incipient sea-breeze front is observed over the Ocean, 10 km from the coastline near the Tejo's mouth and 30 km from the Sines coast. A sea breeze is fully developed along most of the coast line by 14.00 LST (Fig. 3) and persists through 19.00 LST (Fig.4), penetrating inland as far as 25 km in the Sines coast. During this period the sea breeze front evolution follows the coastline curvature (Fig. 6), except over the "Serra de Sintra" where a convergence front is located from 13.00 to 19.00 LST. In the southern part of the domain, a weak sea breeze front is formed approximately 40 km offshore and advances landward, reaching the coast at 15.00 LST. No penetration of the sea breeze front is observed at the southern coast of the "Serra de Arrabida"

Curvature effects introduced by the coastal curvature in the vicinity of the Tejo's mouth combined with the large scale northeasterly flow and onshore/upslope flow near the "Serra de Sintra" form and eddy from 16.00 to 19.00 LST. A similar interaction between the large scale flow and the sea breeze cell form an eddy south of the "Serra de Arrabida" that lasts from 15.00 to 18.00 LST.

Through the analysis of the vertical profile of the atmosphere it is possible to distinguish three sea breezes with different behaviours. On the southern region near Sines the sea breeze is shallow, limited to approximately 150 m AGL. This behaviour is coherent with experimental evidences of sea breeze development in offshore gradient winds (Atkinson, 1981). In the mid-section, along the west coast of the "Serra de Arrabida", the sea breeze cell is better defined, up to 200 m, with a return flow between 200 to 500 m, from 16.00 to 20.00 LST. Finally, over the "Serra de Sintra", the conjugation between the sea breeze and the upslope flow creates a very strong convergence region. This front is mostly active from 14.00 to 19.00 LST. Upward vertical velocity reaching 0.3 m/s were calculated on a layer 250 to 720 m AGL.

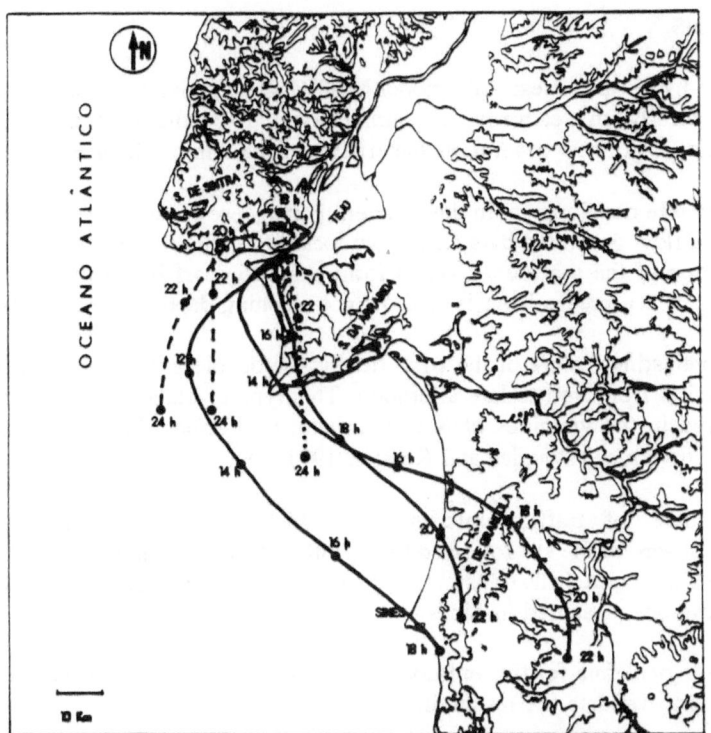

Figure 7. Pathlines with starting point in Lisbon at 30 m AGL. Solid lines represent pathlines initiated at 8.00, 10.00 and 12.00 LST; dashed lines represent pathlines initiated at 14.00 and 16.00 LST; dotted line represents the pathline initiated at 18.00 LST.

During the evening of 4 August 1992, onshore flow at the surface was gradually replaced by nocturnal drainage flow (Fig. 5). Apparently this offshore directed flow is significantly weaker in the southern part of the modeling domain and is not able to overcome the large scale forcing from west and creates a relatively stagnant region during the night.

Transport patterns

Pathlines with starting points in Lisbon at 30 m AGL were calculated every 2 hours, to examine the transport of atmospheric pollutants within the modeling region. This set of pathlines, shown in Figure 7, represents the trajectories of pollutants emitted by traffic in Lisbon downtown. Special care should be drawn in the analysis of Figure 7, because the three-dimensional pathline which incorporates vertical motion is not represented.

Pathlines developed for Lisbon reveal the existence of three different transport regimens. From early morning to 13.00 LST pollutants are carried offshore and returned to shore 50-70 km South, with the sea breeze. Under these patterns pollutants emitted in Lisbon during the morning would reach the Sines coast, via an overwater route, between 18.00 and 22.00 LST. This observation is consistent with conclusions of a statistical comparison of ozone concentration values between two monitoring stations located in Lisbon and Sines (Coutinho e Borrego, 1991b) which showed higher and later ozone peak values in Sines. A different regimen is initiated at 14.00 and lasts till 17.00 LST. During this period polluted air masses are transported upslope and slowly to the west reaching the convergence region over the "Serra de Sintra" described earlier. Around 20.00 LST these air masses have reached aproximately 200 m and are transported by the large scale flow to south over the sea. Finally, after 17.00 LST pollutant emitted in Lisbon are carried south at ground level by the sea breeze developed at the northern coast.

CONCLUSIONS

The non-hydrostatic mesoscale model MEMO was applied to simulate the mesoscale flow in the greater Lisbon area for typical summer synoptic conditions. Simularion results for the 4th of August 1992 show several characteristics of the wind field at this site:

→ sea breeze development in the southern Portuguese coast is affected by a northeast opposing flow generated by a strong but local pressure gradient. The extent of inland penetration of the sea breeze varies from coast to coast but is limited to 20-25 km inland. The sea-breeze layer is shallow not reaching depths over 250-300 m.

→ an interesting feature of the wind field is the devolopment of two different eddies that persists from mid to late afternoon. This type of eddies has been observed and monitored in the vicinities of Santa Barbara, California where the coastline shows similar curvatures (Douglas and Kessler, 1991).

→ pathlines of traffic emission from Lisbon indicate a morning transport southwestward to the sea followed by a southeastward transport within the sea breeze into the Sines coast. This transport mechanism could be responsible for the high ozone concentrations measured in Sines.

Meteorological data for this episode measured in several locations has been collected and will be used to validate the results in the near future.

REFERENCES

Atkinson B.W., 1981, "Meso-scale Atmospheric Circulations", Academic Press, London.

Borrego, C., Coutinho M. and Rua, J., 1991, Eulerian modeling of atmospheric dispersion over Portugal: background concentrations and emission data preparation, Air Pollution Modeling and its Applications VIII, Eds. H. van Dop and D.G. Steyn, NATO-CCMS Vol.15, Plenum Press, New York.

Coutinho e Borrego, 1991, Características dos episódios de produção fotoquímica em Portugal, 3ª Conferência Nacional de Qualidade do Ambiente, Lisbon.

Coutinho , M., Borrego, C., Rua J. and Costa M.J., 1989, Application and implementation of an atmospheric pollution interregional model to Portugal, Man and his Ecosystem, Eds. L.J. Brasser and W.C. Mulder, Elsevier, Amsterdam.

Douglas, S.G. and Kessler, R.C., 1991, Analysis of mesoscale airflow patterns in the south-central coast air basin during the SCCCAMP 1985 intensive measurement periods, *J.Appl.Meteor.* Vol.30-5, 607-631.

Ferreira, D.B., 1984, Le Systeme Climatique de l' Upwelling Ouest Iberique, Linha de Acção de Geografia Física - Relatório nº19, Centro de Estudos Geográficos - INIC, Lisbon.

Flassak, T., 1990, Ein Nicht-Hydrostatisches Mesoskaliges Modell zur Beschreibung der Dynamik der Planetaren Grenzschicht, VDI Verlag, Dusseldorf

Millan, M.M., Artinano, B., Alonso, L., Castro, M., Fernandez-Patier, R. and Goberna, J., 1992, Mesometeorological Cycles of Air Pollution in the Iberian Peninsula, Air Pollution Research Report 44, Commission of the European Comunities, Brussels.

Moussiopoulos, N., 1987, An efficient scheme to calculate radiative transfer in mesoscale models, *Environmental Software* 2, 172-191

Nadal, J.M., 1980, El Clima de Baleares, PhD Thesis, University of Barcelona, Barcelona.

DISCUSSION

J. WALMSLEY

Regarding the winter sea breeze, what mechanism is forcing it? What are typical land and sea temperatures in February and were those temperatures used in the model application?

M. COUTINHO

During winter, land breezes play an important role in the transport of air pollutants. Sea temperatures stabilize around 14-16° C. Meanwhile, during clear sky nights, land temperatures might drop to 5-7° C. During mid afternoon, a 2-3° C gradient is formed between land and sea temperature which enables the formation of a weak sea breeze, strongly catalyzed by upslope flows.

R. BORNSTEIN

Winter sea breezes have also been found in New York City and off of Lake Michigan. Do you think that Lisbon slowed the sea breeze penetration, as has been found in N.Y. C.

M. COUTINHO

Till present the land use data-base used for the simulation does not represent adequately the urban area of Lisbon. As a result of this, simulations cannot show any urban effect.

G. KALLOS

Sea breezes have been reported during winter also in the area of Athens. As we found from several winter simulations, sometimes the flow behaviour is similar to the sea breeze but it is not a real breeze because there is no strong temperature gradients between the land and the sea. In your case, sometimes the flow pattern must be due to the orography and the gaps between main orographic features or the land-water distribution.

M. COUTINHO

You are correct. Winter wind fields are strongly influenced by the orography.

PROBLEMS OF MODELLING THE LONG-RANGE TRANSPORT OF REDUCED NITROGEN

Helen M. ApSimon, Brian M. Barker, and Serpil Kayin

Imperial College Centre for Environmental Technology
48, Princes Gardens
London SW7 2PE, United Kingdom

INTRODUCTION

As negotiations on a new sulphur protocol under the UN Economic Commission for Europe reach an advanced stage, increasing attention is being paid to the contribution of nitrogen compounds to acidification. In this context European emissions of nitrogen as ammonia (estimated at almost 8 million tonnes of N over the EMEP map area- EMEP 1993), originating mainly from livestock, are comparable with those as NOx (7 million tonnes as N over the same map area). However as a reactive gas emitted at or near ground level from a distribution of sources, there are special problems in modelling the subsequent dispersion and atmospheric transport out to long distances. There are also large uncertainties about the nature and magnitude of emissions, and the interaction of ammonia with other atmospheric species. This paper illustrates some of these difficulties by application of a Lagrangian model TERN (Transport over Europe of Reduced Nitorgen) to selected episodes.

EMISSIONS OF AMMONIA

Ammonia emissions come primarily from agricultural sources, particularly from livestock and to a lesser extent from fertilizers and arable crops; although there is some evidence for potentially significant emissions from a combination of humans, pets, sewage, traffic and combustion, and miscellaneous other small sources (ApSimon et al 1987, Kruse et al 1989). The livestock emissions originate from grazing on pastures, animal housing and storage of wastes, and from subsequent disposal and spreading of wastes. There are large uncertainties in such emissions. For example emissions from grazing depend on such factors as meteorological conditions, the state and characteristics of the soil and vegetation, and farming practice and fertilizer applications, as well as the numbers of animals of each type. As emission takes place there is competition between mixing upwards by turbulent diffusion and local redeposition. Measurements of emissions over pastures are often derived from concentration profiles, which are frequently fluctuating, sometimes giving a net downwards flux as well as an emission. Such measurements really represent a net flux of emissions diffusing upwards and deposition back down to the surface; they may also be influenced by chemical reactions. However models generally treat emission and dry deposition as two different processes, and hence it is necessary to be careful about what the estimated emissions represent.

Emission from animal housing, storage of manures and slurries, and spreading of wastes are subject to different but equally large variability. Moreover, although there are possible ways of reducing emissions of ammonia, care has to be taken that these do not have other adverse environmental effects. In addition emissions are subject to large seasonal and diurnal variations, so that for example emission is greater during the day when vertical mixing tends to be greater.

In the calculations below emission have been based on annual emissions as used within the EMEP programme.

The TERN model for atmospheric transport of sulphur and nitrogen species

For long-term average estimates of transboundary fluxes over several years, relatively simple Lagrangian models are generally used, such as the EMEP model. However, the rapid increase in computing capacity has led to the development of complex Eulerian models; such models include 3-dimensional windfields and chemical reactions involving a large number of species, but are still limited by such factors as the grid resolution and assumptions about mixing. Moreover, the treatment of ammonia as a reactive gas released at ground level requires careful treatment of vertical mixing and concentration profiles within the lowest few hundred metres.

Thus in the TERN model we have incorporated sub-models of the boundary layer, with schemes to derive mixing layer depth and diffusivity profiles according to meterological conditions and underlying surface. The specification of vertical spacing is flexible, usually with thin layers of the order of 1 metre thick near the ground, and thicker layers up to 100 metres deep aloft. The diffusion equation is integrated with a straightforward 4th-order Runge-Kutta treatment, allowing flexibility in the boundary conditions to treat different assumptions about emissions and deposition. We have also allowed for temporal as well as geographical variations in emissions, and investigation of compensation points (see below) as a controlling factor for NH3 emissions.

The chemistry has been kept simple allowing direct comparison with the EMEP model, and simple investigation of diurnally variable factors, such as oxidation rates of SO2 to H2SO4; it includes SOx ,NOx and NHx plus derivative species such as PAN, but does not attempt to represent detailed hydrocarbon chemistry and photchemistry. There is an option to include cloud chemistry allowing for SOx, NOx and NHx species plus simple specifications of O3, H2O2 and HCHO according to season; cloud water profiles and other parametrisations have been taken from our detailed storm model, DROPS. The uptake of gaseous species is determined by equilibria in accordance with Henry's law, allowing for pH and temperature-dependent oxidation rates by O3 and H2O2. Trajectories and meteorological data along them have been taken from our meteorological databases spanning Europe through 1982 and 1983, but are specified as a formatted input file which can also be taken from standard forecasting model output.

The model has been described in more detail in earlier work, one paper applying it to episodes of high aerosol concentrations and poor air quality in easterly air streams arriving in the UK (ApSimon et al - in press); and the cloud chemistry in Kayin (1993) on the potential influence of NH3 on cloud chemistry and deposition of sulphur in convective showers.

Results and interpretation

The TERN model has been applied to a wide variety of episodes giving high wet deposition of NH4. These can be separated into different classes- those giving high aerosol concentrations and/or high SO4 and NH4 concentrations in subsequent rain, those with high rainfall, and those with a combination of concentration and rainfall which still yield an event with high wet deposition of NH4. The different types of episode tend to be correlated with meteorological situations, and illustrative results are given below accordingly.

Application of TERN to episodes with high aerosol concentrations arriving in the UK in anticyclonic conditions from continental Europe

These episodes included trajectories crossing high SO2 emissions in central Europe and subsequent transport over high ammonia emitting regions such as the Benelux countries. They yield high burdens of aeorosols resulting from photochemical oxidation in sunny conditions, whose acidity is only partly neutralised by NH3, and can subsequently provide high deposition in convective showers.

Studies of such episodes illustrate the effects of diurnal variations in emissions, turbulence and chemistry. The coincidence of higher NH3 emissions during the day when they can disperse upwards away from the surface more readily, and the enhanced photochemical production of H2SO4 and HNO3 which react with NH3 to form the less readily deposited NH4 aerosol from, combine to increase the long-range transport of NHx: diurnal variations in dry deposition velocity may also affect budgets of species transported, reducing deposition at night and hence helping to maintain higher air concentrations of NH3 (unless there is dew formation or surface wetting). The importance of vertical diffusion, especially near the surface, has been illustrated by comparison with simpler approaches taken in other models such as the EMEP model. It was also shown how control of SO2 emissions could reduce the proportion of NHx transported over longer distances.

Figure 1 illustrates comparison of calculated profiles of the major species in the TERN model with characteristic profiles observed at the Cabauw mast in the Netherlands. The general characteristics are reproduced very well. However the ammonia emissions in this region are very high and it would be interesting to have equivalent observations elsewhere for similar comparisons.

An interesting aspect of this was the sensitivity to the reaction rates of NH3 with H2SO4 and HNO3, which we concluded would depend not only on the chemical reaction rates but on the inhomogeneous mixing of rural air and urban and power station plumes. This is a sub-grid scale phenomenon given little attention in most modelling studies. The inhomogenoeus mixing also affects the equilibrium between NH4NO3 aerosol and the gaseous NH3 and HNO3, and could be a factor in reconciling observations in conflict with theoretical prescriptions for such equilibrium. The modelled profiles of NH4NO3 can be quite complex as a result of temperature profiles affecting the equilibrium constant and favouring formation at higher altitudes, whereas concentration profiles of ammonia have higher values near the ground.

As indicated above one matter of concern is that for many types of agricultural area emissions have been determined by flux measurements, which really represent a net effect of both emissions and local dry deposition. Yet modelling studies generally include a full treatment of dry deposition, which may hence be allowed for twice over in emitting areas. Over some areas, particularly crops, a compensation point mechanism has been proposed whereby an area may either emit or become a net sink at different times according to a balance between concentrations in the air and the stomata. It has been shown with TERN that using compensation points to drive emissions can produce similar overall emission to traditional inventories, although the local re-deposition can be sensitive to local mixes of land-use when emitting areas are intermingled with areas which readily take up NH3. This raises further very complex sub-grid scale problems, which could be explored by smaller scale studies.

The TERN model was also used to estimate concentrations in precipitation based on a simple wash-out model. These were compared with more detailed modelling for individual convective showers based on the DROPS model (Kayin 1993).

Cloudy trajectories and wet deposition episodes
The dry anticyclonic episodes considered above are relatively simple in that they did not involve transport in cloud until convective clouds formed yielding precipitation at the end of the trajectory. Other episodes considered were quite different in nature with significant influence of cloud chemistry and precipitation en route. All were treated using TERN with and without inclusion of cloud chemistry and precipitation, and with straightforward wash-out modelling of wet deposition; rainfall was extracted from our 3 hourly meteorological database for Europe to coorespond to regional conditions along the air mass trajectory (that is patchy precipitation is averaged out over grid-squares approximately 150 by 150 km). The wash-out coefficients used in TERN were consistent with those in the EMEP model, and give a linear relationship between air concentrations and wet deposition for each species. Model results were compared with monitoring data on precipitation at the end point of the trajectories.

At the start of trajectories before passing over emitting areas, allowance was made for initial background concentrations, again based on data from the EMEP model. To examine the contribution of this background to the concentrations in precipitation, runs were also performed omitting this contribution. This clearly showed for example that wet deposition of NH4 in eastern England has a large contribution from strong westerly winds, and that background concentrations from the Atlantic are still responsible for much of this even after passing across the UK emissions to reach eastern England.

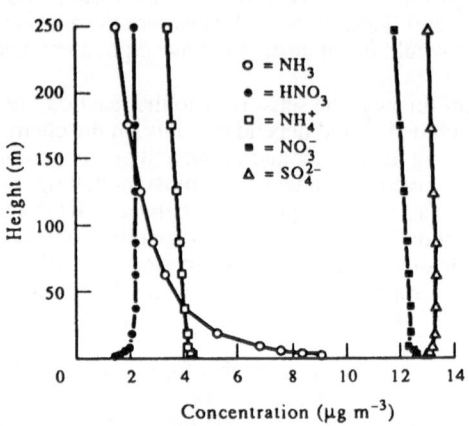

a) as calculated for case 11 in table 2

Vertical distribution of gases and aerosols

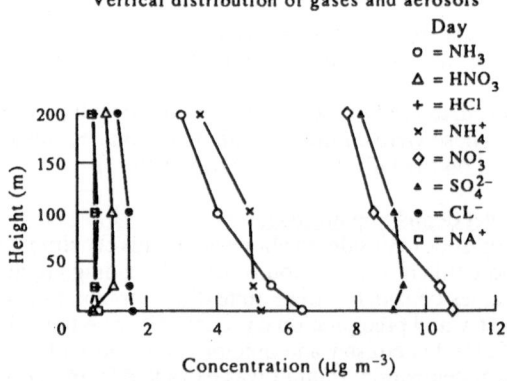

b) reproduced from measurements on the Cabauw mast

Figure 1   Concentration profiles over the Netherlands (taken from ApSimon *et al* 1994)

It was apparent that, in many cases examined, a large proportion of the material had often already been scavenged from the air mass before arrival at the observation site. As NH4 concentrations were frequently underpredicted, especially in strong westerly winds, we wished to investigate whether the overall emissions were sufficient, and the underestimates of wet deposition due to too much depletion in the model en route. To test this we also undertook calculations omitting the rain along the trajectory until the end destination point, thus giving an upper limit on the rainwater concentrations there. This could be reasonably realistic with showery precipitation where, although some of the air mass is drawn through the shower columns, the remaining air has not been cleansed.

The calculations were undertaken with different amounts of H2O2 available to investigate the effects on oxidation and wet deposition of SO2 (ie cases were analysed assuming a) no H2O2, (b) a low initial H2O2 concentration of 0.1ppb in the air mass and low production rate equivalent to 0.1 ppb per daytime period, and (c) a higher initial H2O2 of 1ppb with a production rate of 1ppb per daytime period). The H2O2 is readily taken up in clouds with rapid oxidation of SO2 and lowering of pH.

Illustrative results are given below for different types of episode, taking just one example of each from those analysed.

(i) episodes with southerly trajectories giving high precipitation on arrival at Stoke Ferry in eastern UK

An example of this type of episode is 31 May with 21.4 mm of rain of pH 4.5 yielding the highest daily wet deposition of both SO4 and NH4 at Stoke Ferry in 1983. Table 1a shows the observed concentrations of SO4 (sea salt corrected),NO3 and NH4 compared with calculations, and figures in brackets correspond to values derived assuming no wet deposition along the trajectory until the final rain (clearly an overestimate here). Background concentrations play only a small role in this trajectory passing over France.

The wet deposition of SO4 has a big contribution from quite high SO2 concentrations after passing over southern England and the London area, and judging by subsequent calculations of wet removal in convective storms described below, would be overestimated with the wash-out rate assumed. NO3 may also be overestimated as the trajectory picked up NOx emissions over major traffic areas at night when they would probably have been lower than assumed. However such high rainfall suggest a vigorous and deep storm circulation which would remove a greater proportion of the water content in the air drawn-through it, giving greater dilution of pollutants than assumed in the modelling. This would reduce the concentrations of all species, including the NH4.

The cases with different availability of H2O2 illustrates the competition between different oxidation mechanisms very well. With low H2O2 in-cloud oxidation by O3 contributes 3 times as much SO2 oxidation in clouds as H2O2; whereas with high H2O2, the H2O2 oxidises 5 times as much as the O3; with greater in-cloud oxidation the dry oxidation of SO2 is also lower.

Other episodes of this type produced comparable agreement and conclusions.

Table 1a. 31 May

(mg/l)	SO4	NO3	NH4
Observed	1.75	0.71	1.16
Calculated			
a)no cloud chemistry	3.80 (9.03)	2.01 (3.56)	1.08 (2.64)
b)with cloud chemistry but no H2O2	3.84 (9.63)	2.06 (4.36)	1.16 (3.35)
c)with cloud chemistry and high H2O2	4.54 (10.54)	2.04 (4.34)	1.16 (3.55)

(ii) trajectories drawn into frontal systems

Trajectories drawn into frontal systems can give very bad results, illustrating the limitations of using 2-dimensional trajectories in such analysis. An example is that of 25 February when just 2.4mm of rain of pH 4.7 fell at Stoke Ferry, giving the highest observed concentration of NH4 in precipitation in 1983 there, and the second highest concentration of SO4. Results are given in table 1b, this time with in-cloud oxidation based on the low availability of H2O2 felt to be more likely in such winter conditions.

Table 1b. 25 Feb.

(mg/l)	SO4	NO3	NH4
Observed	5.59	1.78	4.94
Calculated			
a)no cloud chemistry	0.78 (3.62)	0.02 (1.68)	0.08 (2.51)
b)with cloud chemistry			
and low H2O2	0.87 (4.90)	0.03 (1.66)	0.08 (2.46)

Evidently the wash-out along the trajectory is vastly overestimated, removing all the NO3 and NHx. Even excluding any rain along the trajectory concentrations of NH4 are still only half those observed. However in a frontal system the concentrations in rain are likely to be generated by air carried aloft in the frontal conveyor belts, and of quite different origin from that indicated by surface back trajectories.

(iii) Strong westerlies
However the most puzzling trajectories were those with strong westerlies bringing air in from the Atlantic across the UK. These accounted for 40% of the 20 episodes giving the highest wet deposition at Stoke Ferry in 1983. Since they passed rapidly over the UK, with little time to pick up large emissions, the background concentrations assumed over the Atlantic make a significant contribution. But the concentrations of NH4 in precipitation were consistently underestimated, even ignoring precipitation en route, although there seemed to be sufficient SO4. The episode on 14 March generating 5.9mm of rain of pH 5.5 after crossing the British Isles in 18 hours, is a good example- see table 1c. Another aspect of these trajectories is that the rain is of high pH, and hence the O3 tends to dominate the in-cloud oxidation even with high availability of H2O2, and a relatively large proportion of the SO2 emissions is oxidised to SO4 despite the short time available, increasing the removal efficiency in rain.

Table 1c. 14 March

(mg/l)	SO4	NO3	NH4
Observed	4.16	0.66	3.61
Calculated			
a)no cloud chemistry	0.94 (3.80)	~0.0 (0.73)	0.09 (1.78)
b)with cloud chemistry			
but no H2O2	0.93 (4.09)	~0.0 (0.76)	0.09 (1.82)
c)with cloud chemistry			
and high H2O2	0.94 (4.16)	1.03 (0.81)	0.09 (1.83)

Again this scenario suggests that the air scavenged has avoided previous rain along the trajectory. Even then the NH4 concentrations are too low, and in other similar scenarios the underprediction is worse. Various reasons have been considered for this. There could be a higher NH4 aerosol background at the start of the trajectory, possibly carried in elevated layers over the Atlantic. It could be that ammonia emissions across the UK are higher than those assumed, but if so this is consistent for all the westerlies analysed and not just for an isolated case. The wet removal efficiency of NHx could be underestimated, but since nearly all the wet NHx deposition comes from NH4 aerosol and very little from NH3, this would mean a higher wet removal for the NH4 than for the associated SO4 and NO3 in aerosols. The other possibility is that there are some unidentified local sources of NH3 to the west of the Stoke Ferry site. However this phenomenon also seems to occur at other sites too.

The effect of ammonia on wet deposition of SO2
The analysis of wet deposition episodes showed a tendency to overestimate the wet deposition of SO4 in air masses with higher concentrations of SO2. This is consistent with a non-linear removal efficiency of SO2 dependent on cloud chemistry. Accordingly this was investigated by a PhD student, Serpil Kayin, who coupled the cloud chemistry model in

Table 2.

Concentrations of the species used in washout ratio calculations.

Case No	SO_2 ($\mu g.m^{-3}$)	NH_3 ($\mu g.m^{-3}$)	SO_4 ($\mu g.m^{-3}$)	$(NH_4)_2SO_4$ ($\mu g.m^{-3}$)	HNO_3 ($\mu g.m^{-3}$)	O_3 (ppb)	H_2O_2 (ppb)
1	2	0	1	1	1	30	1
2	2	1	1	1	1	30	1
3	10	1	1	1	1	30	1
4	10	2	1	1	1	30	1
5	20	0	1	1	1	30	1
6	20	1	1	1	1	30	1
7	20	2	1	1	1	30	1
8	50	1	1	1	1	30	1
9	50	2	1	1	1	30	1
10	100	1	1	1	1	30	1
11	100	3	1	1	1	30	1

Variation in washout ratios in storm column simulations.

Case No	Base Case	Lower U	Increased R	Higher CCN
1	4.0×10^5	5.2×10^5	3.6×10^5	8.6×10^5
2	4.7×10^5	6.0×10^5	4.2×10^5	9.8×10^5
3	1.6×10^5	2.0×10^5	1.5×10^5	3.3×10^5
4	2.1×10^5	2.6×10^5	1.9×10^5	4.4×10^5
5	7.6×10^4	9.4×10^4	7.0×10^4	1.6×10^5
6	8.7×10^4	1.1×10^5	8.0×10^4	1.8×10^5
7	1.1×10^5	1.4×10^5	1.0×10^5	2.3×10^5
8	4.0×10^4	4.9×10^4	3.8×10^4	8.4×10^4
9	4.9×10^4	6.1×10^4	4.6×10^4	1.0×10^5
10	2.3×10^4	2.8×10^4	2.2×10^4	4.8×10^4
11	3.2×10^4	4.0×10^4	3.0×10^4	6.8×10^4

TERN with the DROPS model of the life cycle of a convective shower (Kayin 1993). Air parcels were followed through the developing storm column and the fate of the pollutants traced to determine the amounts deposited, and the amounts dispersed into the free troposphere. Concentration of different species in the air entering the storm from the mixing layer were compared with the contributions they subsequently made to concentrations in rain to determine wash-out ratios. Some assumptions had to be made about chemical reactions in the ice phase, and it was decided to assume negligible reactions therein; this was less important in the summer storms considered where most of the chemistry had already taken place at lower levels in the storm updraught.

Table 2 shows a selection of the results, showing how even a modest amount of NH3 can influence the wash-out ratio by $\pm$ 50%; the range of NH3 concentrations assumed represent average levels over the depth of the mixing layer rather than at the ground. The 11 different cases have a wide range of SO2 concentrations corresponding to very clean to polluted air, as indicated with the different concentrations of accompanying NH3 in the air entering the storm, in the upper part of the table. It is assumed that this air will be drawn into the base of the storm from the whole mixing layer, and that the same H2O2 levels pertain in air subsequently entrained into the head of the rising storm updraught (they may well be higher than this). The effective wash-out ratios relating the concentrations of sulphur in precipitation to that in the ingoing air are given in the lower part of the table for a standard case of a shower giving about 5mm of precipitation, together with some variations to give a weaker or stronger storm (lower maximum updraught velocity U, or increased radius R, respectively), or alter the internal physics of the storm by changing the CCN number density. Other results have been derived assuming different oxidant concentrations, and illustrating the dependence on the amount of H2O2 or O3.

By comparison the EMEP model (from which the wash-out rates in TERN were derived) uses a wash-out ratio of 3.10^5 for SO2 and 10^6 for SO4. But clearly there are very large variations, with efficient scavenging in clean air masses and high proportions vented to the free troposphere in polluted air. This is consistent with the limited increase in SO4 in precipitation over urban areas with higher SO2 concentrations from local sources. However, it is clear that even small amounts of NH3 can have some effect on the capacity for wet removal of SO2.

Summary and conclusions

The TERN model has been used to illustrate some of the complexities of modelling the emission and long-range transport of ammonia. Whereas some of these difficulties may be attributed to the nature of the 2-dimensional Lagrangian approach (eg in frontal systems), many of them are associated with the reactive nature of NH3 and the manner of its emission, and are not resolved by 3-dimensional Eulerian models. The large uncertainties about emissions also need to be remembered.

Acknowledgements
We are grateful to the Natural Environment Research Council for funding this work.

References
ApSimon. Kruse M and Bell JNB (1987) Ammonia emissions and their role in acid deposition. Atmospheric Environment 19(1), p 99-111

ApSimon HM, Barker B M and Kayin S Modelling studies of the atmospheric release and transport of ammonia - applications of the TERN model to an EMEP site in eastern Engand in anticyclonic episodes.Atmospheric Environment- in press.

EMEP (1993)Calculated budgets for air borne acidifying components in Europe. EMEP/MSC-W Report 1/93

Kayin S (1993) Wet deposition in convective storms and effects on transboundary air pollution. Ph D thesis University of London.

Kruse M, ApSimon H.M, and Bell J.N.B (1989) Validity and uncertainty in the calculation of an emission inventory for ammonia arising from agriculture in Great Britain. Environmental Pollution 56, p 237-257

Kruse-Plass M, Barker B M and ApSimon HM (1993) A modelling study of the effect of ammonia on in-cloud oxidation and deposition of sulphur. Atmospheric Environment 27A No2 p223-234.

DISCUSSION

P. SEIBERT: What is the behaviour of the other species in the cases of anomalously high NH_4 during westerly winds?

H. APSIMON: If allowance is made for possible overestimates of wet scavenging before reaching the receptor then estimates for other species are consistent with calculated levels. However even if no depletion by wet deposition is assumed NH_4 is still underestimated at the recepter point in the strong westerly episodes coming from Ireland aims the UK.

COUPLING THE PHOTOCHEMICAL, EULERIAN TRANSPORT AND 'BIG-LEAF' DEPOSITION MODELLING IN A THREE DIMENSIONAL MESOSCALE CONTEXT

Roberto San José, Luis Rodríguez, Magdalena Palacios and Javier Moreno

Group of Environmental Software and Modelling
Computer Science School - Technical University of Madrid
Boadilla del Monte - 28660 (Madrid, Spain)

INTRODUCTION

The purpose of this paper is to show the most important features of the air dispersion modelling focused on the photochemical, Eulerian transport and deposition modelling. The model we describe will take into account most recent advances on atmospheric photochemistry, it will account with the source and receptor oriented approaches and finally will focus on the deposition processes based on a landuse classification for accounting the canopy resistance effects.

In the last decades many efforts have been devoted to investigate the possibility of modelling the complex dispersion processes, Pielke (1984). Because of lack of computing power only in the last decade, we have been able to start to model the different parts on the pollution modelling. The coupling of the different modules is a important question and is under intensive research, (Cubasch, 1991) We will show in this paper the different modules of complete air dispersion package and we will show some details about them.

THE MESOSCALE METEOROLOGICAL MODEL

The models described above are going to be applied into a three dimensional grid.

Air Pollution Modeling and Its Application X, Edited by S-V. Gryning
and M. M. Millán, Plenum Press, New York, 1994

One of the different approaches to this problem is to use a non-hydrostatic mesoscale meteorological model to provide the adequate wind, temperature and humidity fields. This model is based on the numerical solution of the following conservation equations:

$$\frac{\partial(\rho u)}{\partial t} + \frac{\partial(\rho uu)}{\partial x'} + \frac{\partial(\rho uv)}{\partial y'} + \frac{\partial(\rho uw)}{\partial z'} = -\frac{\partial p'}{\partial x'} + R_u \tag{1}$$

$$\frac{\partial(\rho v)}{\partial t} + \frac{\partial(\rho vu)}{\partial x'} + \frac{\partial(\rho vv)}{\partial y'} + \frac{\partial(\rho vw)}{\partial z'} = -\frac{\partial p'}{\partial y'} + R_v \tag{2}$$

$$\frac{\partial(\rho w)}{\partial t} + \frac{\partial(\rho wu)}{\partial x'} + \frac{\partial(\rho wv)}{\partial y'} + \frac{\partial(\rho ww)}{\partial z'} = -\frac{\partial p'}{\partial z'} + R_w - \rho'g \tag{3}$$

$$\frac{\partial(\rho u)}{\partial x'} + \frac{\partial(\rho v)}{\partial y'} + \frac{\partial(\rho w)}{\partial z'} = 0 \tag{4}$$

$$\frac{\partial(\rho\theta)}{\partial t} + \frac{\partial(\rho u\theta)}{\partial x'} + \frac{\partial(\rho v\theta)}{\partial y'} + \frac{\partial(\rho w\theta)}{\partial z'} = R_\theta \tag{5}$$

where, u, v and w are the components of the velocity in the Cartesian coordinates (x',y',z'), respectively. R_u, R_v and R_w include turbulent diffusion and Coriolis terms. R_θ is composed by a turbulent diffusion term and a solar and infrared radiation term (Moussiopoulos N. et al. 1991).

Because of the irregular lower boundary the governing equations are transformed from Cartesian coordinates (x',y',z') in terrain-following coordinates. For the vertical coordinate η, the transformation

$$\eta = H\left[\frac{z' - z_s(x',y')}{H - z_s(x',y')}\right] \tag{6}$$

is performed, where H and $z_s(x',y')$ are the height of the upper and lower boundary, respectively. To allow non-equidistant mesh size, for example, to have better resolution near the ground, the additional transformation $x(x')$, $y(y')$ and $z(\eta)$ is employed, with arbitrary monotonic functions $x(x')$, $y(y')$ and $z(\eta)$. For $z(\eta)$ the function

$$\eta = \frac{(1+\alpha)^{z-1/2} - 1}{\alpha}\Delta\eta_{min} \tag{7}$$

is applied (Flassak and Moussiopoulos N, 1987) where α is a parameter to be determined. $\Delta\eta_{min}$ is the minimum grid spacing in the lowermost computational grid-cell.

This model have three different main modules: the solar radiation module, which describes the different light intensities for different wave lengths, the numerical module, which basically is the solution of the Helmholtz equation by using finite differences or Fast Fourier Transformations and the Physical Parameterizations which are based on the Similarity Theory. The solar radiation model is focused on the evaluation of the averaged radiation flux divergence. This flux is usually splitted into the reflected, transmitted, emitted and diffusive parts. The numerical part is solved by using the Neuman and Dirichlet boundary conditions discretized on a staggered grid. In addition, the three dimensional problem is usually converted into a X x Y one-dimensional problems. The non-hydrostatic option allows to simulate the pressure field taking into account the inhomogeneities of the terrain. So that, this option is ideal for complex terrain applications with a small cost measuring network because with only one vertical sounding we can perform a 2-3 days simulation without significant problems. However, the computing requirements are high because the resolution should be high for quality predictions of the wind, temperature and humidity fields, (Flassak, 1990).

THE PHOTOCHEMICAL AND TRANSPORT MODEL

The Eulerian grid dispersion models are given by a linear advection-diffusion equation using K-theory which describe the dispersion of a single pollutant in the spatial domain. We can express the transport equation as:

$$\underbrace{\frac{\partial c_i}{\partial t}}_{(A)} + \underbrace{\frac{\partial \left(u_j c_i\right)}{\partial x_j}}_{(B)} = \underbrace{\frac{\partial}{\partial x}\left[K_{jj}\frac{\partial c_i}{\partial x_j}\right]}_{(C)} + \underbrace{R_i}_{(D)} + \underbrace{E_i}_{(E)} + \underbrace{PH_i}_{F} \qquad (8)$$

$$i = 1,...Number\ of\ species$$

In case of cloud, rain and snow phases:

$$\frac{\partial \left(s_k c_{ik}\right)}{\partial t} + \frac{\partial}{\partial x_j}\left(u_j - w_{sk}\right)s_k c_{ik} = \frac{\partial}{\partial x_j}\left(K_{jjk}c_{ik}\frac{\partial s_k}{\partial x_j}\right) + R_{ik} + PH_{ik} \qquad (9)$$

$$i = 1,...Number\ of\ species$$

where, c_i denotes the gas phase concentrations, c_{ik} denotes the liquid phase concentrations, u_j are the velocity components, x_j is the spatial coordinates, s_k is the liquid water content, w_{sk} are the settling velocities for the water droplets, k represents the values 1, 2 and 3 which correspond with the cloud, rain and snow, K_{jj} is the eddy diffusivity tensor, R_i is the rate of the chemical reaction, E_i is the rate of emissions and PH_i is the rate of mass transfer, (Moussiopoulos N. 1990). The term A represents the unsteady accumulation of mass, B the effect of the advection fluxes on the mass, C the effect of the diffusive fluxes on the mass, D the chemical reaction production/destruction term, E the source term and F the effect of phase change on the mass. The photochemical cycle is usually addressed as the NO, NO_2 and the O_3 and the conversion rate constants, k_1, k_2, k_3 and k_4 are playing a key role.

THE DEPOSITION MODEL

Modelling the dry and wet deposition processes is taking much attention from the scientific community in the last years because (1) it affects to the lower boundary condition of the numerical transport model and (2) it is strongly related with the biological and biochemical effects over the terrain. The wet deposition processes have been modelled by using appropriate scavenging coefficients Λ which varies with the rainfall type and rate, saturation conditions, and contaminant characteristics. (Slinn 1976, Hanna et al. 1982) The wet deposition is obtained as,

$$W_{ijk}(x,z) = g\,t \int_0^z \Lambda c_{ijk}(x,z)\,dz \qquad (10)$$

where, g is the acceleration of the gravity, t is the time and z is the depth of the wetted plume layer.

The dry deposition processes have a different approach. The 'Big-Leaf' method is widely used. This techniques assumes that the fluxes are constant along the surface boundary layer which is a common assumption for the Theory of Similarity when applied to the Atmospheric Boundary Layer, however, it heads some problems when dealing with highly reactive gases. The method is based on the well known "resistance law" which divides the biosphere/atmosphere layer into the Canopy, Bulk and Aerodynamic layers which are parameterized by using very different background information. So that, the aerodynamic resistance layer is found just over the terrain or cover of the terrestrial surface. The Monin-Obukhov length and the friction velocity are used to obtain values of the r_a. The Bulk resistance represents the very thin interface between the solid surface (vegetation, grass, trees, etc.) and the air ambient. This thin layer is governed by conduction and thermal diffusivity processes and the friction velocity, roughness length thermal diffusivity of the air and pollutant are the control parameters. Finally, the canopy resistance takes into account the different physical, chemical and biological properties of the surface. Wesely (1989) showed a parameterization of the resistance by using seven different types of resistances and eleven landuse types. The resistances are due to stomatal, mesophyll and cuticular resistances produced by the leaves (in case of existing) and effects of lower leaves of the canopy and soil resistance at ground surface. The parameterization os these resistances account for the total incoming solar radiation, surface temperature, Henry's law constant and reactivity of the pollutant, (San José et al. 1985, 1988). The final resistance expression for the canopy resistance can be expressed as:

$$r_c = \left[\cfrac{1}{\cfrac{1}{\left[r_s \left(\frac{D_{H_2O}}{D_t} \right) + r_m \right]} + \frac{1}{r_{lu}} + \frac{1}{(r_{dc} + r_{cl})} + \frac{1}{(r_{ac} + r_{gs})}} \right] \qquad (11)$$

LAND USE CLASSIFICATION AND FIELD EXPERIMENTS

In order to accomplish with a detailed land use classification we have developed a computer software code to adapt the handmade land use classification to any other classification. The domain is 71000 m x 97000 m with the Madrid Metropolitan Area located approximately in the geometric center of this area. The initial land use map was elaborated by using fifteen different land use types associated with the most characteristic types found in the area. Afterwards, a matrix transformation was used to prepare a land use classification of seven types of terrain: water, arid, few vegetation, farm land, forest, suburban and urban which has been widely used in different mesoscale applications in Europe in the past years (Moussiopoulos et al. 1991). Figure 1 shows this handmade classification with a resolution of 250 m. It is clearly distinguished the urban areas formed by the Metropolitan Area of Madrid (clear grey type) and the different lakes of the region (darkest areas).

In addition, data from a field experiment performed in the Low Atmospheric Research Center (C.I.B.A.) in the Valladolid Area (city located at 200 km at the northwest of Madrid) in 1991 has been used to validate the Deposition Model taken into account this land use classification. Figure 2 shows the validation of the Deposition Model by using the MBR (Modified Bowen Ratio Technique) and the Big Leaf Model for different landuse types (land use type 8 correspond with Wesely (1989) classification for eleven types of terrain). This complex deposition model is implemented into the Transport Model described above.

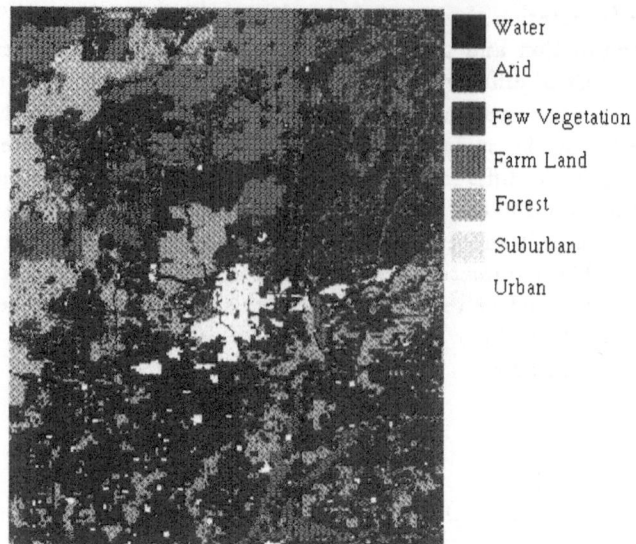

Figure 1 Land-use Classification of the Madrid urban and environ areas. Classification is handmade prepared for fifteen different types of terrain and transformed into seven types showed in the figure.

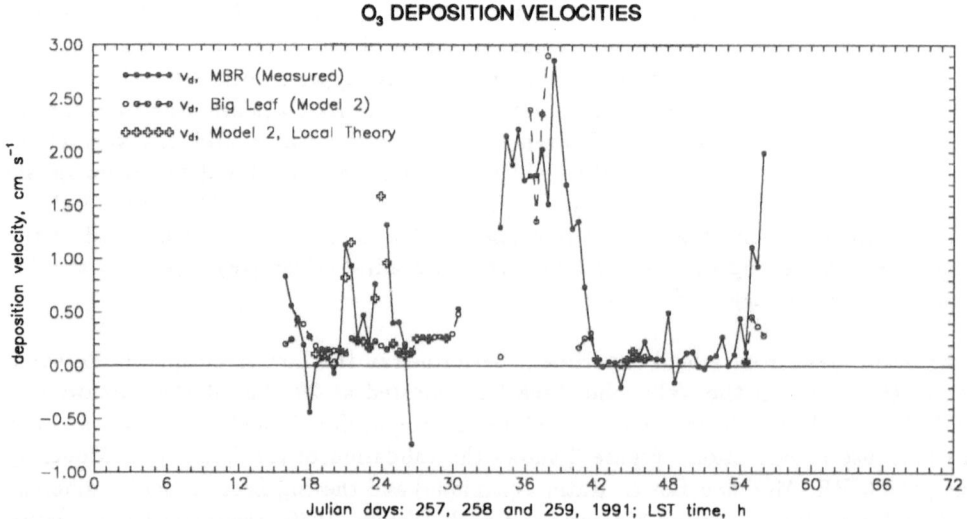

Figure 2 Deposition velocities measured during the Field Experiment carried out at the CIBA site (Valladolid, Spain) during September, 1991. The Big Leaf Model corresponds with the landuse classification due to Wesely (1989).

CONCLUSIONS

A non-hydrostatic mesoscale model, a photochemical model, an Eulerian transport model, an automatic land use classification and a deposition model have been described in this paper for the Madrid urban and environ Area. We have prepared a handmade land use classification and applied data from a Deposition Field Experiment to validate the deposition values for ozone.

ACKNOWLEDGEMENTS

The authors are grateful to Professor N. Nicolas Moussiopoulos for providing most valuable information on the non-hydrostatic mesoscale model MEMO and photochemical model MARS and the code of MEMO which has been used to make initial runs for the Madrid Area.

REFERENCES

Cubasch U. 1991. Preliminary Assessment of the Performance of a Global Coupled Atmosphere-Ocean Model. pp. 137-150. 'Greenhouse-Gas-Induced climate change: A critical appraisal of simulations and observations', M.E. Schlesinger (Ed.) ISBN: 0-444-88351-7.

Flassak T and Moussiopoulos N. 1987. An application of an efficient non-hydrostatic mesoscale model. Boundary-Layer Meteorology, 41, 135-147.

Flassak T. (1990). Ein nicht-hydrostatisches mesoskaliges Modell zur Beschreibung der Dynamik der planetaren Grenzschicht. Fortsch. Ber. VDI Reihe 15 Nr. 64, VDI Verlag GmbH. Dusseldorf 1990. ISBN: 3-18-147415-0.

Hanna S.R., Briggs G.A. and Hosker R.P. Jr. 1982. Handbook of Atmospheric Diffusion. DOE/TIC-11223, U.S. Department of Energy, Washington, D.C.

Moussiopoulos N. 1990. Air pollution levels in Athens: A test case for Environmental Software, in: 'Computer techniques in Environmental Studies III' Ed: P. Zannetti, Computational Mechanics Publications, Southampton Boston.

Moussiopoulos N., Flassak T and Kessler C. 1991. Modelling of photosmog formation in Athens, in: 'Air Pollution Modelling and it Application VOL IX'. H. van Dop and G. Kallos, ed, Plenum Publishing Corp., New York.

Pielke R.A. 1984. Mesoscale meteorological modelling. Academic Press Inc. (London).

San José R., Casanova J.L., Viloria R.E. and Casanova J. (1985). Evaluation of the turbulent parameters of the unstable surface boundary layer outside Businger's range. Atmospheric Environment, 19, 1555-1561.

San José R. and J. Casanova (1988). An empirical method to evaluate the height of the convective boundary layer by using small mast measurements. Atmospheric Research, 22, 265-273.

Slinn, W.G.N. 1976. Dry deposition and resuspension of aerosol particles - A new look at some old problems, in 'Atmosphere-Surface Exchange of Particulate and gaseous Pollutants, ERDA Symposium Series 38, Energy Research and Development Administration, pp. 1-40.

Wesely M.L. 1989. Parameterization of surface resistances to gaseous dry deposition in regional-scale numerical models. Atmospheric Environment 23, 6, 1293-1304.

DISCUSSION

G. SCHAYES   Can you explain the values of the deposition velocity as high as 2 cm/s and more, shown in your documents.

R. SAN JOSE   The canopy resistance, following Wesely parameterization, sometime can be very low and on the other hand the strong unstable conditions together with very dry situation in the central part of Spain allow very low aerodynamic resistances. This facts together can explain deposition velocities found in that part of Spain.

GEM: A LAGRANGIAN PARTICLE MODEL FOR THE DISPERSION OF PRIMARY POLLUTANTS IN URBAN CANYONS-SENSITIVITY ANALYSIS AND FIRST VALIDATION TRIALS

Guido Lanzani[1], Matteo Tamponi[2]

[1] Province of Como
   Como, IT
[2] PMIP/USSL 16
   Lecco, IT

INTRODUCTION

The objective of the work is to develop a model able to describe the microscale concentration field of primary pollutants in urban topography and particulary within street canyons. This kind of topography influences wind and turbulence fields that become very irregular (De Paul and Sheih, 1985; Nakamura and Oke, 1988). So, concentration field presents complex patterns (Hoydysh and Dabberdt, 1988).

It is useful to develop a microscale dispersion model because then it is possible to find the position of the concentration peaks. Moreover it can be used to understand the spatial representativeness of fixed points of measurement, such as those of monitoring network of air quality.

Some models for studying the dispersion within an urban canyon already exist. For example the CANYON model (Johnson et al., 1973) and the CPBM model (Yamartino and Wiegand, 1986) may be considered. The first one is a classical empirical box model; the second one is a Plume-Box model.

This paper presents a new Lagrangian particle model, in order to study the microscale dispersion of primary pollutants. The reason of this choice is that trajectory models describe the diffusion of pollutants in atmosphere in a natural way, even over a complex terrain, as in the case of an urban street canyon. Moreover, the developed model is able to work out the concentration field in any urban geometry, if the mean wind and the turbulence fields are known. On the contrary the previous microscale models are suitable just for street canyon geometries.

Air Pollution Modeling and Its Application X, Edited by S-V. Gryning
and M. M. Millán, Plenum Press, New York, 1994

STRUCTURE OF THE MODEL

The model has been developed following the approach derived by Thomson (1985). The temporal evolution of the each particle's position is described by the simple relation:

$$x_{n+1} = x_n + [U(x_n) + s_n] * \Delta t \qquad (1)$$

The variable $U(x_n)$ describes the mean velocity field and its value is a function only of the position assumed by the particle at the nth temporal step and, in a stationary model like the one developed, it does not depend on time. The fluctuation s_n is determined by the following relations:

$$s_n^i = s_n'^i - \Delta t * T^{ij} * s_n'^j + \mu_n^i \qquad (2)$$

and alternatively, at subsequent time steps:

$$s_n'^i = s_{n-1}^i + \Delta t * \left[\frac{\partial \sigma^{i1}}{\partial x^1} + \left(2 * \frac{\sigma^{i1}}{\rho} \right) * \frac{\partial \rho}{\partial x^1} \right] \qquad (3)$$

$$\chi^{ij}(x_n) * s_n'^j = \chi^{ij}(x_{n-1}) * s_{n-1}^j \qquad (4)$$

where $\underline{\mu}$ is a casual vector extracted from a normal multivaried distribution with average zero and second moments given by:

$$\overline{\mu^i \mu^j} = \Delta t * (T^{ik} \sigma^{kj} + T^{jk} \sigma^{ki}) \qquad (5)$$

for i,j = 1, 2, 3 and with the summed convention on the repeated indexes. $\sigma_{i,j}(x)$ are the components of the covariance tensor of the velocity and $\chi^{ij}(x)$ are the components of the matrix inverse of the covariance tensor at x position. $T^{ij}(x)$ are the components of the inverse of the matrix of the local Lagrangian scale times.

The components of the mean velocity perpendicular to the canyon street axis are described by means of the analytical solution, found by Hotchkiss and Harlow (1973), of the Navier-Stokes equations, linearized and relative to an incompressible fluid:

$$u = \frac{A}{K} * [e^{K*(z-H)} * (1 + K*(z-H)) - \beta * e^{-K*(z-H)} * (1 - K*(z-H))] * \sin(K*x) \quad (6)$$

$$w = -A * (z-H) * (e^{K*(z-H)} - \beta * e^{-K*(z-H)}) * \cos(K*x) \qquad (7)$$

where:

$$\beta = e^{-2*K*H} \qquad A = \frac{K*u_o}{1-\beta} \qquad K = \frac{\pi}{L} \qquad\qquad (8)$$

and L and H are respectively the width and the height of the canyon, and u_o is the wind speed measured at height H at the center of the canyon. The component of the wind velocity parallel to the street axis is described by a logarithmic profile, suggested by Yamartino and Wiegand (1986), according to the relation:

$$v(z) = v_r * \frac{\log[(z+z_o)/z_o]}{\log[(z_r+z_o)/z_o]} \qquad\qquad (9)$$

where v_r is the component of the wind velocity parallel to the street axis measured at a reference altitude z_r, and z_o represents a roughness parameter. Yamartino and Wiegand suggest the following values of z_o for the canyon street:
z_o = 0.4 m when the vortex does not develop; z_o = 0.04 m if the vortex develops; z_o= 400 m (if one of the canyon street buildings has dimensions much bigger than the others, it perturbs the wind flow above the canyon).

The covariance tensor of the velocities has been described only by the diagonal components which were described by Yamartino and Wiegand (1986) on the basis of fitting the experimental data (Leisen and Sobottka, 1980; Builtjes, 1984) as a function of the parameters influencing the structure of the atmospheric turbulence: the total solar radiation; the wind velocity at roof level; the vehicle flow and the travelling speed of the vehicle passing through; the dimensions of the canyon; the position inside the canyon.

On the contrary, with regard to our knowledge, there is no information about the tensor of the inverse of the Lagrangian times in a street canyon. Thus it was decided to define the value of the diagonal components of the tensor of the inverse of the Lagrangian times according to the following relations:

$$\frac{1}{T^{x\,x}} = \frac{k*L}{\sigma_u} \qquad \frac{1}{T^{y\,y}} = \frac{k*D}{\sigma_v} \qquad \frac{1}{T^{z\,z}} = \frac{k*H}{\sigma_w} \qquad\qquad (10)$$

where k is the von Karman constant and its value is 0.4, and D is the minimum between H and L. In this way it is supposed that the length of the turbulence eddy is comprised within the geometric size of the canyon, and the velocity characteristic of the turbulence process is the standard deviation of the wind velocity. In fact the physical meaning of the Lagrangian times is the duration of the turbulence eddy.

SOME RESULTS OF SENSITIVITY ANALYSIS

Model answers to the variation of input parameters are investigated, with respect to a standard situation. In particular, the influence of the change of emission rate, of the number of traffic lanes, of total solar radiation and of street canyon dimensions are studied. It has been found out that model answers are physically corrected with regard to an input parameter change.

The standard situation is characterized by a 180 m. long, 15 m. wide, 15 m. high canyon. The mean wind field component perpendicular to canyon axis at roof level is 2.5 m/s, while the parallel one is 0.0 m/s . The total solar radiation is 0.12 kW/m2. The source strength is described with 4 traffic lanes each one of 900 vehicles per hour, with a composite emission factor of 40 gr/(km*vehicle). In fig. 1 the predicted concentration contours of carbon monoxide can be seen for the median section of the street, normal to the canyon axis in the situation of reference of the sensitivity analysis. The concentration contours reproduce qualitatively well experimantal results. It can be noticed that leeward concentration values are greater than windward ones. Besides windward and, above all, leeward concentration values decrease with height.

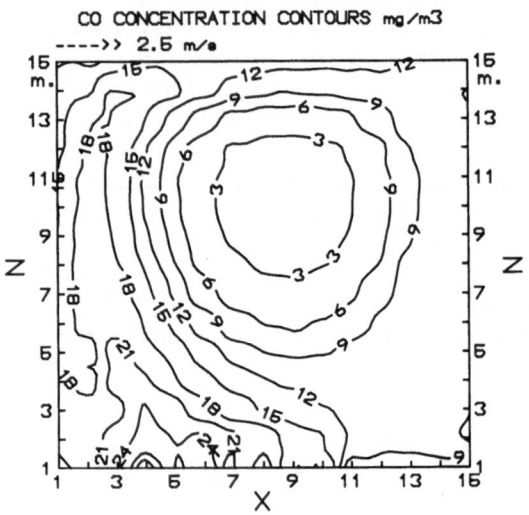

Figure 1. Predicted concentration contours of carbon monoxide for the median section normal to the street canyon axis in the situation of reference of the sensitivity analysis.

In the sensitivity analysis only one parameter is changed each time, with respect to the situation of reference. In fig. 2 the results obtained changing the number of lines, but not the total emissions, can be seen. The horizontal concentration profiles at 3 different heights from the ground with 2 and 4 lanes are shown. Only the profiles at the bottom of the canyon are different, and 2 and 4 peaks may be recognized respectively. On the contrary the profiles for 2 and 4 lanes are not very different in the center and in the top of canyon.

In figure 3 the results of the change of solar radiation are shown. Vertical concentration profiles of carbon monoxide leeward and in the middle of the street canyon are rappresented for 2 different values of total solar radiation (0.120 kW/m2 and 0.5 kW/m2) . The leeward values of 0.12 kW/m2 profile are higher than those of 0.5 kW/m2 profile, since at leeward side, if the total solar radiation is lower, the dispersion is lower and the values of concentration are higher. On the contrary in the middle of the canyon values are higher if total solar radiation is higher. In fact in the middle of the canyon concentration field is highly influenced by turbulence. So if total solar radiation is higher, turbulence is higher and values are higher.

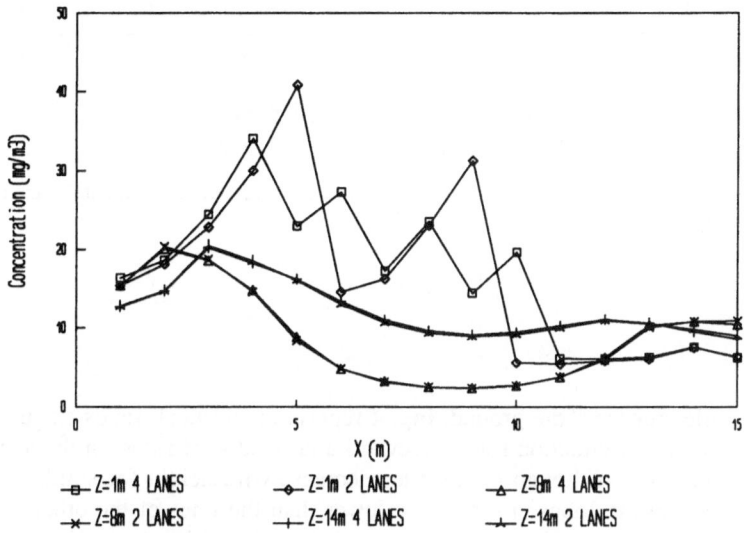

Figure 2. Horizontal concentration profiles of carbon monoxide at 3 different heights from the ground with 2 and 4 lanes

MODEL INTERCOMPARISON

The concentration profiles calculated by GEM model are compared with those calculated by other models, specifically developed for street canyons, in the situation of reference of the sensitivity analysis. The classic model of Johnson (et al., 1973), implemented in the subroutine CANYON of the code APRAC3 (Air Pollution Research Advisory Committee Version 3) (Simmon et al., 1981), and the model CPBM (Canyon Plume Box Model), more recently developed by Yamartino and Wiegand (1986), in version 3 (Yamartino et al.,1989) are considered.

The vertical leeward and horizontal trends calculated near the ground turned out to be similar.

On the contrary significant differences were observed for the vertical windward and

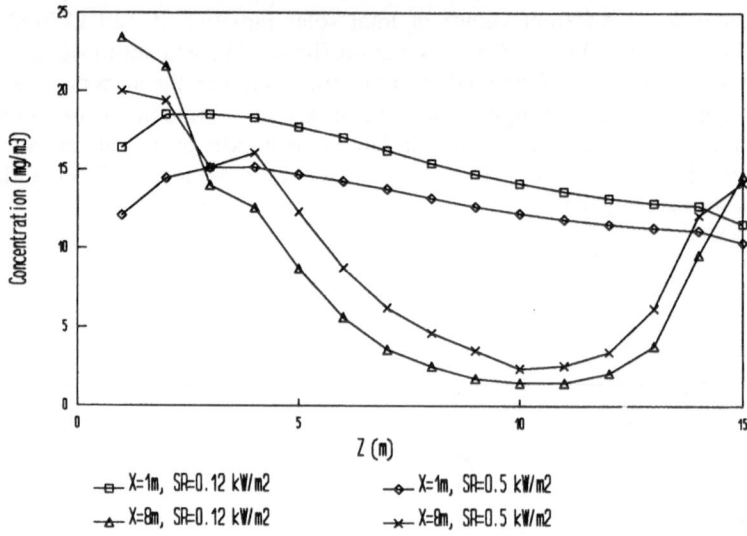

Figure 3. Vertical concentration profile of carbon monoxide leeward and in the middle of the canyon for 2 different values of total solar radiation: 0.12 kW/m2 and 0.5 kW/m2

horizontal profiles far from the ground. Fig. 4 reports the vertical trends calculated by the models in the reference situation for the leeward and windward sides. In the leeward case, it should be noted that although the profiles decrease with height for similar slopes, the values calculated by the Johnson model are lower than the ones of the other two models. While the GEM and CANYON models supply windward profiles decreasing with height, the CPBM model gives a profile increasing with height.

PRELIMINARY VALIDATION

Preliminary validation trials of the model were developed. With this aim two experimental campaigns were considered: the one carried out by P. Leisen and H. Sobottka's group (1980) in two canyon streets in Cologne, and the one made by R. Joumard's group (Joumard and al., 1980a; Joumard and al., 1980b) in a canyon street in Lyon.

Table 1 reports the characteristics of the 2 experiments and the values of the statistical indexes considered for the application of GEM to the two campaigns.

These first validation trials cannot be considered exhaustive as only a limited number of experimental data were available. In fact for none of the experimental campaigns, Lyon and Cologne, it was possible to use full data sets but only data extracted from the above cited references. The intention was to give only a first approximate evaluation of the model performances.

Therefore it would be useful to subject the model to a thorough validation procedure, based on wider and temporarily more detailed sets of experimental data.

Table 1. Parameters characterizing the validation trials.

	Cologne Experiment	Lyon Experiment
Type of experiment	Wind tunnel and 2 street canyons	1 street canyon
Period	2 years (1977-79)	May 1977
Concentration data	Normalized	mg/m3
Background Concentration	Not considered	Considered
NMSE	0.140	0.062
BIAS	-0.277	0.254

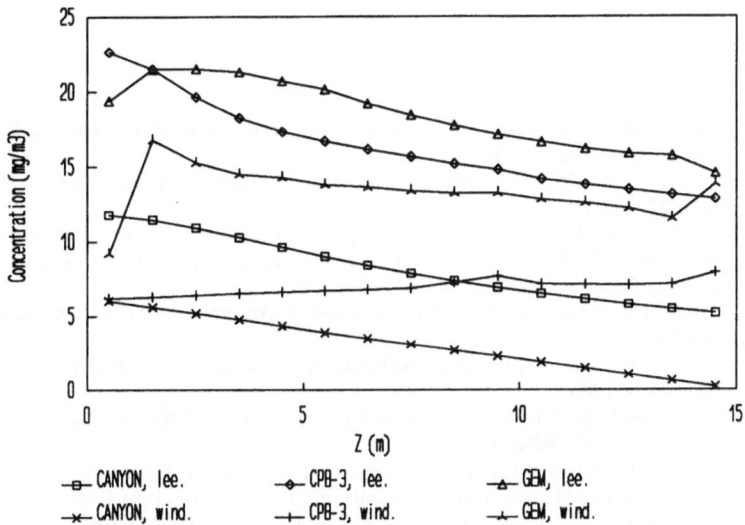

Figure 4. Leeward and windward concentration profile of carbon monoxide calculated by three model, GEM, CPBM and CANYON, in the reference situation of the sensity analysis.

CONCLUSION

It is presented a new model for the dispersion of primary pollutants within a street canyon.

The sensitivity analysis shows that the model indicates, in a physically correct way, the variation of the main input parameters. Furthermore the concentration profiles calculated by the model are quite similar to those calculated with two models designed for the specific case of the canyon street. In the few cases where GEM has a quite different behaviour from the ones of the other two models, it shows physically verosimilar profiles.

The results of the preliminary validation trials, relative to the two considered measuring campaigns, also confirm the positive performances given by the model. Hence, on the whole, the evaluation processes carried out are very encouraging and should stimulate the continuation of the work. In fact the model is applicable to generic topographic situations, quite different from the ones of the canyon street, provided the characteristics of the turbulence and wind field in the considered geometries are known.

ACKNOWLEDGEMENTS

The authors thank the Province Administration and IV U.O. of P.M.I.P.-U.S.S.L. 75/III, Milan for assistance and information. Special thanks go to Dr. Gualdi for promoting the study, Dr. Angelino and Ms. Zappa for useful suggestions, Prof. Yamartino for the model CPB-3 and to Drs. Graziani, Joumard and Pankrath for bibliographical material.

REFERENCES

Dabberdt W.F. e Hoydysh (1990) Street canyon dispersion: sensitivity to block shape and entrainment. *Atmospheric Environment* **24**, 1143-1153.

De Paul F.T. e Sheih C.M. (1986) Measurements of wind velocities in a street canyon. *Atmospheric Environment* **20**, 455-459.

Hotchkiss R.S. e Harlow F.H. (1973) Air pollution transport in street canyons. EPA-R4-73-029.

Hoydysh W.G. e Dabberdt W.F. (1988) Kinematics and dispersion characteristics of flows in asymmetric street canyons. *Atmospheric Environment* **22**, 2677-2689.

Johnson W.B., Ludwig F.L., Dabberdt W.F., Allen R.J. (1973) An urban diffusion simulation model for carbon. *JAPCA* **23**

Joumard R. e Vidon R. (1980a) Dispersion de polluants dans une rue en U. *Pollution Atmospherique*, Octobre-Decembre 1980

Joumard R e Vidon R. (1980b) Distribution de polluants dans une rue en U: 3-résultats statistiques des teneurs et trafics Rapport IRT-CERNE, Février 1980.

Leisen P. e Sobottka H. (1980) Simulation of the dispersion of vehicle exhaust gases in street canyons: comparison of wind tunnel investigations and full scale measurements. IMA Conference on Modeling of Dispersion in Transport Pollution, Southend-on-Sea, England, 17-18 March.

Nakamura Y., Oke T.R.(1988) Wind, temperature and stability conditions in an East-West oriented urban canyon. *Atmospheric Environment* **22**

Simmon P.B., Patterson R.M., Ludwing F.L. and Jones L.B. (1981) The APRAC3-Mobil emission and diffusion modelling package. EPA 909-9-81-002. 215 Fremont Street, San   Francisco, CA.

Thomson D.J.(1985) A random walk model of dispersion in turbulent flows and its application to dispersion in a valley. *Quart.J.R.Met.Soc.* **112**, 511-530.

Yamartino R.J. e Wiegand G. (1986) Development and evaluation of simple models for the flow, turbolence and pollutant concentration fields within an urban street canyon. *Atmospheric Environment* **20**, 2137-2156.

Yamartino R.J., Strimaitis D.G., Messier T.A. (1989) Modification of highway air pollution models for complex site geometries. Report No. FHWA-RD-89-112. Draft.

DISCUSSION

HAN VAN DOP

Can your model also handle the general case where the wind vector makes an arbitrary angle with the street canyon-axis? Should there be homogeneity in the canyon axis direction?

G. LANZANI

Yes, GEM model also handle the general case. The homogeneity in the canyon axis direction depends on concentration values in cross-sections and on canyon length. For examples if canyon length is not finite, concentration calculated by model are homogeneous in the canyon axis direction and the presence of a component of wind parallel to canyon axis affects only mechanical turbulence. Instead if at the begining (with respect to wind direction) of the canyon there is a cross-section with a background concentration equal to zero, the concentration calculated by model increases in the canyon axis direction. In this case homogeneity is reached only after 100 - 200 m from the cross-section.

J. KUKKONEN

Chemical transformation is very important in street canyon conditions. Does your model deal with carbon monoxide, and not, for instance, nitrogen oxides?

G. LANZANI

GEM model deals only with not reactive pollutants.

R. BORNSTEIN

Can you modify your model to simulate non-noon (i.e. early a.m. or late afternoon) conditions when only parts of the canyon are illuminated? If so, we have a canyon data set from Sacramento, CA, for you to simulate.

G. LANZANI

GEM model is able to simulate every situation if turbulence field and wind field are known. So it is possible to modify it if a new data-base is avaiable. Actually, GEM model uses a turbulence field described by Yamartino and Weigand (1986): now GEM depends on total solar radiation but not on the position of sun above canyon.

VERIFICATION OF URBAN SCALE TIME-DEPENDENT DISPERSION MODEL WITH SUBGRID ELEMENTS, IN OSLO, NORWAY

Steinar Larssen, Knut Erik Grønskei, Frederic Gram, Leif Otto Hagen and Sam-Erik Walker

Norwegian Institute for Air Research
P.O. Box 64
2001 Lillestrom, Norway

1 INTRODUCTION

Results from monitoring of air pollution concentrations in cities in Norway have shown that nitrogen dioxide (NO_2) is one of the compounds which most often, and to the largest extent, exceeds current air quality guidelines (Hagen, 1992; Larssen, 1993). This is the case both in city streets and in the urban atmosphere in general. In Norway, the highest NO_2 concentrations occur during the winter months, in connection with "episodes" with poor dispersion. In the general urban atmosphere, high 24-hour average values are of greatest concern relative to Air Quality Guideline (AQG), while in the street atmosphere, very high peak (hourly) concentrations may be the most important problem.

This paper describes the testing of an urban scale, time varying dispersion model under development at the Norwegian Institute for Air Research (NILU), to calculate, as a function of time and space, NO_x and NO_2 concentrations in a grid (length 500 m-1 km) and in receptor points within the grid by means of integrated sub grid models. Input to the model are time-varying gridded fields of emissions and dispersion parameters.

The urban scale time dependent model has been used to describe the spatial distribution of air pollution in other urbanized areas (Grønskei et al. 1993), and the model has been further developed to account for subgrid variations as a result of emissions along roads and streets in Oslo. The subgrid model is based on HIWAY-2 (Petersen, W.B. 1980). This model has been modified to account for emission and dispersion conditions in Oslo by Larssen et al. 1990.

The photochemical reaction scheme developed by Hov et al. (1993) will be used to describe photochemistry on the urban and regional scale in Norway.

The model evaluation is based on measurements of air quality and meteorological conditions in Oslo October 1991-June 1992 (Hagen et al. 1993).

The research carried out during the last 2 years, to develop and test the model with sub-grid elements, includes the following elements:

- Development of a data base for testing of the model for the city of Oslo, including emission inventorying, and continuous measurement of NO, NO_2, O_3 and meteorological and dispersion parameters at a number of locations in Oslo during winter and summer periods (described further in chapter 2).
- Model development (described further in chapter 3).
- Km-scale tracer gas dispersion experiments in various regions in Oslo, representing fully and partly built-up areas. These experiment are not further treated in this paper.
- Model testing and modification (chapter 4).

2 THE ESTABLISHMENT OF THE TEST DATA BASE

Description of the test city in Oslo

Oslo is situated at the end of the 100 km long Oslofjord (Figure 1). Surrounded by hills of height 200-500 meters, the city topography is bowl-like with valleys protuding between hills, the main valley rising from the fjord and city centre towards the northeast. The area has a continental-type climate, normally with relatively cold winters (mean winter temperature: -3.9°C) and local drainage winds from the hill valleys, dominated in the city centre by the drainage down the main northeast valley.

Oslo is largely a commercial city. There are only a few minor industrial emission sources and power plants. Car traffic is by far the dominating source of air pollution emissions, while space heating by oil (low-sulphur) and, in cold periods, wood burning, also contributes notably to the air pollution.

Emission inventory

The emission survey covers the urban part of Oslo, the eastern part of Bærum and the northern part of Nesodden, within a grid of 22x18 km^2, as shown in figure 1. As the model is using data on area sources as well as subgrid point and line sources, it is important to present the emission data on different levels for the dispersion model. The main source of nitrogen oxides in the Oslo area is road traffic, so the major work was performed on the calculation of traffic emissions.

Road traffic. Data for traffic intensity for the main roads were available for the morning and afternoon rush, plus for "low traffic" conditions. The data were provided by the traffic authorities in Oslo by using the TRIPS model. (The "low traffic" corresponds to the period at noon, not night traffic.) The average daily traffic is calculated as (2*morning+2*afternoon+14*low), assuming low traffic during 6 night hours. From these 3 data sets fields with emissions of CO, NO_x, NO_2 and VOC within each km^2 were calculated, using routines with emission as a function of the driving conditions. Table 1 shows emission data for the main roads in Oslo 1991. The total length of the main roads within the area was 458 km.

In addition to the emissions from the main roads, emissions were estimated for 954 km local roads to 1160 kg CO/h, 100 kg NO_x/h, 7 kg NO_2/h and 112 kg VOC/h. The emission of the nitrogen oxides from the small roads corresponds to the fraction of the traffic work, 10-12 %, but due to a lower speed on the local roads, the emissions of CO and VOC were about 25 % of the total.

Met.stations
A. Fornebu Airport
B. Bygdøy
C. Blindern (Norw. Met. Inst.)
D. Bjølsen
E. Nordahl Bruns street (City centre)
F. Hovin
G. Skøyen

Air quality stations
1. Skøyen   (City regional)
2. Nordahl Bruns street (City regional)
3. Pilestredet   (Street)
4. Hovin   (City regional)
5. Strømsveien   (Street)
6. Fyrstikkalleen   (50 m from street)
7. Etterstadsletta   (City regional)
8. Holmlia   (Surburban residential)

Figure 1. The Oslo area.

Table 1. Emission data for the main roads in Oslo 1991.

	Traffic work 10^3 car-km/h	CO kg/h	NO_x kg/h	NO_2 kg/h	VOC kg/h
Morning	591	11 277	1 979	154	768
Afternoon	596	12 319	2 180	150	776
Low traffic	280	4 167	949	65	364
Average	265	3 245	799	70	330

Other traffic emissions. The calculation of emissions from the harbour traffic and from Fornebu Airport was based upon detailed data about the traffic intensity, specially for the airport.

The harbour traffic is dominated by large ferries with a mean emission of 84 kg NO_x/h. The emissions from Fornebu Airport was estimated in 1989 to 39.3 kg NO_x/h as a daily mean value. The airport is closed during the night giving a mean emission of 58.9 kg NO_x/h from 07-23.

Heating. The emission from industry and the consumption of oil for heating purposes in Oslo are reduced substantially during the last decades, mainly due to low electricity prices, central heating and a fall in the industrial activity. The emission from point sources in Oslo 1991 was estimated to 93 kg NO_x/h, mainly from three incineration plants.

The emission from domestic use of oil and solid fuels was estimated to about 120 kg NO_x/h.

Table 2 shows mean hourly emissions from the Oslo-area for the winter 1991-92.

Table 2. Average hourly emissions of nitrogen oxides from traffic and heating for the winter 1991-92. Unit: kg/h as NO_2.

Traffic		Heating	
Main roads	798.7		
Local roads	99.6	Point sources	93.1
Harbour	84.0	Area oil heating	111.5
Airport	39.3	Solid fuels	8.2
Sum traffic	1 021.6	Sum heating	212.8

Measurement program, NO_x, NO_2, O_3.

A total of 10 measurement stations for nitrogen oxides and ozone were operated during the measurement period starting in October 1991 and ending in July 1992. All stations were not operated simultaneously. The operational schedule is shown in Table 1. The location of the stations within Oslo city boundaries are shown in Figure 1. In addition to these a regional background station was operated, situated in a rural setting on the east coast of the Oslofjord, some 30 km south of Oslo.

The three main stations for testing of the model are the following:

1   Skøyen. Located in a park. Distance to surrounding   City regional station
main streets: about 1 km.

2   Nordahl Bruns street. The location is inside a city   City centre station
block, in a fully built-up area, 4-8 stories buildings.
Distance to streets surrounding the block: 30-50
meters.

3   Hovin. Located in residential area with dispersed 4-   City regional station
stories apartment blocks. Distance to main roads:
250 m to the East, 500 m to the West.

3 MODEL DESCRIPTION

General description

The procedure for calculation and for specification of boundary are described by Grønskei et al. in 1993.

The computer codes are developed to include chemical reactions between different compounds, but in this presentation such reactions were not taken into account.

The vertical structure is shown in Figure 2.

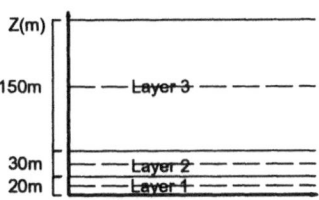

Figure 2. The vertical structure of the three-level model. Emissions from area sources are mixed in layer 1 and layer 2. During inversion situations area source emissions are mixed in layer 1.

When sodar measurements were not available in the current study, measurements from station F were used for wind direction in level three. Plume rise above layer three was not taken into account.

The initial distributions of concentrations are specified by measurements or given the homogeneous value of zero, when distributions are not available.

Subgrid model

Close to point sources and close to roads with high traffic intencity, data on subgrid gradients are needed to describe pollution concentrations at measuring stations and for estimating exposure.

A highway model, corresponding to US-EPA HIWAY (Peterson, 1980) is used to estimate concentrations close to set of roads with high traffic intensity within the square km² ($\Delta C'$). A puff trajectory-model is used to calculate the influence of point sources (ΔC^P). Local area sources are reduced accordningly.

4 MODEL EVALUATION

Figure 3 show observed and calculated NO_x-concentrations as a function of time at the stations number 1, 2, and 4 in Figure 1.

Table 3 show the model evaluation parameters for November and December 1991. The contributions from the subgrid-model are shown in Figure 3 and 4.

Table 3. Model evaluation parameters for November and December 1991.

Station	$(\overline{C})_o (\sigma_o)$ $\mu g\, NO_2 / m^3$		$\overline{C}_B(\sigma_B)$ $\mu g\, NO_2 / m^3$		r	$\overline{\Delta C^2}$ $\mu g / m^3$	$\overline{\Delta C_s^2} / \overline{\Delta C^2}$ %	Index of agreement
1. Skøyen	195	(239)	165	(199)	0.77	24 863	34	0.86
2. Nordahl Brun	155	(181)	227	(313)	0.61	67 053	8	0.69
4. Hovin	280	(391)	262	(350)	0.72	78 616	25	0.84
8. Holmlia	94	(98)	117	(120)	0.51	12 476	15	0.69

$(\overline{C})_o$	: mean value of observed concentrations
σ_o	: standard deviation of observed concentrations
$(\overline{C})_B$	: mean value of calculated concentrations
σ_B	: standard deviation of calculated concentrations
r	: correlation coefficient
$\overline{\Delta C^2}$	: mean square difference between observed and calculated concentration values
$\overline{\Delta C_s^2} / \overline{\Delta C^2}$	: the systematic part of the mean square difference.

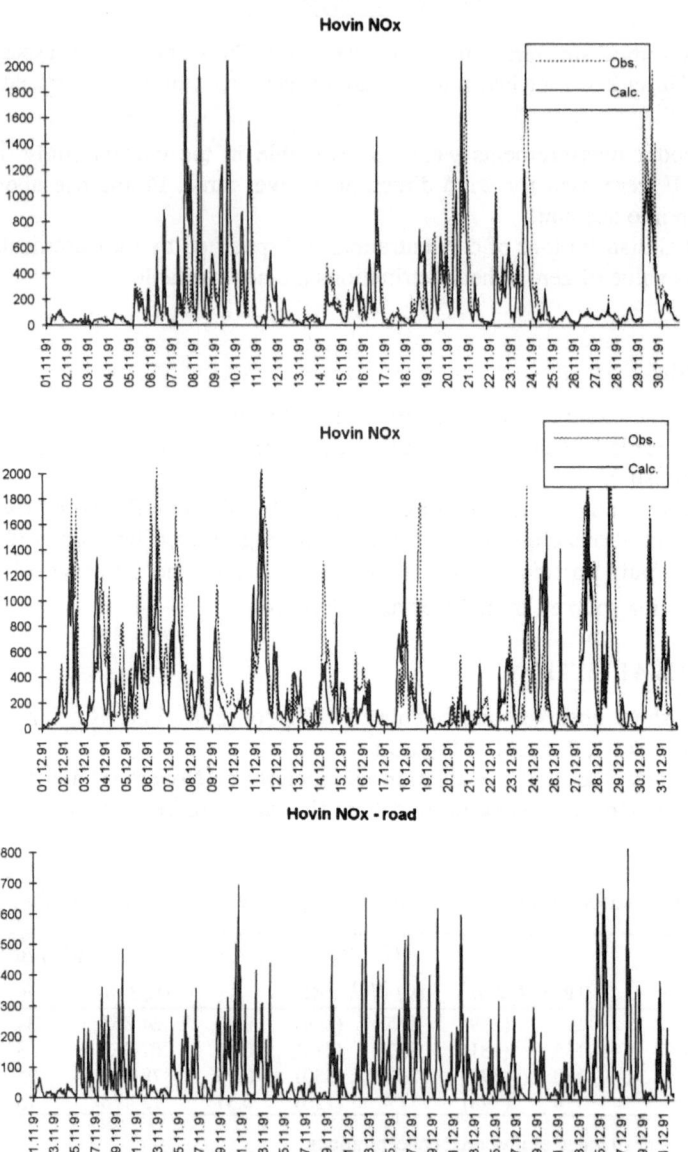

Figure 3. Observed and calculated NO_x-concentrations at station number 1 (Skøyen). The contributions from the subgrid model are marked "Skøyen NO_x-road". Unit: $\mu g/m^3$ as NO_2.

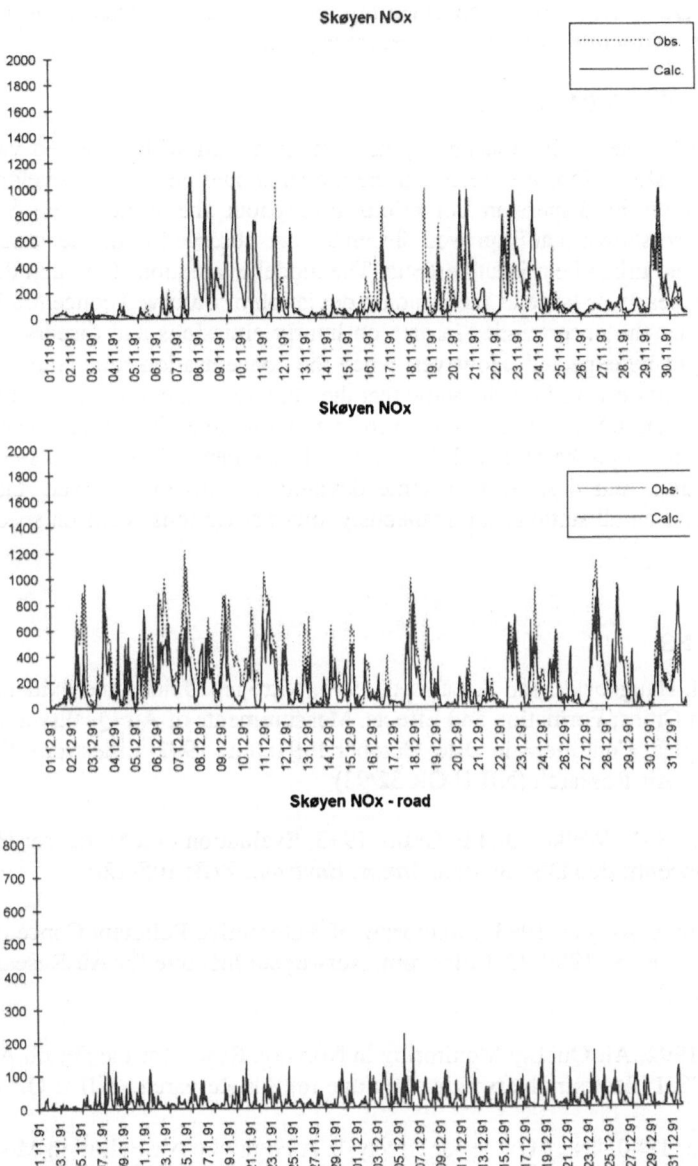

Figure 4. Observed and calculated NO$_x$-oncentrations at station number 4 (Hovin). The contributions from the subgrid model are marked "Hovin NO$_x$-road". Unit: μg/m³ as NO$_2$.

The hourly fluctuations in concentration values are simulated well at all stations and the following deviations should be mentioned.

- The maximum values in the center of the city (station number 2) are overestimated even when the subgrid concentrations are not taken into account.
- Hourly values may deviate considerably and stochastic concentration contributions have to be taken into account in all receptor points.

5 DISCUSSION OF RESULTS

Hourly calculated NO_x-concentrations compared well with observed values at four stations in the Oslo region. It is seen that the model should be further developed to better describe variation in dispersion conditions throughout the urban area. Several other studies have also shown that improved dispersion are observed in the center as a result of tall buildings and urban heat island effects. The model evaluation show that the root mean square error is nearly as large as the standard deviation of observed concentrations.

Considering the uncertainties in data on hourly emissions and dispersion conditions this result might be expected. However, when the concentration fluctuations are studied, the calculated and observed results show that the observed concentration "signals" are well simulated on all stations, and additional information on the emission and the dispersion of urban air pollution may be obtained, by statistical treatment of the time series. A detailed study of Figure 3 and 4 show that some deviations between observed and calculated values are found on all stations simultaneously, other deviations occur on specific stations only.

REFERENCES

L.O. Hagen, I. Haugsbakk, and S. Larssen, 1993, Nitrogen Oxides and oxidants in Norwegian Cities; Formation and Effects. Measurements of Air Quality and Meteorological Conditions in Oslo, October 1991-June 1992. Lillestrøm, Norwegian Institute for Air Research (NILU OR 32/93)

K.E. Grønskei, S.-E. Walker, and F. Gram, 1993, Evaluation of a Model for Hourly Spatial Concentration Distribution. *Atmos. Environ.*, 27B: 105-120

S. Larssen, and A. Røstad, 1993, Monitoring of Automotive Pollutant Concentrations in Oslo for the Period 1989-92. Lillestrøm, Norwegian Institute for Air Research,Norway (NILU OR 7/93)

L.O. Hagen, 1992, Air Quality Monitoring in Norway. Result for the Period April 1991-March 1992. Lillestrøm, Norwegian Institute for Air Research, (NILU OR 66/92).

W.B. Petersen, 1980, Users Guide for Hiway-2: A Highway Air Pollution Model. Research Triangle Park, NC., U.S. Environmental Protection Agency (EPA-600/8-80-018).

S. Larssen, D.A. Tønnesen, and M. Johnsrud, 1990, Calculation of Air Pollution Levels in Vålerenga/Gamlebyen, an Area in Oslo with a Large Traffic Burden. Lillestrøm, Norwegian Institute for Air Research (NILU OR 19/90).

A. Strand and Ø. Hov, 1993, Two diemnsional global study of the troposheric ozone production. (Submitted for publication in J. of Geophysical Res.)

DISCUSSION

N. MOUSSIOPOULOS   You seem to have an increasing underprediction in the course of November. Could this be related to the treatment of cold-start emissions (which, most probably, are important in Norway)?

K.E. GRØNSKEI   The cold-start emission is taken into account defining emission factors for car traffic. However, we will consider your suggestion as one of several reasons for discrepancies between observed and calculated concentrations.

OBSERVATION AND SIMULATION OF URBAN-TOPOGRAPHY BARRIER EFFECTS ON BOUNDARY LAYER STRUCTURE USING THE THREE-DIMENSIONAL TVM/URBMET MODEL

R. Bornstein[1], P. Thunis[2], and G. Schayes[3]

[1]Department of Meteorology
San Jose State University
San Jose, California, USA

[2]Environmental Institute
European Community Joint Research Center
Ispra, Italy

[3]Institute of Astronomy and Geophysics
Catholic University
Louvain La Neuve, Belgium

INTRODUCTION

Urban areas affect prevailing mesoscale and synoptic flow patterns due to a variety of physical processes, including urban heat island induced accelerations, surface roughness induced decelerations, and building barrier effects. Analysis of data collected over New York City (NYC) has shown that the city is capable of significantly altering the speed and/or direction of movement of thunderstorm cells, sea breeze fronts, and synoptic fronts. Analyses of the above effects, plus additional analysis (during periods without such features) of surface flow patterns, surface convergence fields, tetroon vertical velocities, and double theodolite velocity fields all point to the urban barrier effect as the most significant factor altering mesoscale flow over NYC (Bornstein and LeRoy[1]).

Results from simulations using the URBMET/TVM mesoscale model are presented to illustrate the different contributions to the observed urban barrier effect over NYC arising from the various urban and coastal influences described above.

FORMULATION

The TVM three-dimensional mesoscale vorticity-mode numerical model of Schayes and Thunis[2] originated from the PBL URBMET model of Bornstein et al[3] and is hydrostatic and Boussinesq (and hence incompressible). The model contains a soil sub-surface layer and an atmospheric layer, which is divided into two sub-layers: a constant flux surface layer in which time dependent meteorological profiles are calculated from analytical stability functions dependent only on height, and a transition layer in which the hydrodynamic and thermodynamic equations are solved numerically with finite differences. Surface temperature and moisture values are computed using the surface energy and moisture balance equations, respectively.

The transition-layer equations are derived from the exact equations of motion for a Reynolds averaged viscous Newtonian fluid in a rotating coordinate system by assuming that the atmosphere is Boussinesq and hydrostatic. The basic Reynolds averaged equation of motion in the transition layer for the URBMET were given by Bornstein et al[3]. Both the infrared radiative flux divergence term in the energy equation and the new higher order turbulence closure scheme (obtain vertical eddy diffusivities) are discussed in Schayes and Thunis[2].

The basic equations of motion are rewritten in vorticity form, so that pressure can be eliminated from the equations, as the upper boundary condition on this variable is not well posed in primitive equation meso-scale models. It is also frequently a source of instability, generating waves extraneous to the desired solution. On the other hand, the vorticity approach requires additional integrations to recover velocity components from the required stream functions.

The equations of Bornstein et al[3] were transformed in TVM to new coordinates to account for topographic influences. In this new system, only the vertical coordinate is transformed following Pielke[4]. Vertical grid spacing is thus a function of horizontal location, and only the surface level is terrain following. To obtain these equations, it was assumed that the horizontal gradient of the new vertical coordinate can be neglected versus its vertical gradient, which requires terrain slope angles much less than 45^0 (Pielke[4]).

The original URBMET model used two different turbulence closure formulations, i.e., the first order closure of O'Brien in which vertical eddy diffusivities are specified as third degree polynomials depending only on SBL characteristics and a higher order integral length scale formulation based on Mellor and Yamada[5]. TVM uses a 1.5 turbulent kinetic energy (TKE) closure. The dissipation and diffusion mixing length formulation of Therry and Lacarrere[6] is able to reproduce features normally only found with higher order models.

The current SBL formulation uses the Businger, et al[7] forced and mixed convective SBL stability functions. For the very stable case, a modified Webb[8] formulation is used. The Webb formulation slows the rate at which SBL fluxes approach zero in very stable conditions, as z/L exceeds unity.

The water vapor and carbon dioxide infra-red flux divergence term of is calculated using the Sasamori[9] scheme, rather than with the emissivities of Atwater[10] used in Bornstein et al[3]. As this process generally dominates that of the solar flux, this latter effect is currently neglected. Incoming surface insolation is calculated for a horizontal surface following the method of Schayes and Thunis[2], while surface inclination effects are included following Pielke[4].

Boundary conditions are specified at each of the six external model boundaries and at the two internal boundaries (i.e., the surface and SBL top). TVM imposes the following:
> Open lateral boundary conditions, which normally permit perturbations to cross a lateral boundary; however, the

variable grid formulation of the current application uses stretched horizontal grid spacings near lateral boundaries to move them away from the region of activity. This stretching, in combination with a complex terrain can allow new perturbations to grow in these regions. This problem is minimized by use of horizontal topography at the outer four grid points at each lateral boundary.

> The model top wind is geostrophic and its vertical temperature and humidity values match those of the synoptic scale. No constraints, in addition to a zero surface-value, are imposed on the vertical velocity component w, which is computed to satisfy continuity. Note that imposition of a zero w at the model top would be an over-specification. Complex topography in a hydrostatic PBL model also allows for formation and propagation of (almost) vertically propagating gravity waves in stable conditions. As these waves can reflect at the upper boundary, a filter is used in a damping layer (consisting of the five uppermost domain levels) that gradually attenuates rising perturbations. The filter smooths all prognostic variables (except TKE) at each time step by use of their four neighboring values. As real topography contains unresolvable features (smaller than two horizontal grid spacings), the same filter is used on observed topographic height values to prevent propagation of "two delta x" waves.

> Time and space varying surface temperature and humidity values are calculated by a soil sub-model using soil heat and moisture fluxes. While water surface-temperature is assumed constant, soil surface temperature is calculated from the prognostic "force-restore" equation of Deardorff[11], in which vegetation is accounted for indirectly via specified values of soil specific heat and density. Surface soil heat flux is obtained using a residual method with the surface energy balance equation. Soil moisture was treated by Deardorff[11] in a similar manner to soil temperature; however, the approach produces unrealistic latent heat fluxes unless a full vegetation model is simultaneously implemented. The simple Pennmann-Monteith[12] formulation is thus used. It uses a similarity theory based aerodynamic resistance from Thom and Oliver[13] and a specified constant surface resistance (that includes vegetative effects) to compute the surface latent heat flux. The URBMET code of Bornstein et al[3] used a double iterative technique to obtain both surface temperature and specific humidity from simultaneous solutions of the surface energy and moisture balance equations.

> SBL turbulent fluxes are imposed at the internal boundary at the bottom of the transition layer (i.e., at the first numerical grid point beyond h_s). Following Therry and Lacarrere[6], turbulent energy is set equal to $4u_*$ both in the SBL and at the lowest transition layer numerical grid point.

RESULTS

A series of 30 hour simulations were carried out for an urban heat island (maximum surface intensity of about 6 K) sea breeze period. A constant geostrophic wind speed was specified as 3 m/s from the NW. Surface roughness values ranged from 1 cm for water areas, 0.5 m for rural areas, 3 m for urban areas, to 4 m for six "super' urban grid areas around the tip of Manhattan. Topography was limited to 20 m for urban areas and

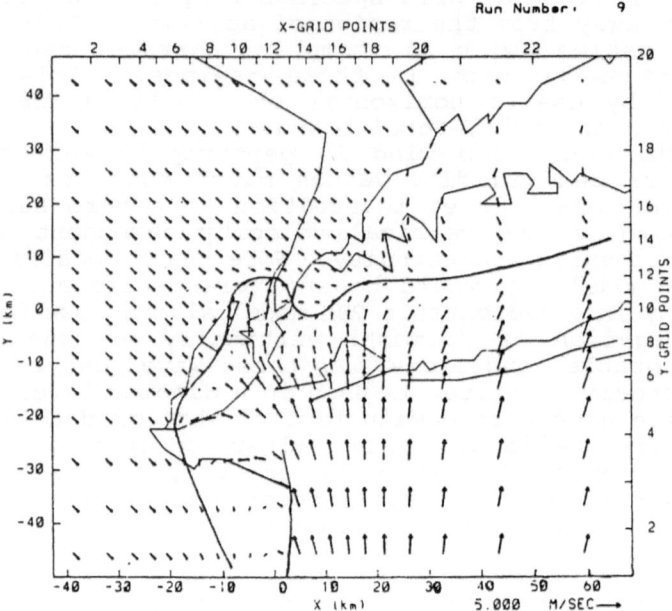

Figure 1a. Simulated 25 m level "base case" flow field over NYC at 1500 LST.

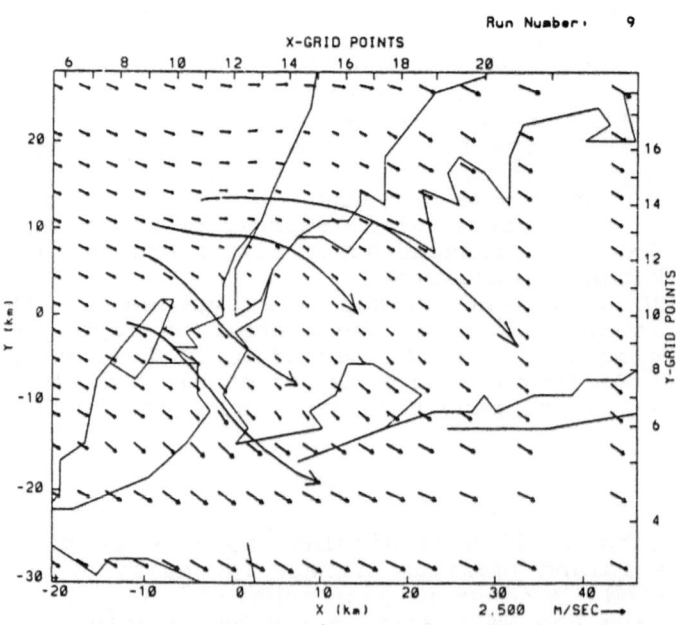

Figure 1b. Simulated 25 m level "base case" flow field over NYC at 0600 LST of the second simulation day.

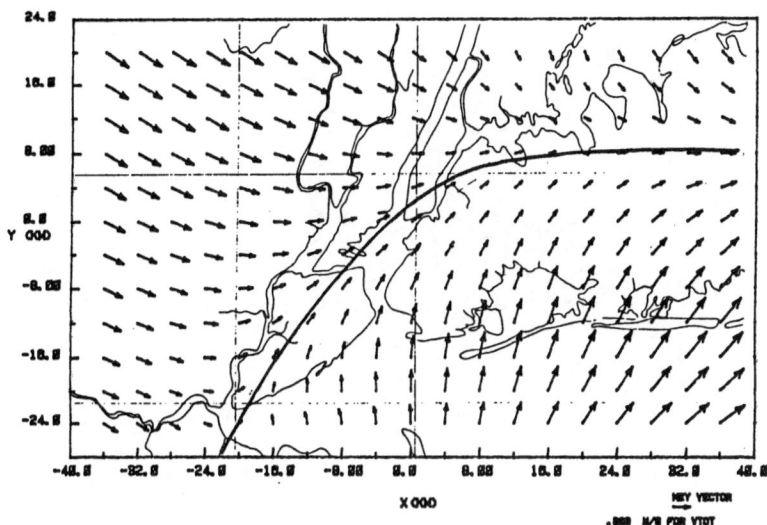

Figure 2. Simulated 37.5 m level flow field over NYC at 0600 LST of the second simulation day for case with no urban topography.

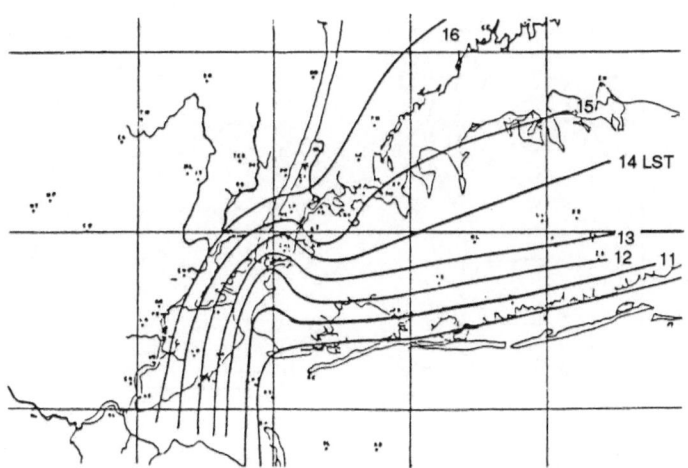

Figure 3. Observed sea breeze frontal isochrones, where 16 is 1600 LST.

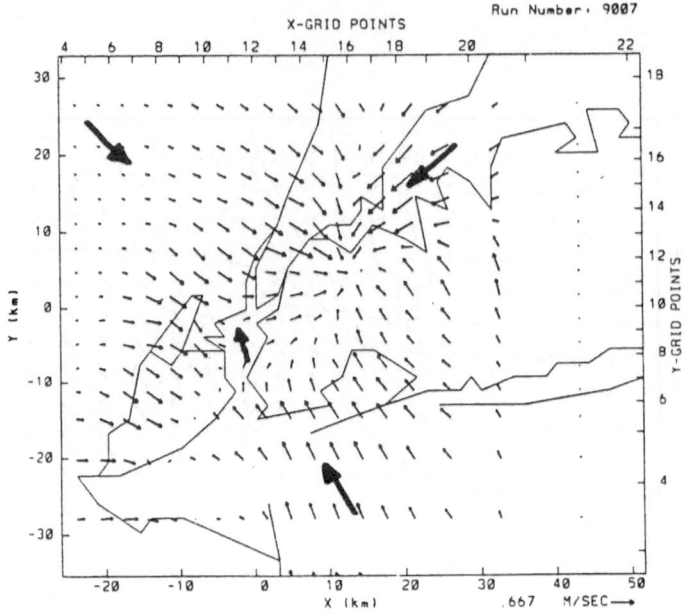

Figure 4a. Simulated 25 m level "difference" flow field over NYC at 1200 LST, with corresponding "base" case flow vectors shown.

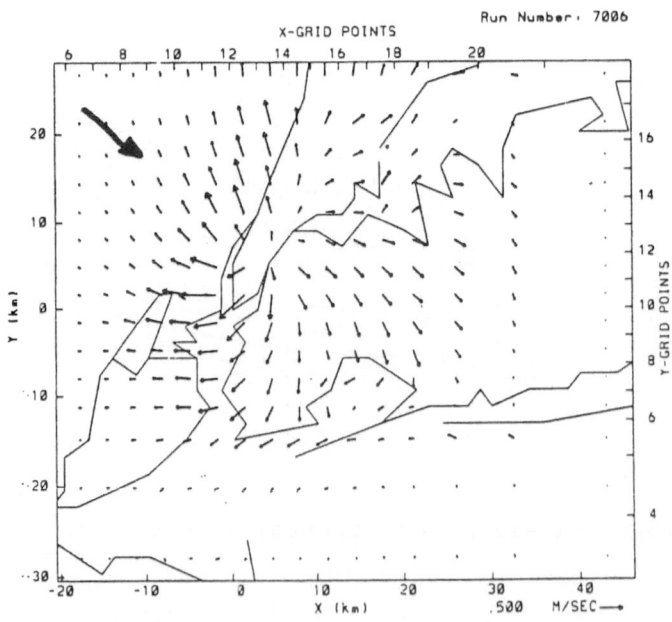

Figure 4b. Simulated 25 m level "difference" flow field over NYC at 0300 LST of second simulation day, with corresponding "base" case flow vectors shown.

70 m for the super urban areas. The grid was 24 by 24 in the horizontal (Figure 1a) and 16 levels to 2 km in the vertical.

The horizontal distribution of daytime horizontal wind velocity at the 25 m level after 18 hours of the "base" simulation (including urban topographic, roughness, and anthropogenic heating effects) at the simulated time of 1500 LST (Figure 1b) shows that sea breeze frontal movement has been retarded over the city center, as seen in the wave like shape of the front over the city.

Corresponding results from a similar simulation from Bornstein et al[3], that did not include urban topographic effects, does not show a wave like frontal shape over the city (Figure 2). The corresponding observations do in fact show such a wave like perturbation (Figure 3), indicating that it arises due to urban topographic effects.

The new "base" simulation nighttime results after 33 hours (at 0600 LST) show a diffluent flow around the warm city and a local minimum of speed over the city (Figure 1b). Diffluence from the urban topographic barrier effect was not present in the old nighttime results (not shown). The local urban minimum wind speed area, however, is due mostly to the large urban roughness effect, as it was present in the old results.

Urban topographic effects can be quantified by subtracting (from the new base fields) the corresponding values from a second new simulation that is identical to the first new results, except that all urban and super urban topographic heights are set to zero. The resulting daytime "difference" field at 1200 LST (Figure 4a) shows a convergent upslope flow over the city, as expected with a raised topographic feature.

These perturbation flows generally are in the same direction as the "base" case flows in each local domain area, i.e., see the NW synoptic flow NW of the city, the SSE sea breeze flow S of the city, and the NE Long Island Sound sea breeze flow NE of the city. Concurrence of each of the local flow directions produce the convergence seen in the figure.

The corresponding nighttime "difference" field at 0300 LST (Figure 4b) shows the expected divergent downslope flow from the city. Given the generally northwesterly offshore flow during this period, the following effects are noted: along streamline deceleration NW of the city, along streamline acceleration SW of the city, and diffluence NE and SW of the city. Both the along streamline speed changes and the diffluence contribute to a divergence effect (as shown in the natural coordinate form of the horizontal divergence equation).

REFERENCES

1. R. Bornstein and M. LeRoy, Urban barrier effects on convective and frontal thunderstorms, in: "*Preprint Volume 4th AMS Conference on Mesoscale Processes*", AMS, Boston, MA (1990).

2. G. Schayes and P. Thunis. "A Three-Dimensional Mesoscale Model in Vorticity Mode," Contribution No. 60, *Institute of Astron. and Geophysics*, Catholic University of Louvain, (1990).

3. R. Bornstein et al, Application of linked three-dimensional PBL and dispersion models to New York City, in: "*Air*

Pollution Modeling and its Application V", ed. D.
Wispelaere et al, pp. 543-564 (1986).

4. R. Pielke. *"Mesoscale Meteorological Modelling"*, Academic
   Press, New York (1984).

5. G. Mellor and T. Yamada, Development of a turbulence closure
   model for geophysical fluid problems, *Rev. Geophy. Space
   Sci.* 4:851 (1982).

6. G. Terry and P. Lacarrere, Improving the eddy kinetic energy
   model for planetary boundary layer description, *Bound.
   Layer Meteor.* 25:63 (1983).

7. J. Businger et al, Flux-profile relationships in the
   atmospheric surface layer, *J. Atmos. Sci.* 28:181 (1971).

8. E. Webb, Profile relationships: the log-linear range and
   extension to strong stability, *Quart. J. Roy. Meteor. Soc.*
   96;67 (1970).

9. T. Sasamori, The Radiative Cooling Calculation for
   Application to General Circulation Experiments, *J. Appl.
   Meteor.* 7:721 (1968).

10. M. Atwater and P. Brown, Numerical computation of the
    latitudinal variation of solar radiation for an atmosphere
    of varying opacity, *J. Appl. Meteor.* 13:289 (1974).

11. J. Deardorff, Efficient prediction of ground surface
    temperature and moisture, with inclusion of a layer of
    vegetation, *J. Geophs. Res.* 83:1198 (1978).

12. J. Monteith, Evaporation and surface temperature, *Quart. J.
    Roy. Meteor. Soc.* 107:1-27 (1981).

13. P. Thom and T. Oliver, On Pennmann's equation for
    estimating regional evaporation, *Quart. J. Roy. Meteor.
    Soc.* 103:345 (1977).

DISCUSSION

R. SAN JOSE: How are you planning to apply satellite data to the simulations?

B. BORNSTEIN: He is planning to use data from the NASA similar resolution to LANDSAT-7 and this is an important future task.

G. ADRIAN: In the vertical cross section of the flow over a city represented by a roughness - and heat island gravity waves should be expected. Why are they not visible in your results?

R. BORNSTEIN: There is no topography and no elevated stable layers, and hence no gravity waves in that particular simulation.

WIND FLOW AND PHOTOCHEMICAL SMOG IN THESSALONIKI:
MODEL RESULTS COMPARED WITH OBSERVATIONS

Nicolas Moussiopoulos, Athena Proyou and Peter Sahm

Laboratory of Heat Transfer and Environmental Engineering
Aristotle University, 54006 Thessaloniki, Greece

ABSTRACT

The EUMAC Zooming Model is applied to simulate wind flow and pollutant transport and transformation in the Greater Thessaloniki Area (GTA). The simulation results agree fairly well with observations during the Thessaloniki '91 Field Measurement Campaign. Both the calculation and the observation show that the ozone levels in the GTA are decisively affected by the sea breeze circulation and the nighttime inversion.

INTRODUCTION

With more than one million inhabitants, Thessaloniki, the capital of Macedonia, is the second largest city in Greece. It stretches over twenty kilometers in a bowl formed by low hills facing a bay that opens into Thermaikos Gulf. Industrial activities and road traffic cause significant air pollutant emissions. In order to analyse the photochemical smog levels associated with these emissions, a comprehensive field measurement campaign was carried out in the Greater Thessaloniki Area (GTA) in 1991. The objective of this campaign was to obtain all observational evidence needed to verify mathematical models describing dispersion and chemical transformation of atmospheric pollutants.

In this paper results of the EUMAC Zooming Model (EZM) for the wind flow and the transport and transformation of air pollutants in the GTA are compared with observations during the Thessaloniki '91 Field Measurement Campaign. Brief information on this campaign is given in the next section. Subsequent sections contain a short description of the EZM and details on the case studied. Selected simulation results are then presented and extensively discussed in comparison with corresponding results of the field measurement campaign.

THE THESSALONIKI '91 CAMPAIGN

The Thessaloniki '91 Field Measurement Campaign was planned and executed by scientists of the Institute of Meteorology and Climatology of the Karlsruhe Nuclear Research Centre and the Laboratory of Heat Transfer and Environmental Engineering of the Aristotle University Thessaloniki. The main aim of the campaign was to perform simultaneous measurements of the ground level ozone concentrations and meteorological quantities in the periphery of the GTA. To complete the picture of transport and transformation mechanisms in the GTA, additional meteorological measurements were performed, including the vertical variation of wind velocity, air temperature and ozone concentration by

Air Pollution Modeling and Its Application X, Edited by S-V. Gryning
and M. M. Millán, Plenum Press, New York, 1994

tethersonde ascents and the diurnal variation of the atmospheric boundary layer structure by operating an acoustic sounder. These measurements were set up and performed by scientists of the University of Athens and the National Observatory of Athens.

The results of the Thessaloniki '91 Field Measurement Campaign elucidate the mesoscale wind flow characteristics and the major mechanisms of air pollutant transport and transformation in the GTA. A complete presentation of the campaign results can be found elsewhere (Moussiopoulos and Kaiser, 1993).

THE EUMAC ZOOMING MODEL (EZM)

The EUMAC Zooming Model (EZM) was designed for the refined modelling of transport and chemical transformation of pollutants in selected European regions in the frame of EUROTRAC's subproject EUMAC. The EZM has been already successfully applied and verified for various European airsheds including the Upper Rhine Valley and the areas of Zurich, Graz, Barcelona, Lisbon and Athens. EZM's core are the prognostic mesoscale model MEMO and the photochemical dispersion model MARS.

The nonhydrostatic model MEMO

MEMO is a fully vectorized physically complete nonhydrostatic mesoscale model (Moussiopoulos et al., 1993). Within MEMO, the conservation equations for mass, momentum and scalar quantities as potential temperature, turbulent kinetic energy and specific humidity are solved. The model uses the anelastic approximation to filter sound waves. In the present case the model version using the Boussinesq approximation is applied. The governing equations are solved in terrain-following coordinates. Non-equidistant grid spacing is allowed in all directions. The numerical solution is based on second-order discretization applied on a staggered grid. As an important feature of MEMO, conservative properties are fully preserved within the discrete model equations. The discrete pressure equation is solved with a direct elliptic solver in conjunction with a generalized conjugate gradient method.

Advective terms are treated with the TVD scheme. Turbulent diffusion is described with an one-equation turbulence model (conservation equation for the turbulent kinetic energy and algebraic equation for the mixing length). At roughness height similarity theory is applied. The radiative heating/cooling rate in the atmosphere is calculated with an efficient scheme based on the emissivity method for longwave radiation and an implicit multilayer method for shortwave radiation. The surface temperature over land is computed from the surface heat budget equation. The soil temperature is calculated by solving an one dimensional heat conduction equation for the soil. At lateral boundaries generalized radiation conditions are implemented. In its latest version, MEMO allows performing multiple nested grid simulations.

The photochemical dispersion model MARS

MARS uses a fully implicit selfadaptive method, i.e. an implicit algorithm characterized by a variable time step and a variable order (Moussiopoulos, 1989). The overall solution procedure is controlled by the local error (the additional error of the algorithm made in a time step), which is not allowed to exceed a given tolerance.

Differently than in other proposed photochemical dispersion models, MARS allows describing the combined effects of vertical diffusive transport and chemical transformation of pollutants (Graf and Moussiopoulos, 1991). By this approach the feasible error caused by splitting the operators associated with vertical diffusion and chemistry is avoided. Operator splitting is thus restricted to the justifiable separate treatment of advective transport.

Generally, chemical kinetics may be taken into account by the aid of any suitable reaction mechanism. In several applications KOREM, a modified version of the compact mechanism of Bottenheim and Strausz (1982), was found capable producing ozone diurnal cycles which are in good agreement with those obtained with far more sophisticated contemporary reaction mechanisms.

CASE SPECIFICATION

Simulations with the EZM were performed for September 30, 1991, a typical sea breeze day in the GTA. For a proper description of the transport phenomena in the GTA, MEMO was first applied to calculate the wind field in a 500×600 km^2 area comprising practically the whole of Northern Greece and Bulgaria (frame 'A' in Fig. 1a). This 'coarse grid' simulation was performed at a spatial resolution of 10 km. As an example, Fig. 1 shows the computed ground level wind field in the SW section of the coarse grid domain (frame 'B' in Fig. 1a) at 6 p.m. of September 30, 1991. The displayed result shows that MEMO successfully reproduces the extended sea breeze observed on that day in the GTA.

Using the coarse grid results to define appropriate boundary conditions, MEMO was then applied to describe the wind flow in a 60×120 km^2 area around the GTA (domain '1' in Fig. 1a) at a horizontal resolution of 2.5 km. Pollutant dispersion and photosmog formation in the GTA were studied with MARS at the same resolution for a 50×70 km^2 area (domain '2' in Fig. 1a, shown enlarged in Fig. 1b). In the latter simulation nineteen non-equidistantly distributed layers were considered in the vertical direction, the minimum spacing at ground level not exceeding 20 m. The numerical simulations were performed over a time period of three days. The results for the third day were proved to be independent of the assumed initial concentrations.

Chemical transformations were modelled on the basis of the KOREM mechanism. Photolysis rates were computed as functions of the solar zenith angle. Deposition mechanisms were neglected, as for the given length scale deposition fluxes are negligible compared to the advective ones.

Available emission data were used to compile an emission inventory with a spatial resolution of 1 km. The diurnal variation of the emission rates was described by the aid of appropriate assumptions for the major emission sources in Thessaloniki. In the case of traffic, emissions were derived from the driving patterns in the main roads of the city.

MODEL RESULTS

Figure 2 shows the calculated diurnal variation of the surface level wind, NO, NO$_2$ and ozone fields in the GTA. The model results for the wind velocities at the stations

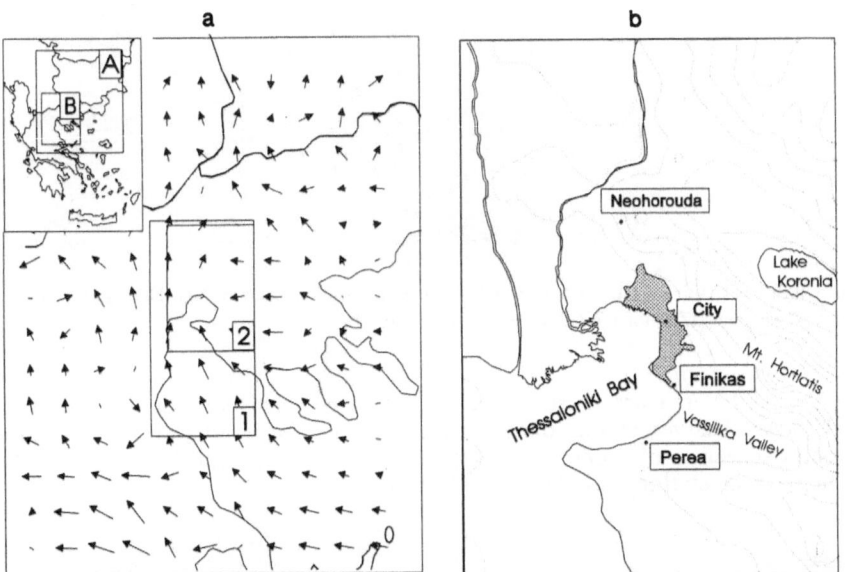

Figure 1. (a) Result of MEMO for the wind field at 6 p.m. of September 30, 1991 in the subregion 'B' of the coarse grid domain 'A'. The frames '1' and '2' correspond with the fine grid domains of MEMO and MARS, respectively. (b) Enlargement of frame '2'. The area occupied by the city of Thessaloniki is stippled.

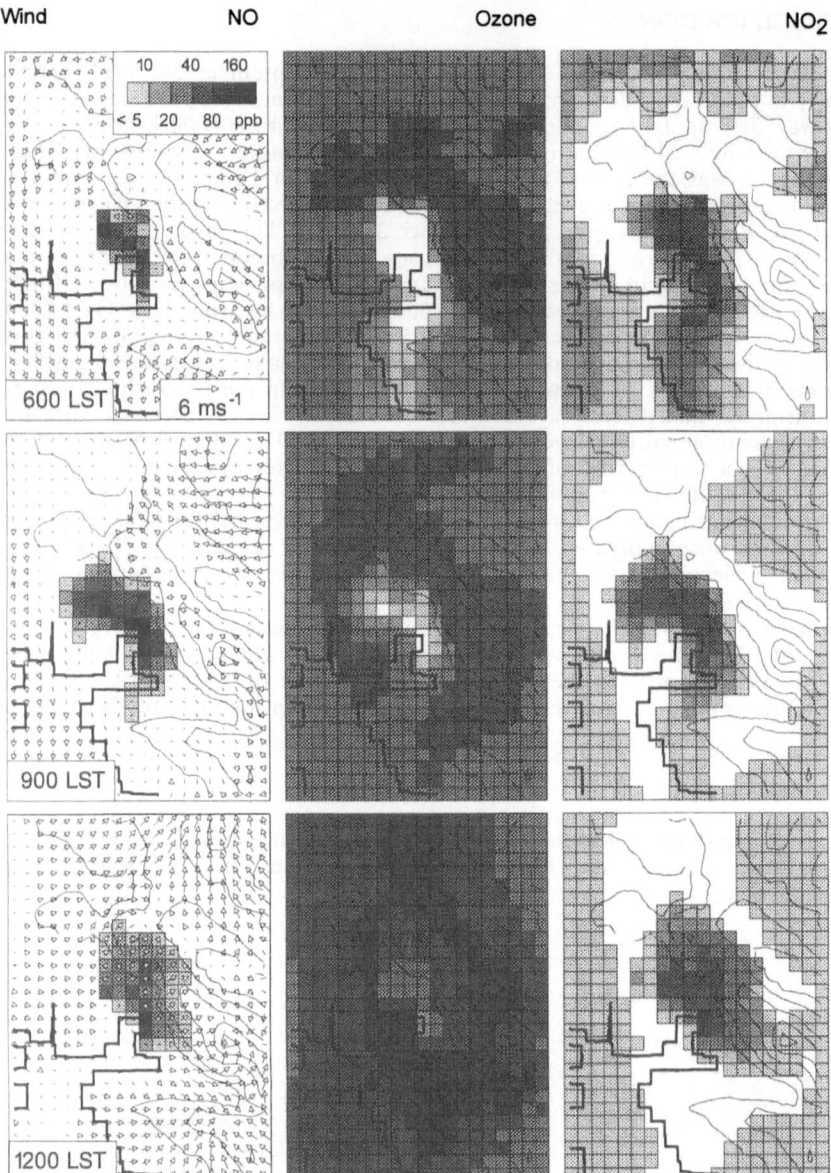

Figure 2. Diurnal variation of the wind field and of the concentration fields of NO, NO_2 and ozone at surface level predicted with the EZM for the GTA on September 30, 1991.

Neohorouda and Perea are illustrated in Fig. 3 in comparison with the observed ones. Figure 4 shows observed and predicted concentrations of primary pollutants in the city of Thessaloniki as well as of ozone at the stations Neohorouda, Finikas and Perea. The location of these three stations is indicated in Fig. 1b.

Wind flow

The nighttime wind pattern reflects a weak offshore air motion related to both land breeze and downslope winds. It should be noted that this wind pattern results in air mass

Wind   NO   Ozone   NO$_2$

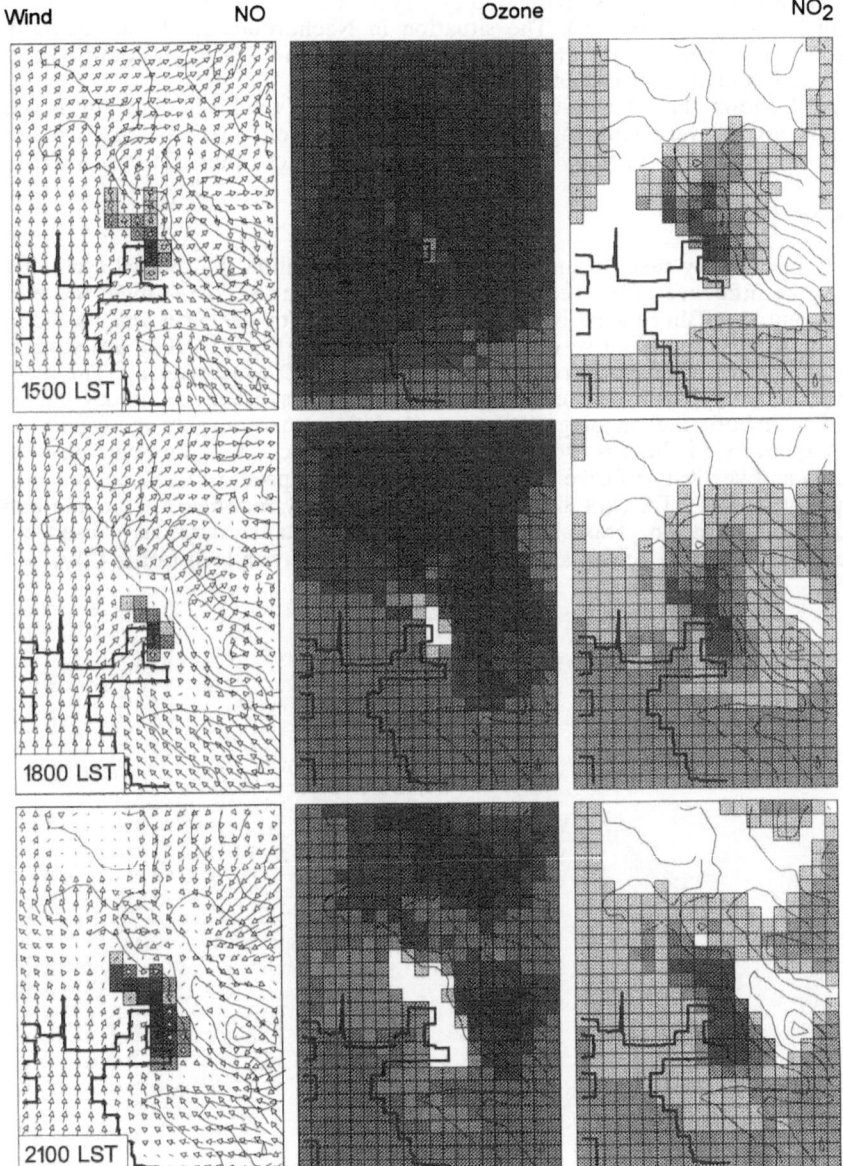

Figure 2. Continued.

stagnation above the city of Thessaloniki which persists until the sea breeze is developed late in the morning. The sea breeze circulation is first characterized by low wind speeds and directions perpendicular to the shore. In the afternoon, however, it evolves to an extended sea breeze originating from the Thermaikos Gulf and having higher speeds and a practically uniform direction from the South. This transition after 2 p.m. is nicely reflected in both the observed and predicted diurnal variation at Perea (Fig. 3). The nocturnal wind flow at Perea is dominated by winds from ESE directions originating from the Vassilika Valley (cf. Fig. 1b). The situation in Neohorouda is characterized by sea breeze from a uniform direction (SSW) between 9 a.m. and 7 p.m. and downslope winds from NE in the remaining hours of the day.

Figure 3 shows that the results of MEMO for the wind velocities in Perea and Neohorouda are in very good agreement with observations. An analogous agreement between predicted and observed wind velocities was found for other stations as well.

Pollutant concentrations

CO is only slightly reactive and thus it essentially exhibits the diurnal cycle of transport under the influence of the sea breeze circulation in the GTA. At night the emission level is low and an offshore wind is blowing over the city. For these reasons the downtown CO concentration before sunrise does not exceed 4 ppm (Fig. 4). High emissions in the early morning lead to a sharp increase of the downtown CO concentration. The maximum of approximately 7 ppm is reached at about 8 a.m.. The subsequent midday decrease is associated with both the ongoing sea breeze and the monotonically increasing vertical mixing. A recovery of the downtown CO concentration occurs in the evening when the wind ceases and vertical diffusive transport weakens. The predicted CO concentration in the periphery of the GTA (not shown here) deviates only marginally from the assumed background concentration. This shows that primary pollutant emissions caused by road traffic have a direct influence on air quality mainly in the urban area of the GTA.

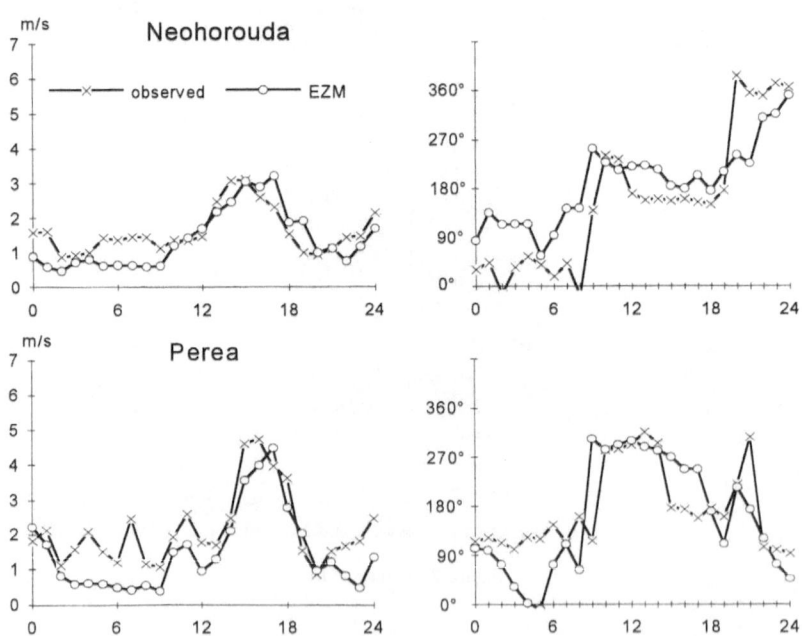

Figure 3. Diurnal variation of the wind velocity at Neohorouda and Perea on September 30, 1991: Predictions with the EZM (MEMO) compared with observations.

The predicted diurnal cycle of the surface level NO pattern reveals that NO emitted in the GTA essentially stays inside the urban area, in the daytime because of rapid oxidation to NO_2, during the night because of the weakness of the transport processes. The relatively high daytime values of NO_2 in the urban area are associated with the intense photochemical production of NO_2, which obviously counterbalances the dilution due to advection and vertical diffusion. It is worthy of notice that the nighttime offshore air motion leads to the accumulation of nitrogen oxides above the Thessaloniki Bay close to the shore. Apparently, pollutants transported during the night onto the sea may be transported back to the urban area by the daytime sea breeze thus leading to a further increase in peak concentrations and also affecting photosmog formation during the day.

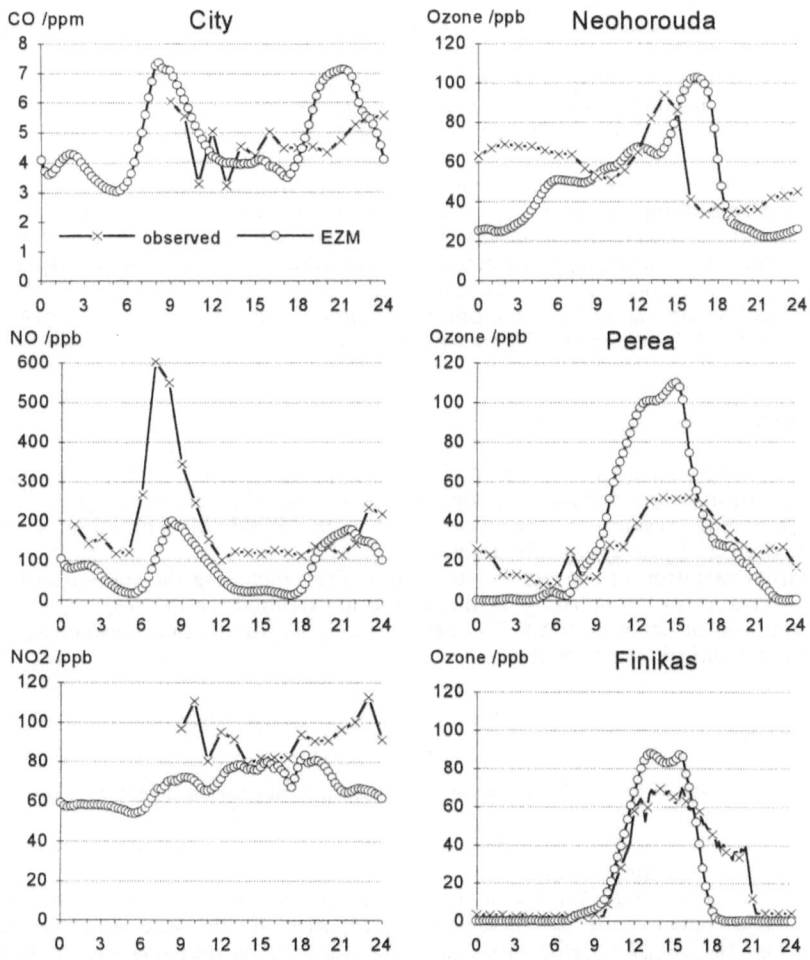

Figure 4. Diurnal variation of primary pollutant concentrations in the centre of Thessaloniki and of the ozone concentration at three locations in the periphery of Thessaloniki on September 30, 1991: Predictions with the EZM (MARS) compared with observations.

Ozone formation starts first above the Thessaloniki Bay, namely in polluted air masses which were advected onto the sea by the nighttime land breeze (9 a.m., cf. Fig. 2). Later during the day, air mass with a high photochemical oxidant content is advected by the sea breeze to the city of Thessaloniki. In the city centre surface level ozone formation is considerably dampened by the high local NO emissions. Highest ozone levels are reached at about 3 p.m.: With the exception of the city centre, the ozone concentration exceeds 80 ppb over the entire GTA. In the evening, NO and VOC emissions serve as fast-acting sinks for ozone. Consequently, the ozone levels gradually decrease: The nocturnal ozone levels are close to zero in the city of Thessaloniki and in the industrial area to the NW of the city. On the contrary, the nocturnal ozone concentrations clearly exceed the assumed background level (30 ppb) in the hilly suburban area to the North and to the East of Thessaloniki. This behaviour can be explained by the fact that at locations with an altitude exceeding the level of the nighttime inversion (observed height: 50 to 100m; Helmis et al., 1993) high ozone concentrations may be preserved as they are separated from the surface level emissions.

The results of MARS for the primary pollutant concentrations in the city of Thessaloniki agree fairly well with the corresponding observations from a qualitative point of view. The underprediction, especially in the case of NO, is not surprising: The measurements in the centre of the city are to a large extent affected by local influences which cannot be resolved in the simulation. Moreover, the ozone concentrations predicted with MARS generally correspond to the observed behaviour. The systematic overprediction of the ozone maximum values could indicate an overestimation of the assumed VOC reactivity. The reasons for the early ozone decrease at Finikas in the afternoon remain to be analysed.

The underprediction of the nighttime ozone levels at Neohorouda is most probably related to the smoothing of the orography in the model simulation: Given that Neohorouda has an altitude of about 200 m, the measurements are performed above the ground-based inversion. As a consequence of orography smoothing, the grid cell corresponding to the location of Neohorouda has an altitude of less than 100 m. Therefore, in the model simulation Neohorouda is, at least temporarily, below the inversion. So, the ozone levels are affected by the NO content of the transported air mass.

CONCLUSIONS

The presented simulation results of the EZM for the wind flow and the dispersion and transformation of air pollutants in the GTA on September 30, 1991 agree fairly well with corresponding results of the Thessaloniki '91 Field Measurement Campaign. Specifically, the EZM proves capable describing

▶ the diurnal variation of the sea breeze in the GTA, including the transition from the breeze structure in the morning to the afternoon extended sea breeze;
▶ the formation of ozone in the GTA and especially the interaction between sea breeze and photochemical transformations.

REFERENCES

Bottenheim, J.W. and Strausz, O.P., 1982, Modelling study of a chemically reactive power plant plume, Atmos. Environ. 16:85.

Graf, J. and Moussiopoulos, N., 1991, Intercomparison of two models for the dispersion of chemically reacting pollutants, Contr. Phys. Atmos. 61:13.

Helmis, C.G., Asimakopoulos, D.N., Tombrou, M. and Soilemes, A., 1993, The wind field and the atmospheric boundary layer structure over the Greater Thessaloniki Area: II. The atmospheric boundary layer structure, in: Thessaloniki '91 Field Measurement Campaign, N. Moussiopoulos and G. Kaiser, eds, Scientific Series of the International Bureau / Forschungszentrum Juelich GmbH, Vol. 18.

Moussiopoulos, N., 1989, Mathematische Modellierung mesoskaliger Ausbreitung in der Atmosphaere, VDI-Verlag, Duesseldorf.

Moussiopoulos, N., Flassak, Th., Sahm P. and Berlowitz D., 1993, Simulations of the wind field in Athens with the nonhydrostatic mesoscale model MEMO, Environmental Software 8:29.

Moussiopoulos, N. and Kaiser, G., eds, 1993, Thessaloniki '91 Field Measurement Campaign, Scientific Series of the International Bureau / Forschungszentrum Juelich GmbH, Vol. 18.

GLOBAL AND LONG-RANGE TRANSPORT

chairmen: S.E. Gryning

                                         F.A. Schiermeier

rapporteurs: E. Batchvarova

                                       M. Rotach

A THREE DIMENSIONAL HEMISPHERIC AIR POLLUTION MODEL USED FOR THE ARCTIC

Jesper Christensen

National Environmental Research Institute
Frederiksborgvej 399
4000 Roskilde
Denmark

1. INTRODUCTION

There have been a considerably scientific interest of the air pollution in the Arctic. The reason for this is that the arctic area is rather sensitive to the pollution, and the pollution there is a finger print of the air pollution for the whole Northern Hemisphere. In the last 20 years several groups have done measurements in the Arctic (see, for example, AGASP, 1984; Arctic Air Chemistry, 1985, 1989, 1993). All these measurements shows that the air pollution have a great seasonally variation with relative high concentrations during the winter and the early spring with a maximum normally in march and very low concentrations during the summer. The air pollution is also characterized by an episodic nature, because the origin of the pollution is only due to the long-range transport, and a deep vertical distribution, because of the very cold and stable Arctic atmosphere.

In this paper a hemispheric model, that is in progress of development at the National Environmental Research Institute (NERI), will be described. This model is an extension of the regional long-range transport models, that have been developed at NERI, see Zlatev and Christensen, 1989, where a simple sulphur model, covering the whole Europe, is discussed, and Zlatev et al., 1992 and 1993, where a big model, including many photochemical non-linear reactions, is discussed. These models is not directly applicable to describe the transport on a hemispherical scale, because of the vertical structure (only 2-d boundary layer models).

It was decided to use the terrain-following coordinates σ in the model. The grid resolution is 150 km at 60°N and the present version of the model includes only linear sulphur chemistry (SO_2 to SO_4).

The vertical resolution of the meteorological input is rather coarse for the planetary boundary layer. In some models, for example, Iversen (1989), the boundary layer, especially the mixing height is parameterized quite simple. In the model, presented in this paper, the boundary layer is parameterized more advanced by using the surface meteorological input parameters.

Wet deposition is very important for SO_2 and especially for SO_4. Because the meteorological input does not include any information about the precipitation, it is necessary to parameterize the wet deposition, for example by using the humidity directly to estimate the scavenging, see Pudykiewicz, 1991. In the model the precipitation is calculated by using a condensation scheme similar to scheme descibed in Sundqvist et al. (1989).

This work has been partly supported by the Danish Research Academy and the Danish Natural Science Research Council.

2. DESCRIPTION OF THE MODEL

The model is based on a set of coupled continuity equations for each species, where the coupling is through the chemical reactions. Each equation is derived by the assumption that the pure advection cannot change the mass mixing ratio for a species. The total continuity equation (see, for example, Pudykiewicz, 1991) for each species is

$$\frac{\partial q_i}{\partial t} = -u\frac{\partial q_i}{\partial x} - v\frac{\partial q_i}{\partial y} - \dot{\sigma}\frac{\partial q_i}{\partial \sigma}$$

$$+\frac{\partial\left(K_x\frac{\partial q_i}{\partial x}\right)}{\partial x} + \frac{\partial\left(K_y\frac{\partial q_i}{\partial y}\right)}{\partial y} + \frac{\partial\left(\Gamma^2 K_z\frac{\partial q_i}{\partial \sigma}\right)}{\partial \sigma} \tag{1}$$

$$- W_i q_i + E_i + Q(c_1, c_2, \cdots, c_{nq}) \qquad , \quad i=1, nq$$

where x and y is the horizontal coordinates in the east and north direction respectively. The vertical coordinate σ is a terrain following coordinate and is defined by $\sigma = p/p_s$, where p is the pressure and p_s is the surface pressure and nq is the number of species. The first term is the advection term, where u and v is the horizontal wind fields in east and north direction respectively and $\dot{\sigma}$ is the vertical velocity. The second term is the diffusion, where K_x and K_y are the horizontal diffusion coefficient and K_z is the vertical diffusion; $\Gamma = g\sigma/RT$, where g is the acceleration due to the gravity, R is the gas constant and T is the temperature. The third term is the wet deposition term, and the last two terms are the emissions and the chemistry. In the present version nq=2, where the two species are SO_2 and sulphate. Eq. (1) was extended to also include the transport of humidity and cloud water (i. e. nq=4), where the emission is given by the latent heat flux. The equations are coupled through condensation and evaporation of cloud water, and the lost term is the release of precipitation.

3. NUMERICAL SOLUTIONS OF THE TRANSPORT EQUATIONS

The Northern Hemisphere was transformed to a plane, by using the polar stereographic projection system true at 60° North (see, for example, Haltiner and Williams, 1980; Iversen, 1989; Pudykiewicz, 1991). Then the time integration is done by splitting (1) into four sub-models:

1. Advection
2. Horizontal diffusion
3. Vertical diffusion (dry deposition for sulphur and latent heat-flux)
4. Emission, chemistry, condensation-evaporation-precipitation and wet deposition

After this splitting the time integration of (1) is obtained by doing a time integration on each of the four sub-models. This procedure make it much more easy to find a numerical solution for the whole model, because it is possible to find an efficient numerical solution for each sub-model. Splitting is commonly used in air pollution models (see, for example, Mcrae et al., 1982). The horizontal space of the model is define on a regular 96x96 grid (see fig. 1), that covers most of the Northern Hemisphere, with a grid-distant of 150 km at 60°N. The vertical grid is an irregular grid with 12 grid-points, which cover most of the whole troposphere, see table 1.

The horizontal transport equations are solved by using a pseudospectral algorithm (see Bartnicki et al., 1990; Christensen, 1993). The pseudospectral algorithm demands periodical boundary conditions, which are imposed by some artificial sinks on the horizontal boundaries (Zlatev et al., 1992; Christensen, 1993).

The vertical transport equation is solved by applying an finite element algorithm; see Pepper et al., 1979, and Christensen, 1993. This algorithm is also used for the horizontal transport in some air pollution models, see, for example, Carmichael and Peters, 1984, and Piedelievre et al., 1990. The boundary conditions at the lower boundary (at the ground) is defined by that the flux out or in of the model domain is only due to the dry deposition for the sulphur species or the latent heat flux for the vapour. At the upper boundary there are free boundary conditions.

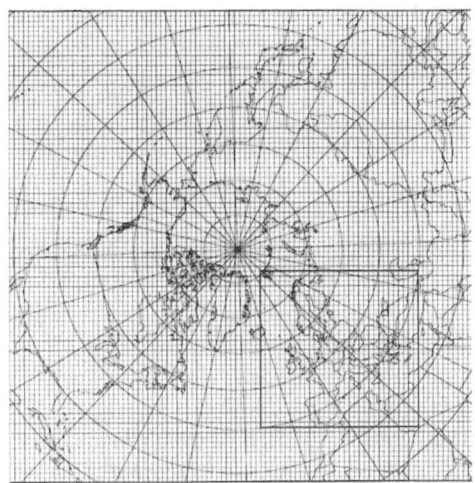

Figure 1. The model area and the horizontal grid.

It should be emphasized that for a long-range transport model on hemispheric scale, used to study the transport to the Arctic, it is very important, that one choose numerical schemes, which are efficient and give small numerical errors, because the atmospheric transport is the only reason for the air pollution in the Arctic. The numerical schemes used in this paper is tested carefully (see Christensen, 1993) and was found to work satisfactory well concerning numerical efficiency and numerical errors.

4. PHYSICAL PARAMETERS

In this section the main physical processes will be described. All the meteorological inputs are obtained from the European Centre for Medium-range Weather Forecasts (ECMWF), ECMWF/TOGA Basic Level III data sets and the supplementary sets on a 2.5°x2.5° grid. All the meteorological data are interpolated to the grid, used in the model. The time resolution is 6 hour for the cloud cover, wind stress and the heat flux (the last two are accumulated fields) and 12 hour for the other, which are 3-d wind, temperature, humidity and geopotential for the upper levels (1000mb, 850mb, 700mb, 500mb, 400mb, 300mb, 250mb, 200mb, 150mb, 100mb, 70mb, 50mb, 30mb, 10mb) and 10 meter wind, 2 m temperature, 2 m dewpoint, surface pressure, surface temperature and surface geopotential for the surface level).

The advection depends on the 3-d wind fields from the meteorological input.

In the model it is assumed that the horizontal diffusion is constant: $K_x=K_y=1\cdot10^5$ m²/s. This diffusion is some artificially compared with implicit diffusion due to the grid-resolution, but it smooths the numerical solution of the advection.

The emissions used in the model are based on the global inventory of sulphur emissions described in Spiro et al., 1992, the EMEP emissions for Europe (Iversen et al., 1989) and the circumpolar emissions survey by Semb, 1985. The emissions are distributed evenly up to 500 m above the surface. Natural emissions of dimethyl sulphide (DMS) from the oceans are also included in the model, and they depend on the concentrations of DMS in the oceans and u_{10}:

Table 1. Vertical structure used in the model

level no.	σ
1	0.400
2	0.500
3	0.570
4	0.640
5	0.705
6	0.787
7	0.851
8	0.902
9	0.935
10	0.960
11	0.980
12	1.000

$$FLUX_{DMS} = Ku_{10}^2c_o \qquad (2)$$

where K=[0.15⁄0.01/3600] s/m and c_o is the concentration of DMS in the ocean (see, for example, Tarrason, 1991, and Hertel et al., 1993). In the model it is assumed, that 44% of the DMS flux are going to SO_2 and 4% are going to SO_4 (see Hertel et al., 1993).

The chemistry is simple linear, where the oxidation rate of SO_2 to SO_4 depends on the time of the year and the latitude (see, for example, Iversen, 1989, and Iversen et al., 1991). This spatial and temporal variation of the oxidant rate try to take into account the variations of OH in the gas phase and H_2O_2 and O_3 in the liquid phase. The maximal oxidant rate is $5\cdot10^{-6}$ s⁻¹ at equator and the minimal rate is $0.3\cdot10^{-6}$ s⁻¹ at the North Pole in January.

The dry deposition for SO_2 have also a temporal and spatial variation due to the low deposition of SO_2 over non-melting snow, the variation of the biochemical uptake of the vegetation and that over the open water the dry deposition is very effective, see Iversen et al.,1991.
The maximal dry deposition velocity is 0.8 cm/s.
The dry deposition for sulphate is 0.1 cm/s over the whole area (see Zlatev et al., 1992).

4.1 Parameterization of the boundary layer

The expressions for the K_z profile in the surface layer is based on Monin-Obukhov similarity theory (see Seinfeld, 1986)

$$K_z = \frac{\kappa u_* z}{\phi(z/L)} \tag{3}$$

where $\phi(z/L)$ is the similarity function, κ is the von Karman constant ($=0.35$), u_* is the friction velocity, z is the height above the surface, and L is the Monin-Obukhov length. u_* and L are calculated from the meteorological input with a time resolution on 6 hours.
The K_z profile for the surface layer is extended to the whole vertical by the following expression which ensures that vertical diffusion is small (0.1 m^2/s) above the mixing-layer

$$K_z = \max\left(\frac{\kappa u_* z}{\phi(z/L)}(1 - z/z_{mix}), 0.1 \ m^2/s\right) \tag{4}$$

The height of the mixing layer, z_{mix}, is calculated by using a parameterization for height of the boundary layer similar to what, there is described in Berkowicz and Olesen, 1990, and Gryning and Batchvarova, 1990. This parameterization is based on a simplified energy balance equation for the internal boundary-layer, which could have following expression:

$$\left(N^2 z_{mix} + 1.9\frac{u_*^2 + w_*^2}{z_{mix}}\right)\frac{dz_{mix}}{dt} = \frac{w_*^3 + 1.25 u_*^3}{z_{mix}} - \frac{u_*^2 + 0.5 w_*^2}{1100s} \tag{5}$$

where N is the Brunt-Vaisala frequency, $N = \sqrt{\gamma g/T}$ (γ is the lapse rate) and $w_* = \left(\frac{g}{T}\max(H_d, 0)z_{mix}\right)^{1/3}$,

where H_d is the positive upward sensible heat flux. The first term on the rhs of (5) are the growth of the mixing height due to convective and mechanically produced energy, and the last term are dissipation of the mechanically and convective energy.

4.2 Precipitation and wet deposition

The precipitation is calculated by using a part of the condensations scheme described in Sundqvist et al, 1989. Every 12 hours a new humidity field is read, and that field is used as an initial field for the condensation scheme. Emissions of humidity into the atmosphere is given by the latent heat flux (6 hours accumulated meteorological input). The transport of vapour and cloud water are the same as for the sulphur species, therefore (1) is extended to also include vapour and cloud water. There is condensation of cloud water if the relative humidity exceeds the condensations level, which depends on the cloud-cover (meteorological input every 6 hour), land-cover and temperature. Evaporation of cloud water or rain water is possible if the relative humidity is lower than the condensations level.
The rate of release of precipitation is described by

$$P = \frac{1}{\tau}m\left[1 - \exp\left(-\left(\frac{m}{cl \ m_r}\right)^2\right)\right] \tag{6}$$

where m is cloud water content, m_r is the threshold, which m/cl (cl=cloud cover) should exceed before there is a release of precipitation and τ is the characteristic timescale. m_r has a typical value from $3 \cdot 10^{-4}$ to $5 \cdot 10^{-4}$ (mixing ratio) and τ from 10^3s to 10^4s, depending on the cloud type and the temperature.

The wet deposition is parameterize by using a simple scavenging ratio formulation (see Tarrason and Iversen, 1992). For SO_2 at a given level i the scavenging coefficient is given by

$$W_{SO_2} = \frac{\Lambda_{bc}}{H} \sum_{j=1}^{i} P_j \tag{7}$$

where P_j is the rate of release of the precipitation at level j (in m/s), $\sum_{j=1}^{i} P_j$ is the total precipitation at level i and H is a typical height (=1000m). Λ_{bc} is a below cloud scavenging ratio and have the value $2.5 \cdot 10^5$. For SO_4 the scavenging coefficient is given by

$$W_{SO_4} = \frac{\Lambda_{bc}}{H} \sum_{j=1}^{i-1} P_j + \frac{\Lambda_c}{H} P_i \tag{8}$$

where Λ_c is the in-cloud scavenging ratio and have the value 10^6, which means that scavenging of SO_4 is much more efficient in the clouds, where the precipitation are released.

5. NUMERICAL RESULTS

The model have been run for July 1990, October 1990, January 1991 and March 1991. Fig 2. and fig 3. shows as an example the precipitation and the mixing height calculated by the model compared with data given by the EMEP Meteorological Synthesizing Centre - West, Norwegian Meteorological Institute average over the Northern part of Central Europe for October 1990.

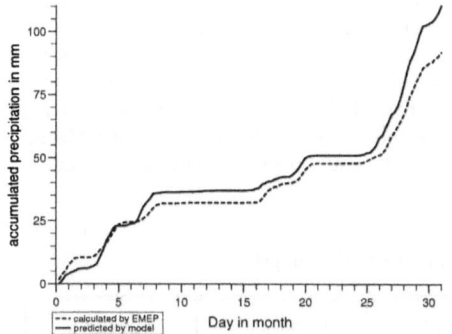
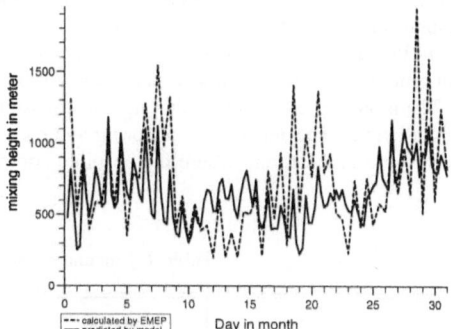

Figure 2. Total accumulated precipitation in mm average over the Northern part of the Central Europe for October 1990.

Figure 3. The mixing heights average over the Northern part of the Central Europe for October 1990.

Initial concentrations of SO_2 and SO_4 is given by the monthly mean values for the particular month. On Fig. 4 and 5 there are shown scatter plots for October 1990, where monthly mean values of SO_2 and SO_4 concentrations are compared with measurements from EMEP stations in Europe (see Schaug et al., 1992 and 1993). The figure shows, that there is a good correlation between observed and calculated concentrations. For the other months there are the same good results.

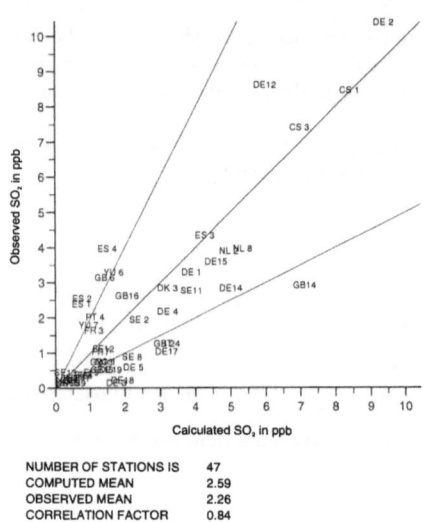

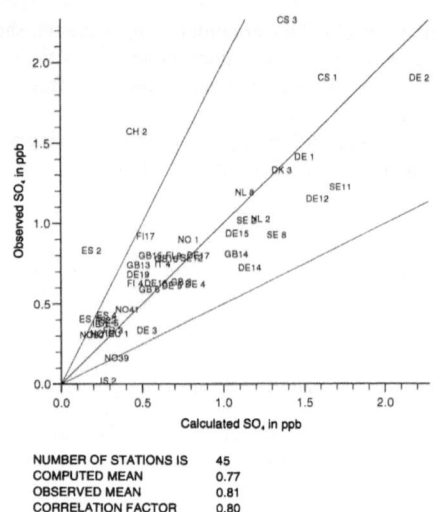

NUMBER OF STATIONS IS	47
COMPUTED MEAN	2.59
OBSERVED MEAN	2.26
CORRELATION FACTOR	0.84

NUMBER OF STATIONS IS	45
COMPUTED MEAN	0.77
OBSERVED MEAN	0.81
CORRELATION FACTOR	0.80

Figure 4. Comparison of calculated SO$_2$ concentrations with measurements in Europe for October 1990.

Figure 5. Comparison of calculated SO$_4$ concentrations with measurements in Europe for October 1990.

For the Arctic areas the model have been compared with measurements from two sites: Station Nord in northeastern Greenland (81° 36' N, 16° 40' W), see Wåhlin in Arctic Air Chemistry, 1993, and Ny-Ålesund at Spitzbergen (78° 54' N, 11° 53' E), see Schaug et al, 1992 and 1993. The model gives also for these two stations reasonable well results on monthly basis for the four months. In table 2 the monthly mean values of SO$_2$ and SO$_4$ concentrations are shown for the four months.

The correlation between calculated and measured concentrations is more poor on shorter time periods for the Arctic stations than for the EMEP stations in Europe, mainly because it is long-range transport and the poor time resolution (12 h) of the windfields.

The model have also been used to estimate the contribution of different sources on the Northern Hemisphere to the air pollution in the Arctic areas. The emissions are break up in European, Asia, North-America, natural (DMS) and Norilsk (very important single source for the Arctic). In Table 3 the results for the total mean concentration of sulphur for the lowest 500m above the ground, the total deposition of sulphur and the contributions from the different sources north for 75° N are summarized.

In January and March the initial concentrations have a rather large contribution to the mean air concentrations, which indicate that in the Arctic winter the life time of the sulphur species is large.

The model results indicate that natural emissions of DMS from the oceans could be an inportant source of the sulphur species in the late spring and early summer period, where the concentration of DMS in the oceans are highest, and where the absolute contribution from the anthropogenic emissions is small.

Table 2. Calculated and measured concentrations in ppb.

		July 90		October 90		January 91		March 90	
		SO$_2$	SO$_4$	SO$_2$	SO$_4$	SO$_2$	SO$_4$	SO$_2$	SO$_4$
Station Nord	model	0.02	0.06	0.02	0.02	0.9	0.2	0.2	0.2
	obs	-	-	0.02	0.05	0.6	0.2	0.2	0.2
Spitzbergen	model	0.05	0.10	0.07	0.10	0.9	0.2	0.4	0.4
	obs	0.05	0.15	0.04	0.04	0.6	0.2	0.4	0.3

Table 3. Total concentrations of S in air (ppb) up to 500 m above surface, the total depositions of S (mg/m^2) and the contributions from different sources (%).

	July 90		October 90		January 91		March 91	
	air	dep	air	dep	air	dep	air	dep
Total value	0.13	2.5	0.17	5.0	1.4	12.2	0.7	7.3
Norilsk	9 %	16 %	59 %	31 %	38 %	23 %	54 %	30 %
Europe	42 %	30 %	10 %	41 %	8 %	25 %	7 %	28 %
Asia	15 %	13 %	24 %	18 %	17 %	17 %	16 %	17 %
North America	9 %	6 %	2 %	4 %	0 %	2 %	3 %	7 %
DMS	18 %	9 %	1 %	1 %	0 %	1 %	1 %	3 %
Initial	7 %	26 %	5 %	6 %	36 %	32 %	20 %	15 %

6. CONCLUDING REMARKS AND PLANS FOR THE FUTURE

An Eulerian Long-Range model, covering most of the Northern Hemisphere, was presented here. The chemistry, used in the model, is a simple linear SO_2-SO_4 chemistry. The model is a three dimensional model with new way to parameterize the height of the mixing layer. The model has also a condensation scheme to calculate the rate of the release of precipitation, which is very important for the deposition of SO_2 and SO_4 and therefore also for the life time of the sulphur species in the troposphere.

The model have been run for four months: July 1990, October 1990, January 1991 and March 1991. The calculated mixing heights and precipitations for Europe have been compared with data given by EMEP, and the model reproduces reasonably well these data.

The model calculations of sulphur species have been compared with measurements in Europe (EMEP stations) and with measurements in the arctic areas, Station Nord by NERI and Spitzbergen by NILU in Norway. Again the model reproduced reasonably well the measured concentrations, especially the large seasonally variation of the sulphur concentrations in the arctic areas.

The model have been used to study the contribution to the sulphur concentrations and depositions in the Arctic from different sources in the Northern Hemisphere. The results shows, that Norilsk, Europe and Asia are the most important sources.

Plans for the future to run the model for a longer period (one year) and do a larger evaluation of the model, especially the parameterization of physical processes. The next step would be to improved the chemistry, both the gas-phase and the aqueous-phase.

The model development will be an important part of the danish part of the international Arctic Monitoring and Assessment Programme, AMAP.

REFERENCES

AGASP (1984): Special issue edited by R. C. Schnell. Geophys. Res. Lett., 11, 359-472.
Arctic Air Chemistry (1985): Special issue edited by K. A. Rahn. Atmos. Env., 19, 1987-2208.
Arctic Air Chemistry (1989): Special issue edited by K. A. Rahn. Atmos. Env., 23, 2345-2638.
Arctic Air Chemistry (1993): Proceedings of the 5th International Symposium on Arctic Air Chemistry, Copenhagen, Denmark, Septemper 8-10, 1992, edited by N. Z. Heidam. NERI Tecnical Report NO. 70, National Environmental Research Institute, P. O. Box 358, Frederiksborgvej 399, DK-4000 Roskilde, Denmark.
Bartnicki, J., K. Olendrzynski, K. Abert, P. Seibert and B. Morariu (1990): Numerical Approximation of the Transport Equation: Comparison of Five Positive Definite Algorithms. IIASA Working Paper WP-90-10, International Institute for Applied System Analysis, Laxenburg, Austria.
Berkowicz, R. and H. R. Olsen (1990): Modelling the Internal Boundary Layer at a Coastal Site. In: "Preprint from Ninth Symposium on Turbulence and Diffusion, April 30 - May 3, 1990, Risø, Denmark".

Carmichael, G. R. and L. K. Peters (1984): An Eulerian Transport/transformation/removal Model for SO_2 and Sulphate - I. Model Development. Atmos. Env., 18, 937-952.

Christensen, J. (1993): Testing Advection schemes in a Three Dimensional Air Pollution Model. Accepted by Mathl. Comput. Modelling.

Gryning, S. E. and E. Batchvarova (1990): Analytical for the Growth of the Coastal Internal Boundary Layer During Onshore Flow. Q. J. Meteorol. Soc., 116, 187-203.

Haltiner, G. J. and R. T. Williams (1980): Numerical Prediction and Dynamic Meteorology. John Wiley & Sons, New York.

Hertel, O., J. Christensen and Ø. Hov (1993): Modelling of the Endproducts of the Chemical Decomposition of DMS in the Marine Boundary Layer. Submitted to Atm. Env.

Iversen, T (1989): Numerical Modelling of the Long Range Atmospheric Transport of Sulphur Dioxide and Particulate Sulphate to the Arctic. Atmos. Env., 23, 2571-2595.

Iversen, T., N. Halvorsen, S. Mylona and H. Sandnes (1991): Calculated Budgets for Airborne Acidifying Components in Europe, 1985, 1987, 1988, 1989 and 1990. Report No 1/91. EMEP Meteorological Synthesizing Centre - West, Norwegian Meteorological Institute, PO Box 43, Blindern, N-0313 Oslo 3, Norway.

Iversen, T., J. Saltbones, H. Sandnes, A. Eliassen and Ø. Hov (1989): Airborne Transboundary Transport of Sulphur and Nitrogen over Europe - Model Description and Calculations, Report No 2/89. EMEP Meteorological Synthesizing Centre - West, Norwegian Meteorological Institute, PO Box 43, Blindern, N-0313 Oslo 3, Norway.

Mcrae, J., W. R. Goodin and J. H. Seinfeld (1982): Numerical Solution of the Atmospheric Diffusion Equation for Chemical Reaction Flows. J. Comput. Phys., 45. 1-42.

Pepper, D. W., C. D. Kern and P. E. Long, Jr, (1979): Modelling The Dispersion of Atmospheric Pollution Using Cubic Spline and Chapeau Functions. Atmos. Env., 13, 223-237.

Phillips, N. A. (1957): A Coordinate System Having some Special Advantages for Numerical Forecasting. J. Meteorol., 14, 184-185.

Piedelievre, J., L. Musson-Genon and F. Bombay (1990): MEDIA - An Eulerian Model of Atmospheric Dispersion: First Validation on the Chernobyl Release. Mon. Wea. Rev., 29, 1205-1220.

Pudykiewicz, J. (1991): Environmental Prediction Systems: Design, Implementation Aspects and Operational Experience with Application to Accidental releases. In: "Air Pollution Modelling and its Applications VIII". (H. van Dop and D. G. Stein, eds), Plenum Press, New Yor 145-159.

Schaug, J., U. Pedersen, J. E. Skjelmoen and I Kvalvågnes (1992): Data Report 1990. Part 2: Monthly and Seasonal summaries, EMEP/CCC-Report 3/92. Norwegian Institute For Air Research, PO Box 64, N-2001 Lillestrøm, Norway.

Schaug, J., U. Pedersen, J. E. Skjelmoen and I Kvalvågnes (1993): Data Report 1991. Part 2: Monthly and Seasonal summaries, EMEP/CCC-Report 5/93. Norwegian Institute For Air Research, PO Box 64, N-2001 Lillestrøm, Norway.

Seinfeld, J. H. (1986): Atmospheric Chemistry and Physics of Air Pollution. John Wiley & Sons, New-York.

Semb, A. (1985): Circumpolar SO_2 Emission Survey. Proj. No. 0-8516, Norwegian Institute for Air Research, Lillestrøm, Norway.

Spiro, P. A., D. J. Jacob and J. A. Logan (1992): Global Inventory of Sulphur Emissions With 1° x 1° Resolution. J. Geophys. Res., 97, 6023-6036.

Sundqvist, H., E. Berge and J. E. Kristjansson (1989): Condensation and Cloud Parametrization Studies with a Mesoscale Numerical Weather Prediction Model. Mon. Wea. Rev., 117, 1641-1657.

Tarrason, L. (1991): Biogenic Sulphur Emissions from the North Atlantic Ocean, Note No 3/91. EMEP Meteorological Synthesizing Centre - West, Norwegian Meteorological Institute, PO Box 43, Blindern, N-0313 Oslo 3, Norway.

Tarrason, L. and T. Iversen (1992): The Influence of North American Anthropogenic Sulphur Emissions over Western Europe. Tellus, 44B, 114-132.

Zlatev, Z. and J. Christensen (1989): Studying the Sulphur and Nitrogen Pollution over Europe. In: "Air Pollution Modelling and its Applications VII". (H. van Dop, ed), Plenum Press, New York, 1989, 351-360.

Zlatev, Z., J. Christensen and Ø. Hov (1992): An Eulerian Model for Europe. J. Atmos. Chem., 15, 1-37.

Zlatev, Z., J. Christensen and A. Eliassen (1993): Studying High Ozone Concentrations by Using the Danish Eulerian Model. Atmos. Env., 27A, 845-865.

126

DISCUSSION

H. VAN DOP: How did you model the stably stratified boundary layer? If you used the same formula as for the CBL, what are representative heights over land? Are they stationary during arctic continental winter?

J. CHRISTENSEN: The model for the boundary height combines both the CBL and the stably stratified boundary layer. In the pure stable case the height of the boundary height are propotional with u_*. The heigth of a stable boundary layer is given by Hstable$=1375$ u_*, i. e. the heights are not stationary, but depends on u_*.

J. LANGNER: An important observational feature of Arctic pollution is the layered structure of the pollution at different levels in the troposphere. Can your model decribe those features?

J. CHRISTENSEN: In principle the model could describe the Arctice haze. But I have not yet tried to look at the vertical distribution to see, if there is this Arctic haze. In that connection I think it is very important to have the correctly vertical distribution of the emissions for the sources close to the Arctic, because of the stable Arctic atmosphere. The emissions are distributed evently up to 500 m above the ground in the present version of the model.

T. IVERSEN: I congratulate with another model aimed at Arctic pollution studies (there are not very many). Have you tried to estimate how large part of the Arctic Haze is coming from different main source areas (Europe, America, etc)?

J. CHRISTENSEN: Yes, I have. Table 3 in the paper shows how important the different source areas are for the Artic pollution.

MODEL SIMULATION OF THE ATMOSPHERIC INPUT OF TRACE METALS INTO THE NORTH, BALTIC, MEDITERRANEAN AND BLACK SEAS

Joseph Alcamo,[1] László Bozó[2] and Jerzy Bartnicki[3]

[1]RIVM Environmental Forecasting Division
A. von Leeuwenhoeklaan
3720 BA Bilthoven P.O. Box 1
The Netherlands
[2]Institute for Atmospheric Physics
H-1675 Budapest P.O. Box 39
Hungary
[3]IBM Bergen Environmental Sciences and Solutions Center
N-5008 Bergen, Norway

INTRODUCTION

Since the middle of this century, energy generation, industrial production and transportation have caused serious environmental contamination by trace elements including heavy metals. The rate of contamination can vary from place to place as a function of source densities and intensities of heavy metals flux as well as meteorological conditions. Aerosols containing heavy metals can be transported far away from their sources by advection before being deposited.

The TRACE model developed at IIASA (Alcamo et al., 1992) focuses on European -scale emission, transport and deposition of As, Cd, Pb and Zn. An application of this model to the estimation of total deposition and budgets of heavy metals over Eastern Europe is given in Bozó et al. (1992). On the basis of TRACE model runs this paper presents calculations of long-term heavy metal deposition into the Regional Seas (North, Baltic, Mediterranean and Black) around the European continent. The country-to-sea contributions of the European countries to the Regional Seas are also estimated by means of TRACE model.

COMPUTATION OF WET AND DRY DEPOSITION

First the concentration of toxic metals at each receptor points (EMEP 150 X 150 km² grid) are calculated using a loss term (Alcamo et al., 1992). This equation gives the air concentration of a pollutant at a receptor located any distance downwind from a source. In the second step of the calculation wet and dry deposition of toxic metals considered are computed from the air concentration. Wet deposition is computed by means of a scavenging ratio:

$$d_w = c(x_r, y_r) \, W_q \, P, \tag{1}$$

where P is annual precipitation, W_q is the scavenging ratio (the ratio of the concentration of heavy metals in precipitation in $\mu g \, L^{-1}$ to their concentration in air in $ng \, m^{-3}$) and $c(x_r, y_r)$ is the concentration of pollutant at the receptor point. The scavenging ratio is set to 500,000 for As, Cd and Zn, and 200,000 for Pb. These values are based on the measurements carried out by Chan et al. (1982) who performed a spatialy and temporaly thorough study of the scavenging ratio of metals. Larger scavenging ratio for As, Cd and Zn is consistent with experimental (Lindberg, 1982) and theoretical evidences (Slinn, 1984) that the scavenging ratio increases with increasing particle diameter. Sensitivity of model calculations to the uncertainty of scavenging ratios is given in Alcamo et al., 1992).

Dry deposition is computed from:

$$d_d = c(x_r, y_r) \, v_d. \tag{2}$$

The dry deposition velocity v_d is computed from the semi-empirical model of Sehmel (1980). This model is based on wind-tunnel experiments and theoretical removal rates via Brownian diffusion and gravitational settling. Dry deposition velocities are computed as a function of particle size (D), surface roughness (z_0) and friction velocity (u_*):

$$v_d = \sum_{i=1}^{6} v_d(D_i, u_*, z_0) \, f(D_i), \tag{3}$$

where $v_d(D_i, u_*, z_0)$ is taken from the curves presented in Sehmel (1980). Data for u_* and z_0 were obtained on a European grid with a spatial resolution of 150 x 150 km^2 from the EMEP Synthesizing Center West (J. Saltbones, personal communication).

The variable $f(D_i)$ is the fraction of mass with diameter D_i in each of 6 classes. For these data we have provisionally used the Mediterranean measurements by Dulac et al. (1989) to represent Southern European conditions. To represent Northern European conditions, we used particle data measured in Norway by Cornille et al. (1991). These data were used (1) because they cover relatively long periods of measurements (3-12 months) rather than only short field campaigns; (2) they were collected at several sites and represent a wide geographical area rather than a single station; (3) the measurement sites were probably not significantly affected by local sources but nevertheless were influenced by distant anthropogenic sources. These characteristics are consistent with the assumptions of the TRACE model.

Comparisons of measured and computed air concentrations and wet depositions over the European continent are given in details in Alcamo et al. (1992). Briefly it should be noted that fairly good agreement was calculated for As and Pb air concentrations as well as wet depositions. The TRACE model systematically underestimated the Cd and Zn air concentrations and wet depositions (significant correlation between the measurements and computations but high bias occured between them). The reason for this underestimation could be that (1) measurements of Cd, Cu and Zn can be greatly overestimated if the collectors are not acid-washed (Ross, 1986), and (2) the emissions are underestimated or - in the case of Zn - significant source of Zn has not been taken into consideration.

DEPOSITION OF TRACE METALS OVER THE REGIONAL SEAS

The annual total deposition of heavy metals is an important indicator of their long-term impact on the environment because many metals are known to gradually accumulate among others in living organisms of surface waters. Figs. 1(a) through (d) and Figs. 2(a) through (d) present the computed country-to-sea total deposition (wet and dry) of lead and arsenic for the North, Baltic, Mediterranean and Black Seas. The calculations refer to 1982 for As and 1985 for Pb. These emission data were the most recent ones available for gridded emissions covering all of Europe.

Concerning the North Sea the largest contributions to the total Pb deposition are from the United Kingdom, West-Germany and The Netherlands. It is interesting that due to the high As emission the former Soviet Union contributed a lot to the total As deposition over the North Sea. In the case of the Baltic Sea the most of the origin of Pb and As deposited over the Sea is from the Soviet Union. On the basis of our estimation around one quarter of the Pb deposition over the Mediterranean Sea comes from Italian sources. In the case of Black Sea the countries surrounding this Sea (Soviet Union, Bulgaria and Romania) contribute the most to the Pb and As deposition.

The total deposition of As, Cd, Pb and Zn to the North, Baltic, Mediterranean and Black Seas are summarized in Table 1.

Table 1. Total deposition of trace metals computed by TRACE model (t yr^1)

	As	Cd	Pb	Zn
North Sea	81.9	16.6	2247	821
Baltic Sea	138.4	20.1	1567	706
Mediterranean Sea	218.1	58.9	3549	2093
Black Sea	57.9	12.5	1101	493

Total deposition calculations for As, Cd and Zn refer to 1982 while those for Pb are based on 1985 gridded emission and meteorological database. Several other model simulations have been performed to estimate the total deposition of trace metals into the Regional Seas (e.g. Van Jaarsveld et al., 1986; Krell and Roeckner, 1988; Grassl et al., 1989; Van Jaarsveld and De Leeuw, 1993). Most of the model estimations concerning the total deposition of trace metals over the Regional Seas around Europe refers to the North Sea. This is not surprising since the North Sea is surrounded by highly industrialized countries which pollute it not only via the rivers and direct dumping but also the atmosphere. Total lead depositions calculated by different models are in the range of 1531 - 2878 t yr^1 (Krell and Roeckner, 1988; Grassl et al. 1989). The corresponding modelled range for Cd deposition is 14 - 20 t yr^1 (Krell and Roeckner, 1988; Warmenhoven et al., 1989). Petersen et al. (1989) estimated the Pb flux into the Baltic Sea to be 1807 t yr^1. Concerning the Mediterranean Sea (more exactly the Western part of it), total atmospheric flux estimation for Cd, Pb and Zn are available in the literature (Arnold et al., 1982).

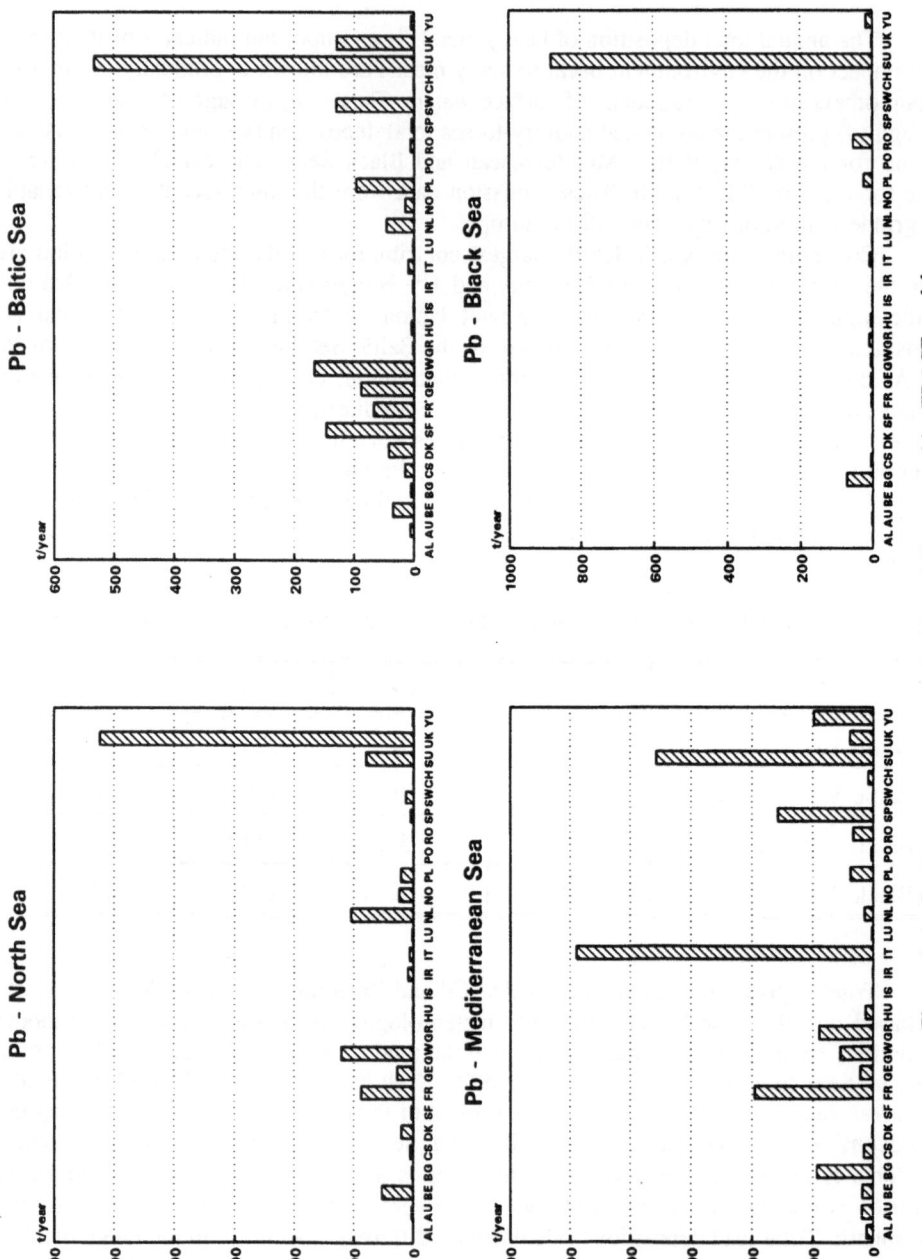

Figure 1. (a–d). Country-to-Sea total deposition of Pb computed by TRACE model.

132

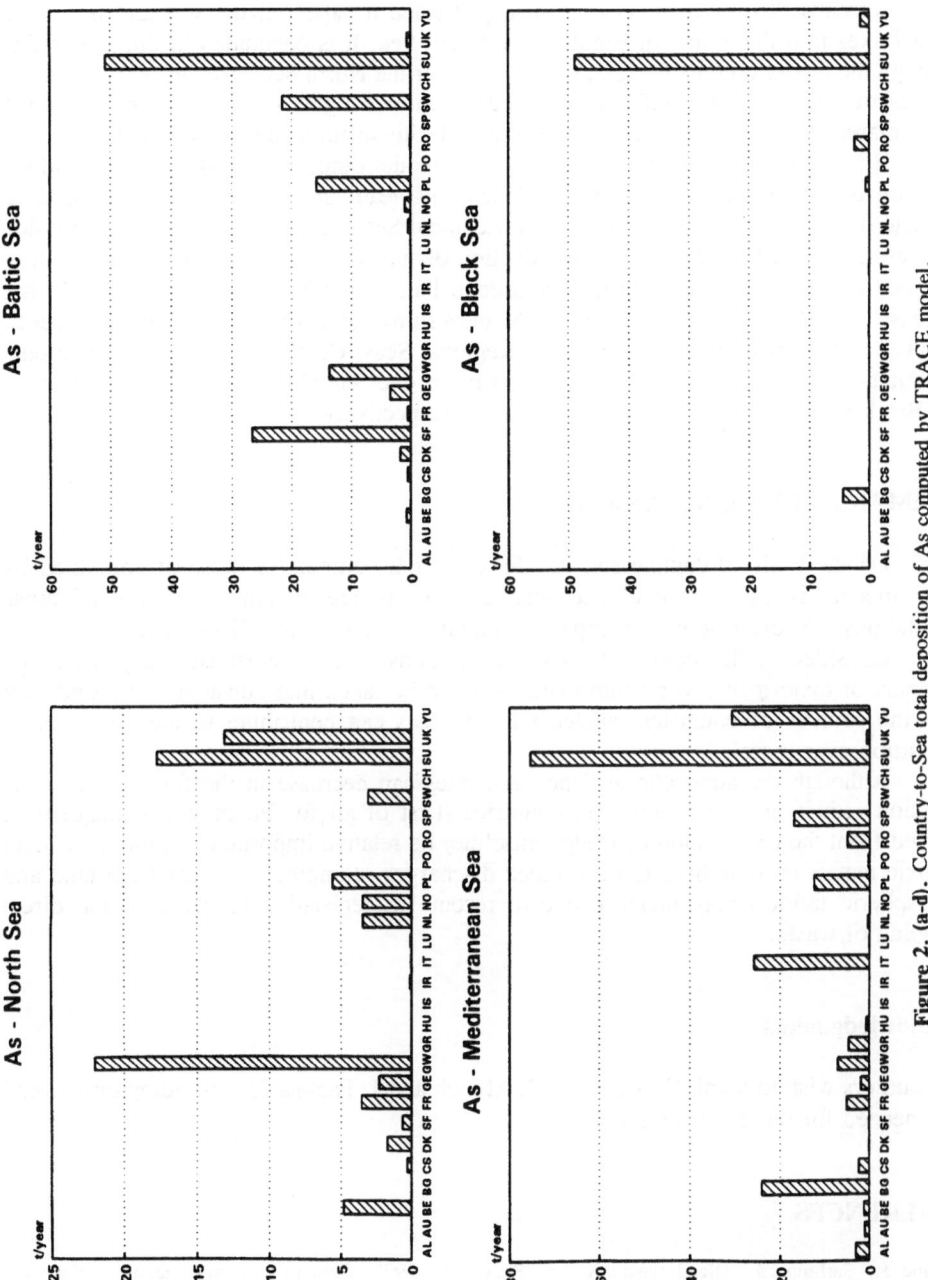

Figure 2. (a-d). Country-to-Sea total deposition of As computed by TRACE model.

Their estimations are 0.13, 10.5 and 11 kg km^{-2} yr^{-1}, respectively. Fluxes for Cd, Pb and Zn computed by means of TRACE model are 0.13, 7.83 and 4.62 kg km^{-2} yr^{-1}, respectively. It should be noted, however, that in calculations performed by TRACE almost the whole Eastern Basin of the Mediterranean Sea (as far as it is covered by EMEP-grids) is included.

Recently, Rojas et al. (1993) have published a paper on the wet and dry trace metal fluxes into the Southern Bight of the North Sea. It is concluded in this paper that although the model outputs for total deposition over the North Sea agree very well, on the basis of measurements the authors estimated 8-10 times higher deposition for Zn and Cd as compared to the model results whereas for Pb this difference is only a factor of 2.

It has been mentioned above that all the model outputs are based on the gridded emission data valid for the mid-80's. Concerning lead, an attempt has been made to estimate the total deposition of Pb into the Regional Seas based on country-total emission data referring to 1989. Due to the introduction of fuel with lower lead content and even unleaded gasoline in Europe during the second half of the 80's the emission of Pb has decreased significantly, first of all in the Western European countries. It should also result in lower total depositions of Pb over the Regional Seas. On the basis of TRACE model computations the 1989/1985 total deposition ratios over the North, Baltic, Mediterranean and Black Seas are 0.58, 0.80, 0.74 and 0.98, respectively.

SUMMARY AND CONCLUSIONS

On the basis of comparison of TRACE model outputs to those of computed by other models it can be concluded that our results are within the range of those calculations. Unfortunately no comparison could be done for the Black Sea.

Considering the country-to-sea contributions it can be stated that long-range transport of toxic metals is an important factor to be taken into consideration, since not only the sources of countries, adjacent to the seas can contribute to the atmospheric deposition over there.

Although the atmospheric deposition rates can decrease in the future due to the emission reduction in the European countries (first of all for Pb of which majority is emitted from the combustion of leaded gasoline) its relative importance in the total input of toxic metals into the Seas (rivers, piped discharges, dumpings, run-off from land and atmospheric fallout) may increase due to recent EC legislation to decrease the direct dumping of waste.

Acknowledgments

The authors wish to thank F. Axenfeld, J. Münch and J. Pacyna for providing emissions' data needed for these calculations.

REFERENCES

Alcamo J., Bartnicki J., Olendrzynski K. and Pacyna J., 1992, Computing heavy metals in Europe's atmosphere-I. Model development and testing, *Atmos. Environ.* 26A: 3355-3369.

Arnold M., Seghaier A., Martin D., Buat-Menard P. and Chesselet R., 1982, Geochemie de l'aerosol au-dessus de la Mediterranee occidentale. *Workshop on Pollution of the Mediterranean*, Cannes, France.

Bozó L., Alcamo J., Bartnicki J. and Olendrzynski K., 1992, Total deposition and budgets of heavy metals over Eastern Europe, *Idöjárás* 96:61-80.

Chan W.H., Tang A.J.S., Chung D.H.S. and Lusis M.A., 1986, Concentration and deposition of trace metals in Ontario-1982, *Wat. Air Soil Pollut.* 29:373-389.

Cornille P., Maenhaut W. and Pacyna J., 1991, PIXE analysis of size-fractionated aerosol samples collected at Birkenes during spring 1987, *Nucl. Instr. Meth. Phys. Res.* B50:310-316.

Dulac F., Buat-Menard P., Ezat U., Melki S. and Bergametti G., 1989, Atmospheric input of trace metals to the western Mediterranean: uncertainties in modelling dry deposition from cascade impactor data, *Tellus* 41B:362-378.

Grassl H., Eppel D., Petersen G., Schneider B., Weber H., Gandrass J.G., Reinhardt K.H., Wodarg D. and Fliess J., 1989, Stofeintrag in Nord und Ostee über die Atmosphare. *GKSS Report 89/E/8*, GKSS, Geesthacht, Germany.

Krell U. and Roeckner E., 1988, Model simulation of the atmospheric input of lead and cadmium into the North Sea, *Atmos. Environ.* 22:375-381.

Lindberg S., 1982, Factors influencing trace metal, sulfate and hydrogen ion concentrations in rain, *Atmos. Environ.* 16:1701-1709.

Rojas C.M., Injuk J., Van Grieken R.E. and Laane R.W., 1993, Dry and Wet deposition fluxes of Cd, Cu, Pb and Zn into the southern bight of the North Sea, *Atmos. Envir.* 27A:251-259.

Ross H., 1986, The importance of reducing sample contamination in routine monitoring of trace metals in atmospheric precipitation, *Atmos. Environ.* 20:401-405.

Sehmel G.A., 1980, Particle and gas deposition: a review, *Atmos. Environ.* 14:983-1011.

Slinn W.G.N., 1984, Precipitation scavenging, *in*: Atmospheric Science and Power Production, ed. Randerson D, Technical Information Center, Office of Scientific and Technical Information, United States Department of Energy.

Van Jaarsveld J.A. and De Leeuw F.A.A.M., 1993, Source receptor relations for the calculation of atmospheric deposition to the North Sea: Nitrogen and Cadmium, *RIVM Report* No.222401002, Bilthoven, The Netherlands.

Van Jaarsveld J.A., Van Aalst R.M. and Onderdelinden D., 1986, Deposition of metals from the atmosphere into the North Sea: model calculations. *RIVM Report* 842015002, Bilthoven, The Netherlands.

Warmenhoven J., Duisen J., de Len L. and Veldt C., 1989, The contribution of the input from the atmosphere to the contamination of the North Sea and the Dutch Wadden Sea. *TNO Report* R89/349A.

DISCUSSION

T. IVERSEN   What are the physico-chemical reasons behind choosing different scavenging ratios?

L. BOZO   The use of a larger svavenging ratio for As, Cd and Zn than for Pb is consistent with experimental evidence and the theoretical argument that above a certain minimum diameter the scavenging ration increases with increasing particle diameter. The mass median diameters of aerosol particles containing As, Cd and Zn in the atmosphere is usually larger than that of Pb.

T. IVERSEN   How do these atmospheric contributions compare with contributions from river run-off?

L. BOZO   The atmospheric input into the Seas around Europe gives a significant contribution to the total (atmospheric deposition, river run-off, dumping, discharges) input of trace metals. The contribution of atmospheric input depends on which toxic metal and Sea is considered. For example the atmospheric input of Pb and Cd into the North Sea has been estimated to be approximately one third of the total.

H. SCHLÜNZEN   Did you include non-European emission data?

L. BOZO   The TRACE model focuses on European-scale emission, transport and deposition so non-European sources were not considered.

H. SCHLÜNZEN   Are the results for Black and Mediterranean Seas reliable?

L. BOZO   The TRACE model results could be compared to other model outputs. Good agreements were found in the case of North and Baltic Seas. No model comparison was made for the Mediterranean and Black Seas due to the lack of other model results. Cd, Pb and Zn total atmospheric fluxes into the Mediterranean Sea computed by TRACE were compared to data measured by Arnold et al. (1982) which is summarized in our paper.

S.J. ROBERTO   Have you made any estimation about the importance of the wet and dry deposition respecting the total deposition?

L. BOZO   Yes, we have. The contribution of wet deposition to the total deposition of trace metals considered is between 70 - 80% over these Seas.

MODEL OF LONG-RANGE POLLUTANTS TRANSPORT AND ACIDITY
OF PRECIPITATION FOR BALTIC REGION

D. Syrakov, M. Kolarova

Hydrometeorological Service
Tsarigradsko chousse 66
1184 Sofia, Bulgaria

D. Perkauskas, K. Senuta and A. Mikelinskene

Institute of physics
Savanoriu 231
2028 Vilnius, Lithuania

In the present paper the investigation of the long-range pollutants transport includes the formation of the acid precipitation. For this purpose a proper combination of models which describe the different stages of this complex phenomenon is realised [1,2].

In the paper a combination between the Eulerian and Lagrangian approaches for investigation of hydrodynamical problems is presented. Any volume of polluted air is identified by the trajectory of its center. The evolution of this object (its diffusion and the physical-chemical processes) is investigated on the basis of an analytical solution of the appropriate differential equation in Eulerian coordinates:

$$\vec{r}_c(t+\Delta t) = \vec{r}_c(t) + \vec{c}(\vec{r}_c)\,\Delta t \tag{1}$$

where Δt is properly defined time step which secures the linearization of the process. The vector $\vec{c}$ is the external velocity of the flow in the point (x^c, y^c, z^c) in the moment t which has components u,v,w.

The theoretical and experimental investigations reveal that the diffusion process in the atmospheric PBL can be separately investigated in horizontal and vertical directions. This fact permits the concentration in a point (x,y,z) and moment t due to a particular volume to be written as:

$$C^k_{ij}(x,y,z,) = Q^k_{ij}(t_{ij})\,q_h(x,x^c_{ij},y,y^c_{ij},t_{ij})\,q_z(z,z^c_{ij},t_{ij})\,q_w(t_{ij}) \tag{2}$$

where Q^k_{ij} is the pollutants quantity in the j-th puff emitted by the i-th source, q_h and q_z are horizontal and vertical distribution, q_w is the wash-out function, t_{ij} is the puff lifetime.

For the determination of horizontal distribution function a Cartesian coordinate system with its origin in the puff center is chosen. The diffusion field in the case of horizontally homogeneous turbulence satisfies the following equation:

$$\frac{\partial q_h}{\partial t} = K_h \left(\frac{\partial^2 q_h}{\partial \xi^2_1} + \frac{\partial^2 q_h}{\partial \xi^2_2} \right) \tag{3}$$

where K_h is the horizontal turbulent exchange coefficient.

The initial conditions for the solution of the eq.(3) are:

$$q_h(r,o) = \begin{cases} A\exp(-r^2/a_0^2), & |r| \leq r_0 \\ 0, & r > r_0 \end{cases} \tag{4}$$

where $r^2 = \xi^2_1 + \xi^2_2$, r_0 is radius of the surface source, a_0 is variable proportional to the dispersion, A is constant, determined on the basis of mass conservation law.

If the coordinate system is taken on the earth surface in point (x,y) then concentration due to a source with center (x^c, y^c) will be:

$$q_h(x, x^c, y, y^c, t) = \frac{1}{\pi a^2(t)} \exp - \frac{(x-x^c)^2 + (y-y^c)^2}{a^2(t)} \tag{5}$$

The function q_z which determines the vertical distribution of the pollutants can be obtained on the basis of the semi-empirical turbulent diffusion equation:

$$\frac{\partial q_z}{\partial t} = K_z \frac{\partial^2 q_z}{\partial z^2}, \, K_z = const \tag{6}$$

subject to the following general boundary and initial conditions:

$$q_z = \delta(z - z_c), \, t=0$$

$$K_z \frac{\partial q_z}{\partial z} = \beta q_z, \, z=0 \tag{7}$$

$$K_z \frac{\partial q_z}{\partial z} = 0, \, z = z_i$$

In eq.(6) β is constant which determines the interaction between pollutants and underlying surface, z_i is the mixing height.

The system of equations (6)-(7) describes the different diffusion conditions which are possible in the PBL and the solution is presented in [3].

The most important mechanism for the formation of acid precipitation is the direct transformation of sulphur dioxides:

$$\frac{d}{dt} [SO_2] = -K_1 [SO_4^{2-}] \tag{8}$$

$$\frac{d}{dt} [SO_4^{2-}] = K_1 [SO_2]$$

The symbol [ ] denotes concentration (or quantity) of a given pollutant. When using eq.(8) the sulphur balance is only secured.

Chemical transformations which undergo the two basic NO and NO_x pollutants are rather complicated. Smog chamber measurements of the typical chemical reactions of the NO compounds show that 90% of them are realised among NO, NO_2, PAN (peroxyacetil nitrate) and NO_3^-. This allows to include in the model only the chemical transformations of these five components [4]:

$$\frac{d}{dt} [NO_2] = -K_2 [NO_2] + K_4 [PAN] \tag{9}$$

$$\frac{d}{dt} [PAN] = 0.5 K_2 [NO_2] - (0.5 K_3 + K_4) [PAN]$$

$$\frac{d}{dt} [NO_3^-] = 0.1 (K_2 [NO_2] + K_3 [PAN])$$

$$\frac{d}{dt} [HNO_3] = 0.4 (K_2 [NO_2] + K_3 [PAN])$$

$$0 = [NO_2] + K_5 [NO]$$

The eqs.(9) are linear and rate of chemical transformations are functions of time and space.

The interaction of the pollutants with the precipitation causes a decrease of their quantities in the atmosphere because of wash-out process which in turn increases the acidity of precipitation. The wash-out process is described by the function $q_w(t_{ij})$ in eq.(2) which is solution of the equation:

$$\frac{d}{dt} q_w{}^k = -\lambda_w{}^k q_w{}^k, \; q_w{}^k(0) = 1 \tag{10}$$

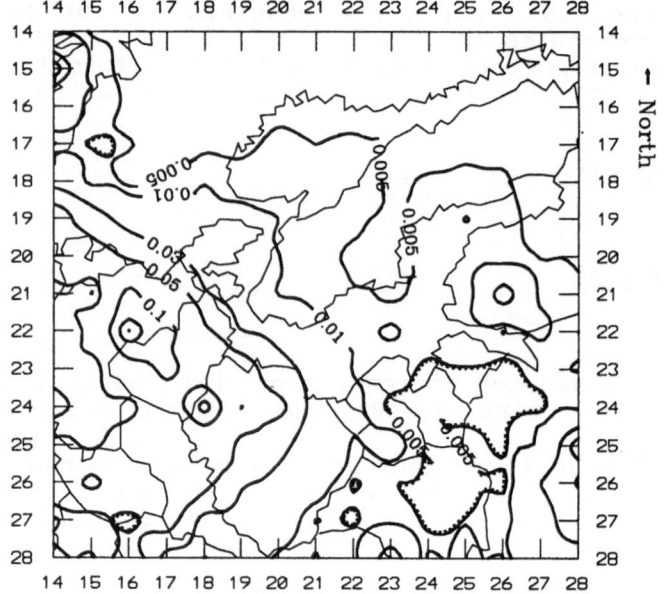

Figure1. Mean SO$_2$ concentration (mg/m^3)

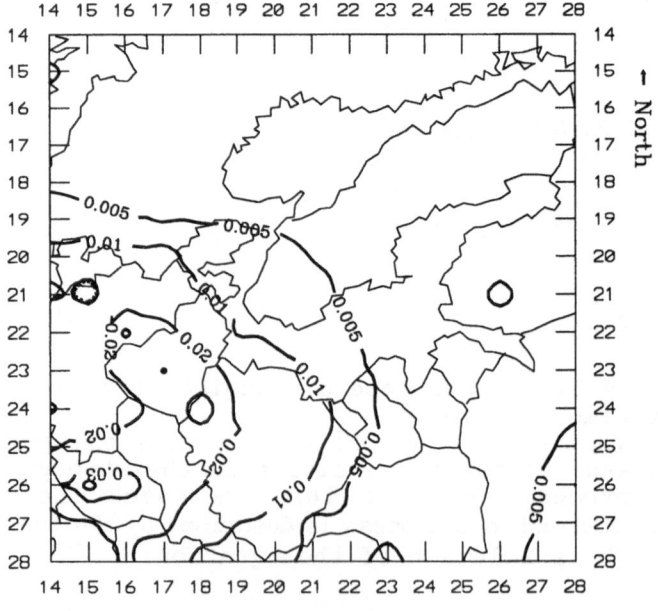

Figure2. Mean NO$_2$ concentration (mg/m^3)

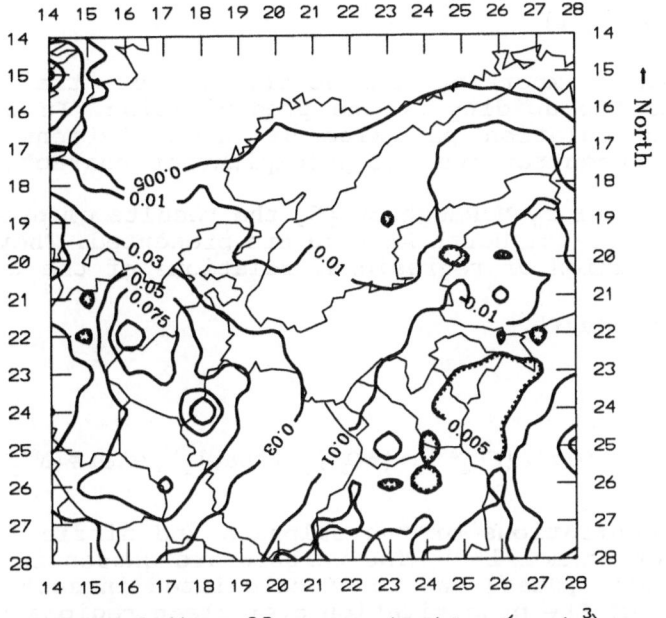

Figure3. Mean SO₂ concentration (mg/m³)

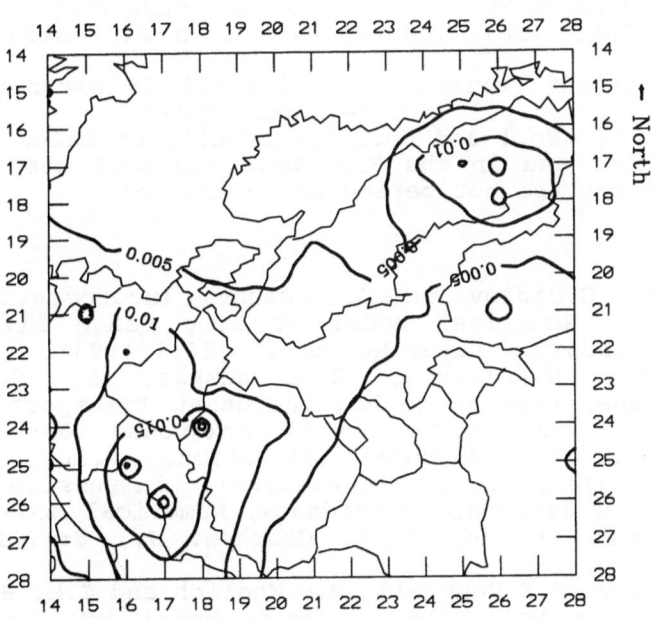

Figure4. Mean NO₂ concentration (mg/m³)

where $\lambda_w{}^k$ is the wet deposition coefficient for the k-th pollutant. Apparently

$$q_w{}^k(t) = \delta_w \exp(-\lambda_w{}^k t)$$

The most important characteristic of the long-range transport is the acidity of the precipitation. In the study a model for calculation pH value of precipitations using the background concentrations and precipitation characteristics is proposed.

In the paper by McNaughton [5] the results of precipitation acidity analysis for North America are presented. They were used for the derivation of regresional relations of the type:

$$pH = a_1 p\sum (anions) + b_1 \qquad\qquad (11)$$

where

$$p\sum [anions]_r = -\log(2[SO_4{}^{2-}]_r + [NO_3{}^-]_r + [base]_r), \ where [NO_3{}^-], [SO_4{}^{2-}]$$

are the concentrations of the sulphate and nitric ions in the precipitation in mol/l . The weights are chosen in accordance to the valency. [Base] has to be calculated in such a way that the pH value of the precipitation over clean regions to have the characteristic contemporary value of 5.6.

The numerical model for long range pollutants transport and acidity of the precipitation is applied to the region of Baltic sea and part of the Central and East Europe which comprises Baltic republics, Belorusia, Poland, Germany, Denmark, Sweden, Finland, part of Russia, Hungary, Chechoslovakia, Austria. The square net consists of 15x15 points and space step 150km. It is part of the EMEP grid. The emission data was taken from EMEP reports.

Two different periods from 23 until 29 January 1986 and from 18 until 24 July 1986 were investigated.

In the Fig. No.1 and No.2 the results of calculations for the cold period and in the Fig. No.4 and No.5 the results of calculations for the hot period are presented.

LITERATURE

1. D.Syrakov, G.Djolov and D.Yordanov, Incorporation of PBL Dynamics in a Numerical Model of Long Range Air-Pollutant Transport, Boundary - Layer Meteorol., 26:1(1983).
2. D.Syrakov, M.Kolarova, D.Perkauskas, K. Senuta and A.Mikelinskiene, Long-Range Air Pollutant Transport Model for Baltic Region, in: Problems of Background Monitoring and Assessment of Environmental Situation, F.Rovinsky, ed., Hydrometeorological Pub., Moscow,(1989). (in russian)
3. D.Syrakov, G.Jolov and D.Yordanov, Numerical Model of Long-Range Transport in PBL, Bulg. Geophys. J., 8:1,(1982). (in russian)
4. R.Brodzinsky, B.K.Cantrell, R.M.Endlich and C.M. Bhumvalkar, A Long-Range Air Pollution Transport Model for Eastern-North America - II. Nitrogen Oxides., Atmos. Environ., 18:2360(1984).
5. D.J.McNaughton, Relationships between Sulphate and Nitrate Ion Concentrations and Rainfall pH for Use in Modelling Applications, Atmos. Environ., 15:1075(1981).

MODELLING THE LONG-RANGE TRANSPORT AND DEPOSITION OF PERSISTENT ORGANIC POLLUTANTS OVER EUROPE AND ITS SURROUNDING MARINE AREAS

J.A. van Jaarsveld, W.A.J. van Pul, and F.A.A.M. de Leeuw

National Institute of Public Health and Environmental Protection (RIVM)
P.O. Box 1
3720 BA Bilthoven
The Netherlands

INTRODUCTION

For a long time it has been recognized that the atmosphere is an important pathway for the transport of pollutants from industrialized and densely populated areas to ecosystems near to or far from these areas. In particular, the important role of the atmospheric pathway in the pollution of Europe's marginal seas by land-based sources was identified more than a decade ago (Cambray et al., 1979; Van Aalst et al., 1982). In case of persistent organic pollutants (POPs) transport through the atmosphere may even lead to hemispheric or global distribution of the pollutants.

POPs have not yet been included in Europe-wide monitoring networks such as are run by the ECE-EMEP programme. Only in the PARCOM-ATMOS network (mainly around the North Sea) some pesticides are part of the (voluntary) monitoring programme.

The lack of measurements with sufficient geographical coverage is the main reason for using an emission-based modelling approach. Modelling, however, also has the important advantage of being able, directly or indirectly, to relate emissions resulting from human (economic) activities to pollutant levels in the environment; it thus provides the opportunity of predicting the levels to be expected in association with specific economic scenarios.

This paper describes how the atmospheric deposition of benzo(a)pyrene (B(a)P) and lindane (γ-HCH) to land and sea surfaces in Europe is calculated on the basis of their emissions in Europe. B(a)P and γ-HCH are selected as representatives of organic compounds mainly in particle-phase and mainly in gas-phase respectively. The work is being carried out as a part of the so-called ESQUAD project (Van den Hout, 1994), a pilot study aimed at the development of a modelling system for non-acidifying pollutants analoguous to similar systems for acidification analysis. The work presented here is described in more detail in Van Jaarsveld (1994) together with similar calculations for Cd, Cu and Pb.

Air Pollution Modeling and Its Application X, Edited by S-V. Gryning
and M. M. Millán, Plenum Press, New York, 1994

EMISSIONS

The 1990 emission data used in this study were inventoried as a part of the ESQUAD-project by TNO (Berdowski and Veldt, 1994). Emissions are given for each 1^0 longitude x 0.5^0 latitude grid square. In addition, a number of large point sources are included which are identified by their coordinates.

B(a)P emissions are dominated by domestic (wood) combustion. It may be expected that in wintertime emissions are high and also ambient air concentrations. The effect of higher emissions in situations with lower mixing volumes as frequently occur in wintertime is accounted for in the model by taking emissions as a function of the degree-day value.

Lindane is used as an insecticide, on crops as well as in soil. Highest levels are usually found in spring and early summer. For this study it is assumed that emissions of lindane are constant in time. It is not expected that this will systematically affect annual mean concentrations. In Figure 1 annual variations are given as measured in the Netherlands. The γ-HCH pattern is very similar to those measured at other PARCOM-ATMOS stations.

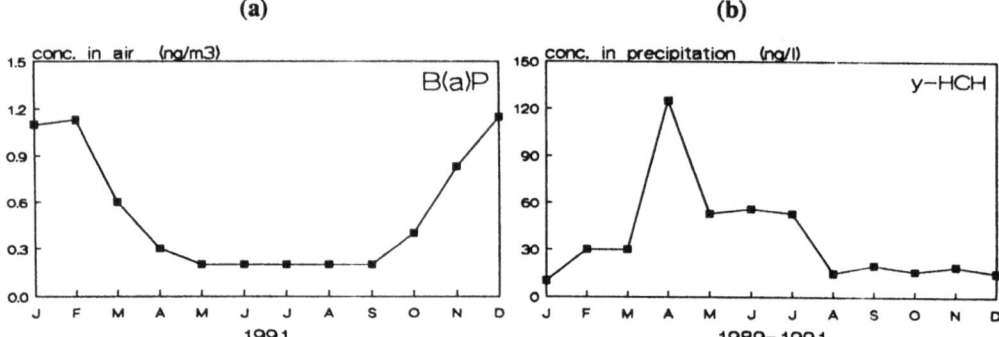

(a)   **(b)**

Figure 1. (a): Monthly mean concentrations of benzo(a)pyrene in air in 1991 for an urban site (Rotterdam). Mean concentrations based on 5-6 samples/month (DCMR, 1992). (b): Monthly mean concentrations of γ-HCH in precipitation for the period 1989-1990 at the PARCOM-ATMOS site in the Netherlands.

THE TRANSPORT AND DEPOSITION MODEL

The model used for the calculations is the so-called TREND model (Van Jaarsveld and Onderdelinden, 1994) which has been used in previous studies of deposition to the North Sea (Van Jaarsveld *et al.* 1986, Warmenhoven *et al.*, 1989) and the Rhine catchment area (Baart and Diederen, 1991). The model was originally developed for the calculation of transport and deposition of acidifying compounds such as ammonia (Asman and Van Jaarsveld, 1992). In these studies also the more general model concept was validated by comparing model results with measurements of concentrations in air and in precipitation (Van Jaarsveld, 1989).

The version used here (EUTREND v1.11) has recently been upgraded to cover the entire European continent with its marginal seas and to calculate deposition as a function of surface characteristics. This Lagrangian-type model is driven by meteorological data obtained through The Netherlands Meteorological Institute (KNMI) from the European Centre for Medium range Weather Forecasts (ECMWF) in Reading, England in combination with synoptic surface observations from more than 1500 stations in Europe, also obtained from ECMWF.

Atmospheric processes included in the model are: emission, dispersion, advection, chemical conversion and wet and dry deposition. With respect to deposition, it is important to mention here that the model describes the behaviour of pollutants attached to particles as a function of particle size. Chemical reaction rates are used in the model independently of concentration levels, which means that only linear chemistry can be described. The model has variable spatial resolution, although it is applied here using a fixed receptor grid with a 1^0 long. x 0.5^0 lat. resolution.

DEPOSITION AND CONVERSION PARAMETERS

Wet deposition

Although the TREND model discriminates between washout and rainout, in the following it is described as a bulk process. In such a case it is implicitly assumed that the layer below the cloud is well mixed so that the concentration at ground level is equal to the concentration at the cloud base. The wet deposition flux F_w (g m^{-2} h^{-1}) during a precipitation event can be calculated as:

$$F_w = C_p R_i \tag{1}$$

where R_i is the average rain intensity (m h^{-1}) and C_p the average concentration in precipitation during the event (g m^{-3}). C_p is related to the (local) average concentration in air C_a by means of a scavenging ratio $W = C_p / C_a$. The wet scavenging rate during precipitation Λ_w [h^{-1}] is related to the scavenging ratio W through: $\Lambda_w = W R_i / z_i$, where z_i is the mixing layer height. In case of gases the (equilibrium) ratio of the concentration of a substance in water and air can also be related to the Henry coefficient for a specific substance using: $W = R T / H$, where R is the gas constant (8.3 Pa m^3 mol^{-1} K^{-1}), T the temperature (K) and H the Henry coefficient (Pa m^3 mol^{-1}). The latter relation assumes that the gases are reversibly soluble in water i.e. do not react and/or ionize in the rain droplets and do not attach to particles in the droplets.

Dry deposition

The dry deposition flux (g m^{-2} h^{-1}) can be calculated from:

$$F_d = v_d(z) (C_a(z) - C_s) \tag{2}$$

where $v_d(z)$ is the dry deposition velocity (m s^{-1}) at height z, $C_a(z)$ is the average air concentration at height z [g m^{-3}] and C_s the average air concentration at the absorbing surface (soil, vegetation or water). C_s may be considered zero for substances which immediately react at the surface with other substances or for substances attached to particles. The dry deposition velocity $v_d(z)$ is described in the model as the inverse of three resistances: $v_d(z) = (r_a(z) + r_b + r_c)^{-1}$, where $r_a(z)$ presents the resistance of the turbulent layer with height z, r_b the resistance of the quasi laminar layer immediately adjacent to the surface and r_c the (internal) resistance of the receiving surface. r_a and r_b are calculated in the model according to formulations given by Wesely and Hicks (1977) as functions of roughness length, friction velocity and Monin Obukhov length. The latter parameters are calculated from routinely measured meteorological parameters using the scheme of Beljaars and Holtslag (1990).

Gas/particle partitioning

Important for removal and deposition processes is the physical state of pollutants in the atmosphere. Mainly depending on the vapour pressure pollutants may be in the gas phase or in the particle phase or occur in both phases. Junge (1977) proposes the following model for the gas-particle partitioning of semi-volatile organic compounds in the atmosphere:

$$\phi = c \theta (p^0_L + c \theta)^{-1} \tag{3}$$

where ϕ is the ratio of adsorbed organic vapour on aerosol to the total amount of vapour in air, θ the aerosol surface area (m^2 m^{-3} air), p^0 the solute saturation vapour pressure (Pa) and c a constant that depends on heat of condensation and molecular weight. Junge assumed $c \sim 0.17$ Pa m for high molecular weight organics. Since vapour pressures are strongly temperature dependent, the fraction of a substance absorbed to particles will also be temperature dependent. For certain organics this may mean that in tropical regions the organic is mainly in vapour phase while for arctic regions the particle phase is dominant.

The particle phase fractions calculated are 0.002 and 0.83 for γ-HCH and B(a)P respectively. These

145

figures are roughly in agreement with observations. On the basis of these results it was decided to model γ-HCH as a gas exclusively and B(a)P in particle phase exclusively. This choice is for B(a)P further justified by considering average European temperatures being closer to 10^0C than the 20 ^{0}C at which vapour pressures are specified.

Deposition parameters for B(a)P

Particle size distribution. With respect to particle-size distributions is essential to know the sizes of the particles as they are when released to the atmosphere. Another problem is that pollutants may not be distributed over (carrier) particles according to their mass or volume but - in case of pollutants which appear in both gas and vapour phase - according to their surface area. In fact both production process parameters play a role in this as well as the way exhaust gases are treated by cleaning equipment. Little quantitative information has become available about the processes in which particles are formed and evolve and the processes in which pollutants are distributed over or incorporated in carrier particles. With respect to sizes and morphological properties of emitted particles some general aspects may be mentioned. First, most filter equipment remove larger particles more efficiently than smaller particles. Second, some combustion sources such as coal-fired power plants produce fly ash predominately consisting of nonporous spheres (Schure *et al.*,1985) while other such as refuse incinerators may produce irregularly shaped particles with a high specific surface area (Taylor *et al.*, 1982). Third, it has been shown that large specific surface areas are attributable to carbonaceous particles of highly porous character (Schure *et al.*, 1985).

Table 1. Properties of the particle size classes with respect to dry and wet deposition for land surfaces. Dry deposition velocities are representative for a roughness length of 0.15 m

size class	median aerodyn. diam.	initial mass distribution for B(a)P	scavenging ratio W [a])	mean scavenging rate Λ_{eff}	mean dry dep. velocity v_d [c])	mean atm. residence time $T_{1/2}$ [b])
μm	μm	%		s^{-1}	m s^{-1}	h
< 0.95	0.2	14	1.2×10^5	2.0×10^{-6}	0.00065	63
0.95 - 4	1.5	9	10^6	1.5×10^{-5}	0.0025	11
4 - 10	6	10	10^6	1.5×10^{-5}	0.0071	8
10 - 20	14	17	10^6	1.5×10^{-5}	0.0132	6
> 20	40	50	10^6	1.5×10^{-5}	0.067	2

[a]) scavenging ratio during precipitation
[b]) $T_{1/2} = \ln 2/(v_d/zi + \Lambda_{eff})$ where zi is the mixing layer height
[c]) dry deposition velocity for z= 50 m

The particle-size distribution for B(a)P used in the calculations is given in Table 1. For this substance it is assumed that it is mainly attached to soot particles which originate from domestic open stoves not equipped with emission reduction devices. The particle-size distribution is assumed to be similar to that of an (old) municipal waste incinerator as recently measured in the Netherlands (Slob *et al.*,1992). From these measurements it followed that a large part of the emitted dust consists of fast settling particles. The effect of the size distribution on the calculated deposition is investigated by means of a few sensitivity runs by Van Jaarsveld (1994).

Deposition parameters for γ-HCH

Wet deposition. Empirical W values for γ-HCH reported by Atlas and Giam (1988) range from 14000-41000. Similar W values can be calculated on the basis of Henry constant values reported in the literature. For the model calculations a (fixed) W value of 34000 is used (corresponding H = 0.073 Pa at 10 ^{0}C). This choice neglects the fact that W values may geographically differ in Europe due to temperature effects. Still the effect of the temperature range on W falls within the range of literature values of the Henry constant.

Dry deposition velocity for soils. Relatively little is known about dry deposition velocities for the more persistent organics such as γ-HCH. We assume that the initial dry deposition velocity for uncontaminated soil ($C_s = 0$) will be mainly determined by atmospheric resistances. This assumption leads to deposition velocities in the range of 0.002-0.005 m s^{-1}.

Dry deposition velocity for water surfaces. The relatively high water solubility of γ-HCH and the relatively rapid mixing in surface waters will lead to negligible resistances in water layers compared to resistances in the atmosphere (e.g. Liss and Slater, 1974). Therefore the standard resistance model is applied with a variable r_a and r_b, and r_c set to zero. This gives an concentration weighted yearly averaged v_d for sea surfaces of approximately 0.004 m s^{-1}.

Flux of γ-HCH across interfaces; re-emission

Fluxes to soils; air/soil exchange. Jury *et al.* (1983) assume a half-life time for γ-HCH in soil in the order of a year. With such a half-life time, the concentration in the upper soil layer will soon limit the dry deposition flux because the surface concentration C_s will soon approach C_a. In a steady state situation the time averaged total deposition flux becomes equal to the sum of degradation, uptake by plants and leaching to ground water. Preliminary calculations with an air-soil exchange model similar to the model of Jury *et al.* (1983), reveals that (near) saturation of the upper layer will take place within days. The calculations also show a diurnal cycle of deposition and (re-) emissions, mainly driven by temperature and moist evaporation cycles.

In the current version of the TREND model it is not possible to calculate and administrate soil concentrations dynamically. The effect of saturation and the possible re-emission of previously deposited material is taken into account by introducing an effective dry deposition velocity v_d eff such that:

$$F_d = v_d(z) \; (C_a(z) - C_s) = v_d \, eff \; C_a(z) \tag{4}$$

It might be clear that such an approach is only valid for steady state situations which means in our situation for calculating of long term average deposition fluxes. An effective dry deposition velocity of 5×10^{-5} m s^{-1} is taken as suggested by Bakker *et al.* (1994) for forest soils. This deposition velocity results in a deposition flux which is almost negligible when compared to the average wet deposition flux. The calculations with the air-soil exchange model indicate that an important fraction of the wet deposited γ-HCH will volatilize after a shower. Therefore, the total deposition fluxes to land surfaces as presented here are probably overestimated.

Fluxes to sea surfaces; air-sea exchange. Vertical mixing in the upper layer of seas and oceans is high but usually limited to a layer of 10-100 m, the so called mixing layer. When the degradation rate in the water is low as is the case for γ-HCH and horizontal advection is neglected, then after some time the concentration in the water will become into equilibrium with the air above the water surface. The characteristic time τ_e in which this equilibrium will be reached can be calculated as:

$$\tau_e = \frac{h}{v_d \dfrac{H}{RT}} \tag{5}$$

where h is the thickness of the mixed layer (m). This approach suggests that the concentration in the water develops with time t according to $(1 - \exp(-t / \tau_e))$. For $h = 50$ m, $v_d = 0.004$ m s^{-1} and H taken at 10 ^{0}C, this time will be in the order of 10 years. For comparison: in case of CO_2 which is much less water soluble a characteristic exchange time constant is calculated in the order of 10 days. Exchange time constants for other organics such as PCBs may be in the order of months (GESAMP, 1989). With a τ_e of 10 years, processes in the water other than deposition become important, such as hydrolysis, photolysis, sorption-desorption and sedimentation (Saleh *et al.*, 1982). The conclusion is that a water to air flux is only temporally possible in Europes marginal seas. The atmosphere is, however, not the only source of γ-HCH. At places where rivers deliver their inputs the surface water, concentrations may be much higher and this may be extended over significant areas because of the poor mixing of fresh and salt water. Another conclusion from the above is that oceans and seas provide an enormous buffering capacity for γ-HCH or in general for organics with a low Henry's law constant. Due to this buffering capacity re-emission to the atmosphere will only take place when atmospheric concentrations become (temporally) very low. Transport to remote areas via repeated deposition/emission processes as sometimes suggested is therefore not likely an important pathway.

Saturation of surface water. Measurements of γ-HCH concentrations in sea surface waters with some geographical coverage are not available. Also reliable information about seasonal variations is lacking. However, on the basis of an exchange time constant of 10 years it can be expected that the seasonal variation is small.

The average concentration in surface water of the North Sea is in the order of 1 μg m^{-3} (RIZA/DGW, 1992). The corresponding equilibrium concentration in the atmosphere is in the order of 30 ng m^{-3} From the calculated values of C_a for the same area and estimated variability in time due to changing meteorological conditions and variations in emissions it is estimated that this equilibrium concentration is exceeded in at least 90% of the time. So a flux from the sea to the atmosphere will not take place in more than 10% of the time. It is expected that this is for other regional seas not significantly different.

Similar to the method followed for the calculation of dry deposition fluxes to soils the effect of saturation of the water and possible temporarily changes in flux directions have been accounted for by taking an effective dry deposition velocity v_d *eff* of 0.002 m s^{-1}. The full effect from (temporal) saturation can only be taken into account when an atmospheric model can be dynamically coupled to a water quality model.

Reaction rates for B(a)P and γ-HCH

B(a)P. A number of experimental studies have demonstrated that many PAH are susceptible to photochemical and/or chemical oxidation under simulated atmospheric conditions e.g. Thomas *et al.*, (1968). B(a)P may be considered as one of the more reactive PAHs (Masclet *et al.*, 1986). Although the degradation rates by photolysis of different PAH show wide variation, it has become apparent that the extent of photochemical decay is strongly influenced by the nature of the substrate on which they are absorbed. Baek *et al.* (1991) give a review of results obtained by different studies on PAH behaviour. From their work it can be concluded that for atmospheric conditions no conversion rates have been quantified. High conversion rates are, however, unlikely because there are strong indications that B(a)P is long range transported (Lunde and Bjorseth, 1977).

For this study the conversion rate is set to zero as a first approach mainly because it is expected that B(a)P carrying fly ash particles have a high carbon content and the corresponding photolytic half-life time would be in the order of hundreds of hours as suggested by Behymer and Hites (1988). The impact of applying the full range of possible conversion rates is subject to further investigation.

γ-HCH. About conversion of γ-HCH to other isomers in the atmosphere is not much known. Steinwandler (1976) carried out some laboratory test in which he radiated γ-HCH in different solutions with UV light. The reaction product was mainly α-HCH. Two hour illumination of γ-HCH on a glass surface resulted in a α-HCH conversion of about 2%. Results of this study cannot be translated into reaction rates for atmospheric conditions. It is, however, generally accepted that γ-HCH is quite stable in the atmosphere.
The conversion rate of 2.5 x 10^{-7} s^{-1} used in this study is taken from an earlier model study carried out by Warmenhoven *et al.*(1989).

BACKGROUND CONCENTRATIONS

Influences of sources outside the emission area are taken into account by adding background concentrations to both concentrations in air and rain. The values of these concentrations are estimated from reported measurements in remote areas such as the Pacific Ocean and the Arctic. Background air concentrations for B(a)P and γ-HCH were taken as 0.001 and 0.015 ng m^{-3} respectively. Corresponding values for concentrations in precipitation were 0.1 and 0.3 ng l^{-1} respectively. For both substances considered in this study the background concentrations play only a role close to the borders of the receptor area.

RESULTS

Maps of calculated total deposition for the 2 substances are given in Figure 2 and 3. The deposition pattern for B(a)P reflects rather closely the emission densities of the substance in the different countries. High B(a)P deposition is found in Germany and Eastern European countries mainly because of the relatively large amount of wood that is burned in these areas. The deposition pattern of the gaseous γ-HCH deviates from the other (particle phase) substances in a way that the deposition is relatively high for the regional and marginal seas compared to the deposition to land surfaces. This is mainly due to the fact that the dry deposition velocity of γ-HCH is much more surface type dependent than that of substances attached to particles. For B(a)P dry and wet deposition processes result in deposition loads which are of comparable magnitude over land areas. For the marginal seas and also for remote land areas wet deposition is almost twice the dry deposition load. The situation for γ-HCH is different. Here, for land areas dry deposition is in the order of 10% of wet deposition, while for water surfaces dry deposition is twice the wet deposition.

From mass balance calculations it appears that more than 80% of the emitted B(a)P is deposited in the chosen receptor area and only 45 % of the emitted γ-HCH. With the removal rates used in this study, atmospheric half-life times can be calculated at 3-7 days. Such half-life times easily enable direct transport from areas in Central Europe to the Arctic.

Effective dry deposition velocities, wet deposition rates and degradation rates as they follow from the calculations are summarized in Table 2. The difference in dry deposition velocities for land and sea surfaces is partly due to the difference in roughness length and partly due to the different atmospheric stability regimes. The low effective dry deposition velocity of γ-HCH to land surfaces is essentially the result of surface saturation caused by the limited diffusive transport in soils as discussed earlier.

Table 2. Effective yearly average dry deposition velocities and wet deposition rates.

	dry deposition velocity mean (min.-max.)		mean wet deposition rate	mean degradation rate
	land surface	sea surface		
	m s^{-1}	m s^{-1}	s^{-1}	s^{-1}
B(a)p	0.0063 (0.002-0.02)	0.0032 (0.0005-0.0045)	5 x 10^{-6}	0
γ-HCH	5 x 10^{-5}	0.0019 (0.0016-0.0020)	9 x 10^{-7}	2.5 x 10^{-7}

DISCUSSION

Comparison with measurements

For the pollutants considered in this study there are not many observations available. Within the monitoring activities of PARCOM-ATMOS γ-HCH is measured at a number of coastal stations. A comparison of model results with these measurements on a yearly average basis is given in Figure 4.

B(a)P is not part of any monitoring programme in Europe. The few data reported in the literature concern either old data or measurements over very short periods. Comparing modelled B(a)P concentrations in air with observations in the Netherlands, Norway and the Mediterranean shows that air concentrations for the Netherlands compare fairly well. Concentrations for the Mediterranean are

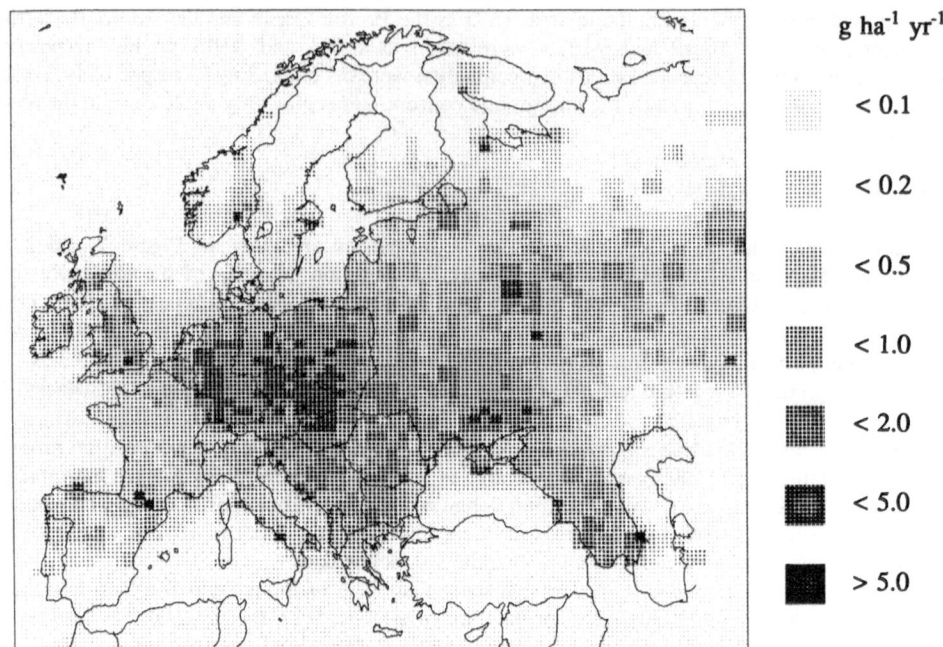

g ha^{-1} yr^{-1}

< 0.1
< 0.2
< 0.5
< 1.0
< 2.0
< 5.0
> 5.0

Figure 2. Calculated total deposition of B(a)P

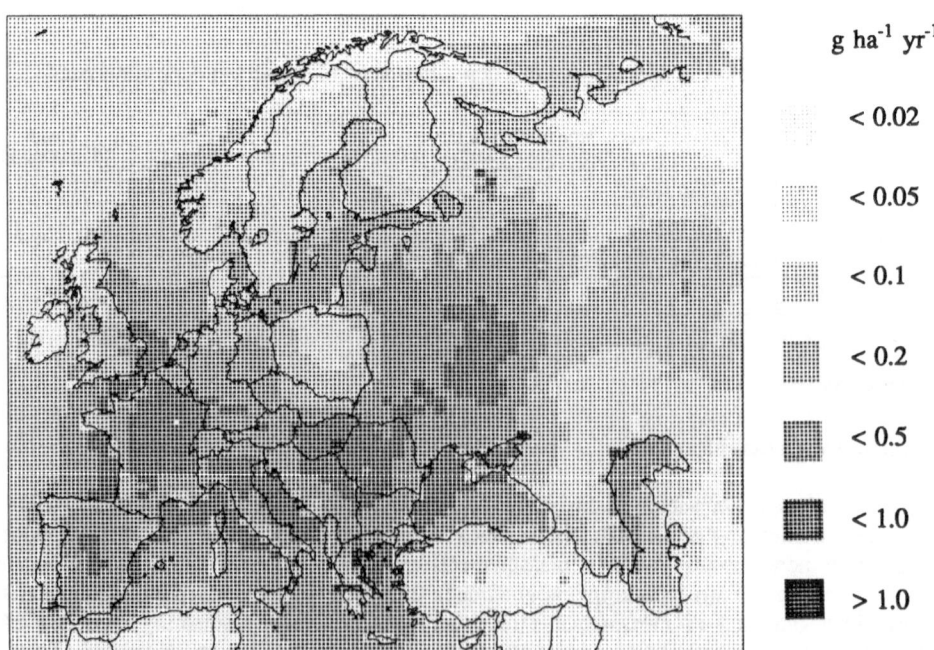

g ha^{-1} yr^{-1}

< 0.02
< 0.05
< 0.1
< 0.2
< 0.5
< 1.0
> 1.0

Figure 3. Calculated total deposition of γ-HCH

over predicted (factor ~ 4) and for Norway under predicted (factor ~ 2); however, the measurements cover only a few episodes so no firm conclusions can be drawn.

In Table 3, calculated γ-HCH deposition loads are compared with deposition loads inferred from measurements. If one considers that the current approach underestimates wet deposition by a factor of almost 3 (see Figure 4), then the calculated total depositions come close to the GESAMP (1989) estimates for both the North Sea and the Baltic Sea.

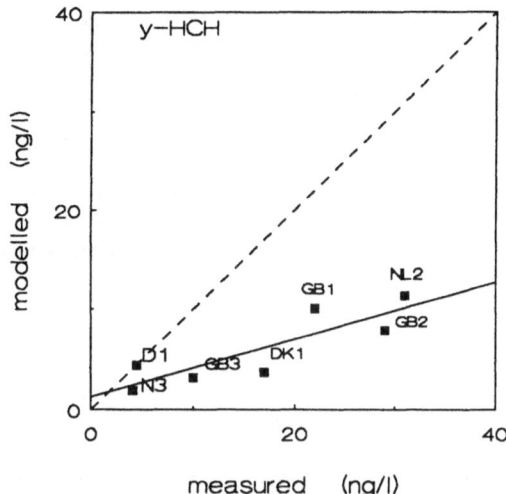

Figure 4. Comparison of calculated annual mean γ-HCH concentrations in precipitation with measurements from the PARCOM-ATMOS network around the North Sea (PARCOM, 1991).

Table 3. Comparison of calculated γ-HCH deposition loads with other estimates.

Area	This study	GESAMP (1989) (data from 1974-1987)	other estimates
	tons yr^{-1}	tons yr^{-1}	tons yr^{-1}
North Sea	6.2	17.6	5.3-8.1[c]
Baltic Sea	4.3	12.8	
Mediterranean Sea	29[a]	5.6[b]	2.6[d]

[a] area north of 36^0 N only (20×10^5 km^2)
[b] northwestern Mediterranean basin only (5×10^5 km^2)
[c] PARCOM (1991); wet deposition only
[d] Villeneuve and Cattini (1986); western Mediterranean basin only (8.1×10^5 km^2)

Uncertainties

Large uncertainties may be involved in emission based deposition calculations. The total uncertainty may be considered as the sum of:

a. uncertainties introduced by the model concept,
b. uncertainties in substance specific parameters and
c. uncertainties in emissions.

a. This includes also meteorological aspects of the model. Such general aspects can be tested using the model for substances such as SO_2, for which much more reliable data (emissions and measurements) are available. This type of uncertainty is therefore relatively small (of the order of ± 30% for total deposition on a yearly basis).

<u>b</u>. The choice of deposition and degradation parameters has a great impact on the calculated deposition. The uncertainty in calculated B(a)P deposition is probably much higher than for the other substance; the literature contains contradicting data regarding photolytic degradation and also for particle size distributions. A simple sensitivity analysis shows that this error is highest at large distances from the main emission areas and can reach ± 200% for total deposition on a yearly basis. Within (high) emission areas this kind of uncertainty has a magnitude of only ± 30 %. A remark must also be made about uncertainty in deposition due to sub-grid effects. Variations in dry deposition velocities within a grid cell can easily reach a factor two because only grid cell averaged surface roughness was taken into account. Deposition to specific ecosystems such as forests may therefore be underestimated.

<u>c</u>. The uncertainties in emissions can be very high, especially for organic compounds (Berdowski and Veldt, 1994). The discrepancies found in this study between model predicted concentrations and measurements for γ-HCH may point to systematically underestimated emissions.

CONCLUSIONS

Atmospheric deposition over Europe of B(a)P and γ-HCH was calculated using a new version of the RIVM-TREND model on the basis of a new high resolution emission inventory for the year 1990. From these calculations the following conclusions may be drawn:

- The computed levels around the North Sea of γ-HCH are a factor of about 3 lower than measurements. This points to underestimated emissions. Insufficient information was available to test the B(a)P results.
- More than 80% of the B(a)P in Europe which is emitted and transported in the form of particles will deposit within the chosen model area. Only 45% of the emitted (gaseous) γ-HCH is estimated to be deposited or destroyed in the model area, the rest will be distributed over neighbouring continents.
- The characteristic air/sea exchange time constant for γ-HCH is in the order of 10 years. Re-emission of previous deposited γ-HCH will seldom occur and does not play an important role in the hemispheric distribution of γ-HCH.
- Large uncertainties still exists in atmospheric degradation rates for B(a)P and γ-HCH. Furthermore the exchange of γ-HCH at the air/soil and air/sea interfaces is strongly parameterized in the present calculations. It is, however, not believed that these uncertainties and simplifications lead to large systematic deviations in the results presented here.

REFERENCES

Asman, W.A.H., and van Jaarsveld, J.A., 1992, A variable-resolution transport model applied for NH$_x$ in Europe. *Atmospheric Environment* 26A:445-464.

Atlas, E., and Giam, C.S., 1988, Ambient concentration and precipitation scavenging of atmospheric organic pollutants. *Water, Air, and Soil Pollut.*, 38:19-36.

Baart, A.C., and Diederen, H.S.M.A., 1991, Calculation of the atmospheric deposition of 29 contaminants to the Rhine catchment area. TNO Report R91/219, Delft, The Netherlands.

Baek, S.O., Field, R.A., Goldstone, M.E., Kirk, P.W., Lester, J.N., and Perry, R., 1991, A Review of atmospheric polycylclic aromatic hydrocarbons: Sources, fate, and behaviour. *Water, Air, Soil Pollut.* 60:279-300.

Bakker, D.J., van den Hout, K.D., Reinds, G.J., and de Vries, W., 1994, Critical and present loads of lindane and benzo(a)pyrene for European forest soils. Background report of the ESQUAD project, IMW-TNO report, Delft, the Netherlands. (in prep.)

Behymer, T.D., and Hites, R.A., 1988, Photolysis of polycyclic aromatic hydrocarbons adsorbed on fly ash. *Environ. Sci. Technol.* 22:1311-1319.

Beljaars, A.C.M., and Holtslag, A.A.M., 1990, A software library for the calculation of surface fluxes over land and sea. *Environmental Software* 5:60-68.

Berdowski, J.J.M., and Veldt, C., 1994, Emission of cadmium, copper, lead, benzo(a)pyrene and lindane to air in Europe in 1990. Background report of the ESQUAD project. TNO-report 94-..., Delft, the Netherlands (in prep.)

Cambray, R.S., Jeffries, D.F., and Topping, G., 1979, The atmospheric input of trace elements to the North Sea. *Mar. Sci. Communic.* 5:175-194.

DCMR, 1992, Measurements carried out by Dienst Centraal Milieubeheer Rijnmond (Pers. comm.), Schiedam, The Netherlands.

GESAMP, 1989, IMO/FAO/UNESCO/WMO/WHO/IAEA/UN/UNEP Joint group of experts on the scientific aspects of marine pollution. The atmospheric input of trace species to the world oceans. Reports and studies No. 38, WMO, Geneva.

Junge, G.E., 1977, Basic considerations about trace constituents in the atmosphere is related to the fate of global pollutants. In: *Fate of pollutants in the air and water environment*. Part I, I.H. Suffet (ed.) (Advances in environmental science and technology, Vol. 8), Wiley-Interscience, New York.

Jury, W.A., Spencer, W.F., and Farmer, W.J., 1983, Behaviour assessmnet model for trace organics in soil: I. Model description. *J. Environ. Qual.* 12, 558-564

Liss, P.S., and Slater, P.G., 1974, Fluxes of gases across the air-sea interface. *Nature* 247:181.

Lunde, G., and Bjorseth, A., 1977, Polycyclic aromatic hydrocarbons in long-range transported aerosols. *Nature* 268:518-519.

Masclet, P., Mouvier, G., and Nikolaou K., 1986, Relative dcay index and sources of polycyclic aromatic hydrocarbons. *Atmospheric Environment* 20:439-446.

PARCOM (1991) Measurements and calculations of atmospheric input to the North Sea in 1990. Annex to the Summary Record of the Ninth Meeting of the Working Group on the Atmospheric Input of Pollutants to Convention Waters, London: 5-8 November 1991.

RIZA/DGW (1992) Speuren naar sporen I. Report RIZA 92.057/DGW 92.040, Rijkswaterstaat, The Hague, The Netherlands

Saleh, F.Y., Dickson, K.L., and Rodgers, J.H. Jr., 1982, Fate of lindane in the aquatic environment: Rate constants of physical and chemical processes. *Environ. Tox. Chem.* 1:289-297.

Schure, M.R., Soltys, P.A., Natush, D.F.S., and Mauney T., 1985, Surface area and porosity of coal fly ash. *Environ. Sci. Technol.* 19:82-86.

Slob, W., Troost, L.M., Krijgsman, M., de Koning, J., and Sein, A.A., 1993, Combustion of municipal waste in the Netherlands. Report no. 730501052, RIVM/TNO/VROM, The Netherlands

Steinwandler H., 1976, Beitrage zur umwandlung der HCH - Isomere durch einwirkung von UV - Strahlen I. Isomiering des Lindans in α-HCH. *Chemosphere*, **4**, 245-248

Taylor, D.R., Tompkins, M.A., Kirton, S.E., Mauney, T. Natusch, D.F.S., and Hopke, P.K., 1982 Analysis of fly ash produced from combustion of refuse-derived fuel and coal mixtures. *Environ. Sci. Technol.* 16:148-154.

Thomas, J.F., Mukai, M., and Tebbens, B.D., 1968, Fate of Airborne benzo[a]pyrene. *Environ. Sci. Technol.* 2: 33

van Aalst, R.M., van Ardenne, R.A.M., de Kruk, F.J., and Lems, T., 1982, Pollution of the North Sea from the atmosphere. TNO Report CL 82/152, Delft, The Netherlands.

van den Hout, K.D. (ed.), 1994, The impact of atmospheric deposition of non-acidifying compounds on the quality of European forest soils and the North Sea. Mean report of the ESQUAD project. TNO report R93/329, Delft, the Netherlands.

van Jaarsveld, J.A., 1989, A model approach for assessing transport and deposition of acidifying components on different spatial scales. In: *Changing Composition of the Troposphere*. Special environmental report no. 17, (WMO- no. 724), 197-204, Geneva, Switzerland.

van Jaarsveld, J.A., 1992, Estimating atmospheric inputs of trace constituents to the North Sea; methods and results. In: *Air Pollution Modelling and its Application IX*, Edited by H. van Dop and G. Kallos, 249-258, Plenum Press, New York.

van Jaarsveld, J.A., and Onderdelinden, D., 1994, TREND; An analytical long-term deposition model for multi-scale applications. RIVM Report no. 228603009, Bilthoven (in prep.).

van Jaarsveld, J.A., 1994, Atmospheric deposition of cadmium, copper, lead, benzo(a)pyrene and lindane over Europe and its surrounding marine areas. RIVM Report no. 722401002, Bilthoven (in prep.).

Villeneuve, J.P., and Cattini, C., 1986, Input of chlorinated hydrocarbons through dry and wet deposition to the Western Mediterranean, *Chemosphere* 15:115-120

Warmenhoven, J.P., Duiser, J.A., de Leu, L.Th. and Veldt, C., 1989, The contribution of the input from the atmosphere to the contamination of the North Sea and the Dutch Wadden Sea. TNO Report R 89/349A, Delft, The Netherlands.

Wesely, M.L., and Hicks, B.B., 1977, Some factors that affect the deposition rates of sulphur dioxide and similar gases on vegetation. *J. Air Pollut. Contr. Assoc.* 27:1110-1116.

DISCUSSION

K.E. GRØNSKI: Considering the process of re-emission, could you describe the parameters that are used?

J.A. VAN JAARSVELD: Re-emission of previous deposited γ-HCH could not be taken into account dynamically, because this would require dynamic modelling of the soil and water compartments as well. Instead we used effective deposition velocities to describe the net deposition flux to both soil and water surfaces. These velocities are estimated using separate one-dimensional air-soil and air-water exchange models. Current concentration levels of γ-HCH in the North Sea are such that re-emission is estimated to take place not more than 10% of the time. In contrast to the relatively well mixed surface water the top soil layer is found to be near saturated resulting in very low net deposition fluxes. The effective dry deposition velocity taken for water surfaces is 2 x 10-3 m s^{-1} and for land surfaces 5 x 10-5 m s^{-1}.

G. GRAZIANI: Results of LRT statistical models are affected both by the lack of information on emissions and by model approximations. Where is it more effective to invest in the future?

J.A. VAN JAARSVELD: I think that model results for persistent organic compounds are most affected by poorly known emissions. General aspects of LRT models can be tested using well known compounds such as sulphur for which huge databases of both emissions and measurements are available. The influence of compound specific parameters, especially the removal parameters can be evaluated through sensitivity studies. Emissions, however, directly affect model results. This study reveals that with the current estimates of γ-HCH emissions the concentrations in precipitation are underpredicted by a factor of 3, though γ-HCH is one of the best known pesticides. Further work in this field should include research on emissions as a priority, but also measurements of ambient levels for model checking purposes.

P. SEIBERT: How did you obtain the precipitation fields, especially over the sea?

J.A. VAN JAARSVELD: Precipitation data is derived from 6- hourly synoptic observations of more than 1500 stations in Europe. These observations are obtained from ECMWF in Reading, England. Precipitation fields over the sea are obtained by inter- and extrapolation of data of the nearest stations. Although few, there are some stations within the Atlantic area. Of course the uncertainty in precipitation amounts is much larger over sea than over land.

K. JUDA-REZLER: 1) Are the given values for T1/2 in the table accounted for both dry and wet deposition?

2) Does B(a)P and γ-HCH undergo chemical reactions, and is it an important removal mechanism for them?

J.A. VAN JAARSVELD: The calculated atmospheric residence times include dry deposition, wet deposition as well as chemical degradation. Both B(a)P and γ-HCH undergo chemical reactions in the atmosphere. For γ-HCH almost no data is available. Chemical removal mechanisms are generally considered as less important than deposition processes. Most published data on reaction rates are, however, for laboratory circumstances and cannot be extrapolated to atmospheric conditions where B(a)P is attached to or even is incorporated in particles. Since cases of long range transport of B(a)P have been documented, it is assumed that chemical degradation under atmospheric conditions is not very important. In the present study the degradation rate has been set to zero as a first approach. Chemical degradation of POPs in the atmosphere has certainly to be subject of further research.

A NEW MODEL FOR ROUTINE CALCULATION OF LONG-RANGE TRANSPORT OF AIR POLLUTANTS

Trond Iversen[1] and Erik Berge[2]

[1]Institute of Geophysics, University of Oslo
P.O. Box 1022 - Blindern, 0315 Oslo, Norway
[2]The Norwegian Meteorological Institute
P.O. Box 43 - Blindern, 0313 Oslo, Norway

INTRODUCTION

At the Western Meteorological Centre (MSC-W) for the European Monitoring and Evaluation Programme (EMEP), LRT-models have been applied for many years based on official estimates of emissions and real-time meteorological data (Eliassen and Saltbones, 1983; Hov et al., 1988; Iversen, 1990 and 1993; Sandnes, 1993; Simpson, 1993). So far, only "1.5-level" models have been used with a Lagrangian estimation of horizontal transport. Although showing considerable skill for long-term averages, several shortcomings connected with the crude vertical treatment and coarse horizontal mesh width of 150 km are quite evident. As the work under the UN/ECE Convention for Long-Range Transboundary Air Pollution increases the demand for detailed information about atmospheric transport, these model weaknesses should be diminished as far as possible, without destroying the balance between the level of detail and the need for long-term calculations covering several years. Effects introduced by steep topography such as the Alps are only coarsely resolved by the 150 km grid, and calculations of ecosystem-specific dry deposition also require as good a horizontal resolution as possible. Furthermore, EMEP is supposed to keep budgets of transboundary transport, and these are quite uncertain for countries only resolved by very few grid-squares. A calculation of the subgrid-scale deposition after emission release involving time-resolved characteristics of the atmospheric boundary layer (ABL), is best taken care of by a certain minimum vertical resolution of the ABL. A significant part of the wet deposition in regions about 1000 km away from the main emission sources probably stems from free troposphere just above the ABL, hence some vertical resolution of the lowermost portion of the free troposphere above ABL is also required. This is even more important for models for ozone for which the free troposphere in some regions acts as a reservoir. A scale analysis of the budget-equations of atmospheric constituents on regional scales (L ~ 250-2500 km) shows that horizontal and vertical advection terms are of the same order of magnitude.

Air Pollution Modeling and Its Application X, Edited by S-V. Gryning
and M. M. Millán, Plenum Press, New York, 1994

In recent years the routine weather prediction model at The Norwegian Meteorological Institute (DNMI) has been run with full data assimilation in a grid with mesh width 50 km (LAM50). As a consequence of this and the availability of faster computers (CRAY Y-MP), development of a multilayer model with Eulerian coordinates has been initiated. Today there are several comprehensive, multilayer models available; RADM (Chang et al., 1987) and STEM (Charmichael et al., 1991) in the U.S.A. and ADOM (Venkatram et al., 1988) in Canada. Modified versions of some of these models have also been applied for episodic studies in Europe, e.g. the EURAD model (Hass et al., 1990), but so far they are to resource-demanding to run for routine, multiannual calculations over regional area covering Europe. It is therefore necessary to compromise between the need of details and the need of a widely applicable tool.

As a consequence of the above mentioned need for economy, we wish to keep the numbers of layers in the vertical to a minimum which is strictly needed. The model-equations are therefore formulated with coordinate surfaces which do not intersect neither the ground nor the estimated surface defining the top of the atmospheric boundary layer (ABL). Thus we are able to optimally utilize the vertical resolution by selecting a number of layers inside the ABL and a number of layers above, and we are calculating explicitly the flux through the top of the ABL. Presently we are aiming at three layers inside, and two layers immediately above the ABL. The horizontal advective transport is calculated with a 4th. order version of the Bott scheme, while the vertical is a 2nd. order Bott scheme formulated with a variable mesh-width (Strand and Hov, 1993) combined with an implicit integration of the vertical diffusion. The first version of the model will be run for simplified SO_2-sulphate chemistry only. Presently (October - 1993) there are still no results available with this five-layer model with new vertical coordinate. A version with the 20 sigma-layers of the LAM50 in which the meteorological data are calculated, has been used as a first step. Results for a winter-month (January 1992) and a summer month (July 1992) are available for presentation.

BASIC EQUATIONS IN s-COORDINATES

Even though results are not yet ready with the model version with the new type of vertical coordinate (s-coordinate), we present here the basic equations formulated with these coordinate. Let s(x,y,z,t) be a quantity which can be expressed as an invertible function of the vertical coordinate z; thus we may write

$$z = z(x,y,s,t) \tag{1}$$

A scalar (or vector) function q may either be expressed with z as vertical coordinate: q=q(x,y,z,t), or alternatively with s: q=q(x,y,s,t). If ξ signifies either x, y or t we arrive at the following relationships:

$$\left(\frac{\partial q}{\partial \xi}\right)_z = \frac{\partial q}{\partial \xi} - \frac{\partial z}{\partial \xi}\frac{q_s}{z_s} \tag{2}$$

$$\frac{\partial q}{\partial z} = \frac{q_s}{z_s} \tag{3}$$

where a subscript z signifies derivatives taken along surfaces of constant z, while no subscript on a differentiation symbol signifies constant s. Subscript s on a variable signifies derivatives with respect to s, and subscripts x, y and t similarly express partial derivatives taken with constant s. A material derivative can now be expressed by

$$\frac{Dq}{dt} = q_t + \boldsymbol{v} \cdot \nabla q + \dot{s} q_s \tag{4}$$

where $\boldsymbol{v}$ and ∇ are the horizontal wind and gradient (taken at constant s) vectors respectively, and the "vertical velocity" is

$$\dot{s} = \frac{Ds}{dt} = \frac{w_{rel}}{z_s} = \frac{1}{z_s}[w - z_t - \boldsymbol{v} \cdot \nabla z] \tag{5}$$

where $w = Dz/dt$ is the vertical velocity in z-coordinates and w_{rel} is the vertical velocity relative to the s-surfaces (viz. the volume-flux density through s-surfaces). It is straightforward to show that the vertical advection in s-coordinates is

$$VADV = \dot{s} q_s = w_{rel} \frac{\partial q}{\partial z} \tag{6}$$

It may be beneficial to apply the second expression in practice if the s-coordinate is chosen with a discontinuous derivative with respect to z. In that case q_s and s are not uniquely defined for all s even though the vertical advection of course is well defined. Using the second expression therefore requires less restrictions on the choice of the s-coordinate.

If now q is the mixing ratio of a gravitationally passive constituent ($q = c/\rho$ where c is the concentration and ρ is density of air), the continuity equation for the component can be written

$$q_t + m\boldsymbol{v} \cdot \nabla q + w_{rel} \frac{\partial q}{\partial z} = \frac{\partial}{\partial z}(K \frac{\partial q}{\partial z}) + \frac{S}{\rho} \tag{7}$$

where m is a map-factor, and S is the sum of source- and sink-terms. The coefficient K is the vertical exchange coefficient for turbulent diffusion; horizontal diffusion is neglected. Notice that $\boldsymbol{v}$ is the horizontal wind, and not any vector parallel to the constant-s surfaces which probably is a quite common misunderstanding. It is also necessary to be aware of the pitfall which may arise when the curvature of the s-coordinate surfaces becomes considerable, in which case the transformation formulas should be generalized (e.g. Dutton, 1976). As long as the coordinate surfaces inclines much less than 45° with the horizontal, this effect can be neglected (Pielke and Martin, 1981). As an alternative to eq.(7) one may apply an equation on flux form:

$$(\rho z_s q)_t + m^2 \nabla \cdot (\frac{\rho z_s}{m} q\mathbf{v}) + (\rho q w_{rel})_s = (K\frac{\partial q}{\partial z})_s + z_s S \qquad (8)$$

This version is applicable when using Bott's advection scheme (Bott, 1989 a and b). It is necessary in this case to calculate q and its derivatives in levels in the model where z_s is well defined.

CHOICE OF COORDINATE

As earlier mentioned we wish to keep the number of layers at a minimum without destroying too much of the resulting solution. Earlier EMEP models in routine have consisted of a one-layer description of the ABL or "mixing layer", the free troposphere being a constant background. Now we wish to also resolve and calculate budgets for the lowermost layers of the free troposphere above ABL, as well as to increase the horizontal resolution with a factor three. The latter should be combined with also a division of the ABL into a few layers, so that a thorough vertical mixing inside ABL within 50km grid-squares is not implicitly presumed. This also permits a different treatment of high-level emission sources (viz. tall stacks) than of those at ground level.

It is quite evident that the height of the ABL above the ground is a crucial parameter for the regional scale dispersion of pollutants. We therefore choose a vertical coordinate which has the top of the ABL as a coordinate surface as well as the ground topography. The transport through the top of ABL may then be an explicit parameter in the model formulation. Thus we choose

$$s = \begin{cases} \dfrac{z - h}{H - h} & for\ h \le z < H \\[2ex] \dfrac{z + z_T - 2H}{z_T - H} & for\ H \le z \le z_T \end{cases} \qquad (9)$$

where h is the topography height above sea level, H is height above sea level of the ABL-top, and z_T is a chosen upper bound of the vertical domain presumed to be a constant. It is seen that $s \in [0,1>$ inside ABL, and $s \in [1,2]$ above. Furthermore, z_s is in general not defined for s=1, since z_s are given as

$$z_s = \begin{cases} (H - h) & for\ s \in [0,1> \\ z_T - H & for\ s \in <1,2] \end{cases} \qquad (10)$$

160

The most serious uncertainty connected with this choice for s is the problem of estimating the ABL-height H. As a first attempt we base ourselves on the boundary-layer parameters from the meteorological model which is run with 6-hourly, intermittent data-assimilation. For the surface layer (Prandtl layer) we have turbulent stress and the flux densities of heat and humidity. In the unstable case when the heat-flux is directed upwards, H is determined by distributing vertically the heat input over the model's time step with a new adiabatic sounding. The vertical exchange coefficient, K, is determined by the O'Brien (1970) - profile. In the stable case, K is estimated by the Blackadar (1979) formulas, and H is estimated at the level where there is a significant drop in the K-values going upwards. In the relatively sparse s-surfaces K is estimated by layer averages based on calculations in the dense vertical resolution of the LAM50 model. Both K and H are smoothed with a second order Shapiro-filter in space, and a time-filter.

The vertical and horizontal velocities as well as K are given in staggered points relative to the grid-points keeping q, according to the Arakawa C-grid (Fig. 1).

It is planned to run a five-layer model in s-coordinates in routine, with three layers in the ABL (s<1) and two layers above (s>1); see Fig. 1 for illustration. With the staggering of variables, we use the vertical transport through the top of ABL explicitly in the model. The vertical and horizontal velocities as well as K are given in staggered points relative to the grid-points keeping q, according to the Arakawa C-grid.

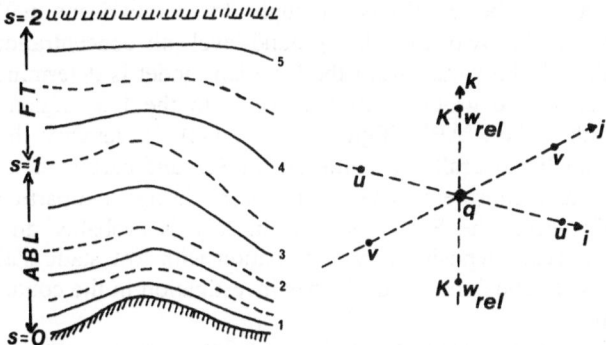

Figure 1. Vertical and horizontal staggering of variables.

SOME RESULTS FROM PRELIMINARY CALCULATIONS

Unfortunately, we are presently unable to show results with the 5-level version of the model. A preliminary experimental version has, however, been made, and calculation results for a winter month (January, 1992) and a summer month (July, 1992) can be presented. Calculations are made in the 20 sigma-surfaces of the LAM50-model, 10 of which are below about 2000 m above the topography. The main reasons for making the calculations were to check the quality and stability of the new LAM50 data, to formulate the Bott-scheme with a variable grid resolution vertically, and to reveal if apparent systematic errors in the routine one-layer Lagrangian models would be influenced beneficially with a multilayer model. Since official EMEP measurements still are not available for 1992, regrettably no comparisons with measurements can be shown.

The model was run with simplified SO_2 - sulphate chemistry with a linear but variable oxidation rate, using different scavenging ratios for in-cloud and sub-cloud washout, applying Blackadar (1979) formulas for vertical turbulent exchange, and taking into account surface-layer aerodynamic resistance towards dry deposition. The different terms in the continuity equations were integrated with time-splitting (McRae et al., 1982). Boundary conditions for the concentrations were taken from a hemispheric-scale model for SO_2-sulphate transport (Tarrason, 1992; Tarrason and Iversen, 1992).

Appreciating the considerable deviations in some points, the LAM50 precipitation fields are quite similar to those used in the Lagrangian model, which over land are Cressman-analysed measurements. Fig. 2 reveals, however, a tendency towards more precipitation in the Eulerian model in the two selected months, in particular in areas with small amounts in January. Fig.3 shows that there also is a considerable spread when comparing the results for wet deposition of sulphur. It is quite clear that in areas where the wet deposition is smaller than average, the deposition is larger in the Eulerian calculations, while the opposite is the case in areas with large depositions. The trend is clearer in July than in January, and it is confirmed when comparing the overall picture, not only the results at some measurement sites. Thus there is an increased optimism for better budgets in semi-remote areas, such as southern Scandinavia, with smaller relative parts of indeterminate origins. All this has to be confirmed, though, by comparisons with measurements. Comparison between the two models' ground-level air-concentrations reveals good agreement, even though the values from the Eulerian model is determined from a 60-100 m thick layer close to the ground, while those from the Lagrangian model are more representative for the whole ABL (figures not shown). In January the Eulerian model estimated slightly smaller overall concentrations of SO_2 and particulate sulphate, whilst in July SO_2 were estimated considerably larger in central Europe and particulate sulphate had no systematic difference. The SO_2 behaviour in July is probably due to the different treatment of subgrid-scale deposition in combination with low static stability in summer. The treatment of local deposition needs further investigation in the context of the Eulerian model formulation.

All in all results produced so far, do not point to serious problems with the input meteorological data. A quite straightforward way to technically run monthly production calculations with the rather comprehensive set of meteorological data has been established. The results are encouraging, even though there are clear needs for improvements of some parts of the model-formulations. In the future it is planned that the new model-system is to be used for research on processes which eventually will be taken into the routine calculations if seen necessary.

ACKNOWLEDGEMENTS

We appreciate the close cooperation with the Numerical Weather Prediction-group at DNMI. In particular we are grateful to Mr. Anstein Foss for technically establishing the LAM50 routine. Helge Styve is contributing to the programming and running of the new Eulerian model.

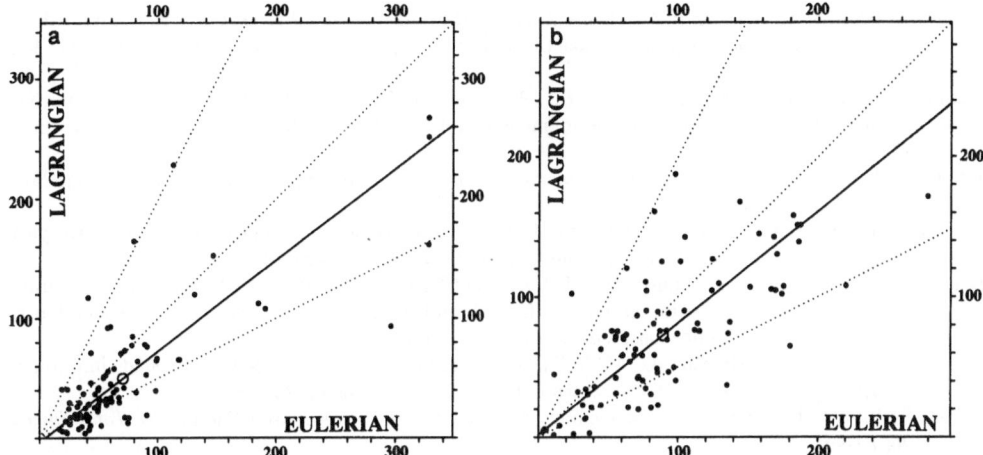

Figure 2. Monthly accumulated precipitation amounts (mm) at EMEP measurement sites. Data from Lagrangian model (ordinate) versus data from Eulerian model (abscissa). Dotted lines denotes perfect agreement and disagreement with a factor 2. The full line is optimal regression. **a)** January 1992. **b)** July 1992.

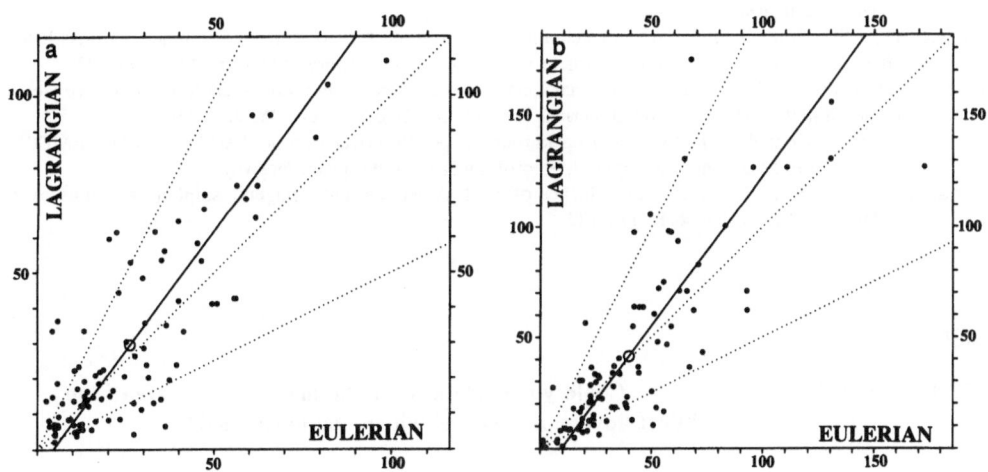

Figure 3. Same as Fig. 2, but for monthly accumulated sulphur wet deposition (mg(S) m^{-2}).

REFERENCES

Bott, A., 1989a, A positive definite advection scheme obtained by non-linear renormalization of the advective fluxes. *Mon. Wea. Rev.*, **117**, 1006-1015.

Bott, A., 1989b, Reply. *Mon. Wea. Rev.*, **117**, 2633-2636.

Chang, J.S., Brost, R.A., Isaksen, I.S.A., Madronich, P., Stockwell, W.R. and Walcek, C.J., 1987, A three-dimensional Eulerian acid deposition model: Physical concepts and formulation. *J. Geophys. Res.*, **92**, 14681-14700.

Charmichael, G.R., Peters, L.K. and Saylor, R.D., 1991, The STEM-II regional acid deposition and photochemical oxidant model - 1. An overview of model development and applications. *Atmos. Environ.*, **25A**, 2077-2090.

Dutton, J., 1976, *The ceaseless wind. An introduction to the theory of atmospheric motion.* Ch. 7.4, pp 248-254. Dover Publ. Inc., New York.

Eliassen, A. and Saltbones, J., 1983, Modelling of long-range transport of sulphur over Europe: a two year model run and some model experiments. *Atmos. Environ.*, **22**, 1457-1473.

Hass, H., Jakobs, H.J., Memmesheilmer, M., Ebel, A., and Chang, J.S., 1990, Simulation of a wet deposition case in Europe using the European Acid Deposition Model (EURAD). 18th ITM on Air Pollution Modelling and its Applications, Vol. 1, 153- 160.

Hov, Ø., Eliassen, A., and Simpson, D., 1988, Calculation of the distribution of NOx compounds in Europe. In: *Tropospheric Ozone.*, I.S.A. Isaksen (ed.), Reidel Publ. Comp., 239-261.

Iversen, T., 1990, Calculations of long-range transported sulphur and nitrogen over Europe. *Sci. Total Envir.*, **96**, 87-99.

Iversen, T., 1993, Modelled and measured transboundary acidifying pollution in Europe - verification and trends. *Atmos. Environ.*, **27A**, 889-920.

McRae, G.J., Goodin, W.R. and Seinfeld, J.H., 1982, Numerical solution of the atmospheric diffusion equation for chemically reacting flows. *J. Comp. Phys.*, **45**, 1-42.

O'Brien, J.J., 1970, A note on the vertical structure of the eddy exchange coefficient in the planetary boundary layer. *J. Atmos. Sci.*, **27**, 1213-1215.

Pielke, R.A. and Martin, C.L., 1981, The derivation of a terrain-following coordinate system for use in a hydrostatic model. *J. Atmos. Sci.*, **38**, 1707-1713.

Sandnes, H., 1993, Calculated budgets for airborne acidifying components in Europe, 1985, 1987, 1988, 1989, 1990, 1991 and 1992. EMEP/MSC-W Report 1/93. The Norwegian Meteorological Institute, Oslo, Norway, Tech. Rep. No. 109.

Simpson, D., 1993, Photochemical model calculations over Europe for two extended summer periods: 1985 and 1989. Model results and comparison with observations. *Atmos. Environ.*, **27A**, 921-943.

Strand, A. and Hov, Ø., 1993, A two-dimensional zonally averaged transport model including convective motions and a new strategy for the numerical solution. *J. Geophys. Res.*, **98**, 9023-9037.

Tarrason, L., 1992, Contributions to sulphur background deposition over Europe: Results for 1988. EMEP/-MSC-W Note 5/92. The Norwegian Meteorological Institute, Oslo, Norway.

Tarrason, L. and Iversen, T., 1992, The influence of north American anthropogenic sulphur emissions over western Europe. *Tellus*, **44B**, 114-132.

DISCUSSION

J. MATTHIJSEN: Could you explain what the main features are of the "Bott advection-scheme" which you mentioned?

T. IVERSEN: The Bott scheme is a generalization of the well known upwind advection scheme which is too diffusive for application in transport models. The generalization consists of better ways to calculate mass-fluxes into and out of grid-boxes. The advection equation in one dimension is written in flux-form, and wind and mass- concentrations are stored in staggered grid-points. In order to calculate the fluxes between grid-boxes, the concentrations are interpolated from their staggered position to the wind-points. With the upwind-scheme, a zero order interpolation (i.e. step function) is used. Bott propose to use higher order polynomials, and we have used 4th order in the horizontal, and second order in the vertical where the grid-spacing is uneven. The scheme is automatically globally mass-conservative, and additional constraints are fulfilled by adjusting the interpoalation procedure, so that the scheme becomes locally mass-consitent and positive definit. All in all the scheme is highly applicable in pollution transport models.

H. VAN JAARSVELD: 1) You describe a high resolution, multi-layer model. This will require much more computer resources than for the Lagrangian model. Can you give an indication of computing time needed?

I assume you are currently using the same chemical scheme as in the lagrangian model. Is that true.

T. IVERSEN: 1) You are right: but I hesitate to give a specific number for required CPU-time at present. The model still is under development, and we have so far only experimented with a 18-level version whilst the aim is to use a 5-level model as a new routine. Furthermore, we co-operate with experts in parallel coding and other CRAY-specific programming working at the CRAY-centre in Trondheim, Norway, in order to speed up the final model-version.

2) At present only two-component chemistry for ox-idised sulphur is included under the development phase. Later on more general chemistry, including nitrogen and photo-oxidants will be included. But we prefer a stepwise procedure in order to gain confidence in the 3D transport calculations.

K. JUDA-REZLER: 1) Is the modelling grid for the lagrangian EMEP model a part of the new model grid?

2) When the new Eulerian model will be working operationally are you going to use only the Eulerian one, or both for routine calculations?

T. IVERSEN: 1) Yes, the well known "EMEP-grid" covers a subdomain of the grid planned for use by the new model. Inside the present EMEP-domain, there will be exactly 9 full new grid-squares inside each present EMEP grid-square.

2) This is not for me to decide, since I am not project leader of EMEP/MSC-W. However, I would assume that there will be some years with overlap where results from both models will be available simultaneously, but that only

one set of calculations will at any time be accredited for official use in decission-making.

B. FISHER: Have you plans to improve the N chemistry in the model? Why put extra computing resources into better vertical resolution rather than improved chemistry?

T. IVERSEN: Again I must answer with caution since I am not responsible for all decisions taken at EMEP/MSC-W. I do not believe that there are immediate plans to improve the nitrogen chemistry in the model for acid deposition. While modelling NO_2 in fully non-linear photochemistry models on the large regional scale us a general problem, we have very good results for NO_2 as well as nitrate in precipitation in our acid deposition model. They are better than for sulphur in precipitation when comparing with official EMEP measurements. However, for nitrogen as well as for sulphur we have some systematic deviations which we believe are due to clear limitations imposed by the one-layer formulation of the routine models. Too much pollution-mass is lost to the free troposphere, and using land-use specific dry deposition velocities and an improved calculation of deposition in the first grid-square require a resolution of the ABL. Furthermore, several countries in Europe are very coarsly resolved by the present 150km-grid, and a grid resolution of 50km should be balanced with a subresolution of the ABL for reasons of consistency. It does not help us if we have the world's best chemistry, if we do not have a reasonable resolution of airmasses and their relative motion.

I feel that there is a continuous internal competition between the need for improvements of chemistry and improvements of meteorological information and transport. In a routine model like those we will have to use for EMEP there will have to be a balanced approach, so that no link in the chain is considerably weaker than others. To be honest I presently think that the weakest link in the chain is the emission numbers. I therefore sincerely hope that the outcome of the UN/ECE Task Force on Emission Inventories hosted by UK will be successful in designing guidelines which the countries are able to follow in practice. Then I think that some of the limitations of the one-layer formulation lead to more serious weaknesses than the presently used scheme. The input meteorological data is probably the strongest link in the present model formulation.

ATMOSPHERIC MERCURY SPECIES OVER EUROPE.
MODEL CALCULATIONS AND COMPARISON WITH OBSERVATIONS FROM
THE NORDIC AIR AND PRECIPITATION NETWORK FOR 1987 AND 1988

G. Petersen[1], Å. Iverfeldt[2], J. Munthe[2]

[1]GKSS Research Centre, Institute of Physics
 Max-Planck-Str. 1, D-21502 Geesthacht, Germany
[2]Swedish Environmental Research Institute (IVL)
 P.O. Box 47086, S-402 58 Göteborg, Sweden

INTRODUCTION

Mercury exists in the atmosphere in different physical and chemical forms, whose properties and interactions with their surroundings determine its transport and transformations, as well as its removal mechanisms such as wet and dry deposition to the earth's surface. It is now generally accepted, that more than 90 % of atmospheric mercury is in the vapor phase while the remainder is associated with particulate matter. The chemical speciation of mercury in the atmosphere is still under discussion although there is evidence for the total predominance of elemental mercury. Recent data indicate that, besides elemental mercury, methylmercury species are present in ambient air in minor quantities. In the emissions from high-temperature combustion facilities without flue gas cleaning systems divalent inorganic mercury compounds in gaseous or particulate form have been identified, but far from sources these species are at very low or undetectable levels.

According to European emission surveys for mercury based upon recently available data, the dominating sources in 1987 and 1988 were chlor-alkali factories in East Germany and coal combustion units without flue gas cleaning systems in East Germany and Czechoslovakia accounting for nearly 40 % of the total anthropogenic mercury emissions in Europe (Axenfeld et al., 1991). These emissions are assumed to be made up of gaseous elemental mercury (Hg^0), particle associated mercury (Hg(part.)) and divalent inorganic mercury compounds (Hg(II)(g). This speciation is considered to be a first approach and the uncertainties involved are hard to estimate because of a lack of reliable data on this subject.

MERCURY MODELLING

The long-period model for sulphur used under the European Monitoring and Evaluation Programme (EMEP) (Eliassen and Saltbones, 1983) has been modified for transport, chemical transformations and deposition of mercury . The chemical scheme in the model is mainly based on results obtained during recent years in various research programmes in

Sweden (Iverfeldt and Lindqvist, 1986; Lindqvist et al., 1991; Munthe et al. 1991; Munthe and McElroy, 1992; Munthe, 1992) and Canada (Schroeder and Jackson, 1987). The results indicate, that gas phase chemistry of elemental mercury is probably of minor importance. Further, there is no evidence that mercury associated with particles and divalent inorganic mercury compounds undergo physical or chemical transformation during transport and diffusion through the atmosphere. Hence, in the present model chemical processes of Hg^0 are restricted to the aqueous phase and it is assumed that partitioning between gas and particulate phase and chemical reactions are negligible for Hg(part.) and Hg(II)(g).

In the case of elemental mercury investigations of redox processes that are relevant to atmospheric conditions have shown, that in the presence of liquid water ozone is a major oxidant for this species:

$$Hg^0 + O_3 + H_2O \rightleftharpoons Hg^{2+} + 2OH^- + O_2 \tag{1}$$

The rate of the reaction can be described by the following equation:

$$\frac{d[Hg^{2+}]}{dt} = k_1 \cdot [Hg^0]_{aq} \cdot [O_3]_{aq} \tag{2}$$

with a second-order rate constant $k_1 = 4.7 \cdot 10^7$ [$M^{-1}s^{-1}$] (Munthe, 1992).

It has also been shown, that sulfite is capable of reducing Hg(II)(g) to Hg^0 in aqueous solutions (Munthe et al., 1991). As a first step the proposed mechanism involves the complexation of Hg^{2+} by sulfite ions

$$Hg^{2+} + 2SO_3^{2-} \rightleftharpoons \left[Hg(SO_3)_2^{2-} \right] \tag{3}$$

As the complexation reaction is very fast, it is assumed that it occurs spontaneously. Subsequently, the complex decomposes to produce Hg^+ which in turn is rapidly reduced to Hg^0. In the present model, the decomposition has been treated as a first order reaction:

$$\frac{d}{dt}\left[Hg(SO_3)_2^{2-} \right] = -k_2 \cdot \left[Hg(SO_3)_2^{2-} \right] \tag{4}$$

with a maximum value of the rate constant $k_2 = 4 \cdot 10^{-4}$ [s^{-1}] (Munthe et al., 1991).

It is assumed that Hg^{2+} is completely complexed, as a large excess of sulfite ions with respect to Hg^{2+} can be expected in atmospheric water droplets. Hence, the reduction rate of divalent mercury can be expressed as:

$$\frac{d[Hg(II)]}{dt} = -k_2 \cdot [Hg(II)] \tag{5}$$

where [Hg(II)] denotes the total dissolved concentration in the water droplets.

As the reductive processes occur simultaneously with the oxidation of elemental mercury by ozone, steady state concentrations of Hg(II) are built up in the water droplets depending on the reaction rates in equation (2) and (5). If those equations are assumed to represent the dominant reactions, the steady state concentration of dissolved oxidized mercury in atmospheric water droplets can be calculated as a function of the rate constants k_1 and k_2, ozone concentration in the aqueous phase and elemental mercury concentration in ambient air:

$$[Hg(II)]_{aq} = \frac{k_1}{k_2 \cdot H_{Hg}} \cdot [O_3]_{aq} \cdot [Hg^0]_{gas} \tag{6}$$

implying net zero flux of Hg^0 across the air/water interface for steady state conditions and hence

$$[Hg^0]_{gas} = H_{Hg} \cdot [Hg^0]_{aq} \tag{7}$$

where H_{Hg} is the Henry's law constant for Hg^0.

The reduction of Hg(II) to Hg^0 by sulfite is assumed to occur independently of the SO_2 concentration in air. This is equivalent to ignoring other complexes than $Hg(SO_3)_2^{2-}$. In reality, other complexes such as $HgCl_2$ are probably as important as the sulfite complex. However, due to the predominance of particulate mercury in precipitation, this simplification will probably not affect the model results at the present degree of detail.

Analysis of mercury in precipitation in polluted areas has shown, that a significant fraction of the total concentration is present in the particulate phase (Iverfeldt, 1991a). There is evidence, that divalent aqueous Hg is adsorbed on soot particles, so that the total Hg concentration in the water droplets exceeds the steady state value of Hg(II) (Iverfeldt, 1991a; Brosset and Lord, 1991; Lindqvist et al., 1991). The adsorption can be described by a mechanism which is known as the Langmuir isotherm:

$$\frac{[Hg(II)]_{ad}}{F} = [Hg(II)]_{aq} \cdot K_{ad} \tag{8}$$

where F is the total soot particle surface area per unit volume of water. K_{ad} represents the adsorption equilibrium constant.

Assuming a mean radius r for the soot particles and substituting F by the soot particle concentration c_{soot}, it can be shown (Petersen, 1992) that the concentration of mercury adsorbed on soot particles with the density ρ is given by:

$$[Hg(II)]_{ad} = [Hg(II)]_{aq} \cdot K_{ad} \cdot \frac{3}{\rho} \cdot c_{soot} \cdot \frac{1}{r} \tag{9}$$

Expressing the constant values of equation (9) as a model specific adsorption equilibrium constant K_3:

$$k_3 = k_{ad} \cdot \frac{3}{\rho} \tag{10}$$

leads to

$$[Hg(II)]_{ad} = [Hg(II)]_{aq} \cdot c_{soot} \cdot \frac{1}{r} \cdot K_3 \tag{11}$$

The total concentration of elemental mercury oxidized by ozone in the aqueous phase $[Hg(II)]_{tot}$ is the sum of the dissolved and the adsorbed fraction:

$$[Hg(II)]_{tot} = [Hg(II)]_{aq} + [Hg(II)]_{ad} \tag{12}$$

By substituting (6) and (11) into (12) one obtains an equation for the total concentration of elemental mercury oxidized by ozone in the aqueous phase as a function of the concentration of elemental mercury in ambient air:

$$[Hg(II)]_{tot} = \frac{k_1}{k_2} \cdot \frac{1}{H_{Hg}} \cdot [O_3]_{aq} \cdot \left(1 + K_3 \cdot \frac{c_{soot}}{r}\right) \cdot [Hg^0]_{gas} \qquad (13)$$

The wet deposition rate is parameterized in terms of the precipitation rate P, the height of the mixing layer h and a scavenging ratio W, which is a proportionality factor relating pollutant concentrations in precipitation to the respective concentrations in air. For elemental mercury the scavenging ratio is given by

$$W(Hg^0) = \frac{[Hg(II)]_{tot}}{[Hg^0]_{gas}} \qquad (14)$$

Here, a negligibly small fraction of dissolved Hg^0 is assumed, since this species most probably accounts for a one percent or less of the total mercury in precipitation.

By rearranging equation (13) the scavenging ratio of Hg^0 can be expressed in terms of ozone and soot carbon concentration, three rate constants and Henry's law coefficient of Hg^0:

$$W(Hg^0) = \frac{[Hg(II)]_{tot}}{[Hg^0]_{gas}} = \frac{k_1}{k_2} \cdot \frac{1}{H_{Hg}} \cdot [O_3]_{aq} \cdot \left(1 + K_3 \cdot \frac{c_{soot}}{r}\right) \qquad (15)$$

The rate constants k_1, k_2 and K_3 have been determined by laboratory studies and regression analyses (Munthe et al., 1991; Munthe, 1992; Petersen, 1992). For the calculation of the scavenging ratio of Hg^0 a mean soot particle radius of 0.5 µm has been assumed. For Henry's law coefficient of Hg^0 the value of 0.29 at 10 °C is used (Sanemasa, 1975).

Besides the above mentioned constant values, input data of 6-hourly ozone and soot carbon concentrations for each grid element are required to calculate time and space dependent scavenging ratios over the entire model domain. For ozone, these data are taken from the output of the EMEP MSC-W ozone model (Simpson, 1992) and from observed data of the OECD-OXIDATE project (Grennfelt et al., 1987). The soot carbon concentrations are derived from calculations of sulfur dioxide concentrations performed with the EMEP-model for acidifying components, assuming constant soot carbon/SO_2 ratios for different regions of the European continent (Penner et al., 1993). Over the North Atlantic, a constant soot carbon background concentration of 0.2 µgm^{-3} has been applied (Cachier et al., 1990).

For mercury associated with particles (Hg(part.)) and for gaseous inorganic mercury compounds Hg(II)(g), scavenging ratios are assumed to be constant in space and time. For Hg(part.), a scavenging ratio of $5 \cdot 10^5$ has been adopted from a long-range transport model for lead, as field measurements have shown a similar size distribution for lead and mercury in particulate form. The gaseous nitric acid scavenging ratio of $1.6 \cdot 10^6$, used with the EMEP-model for acidifying pollutants, has been applied for Hg(II)(g), since the water solubility of the two species is comparably high.

Dry deposition rates of Hg^0 should be negligibly small since the substance is almost insoluble. Therefore, the dry deposition velocity of Hg^0 is zero in the model, although recent studies indicate, that there are measurable deposition rates of vapor-phase Hg to forest surfaces (Iverfeldt, 1991b; Lindberg et al., 1991). Dry deposition velocities of 0.2 cms^{-1} for Hg(part.) and 4 cms^{-1} for Hg(II)(g) at 1 m height have been adopted from the above mentioned models for lead and for nitric acid, respectively. For Hg(II)(g), the dry deposition velocity at 1 m height is corrected to a 50 m height value, which is considered to be representative for the average concentration in the well mixed layer assumed by the model. This correction is done by applying the similarity theory for the constant flux layer which is about 50 m thick (Iversen et al., 1991).

RESULTS

The model has been run for two full years, 1987 and 1988. The general objective of the model simulations was to quantify the atmospheric long-range transport of mercury from the main emission areas in Europe to Scandinavia and the adjacent North Sea and Baltic Sea. A focus of the study was to confirm one of the most important results from the Nordic network for atmospheric mercury, i.e. a relatively weak south-to-north gradient of mercury concentrations in ambient air and a more pronounced one in precipitation over Scandinavia, indicating a substantial impact from Central European anthropogenic mercury sources through long-range transport. To this end, four sites in eastern Germany and four stations of the Nordic network forming a longitudinal corridor extending from the high emission region of Halle/Leipzig/Bitterfeld in Germany to Överbygd in northern Norway (Fig. 1) were selected for an interpretative analysis of comparable field observations and model simulations. The data set of atmospheric mercury measurements performed at the German sites is still much smaller than that from the Nordic network and can hence only be used for a crude evaluation of the model performance. However, the measurements at the Scandinavian sites are most suitable for comparison with model results, since they re-

Figure 1. Location of the stations

present the longest continuous time series reported to date of atmospheric mercury concentration data in Europe. An extensive treatment of the data generated within the Nordic network can be found in Iverfeldt (1991a).

Two different scenarios have been investigated by model calculations:

1. Mercury in the European atmospheric environment is restricted to species, which have been definitely observed in the Nordic network, i.e. Hg^0 and Hg(part.).
2. Scenario 1 plus $HgCl_2$. This species most likely emitted by waste incinerators and other high-temperature combustion processes is very water-soluble and would thus be readily dry and wet deposited in the vicinity of the sources.

Model results from scenario 1 for the 8 stations in Europe depicted in Fig. 1 and observations from one station in Germany and 4 stations of the Nordic network show a relatively weak south-to-north decrease for Hg^0 concentrations in air and a more pronounced one for total mercury concentrations in precipitation (Table 1) indicating high amounts of soot particle associated mercury in precipitation at the central European stations. The correlation between soot carbon and Hg in rainwater becomes evident even for Aspvreten in central Scandinavia when time series of calculated and observed values of the two species are compared (Fig. 2).

Table 1. Atmospheric mercury concentrations in Central and Northern Europe
- annual mean values 1988
(observations: IVL, Göteborg, Sweden and GKSS, Geesthacht, Germany)

	elemental mercury in air $[ngm^{-3}]$		particulate mercury in air $[ngm^{-3}]$		reactive mercury in precipitation $[ngl^{-1}]$		total mercury in precipitation $[ngl^{-1}]$	
	modelled	observed	modelled	observed	modelled	observed	modelled	observed
Halle/Leipzig/ Bitterfeld	10.1		0.279		9.2		330.5	
Neuglobsow	4.9		0.208		6.6		233.9	
Langenbrügge	4.1	4.2	0.111		4.3		65.8	52.0
Kap Arkona	3.6		0.107		4.0		61.6	
Rörvik	2.5	2.8	0.025	0.06	2.3	1.9	17.4	35.0
Aspvreten	2.5		0.019		3.3	2.4	16.5	18.4
Vindeln	2.1	2.5	0.005	0.05	1.9	1.4	7.0	11.2
Overbygd	2.1	2.6	0.002		2.3	1.0	3.9	9.2

Scenario 2 gives about 100 % higher wet deposition fluxes over Southern Scandinavia and over the southern parts of the Baltic Sea. For this scenario it is not possible to evaluate the model performance because of a total lack of $HgCl_2$-measurements in air in that area as well as the fact that a possible contribution of $HgCl_2$ to the total mercury concentration in precipitation is not discernible from other oxidized mercury species.

Both scenarios have been applied to estimate the total atmospheric input of mercury to the Baltic Sea for 1988 (Fig. 3). Comparison between the lower full line (scenario 1) and the upper chaindashed line (scenario 2) reveals the influence of $HgCl_2$. According to the speciation scheme of scenario 1, for which reliable data have been obtained in the Nordic network, 6.1 tonnes per year has to be considered as a realistic estimate. Measurements for mercury species identification in the central European area would be necessary to confirm the existence of divalent inorganic mercury and the significance of this species for the mercury deposition fluxes over central and Northern Europe.

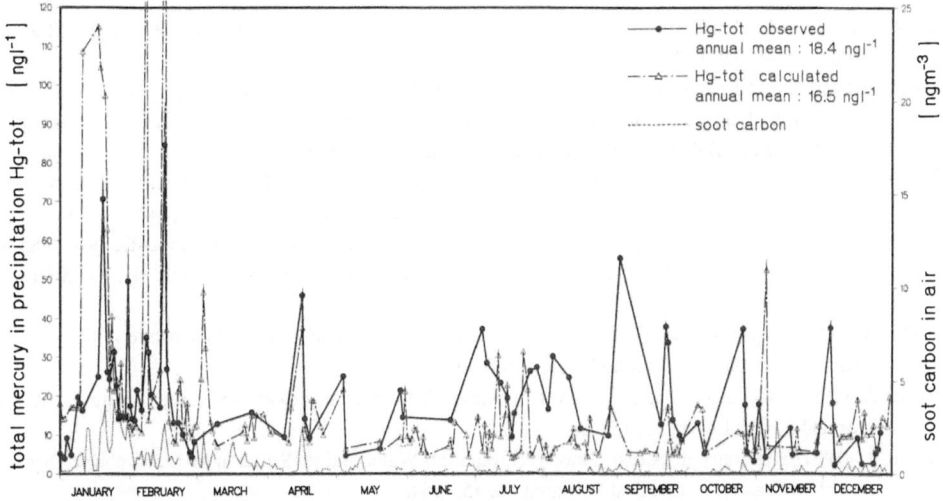

Figure 2. Daily means of modelled and observed mercury concentrations in precipitation and soot carbon concentrations in air - Aspvreten 1988

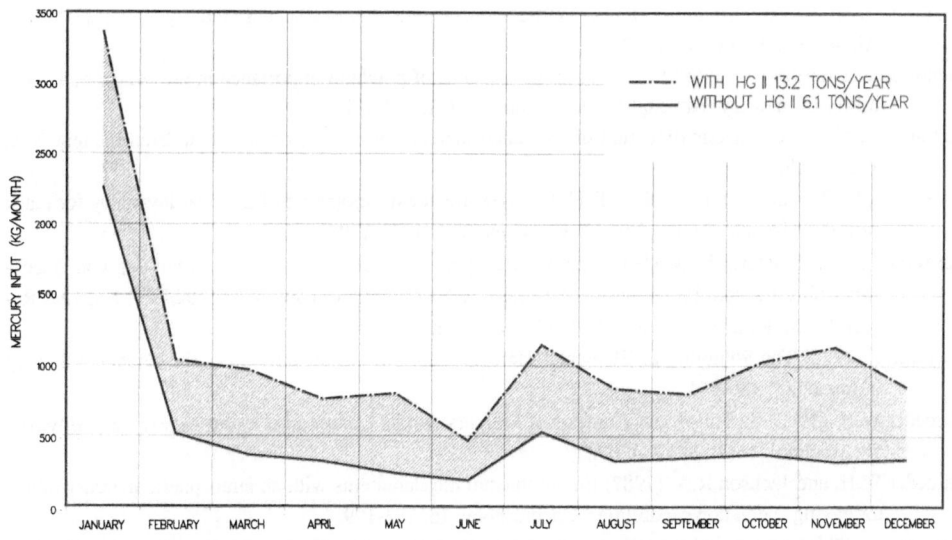

Figure 3. Atmospheric input of mercury to the Baltic Sea - 1988

REFERENCES

Axenfeld F., Münch J. and Pacyna J. (1991) Europäische Test-Emissionsdatenbasis von Quecksilberkomponenten für Modellrechnungen. Umweltforschungsplan des Bundesministers für Umwelt, Naturschutz und Reaktorsicherheit - Luftreinhaltung - 104 02 726. In: Petersen G. (1992) Belastung von Nord- und Ostsee durch ökologisch gefährliche Stoffe am Beispiel atmosphärischer Quecksilberverbindungen. GKSS 92/E/111 External Report, GKSS Research Centre, D-21502 Geesthacht, Germany.

Brosset C. and Lord E. (1991) Hg in precipitation and ambient air. A new scenario. Water, Air and Soil Pollution 56, 493-506.

Cachier H., Bremond M.P. and Buat-Menard P. (1990) Organic and Black Carbon Aerosols over Marine Regions of the Northern Hemisphere. In: Newman and Kian (eds.): Proceedings of the International Conference on Global Atmospheric Chemistry (Beijing, 1989), 249-261.

Eliassen A. and Saltbones J. (1983) Modelling of long-range transport of sulphur over Europe: A two-year model run and some model experiments. Atmospheric Environment 17 No. 8, 1457–1473.

Grennfelt P., Saltbones J. and Schjoldager J. (1987) Oxidant Data Collection in OECD-Europe 1985-87 (OXIDATE). NILU OR: 22/87. Norwegian Institute for Air Research, Postboks 64, N-2001 Lillestrom, Norway.

Iverfeldt Å. and Lindqvist O. (1986) Atmospheric oxidation of elemental mercury by ozone in the aqueous phase. Atmospheric Environment 20, 1567-1573.

Iverfeldt Å. (1991a) Occurrence and Turnover of Atmospheric Mercury over the Nordic countries. Water, Air and Soil Pollution 56, 151-165.

Iverfeldt Å. (1991b) Mercury in Forest Canopy Throughfall Water and its Relation to Atmospheric Dry Deposition. Water, Air and Soil Pollution 56, 553-564.

Iversen T., Halvorsen N.E., Mylona S. and Sandnes H. (1991) Calculated Budgets for Airborne Acidifying Components in Europe 1985, 1987, 1988, 1989 and 1990. EMEP/MSC-W Report 1/91.

Lindberg S.E., Turner R.R., Meyers T.P. and Schroeder W.H. (1991) Atmospheric Concentrations and Deposition of Hg to a Deciduous Forest at Walker Branch Watershed, Tennessee, USA. Water, Air and Soil Pollution 56., 577-594.

Lindqvist O., Johansson K., Aastrup M., Andersson A., Bringmark L., Hovsenius G., Håkanson L., Iverfeldt Å., Meili M. and Timm B. (1991) Mercury in the Swedish Environment. Recent Research on Causes, Consequences and Corrective Methods. Water, Air and Soil Pollution 55.

Munthe J., Xiao Z.F. and Lindqvist O. (1991) The aqueous reduction of divalent mercury by sulfite. Water, Air and Soil Pollution 56, 621-630.

Munthe J. and McElroy W. (1992) Some aqueous reactions of potential importance in the atmospheric chemistry of mercury. Atmospheric Environment 26A, 553-557.

Munthe J. (1992) The aqueous oxidation of elemental mercury by ozone. Atmospheric Environment 26A, 1461-1468.

Penner J.E., Eddleman H. and Novakov T. (1993) Towards the development of a global inventory for black carbon emissions. Atmospheric Environment 27A, 1277-1295.

Petersen G. (1992) Numerische Modellierung des Transports und der chemischen Umwandlung von Quecksilber über Europa. Ph.D. thesis, University of Hamburg, Germany. GKSS 92/E/51 External Report, GKSS Research Centre, D-21502 Geesthacht, Germany.

Sanemasa I. (1975) The Solubility of Elemental Mercury Vapor in Water. Bulletin of the Chemical Society of Japan, 48, 480-484.

Schroeder W.H. (1982) Sampling and Analysis of Mercury and its Compounds in the Atmosphere. Environ. Sci. Technol. 16(7), 394A-400A.

Schroeder W.H. and Jackson R.A. (1987) Environmental measurements with an atmospheric mercury monitor having speciation capabilities. Chemosphere 16, 183-199.

Simpson D. (1992) Long period modelling of photochemical oxidants in Europe. Model calculations for July 1985. Atmospheric Environment 26A, 1609-1634.

DISCUSSION

J. MATTHIJSEN   Does the Hg-redox cycle have an important impact on the oxidation of S(IV)?

G. PETERSEN: The impact of the Hg-redox cycle on the oxidation of S(IV) should be very small, as a large excess of S(IV) with respect to Hg2+ ions in water droplets in the lower troposphere over Europe can be expected.

H. SCHLÜNZEN: How can you calculate daily values with the EMEP model and compare them with measured values which are mean values for several days?

G. PETERSEN: Actually, the measured values for both the concentrations of total gaseous mercury in air and the total mercury concentrations in precipitation, which have been compared against model predicted values at two Scandinavian sites, are daily mean values as well. However, it should be clearly stated that the model has not been designed for a time resolution of one day or less. The intension of the daily comparison of high observed and model predicted peak concentrations was to demonstrate more or less qualitatively, that the emission inventory used with the model is based on realistic estimates for East Germany in 1987 and 1988.

R. KUNZ: Is there any preferred height, where the ozone-mercury-reaction takes place?

G. PETERSEN: The available experimental results are difficult to extrapolate to atmospheric conditions but so far there is no evidence that this reaction takes place at any preferred height within the lower troposphere.

A BOUNDARY LAYER PARAMETERIZATION

FOR GLOBAL DISPERSION MODELS

Carmen J. Nappo[1] and Han van Dop[2]

[1]National Oceanic and Atmospheric Administration, Air Resources Laboratory, Atmospheric Turbulence and Diffusion Division
Oak Ridge, Tennessee 37830
U.S.A.

[2]Institute for Marine and Atmospheric Research
Utrecht University
Dept. of Physics and Astronomy
3508 TA Utrecht
The Netherlands

INTRODUCTION

In this note we present the essentials of a simple algorithm which simulates the main features of a dry, horizontally homogeneous atmospheric boundary layer, a more comprehensive presentation of these ideas are found in Nappo and van Dop (1993). The algorithm or parameterization scheme is designed only for use in global-scale air chemistry and transport models. Details of dispersion in the convective and stable boundary layers are not addressed; instead, we specify the concentration profiles for these PBL conditions. It is this simplicity that makes this algorithm attractive for parameterizing subgrid-scale boundary-layer fluxes in numerical global-scale transport/chemistry models. The algorithm was developed for the Moguntia (Latin word for Mainz, FRG) global chemistry model (Zimmerman, 1987) which uses monthly-averaged meteorological input data. The algorithm can be used over land and sea surfaces provided grid-cell averaged convective mixing heights and stable PBL heights are given. Trace-gas removal and surface concentrations, including diurnal variations, are satisfactorily described. However, the effects of isolated thunderstorms and frontal systems are not parameterized.

Air Pollution Modeling and Its Application X, Edited by S-V. Gryning
and M. M. Millán, Plenum Press, New York, 1994

MODEL FORMULATION

Convective Boundary Layer

We consider a well-mixed layer of average depth h_c above the ground surface within which the concentration $C_m(t)$ of an arbitrary substance is vertically uniform (see Figure 1). Above this layer, concentration, C, decreases uniformly up to the height of the tropopause, H, but, this is not crucial; a vertically uniform distribution would be equally acceptable. Mixed-layer concentrations may change due to the presence of sources, dry deposition, and entrainment at the top of the CBL. Two types of sources are considered, a surface source, q_s (g m^{-2} s^{-1}), and a volume source, q_v (g m^{-3} s^{-1}). The volume source represents chemical processes, for example, photochemical production of ozone. For simplicity, we neglect chemistry and washout by precipitation, but these can be included in a straightforward way. The dry deposition flux at the ground surface is given by

$$\phi_d = v_d \, C_m \tag{1}$$

where v_d is the deposition velocity. The entrainment flux at the top of the CBL is given by

$$\phi_{h_c} = w_e \, (C_{h_c}^{\;+} - C_m) \tag{2}$$

where w_e is the entrainment velocity and $C_{h_c}^{\;\cdot}$ is the concentration just above the mixed layer (see Figure 1).

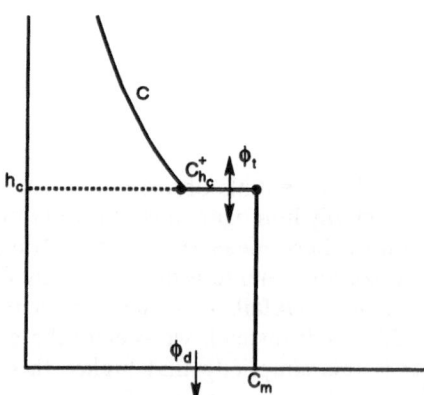

Figure 1. Schematic diagram of the convective boundary layer: h_c is the depth of the convective boundary layer, C_m is the mixed layer concentration, C is the free tropospheric concentration, ϕ_d is the deposition flux, ϕ_t is the flux due to entrainment, and $C_{h_c}^{\;\cdot}$ is the concentration just above the mixed layer.

We assume that the entrainment velocity is equal to the time-averaged rate of growth of the CBL,

$$w_e = \frac{\overline{\partial z_i}}{\partial t} \tag{3}$$

where $z_i(t)$ is the height of the elevated inversion base.

To calculate an average entrainment velocity, we replace the continuous derivative in (3) by its finite difference, i.e.

$$w_e = \frac{\overline{\Delta z}}{\Delta t} = \frac{\overline{z_{max}} - h_s}{t_{SS} - t_{SR}} = \frac{h_c - h_s}{t_{SS} - t_{SR}} . \tag{4}$$

Consider the one-dimensional concentration equation

$$\frac{\partial C}{\partial t} = -\frac{\partial F_c}{\partial z} \tag{5}$$

where F_c is the vertical turbulent flux of concentration, C. Vertically integrating (5) between the ground surface and height h_c, and noting that the concentration in the mixed layer is vertically uniform yields

$$\frac{\partial C_m}{\partial t} = \frac{\phi_{h_c} - \phi_d}{h_c} . \tag{6}$$

Using (1) and (2) in (6) and including source terms q_s and q_v we get

$$\frac{\partial C_m}{\partial t} = -\left(\frac{W_e + V_d}{h_c} \right) C_m + \frac{W_e C_h^+ + q_s}{h_c} + q_v \tag{7}$$

which has the general solution

$$C_m(t) = C_m(0) e^{-\frac{w_e + v_d}{h_c}t} + e^{-\frac{w_e + v_d}{h_c}t} \int_0^t e^{\frac{w_e + v_d}{h_c}t'} \tag{8}$$

$$\times \left(\frac{w_e C_{h_c}^+ + q_s(t')}{h_c} \right) dt' + q_v t$$

where $C_m(0)$ is the initial concentration in the mixed layer at the start of the convective period.

For this illustration of the parameterization, we assume Fickian diffusion above the mixed-layer:

$$\frac{\partial C}{\partial t} = K \frac{\partial^2 C}{\partial z^2} , \tag{9}$$

with the boundary conditions

$$K \frac{\partial C}{\partial z}\Big|_{z=h_c} = \phi_t \tag{10}$$

and

$$K \frac{\partial C}{\partial z}\Big|_{z=H} = 0. \tag{11}$$

The concentration just above the mixed layer, $C_{h_c}^*$, can be obtained from (9) by noting that $C_{h_c}^* \equiv C(z=h_c)$. The concentrations $C_m(t)$ and $C(z,t)$ are the solutions of the (coupled) equations (8) and (9).

Stable Boundary Layer

We assume in this case that the troposphere is uncoupled from the boundary layer, i.e., there is no flux across the top of the SBL. We assume a linear concentration profile (see Figure 2) given by

$$C(z,t) = \frac{C(0,t)-C(0,0)}{h_s}(h_s-z) + C(0,0) \tag{12}$$

where $C(0,0) = C(z,0)$ is the initial profile which is vertically uniform at sunset. Modeling the dry deposition flux as $\phi_d = v_d C(0,t)$, keeping the concentration at $z=h_s$ fixed to the initial value, and using (12), the vertical integral of (5) between the ground surface and h_s is

$$\frac{\partial C_0}{\partial t} = \frac{(q_s-v_d C_0)}{h_s/2} + q_v \tag{13}$$

where $C_o(t) = C(0,t)$. The solution to (13) is

$$C_b(z,t) = C_0(t)\left(1-\frac{z}{h_s}\right) + C_{h_s}\left(\frac{z}{h_s}\right). \tag{14}$$

where $C_{h_s} = C(z,0)$.

The boundary-layer profile is then given simply by

$$C_0(t) = C_{h_s} e^{-\frac{2v_d}{h_s}t} + \frac{e^{-\frac{2v_d}{h_s}t}}{h_s/2}\int_0^t e^{\frac{2v_d t'}{h_s}} q_s(t') \, dt' + q_v t \tag{15}$$

Above the SBL we again assume that C(z,t) is a solution of (9) but now with the lower boundary condition

$$K \frac{\partial C}{\partial z} \Big|_{z=h_s} = 0 \qquad (16)$$

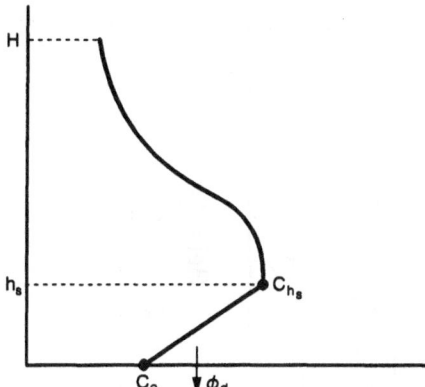

Figure 2. Schematic diagram of the stable boundary layer.

TRANSITION

It is assumed that the concentration profile at the end of a convective episode serves as the initial profile for the subsequent stable episode. This is a straightforward procedure when the SBL height is always smaller than the CBL height. The change from stable to unstable is slightly more complicated. In this case, we "reshape" the stable concentration profile such that it has the required shape for an initial profile in the CBL, and the mass within a unit column is conserved.

RESULTS

A preliminary test run of the PBL parameterization was carried out using a simple one-dimensional model, with a tropopause height of 10000 m, a CBL maximum height, h_C = 1900 m, and a SBL height, h_s, of 300 m. The eddy diffusivity was arbitrarily set at 10 m^2s^{-1}, and the deposition velocity at 0.01 ms^{-1}. However, the parameterization of vertical transport above the boundary layer is not critical here; in an actual application of our parameterization in a global air chemistry model, the details of the vertical transport above the PBL would be given by that model. The model time step was one hour, and the numerical integration of the Fickian diffusion equation (9) was done with a time step of 30 s. All test runs begin at 00:00 with sunrise at 06:00 and sunset at 18:00.

The run begins with a uniform vertical distribution of a passive contaminant with concentration arbitrarily set to 1 g m^{-3}. The surface and volume sources are set to zero. For the first six days of the model run, Figure 3 shows the concentration profiles at 12:00 and Figure 4 shows the concentration profiles at 24:00. Concentration values below 300 m fluctuate greatly with high values during the day and low values at night. Between 300 and about 1000 m concentrations decrease monotonically with time. However, between 1000 and 2100 m concentrations are higher during stable conditions than during convective conditions. These diurnal changes in concentration are due to the combined effects of convective mixing, stable stratification, and surface deposition. The effects of chemistry and advection are expected to considerably change these results.

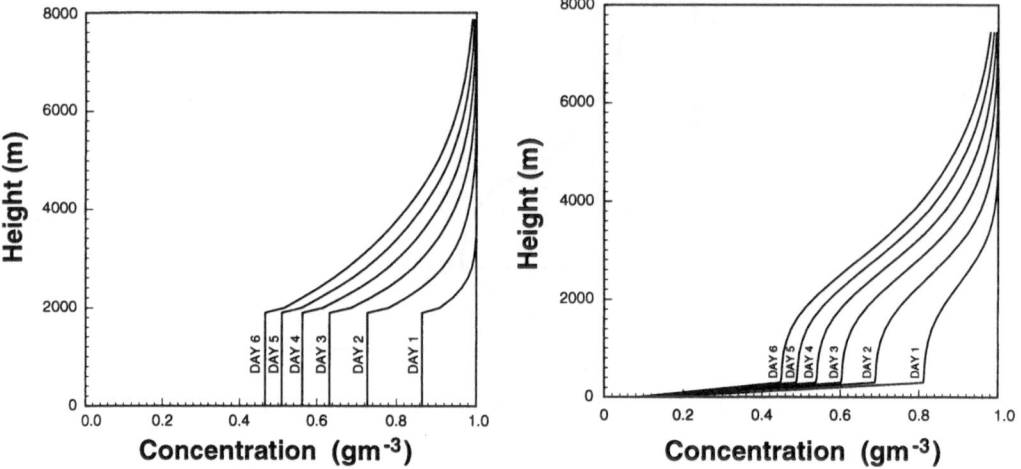

Figure 3. Vertical profiles of concentration at 12:00 for the first six days of simulation. Initial concentration, C_m, is 1 g m^{-3}, deposition velocity, V_d, is 0.01 ms^{-1}, and there are no sources.

Figure 4. Same as Figure 5, but at 24:00.

Figure 5 shows the time variations of surface concentration and total deposited mass. The amplitude of the diurnal variation of surface concentration decreases nearly exponentially with time. The surface concentration minima, which occur at sunrise, decrease rapidly during the first two simulation days, and then decrease much more slowly with time after that. Concentration maxima occur during the first CBL time step when vertical mixing takes place. After this mixing, CBL concentrations decrease with time. For this run, the deposition flux is slightly greater than the entrainment flux. A shorter convection time would result in a larger entrainment velocity and a corresponding larger entrainment flux into the CBL. In this event, CBL concentrations would grow with time rather than decrease with time. Also shown in Figure 5 is the mass deposited to the surface. The deposition rate is much greater during convective conditions than during stable conditions. Figure 6 shows the mass balance for the first six simulation days. The total mass is equal to the sum of the mass in the atmosphere plus the mass deposited to the ground surface. After the first hour of the run, the total mass remains essentially constant with only small diurnal fluctuations. During the first simulation hour, there is a mass loss of about 3 percent of the total initial mass. We believe this to be due to the initial effects of the numerical diffusion scheme above the SBL.

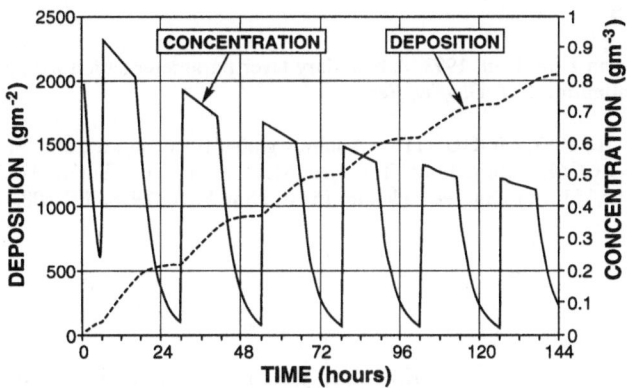

Figure 5. Surface concentrations and deposition for the case shown in Figures 5 and 6.

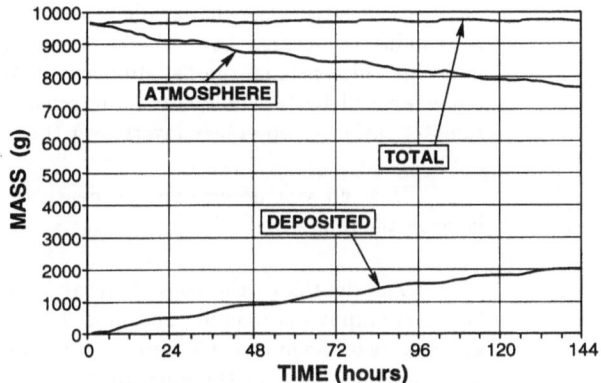

Figure 6. Mass balance for the case shown in Figure 5. TOTAL is the sum of the mass contained in the ATMOSPHERE and mass DEPOSITED at the ground surface.

CONCLUSIONS

We have developed a simple parameterization of the diurnally varying atmospheric boundary layer for use in global dispersion and air chemistry models. The parameterization captures the essential characteristics of the stable and convective boundary layer, and includes the effects of surface and volumetric source terms; however, other features such as the effects of isolated thunderstorms and frontal systems are not taken into account. Initial tests using a one-dimensional model show that the parameterization is realistic and mass conserving. However, these tests can not determine the accuracy of the parameterization. The validity of the parameterization can be determined only by comparison with observations of a not too reactive chemical tracer. Still, the success of these tests encourages us to try this parameterization in a more general air chemistry model and compare the results with the standard boundary-layer parameterization.

REFERENCES

Nappo, C.J. and van Dop, Han, 1993, A boundary layer parameterization for global dispersion models, Submitted to *J. Geophy. Res.*

Zimmermann, P. H., 1987, MOGUNTIA: A handy global tracer model, in: "Proceedings of the 16th NATO/CCMS International Technical Meeting on Air Pollution Modelling and its Applications," H. van Dop, ed., Plenum Press, New York, 1988, 593-608.

DISCUSSION

R.C. EASTER: How does your parameterization treat the boundary layer over sea surfaces?

C. NAPPO: The parameterization scheme requires estimates of monthly-averaged convective and stable boundary-layer heights. Over the sea surface, these values are poorly known, and consequently somewhat crude estimates must be made. It remains to be seen how the parameterization performs in a three-dimensional, global simulation. When these are done, we will better understand the sensitivity of the parameterization to boundary layers over the oceans.

Z. KLAIC: How do you determine the depth of the boundary layer in the model?

C. NAPPO: The depths of the convective and stable boundary-layers are input parameters, and must be specified for each grid square of the model. Climotological values of the mixing depths can be used for these parameters.

J. VILA-GUERAU DE ARELLANO: Could you provide more details on the transition between the CBL and the SBL?

C. NAPPO: The evening transition between convective and stable conditions, and the morning transition between stable and convective conditions are parameterized to occur instantaneously. At the evening transition, the concentration at the top of the SBL is held constant during the night, and a linear profile is assumed to exist between the ground surface and the top of the SBL. Above the SBL, the time varying concentrations are calculated by the tropospheric model. During the morning transition, the concentrations below and above the SBL are averaged in a mass-conserving way up through the depth of the CBL. The resulting vertically-uniform concentration is the initial value for the CBL parameterization.

ANALYSIS OF MID-TROPOSPHERIC CARBON MONOXIDE DATA USING A THREE-DIMENSIONAL GLOBAL ATMOSPHERIC CHEMISTRY NUMERICAL MODEL

Richard C. Easter, Rick D. Saylor, and Elaine G. Chapman

Earth and Environmental Sciences Center
Pacific Northwest Laboratory
Richland, WA 99352

INTRODUCTION

Carbon monoxide is an important atmospheric trace species. It has long been recognized as a major contributor to urban air quality and in high concentrations is known to adversely affect health (Seinfeld, 1986). CO is the third most abundant carbon-containing species in the atmosphere and its reaction with hydroxyl radical (OH) represents a 2000-3000 Tg/yr (1 Tg = 10^{12} g) source of carbon dioxide. On a global basis, through its reaction with OH, CO plays a significant role in the troposphere's overall oxidative capacity (Crutzen and Zimmerman, 1991). Furthermore, depending on the local abundance of nitrogen oxides, CO can participate in reactions that either increase or decrease the formation of tropospheric ozone (Logan et al., 1981).

Several three-dimensional modeling studies have been attempted in the past to better understand the global distribution of CO. The earliest attempt was made by Kwok et al. (1971). In their simulation, CO was transported within a GCM as an inert tracer and allowed to advect within the troposphere over 30 days during a simulated November-December time period. They performed experiments with and without vertical CO transport by cumulus convection and found that around 10 percent of their model's global-mean vertical transport occurred through upward cumulus motion. Peters and Jouvanis (1979) developed a global CO and CH_4 transport model that included chemical kinetics calculations and was separated from a parent GCM such that observed meteorological data for a specified time period could be used to drive a simulation. From their simulations they estimated a residence time for CO of 0.21 yr. The same model was used by Kitada and Peters (1982) to generate global estimates of the distribution of OH concentrations which compared favorably with the limited measurements available at that time. Pinto et al. (1983) used a GCM tracer model with simple chemistry to study the climatological CO budget and its global distribution. They suggested that, in addition to fossil fuel combustion and CH_4 oxidation sources of CO, a large low latitude source of 1300 Tg/yr was necessary to account for observed CO magnitudes and isotopic abundances. Crutzen and Zimmerman (1991) performed a set of global three-dimensional simulations to study the impact of changing anthropogenic emissions of CO, CH_4, and NO_x from pre-industrial values to present-day levels on the atmosphere's total oxidative capacity. They concluded that since the beginning of the industrial revolution a large increase in CO concentrations has occurred, predominantly in the northern hemisphere, which has resulted in a substantial decrease of global OH concentrations. In Saylor and Peters (1990, 1991), the original model of Peters and Jouvanis (1979) was used to analyze CO data from the November 1981 mission of NASA's Measurement of Air Pollution from Satellites (MAPS) program. A three-dimensional simulation was performed for the November 1981 time period and the model's results were compared with the mid- to upper-tropospheric data of Reichle et al. (1986). They found good agreement between the model's results and the observations of the MAPS experiment. Simulations performed without the inclusion of fossil fuel combustion emissions indicated that emissions from biomass burning may be exerting a significant role in global CO distributions.

The purpose of the current work is to extend the results of Saylor and Peters (1990, 1991) to the October 1984 MAPS CO dataset. The data obtained during the second MAPS mission was much more complete in global coverage and thus provides an excellent opportunity for comparison with model simulations. The primary objective of this study is to examine the data of the 1984 MAPS mission, using a three-dimensional model as an interpretive tool. We have performed global, three-dimensional simulations of CO transport and chemistry for September-October, 1984, the time period during which the MAPS experiment was conducted.

Air Pollution Modeling and Its Application X, Edited by S-V. Gryning
and M. M. Millán, Plenum Press, New York, 1994

MAPS MEASUREMENTS OF TROPOSPHERIC CARBON MONOXIDE

The Measurement of Air Pollution from Satellites (MAPS) experiment was designed by NASA to investigate the global distribution of tropospheric CO (Reichle *et al.*, 1986; Reichle *et al.*, 1990). Two experiments have been conducted in this program - the first during November 1981 and the second during October 1984. In each experiment a nadir-viewing gas filter radiometer was placed onboard the U.S. space shuttle and mid- to upper-tropospheric CO data were obtained. Details of the measurement technique and instrumentation used during these experiments can be found in Reichle *et al.* (1986, 1990). Carbon monoxide measurements were made between 38° N and 38° S during the 1981 experiment, while a broader coverage of 57° N to 57° S was obtained during the 1984 experiment. The MAPS measurements represent the first nearly simultaneous set of trace tropospheric species data obtained with such widespread global coverage. In this work we concentrate our analysis on the second MAPS experiment, although some of the major features of the 1984 dataset are also common to the 1981 data.

The second MAPS experiment was performed during October 5-13, 1984; Reichle *et al.* (1990) present results from this dataset. In comparing remotely sensed CO mixing ratios from the MAPS experiment with *in situ* CO measurements taken in the mid-troposphere by aircraft sampling, Reichle et al. (1990) concluded that the MAPS CO data are systematically too low by 20-40 percent. Taking this bias into account, several large-scale features of the MAPS dataset are described in the following. Significant areas of elevated CO mixing ratios (> 80 ppb) occur over southern Africa, central South America and central and eastern Asia. Smaller areas of high CO values occur over the southern Indian ocean, southern Australia and New Zealand, the North Atlantic, and the far northern and southern reaches of the Pacific ocean. Just as striking as these areas of high CO values are the vast areas of very low CO that occur over the central Pacific and in a band stretching from Central America across the Atlantic, over the Sahara, and into western Asia. Some these features are readily explained by correlation with major source areas of CO emissions. For example, the maxima which occur over southern Africa and central South America have been linked to biomass burning activities (Watson *et al.*, 1990 and Connors *et al.*, 1991). However, other features, such as the high values over the southern Indian ocean and over the northeastern portion of Asia, are not as easily linked to known areas of large CO emissions. Also puzzling is the lack of a well-defined plume associated with North America, which is known to be a significant source of anthropogenic CO.

MODEL DESCRIPTION

The Global atmospheric Chemistry Model (GChM) is an Eulerian, three-dimensional model that simulates the emission, transport, chemical transformation and surface interactions of tropospheric trace species. It has been used previously to investigate the impact of fair weather cloud venting processes on the long-range transport of CO and CH_4 (Luecken *et al.*, 1989) and to estimate the long-range transport of sulfur from North America and Europe to the North Atlantic (Luecken *et al.*, 1991).

Governing Equation

The mass conservation equation for a gas-phase atmospheric species as simulated by GChM is given by

$$\frac{\partial r_i}{\partial t} + \frac{u}{a\cos\theta}\frac{\partial r_i}{\partial \phi} + \frac{v}{a}\frac{\partial r_i}{\partial \theta} + w_\sigma \frac{\partial r_i}{\partial \sigma} =$$

$$\frac{1}{p_s a^2 \cos^2\theta}\frac{\partial}{\partial \phi}\left(p_s K_h \frac{\partial r_i}{\partial \phi}\right) + \frac{1}{p_s a^2 \cos\theta}\frac{\partial}{\partial \theta}\left(p_s \cos\theta\, K_h \frac{\partial r_i}{\partial \theta}\right) + \left(\frac{M_a g}{p_s}\right)^2 \frac{\partial}{\partial \sigma}\left(c^2 K_z \frac{\partial r_i}{\partial \sigma}\right) + R_i + CT_i$$

where ϕ and θ are longitude and latitude respectively, σ is the vertical coordinate expressed in terms of normalized pressure

$$\sigma = p/p_s \ ,$$

p is the pressure at a given location, and p_s is the corresponding pressure at the surface. Further, r_i is the mixing ratio of species i, u and v are the curvilinear velocity components toward east and north, respectively, w_σ is the vertical velocity component in the σ coordinate system (dσ/dt), a is the mean radius of the earth, g is the gravitational acceleration, M_a is the molecular weight of air, K_h and K_z are the horizontal and vertical eddy diffusivities, R_i represents chemical production (or loss) of species i, and CT_i is the change of species i per unit time resulting from convective cloud transport. The boundary conditions applied to these equations account for the inflow and outflow of material from the model domain through surface exchange, emission of trace species at the surface, and advection across the model's upper boundary.

Meteorology and Model Transport

Meteorological data from the European Centre for Medium-Range Weather Forecasting (ECMWF) for September 1 through October 15, 1984, were used to drive the GChM simulations reported here. The ECMWF data for this time period are provided on a 1.875° latitude-longitude grid with 16 vertical levels and at 6-h time intervals. The data include horizontal winds, vertical velocity, temperature, humidity, cloud cover, and several surface parameters. The current GChM simulations used the lowest 14 levels (surface to 100 hPa) and a 3.75° horizontal grid resolution.

Advective transport in the model is calculated using a version of the Bott algorithm (Easter, 1993) with 1-h time steps. Vertical turbulent transport is calculated with an implicit in time, centered in space finite difference algorithm. Spatially and temporally varying eddy diffusivity coefficients are calculated according to the procedure of Carmichael (1979). Horizontal turbulent transport is neglected in these simulations.

The parameterization of convective clouds is based on the model of Walcek and Taylor (1986) and is similar to that used in the Regional Acid Deposition Model (Chang et al., 1987). When convective clouds are present over a grid cell, the convective cloud volume (determined by cloud base, cloud top, and areal coverage) is filled with air from below cloud base plus air entrained from the cloud top model level. This takes place over the convective cloud lifetime (assumed 1 h), and then the cloud dissipates. The process of filling the cloud volume transports trace species from below cloud base to higher altitudes. The amount of vertical transport is dependent on the cloud volume and thus the cloud areal coverage (cloud base and cloud top are determined by the temperature and moisture vertical profiles). In the simulations presented here, the areal coverage for convective clouds was taken to be the total cloud coverage from the ECMWF analyses when that coverage was below 35%. As discussed below, convective transport has a major impact on the simulated CO concentrations at upper model levels. We plan to investigate other methods for parameterizing convective transport.

Chemistry

The version of GChM used for this work accounts for the kinetic reactions of CO and CH_4 that occur in the remote troposphere. The gas-phase mechanism of Lelieveld and Crutzen (1990) has been used as the basis for GChM chemical transformations. To reduce the computational burden of the chemical calculations within the model, several simplifying assumptions and techniques have been implemented. First, a latitude-altitude distribution of ozone mixing ratios is specified within the model, based on the climatological distribution of Fishman and Crutzen (1978). Further, NO_x vertical profiles for each grid column are calculated during the simulation based on the methodology of Peters and Jouvanis (1979) and Kitada and Peters (1982). In their formulation, a NO_x vertical profile is specified for each vertical column based on the CO emissions that are specified for that surface grid cell. Since CO emissions in each grid cell are time varying, the NO_x profiles generated by this methodology change dynamically during simulation runtime. Finally, the pseudo-steady-state approximation (PSSA) is imposed at each grid point for all species in the chemistry mechanism except for CO and CH_4. The combination of prescribed NO_x and O_3 distributions and the application of PSSA allows a set of algebraic relationships to be formulated for the calculation of OH radical concentrations. A simple iterative procedure is used to solve the algebraic equations for OH, which is then used to integrate the chemical rate equations of CO and CH_4. To test the accuracy of this simplification procedure, a series of box model chemistry simulations were performed in which the simplified chemical scheme was compared to a full integration of the rate equations with no PSSA applied. The error in CO and CH_4 mixing ratios calculated by the simplified procedure as compared to the full integration was less than 1 percent for most NO_x and O_3 conditions.

CO and CH_4 Emissions

Due to the simplifications made to the chemical transformation calculations, only the emissions of CO and CH_4 were required for the current GChM simulations. Both CO and CH_4 emissions data were taken from the work of DeHaven (1980) and updated based on current knowledge. Table 1 presents a breakdown by category of the emissions used in the current study and a comparison with other investigations.

The major sources of CO are well known, although the exact magnitudes of these sources are uncertain. Emissions from fossil fuel combustion, mainly from automobile exhaust, are the best known and have been estimated by Logan et al. (1981) to lie in the range 400-1000 Tg/yr, while Seiler and Conrad (1987) have estimated 440-840 Tg/yr. CO emissions from biomass burning are much less certain and have been estimated at 310-1250 Tg/yr by Logan et al. (1981) and at 400-1600 Tg/yr by Seiler and Conrad (1987). The large range in these values reflects uncertainties in both the total amount of biomass consumed and in the amount of CO emitted per kg of biomass combusted. Another significant source of CO is the atmospheric oxidation of anthropogenic and natural nonmethane hydrocarbons. Although the source of CO resulting from the oxidation of anthropogenic hydrocarbons is probably small (Logan et al., 1981), the amount of CO generated from natural hydrocarbons has been variously estimated at 280-1200 Tg/yr by Logan et al. (1981), 400-1400 Tg/yr by Seiler and Conrad (1987) and 774 Tg/yr by Warneck (1988). Oxidation of CH_4 represents a further 300-900

187

Table 1. Carbon Monoxide and Methane Emissions (Tg/yr)

Carbon monoxide	Logan et al. (1981)	Seiler and Conrad (1987)	Present study
Anthropogenic fuel use	500	640	570
Oxidation of hydrocarbons	780	975	878
Biomass burning	605	1000	803
Miscellaneous	40	117	0
TOTAL	1925	2732	2251

Methane	Fung et al. (1991)	Present study
Enteric fermentation	80	99
Wetlands, rice patties, etc.	255	256
Biomass burning	55	55
Industrial	75	75
Miscellaneous	35	0
TOTAL	500	485

Tg/yr source (Seiler and Conrad, 1987) of atmospheric CO. The total amount of CO generated from these major sources plus several minor ones (direct emission by plants, from the ocean, and from soils) is thought to be on the order of 3000 Tg/yr (Warneck, 1988). CH_4 emissions for the current simulations have been scaled from the work of DeHaven (1980) to correspond approximately to the results of Fung et al. (1991).

RESULTS AND DISCUSSION

A total of 14 GChM simulations have been performed with a horizontal resolution of 3.75°. The "base case" simulation was performed using GChM's full representation of chemical and physical processes and our best estimates for emissions. To gauge the sensitivity of the model's output when various model components are perturbed from their base case values, thirteen sensitivity simulations were performed. Results from these sensitivity simulations are presented in Table 2 as changes in the average CO mixing ratio in several model sub-domains over the last 15 days (October 1-15, 1984) of the specified simulation. The sensitivity results of Table 2 indicate that the simulated CO is not greatly sensitive to the assumed NO_x or O_3 distributions, changing no more than 5-10 percent for doubled and halved NO_x and O_3 values. Further, the simulated CO is rather insensitive to the formulation of the boundary layer parameterization. The base case simulation contains a boundary layer eddy diffusivity parameterization that allows for the diurnal growth and collapse of the mixed layer based on local stability and surface characteristics. In the constant mixed layer simulation, the mixed layer depth was fixed at the maximum depth of the base case simulation (1740 m, the top of model layer 5) and had no diurnal variation. As evidenced by the small changes in average CO mixing ratio for this simulation, the lack of a diurnally varying mixed layer had little impact on CO distributions. Likewise, in sensitivity simulations where the maximum mixed layer depth was forced to 1050 m and 2620 m, little change was observed in the average CO mixing ratio.

Two simulations were performed where total CO emissions were increased and decreased by 50 percent. As seen in Table 2, increasing total CO emissions increased the global average CO mixing ratio 27 percent, while halving total emissions decreased the average mixing ratio by 25 percent. In the base simulation, CO production from CH_4 oxidation was equal to 56 percent of the total CO emissions. Thus, increasing (decreasing) the CO emissions by 50 percent changes the overall CO source strength by +32 percent (or -32 percent). Due to the nonlinear effects of chemical feedbacks, the changes in global average CO mixing ratio do not respond linearly to changes in CO emissions. Further, three simulations were performed where each of the major CO emissions categories were omitted. The response of the global average CO mixing ratio resulting from these simulations generally follows that expected from the relative magnitude of emissions from each category, although some degree of nonlinearity is present. The largest source category is CO emissions from nonmethane hydrocarbon oxidation; omitting this source reduced the global average mixing ratio by 20 percent. Omitting anthropogenic and biomass burning CO emissions resulted in a global average mixing ratio reduction of 14 percent and 15 percent, respectively.

Figure 1 presents the CO mixing ratio field in the first model layer for the base case simulation, averaged over October 1-15, 1984, corresponding to the time during which the MAPS CO measurements were obtained. The surface CO distribution is heavily dominated by major emission sources. Mixing ratio maxima occur near the major emission sources of eastern North America, Europe, central South America, and central and southern Africa. Peak mixing ratios > 300 ppb occur over eastern North America, central South America and central and southern Africa, while remote regions (e.g., southern Pacific) exhibit mixing ratios < 45 ppb. The model produces a significant hemispheric asymmetry of CO mixing ratios, ranging from 75-150 ppb in the remote northern hemisphere to < 45 ppb in the remote southern hemisphere.

Table 2. Summary of GChM Simulations for October 1-15, 1994. For base case run, given values are average CO (ppbv) for indicated model domain. For other simulations, given values are percent changes relative to the base case.

	Model Sub-Domain						
Domain	Global	Global	Global	Cen Africa	E NA	E Europe	NE Siberia
Model Layer	1-14	1	11	11	11	11	11
Simulation							
Base	66.5	87.1	62.4	110.1	79.4	85.8	94.9
CO Emissions x 1.5	26.8	31.8	26.1	39.9	31.5	32.9	34.0
CO Emissions x 0.5	-25.2	-30.6	-24.3	-36.5	-30.1	-31.5	-33.0
Anthropogenic CO Emissions Removed	-13.8	-18.8	-12.0	-1.2	-34.1	-30.9	-31.0
Biomass Burning CO Emissions Removed	-14.6	-17.4	-15.7	-39.8	-6.3	-6.8	-5.4
CO Emissions from NMHC Oxidation Removed	-20.2	-23.4	-19.0	-27.6	-18.1	-23.8	-28.2
NO_x mixing ratios x 0.5	6.3	4.9	7.2	7.8	6.0	5.6	4.3
NO_x mixing ratios x 2.0	-6.4	-5.0	-7.4	-7.2	-6.0	-5.6	-4.4
O_3 mixing ratios x 0.5	4.2	3.4	4.7	5.8	3.9	3.6	2.3
O_3 mixing ratios x 2.0	-5.0	-4.1	-5.6	-6.6	-4.8	-4.5	-3.4
Constant mixed layer depth=1740m	0.2	-2.6	-0.2	0.5	-0.8	-1.0	-1.7
Maximum mixed layer depth=2620m	-0.1	-2.7	0.5	0.2	0.9	1.4	2.0
Maximum mixed layer depth=1050m	0.4	3.9	0.0	0.4	-0.3	-0.4	-0.9
No convective cloud transport	-3.7	19.3	-26.7	-62.1	-35.3	-27.3	-15.8

Sub-Domain	Description	Longitude Range	Latitude Range
Cen Africa	Central Africa	13°E to 36°E	21°S to 9°N
E NA	Eastern North America	92°W to 69°W	32°N to 51°N
E Europe	Eastern Europe	2°E to 43°E	43°N to 69°N
NE Siberia	Northeastern Siberia	84°E to 126°E	54°N to 77°N

Gas-filter radiometer measurements made during the MAPS experiment correspond to a vertically integrated CO mixing ratio where the strength of the radiometer signal is a function of altitude. Since the peak signal from the CO instrument occurs at an altitude corresponding to level 11 (~350 hPa) in the model domain, the CO mixing ratio distribution in the mid-troposphere at model level 11 is presented in Figure 2 for the base case simulation, averaged over October 1-15. Accounting for the systematic 20-40 percent low bias in the MAPS data, the distribution of CO produced by the model compares favorably with the MAPS data during this time period. Major features of the distribution that are common with the satellite data include: (i) peak CO mixing ratios occur over central South America and central and southern Africa; (ii) an extensive area of elevated CO occurs over northern Asia; (iii) no well-defined plume of elevated CO from North America is apparent; (iv) the lowest mixing ratios occur over the southern Pacific and the far southern Atlantic; and, (v) decreased CO values occur in a band extending from Central America, over northern Africa, and over southern Asia.

Although the model result and the MAPS satellite data agree very well, there are some discrepancies. For example, the location of the maxima over central South America and Africa occur farther southward in the satellite data than in the model results, indicating either greater horizontal transport or a spatial difference between the model's assumed emission sources and the actual location of biomass fires during October 1984. Further, the MAPS data show indications of a CO plume extending from southern Africa across the far southern Indian ocean and possibly swinging northward across New Zealand and southern Australia. The model distribution also exhibits a plume originating in Africa and extending across the Indian ocean; however, the model-produced plume occurs farther north and contains 10-20 ppb lower CO values than does the satellite data. Lastly, a band of lower CO values extending from northern Africa across eastern Europe and into central Asia is

189

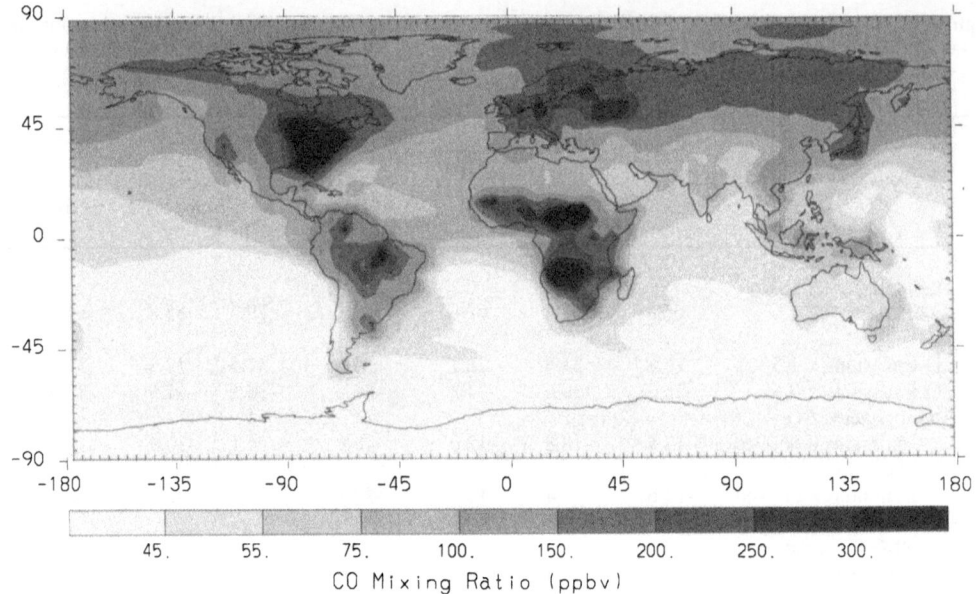

Figure 1. CO mixing ratio distribution from GChM model level 1 (~1010 hPa) averaged over October 1-15, 1984 for the base case simulation.

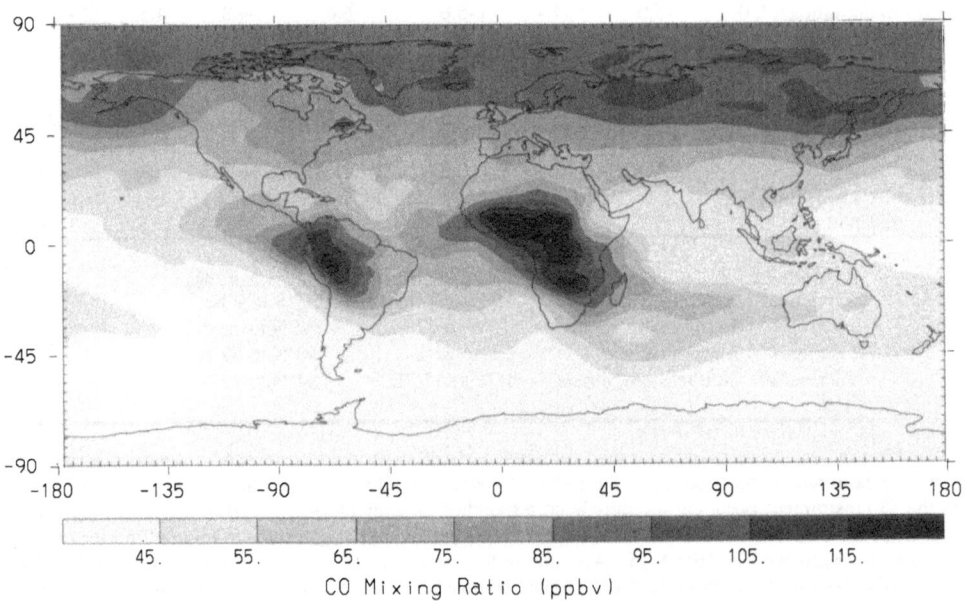

Figure 2. CO mixing ratio distribution at GChM model level 11 (~350 hPa) averaged over October 1-15, 1984 for the base case simulation. Model level 11 corresponds to the altitude of maximum signal strength from the MAPS gas-filter radiometer.

clearly indicated in the MAPS data; the model result does exhibit lower mixing ratios in this region as compared to the areas of peak values, however, the CO values range from 55-75 ppb as compared to < 60 ppb observed in the MAPS experiment.

Figure 3 presents the model produced CO mixing ratio distribution at model level 11 for the sensitivity simulation where the convective cloud transport parameterization was disabled within the model. The results of Figure 3 are presented with the same shading contours as Figure 2 for ease of comparison. As is readily observed, convective cloud transport plays a significant role in the amount of CO that is transported to model layer 11, implying that the CO distribution as observed during the MAPS experiment is largely the product of surface emissions vented rapidly to the mid-troposphere by convective activity associated with clouds. These

190

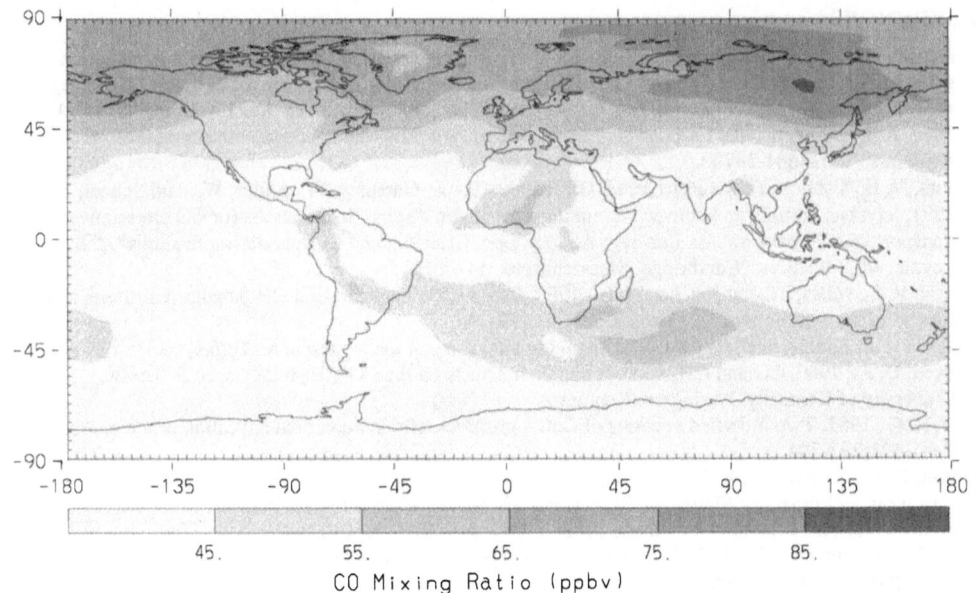

CO Mixing Ratio (ppbv)

Figure 3. CO mixing ratio distribution at GChM model level 11 (~350 hPa) averaged over October 1-15, 1984 for the no cloud convection simulation. Model level 11 corresponds to the altitude of maximum signal strength from the MAPS gas-filter radiometer.

results support the conclusions of Newell *et al.* (1988), Connors *et al.* (1989), Watson *et al.* (1990), and Connors *et al.* (1991), that cloud convective activity serves to transport boundary layer CO into the free troposphere where it can be transported long distances away from its point of emission. As documented in Table 2, the average CO mixing ratio in model layer 11 is 27 percent lower without the convective cloud parameterization enabled. Likewise, CO mixing ratios in the lowest model layer are 19 percent greater, indicating that in the base case simulation a substantial amount of material within the boundary layer is transported to higher model levels. In the central African sub-domain, a decrease of 62 percent in the domain-averaged CO mixing ratio occurred when the convective cloud parameterization was disabled, implying that a majority of the CO observed over this region in the MAPS data is brought to this level by cloud convective activity.

CONCLUSION

The GChM atmospheric chemistry and transport model has been used to analyze the mid-tropospheric CO dataset obtained from NASA's Measurement of Air Pollution by Satellites (MAPS) program. Fourteen simulations with a 3.75° horizontal resolution have been performed, including a base case and 13 sensitivity runs. The model reproduces many, but not all, of the major features of the MAPS dataset. Locations of peak CO mixing ratios associated with biomass burning as observed in the MAPS experiment are slightly farther south than the model result, indicating either greater horizontal transport than present in the model representation or a spatial difference between the location of modeled biomass fires and actual fires. The current version of GChM was shown to be relatively insensitive to the magnitude of the prescribed NO_x and O_3 global distributions and very insensitive to the depth of the mixed layer as parameterized in the model. Cloud convective transport was shown to play an important role in venting boundary layer CO to the free troposphere. This result agrees with prior meteorological analyses of the MAPS dataset (Newell *et al.*, 1988; Connors *et al.*, 1989; Watson *et al.*, 1990; and Connors *et al.*, 1991) that have indirectly inferred the presence of convective activity through satellite-based information. Work is continuing to analyze the results of these simulations further and to perform more detailed comparisons between model results and MAPS data.

Acknowledgement

This research was funded through the Laboratory Directed Research and Development program of Pacific Northwest Laboratory. Pacific Northwest Laboratory is operated for the U. S. Department of Energy by Battelle Memorial Institute under Contract DE-AC06-76RLO 1830.

REFERENCES

Carmichael, G. R.,1979, Development of a regional transport/transformation/removal model for SO_2 and sulfate in the eastern United States, Ph.D. dissertation, University of Kentucky, Lexington, Kentucky.

Chang, J. S., R. A. Brost, I. S. A. Isaksen, S. Madronich, P. Middleton, W. R. Stockwell, and C. J. Walcek, 1987, A three-dimensional Eulerian acid deposition model: Physical concepts and formulation, *J. Geophys. Res.*, 92:14681-14700.

Connors, V. S., Cahoon, D. R., Reichle, H. G., Brunke, E.-G., Garstang, M., Seiler, W., and Scheel, H. E., 1991, Savanna burning and convective mixing in southern Africa: Implications for CO emissions and transport, in "Global Biomass Burning: Atmospheric, Climatic, and Biospheric Implications", J. S. Levine, ed., MIT Press, Cambridge, Massachusetts, 147-159.

Connors, V. S., Miles, T., and Reichle, H. G., 1989, Large-scale transport of a CO-enhanced air mass from Europe to the Middle East, *J. Atmos. Chem.*, 9:479-496.

Crutzen, P. and Zimmerman, P.,1991, The changing chemistry of the troposphere, *Tellus*, 43AB:136-151.

DeHaven, D. A., 1980, CO and CH_4 sources and their effects on the CO-CH_4 budgets, M.S. Thesis, University of Kentucky, Lexington, Kentucky.

Easter, R. C., 1993, Two modified versions of Bott's positive-definite numerical advection scheme, *Mon. Wea. Rev.*, 121:297-304.

Fishman, J., and Crutzen, P. J., 1978, The origin of ozone in the troposphere, *Nature*, 274:855-858.

Fung, I., John, J., Lerner, J., Matthews, E., Prather, M., Steele, L., and Fraser, P., 1991, Three-dimensional model synthesis of the global methane cycle, J. Geophys. Res., 96:13033-13065.

Kitada, T. and Peters, L. K., 1982, A three-dimensional transport-chemistry analysis of CO and CH_4 in the troposphere, in "AMS Second Symposium on the Composition of the Nonurban Troposphere", Williamsburg, Virginia, 96-101.

Kwok, H. C. W., Langlois, W. E., and Ellefsen, R. A., 1971, Digital simulation of the global transport of carbon monoxide, *IBM J. Res. Develop.*, 15:3-9.

Lelieveld, J., and Crutzen, P.J., 1990, Influence of cloud photochemical processes on tropospheric ozone, *Nature*, 343:227-233.

Logan, J. A., Prather, M. J., Wofsy, S. C., McElroy, M. B., 1981, Tropospheric chemistry: A global perspective, *J. Geophys. Res.*, 86:7210-7254.

Luecken, D. J., Berkowitz, C. M., Bader, D. C., Vukovich, F. M., and Glatzmeir, G. C., 1989, The role of fair weather cumulus clouds in the long range transport of carbon monoxide and methane, American Geophysical Union Spring Meeting, Baltimore, Maryland.

Luecken, D. J., Berkowitz, C. M., and Easter, R. C., 1991, Use of a three-dimensional cloud-chemistry model to study the transatlantic transport of soluble sulfur species, *J. Geophys. Res.*, 96:22477-22490.

Newell, R. S., Connors, V. S., Reichle, H. G., 1988, Regional studies of potential CO sources based on space shuttle and aircraft measurements, *J. Atmos. Chem.*, 6:61-81.

Peters, L. K., and Jouvanis, A. A., 1979, Numerical simulation of the transport and chemistry of CH_4 and CO in the troposphere, *Atmos. Environ.*, 13:1443-1462.

Pinto, J. P., Yung, Y. L., Rind, D., Russell, G. L., Lerner, J. A., Hansen, J. E., and Hameed, S., 1983, A general circulation model study of atmospheric carbon monoxide, *J. Geophys. Res.*, 88:3691-3702.

Reichle, H. G., Connors, V. S., Holland, J. A., Hypes, W. D., Wallio, H. A., Casas, J. C., Gormsen, B.B., Saylor, M. S., and Hesketh, W. D., 1986, Middle and upper tropospheric carbon monoxide mixing ratios as measured by a satellite-borne remote sensor during November 1981, *J. Geophys. Res.*, 91:10865-10887.

Reichle, H. G., Connors, V. S., Holland, J. A., Sherrill, R. T., Wallio, H. A., Casas, J. C., Condon, E.P., Gormsen, B. B., and Seiler, W., 1990, The distribution of middle tropospheric carbon monoxide during early October 1984, *J. Geophys. Res.*, 95:9845-9856.

Saylor, R. D., and Peters, L. K., 1990, The contribution of anthropogenic emissions to the global distribution of CO in the troposphere, Symposium on the Chemistry of the Global Atmosphere, Commission on Atmospheric Chemistry and Global Pollution, Chamrousse, France.

Saylor, R. D., and Peters, L. K., 1991, The global numerical simulation of the distribution of CO in the troposphere, "Air Pollution Modeling and Its Application VIII", H. van Dop and D. G. Steyn, eds., Plenum Press, New York, 485-496.

Seiler, W., and Conrad, R., 1987, Contribution of tropical ecosystems to the global budget of trace gases, especially CH_4, H_2, CO and N_2O, in "The Geophysiology of Amazonia", R.E. Dickinson, ed., John Wiley & Sons, New York.

Seinfeld, J., 1986, "Atmospheric Chemistry and Physics of Air Pollution", John Wiley & Sons, New York.

Walcek, C. J., and G. R. Taylor, 1986, A theoretical method for computing vertical distributions of acidity and sulfate production within cumulus clouds, *J. Atmos. Sci.*,43:339-355.

Warneck, P., 1988, "Chemistry of the Natural Atmosphere", Academic Press, New York.

Watson, C. E., Fishman, J., and Reichle, H. G., 1990, The significance of biomass burning as a source of carbon monoxide and ozone in the southern hemisphere tropics: A satellite analysis, *J. Geophys. Res.*, 95:16443-16450.

DISCUSSION

R. SAN JOSE: Can you say something about the CPU time for your simulations?

R.C. EASTER: A 1.5 month simulation with 3.75 degree latitude-longitude resolution requires 15 hours on a SUN Sparc-10 workstation. Without the simplifications we have made to the photochemistry, the CPU time would be at least an order of magnitude greater.

R. SAN JOSE: How are you implementing the model on the massively parallel computer?

R.C. EASTER: We began with a domain decomposition of the latitude-longitude domain. However, the chemistry takes most of the CPU time and it only occurs on the daylight side of the globe. Thus we are working on a dynamic load balancing technique to improve performance for the chemistry calculations.

M. KROL: What scheme did you use to parameterize transport to the stratosphere?

R.C. EASTER: The upper boundary of the model is at 100 mb. A simple inflow-outflow boundary condition is used at the boundary, and transport across the boundary is determined by the vertical velocities in the ECMWF meteorological data. When the vertical velocity is downwards, into the model domain, the CO mixing ratio in the inflow air is specified at 40 ppb.

M. KROL: Do you include cloud chemistry and reactions of nitrogen species on aerosols?

R.C. EASTER: No. Currently the chemistry is limited to the gas-phase portion of the Lelieveld and Crutzen mechanism. However, due to the fact that CO tropospheric chemistry is not greatly affected by these processes, negligible error in CO distributions is introduced through their omission. Cloud chemistry and aerosol surface reactions would be expected to have a greater impact on the distributions of ozone and nitrogen oxides, but these species are currently specified within the model.

M. VAN WEELE: Did you take into account the radiative effects of clouds on photodissociation rates in the gas phase chemistry?

R.C. EASTER: We have not done that. One difficulty is that the ECMWF meteorological archives for this time period (1984) contain very limited information on cloud extent.

M. VAN WEELE: Can you give a general figure for the accuracy of the CO measurements?

R.C. EASTER: Reichle et al. (1990) estimated that the random error for the CO measurements was 10%. However, they also estimated that the measurements were biased 20-40% low as compared with concurrent aircraft-based measurements.

THE INFLUENCE OF AQUEOUS PHASE CHEMISTRY ON LONG
TERM OZONE CONCENTRATIONS OVER EUROPE

J. Matthijsen[1], P.J.H. Builtjes[2], M.G.M. Roemer[1]

[1]TNO Institute of Environmental Sciences (IMW)
Dept. of Environmental Chemistry
P.O. Box 6011
2600 JA Delft
The Netherlands

[2]TNO-IMW and IMAU
Institute for Marine and Atmospheric Research
University of Utrecht
Princetonplein 5
3584 CC Utrecht
The Netherlands

INTRODUCTION

Clouds affect tropospheric ozone levels through aqueous phase chemical reactions and by altering radiation transfer and mixing. Cloud droplets can act as important sinks for ozone and ozone precursors such as oxidized hydrocarbons and NO_x ($NO+NO_2$) through aqueous oxidation and/or wet removal. Moreover the presence of liquid water separates soluble and insoluble species which again can effect the ozone formation. For instance NO, which is relatively insoluble will be separated in the presence of cloudwater from the relatively soluble HO_2, reducing effectively the formation of ozone in the gasphase.

Recently published models indicate that cloud chemistry can significantly affect tropospheric ozone levels. On a global scale Lelieveld and Crutzen, (1990) show with model calculations that clouds may reduce ozone concentrations by as much as 50%. Dentener (1993) found smaller reductions of the order of 10% with an improved parametrization of the aqueous phase chemistry. For the 30°N-60°N zonal band model studies suggest that ozone levels may be reduced by 10% to 30% by including cloud chemistry (Jonson and Isaksen, 1993).

Air Pollution Modeling and Its Application X, Edited by S-V. Gryning
and M. M. Millán, Plenum Press, New York, 1994

Walcek et. al (1993) calculated the impact of aqueous phase chemistry on the net ozone formation rate as a function of the concentration of NO_x and nonmethane organic species (nmoc). They found that ozone formation rates were reduced by 2 ppb/hour up to 20 ppb/hour for NO_x concentrations between 400 ppt and 4 ppb and nmoc concentrations ranging from 1 to 1000 ppb. For higher and lower NO_x concentrations the ozone formation rate was hardly influenced by aqueous phase chemistry. The influence of aqueous phase chemistry on ozone, therefore depends on the local photochemical regime.

The predicted effects of clouds could be greater if the reactions of dissolved Iron (Fe) and Copper (Cu) are included. The reactions of Fe and Cu play an important role in the cloud water concentrations of oxidizing species like H_2O_2, HO_2/O_2^- and the OH radical (Sedlak and Hoigné, 1993). Moreover, Fe and Cu are believed to affect the oxidation of aqueous S(IV) to S(VI) (e.g.: Jacob et. al, 1989), which can in turn alter aqueous oxidant concentrations.

Regional and global photochemical models calculate tropospheric photo-oxidant and pollutant concentrations for areas with different emission and aerosol conditions. Moreover Fe and Cu are generally more abundant in continental aerosols than in aerosols from marine origin. It is therefore expected that continental clouds have a larger impact on ozone and related species than marine clouds. Also observations of ozone during cloud episodes (Möller et al., 1993) indicate that the ozone removal capacity of clouds increases with increasing pollution.

On a regional scale the LOTOS-model (Long Term Ozone Simulation) calculates ozone and other photo-oxidant concentrations in the lowest 3 km of the troposphere for Europe (Builtjes, 1992). The model region reaches from -10° E to 60° W and from 35° N to 70° N and it therefore includes both marine and continental areas. To account for the effect of aqueous phase chemistry in the LOTOS model a cloud model is developed.

In this paper we investigate the influence of aqueous phase reactions, which include Fe and Cu reactions, on ozone and related species with the cloud model. Results are shown for different photochemical conditions (marine, continental: averaged and polluted) and different initial Fe and Cu concentrations. Since stratiform clouds are the dominant cloud form (> 50%) on northern midlatitudes, model experiments are performed under typical stratus cloud conditions.

CLOUD MODEL

In this study we use a two layer model (layer 1: 0-1000 m., layer 2: 1000-1500 m.). Dry deposition at the surface, exchange between the two layers and exchange at the top with fixed typical marine or continental free tropospheric concentrations are described. For each of three cases the model was initialized with a run of five days under cloud free conditions with air at a specified temperature (288K), pressure (1 atm). Furthermore sunlight conditions for 21 June and northern midlatitudes were used.

After five days of initialization a stratus cloud is defined in layer 2 for the duration of one hour with a fixed liquid water content of 0.3 g/m^3 and mono-disperse cloud droplets with a radius of 10 μm for marine stratus and 5 μm for continental stratus. The cloud is modelled without rainout.

Typical emissions for marine and continental initializations and runs (Matthijsen and Diederen, 1992, and Asman, 1992) are specified in table 1. The polluted runs are initialized with a five days run using the averaged continental emissions, which are at the fourth day of initialization replaced by high emissions of an industrialized and urban area in the Netherlands (Berdowski and van der Most, 1993).

Table 1. Marine and continental emissions (in molecules/cm^2/s).

Species	Marine	Continental	
		averaged	polluted
	$\times 10^7$	$\times 10^{10}$	$\times 10^{11}$
NO_x	1.9	4.8	48.0
SO_2	0.0	5.9	44.0
CO	0.0	76.8	17.0
Ethane	0.0	2.5	0.7
Propane	1.9	1.2	1.2
Butane	1.2	2.0	8.8
Ethene	0.0	1.5	1.7
Propene	45.1	2.0	0.8
Toluene	10.0	0.4	2.1
Ammonia	0.0	12.0	11.0

Gasphase reactions

Gasphase chemistry is described by a consized version of the mechanism according to Hough and Derwent (1987). The mechanism describes over 180 reactions among over 90 chemical species in the troposphere.

Aqueous phase reactions

The aqueous phase chemistry mechanism used is a combination of the mechanism according to Lelieveld and Crutzen (1991) and photo-oxidant relevant Fe and Cu aqueous phase reactions according to Sedlak and Hoigné (1993). For the results presented here we have included over 50 aqueous phase reactions among the dominant short-lived radical species that are affected by the presence of cloudwater. The combined gas and aqueous phase chemical reaction mechanism includes all relevant solubilities and dissociation of species, and treats mass transfer between the gas and the aqueous phase according to Schwartz (1986). NO and NO_2 are modelled to be insoluble. The effect of the presence of the stratus cloud on the photolysis rates is according to van Weele and Duynkerke (1993).

Typical marine and continental initial concentrations are used for SO_4^{2-}, NH_4^+, NO_3^- and Cl^- due to aerosols (Warneck, 1988). The initial pH is 5.0 and the initial Fe and Cu concentrations (Table 2), based on observations are according to Sedlak and Hoigné (1993) and Sedlak (personal communication, 1993). The high polluted continental values are taken to be the upper limit of Fe and Cu concentration representing with the accompanying inital concentrations a 'worst case' for the impact of aqueous phase chemistry on ozone and related species.

Table 2. Marine and continental initial Fe and Cu concentrations (in μmole/l).

Metals	Marine	Continental	
		averaged low - high	polluted low - high
Fe^{2+}	0.04	0.08 - 0.8	0.8 - 8
Fe^{3+}	0.01	0.02 - 0.2	0.2 - 2
Cu^+	0.0	0.0	0.0
Cu^{2+}	0.0005	0.001 - 0.01	0.01 - 0.1

RESULTS

The results are displayed in tables 3 to 7. The effect of the stratus cloud on ozone and related species is expressed as the absolute and relative differences between the after-cloud gasphase concentrations in layer 2 and the concentrations in that layer calculated for the same time with a cloud free run. The results are shown for each case calculated with and without Fe and Cu aqueous phase reactions (+ metals respectively no metals). For the continental cases the + metal-simulation results are shown for low and high initial Fe and Cu concentrations (according table 2). Thus table 3 to 5 give the overall effect of the modelled stratus cloud, lasting one hour, on gasphase concentrations in the cloud layer.

Table 3. Cloud effect on **marine** concentrations of gasphase species in layer two, 10 minutes after cloud evaporation. (ΔC is the difference between the concentration calculated with the model including clouds and the concentration for a cloud free run). The absolute ΔC and concentration of OH and HO_2 are in molecules/cm^3. (averaged pH=4.4)

species	cloud free concentration	ΔC (in ppt)		ΔC (in %)	
		+ metals	no metals	+ metals	no metals
O_3	32 ppb	-206	-252	-0.6	-0.8
NO_x	7 ppt	-1	-1	-13	-16
SO_2	26 ppt	-25	-25	-97	-97
HCHO	356 ppt	-86	-89	-24	-25
H_2O_2	880 ppt	-14	+22	-2	+3
HCOOH	14 ppt	+42	+38	+311	+276
CH_3OOH	811 ppt	+45	+53	+6	+7
OH	4.3×10^6	$+2.2 \times 10^4$	$+2.6 \times 10^4$	+0.5	+0.8
HO_2	4.1×10^8	-1.1×10^7	-7.7×10^6	-3	-2

Table 4. Cloud effect on **continental averaged** concentrations of gasphase species in layer two, 10 minutes after cloud evaporation. (ΔC is the difference between the concentration calculated with the model including clouds and the concentration for a cloud free run). The absolute ΔC and concentration of OH and HO_2 are in molecules/cm^3. (averaged pH$=3.8$)

species	cloud free concentration	ΔC (in ppt)			ΔC (in %)		
		+ metals		no metals	+ metals		no metals
		low	high		low	high	
O_3	35 ppb	-446	-437	-463	-1.3	-1.2	-1.3
NO_x	90 ppt	+17	+22	+15	+19	+25	+17
SO_2	1.3 ppb	-1204	-1187	-1206	-95	-93	-95
HCHO	783 ppt	-134	-52	-162	-17	-7	-21
H_2O_2	1.7 ppb	-731	-967	-692	-42	-56	-40
HCOOH	51 ppt	+38	+78	+32	+74	+153	+64
CH_3OOH	378 ppt	-10	-74	+17	-3	-20	+4
OH	5.0×10^6	$+2.9 \times 10^5$	$+3.1 \times 10^5$	$+2.7 \times 10^5$	+6	+6	+5
HO_2	5.7×10^8	-3.4×10^7	-3.4×10^7	-3.5×10^7	-6	-6	-6

Table 5. Cloud effect on **continental polluted** concentrations of gasphase species in layer two, 10 minutes after cloud evaporation. (ΔC is the difference between the concentration calculated with the model including clouds and the concentration for a cloud free run). The absolute ΔC and concentration of OH and HO_2 are in molecules/cm^3. An absolute ΔC smaller than 0.5 ppt is represented by (-). (averaged pH$=4.4$)

species	cloud free concentration	ΔC (in ppt)			ΔC (in %)		
		+ metals		no metals	+ metals		no metals
		low	high		low	high	
O_3	10 ppb	-238	-229	-220	-2.4	-2.3	-2.2
NO_x	67 ppb	-31	-37	-51	-0.1	-0.1	-0.1
SO_2	47 ppb	-1847	-1753	-1745	-4	-4	-4
HCHO	970 ppt	-921	-921	-921	-95	-95	-95
H_2O_2	368 ppt	-362	-362	-362	-98	-98	-98
HCOOH	59 ppt	-1	-	-	-1.8	-0.1	-0.1
CH_3OOH	76 ppt	-1	-1	-1	-0.8	-0.8	-0.9
OH	4.2×10^5	-1.5×10^5	-1.4×10^5	-1.4×10^5	-35	-34	-33
HO_2	4.3×10^5	-2.0×10^5	-1.9×10^5	-1.9×10^5	-46	-45	-45

Tables 6 and 7 show averaged aqueous phase reaction rates (molecules/cm^3/s) during the presence of the stratus cloud for the marine cases and for the continental averaged cases with high Fe and Cu initial concentrations, respectively, with and without Fe and Cu aqueous phase reactions. The reaction rate unit is 1×10^5 molecules/cm^3/s, which coincides for the cloud period of one hour with a reaction rate of about 14 ppt/hour. To convert the reaction rate unit from molecules/cm^3/s to mole/l/s the reaction rates are to be multiplied with 5.534×10^{-15}.

Table 6. Aqueous phase production, loss and net reaction rates (in x10^5 molecules/cm^3/s) for the **marine** cases.

aqueous species		Prod.	Loss	Net
O$_3$	+ metals	-	3.5	**-3.5**
	no metals	-	5.0	**-5.0**
HCHO	+ metals	0.5	4.5	**-4.0**
	no metals	0.6	4.4	**-3.8**
H$_2$O$_2$	+ metals	8.4	4.6	**+3.8**
	no metals	8.5	3.4	**+5.1**
OH	+ metals	5.8	8.0	**-2.2**
	no metals	5.6	7.9	**-2.3**
HO$_2$/O$_2^-$	+ metals	31.5	43.9	**-12.4**
	no metals	7.6	20.6	**-13.0**

Table 7. Aqueous phase production, loss and net reaction rates (in x10^5 molecules/cm^3/s) for the **continental averaged** cases with and without high initial Fe and Cu concentrations.

aqueous species		Prod.	Loss	Net
O$_3$	+ metals	-	0.9	**-0.9**
	no metals	-	3.5	**-3.5**
HCHO	+ metals	13.7	21.0	**-7.3**
	no metals	7.9	18.9	**-11**
H$_2$O$_2$	+ metals	13.5	44.1	**-30.6**
	no metals	10.9	31.7	**-20.8**
OH	+ metals	14.8	19.7	**-4.9**
	no metals	4.3	10	**-5.8**
HO$_2$/O$_2^-$	+ metals	72.3	106.4	**-34.1**
	no metals	2.9	24.9	**-22**

The net effect of the stratus cloud on concentrations is mainly caused by gas and aqueous phase cloud chemistry in the cloud layer, since the contribution of transport from other layers is relatively small for the stratus cloud period of one hour.

The overall effects of stratus clouds on concentrations show that the relative destruction of O$_3$, HCHO, H$_2$O$_2$, and HO$_2$ increases with increasing pollution. HCOOH, concentrations show a relative increase after cloud evaporation. This increase becomes smaller with increasing pollution. NO$_x$ reacts on the changes in the gasphase concentrations due to transfer of reactants from the gasphase to the aqueous phase and vice versa, aqueous phase chemistry and altered photolysis rates. In the continental averaged case an increase of the NO$_x$ concentration was found with a maximum of 25% due to these cloud processes. OH levels are somewhat increased in the marine and continental averaged cases, whereas in the continental polluted case up to 35% of OH is depleted after cloud evaporation. The overall cloud effect on CH$_3$OOH concentrations is in most cases rather small. Under the continental averaged circumstances with high Fe and Cu initial concentration a maximum destruction of 20% is calculated.

The destruction of ozone in the aqueous phase (see table 6 and 7) is relatively small with a maximum of 30% compared to the overall reduction effect of stratus clouds on ozone concentrations, which was found to be between 200 and 465 ppt/hour (table 3 to 5). Under marine circumstances the destruction of O_3 in the aqueous phase is somewhat larger than under more polluted conditions. For HCHO, H_2O_2, OH and HO_2 the aqueous phase is relatively a much more important sink, especially for H_2O_2 and HO_2.

Including Fe and Cu aqueous phase reactions in our calculations in general was found to have little effect on the gasphase concentrations after cloud evaporation. In the marine and continental polluted photochemical regimes and also in the continental averaged regime with low initial Fe and Cu concentrations the net effect was found to be small. In spite of the small effect in most cases aqueous phase reaction pathways are considerably changed by adding Fe and Cu reactions.

Only in the continental averaged regime with high initial Fe and Cu concentrations we calculated considerable differences between the concentrations of the runs with and without Fe and Cu aqueous phase reactions. In this case including Fe and Cu aqueous phase reactions lead to an increased destruction ($\approx 50\%$) in the aqueous phase of H_2O_2 and HO_2/O_2^-, and to a decreased destruction in the aqueous phase of O_3 (75%), HCHO (35%) and OH (15%).

In all cases including Fe and Cu aqueous phase reactions lead to higher OH and lower HO_2/O_2^- concentrations in the aqueous phase. In the continental averaged case with high initial Fe and Cu concentrations the aqueous concentrations of OH and HO_2/O_2^- are respectively 175% higher and 80% lower compared to the no metal case. These differences may considerably affect the amount of hydrocarbons oxidized in the aqueous phase and therefore on its turn long term O_3.

CONCLUSION

From the simulations of non-raining stratus clouds in three different photo-chemical regimes, marine, continental averaged and polluted, we conclude:

- The direct effect of the modelled stratus clouds on O_3 was found to be limited to a maximum of about 0.5 ppb/hour (for continental averaged circumstances).

- Indirectly the presence of stratus clouds affect ozone levels on the long term by altering concentration levels of ozone related species like NO_x, HCHO and H_2O_2. The effect of stratus clouds on these species depend for a great deal on the ambient photochemical regime.

- Our calculations indicate that the aqueous phase reactions of Iron and Copper have in general a small impact on the concentrations of ozone and related species, except under continental averaged conditions combined with high Fe and Cu concentrations.

Acknowledgments

For providing us with the relevant transition metal chemistry and measurements and for the many helpful discussions we thank David Sedlak.

REFERENCES

Asman, W.A.H., 1992, 'Ammonia Emission in Europe: Updated emission and emission variation', RIVM report, nr. 228471008, Bilthoven, The Netherlands.

Berdowski, J.J.M. and P.F.J van der Most, (ed.), 1993, 'The Netherlands, 5th Emission Inventory, 1990', nr. 13.

Builtjes, P.J.H., 1992, The LOTOS - Long Term Ozone Simulation-project - Summary report, TNO-report nr. IMW-R-92/240, TNO, Delft, The Netherlands.

Hough A.M. and R.G. Derwent, 1987, Computer Modelling studies of the distribution of photochemical ozone production between different hydrocarbons, Atm. Environ., Vol. 21, 9, pp 2015-2033.

Jacob, D.J., E. W. Gottlieb and M.J. Prather, 1989, Chemistry of a Polluted Cloudy Boundary layer, J. of Geophys. Res., Vol. 94, No. D10, pp 12975-13002.

Jonson, J.E. and I.S.A. Isaksen, 1993, Tropospheric ozone chemistry. The impact of cloud chemistry, J. of Atm. Chem., 16, pp. 99-122.

Lelieveld, J. and P.J. Crutzen, 1991, The role of clouds in tropospheric photochemistry, J. of Atm. Chem., 12, pp. 229-267.

Matthijsen, J. and H.S.M.A. Diederen, 1992, The effect of zonal averaging in global modelling of tropospheric ozone distributions, in: Air Pollution Modeling and its Application IX, Eds.: H. van Dop and G. Kallos, Plenum Press, New York.

Möller, D., Wieprecht, W., Acker, K. and G. Mauersberger, 1993, Evidence for Ozone Destruction in Clouds, Fraunhofer Institute for Atmospheric Environmental Research, Air Chemistry, Berlin.

Sedlak, D.L. and J. Hoigné, 1993, The Role of Copper and Oxalate in the redox of Iron in Atmospheric Waters, Atm. Environ., Vol. 27A, No. 14, pp 2173-2185.

Schwartz, 1986, In: Jaeschke (ed.) Chemistry of multiphase atmospheric systems. Springer-Verlag, Berlin, 415-471.

Walcek, C.J., Hong-Hsee Yuan and W.R. Stockwell, 1993, The Influence of Heterogeneous Atmospheric Chemical Reactions on the Formation of Ozone in Polluted Air, presented at 86th Annual Meeting & Exhibition, Denver, Colorado, June 13-18.

Warneck, P., 1988, Chemistry of the Natural Atmosphere, Int. Geophysics series vol. 41, Academic Press, New York, pp 334-339.

van Weele, M. and P.G. Duynkerke, 1993, Effect of clouds on photo-dissociation of NO_2: observations and modelling, J. Atm. Chem, 16, pp. 231-255.

A PARAMETERIZATION OF THE EFFECT OF CLOUDS ON PHOTODISSOCIATION RATES; COMPARISON WITH OBSERVATIONS

Michiel van Weele, Jordi Vilà-Guerau de Arellano, Peter G. Duynkerke
Institute for Marine and Atmospheric Research Utrecht (IMAU)
Utrecht University
P.O. Box 80005, 3508 TA, Utrecht, The Netherlands

1   INTRODUCTION

Many molecules in the atmosphere dissociate under influence of incident sunlight, a process called photodissociation. This photodissociation typically initiates a sequence of chemical reactions (radical formation) and is therefore a key process in simulating the chemistry of the atmosphere.

It is clear that reasonable accuracy is required in the calculation of photodissociation rates in these models in order to simulate chemical transformations properly. Unfortunately, the accuracy that can be achieved is limited by both uncertainties in atmosphere conditions (e.g. clouds effects) and uncertainties in photochemical data. The latter issue was addressed by Thompson and Stewart (1991), who performed a sensitivity analysis in order to assess the effect of uncertainties in chemical kinetics on the concentrations of trace gases that are calculated in a photochemical model. The two most critical reactions turned out to be the photodissociation of ozone and nitrogen dioxide. Their result shows that photochemical models need an accurate calculation of photodissociation rates.

This paper investigates the effect of uncertainties in photodissociation rates due to (the variability of) atmospheric conditions. Photodissociation rates depend on the intensity of sunlight in the atmosphere, more specifically on the actinic flux (Madronich, 1987). The photodissociation reactions that are most critical for tropospheric chemistry are driven by sunlight in the spectral region from 290 - 700 nm. The variations in space and time of the actinic flux are due to spatial inhomogeneities in the scattering and absorption properties of the atmosphere. The physical processes that are important include Rayleigh (molecular) scattering, gaseous absorption (mainly ozone), absorption by aerosols, and Mie scattering by aerosols and cloud droplets. Other important boundary conditions for the actinic flux are the extraterrestrial solar flux, solar zenith angle and ground albedo.

For the radiative properties of the atmosphere we concentrate on clouds because they have a dramatic effect on photodissociation rates and show most clearly the effect of spatial inhomogeneities in the atmosphere. Here, we only consider vertical inhomogeneities, i.e. layered clouds. Thompson (1984) showed that the ozone production rate in the background troposphere is changed significantly by the radiative effect of clouds. However, although it is quite clear that the chemistry is dramatically changed, it is still uncertain to what extent these changes affect the concentration of trace gases.

In section 2 we will address the accuracy of photodissociation rates needed for atmospheric chemistry models and the accuracy that can be achieved. The discussion provides a justification for the development of the parameterization scheme which includes the effect of clouds and ground albedo, presented in section 3a. The parameterization scheme is validated in section 3b against measurements of vertical profiles of actinic fluxes made by Vilà-Guerau de Arellano and Duynkerke (1993).

2   REQUIRED AND ACHIEVABLE PHOTODISSOCIATION RATES

In current tropospheric chemistry models height and time dependent photodissociation rates are often interpolated from tabulated values which have been calculated with a radiative transfer model. In these calculations a mean atmospheric state is assumed for the ozone concentration and aerosol loading under clear sky conditions with standard ground albedo. Therefore, parameterizations are required in order to account for the effect of clouds and ground albedo (e.g. Thompson, 1984; Madronich, 1987; Chang et al., 1987; van Weele and Duynkerke, 1993). The effect of ground albedo, especially important above highly reflecting surfaces (snow and ice), is sometimes neglected. The main advantage of the use of parameterizations is to avoid computer time consuming calculations for all atmospheric conditions. Whereas the atmospheric input is partly based on some crude assumptions and cloud coverage is badly known.

The question remains: "Does the use of tabulated 'standard' values completed with parameterizations yield sufficiently accurate photodissociation rates?" Two problems can be distinguished: 1) What accuracy is *required* in atmospheric chemistry models and 2) what accuracy can be *achieved* in the calculation of photodissociation rates.

The *required* accuracy should be estimated from sensitivity studies with an atmospheric chemistry model. The differences in results for several atmospheric (both meteorological and chemical) conditions should give insight in the required accuracy for photodissociation rates in such models. The radiative effect of clouds was studied by Thompson (1984). This paper suggests that, although the chemistry is highly perturbed, detailed knowledge about photodissociation rates is probably not required, because (e.g. daily) integrated values may be decisive. On the other hand Dvortsov et al. (1992) state that differences between current atmospheric chemistry model results can at least partially be ascribed to different photodissociation rates. Therefore, the question of required accuracy in atmospheric chemistry models can not yet be answered satisfactory.

The accuracy that can be *achieved* is determined by the uncertainties in the calculation of the photodissociation rates itself. The photodissociation rate J_i of a species i is defined as

$$J_i = \int_{\lambda_1}^{\lambda_{2,i}} \sigma_i(\lambda) \, \varphi_i(\lambda) \, F(\lambda) \, d\lambda \tag{1}$$

where λ_1 is the short wavelength limit ($\pm$ 290 nm for the troposphere due to the shielding by the ozone layer), $\lambda_{2,i}$ is the long wavelength limit (determined by the strength of the chemical bonds of the considered species), σ_i is the absorption cross section of the considered molecule (in units of area per molecule), ϕ_i the quantum yield (the amount of dissociated molecules for a given number of incident photons) and F the actinic flux (in units of energy per area per time). The actinic flux at a certain location in the atmosphere is defined as the intensity of incident light from all directions, irrespective of the direction of incidence. See Madronich (1987) for a thorough discussion of the definition of the actinic flux.

The accuracy in the calculation of J_i is limited due to uncertainties in the photochemical data (quantum yield and absorption cross section) and uncertainties in actinic flux levels. The uncertainty in photochemical data is different for each chemical species. Toon et al. (1989) state that, generally, 'the combined uncertainty of cross sections and quantum yields is at least 10%'. Shetter et al. (1992) estimate 10% uncertainty in the data for NO_2. Thompson and Stewart (1992) give uncertainties between 10 and 100% due to chemical kinetics uncertainties for the most important photodissociation rates. Therefore, we take 10% accuracy as the best available general estimate of the minimum uncertainty in the photochemical data.

Secondly, the achievable accuracy is determined by the calculation of the actinic fluxes in space and time as a function of wavelength. Madronich (1987) gave an overview of studies concerning actinic flux calculations. Extensive tables for actinic fluxes under a variety of atmospheric conditions were given by Demerjian (1980) and Finlayson-Pitts and Pitts (1986).

Very detailed radiative transfer models exist that can calculate actinic fluxes very accurate for given atmospheric conditions. However, all radiative transfer models still have enormous difficulties with incorporating horizontal inhomogeneity (e.g. partial cloudiness) and the most sophisticated models are very time consuming.

In the literature a variety of methods is presented that aim to simplify the radiative transfer problem. One of the most widely used methods is the so called Isaksen-Luther method reviewed by Dvortsov et al. (1992). Because only isotropic scattering is assumed, the scheme is inappropriate for tropospheric calculations where clouds and aerosols are important. Therefore, it is preferable to use a more sophisticated and hardly more time-consuming approach, based on a 'two-stream approximation' of the radiative transfer equation. Toon et al. (1989) showed by comparison with more detailed radiative transfer calculations that the use of two-stream techniques yield errors in both albedos and actinic fluxes ('mean intensities' in Toon's paper) that are 'generally less than 10% but may exceed 10% if the solar zenith angle is large or if the quantity being calculated is small'.

We compared a delta-Eddington model (Joseph et al., 1976; van Weele and Duynkerke, 1993) with an 'exact' doubling-adding model (Stammes et al., 1989). Errors are generally less than 10% although actinic fluxes just below cloud top can be underestimated by 20% in the delta-Eddington approximation due to strong anisotropy.

We conclude that for the calculation of photodissociation in atmospheric chemistry models we prefer to use tabulated values for clear sky situations, calculated with the delta-Eddington model. The gain in accuracy by using the delta-Eddington model for cloud effects may be limited compared to the accuracy that can be obtained with parameterizations. Besides, accurate and rather costly calculations are probably not justified in view of the minimum of 10% accuracy in the photochemical data..The use of a parameterization scheme for the effect of clouds and ground albedo is justified as long as the accuracy of this approach is in the order of 10%. A suitable parameterization scheme is presented in the next section.

3   A PARAMETERIZATION SCHEME FOR THE EFFECT OF CLOUDS

The parameterization scheme discussed in this section was already presented in van Weele and Duynkerke (1993). There a full derivation is given. The final equations are given in the Appendix below. A brief discussion of the most important concepts is given in section 3a. In section 3b we will compare the parameterization scheme with observations of actinic fluxes made during the ASTEX (Atlantic Stratocumulus Transition EXperiment) campaign in June 1992. For a complete overview of these observations see Vilà-Guerau de Arellano and Duynkerke (1993).

3a   Parameterization Description

A complete calculation of the effect of clouds on photodissociation rates consists of i) determination of the actinic flux as a function of height and wavelength in the cloudy atmosphere and ii) integration according to equation (1). This requires complete radiative transfer calculations for several wavelengths every time atmospheric conditions change. From the former section we can learn that, generally, the computational costs of such detailed calculations can not be justified due to other limitations in the accuracy. Therefore, in van Weele and Duynkerke (1993) a parameterization scheme was presented for the effect of clouds on the actinic flux. It is recognised that in the spectral region of interest for photodissociation in the troposphere [290 - 700 nm] the optical properties of clouds are almost independent of wavelength. Therefore, the effect of a cloud on photodissociation rates is approximately the same for all relevant wavelengths and we can approximate

$$J_i = C \cdot J_{i, \text{ clear}} \qquad (2)$$

Here C is a dimensionless proportionality factor that contains all parameters that relate the photodissociation rates in a cloudy atmosphere to the clear sky values. This factor is a complex function of height (z), cloud optical thickness (τ_C), solar zenith angle (θ_0), the relative amount of diffuse (F_*) and direct light (F_0) incident at cloud top ($F_* + F_0 = 1$), cloud-base and -top height and spectral domain. Equations for C are given in the Appendix.

It was discussed in van Weele and Duynkerke (1993) that the most critical parameters for the proportionality factor are cloud optical thickness and solar zenith angle. These two parameters determine cloud albedo, for the other parameters a rough estimate is sufficient.

Cloud-base and -top height and spectral dependency are incorporated in the scheme in the parameters F_0 (= 1 - F_*), A_L and A_U. Here A_U is the reflectivity of the atmosphere above the cloud for upward directed light and A_L the reflectivity of the atmosphere below the cloud for downward directed light. Estimations for these parameters as a function of height and spectral region can be obtained with a radiative transfer model.

The effect of ground albedo can be incorporated in the parameter A_L. Therefore, no separate parameterization scheme for the effect of ground albedo is required.

Given the values for the proportionality factor at cloud top (C_A), cloud base (C_B) and in the cloud (C_C), we need to specify the height dependence of the proportionality factor above and below the cloud. Fortunately, the proportionality factor below cloud (C_B) is almost independent of height due to multiple scattering between cloud base and the ground. This is confirmed by the measurements performed by Vilà-Guerau de Arellano and Duynkerke (1993). Above the cloud a decreasing effect of the cloud is seen due to backscattering of the upward directed radiation. A linear decrease of C_A with height should be sufficient to account for this effect. An estimate for the slope is easily determined by radiative transfer calculations, eventually for different wavelength regions.

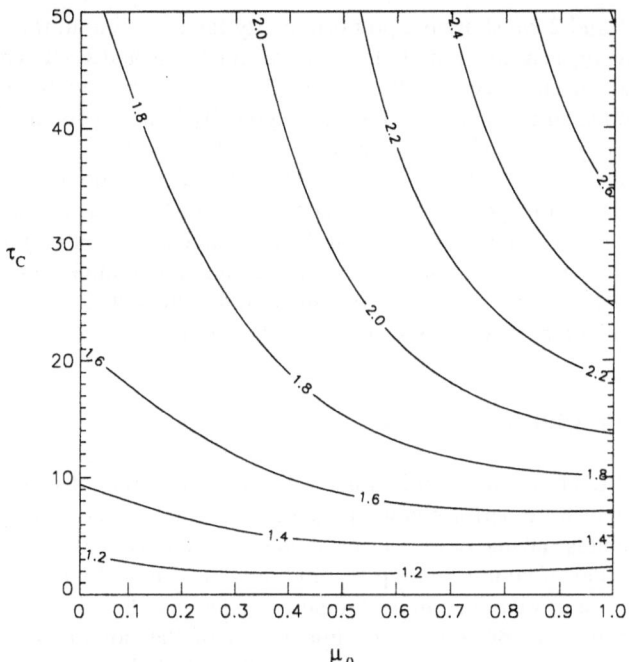

Figure 1. The proportionality factor for the actinic flux at cloud top (C_A) as a function of cloud optical thickness (τ_C) and cosines of solar zenith angle ($\mu_0 = \cos \theta_0$) calculated with the parameterization scheme for $A_L = 0.06$, $A_U = 0.25$ and $F_0 = 1 - F_* = 0.4 \cos\theta_0 + 0.34$.

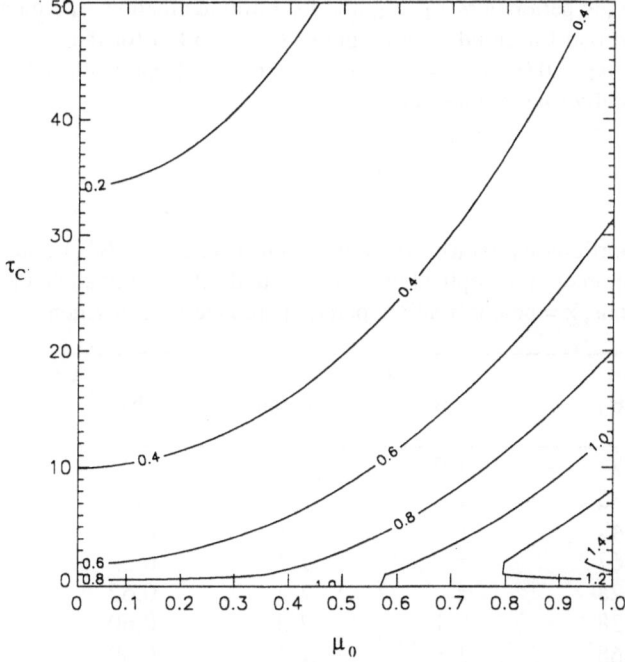

Figure 2. Same as figure 1 for the proportionality factor below cloud (C_B).

In Figures 1 and 2 we show the proportionality factors at cloud top (C_A) and cloud base (C_B), respectively, as a function of cloud optical thickness and solar zenith angle ($\mu_0 = \cos \theta_0$). In the figures we used $A_L = 0.06$, $A_U = 0.25$. The relative amount of direct incident radiation depends on solar zenith angle and varies typically between about 40 and 70%. We used the following expression: $F_0 = 0.4 \cos\theta_0 + 0.34$. These estimates are typical for low boundary layer clouds, for wavelengths in the 330 - 390 nm region above a surface with ground albedo of 0.05 (corresponding with the conditions of the measurements discussed in the next section). More detailed knowledge of these parameters is not required because the critical parameters for the proportionality factor are cloud optical thickness and solar zenith angle. Notice that below cloud actinic fluxes can be higher than clear sky values (Figure 2) due to trapping of radiation between cloud base and the ground.

3b   Comparison with Observations

During the ASTEX campaign in June 1992 measurements of actinic fluxes in the marine boundary layer were performed with a photo-electrical detector (Van de Hage et al., 1993). The instrument is capable of measuring actinic fluxes in the 330 - 390 nm region.

Vertical profiles of actinic flux up to 1500 m were obtained with a tethered balloon under various conditions (Vilà-Guerau de Arellano and Duynkerke, 1993). Typically marine stratocumulus clouds were present in the upper part of the boundary layer during the measurements. The data set was completed with detailed measurements of cloud characteristics (cloud-base and -top height and cloud optical thickness).

In Table 1 we compare the proportionality factors calculated with the parameterization scheme with measured values. The measured actinic flux values at cloud top and cloud base were calculated by comparison with clear sky values for the same zenith angle. Cloud top was not always reached by the balloon. For the calculations with the parameterization scheme we need to specify the parameters A_L, A_U and F_0. Ground albedo at the measuring site was 0.05. We estimated that for clouds at a height of 0.5 - 1.5 km for the 330-390 nm spectral region we can use $A_L = 0.06$, $A_U = 0.25$ and $F_0 = (1-F_* =) 0.4 \cos\theta_0 + 0.34$. The error in these values is typically smaller than about 20%.

Table 1. Proportionality factors for the actinic flux above (C_A) and below (C_B) cloud as a function of cloud optical thickness τ_C and solar zenith angle θ_0. (obs: observations; x = no observation; param: parameterization scheme)

τ_C	θ_0	C_A obs	C_A param	C_B obs	C_B param
6	23	1.9	1.5	0.92	1.2
6	35	1.6	1.5	0.86	1.0
10	49	x	1.7	0.91	0.69
13	67	x	1.7	0.53	0.44
20	28	x	2.2	0.62	0.68
23	38	2.1	2.2	0.60	0.54
23	68	1.8	1.8	0.39	0.32

The measured values should not be interpreted as exact, because especially for the optical thinner clouds the variation in cloud optical thickness and horizontal inhomogeneity (with partial cloudiness as an extreme case) in the atmosphere is even during one sounding extremely large. Further, as one sounding (up and down) takes about one hour also zenith angle varies (typically ± 8 degrees).

However, we can conclude that for zenith angles between about 20 and 70 degrees and a cloud optical thickness up to about 23, the measured proportionality factors lay in the range $0.4 < C_B < 0.9$ and $1.6 < C_A < 2.1$. This range of values compares well with the values that are obtained with the parameterization scheme.

4   CONCLUDING REMARKS

The accuracy of calculated photodissociation rates is limited both due to uncertainties in chemical kinetics and uncertainties in atmospheric conditions. The variability of nature (cloud coverage, morphology, microphysics, etc.) is too difficult to incorporate in any model. It is concluded that detailed radiative transfer calculations of actinic fluxes for specific conditions can not be justified for (global) atmospheric chemistry models.

A parameterization scheme is presented which includes the effects of clouds (and ground albedo) on photodissociation. It can easily be implemented in atmospheric chemistry models. It is shown that the scheme compares well with measurements of actinic fluxes in a cloudy atmosphere. However, more observations, e.g. in the 300 - 330 nm region, are needed to evaluate the applicability of the scheme in more detail. Further, as this paper only concentrates on the effect of vertical inhomogeneities, the effects of partial cloudiness should be assessed.

The scheme will be implemented in a 1-D tropospheric chemistry model. Sensitivity studies will be performed in order to assess the radiative effect of clouds on model results. These studies, together with detailed measurements, should give insight in the accuracy that is required for photodissociation rates in atmospheric chemistry models .

APPENDIX

(see van Weele and Duynkerke (1993) for a full derivation)

The proportionality factor above (C_A) and below (C_B) a cloud are given by

$$C_A = \frac{(F_{A\downarrow} + F_{A\uparrow})}{F_{clr}} \qquad\qquad C_B = \frac{(F_{B\downarrow} + F_{B\uparrow})}{F_{clr}} \qquad\qquad \text{(A1, A2)}$$

where the up- and downward contributions to the actinic flux are given by

$$F_{A\downarrow} = F_0 + F_* + A_U F_{A\uparrow} \qquad\qquad\qquad\qquad \text{(A3)}$$

$$F_{A\uparrow} = 2\mu_0 A_C{}^0 F_0 + A_C{}^*(F_* + A_U F_{A\uparrow}) + (1 - A_C{}^*) F_{B\uparrow} \qquad\qquad \text{(A4)}$$

$$F_{B\downarrow} = 2\mu_0(1 - A_C^0)(1 - e^{-\tau_C/\mu_0})F_0 + F_0\, e^{-\tau_C/\mu_0} +$$
$$(1 - A_C^*)(F_* + A_U\, F_{A\uparrow}) + A_C^* F_{B\uparrow} \tag{A5}$$

$$F_{B\uparrow} = A_L(F_{B\downarrow} - F_0\, e^{-\tau_C/\mu_0}) + 2\mu_0 A_L F_0\, e^{-\tau_C/\mu_0} \tag{A6}$$

$$F_{clr} = \frac{1 + A_L}{1 - A_L A_U} F_* + \left(1 + \frac{2\,\mu_0\, A_L\,(1 + A_U)}{(1 - A_L A_U)}\right) F_0 \tag{A7}$$

Here μ_0 is the cosines of solar zenith angle and τ_C is cloud optical thickness. In the delta-Eddington approximation the cloud albedo for direct (A_C^0) and diffuse incident light (A_C^*) can be written as

$$A_C^0 = 1 - \frac{(2 + 3\mu_0) + (2 - 3\mu_0)\, e^{-\tau_C(1-g^2)/\mu_0}}{(4 + 3\,(1 - g)\,\tau_C)} \tag{A8}$$

$$A_C^* = 1 - \frac{4}{(4 + 3\,(1 - g)\,\tau_C)} \tag{A9}$$

where g is typically 0.87. In a homogeneous cloud layer the in-cloud proportionality factor for the actinic flux as a function of in-cloud optical thickness (τ) is given by

$$C_C\,(\tau) = \frac{4\pi\, I_0(\tau) + F_0\, e^{-\tau(1 - g^2)/\mu_0}}{F_{clr}} \tag{A10}$$

with

$$I_0(\tau) = B_1 - \frac{3}{4\pi}\mu_0^2\, F_0\, e^{-\tau(1-g^2)/\mu_0} - B_2\,(1 - g)\,\tau$$

Boundary conditions at cloud top and at cloud base, provided by equations A1 and A2, yield the constants B_1 and B_2

$$B_1 = \frac{1}{4\pi}\Big[(3\mu_0^2 - 1)F_0 + C_A F_{clr}\Big]$$

$$B_2 = \frac{(C_A - C_B)F_{clr} + F_0(3\mu_0^2 - 1)(1 - e^{-\tau_C(1-g^2)/\mu_0})}{4\pi\,(1 - g)\,\tau_C}$$

while in-cloud optical thickness can be related to in-cloud height z' by

$$\tau\,(z') = \tau_C\,[\,1 - z'^{\,5/3}\,] \qquad\qquad z' = \frac{z - z_b}{z_t - z_b} \quad (0 \le z' \le 1) \tag{A11}$$

where z_b is cloud base and z_t is cloud top.

REFERENCES

Chang, J.S., Brost, R.A., Isaksen, I.S.A., Madronich, S., Middleton, P., Stockwell, W.R., and Walcek, C.J., 1987, A three-dimensional Eulerian acid deposition model: physical concepts and formulation, *J. Geophys. Res.* 92:14681.

Demerjian, K.L., Schere, K.L., and Peterson, J.T., 1980, Theoretical estimates of actinic (spherically integrated) flux and photolytic rate constants of atmospheric species in the lower troposphere, *Adv. Environ. Sci. Technol.*, 10:369.

Dvortsov, V.L., Zvenigorodsky, S.G., Smyslaev, S.P., 1992, On the use of Isaksen-Luther method of computing photodissociation rates in photochemical models, *J. Geophys. Res.* 97:7593.

Finlayson-Pitts, B.J., and Pitts, J.N., Jr., 1986, "Atmospheric Chemistry Fundamentals and Experimental Techniques," Wiley, New York.

Joseph, J.H., Wiscombe, W.J., and Weinman, J.A., 1976, The delta-Eddington approximation for radiative flux transfer, *J. Atmos. Sci.* 33:2452.

Madronich, S., 1987, Photodissociation in the Atmosphere: 1. Actinic flux and the effects of ground reflections and clouds, *J. Geophys. Res.* 92:9740.

Shetter, R.E., McDaniel, A.H, Cantrell, C.A., Madronich, S., and Calvert, J.G., 1992, Actinometer and Eppley radiometer measurements of the NO_2 photolysis rate coefficient during the Mauna Loa Observatory Photochemistry EXperiment, *J. Geophys. Res.* 97:10349.

Stammes, P., de Haan, J.F., and Hovenier, J.W., 1989, The polarized internal radiation field of a planetary atmosphere, *Astron. Astrophys.* 225:239.

Thompson, A.M., 1984, The effect of clouds on photolysis rates and ozone formation in the unpolluted troposphere, *J. Geophys. Res.* 89:1341.

Thompson, A.M., and Stewart, R.W., 1991, Effect of chemical kinetics uncertainties on calculated constituents in a tropospheric photochemical model, *J. Geophys. Res.* 96:13089.

Toon, O.B., Mckay, C.P., and Ackerman, T.P., 1989, Rapid calculation of radiative heating rates and photodissociation rates in inhomogeneous multiple scattering atmospheres, *J. Geophys. Res.* 94:16287.

Van de Hage, J.C.H., Boot, W., van Dop, H., Duynkerke, P.G., and Vilà-Guerau de Arellano, J., 1993, A photoelectric detector suspended under a balloon for actinic flux measurements, accepted in *J. Atmos. Ocean Technol.*

Van Weele, M., and Duynkerke, P.G., 1993, Effect of clouds on the photodissociation of NO_2: observations and modelling, *J. Atmos. Chem.*, 16:231.

Vilà-Guerau de Arellano, J., and Duynkerke, P.G., 1993, Tethered-balloon measurements of actinic flux in a cloud-capped marine boundary layer, submitted to *J. Geophys. Res.*

DISCUSSION

R. EASTER — How do your results compare with previous theoretical studies such as Madronich?

M. VAN WEELE — The presented parameterization is based on the study of Madronich(1987). Here, the developed parameterization scheme is compared with direct observations of actinic flux. These observations compared well with the parameterization scheme and therefore corroborate Madronich theoretical predictions.

EVALUATION OF RADIATIVE FLUX AND TROPOSPHERIC CHEMISTRY UNDER GLOBAL CLIMATE CHANGE SCENARIOS

Kevin C. Crist ,[1] Gregory R. Carmichael,[1] Kuruvilla John [2]

[1]Department of Chemical and Biochemical Engineering and Center for Global and Regional Environmental Research University of Iowa, Iowa City, IA 52242
[2]Department of Atmospheric Sciences State University of New York-Albany Room # 134, NYSDEC Albany, NY 12233

ABSTRACT

A detailed one dimensional radiation model coupled with a comprehensive regional scale photochemical model(STEM-II) is used to evaluate ground level ultraviolet radiation and changes in photochemical oxidant concentrations under potential global change scenarios. The simulations are conducted for a representative metropolitan region in eastern United States. The simulations involve perturbations to radiative flux based on trends of stratospheric ozone loss. The impact of increased radiative flux on tropospheric ozone is analyzed. In addition, the effects of tropospheric pollution levels on ground level radiative flux are studied. For the urban environment increased tropospheric ultraviolet radiation leads to an increase in photochemical production while ground level radiative fluxes are significantly reduced as a result of increased levels of tropospheric ozone production.

INTRODUCTION

Anthropogenic emissions of chloroflurocarbons(CFCs) are the major contributor to destruction of stratospheric ozone documented over the last decade. Stratospheric ozone depletion was originally associated with the polar regions of the southern hemisphere remote from heavily populated areas. However, recent analysis of data obtained from the Nimbus 7 satellite has shown a decline in the total ozone column over a majority of the globe(Herman, et al., 1991; Niu, et al., 1992; Stolarski, et al., 1991) This analysis has shown a significant downward trend (%0.4-%0.8 a year) over the last decade for much of the northern hemisphere including the heavily populated areas of northern United States, Canada, Scandinavia, Europe, and the former Soviet Union.

Stratospheric ozone is a primary filter for ultraviolet-B radiation (UV-B). Depletion of the stratospheric ozone allows more UV-B radiation to reach the earth's surface. This has led to trepidation about the potential impacts to human health, agriculture and the ecological and aquatic systems(EPA, 1987). UV-B has been directly linked to cataracts and skin cancers, and there is concern that UV-B may also act to depress the immune system. Based on dose-response relationships derived from animal experiments and human epidemiological studies, it is estimated that non-melanoma cancers increase at a rate of 3% per 1% decrease in stratospheric ozone, and melanomas increase at 1 to 2% per 1% decrease in stratospheric ozone(Scotto, et al., 1981).

Air Pollution Modeling and Its Application X, Edited by S-V. Gryning
and M. M. Millán, Plenum Press, New York, 1994

The absorption of solar radiation is function of the total atmospheric ozone profile, the stratosphere and troposphere. Recent investigations (Bruhl and Crutzen, 1989) on ground level radiative flux as related to changes in tropospheric and stratospheric ozone have shown intriguing results. While any reductions in stratospheric ozone leads to an increase in radiation transfer to the troposphere, the presence of tropospheric ozone can reduce the radiative transfer to the surface. This mitigation effect of tropospheric ozone is enhanced due to the disproportional role of tropospheric ozone as a filter for solar radiation. Studies conducted by Bruhl and Crutzen(1989) have shown that molecule for molecule tropospheric ozone can absorb more radiation than stratospheric ozone. This greater affinity for absorption of ultraviolet radiation(UVR) is explained by a longer optical path length which is the result of an increase in diffuse(scattered) radiation in the troposphere.

The extinction of UVR by tropospheric constituents can not only lead to a decrease in ground level radiative flux but this interaction can also affect the tropospheric ambient air quality. With stratospheric ozone loss, resulting in an increase flux of solar radiation to the troposphere, an increase in the reactivity of the troposphere is expected(Liu and Trainer, 1987; UNEP, 1989). For locations with elevated NOx emissions the increased flux of UVR can result in an increase in tropospheric ozone.

Ozone in the stratosphere is formed through the photolysis of molecular oxygen. In the troposphere however, ozone is formed through the photolysis of nitrogen dioxide which photo dissociates at longer wavelengths.

$$NO_2 \xrightarrow{j_1(h\upsilon)} NO + O \tag{1}$$

$$O + O_2 + M \longrightarrow O_3 + M \tag{2}$$

The ozone formed then reacts rapidly with NO to set up the following photochemical equilibrium.

$$O_3 + NO \xrightarrow{k_3} NO + O_2 \tag{3}$$

$$[O_3] = j_1 [NO_2] / k_3 [NO] \tag{4}$$

For ozone to build up in the troposphere other processes are required that oxidize NO to NO_2 without consuming ozone. For polluted environments the oxidation of NO to NO_2 can occur via peroxyl radicals(HO_2 and RO_2) which are formed as a result of the oxidation of hydrocarbons(RH) and carbon monoxide.

The decline of stratospheric ozone and the resulting increase of UVR to the troposphere will alter the ozone photochemical equilibrium(4). For tropospheric chemistry the increase flux of UVR will increase the rate photo dissociation of ozone with the formation of O('D).

$$O_3 + h\upsilon \longrightarrow O('D) + O_2 \tag{5}$$

The increase production of O('D) leads to a host of chemical reactions with the result of increasing the chemical reactivity or oxidizing nature of the atmosphere.

For urban areas the increase in photo-dissociation could result in a net increase of surface ozone(UNEP, 1989). The rapid rise of O('D) would lead to an increase in production of odd-hydrogen radicals which are the major oxidizing agents of the atmosphere. The oxidation products of organics such as aldehydes and ketones are then available to oxidize NO to NO_2 without consuming ozone.

This effect has been demonstrated with box model studies conducted by Liu and Trainer(1987). For industrialized nations, with typical high NOx concentrations, an increase of UVR to the troposphere can lead to a net increase in the production of ozone. Conversely, Liu and Trainer(1987) demonstrated that for low NOx regions (i.e. remote oceanic environments) an increase in UVR to the troposphere can lead to a net loss in ozone production.

The focus of this study is two fold: First detailed radiative flux calculations are made utilizing total ozone column trends data obtained from the Nimbus 7 satellite, and tropospheric ozone levels based on future photochemical smog scenarios. Second, the effects of increased UVR to the troposphere on ambient ozone concentrations are evaluated with a three dimensional regional scale photochemical model (STEM-III)(Carmichael, et al., 1991). The study domain covers the eastern United States. Ground level radiative flux calculations are conducted for a representative metropolitan region in the eastern United States.

REVIEW OF MODEL CALCULATIONS

The calculations for the propagation of solar radiation through the atmosphere encompasses an interactive process with atmospheric constituents involving scattering and absorption of solar radiation. For absorption, stratospheric ozone is the major atmospheric constituent involved with radiative transfer, and it is the total column ozone(tropospheric and stratospheric) that determines the quantity of photons which reach the earth's surface.

To model the propagation of radiation throughout the atmosphere a delta-two stream method presented by Zdunkowski et al.(1980) is utilized. The propagation of solar radiation(diffuse and direct) is described via three differential equations; for diffuse upward(F_1) diffuse downward(F_2) and parallel solar radiation(S);

$$\frac{dF_1}{d\tau} = \alpha_1 F_1 - \alpha_2 F_2 - \alpha_3 S/\cos\theta$$

$$\frac{dF_2}{d\tau} = \alpha_2 F_1 - \alpha_1 F_2 + \alpha_4 S/\cos\theta$$

$$\frac{dS}{d\tau} = -(1-\varpi g^2)S/\cos\theta \tag{6}$$

where τ is the optical thickness which includes absorption by ozone and scattering by molecular species, clouds and aerosols and, μ_0 is the cosine of the solar zenith angle. The coefficients of equation (1) are defined as follows;

$$\alpha_1 = 2(1-\varpi(1-\beta_0)) \qquad \alpha_3 = (1-g^2)\varpi\beta(\mu_0)$$

$$\alpha_2 = 2\varpi\beta_0 \qquad \alpha_4 = (1-g^2)\varpi(1-\beta(\mu_0)) \tag{7}$$

ϖ is the albedo for a single scattering event defined as,

$$\varpi = k_s/\{k_s + k_a\} \tag{8}$$

and k_s and k_a are the coefficients for scattering and absorption, respectively. In addition, $\beta(\mu_0)$ and β_0 are the backward scattering coefficients for parallel and diffuse light and g accounts for the anisotropic scattering by aerosols and cloud droplets.

In the above equation $\beta(\mu_0)$ and β_0 are approximated as follows for aerosols and clouds:

$$\beta_0 = \frac{3}{8}(1-g) \qquad \beta(\mu_0) = \frac{1}{2} - \frac{3}{4}\frac{g}{1+g}\cos\theta \tag{9}$$

For molecular scattering both the coefficients are set equal to 0.5.

Equation 1 is solved explicitly in one nm increments and for each hour the sun is above the horizon. The results are then integrated over the wavelengths categorized for UV-B(280-320nm). The majority of the spectral data utilized in the model calculations were obtained from De-More et al.(1982). In addition to UV-B levels, a dose rate for erythemal induction was calculated as described by Madronich(1992);

$$R(u_0) = \int_\lambda E(\lambda, u_0) A(\lambda) d\lambda$$

(10)

where $A(\lambda)$ is the erythemal function, and E is the resulting ultraviolet(UV) spectral irradiance at the earth's surface.

To simulate potential future tropospheric ambient pollution levels and to study the effects of UVR on tropospheric chemistry a three - dimensional regional scale comprehensive photo-chemistry model was used (Carmichael, et al., 1991). The model domain covering the entire eastern United States was broken down into horizontal grid spacing of 80km with the vertical resolution set at 500km. The NAPAP v.2 emissions inventory was used in this study. A detailed description of the emissions, and meteorological inputs is provided by Shin(1990). The evaluated episodes were referenced to a base case simulation for a clear sky summer event on the 7th-11th of June, 1984(Shin, 1990).

RESULTS

Ground Level UV Flux

The sensitivity of ground level UV-B fluxes and erythemal dose rate to changes in the vertical ozone profile is studied under several regional scale scenarios for the eastern half of the United States. Due to limited detailed ozone profile data, a vertical distribution of ozone from ozonesondes at Hohenpeissenburg Germany(47° N) was used as the referenced profile. The profile was then adjusted using the ratio of the integrated profile to the averaged ozone column amounts measured from the Nimbus 7 satellite for the eastern half of the United States. The total ozone column measurements for 40^oN for the spring equinox of 1991 were used.

To evaluate the range of potential UV-B exposures due to changes in both stratospheric and tropospheric ozone levels four scenarios were studied. In the first scenario(case-1) the stratospheric ozone profile was as described above, and the tropospheric ozone profile was that calculated using a three dimensional regional scale photochemical model(STEM-II)(Carmichael, et al., 1991). In this study the tropospheric ozone profile used in case-1 is that calculated for a regional scale ozone episode which occurred in June 1984. From this simulated episode the model calculated profile representative of tropospheric loading for an east coast metropolitan area was chosen.

The remainder of the scenarios represent potential UV-B exposures for the year 2010. For each of these scenarios the stratospheric ozone profile is decreased from the base case(case 1) by projecting the present trends in total column ozone loss to the year 2010. A decrease of 4% per decade was used base on the recent analysis by Stolarski et al.(1992) for the mid latitudes of the northern hemisphere.

Three different tropospheric profiles were used. In Case-2, the tropospheric ozone profile was held constant simulating no change in photochemical smog conditions. Case-3, the tropospheric ozone profile for the year 2010 was calculated using the STEM-II model taking into account estimates of future emissions, changes in water vapor and temperature, and UV flux (for photochemical oxidant calculations) based on projected climate change projected to the year 2010. The perturbations from the base case included a 33% increase in NO_X emissions, a 25% decrease in SO_2 emissions, an approximate 3 K change in surface temperature, and 20-27% increase in the water vapor column for the lower troposphere(John, et al., 1993). The climate changes were consistent with present climate change scenarios(International Panel on Climate Change, 1991), while those for the emissions are estimated from EPA(1991).

The final scenario(case-4) represents the maximum potential exposure to UV-B. The tropospheric profile is that estimated for pre-industrial conditions based on a calculation by Roemer(1991). This condition represents the "best case" air pollution scenario in the sense that anthropogenic emissions are essentially eliminated and tropospheric ozone is reduced to levels associated with a pre-industrial era. These four scenarios are utilized to generate atmospheric ozone profiles, which are depicted in Figures 1 and 2. These profiles were used to calculate the ground level radiative flux.

The results from the simulations are summarized in Table 1. The values for UV-B and UV erythemal weighted induction are integrated noon time values. Comparing scenario 1 and 2 it is found that a decrease in stratospheric ozone of 8%, which is a 6.9% change in the total column increases the noon time UV-B at the surface by 4.6% and increases in the erythemal dose by 8.4%. The resultant RAFs are 0.661 and 1.206 for UV-B and erythemal dose, respectively. A RAF is the percent change in ground level radiative flux resulting from each 1% change in the total ozone column. The erythemal weighted UV results compare closely with spectral measurement obtained by McKenzie(1991) who determined a 1.25 % ± .20% in erythemal induction for every 1% reduction in total ozone at 40°S.

Table 1. Noon time UV-B and UV-erythemal weighted fluxes for the following scenarios: 1) Base case. 2) Stratospheric reduction/tropospheric increase. 3) Stratospheric reduction/tropospheric reduction. DU(Dobson units)=2.69E+16 molecules/cm^2.

Case	Integrated Profile (DU)	UV-B (w/m^2)	Erythemal (w/m^2)
1	328	2.481	0.124
2	305	2.595	0.134
3	311	2.558	0.131
4	279	2.805	0.154

Simulations 2, 3, and 4 reflect the impact of changing tropospheric ozone on the ground level UV flux. For each of these simulations the stratospheric ozone levels are those estimated for the year 2010 and reflect a decrease of 4% per decade from the present values. Comparison of simulations 2 and 3 show the effect of increasing tropospheric ozone on UV-B exposure. It is found that a 13.4% increase in the tropospheric ozone levels, which represents a 1.8% increase in the total column ozone (stratospheric and tropospheric), reduces the ground level UV-B flux by 1.4% and erythemal dose by 2.4%. This produces RAFs of 0.784 and 1.31 for UV-B and erythemal induction, respectively.

Comparison of simulations 2 and 4 shows the changes of radiative exposure under conditions of photochemical smog reduction. For this simulation the total ozone column is reduced by 8.5% resulting in an increase in UV-B flux by 8.1% and an increase in the erythemal dose rate by 14.9%. The RAFs are 0.941 for UV-B and 1.74 for erythemal dose. These RAFs are significantly higher than the RAFs calculated under stratospheric ozone changes(case-1, and 2).

The computed RAFs from these simulations highlight the disproportional impact of tropospheric ozone as a filter of UV-B radiation. With respect to ground level radiative flux (under limiting conditions) this demonstrated disproportional effect can offset stratospheric ozone and total ozone reductions when elevated tropospheric ozone levels exist This effect was demonstrated by Bruhl and Crutzen(1989) from the analysis of the Hohenpeissenberg ozone profile data from 1968 to 1982.

The Effects on Tropospheric Chemistry

Model simulations utilizing a 10% increase in UVR to the troposphere was utilized to study the effects on ambient tropospheric ozone levels for the eastern United States. All other factors i.e. emissions, temperature, and water vapor concentrations were held constant(i.e. to the base case simulation for 1984). A comparison between regional scale ozone levels for the base case and UV perturbed case(Figure 1) showed that regional ozone levels increased from 5-15% throughout the study domain. With the largest net increase along the industrialized corridors. These results compare closely with previous studies(Liu & Trainer, 1987; UNEP, 1989) which identified that for industrialized areas with sufficient quantities of NOx and VOCs the increased energy to the troposphere will have a positive feedback on ozone production. In addition, the increased production of tropospheric ozone will increase the tropospheric filtering of UV-B radiation.

The relationship of an increase in tropospheric ozone as a result of an increase UV flux to the troposphere is further demonstrated by looking at the diurnal variations of ozone for both rural and urban profiles. Ground level diurnal ozone comparisons between the base case and perturbed UV case, Figure 2, for a rural and urban Georgia location show a significant increased in ozone with the largest increase in the urban setting.

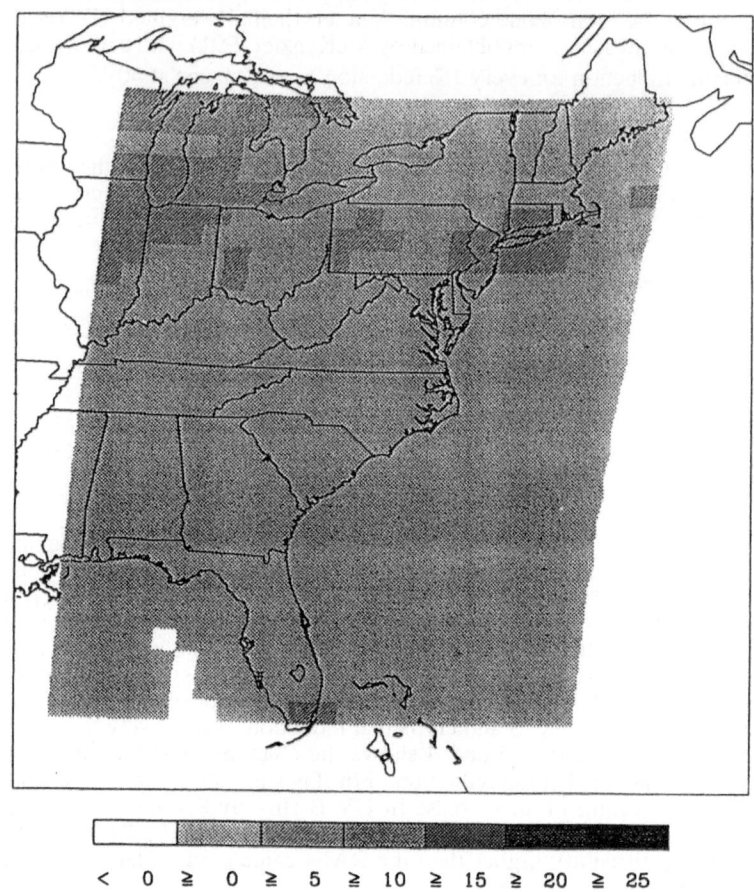

Figure 1. Percent change in ozone concentration: Base case vs peturbed UV flux case.

CONCLUSION

This study demonstrates that tropospheric ambient air quality can significantly effect the potential increases in UV flux at the surface as a result of stratospheric ozone loss. In industrialized areas with elevated tropospheric ozone levels the potential health risk from UV-B exposure becomes masked, and UV-B exposure is reduced due to the role of tropospheric ozone as a filter of UV-B. Additionally, in these regions of high ozone levels with elevated levels of NOx, a potential positive increase in tropospheric ozone is expected further degrading the ambient air quality, which in-turn will provide additional tropospheric UV-B filtering.

The effects of ambient air quality with respect to stratospheric ozone loss could have a significant impact on future emission reduction policies. As air quality is improved the alternate risk of UV-B radiation will carry increasing weight. In addition, the increased flux of UVR to the troposphere will partially offset projected ozone reductions targeted by the emission policies.

An accurate assessment of potential exposure levels to UV-B and changes in ambient air quality will require continued model investigations. The models used in these studies should take into account changes in stratospheric and tropospheric composition. In the troposphere, effects of SO_2, aerosols, and NO_2 on UV-B filtering should be analyzed. In addition, the effects of increased radiation on tropospheric concentrations of NO_x, OH, HO_2, and H_2O_2 should also be investigated.

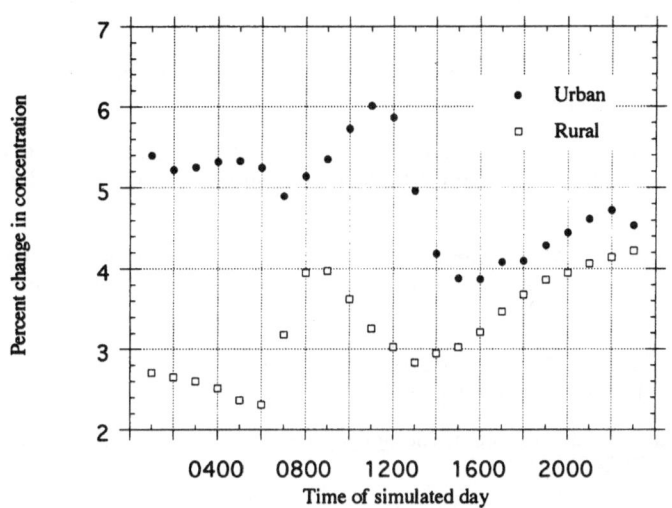

Figure 2. Percent diurnal change rural and urban Gerorgia location: Base case vs peturbed UV.

ACKNOWLEDGMENTS

This research was supported by the Center for Health Effects of Environmental Contamination, University of Iowa, Iowa City Iowa. We thank Drs. C. Bruhl and P. J. Crutzen of the Max Plank Institute, Mainz Germany and Professor Zdunkowski at the University of Mainz for making available their delta-two stream radiation model.

REFERENCES

Bruhl C., and P. J. Crutzen, On the disproportionate role of tropospheric ozone as a filter against solar UV-B radiation, Geophys. Res. L.,16, 703-706,1989

Carmichael G. R., L. K. Peters, and R. D. Saylor, The stem-II regional scale acid deposition and photochemical oxidant applications, Atmospheric Environment, 1991

DeMore W. B., R. T. Watson , D. M. Golden, R. F. Hampson, M. J. Kurylo, C. J. Howard, M. J. Molina, and A. R. Ravishankara, Chemical kinetics and photochemical data for use in stratospheric modeling, Jet Prop. Lab. Publ. 82-57, NASA, Pasadena, CA, 1982.

EPA, National Air Quality and Emissions Trends Report Technical Support Division, Office of Air Quality Planning and Standards, USEPA.(No. EPA-450/4-91-023), 1991

EPA, Ultraviolet Air Quality and Emissions Trends Report 1990, EPA-450 /4-91-023, Technical Support Division, Office of Air Quality Planning and Standards, U.S.E.P.A., Research Triangle Park, N.C., 1991

Herman J. R., R. Hudson, R. McPeters, R. Stolarski, Z. Ahmad, X. Y. Gu, S. Taylor, and C. Wellemeyer, A new self-calibration method applied to TOMS and SBUV back-scattering ultraviolet data to determine long-term global ozone change, J. Geophys. Res., 96, 7531-7545,1991

International Panel on Climate Change. Climate Change- The IPCC Scientific Assessment No. Cambridge University Press, Cambridge), 1991

John K., K. C. Crist, and G. R. Carmichael, An Assessment of the Impacts of Global Climate and Emissions Change on Regional-Scale Tropospheric Chemistry. In AWMA, . Denver, CO, 1993

Liu S. C., and M. Trainer, Responses of the tropospheric ozone and odd hydrogen radicals to column ozone change, J. Atmos. Chem., 6, 221-221, 1987

McKenzie R. L., W. A. Matthews, and P. V. Johnson, The relationship between erythemal UV and ozone, derived from spectral irradiance measurements, Geophy. Res. L.,18, 2269-2272, 1991

Niu X., J. E. Frederick, M. L. Stein, and G. C. Tiao, Trends in column ozone based on TOMS data dependence on month latitude and longitude, J. Geophy. Res.,97, 14,661-14,669, 1992

Roemer M. G., Ozone and the GreenHouse Effect TNO Institute of Environmental Sciences.(TNO-report No. R91/227), 1991

Scotto J., T. Fears, and J. Fraumeni, Incidence of non-melanoma skin cancer in the United States NIH.(No. 82-2433), 1981

Shin W. C., Comprehensive air pollution modeling on multiprocessing environments: Application to regional scale problems. Ph.D., University of Iowa, Iowa City, 1990

Stolarski R. S., P. Bloomfield, R. D. McPeters, and J. R. Herman, Total ozone trends deduced from Nimbus 7 TOMS data, Geophys. Res. L.,18, 1015-1018, 1991

UNEP, Environmental Effects Panel Report United Nations Environmental Programme.(No. ISBN 92-807), 1989

Zdunkowski W. G., R. M. Welch, and G. Korb, An investigation of the structure of typical two stream methods for the calculation of solar fluxes and heating rates in clouds, Atmos. Phys., 53, 147-167, 1980

DISCUSSION

J. VILA-GUERAU DE ARELLANO:	What value did you use for your ground albedo
K. CRIST	The ground albedo was set at 0.05. This value was held constant for all the runs to evaluate the effects of atmospheric ozone.

NEW DEVELOPMENTS

chairmen: M. Millan
                        W. Physick

rapporteurs: R. Salvador
                        D.C. DiCristofaro

SIMILARITY AND SCALING FOR CONVECTIVE BOUNDARY LAYERS (EXTENDED SUMMARY[1])

Sergej Zilitinkevich

Alfred Wegener Institute for Polar and Marine Research
Am Handelshafen 12
27570 Bremerhaven, Germany

1. ITRODUCTION

Convective boundary layers (CBLs) are observed in the atmosphere and ocean, reproduced in wind tunnels, and investigated numerically with large-eddy simulation models. A host of data on vertical distribution of turbulent statistics and mean flow characteristics have been collected. The problem arises as to how we can compare the data representing CBLs of different geometrical scales, in different media, and characterized by different buoyancy forces and velocity shears.

Without velocity shear, this problem is solved with the aid of scalings proposed by Deardorff (1970) and Wyngaard (1983) for the entire shear-free CBL in non-penetrative and penetrative regimes, respectively. With shear, the problem is also solved but only for the region close to the surface (the surface layer) by means of the Monin-Obukhov (1954) similarity theory. Extensive studies dealing with analyzing the experimental/numerical data in terms of these approaches have been carried out. These results make up the basis of our knowledge of convective flows pointed out above. The CBL influenced by the velocity shear by comparison has been neglected, to a great extent because of the lack of proper scaling.

It is hardly surprising that little progress has been made in the general case up to the present. The structure of the flow looks too complex. In particular, the list of governing parameters of turbulence generated by both buoyancy and

---

[1] The paper in full is to be published in "Boundary-Layer Meteorology".

shear is long enough to make straightforward application of dimensional analysis practically useless.

Traditional turbulence closure models do not apply well to these flows, as the models are unable to distinguish between convective and mechanical contributions in turbulence as well as the contributions from updraughts and downdraughts. Indeed, these components show a principal difference in nature, while closure models deal with bulk characteristics, such as kinetic energy, its dissipation rate, its vertical flux etc., assimilating the above contributions without differentiating between them. In a strict sense the turbulence closure problem for convective shear flows should be treated in spectral terms, although this seems too complicated from the practical point of view. Another side of the same problem is non-local nature of turbulence, which manifests itself in counter-gradient transport. These essential features of convective shear flows are taken into account in theoretical concepts and non-traditional closures proposed recently (Berkowicz and Prahm, 1979, 1984; Lenschow and Stephens, 1980; Wyngaard, 1983, 1987; Fiedler, 1984; Stull, 1984, 1988; Hunt, 1984; Hunt, Kaimal and Gaynor, 1988; Schumann, 1988; Holtslag and Moeng, 1991; Lykossov, 1992).

Looking at the problem from the point of view of scaling, an essential difference exists between more or less isotropic movements produced by the velocity shear and basically vertical, buoyancy-driven convective plumes, representing coherent structures rather than usual turbulent, disordered eddies. This suggests the idea underlying a generalized scaling discussed below in Sections 3 and 4, viz., decomposition of the velocity fluctuations into convective (c) and mechanical (m) components with different inherent scales.

2. OVERVIEW

Widely used scaling for nonstationary shear-free CBL was given by Deardorff (1970, 1972, 1974). He introduced the following length, velocity and temperature scales:

$$h, \qquad w_* = (\beta Q_s h)^{1/3}, \qquad \theta_* = Q_s / w_*. \qquad (1)$$

Here, h is the CBL depth, Q_s is the virtual potential temperature flux at the surface and β is the buoyancy parameter. For air, $\beta = g/\theta_s$, where g is the acceleration due to gravity and θ_s is the surface value of virtual potential temperature, θ. One-point moments of convective turbulence, normalized by means of the above scales, were expected to be universal functions of the dimensionless variable

$$\zeta = z/h, \qquad (2)$$

where z is the height. This scaling shows reasonably good agreement with both experimental and large-eddy simulation data (e.g. Moeng and Wyngaard, 1984, 1989; Nieuwstadt et al., 1993).

The effect of entrainment of warmer fluid at the CBL top was not taken into account in the Deardorff scaling. In order to parameterize this effect, Wyngaard (1983, 1987) and Wyngaard and Brost (1984) proposed decomposition of convective turbulence into bottom-up (b) and top-down (t) components, the former associated with updraughts, the latter, with downdraughts. Statistical moments were thus expressed as sums of the contributions from the b-type, t-type and covariance b,t-type terms. The b-terms were scaled using Eq. (1); while the t-terms, in the same way but substituting, instead of Q_s, the modulus of the virtual potential temperature flux due to entrainment, $Q_h \equiv Q(h)$.

This decomposition of convective turbulence was further elaborated by Sorbjan (1988, 1989, 1990, 1991) and Sorbjan and Zilitinkevich (1993). In the spirit of local scalings proposed in general terms by Townsend (1661) and for stable boundary layers by Nieuwstadt (1984), local potential temperature flux, $Q(z)$, was adopted as basic quantity characterizing the turbulence regime at a height z. According to the mixed-layer concept, the latter was expressed as $Q=Q_s(1-z/h)+Q_h z/h$, so that the b- and t-components of the potential temperature flux, Q_b and Q_t were deduced to be

$$Q_b = Q_s(1-z/h), \qquad Q_t = Q_h z/h. \tag{3}$$

Then the following z-dependent length, velocity and temperature scales were introduced: for b-statistics:

$$l_b = \frac{z}{1+C_b z/h}, \qquad W_b = (\beta Q_b l_b)^{1/3}, \qquad \theta_b = Q_b/W_b, \tag{4}$$

and for t-statistics:

$$l_t = \frac{(h+\Delta h - z)}{1+C_t(h+\Delta h - z)/h}, \qquad W_t = |\beta Q_t l_t|^{1/3}, \qquad \theta_t = |Q_t|/W_t, \tag{5}$$

where Δh is the "half-thickness" of the entrainment zone, and C_b and C_t are empirical dimensionless constants ($C_b = C_t \approx 2$).

As a result, any statistical moment, X, was represented as

$$X = X_b + X_t, \qquad \frac{X_b}{l_b^a W_b^b \theta_b^c} = constant, \qquad \frac{X_t}{l_t^a W_t^b \theta_t^c} = constant, \tag{6}$$

with the exponents a, b and c dependent on the dimension of X.

The data of laboratory experiments (Adrian et al., 1986), large-eddy simulation (Moeng and Wyngaard, 1989; Schmidt and Schumann, 1989) and direct numerical simulation (Coleman et al., 1990) are in reasonable agreement with the above scaling predictions, except near the CBL upper boundary.

In all scalings discussed above, the role of the velocity shear was not considered. Geophysical CBLs, however, nearly always develop in the presence of winds, in the atmosphere, or currents, in the ocean. Shear-generated turbulence is almost always significant near the surface. In baroclinic flows,it is significant throughout the CBL and in the entrainment zone. Thus, at least two additional governing parameters of turbulence are to be introduced. These are the values of normalized vertical flux of momentum, τ, at the surface, $\tau_s = \tau(0) \equiv u_*^2$ (u_* is called friction velocity), and at the CBL top, $\tau_h = \tau(h)$.

An extension of the Deardorff scaling, Eqs. (1) and (2), for convective shear flows in the regime, when no shear is observed at the CBL top, was proposed by Zilitinkevich and Deardorff (1974) and then applied to derivation of the resistance and heat/mass transfer laws (e.g. Melgarejo and Deardorff, 1974; Zilitinkevich, 1975). It follows from this scaling that dimensionless statistical moments, are functions of three arguments: the dimensionless height, $\zeta = z/h$, and the following two dimensionless numbers: $\Pi_* \equiv u_*/w_*$ and $\Pi_f \equiv hf/w_*$, where, f is the Coriolis parameter. The scaling is clearly simplified as $\Pi_f \ll 1$, which is typical for geophysical CBLs.

The lower portion ($z \ll h$) of both stably and unstably stratified boundary layers, called the surface layer, was examined comprehensively on the basis of the Monin-Obukhov similarity theory with the following results (Monin and Obukhov, 1954; Monin, Yaglom, 1971). Governing parameters of the surface-layer turbulence, determining the dependence of statistical moments on the height, z, are u_*, Q_s, β. The length, velocity and temperature scales composed from these parameters are

$$L = -u_*^3/\beta Q_s, \qquad u_*, \qquad T_* = -Q_s/u_*. \qquad (7)$$

Statistical moments, nondimensionalized by means of these scales, represent universal functions of the dimensionless height

$$\xi = z/L, \qquad (8)$$

which characterize the effect of stratification on vertical profiles.

At $z/|L| \ll 1$, statistical moments depend only on u_*, Q_s and z, and therefore are easily determined from dimensional analysis. Such a regime, corresponding to the *mechanical turbulence layer* (MTL), is called forced convection.

At $-z/L \gg 1$, statistical moments depend on Q_s, β and z only, and once again are easily determined from dimensional analysis. This results in the free convection formulation. *The free convection layer* (FCL) is thus limited from below and from above: $-L \ll z \ll h$.

In the intermediate, *convective&mechanical layer* (CML) between the MTL and the FCL, the shape of the functions determining vertical profiles of statistical moments cannot be determined from conventional dimensional arguments. An attempt to solve the problem on the basis of generalized (directional) dimensional analysis was made by Zilitinkevich (1971, 1973) and Betchov and Yaglom (1971). Taking into account the fact that convective plumes are driven by vertically oriented buoyancy forces while mechanical turbulence is maintained chiefly by mean horizontal velocity shear, principally different dimensions (units) of length were introduced: $\mathcal{L}_c$ for convective ("vertical") and $\mathcal{L}_m$ for mechanical ("horizontal") distance. Correspondingly, symbolizing the dimension of time as $\mathcal{T}$, principally different generalized dimensions were adopted for the components of the velocity fluctuations: $\mathcal{L}_c/\mathcal{T}$ for convective, and $\mathcal{L}_m/\mathcal{T}$ for mechanical ones.

It was then assumed that in a certain part of the CML, which can be called the *alternative turbulence layer* (ATL), vertical velocity fluctuations, w', basically of convective origin, are to be measured in $\mathcal{L}_c/\mathcal{T}$ units, while horizontal fluctuations, u' and v', basically of mechanical origin, in $\mathcal{L}_m/\mathcal{T}$ units. These assumptions led to the following directional dimensions for the governing parameters of the Monin-Obukhov similarity theory:

$$u_*^2 \sim \mathcal{L}_c\mathcal{L}_m/\mathcal{T}^2, \qquad \beta \sim \mathcal{L}_c/\mathcal{T}^2\Theta, \qquad Q_s \sim \mathcal{L}_c\Theta/\mathcal{T}. \qquad (9)$$

Hence no vertical length scale could be composed in the ATL, and the application of the Π-theorem immediately resulted in simple power-law expressions of vertical profiles of statistical moments.

This model was verified only recently by Kader and Yaglom (1990). Quite high level of correspondence for practically all moments examined was discovered in the layer $(0.1-0.3)<-z/L<(1-3)$, which was reasonably deduced to be the ATL.

A schematic presenting sublayers distinguished in the CBL is shown in Fig. 1.

3. DECOMPOSITION OF TURBULENCE INTO CONVECTIVE & MECHANICAL COMPONENTS. STATISTICAL MOMENTS IN THE SURFACE LAYER

No contribution to vertical velocity fluctuations, w', was assumed in the ATL from shear-generated turbulence, and, vice versa, from buoyancy-driven turbulence to horizontal velocity fluctuations, u', v'. This certainly can be adopted as a more or less reasonable approximation in a limited layer only.

For the general case, Zilitinkevich (1993) proposed decomposition of the vector velocity fluctuation, $\mathbf{u}'$, into convective (c) and mechanical (m) components: $\mathbf{u}' = \mathbf{u}'_c + \mathbf{u}'_m$, the first one, $\mathbf{u}'_c$, being representative of the same convective flow as under consideration but without any shear, the second one, $\mathbf{u}'_m$, expressing the residual, $\mathbf{u}' - \mathbf{u}'_c$. He identified then the velocity fluctuation components of convective and mechanical origin as characterized by different dimensions, $\mathcal{L}_c/\mathcal{T}$ and $\mathcal{L}_m/\mathcal{T}$, as before, but no longer neglect the contributions from convective turbulence into u' and v' and, generally speaking, from mechanical turbulence into w'.

In the *convective&mechanical layer* (CML) between the ATL and FCL, the typical value of $|w'_m|$ was assumed to be small as compared with that of $|w'_c|$ (this statement was actually treated as a condition identifying the CML). Then the

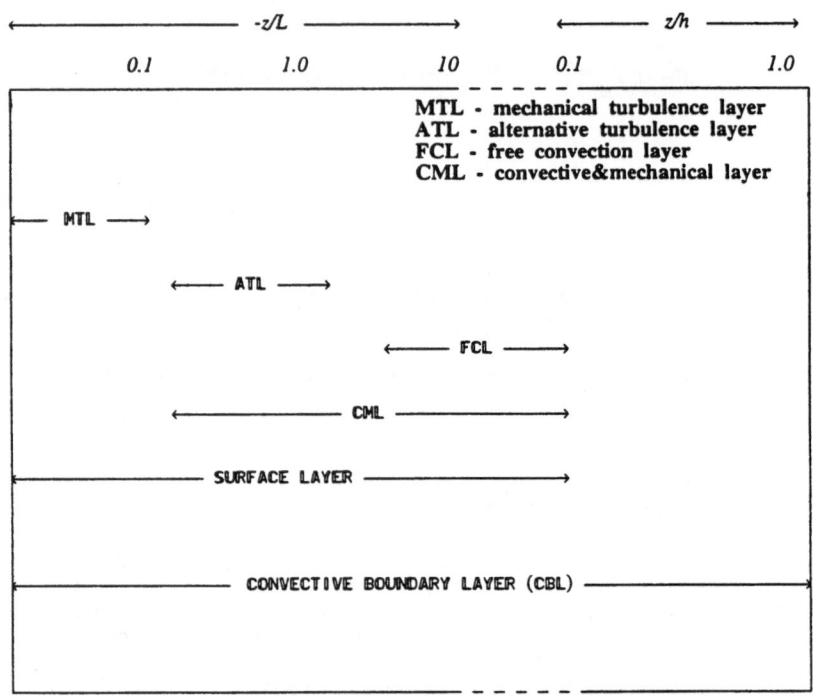

DIMENSIONLESS HEIGHTS (IN LOGARITHMIC SCALES)

Figure 1. A schematic of different sublayers in the CBL.

generalized dimensions of the governing parameters of the Monin-Obukhov similarity theory in the CML were deduced to be determined by Eq. (9). Using the Π-theorem, the following surface-layer scales were uniquely determined for convective length ($\sim \mathcal{L}_c$), convective velocity ($\sim \mathcal{L}_c/\mathcal{T}$), mechanical velocity ($\sim \mathcal{L}_m/\mathcal{T}$), and temperature ($\sim \Theta$):

$$z, \qquad W_c = (\beta Q_s z)^{1/3}, \qquad U_c = u_*^2/W_c, \qquad \theta_c = Q_s/W_c. \tag{10}$$

Correspondingly, in the whole part of the surface layer overlying the MTL, any one of statistical moments, X, was represented as the sum of convective, X_c, mechanical, X_m, and covariance, $X_{cov}^{(k)}$ ($k=1,2,3,..$), components: $X = X_c + X_m + \sum_k X_{cov}^{(k)}$, each one determined from the equation

$$\frac{\{X_c, X_m, X_{cov}^{(k)}\}}{z^a W_c^b U_c^c \theta_c^d} = constant. \tag{11}$$

Certainly, the exponents a, b, c, d here should be different for different components of the moment, being dependent on their generalized dimensions.

The expressions of some important statistical moments derived in this way are

$$e \equiv \tfrac{1}{2}(\overline{u'^2} + \overline{v'^2} + \overline{w'^2}) = C_{ec} W_c^2 + C_{ecm} W_c U_c + C_{em} U_c^2, \tag{12}$$

$$F_{e\&p} \equiv \overline{e'w'} + \overline{\frac{p'w'}{\rho}} = C_{Fe\&pc} W_c^3 + C_{Fe\&pcm} W_c^2 U_c + C_{Fe\&pmc} W_c U_c^2 + C_{Fe\&pm} U_c^3, \tag{13}$$

$$\varepsilon \equiv \frac{\nu}{2}\overline{\left(\frac{\partial u'_i}{\partial x_j} + \frac{\partial u'_j}{\partial x_i}\right)\left(\frac{\partial u'_i}{\partial x_j} + \frac{\partial u'_j}{\partial x_i}\right)} = C_{\varepsilon c} \frac{W_c^3}{z} + C_{\varepsilon cm} \frac{W_c^2 U_c}{z} + C_{\varepsilon m} \frac{W_c U_c^2}{z}, \tag{14}$$

$$\sigma_T^2 \equiv \overline{\theta'^2} = C_{TT} \theta_c^2 = C_{TT} Q_s^2/W_c^2, \tag{15}$$

$$F_T \equiv \overline{\theta'^2 w'} = C_{FTc} \theta_c^2 W_c + C_{FTm} \theta_c^2 U_c, \tag{16}$$

$$\varepsilon_T \equiv \kappa \overline{\frac{\partial \theta'}{\partial x_j} \frac{\partial \theta'}{\partial x_j}} = C_{\varepsilon T} \theta_c^2 W_c/z. \tag{17}$$

Here, e is turbulent kinetic energy; $F_{e\&p}$ is vertical flux of the fluctuations of energy, e', and pressure, p' (ρ is the fluid mean density); ε it the rate of viscous dissipation of turbulent kinetic energy due to molecular viscosity, v; σ_T^2 is the variance of potential temperature; F_T is vertical flux of this variance; ε_T is the rate of decay of this variance due to molecular heat conductivity, κ; C_{ec}, C_{ecm}, C_{em}, etc. are dimensionless constants. Preliminary results of the verification of these scaling predictions were given by Zilitinkevich (1993).

4. STATISTICAL MOMENTS IN THE BOUNDARY LAYER

For convective boundary layer influenced by the velocity shear, Zilitinkevich (1993) adopted linear, mixed-layer approximation for $Q(z)$ and $\tau(z)$ which implies the following decompositions into the bottom-up and top-down components:

$$Q = Q_b + Q_t, \quad \text{where} \quad Q_b = Q_s(1-\zeta), \quad Q_t = Q_h\zeta = R_Q Q_s \zeta, \tag{18}$$

$$\tau = \tau_b + \tau_t, \quad \text{where} \quad \tau_b = u_*^2(1-\zeta), \quad \tau_t = \tau_h\zeta = R_\tau u_*^2\zeta, \tag{19}$$

where $R_Q \equiv Q_h/Q_s$ and $R_\tau \equiv \tau_h/u_*^2$ are the entrainment coefficients for heat and momentum.

Then keeping the same directional dimensions as before for convective and mechanical contributions in the velocity fluctuations ($\mathcal{L}_c/\mathcal{T}$ and $\mathcal{L}_m/\mathcal{T}$, respectively), he introduced the following CBL scales of convective length ($\sim\mathcal{L}_c$), convective velocity ($\sim\mathcal{L}_c/\mathcal{T}$), mechanical velocity ($\sim\mathcal{L}_m/\mathcal{T}$) and potential temperature ($\sim\Theta$): for bottom-up statistics:

$$l_b = z/(1+C_b z/h), \quad W_b = (\beta Q_b l_b)^{1/3}, \quad U_b = \tau_b/W_b, \quad \theta_b = Q_b/W_b, \tag{20}$$

and for top-down statistics:

$$l_t = (h+\Delta h-z)/[1+C_t(h+\Delta h-z)/h], \quad W_t = |\beta Q_t l_t|^{1/3}, \quad U_t = |\tau_t|/W_t, \quad \theta_t = |Q_t|/W_t, \tag{21}$$

where w_* and θ_* are the CBL bulk vertical velocity and potential temperature scales, Eq. (1), U_* is new CBL bulk horizontal velocity scale,

$$U_* = u_*^2/w_* = u_*^2/(\beta Q_s h)^{1/3}. \tag{22}$$

According to the proposed scaling, vertical profile of a statistical moments, X, in the part of the CBL overlying the near-surface MTL is determined as follows. At first, the moment is represented as the sum of convective, mechanical and covariance (convective&mechanical) contributions; each one is then decomposed into bottom-up (b) and top-down (t) components; and finally, both b- or t-components of all (convective, mechanical and covariance) contributions are expressed with the aid of generalized dimensional analysis in terms of the corresponding b- or t-scales, Eqs (20) and (21):

$$X= X_c+X_m+\sum_k X_{cov}^{(k)}= X_{c,b}+X_{m,b}+\sum_k X_{cov,b}^{(k)}+X_{c,t}+X_{m,t}+\sum_k X_{cov,t}^{(k)}, \qquad k=1,2,3,... \qquad (23)$$

Thus the scaling results in the following expressions of the terms on the right-hand side of Eq. (23):

$$\frac{\{X_{c,b}, X_{m,b}, X_{cov,b}^{(k)}\}}{l_b^a W_b^b U_b^c \theta_b^d}= constant, \qquad \frac{\{X_{c,t}, X_{m,t}, X_{cov,t}^{(k)}\}}{l_t^a W_t^b U_t^c \theta_t^d}= constant, \qquad (24)$$

where the exponents a, b, c and d are fully determined by directional dimension of the considering term. This scaling embraces the following particular cases considered above: in shear-free CBLs ($\tau_s=\tau_h=0$), it reduces to Eqs. (3)-(6); in the surface layer, to Eqs. (10) and (11). In both cases it works fairly well. No verification has been done, however, for the general case.

Decomposition of turbulence into convective and mechanical components does not imply the same procedure for mean gradients, $\partial u/\partial z$ and $\partial \theta/\partial z$. The latter are to be determined from the energy budget equations for the fluctuations of velocity and potential temperature:

$$\tau \partial u/\partial z= -\beta Q+\partial(F_e+F_p)/\partial z+\varepsilon, \qquad -Q\partial\theta/\partial z= \frac{1}{2}\partial F_T/\partial z+\varepsilon_T, \qquad (25)$$

where statistical moments in the right hand sides are determined from the same type equations as Eqs. (12)-(17), but generalized for the CBL.

Acknowledgements. This work was carried out at the Alfred Wegener Institute for Polar and Marine Research (AWI Contribution No. 692). It was partially supported by the ICSC - World Laboratory.

REFERENCES

Adrian, R.J., R.T.D.S. Ferriera and T. Boberg: 1986, "Turbulent thermal convection in a wide horizontal fluid layer", *Exp. Fluids* **4**, 121-141.

Berkowicz, R., and L.P. Prahm: 1979, "Generalization of K-theory for turbulent diffusion. Part 1: Spectral turbulent diffusivity concept", *J. Appl. Meteorol.* **18**, 266-272.

Berkowicz, R., and L.P. Prahm: 1984, "Spectral representation of the vertical structure of turbulence in the convective boundary layer", *Quart. J. Roy. Metorol. Soc.* **110**, 35-52.

Betchov, R., and A.M. Yaglom: 1971, "Comments on the theory of similarity as applied to turbulence in an unstably stratified fluid", *Izv. Akad. Nauk SSSR, Ser. Fiz. Atmosf. i Okeana* **7**, 1270-1279.

Coleman, G.N., J.H. Ferzinger and P.R. Spalart: 1990, "A numerical study of the stratified turbulent Ekman layer", *Rept. No. TF-48, Thermosciences Division, Dept of Mechanical Engineering, Stanford University.*

Deardorff, J.W.: 1970, "Convective velocity and temperature scales for the unstable planetary boundary layer", *J. Atmos. Sci.* **27**, 1211-1213.

Deardorff, J.W.: 1972, "Numerical investigation of neutral and unstable planetary boundary layer", *J. Atmos. Sci.* **29**, 91-115.

Deardorff, J.W.: 1974, "Three-dimensional numerical study of the height and mean structure of a heated planetary boundary layer", *Boundary-Layer Meteorol.* **7**, 81-106.

Fiedler, B.H.: "An integral closure model for the vertical turbulent flux of a scalar in a mixed layer", *J. Atmos. Sci.* **41**, 674-680.

Holtslag, A.A.M., and C.-H. Moeng: 1991, "Eddy diffusivity and countergradient transport in the convective atmospheric boundary layer", *J. Atmos. Sci.* **48**, 1690-1698

Hunt, J.C.R.: 1984, "Turbulence structure in thermal convection and shear-free boundary layers", *J. Fluid Mech.* **138**, 161-184.

Hunt, J.C.R., J.C. Kaimal and J.E. Gaynor: 1988, "Eddy structure in the convective boundary layer - new measurements and new concepts", *Quart. J. Roy. Meteorol. Soc.* **114**, 827-858.

Kader, B.A., and A.M. Yaglom: 1990, "Mean fields and fluctuation moments in unstably stratified turbulent boundary layers", *J. Fluid Mech.* **212**, 637-662.

Lenschow, D.H., and P.L. Stephens: 1980, "The role of thermals in the convective boundary layer", *Boundary-Layer Meteorol.* **19**, 509-532.

Lykossov, V.N.: 1992, "The momentum turbulent counter-gradient transport in jet-like flows", *Adv. Atmos. Sci.* **9**, 191-200.

Melgarejo, J.W., and J.W. Deardorff: 1974, "Stability functions for the boundary-layer resistance law based upon observed boundary-layer heights", *J. Atmos. Sci.* **31**, 1324-1333.

Moeng, C.-H., and J.C. Wyngaard: 1984, "Statistics of convective scalars in the convective boundary layer", *J. Atmos. Sci.* **41**, 3161-3169.

Moeng, C.-H., and J.C. Wyngaard: 1989, "Evaluation of turbulent transport and dissipation closures in second-order modeling". *J. Atmos. Sci.* **46**, 2311-2330

Monin, A.S., and A.M. Obukhov: 1954, "Basic laws of turbulent mixing in the atmospheric surface layer", *Trudy Geofiz. Inst. Akad. Nauk SSSR* No 24 (151), 163-187.

Monin, A.S., and A.M. Yaglom: 1971, *Statistical Fluid Mechanics*, vol. 1, MIT Press, Cambridge MA, 769 pp.

Nieuwstadt, F.T.M.: 1984, "The turbulent structure of the stable, nocturnal boundary layer", *J. Atmos. Sci.* **41**, 2202-2216.

Nieuwstadt, F.T.M., P.J. Mason, C.-H. Moeng and U. Schumann: 1993, "Large-eddy simulation of the convective boundary layer: a comparison of four computer codes", *Turbulent Shear Flows 8*, F. Durst, R. Friedrich, F.W. Schmidt, U. Schumann and J.H. Whitelaw, Eds., Springer-Verlag, Berlin, 343-367.

Schmidt, H., and U. Schumann: 1989, "Coherent structure of the convective boundary layer derived from large-eddy simulations", *J. Fluid Mech.* **200**, 511-562.

Schumann, U.: 1988, "Minimum friction velocity and heat transfer in the rough surface layer of a convective boundary layer", *Boundary-Layer Meteorol.* **44**, 311-326.

Sorbjan, Z.: 1988, "Local similarity in the convective boundary layer", *Boundary-Layer Meteorol.* **45**, 237-250.

Sorbjan, Z.: 1989, *Structure of the Atmospheric Boundary Layer*, Prentice Hall, Englewood Cliffs NJ, 317 pp.

Sorbjan, Z.: 1990, "Similarity scales and universal profiles of statistical moments in the convective boundary layer", *J. Appl. Meteorol.* **29**, 762-775.

Sorbjan, Z.: 1991, "Evaluation of local similarity functions in the convective boundary layer", *J. Appl. Meteorol.* **30**, 1565-1583.

Sorbjan, Z., and S. Zilitinkevich: 1993, "Towards parameterization of the convective boundary layer", Unpublished manuscript.

Stull, R.: 1984, "Transilient turbulence theory. Part 1: The concept of eddy-mixing across finite distances", *J. Atmos. Sci.* **41**, 3351-3367.

Stull, R.: 1988, *An Introduction to Boundary Layer Meteorology*, Kluwer Academic Publishers, Dordrecht, 666 pp.

Townsend, A.A.: 1961, "Equilibrium layers and wall turbulence", *J. Fluid Mech.* **11**, 97-120.

Wyngaard, J.C: 1983, "Lectures on the planetary boundary layer", *Mesoscale Meteorology - Theories, Observations and models*, D.K. Lilly and T. Gal-Chen, Eds., Ridel, Dordrecht, 603-650.

Wyngaard, J.C: 1987, "A physical mechanism for the asymmetry in top-down and bottom-up diffusion", *J. Atmos. Sci.* **44**, 1083-1087.

Wyngaard, J.C., and R.A. Brost: 1984, "Top-down and bottom-up diffusion of a scalar in convective boundary layer", *J. Atmos. Sci.* **41**, 102-112.

Zilitinkevich, S.S.: 1971, "On turbulence and diffusion in free convection", *Izv. Akad. Nauk SSSR, Ser. Fiz. Atmosf. i Okeana* **7**, 1263-1269.

Zilitinkevich, S.S.: 1973, "Shear convection", *Boundary-Layer Meteorol.* **3**, 416-423.

Zilitinkevich, S.S.: 1975, "Resistance laws and prediction equations for the depth of the planetary boundary layer", *J. Atmos. Sci.* **32**, 741-752.

Zilitinkevich, S.S.: 1993, "A generalized scaling for convective shear flows", To appear in *Boundary Layer Meteorology*.

Zilitinkevich, S.S., and J.W. Deardorff: 1974, "Similarity theory for the planetary boundary layer of time-dependent height", *J. Atmos. Sci.* **31**, 1449-1451.

DISCUSSION

H. VAN DOP: In order to test your scaling theories you need accurate experimental evidence. For the convective boundary layer (CBL), large-eddy simulation (LES) would be able to provide good enough data since the turbulence is dominated by relatively large scales. For the surface layer, however, one can only hope to have model simulations in the near future, and therefore pure experimental data of high accuracy are required to test the theory. Are sufficiently accurate data available at present?

S. ZILITINKEVICH: Your first statements exactly corresponds to what we are doing. Right now, Dr. Dmitrii Mironov of the St. Petersburg Institute of Limnology works as a visiting scientist at the MMM/NCAR, Boulder with the Chin-Hoh Moeng LES code in order to apply it to the verification of the scaling predictions for the CBL. For the surface layer, further extension of the scaling and concrete outlines of its verification using reliable experimental data is now elaborating in cooperation with different groups. I believe quite accurate data can be found in rich surface-layer-measurements archives collected at the institutions like RISØ, NCAR, Moscow Institute of Atmospheric Physics and some others.

R. SAN JOSE: How the new advances can be applied to passive pollutant profiles in the surface layer.

S. ZILITINKEVICH: The can be deduced from the budget equations for corresponding variances using new scaling expressions of the statistical moments included (in the same way as it is done for the temperature or velocity profiles).

P. ALPERT: In your approach implying decomposition of turbulence into convective and mechanical components, you should face difficulties associated with complexity of covariance contributions in statistical moments. How can you avoid these difficulties?

S. ZILITINKEVICH: The treatment of the covariances really represents one of key points of the new approach. The formal tool for overcoming the difficulties you have mentioned is the generalized dimensional analysis, based on the use of principally different units of length in the dimensions of convective and mechanical velocity fluctuations. This clearly results in the efficient application of the Π-theorem not only to purely convective or purely mechanical contributions but also to covariances. Physical grounds for such expedient lie in the fact that convective and mechanical motions, driven by the forces of principally different nature, are characterized in asymptotic regimes by essentially different scales of length.

A FAST LAGRANGIAN PARTICLE MODEL FOR USE WITH

THREE-DIMENSIONAL MESOSCALE MODELS

W. Physick and P. Hurley

CSIRO Division of Atmospheric Research
Aspendale, Victoria 3195
Australia

INTRODUCTION

In recent years, the air quality modelling field has seen the development of Lagrangian particle dispersion models driven by wind and turbulence predictions from three-dimensional mesoscale models. The Lagrangian models are based on a Langevin equation for the velocity of a fluid particle, and can take a number of different forms, depending mainly on the type of turbulence being simulated. We briefly review these forms, concentrating on the theoretically correct version for simulating dispersion in a convective boundary layer (CBL).

Although computing power has increased rapidly in the last few years, it is still necessary to simplify aspects of Lagrangian theory in order to perform large three-dimensional dispersion studies. We present a simplified model which is fast and is able to reproduce the important features of convective dispersion, but which is still consistent with the tenets of Lagrangian dispersion theory.

THE LANGEVIN EQUATION

The form of the Langevin equation used for modelling the vertical velocity w in *stationary, homogeneous* and *Gaussian* turbulence is

$$dw = -(w / T_{L,w})dt + (2 / T_{L,w})^{1/2} \sigma_w dW \tag{1}$$

where $T_{L,w}$ is the Lagrangian timescale, σ_w is the vertical velocity variance and dW is a Gaussian, white-noise stochastic process with mean zero and variance dt.

The finite-difference form of Eq.(1) used in numerical models is

Air Pollution Modeling and Its Application X, Edited by S-V. Gryning
and M. M. Millán, Plenum Press, New York, 1994

$$w(t + \Delta t) = (1 - \Delta t / T_{L,w})w(t) + (2\Delta t / T_{L,w})^{1/2} \sigma_w r_w, \tag{2}$$

where r_w is a random variable with Gaussian distribution (mean 0, standard deviation 1). Equation (2) is sometimes written in terms of the autocorrelation function $R_w = e^{-\Delta t/T_{L,w}}$ by making use of the relation $n\Delta t / T_{L,w} = (1 - R_w^n) + O(\Delta t^2)$ for $\Delta t \ll T_{L,w}$.

If Eq.(1) is used for *inhomogeneous* convective turbulence, particles accumulate at the top and/or bottom of the mixed layer, i.e. a well-mixed profile does not remain well-mixed with time. Legg and Raupach (1982) attributed this to the mean drift velocity induced by the gradient in vertical velocity variance and derived the following equation, which is equivalent to Eq.(2) with the addition of a further term:

$$w(t + \Delta t) = (1 - \Delta t / T_{L,w})w(t) + (2\Delta t / T_{L,w})^{1/2} \sigma_w r_w + 2\Delta t \sigma_w \partial \sigma_w / \partial z . \tag{3}$$

This equation is currently used in many Lagrangian particle models. However, it was shown by Thomson (1984) that the Legg and Raupach correction term is only applicable in *weakly inhomogeneous* turbulence. Thomson also rigorously derived a new form of the Langevin equation, used earlier by Wilson et al. (1983), and showed that it was correct for *non-stationary, inhomogeneous, Gaussian* turbulence. The equation for Gaussian turbulence derived by Thomson to first order in dt, and referred to here as the w/σ_w-form, is

$$\frac{w(t + \Delta t)}{\sigma_w(z(t + \Delta t))} = \frac{(1 - \Delta t / T_{L,w})w(t)}{\sigma(z(t))} + \frac{\mu(t)}{\sigma_w(z(t + \Delta t))} \tag{4}$$

with $\bar{\mu} = \Delta t \sigma_w \partial \sigma_w / \partial z$ and $\overline{\mu^2} = 2(\Delta t / T_{L,w})\sigma_w^2$.

While Eq.(4) allows the simulation of *non-Gaussian (skewed)* turbulence, and was used for this purpose by Sawford and Guest (1987), the skewness can only be incorporated in an approximate manner. In the skewed case, the third moment of μ is non-zero and an extra term is added to the second moment (see Thomson, 1984).

More recently, Thomson (1987) showed that the two Langevin models (the w-form and the w/σ_w-form) are special cases of a more general form of Langevin equation which is capable of representing *inhomogeneous, non-stationary* conditions and incorporates an exact formulation of *non-Gaussian* turbulence. The general form, under the assumption of independence of the three turbulence components, is written as

$$dw = a dt + (C_0 \varepsilon)^{1/2} dW \tag{5}$$

where ε is the rate of dissipation of turbulent kinetic energy and C_0 is a universal constant. As we shall see, a is a function of σ_w and ε, both measurable quantities, rather than $T_{L,w}$ which is only clearly defined when the turbulence is homogeneous and stationary. Note too that, unlike previous versions of the Langevin equation (for example de Baas et al., 1986, Sawford and Guest, 1987), it is not necessary to incorporate the skewness of the convective boundary layer in the random term, which remains Gaussian.

The finite-difference form of Eq.(5) is

$$w(t + \Delta t) = w(t) + a\Delta t + (C_0 \varepsilon)^{1/2} r_w \Delta t . \tag{6}$$

An expression for the function a is obtained by solving the following form of the Fokker-Planck equation (Thomson, 1987),

$$\frac{\partial(aP_E)}{\partial w} = -\frac{\partial P_E}{\partial t} - \frac{\partial(wP_E)}{\partial z} + \frac{1}{2}C_0\varepsilon\frac{\partial^2 P_E}{\partial w^2} \tag{7}$$

subject to the boundary condition $aP_E \to 0$ as $|w| \to \infty$ (B.L. Sawford, personal communication, 1988). P_E is the probability density function (PDF) describing 1-point Eulerian statistics of the turbulence.

Under *convective* conditions in which the turbulence is assumed to be *stationary*, *inhomogeneous*, and *skewed*, the skewness is incorporated through a PDF made up of two Gaussian functions, one representing the updrafts (+) and the other representing the downdrafts (-) of the convective boundary layer, and written as

$$P_E = pN(m_+, \sigma_+) + (1 - p)N(m_-, \sigma_-) \tag{8}$$

with

$$N(m, \sigma) = \sigma^{-1}(2\pi)^{-0.5}\exp(-(w - m)^2 / 2\sigma^2).$$

Here p is the probability of a particle being in an updraft, m_+ is the mean velocity in an updraft and σ_+ is the velocity standard deviation in an updraft, and similarly for the downdraft terms. The first three moments of P_E are equated to the first three moments of the vertical velocity distribution: $\overline{w} = 0$, $\overline{w^2} = \sigma_w^2$, and $\overline{w^3} = S_w^3$. Then the equations can be solved for the variables p, m_+, and m_- (by making the assumption $\sigma = m$ for both updrafts and downdrafts). The solutions are:

$$p = 0.5(1 - (Sk^2 / (8 + Sk^2))^{0.5})$$
$$m_+^2 = 0.5\sigma_w^2(1 - p) / p$$
$$m_- = -m_+p / (1 - p)$$

where $Sk = (S_w / \sigma_w)^3$ is the degree of skewness of the turbulence. For *stationary*, *inhomogeneous*, *skewed* turbulence, the solution of Eq.(7) is complex and is given by Luhar and Britter (1989).

For *homogeneous, skewed* turbulence, the solution of Eq.(7) is

$$a = -\frac{C_0\varepsilon}{2}\left\{\frac{pN(m_+, \sigma_+)(w - m_+) / \sigma_+^2 + (1 - p)N(m_-, \sigma_-)(w - m_-) / \sigma_-^2}{pN(m_+, \sigma_+) + (1 - p)N(m_-, \sigma_-)}\right\}.$$

For *inhomogeneous, Gaussian* turbulence, the expression for a is

$$a = -\frac{C_0\varepsilon}{2\sigma_w^2}w + \sigma_w\frac{\partial\sigma_w}{\partial z}\left\{1 + \left(\frac{w}{\sigma_w}\right)^2\right\}.$$

Equation (6) with the above form of a reduces to Eq.(3) only in homogeneous turbulence ($\overline{w^2} = \sigma_w^2$), illustrating that the Legg and Raupach drift correction is applicable only in weakly inhomogeneous conditions.

$$a = -\frac{C_0 \varepsilon}{2\sigma_w^2} w \; .$$

Note that in this case, the expression for w (Eq.(6)) is the same as Eq.(2) when use is made of the relation $T_{L,w} = 2\sigma_w^2 / (C_0 \varepsilon)$.

WELL-MIXED EXPERIMENTS

An important characteristic of any dispersion formulation in convective conditions is its ability to maintain a well-mixed profile. We now examine the conditions under which both the simplest and the most complex of the above formulations are able to satisfy this criterion. Firstly, the turbulence parameterizations are presented for each model, both of which use Eq.(6) with the appropriate expression for a.

model 1 - inhomogeneous, skewed turbulence

The vertical velocity moments used in this model are derived from the Willis and Deardorff (1976, 1978, 1981) laboratory experiments (see Sawford and Guest, 1987). A dissipation rate $\varepsilon = 0.6 w_*^3 / z_i$ is used, and skewness is specified via Eq.(8).

model 2 - homogeneous, Gaussian turbulence

The parameterizations used in this model, $\sigma_w = 0.6 w_*$ and $\varepsilon = 0.6 w_*^3 / z_i$ agree with observations in the middle 80% of the atmospheric mixed layer. A value of 2.0 is specified for C_0.

These well-mixed tests were carried out as part of development of a Lagrangian particle model to simulate early-morning fumigation (Hurley and Physick, 1991). After the plume had been totally entrained into the growing mixed layer, all boundary-layer parameters were held constant and the model run for another 100 minutes (about 18 convective timescales). A timestep Δt of 20 seconds was used and an hourly-averaged concentration profile was calculated over the final 60 minutes for each model. From the hourly-averaged steady-state cross-plume integrated concentration profiles (C^y) of Fig.1a,b, it can be seen that only the simpler model 2 is able to maintain a well-mixed profile. Although the central region of model 1 is approximately well-mixed, the profiles near the boundary regions are clearly not.

The reason for these sharp gradients is that the timesteps are not small enough near the boundaries to resolve the scale of turbulence represented by the inhomogeneous turbulence parameterization. The value of the timestep needed to give a well-mixed profile is considered by Thomson (1987) to be related to the parameterizations used:

$$\Delta t = \min\left(0.05 T_{L,w}, 0.1\sigma_w / |a|, 0.01\sigma_w / |w \partial \sigma_w / \partial z|\right) \qquad (9)$$

and is therefore variable in time and space for inhomogeneous turbulence. When the models were modified to use a variable timestep according to Eq.(9), the resulting profiles were both satisfactorily well-mixed (Fig.1c,d). However, the size of the timestep necessary for well-mixedness with model 1 (often less than 1 second) places a severe restriction on the use of the inhomogeneous turbulence formulation for three-dimensional mesoscale dispersion models, in which the number of particles tracked can be very large.

By comparing our numerical simulations to the laboratory experiments of Deardorff and Willis (1982), we have found that for fumigation applications using hourly-averaged concentrations, or more generally, for applications which are concerned with concentrations averaged over several convective timescales, use of Gaussian homogeneous turbulence for the convective boundary layer with the homogeneous Langevin equation (model 2) gives very good results, even for reasonably large timesteps (60 seconds was used) - see Hurley and Physick (1991) for details. In the next section, we examine the performance of models 1 and 2 in simulating dispersion from elevated sources in a fully convective boundary layer.

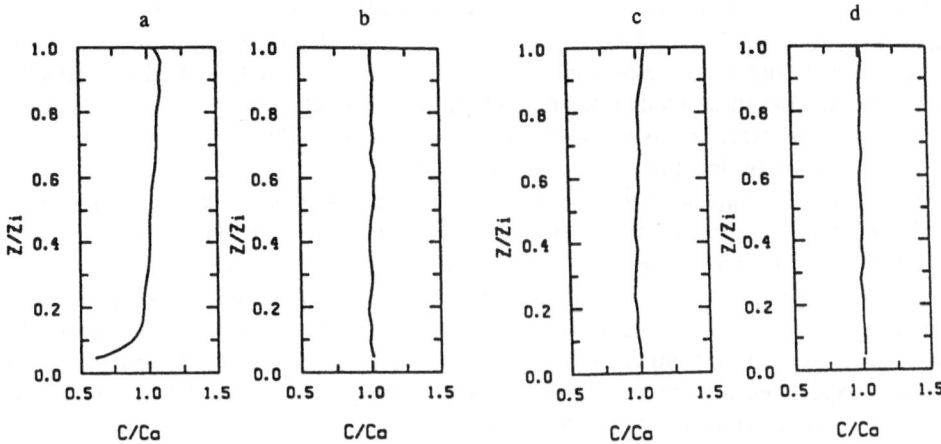

Figure 1. Steady-state cross-plume integrated profiles for (a) model 1 and (b) model 2 with a 20-second timestep, and (c) model 1 and (d) model 2 with a variable timestep. Ca is the analytic solution.

CONVECTIVE DISPERSION FROM ELEVATED SOURCES

The ensemble-averaged concentration fields from the laboratory experiments of Willis and Deardorff (1976, 1978, 1981) and those from various numerical simulations reveal that the plume from an elevated source descends to the ground near the stack and then lifts off again before becoming well-mixed through the depth of the boundary layer further downwind. Maximum concentrations occur at the touchdown point, the position of which is dependent on wind speed, source height and the strength of the turbulence. The descent of the plume is due to the skewed nature of convective turbulence (downdrafts occupy a greater horizontal area than the narrower updrafts). In this section, we investigate the extent to which homogeneous (Gaussian and skewed) models can reproduce this behaviour in convective conditions.

For this purpose we introduce **model 3**, the general form of the Langevin equation (Eq.(5)) using a *skewed homogeneous* turbulence parameterization. This model's homogeneous turbulence is the same as that for model 2 while the skewed PDF is the same as that used for model 1 (Eq.(8)). However the value of skewness used (Sk =0.4) was chosen by performing simulations with Sk = 0.8, 0.6 and 0.4 and comparing the results to the maximum concentrations from the laboratory experiments. It should be noted though that our chosen value, and that of Sawford and Guest (1987) used in model 1, is well within the range of observed values of skewness in the atmospheric CBL.

It is important to note here that the correct boundary condition, consistent with the PDF for w, is one of skewed memory reflection in which w is scaled by the absolute value of the ratio of the mean updraft velocity to the mean downdraft velocity $((1-p)/p)$ when reflecting at the ground, and the inverse of this ratio at the mixing height. This boundary condition gives $C^y = 1$ within 10% at all times at large downwind distances while not affecting maximum values of ground-level concentration (glc) near the source.

Experiment and results

The convective dispersion laboratory experiments of Willis and Deardorff (1976, 1978, 1981) have been simulated by releasing non-buoyant particles at non-dimensional heights of $z_s/z_i = 0.067$, 0.24 and 0.49, where z_i is the depth of the mixed-layer. The release rate is 20 particles per second and a timestep of 60 seconds is used for models 2 and 3. Results from these models are not compared to model 1, but to those from a very similar *skewed inhomogeneous* model of Luhar and Britter (1989).

Details of the experimental set-up and the results can be found in Hurley and Physick (1993), and a brief summary is presented here. When the three simulations and the Willis and Deardorff experiments are presented in the form of dimensionless crosswind integrated concentration contours (C^y), as a function of non-dimensional height and downwind distance X, it is found that all models are able to reproduce the general pattern of plume descent to the ground within $X < 1$ from the source, and the formation of a well-mixed layer further downwind. However, the plume lift-off behaviour of the laboratory experiments is only reproduced by the Luhar and Britter model, indicating that it arises from the inhomogeneity of the turbulence.

A quantitative comparison of the models and observations is made in Table 1 where the maximum ground-level concentration (C_{max}) and its distance from the source (X_{max}) are shown. While there is little difference in X_{max} predictions for the homogeneous models (2 and 3), Table 1 shows that the Gaussian turbulence model 2 underestimates the magnitude of the maximum glc. A larger percentage error occurs for the elevated sources than for the surface source. Most encouragingly, as far as the magnitude and location of C_{max} are concerned, there is little difference between the results from the skewed homogeneous model 3, the skewed inhomogeneous model of Luhar and Britter, and the results from the laboratory experiments.

Table 1. Maximum crosswind integrated ground-level concentrations (C_{max}) and corresponding downwind distances (X_{max}) in parentheses for the three source heights for model 2, model 3, Luhar and Britter (1989) and the laboratory results of Willis and Deardorff.

z/z_i	Model 2	Model 3	L & B	W & D
0.067	6.9 (0.2)	8.6 (0.2)	8.0 (0.1)	7.5 (0.1)
0.24	2.0 (0.6)	3.0 (0.5)	3.0 (0.5)	3.0 (0.5)
0.49	1.0 (1.1)	1.5 (1.2)	1.8 (1.0)	1.8 (0.8)

SUMMARY

A dispersion model using the general form of the Langevin equation derived by Thomson (1987) has been presented. By simulating the laboratory experiments of Willis and Deardorff (1976, 1978 and 1981), we have shown that a model incorporating skewed, *homogeneous* turbulence is able to reproduce the important features of convective dispersion from an elevated source. Through the use of a homogeneous parameterisation we are able to employ a considerably larger timestep (e.g. 20 times) than is allowed in inhomogeneous models. This is an important factor when three-dimensional simulations with a large number of particles are being carried out.

ACKNOWLEDGEMENTS

The authors are appreciative of comments by Brian Sawford on a first draft.

REFERENCES

de Baas, A.F., van Dop, H., and Nieuwstadt, F.T.M., 1986, An application of the Langevin equation for inhomogeneous conditions to dispersion in a convective boundary layer, *Quart. J. Roy. Meteor. Soc.*, 112:165.

Deardorff, J.W., and Willis, G.E., 1982, Ground-level concentrations due to fumigation into an entraining mixed layer, *Atmos. Environ.*, 16:1159.

Hurley, P.J., and Physick, W.L., 1991, A Lagrangian particle model of fumigation by breakdown of the nocturnal inversion, *Atmos. Environ.*, 25A:1313.

Hurley, P.J., and Physick, W.L., 1993, A skewed homogeneous Lagrangian particle model for convective conditions, *Atmos. Environ.*, 27A:619.

Legg, B.J., and Raupach, M.A., 1982, Markov-chain simulation of particle dispersion in inhomogeneous flows: the mean drift velocity induced by a gradient in Eulerian velocity variance, *Bound. Layer Meteor.*, 24:3.

Luhar, A.K., and Britter, R.E., 1989, A random walk model for dispersion in inhomogeneous turbulence in a convective boundary layer, *Atmos. Environ.*, 23:1911.

Sawford, B.L., and Guest, F.M., 1987, Lagrangian stochastic analysis of flux-gradient relationships in the convective boundary layer, *J. Atmos. Sci.*, 44:1152.

Thomson, D.J., 1984, Random walk modelling of diffusion in inhomogeneous turbulence, *Quart. J. Roy. Meteor. Soc.*, 110:1107.

Thomson, D.J., 1987, Criteria for the selection of stochastic models of particle trajectories in turbulent flows, *J. Fluid Mech.*, 180:529.

Willis, G.E., and Deardorff, J.W., 1976, A laboratory model of diffusion into the convective planetary boundary layer, *Quart. J. Roy. Meteor. Soc.*, 102:427.

Willis, G.E., and Deardorff, J.W., 1978, A laboratory study of dispersion from an elevated source within a modeled convective planetary boundary layer, *Atmos. Environ.*, 12:1305.

Willis, G.E., and Deardorff, J.W., 1981, A laboratory study of dispersion from a source in the middle of the convective mixed layer, *Atmos. Environ.*, 15:109.

Wilson, J.D., Legg, B.J., and Thomson, D.J., 1983, Calculation of particle trajectories in the presence of a gradient in turbulent-velocity variance, *Bound. Layer Meteor.*, 27:163.

DISCUSSION

G. GRAZIANI   Dr. Physick, did you validate your skewed model with source release experiments, such as Copenhagen and Karlsruhe data sets?

W. PHYSICK   No, we didn't. We have only simulated the laboratory experiments of Willis and Deardorff.

A RANDOM WALK MODEL FOR
ATMOSPHERIC DISPERSION IN
THE DAYTIME BOUNDARY LAYER

Caterina Tassone, Sven-Erik Gryning and Mathias Rotach[1]

Risø National Laboratory
DK-4000 Roskilde, Denmark

INTRODUCTION

In this paper we present a Random Walk Model (RWM) for atmospheric dispersion in the daytime boundary layer with a continuous transition between neutral and convective conditions. A RWM describes the dispersion of a passive tracer in a flow, whose turbulent structure is known. Particle trajectories are generated using a stochastic model for Lagrangian velocities, and from those the mean concentration field is calculated.

A general formulation of the problem was recently presented by Thomson (1987). The basic assumption in his approach is that the evolution of the joint process (z,w), where z is the height and w the vertical velocity of a particle of tracer, is a continuous Markov process and can therefore be represented by a non-linear stochastic differential equation with Gaussian white noise, which ensures that the Lagrangian velocity will be a continuous function of time. Thomson's model was applied to dispersion in a Convective Boundary Layer (CBL) by Luhar and Britter (1989), who developed a one dimensional model in the z - direction which includes skewness and vertical inhomogeneity.

In this paper we will describe a generalization of Luhar and Britter's model for dispersion in a BL with a continuous transition between neutral and convective conditions. In our model the probability density function (PDF) of vertical velocities is described by the mean, the second, the third and also the fourth order moment of the vertical velocities. The fourth order moment is introduced in order to control the continuous transition from convective to neutral conditions. The model was evaluated against the Copenhagen experiment (Gryning and Lyck, 1984), which represents near neutral and unstable conditions. Good agreement between measurements and model results was found.

---

[1]Permanent affiliation: Swiss Federal Institute of Technology, CH-8092 Zürich, Switzerland

DERIVATION OF THE MODEL EQUATION

General Theory

Following Thomson (1987), and considering only dispersion in the z-direction, the evolution of the joint process (z,w) is described by the following stochastic differential equation:

$$dw = a(z, w, t)dt + b(z, w, t)d\mathcal{W} \qquad (1)$$

$$dz = wdt \qquad (2)$$

where $d\mathcal{W}$ are increments of a Wiener process with mean zero and variance dt. The functions a and b are determined by requiring that the well-mixed condition and a constraint derived from Kolmogorov's theory of local isotropy are fullfilled. If P(z,w) is the PDF of all the tracer particles in the ensemble of flows, the Fokker-Planck equation for eqs. (1) and (2), away from any sources, is:

$$\frac{\partial P}{\partial t} = -w\frac{\partial}{\partial z}P - \frac{\partial}{\partial w}(a(z, w, t)P) + \frac{1}{2}\frac{\partial^2}{\partial w^2}(b^2(z, w, t)P) \ . \qquad (3)$$

From the well-mixed condition it follows that if $P_a(z, w)$ is the PDF of fluid elements, $P_a = P$ should also satisfy eq. (3). The resulting Fokker-Plank equation for P_a can be written in two parts (Thomson, 1987):

$$aP_a = \frac{\partial}{\partial w}(\frac{1}{2}b^2 P_a) + \phi \ , \qquad (4)$$

where ϕ satisfies:

$$\frac{\partial \phi}{\partial w} = -w\frac{\partial P_a}{\partial z} \qquad (5)$$

and $\phi \to 0$ as $|w| \to \infty$. Once an expression for P_a is given, a can be found from eq. (4) if b is known. To derive b, the Lagrangian velocity structure function obtained from eq. (1) is compared to that determined according to Kolmogorov's theory of local isotropy in the inertial subrange. It follows:

$$b = \sqrt{C_0\epsilon} \ , \qquad (6)$$

where C_0 is a universal constant and ϵ is the ensemble average rate of dissipation of turbulent kinetic energy. There is some uncertainty on the value of C_0 (H.C.Rodean,1991). We use $C_0 = 2.0$, in accordance with Luhar and Britter (1989). b can also be expressed as:

$$b = \sqrt{\frac{2\overline{w^2}}{\tau}} \ , \qquad (7)$$

where τ is defined as:

$$\tau = \frac{2\overline{w^2}}{C_0\epsilon} \ . \qquad (8)$$

In homogeneous and stationary turbulence with no mean flow, τ is the Lagrangian integral time scale, while in inhomogeneous turbulence it is a rather loosely defined 'local decorrelation timescale '. It describes the instantaneous rate at which velocity fluctuations are being decorrelated.

Construction of a PDF for a daytime BL

In an unstable BL the PDF of vertical velocity at some height is positevely skewed i.e. the mode is negative and not equal to the mean. Therefore high ground-level concentrations are observed in the case of dispersion from tall stacks (Lamb, 1978). A different situation is observed in neutral stratification, where the wind shear is the only important mechanism generating turbulence. Here the PDF of vertical velocity is essential Gaussian. Small departures from gaussianity were found, but as no quantifiable systematic variation were evident, we will consider the distribution exactly Gaussian.

We construct the PDF by summing two Gaussian distributions:

$$P_a = A(z)P_A(w,z) + B(z)P_B(w,z) \ , \tag{9}$$

where:
- $A(z)$ and $B(z)$ are the probabilities of occurence of updrafts and downdrafts.
- $P_A(w,z) = \frac{1}{\sqrt{2\pi}\sigma_a} \exp\left(-\frac{(w-f\sigma_A)^2}{2\sigma_A^2}\right)$ is a Gaussian PDF of vertical velocities in updrafts with standard deviation σ_A and mean $\overline{w}_A = f\sigma_A$.
- $P_B(w,z) = \frac{1}{\sqrt{2\pi}\sigma_b} \exp\left(-\frac{(w+f\sigma_B)^2}{2\sigma_B^2}\right)$ is a Gaussian PDF of vertical velocities in downdrafts with standard deviation σ_B and mean $\overline{w}_B = f\sigma_B$.

Based on qualitative arguments, Bærentsen and Berkowicz (1984) suggested that for convective conditions $\overline{w}_A = \sigma_A$, and this was later tested by Luhar and Britter (1989). For the general case we have assumed that the mean velocity can be expressed as the product of a function f and the standard deviation. The function f is controlling the transition from a Gaussian PDF, which is obtained for $f = 0$, to a skewed one, for $f > 0$, and it is of the order of 1 for very unstable conditions.

To determine the PDF from eq.(9), 5 parameters have to be specified, i.e. A, B, σ_A, σ_B and and f. Their expressions are derived by solving the system of the 5 moment equations for n=0,1,2,3,4:

$$\overline{w^n}(z) = \int_{-\infty}^{\infty} w^n P_a(w,z)dw \ . \tag{10}$$

Substituting eq. (9) into eq. (10), and remembering that $\overline{w} = 0$, it is possible to find an implicit solution of the system:

$$
\begin{aligned}
\sigma_B &= \frac{-(f^2+1)^2\overline{w^3} + \sqrt{(f^2+1)^4\overline{w^3}^2 + 4f^2(f^2+1)(f^2+3)^2\overline{w^2}^3}}{2f(f^2+1)(f^2+3)\overline{w^2}} \\[2ex]
\sigma_A &= \frac{\overline{w^2}}{(f^2+1)\sigma_B} \\[2ex]
A &= \frac{\sigma_B}{\sigma_A + \sigma_B} \\[2ex]
B &= \frac{\sigma_A}{\sigma_A + \sigma_B} \\[2ex]
f &= \sqrt{-3 + \sqrt{\frac{6A\sigma_A^4 + 6B\sigma_B^4 + \overline{w^4}}{A\sigma_A^4 + B\sigma_B^4}}}
\end{aligned}
\tag{11}
$$

In order to determine A, B, σ_A, σ_B and f, the profiles of the moments $\overline{w^2}$, $\overline{w^3}$ and $\overline{w^4}$ must be known. Furthermore parameterizations for the rate of dissipation of turbulent kinetic energy ϵ and for the wind profile $\overline{u}(z)$ have to be specified.

Final derivation of a and b

Eq. (4) and eq. (5) can be solved using

$$dw = \frac{-(\overline{w^2}/\tau)Q + \phi}{P_a}dt + \left(\frac{2\overline{w^2}}{\tau}dt\right)^{1/2}d\mu$$

$$dz = wdt \tag{12}$$

$$dx = \bar{u}dt$$

where $d\mu$ has mean zero and variance one. Expressions for Q and ϕ are given below:

$$Q = \frac{A(w - f\sigma_A)}{(f\sigma_A)^2}P_A + \frac{B(w + f\sigma_A)}{(f\sigma_A)^2}P_B \tag{13}$$

$$
\begin{aligned}
\phi = {} & -\frac{1}{2}\left(f\sigma_A\frac{\partial A}{\partial z} + A\sigma_A\frac{\partial f}{\partial z} + Af\frac{\partial \sigma_A}{\partial z}\right)erf\left(\frac{w - f\sigma_A}{\sqrt{2}\sigma_A}\right) + \\
& +\sigma_A\left[\sigma_A\frac{\partial A}{\partial z} + A\frac{\partial \sigma_A}{\partial z}(1 + \frac{w^2}{\sigma_A{}^2}) + Aw\frac{\partial f}{\partial z}\right]P_A + \\
& +\frac{1}{2}\left(f\sigma_B\frac{\partial B}{\partial z} + B\sigma_B\frac{\partial f}{\partial z} + Bf\frac{\partial \sigma_B}{\partial z}\right)erf\left(\frac{w + f\sigma_B}{\sqrt{2}\sigma_B}\right) + \\
& +\sigma_B\left[\sigma_B\frac{\partial B}{\partial z} + B\frac{\partial \sigma_B}{\partial z}(1 + \frac{w^2}{\sigma_B{}^2}) - Bw\frac{\partial f}{\partial z}\right]P_B
\end{aligned}
\tag{14}
$$

The function f

The expression for f is determined by solving numerically the system (11) and it depends on the velocity moments $\overline{w^2}$, $\overline{w^3}$ and $\overline{w^4}$. Some additional considerations on the behaviour of f have to be made for the limits $\frac{z}{z_i} \to 0$ and $\frac{z_i}{L} \to 0$. In both of these limits, we require the PDF to be Gaussian, i.e. the 6 parameters in P_a have the following values:

$$\sigma_a = \sigma_B = \sqrt{\overline{w^2}} , \quad \overline{w}_A = \overline{w}_B = 0 , \quad A = B = \frac{1}{2} . \tag{15}$$

This is achieved by requiring that the two following limits for f are fullfilled:

$$f \to 0 \qquad \frac{z}{z_i} \to 0, \frac{z_i}{L} \to 0 \tag{16}$$

$$\frac{f}{\overline{w^3}/\overline{w^2}^{3/2}} \to \infty \qquad \frac{z}{z_i} \to 0, \frac{z_i}{L} \to 0 \tag{17}$$

Not all the choices of the set of parameterization for the vertical velocities moments will lead to the fullfillment of the above required limits. When the two limits for f are fullfilled, then the equation of the model reduces to the form:

$$dw = (-\frac{w}{\tau} + \frac{1}{2}(\frac{w^2}{\overline{w^2}} + 1)\frac{\partial \overline{w^2}}{\partial z})dt + (\frac{2\overline{w^2}}{\tau}dt)^{1/2}d\mu \tag{18}$$

which is the same as given by Thomson(1987) for vertical Gaussian turbulence with no mean flow. Figure 1 shows the function f for 3 different stabilities, $\frac{z_i}{L} = 0.4$, $\frac{z_i}{L} = 10$,

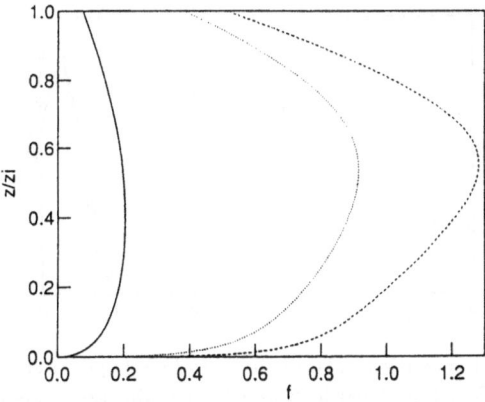

Figure 1 The function f for three different stabilities: $\frac{z_i}{L} = 0.4$ (solid line), $\frac{z_i}{L} = 10$ (dotted line) and $\frac{z_i}{L} = 50$ (dashed line).

$\frac{z_i}{L} = 50$, constructed using the profiles of the vertical velocity moments given in the next section.

PARAMETERIZATIONS OF ATMOSPHERIC TURBULENCE

To determine the model parameters, profiles of the vertical velocity moments $\overline{w^n}$, the rate of dissipation of turbulent kinetic energy ϵ and the wind profile $\overline{u}(z)$ have to be supplied. This is done by considering a set of parameterizations, which can ensure a continuous transition with stability.

For the vertical velocity variance the profile proposed in Gryning et al.(1987) is used:

$$\frac{\overline{w^2}}{w_*^2} = 1.5(\frac{z}{z_i})^{2/3}\exp(-2\frac{z}{z_i}) + (1.7 - \frac{z}{z_i})(\frac{u_*}{w_*})^2 \tag{19}$$

which is the sum of buoyancy and shear produced variances.

As the mixed layer scaling is appropriate for $\overline{w^3}$ even in the surface layer (Hunt et al., 1988), the following third order polynomial is used:

$$\frac{\overline{w^3}}{w_*^3} = 1.3(\frac{z}{z_i})(1 - 0.8\frac{z}{z_i})^2 \tag{20}$$

The coefficients of the polynomial are determined by fitting data for CBLs (Chou et al., 1986, Lenschow et al., 1980, and Willis, cited after Bærentsen and Berkowicz, 1984).

The introduction of a fourth order moment in the model is necessary in order to close the system (11). As very few measurements for $\overline{w^4}$ are available, the following expression is used:

$$\overline{w^4} = K(\frac{z}{z_i}, \frac{z_i}{L})\overline{w^2}^2 \tag{21}$$

Since $\overline{w^4} = 3\overline{w^2}^2$ for a Gaussian distribution, we require that $K(\frac{z}{z_i}, \frac{z_i}{L}) \to 3$ for $\frac{z}{z_i}$ and $\frac{z_i}{L} \to 0$. The limit for convective condition is derived from measurements of the mixed layer height and turbulence carried out during an experimental campaign at the island

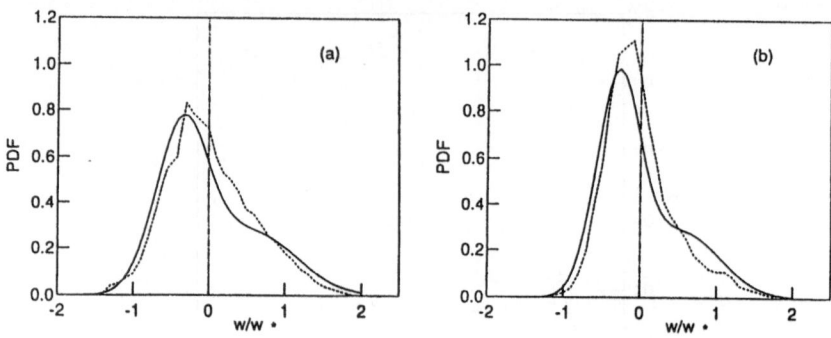

Figure 2 PDF for vertical velocities at two heights (z/z_i): (a) 0.21 (b) 0.79. Solid line: model results for $z_i/L = 50$, dashed line: Deardorff and Willis laboratory measurements.

Anholt in Kattegat, the sea between Denmark and Sweden. From this data set we use only measurements for which $\frac{w_*}{u_*} \geq 4.$, This value for $\frac{w_*}{u_*}$ lies between two typical limits, the lower limit $\frac{w_*}{u_*} \leq 3$ is typical for shear influenced CBL, while the upper limit $\frac{w_*}{u_*} \geq 5$ characterizes very strongly convective conditions (Weil, 1988). An average value for K of 3.5 is found, in good agreement with the convective value for the kurtosis of 3.46, derived from a collection of data relative to several different experiments and covering the entire BL (Lenschow et al., 1993). The resulting expression for $\overline{w^4}$ reads:

$$\overline{w^4} = (3 + \frac{1}{\pi}artan(\frac{w_*}{u_*})(\frac{z}{z_i}))\overline{w^2}^2 \tag{22}$$

which was chosen in order to fullfill the limit (17) for f.

The rate of dissipation of turbulent kinetic energy ϵ is expressed as the sum of two terms, one for convective stability (Luhar and Britter, 1989) and the other one for neutral stability (Stull, 1988):

$$\epsilon = (1.5 - 1.2(\frac{z}{z_i})^{1/3})(\frac{w_*^3}{z_i}) + \frac{u_*^3}{kz}(1 - \frac{z}{z_i})^{3/2} \tag{23}$$

Finally for the profile of mean wind speed we use the expression suggested by Sorbjan (1986):

$$\frac{\partial \overline{u}(z)}{\partial z}\frac{kz}{u_*} = \Phi_m(z/L)(1 - \frac{z}{z_i})^{2/3} \tag{24}$$

where $\Phi_m = (1 - 15z/L)^{-1/4}$ (Stull, 1988) is the surface layer expression for the non-dimensional wind.

APPLICATION AND RESULTS

As a first step PDF for vertical velocities for different stabilities and heights were constructed. PDFs for convective conditions were compared with those observed in laboratory by Deardorff and Willis (1985) and showed qualitatively a good agreement (fig.2). Both for $z \to 0$ and for $L \to \infty$ the PDFs become Gaussian. A comparison

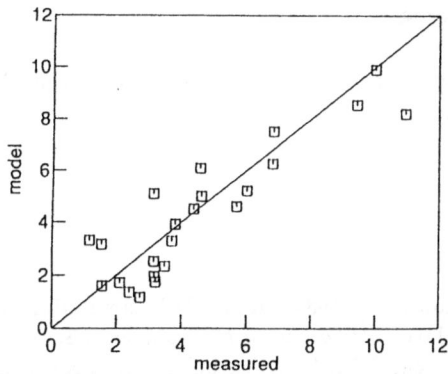

Figure 3 Comparison between concentration predicted from the model and measured ones from the Copenhagen data.

between the contour plots of the dimensionless crosswind integrated concentration for unstable conditions obtained from the model and the ones obtained from Willis and Deardorff (1978, 1981) in laboratory simulations showed that the model could well reproduce the experimental concentrations.

The model was then evaluated against the Copenhagen data, which were collected during atmospheric dispersion experiments carried out in the Copenhagen area under neutral and unstable conditions. The tracer was released without buoyancy from a tower at a height of 115 m and then collected at ground-level positions in up to three crosswind series of tracer sampling units, positioned 2-6 km from the point of release. The site was mainly residential. The meteorological measurements performed during the experiments included the three-dimensional wind velocity fluctuations at the height of release (Gryning and Lyck, 1983). Calculations are performed by releasing 10000 particles at a height of $z_s = 115m$ with an initial vertical velocity of 0. Particles are reflected on two imaginary surfaces, at $\frac{z}{z_i} = 0.001$ and at $\frac{z_i - z}{z_i} = 0.001$.

Dimensionless crosswind concentrations are calculated on a grid with a fixed vertical resolution of 10 m and an horizontal one chosen such that in one time step particles can not travel more than one grid cell.

The profiles for the velocity moments given by eq. (20), (21) and eq. (22) are modified so that the parametrized second order moment at the height of 115 m has the same value as the measured one. This is achieved by defining a new set of parametrizations for each run as:

$$\overline{w^2}_{new} = \overline{w^2}_{old} corr \ , \ \ \overline{w^3}_{new} = \overline{w^3}_{old} corr^{3/2} \ , \ \ \overline{w^4}_{new} = K \overline{w^2}_{new}^2 \ . \qquad (25)$$

where $\overline{w^n}_{old}$ are the profiles defined in eq. (20), (21), (22) and $corr = \frac{\overline{w^2}(115)_{meas}}{\overline{w^2}(115)_{old}}$, with the typical value $corr = 0.7$.

A constant time step of $\Delta t = 1sec$ was used for all the runs but one, for which it was necessary to reduce the time step to $\Delta t = 0.5sec$. For each run, measured concentrations at a height of z=2m at one or more assigned distances from the source were compared with those calculated from the model. Values of both measured and calculated concentration are plotted in figure 3. Good agreement is found, although a certain tendency of underestimating is observed.

ACKNOWLEDGMENTS

This study was supported by the Commission of the European Communities under the STEP Program.

REFERENCES

Bærentsen, J. H., and Berkowicz R., 1984, Monte Carlo simulation of plume dispersion in the convective boundary layer, *Atmospheric Environment* **18**, 701:712.

Chou, S-H., Atlas, D., and Yeh, E-N., 1986, Turbulence in a convective marine atmospheric boundary layer. *J. Atmos. Sci.* **43**, 547:564.

Deardorff, J. W., and Willis, G. E., 1985, Further results from a laboratory model of the convective planetary boundary layer. *Bound.-Layer Meteor.* **32**, 205:236.

Gryning, S.E., and Lyck E., 1984, Atmospheric dispersion from elevated sources in an urban area: comparison between tracer experiments and model calculations, *J. Climate appl. Met.* **23**, 651:660.

Gryning, S. E., Holtslag, A. A. M., Irwin, J. S., and Sivertsen, B., 1987, Applied dispersion modelling based on meteorological scaling parameters. *Atmospheric Environment* **21**, 79:89.

Hunt, J. C. R.,Kaimal, J. .C., and Gaynor J. E., 1988, Eddy structure in the convective boundary layer-new measurements and new concepts. *Q. J. R. Meteorol. Soc.* **114**, 827:858.

Lamb, R. G., 1978, A numerical simulation of dispersion from an elevated point source in the convective planetary boundary layer. *Atmospheric Environment* **12**, 1297:1304.

Lenschow, D. H.,Wyngaard, J. C., and Pennell, W. T., 1980, Mean-field and second-moment budgets in a baroclinic, convective boundary layer. *J. Atmos. Sci.* **37**, 1313:1326.

Lenschow, D. J., Mann, J., and Kristensen, L., 1993, How long is long enough when measuring fluxes and other turbulence statistics? *NCAR Technical Note* NCAR/TN-389+STR, Boulder, Colorado.

Luhar, A. K., and Britter, R. E., 1989, A random walk model for dispersion in inhomogeneous turbulence in a convective boundary layer. *Atmospheric Environment* **23**, 1911:1924.

Rodean, H. C., 1991, The universal constant for the Lagrangian structure function Phys.Fluids A 3(6), 1479-1480.

Sorbjan, Z. , 1986, On similarity in the atmospheric boundary layer. *Bound.-Layer Meteor.* **34**, 377:397.

Stull, R. B., 1988, "An Introduction to Boundary Layer Meteorology," Kluwer Academic Publishers, Dordrecht.

Thomson, D.J., 1987, Criteria for the selection of the stochastic models for particle trajectories in turbulent flows. *J.Fluid Mech.* **180**, 529:556.

Weil, J. C., 1988, Dispersion in the convective boundary layer, *in* : "Lectures on Air Pollution Modeling,"A. Venkatram and J. C. Wyngaard, ed., American Meteorological Society, Boston.

Willis, G. E., and Deardorff, J. W., 1978, A laboratory study of dispersion from an elevated source within a modeled convective planetary boundary layer, *Atmospheric Environment* **12**, 1305:1311.

Willis, G. E., and Deardorff, J. W., 1981, A laboratory study of dispersion from a source in the middle of the convectively mixed layer *Atmospheric Environment* **15**, 109:117.

DISCUSSION

B. FISHER   Have you considered trying to put buoyant source(with plume rise) into your random walk model? I realise that this is complicated, but would be well worthwhile in order to be able to treat more realistic situations.

C. TASSONE   No, I have not considered it yet.

G. GRAZIANI   From where are coming all these observations from Copenhagen data set?

C. TASSONE   The Copenghagen data set includes data from 9 different experiments, carried out under different stability conditions. For each of these experiments the tracer was collected in up to 3 ground-level positions. Therefore we have 23 different observations to compare model results with.

APPLIED MODEL OF THE HEIGHT OF THE DAYTIME MIXED LAYER INCLUDING THE CAPPING ENTRAINMENT ZONE

Ekaterina Batchvarova[1] and Sven-Erik Gryning

Risø National Laboratory
DK-4000 Roskilde, Denmark

INTRODUCTION

Dilution of air pollutants by vertical mixing is mainly confined within the atmospheric boundary layer, which consists of the fully mixed layer and the capping entrainment zone. A fairly accurate estimate of the height to which pollutants are mixed is of main importance in air pollution monitoring and assessment studies.

The turbulent structure of the whole boundary layer can be described with three length scales, being the height above the surface, z, the Monin-Obukhov length, L, and the mixed layer height, h. The height z limits the eddy size to the ground, h limits the vertical extent of the eddies and the Monin-Obukhov length reflects the height at which the buoyancy and shear stress contributions to production of turbulent kinetic energy are comparable. There exist a rich literature on the determination of L from routine meteorological measurements, while the literature on h is less and usually confined to so-called zero order models. These models of mixed layer growth assume that turbulence is sufficient to maintain a layer of nearly uniform potential temperature distribution, capped by an entrainment zone which is represented by a step-like potential temperature jump. The assumption of uniform potential temperature is well in accordance with measurements, whereas the step-like jump in potential temperature is not. The thickness of the entrainment zone is typically 30% of the mixed layer, but can reach a depth comparable to the mixed layer itself. From a practical point of view, the top of the entrainment zone represent the maximum height that pollutants can reach, being typically 10 to 50% higher than the height of the mixed layer predicted by zero-order slab models. Although the pollution concentration at the top of the entrainment zone is smaller than in the bulk of the mixed layer it is important in a number of cases, i.e. the chemical composition of the clouds and consequently long-range transport of pollutants.

In this paper we first shortly present a recently developed zero-order model of the mixed layer height. Then we describe a new parameterization of the depth of the daytime entrainment zone. It is derived from a simple energy budget, in which the eddy

---

[1]Permanent affiliation: National Institute of Meteorology and Hydrology, Bulgarian Academy of Sciences, Sofia 1784, Bulgaria

Air Pollution Modeling and Its Application X, Edited by S-V. Gryning
and M. M. Millán, Plenum Press, New York, 1994

velocity in the mixed layer penetrating the overlying stable entrainment zone is taken to be proportional to the characteristic top-down velocity scale. It gives a better ordering of experimental data, as compared to traditional parcel theory based on bottom-up velocity scale. The entrainment zone model is added to the zero-order model of the height of the mixed layer without any need for additional meteorological information.

MODEL

Mixed-layer height

As the governing equations for the zero-order model for mixed layer growth are already given in Batchvarova and Gryning (1991) the model will be presented here without derivation. The differential equation for h reads:

$$\left\{ \left(\frac{h^2}{(1+2A)\,h - 2B\kappa L} \right) + \frac{Cu_*^2 T}{\gamma g\left[(1+A)\,h - B\kappa L\right]} \right\} \left(\frac{dh}{dt} - w_s \right) = \frac{(\overline{w'\theta'})_s}{\gamma} \quad , \qquad (1)$$

where κ is the von Karman constant, u_* - friction velocity, γ - lapse rate in the free atmosphere, g/T - buoyancy parameter, dh/dt - growth rate of the mixed layer including the effect of the subsidence velocity, w_s, $(\overline{w'\theta'})_s$ - vertical kinematic heat flux at the surface and A, B and C are parameterization constants. The usual value of A is 0.2 (Tennekes, 1973; Stull, 1976a). Based on laboratory experiments Kato and Phillips (1969) reports B=2.5 and Kanta et al. (1977) found B=5. Tennekes (1973) adopted the value of 2.5, Driedonks (1982) found slightly better agreement with experimental data using B=5. Based on Zilitinkevich (1975) and Tennekes and Driedonks (1981) the value of C can be estimated as 8. In this paper we use A=0.2, B=2.5 and C=8.

The inversion strength at the top of the mixed layer, $\Delta\theta$, is given as

$$\Delta\theta = \frac{Ah - B\kappa L}{(1+2A)h - 2B\kappa L}\gamma h \quad . \qquad (2)$$

The relative contribution to the growth of the mixed layer from the mechanical and convective turbulence and the spin-up term can be deduced from Eq. (1). On the left-hand-side the first term stems from the combined effect of mechanical and convective turbulence, the second term is due to the spin-up effect. By comparing the two terms, Gryning and Batchvarova (1990) showed that the spin-up effect is important only near the ground or when the air is nearly neutrally stratified. For typical values of the meteorological parameters in the model, the contribution to the growth rate of the mixed layer due to the spin-up effect equals that from mechanical and convective turbulence at a height of approximately 50 meters. The height increases as γ decreases. In the limit $\gamma = 0$ the growth is controlled entirely by the spin-up term. The present formulation, however, is inappropriate in this limit because it disregards the effect of the Coriolis parameter, which limits the growth of the mixed layer beyond the surface layer. The contributions from mechanical and convective turbulence are equal when

$$(1+2A)h = -2B\kappa L \quad . \qquad (3)$$

With the values of A and B used here this corresponds to

$$h \simeq -1.4\,L \quad . \qquad (4)$$

The growth of the mixed layer is controlled mainly by convective turbulence when its height is larger than $-1.4\,L$, and by mechanical turbulence when it is smaller.

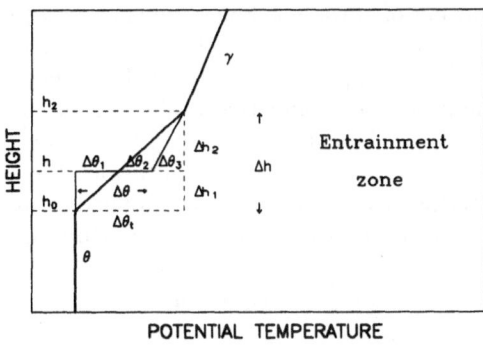

Figure 1. Idealized vertical profiles of the potential temperature used in the models. The zero-order model is illustrated by a thin and the first-order model by a thick solid line.

By setting A=0, B=0, C=0 and w_s=0 Eq.(1) reduces to the encroachment model for the growth of the mixed-layer

$$\frac{dh}{dt} = \frac{(\overline{w'\theta'})_s}{h\gamma} \qquad . \qquad (5)$$

Entrainment zone

The entrainment zone is confined between the well-mixed layer and the stably stratified free atmosphere aloft. It is defined in a horizontally averaged sense. The layer is a result of the interaction of turbulent eddies at the top of the mixed layer with the stably stratified air above. Air in the entrainment zone is composed of basically three types: unmixed free atmosphere air, unmixed air from the mixed layer and a mixture of the two (Crum and Stull, 1987). Overshooting of thermals causes entrainment of parcels of free air between the eddies into the mixed layer. Atop the thermal the entrainment process has been related to Kelvin-Helmholtz waves (Rayment and Readings, 1974). Lateral entrainment along the sides of the thermal was observed by Crum *et al.* (1987).

Clouds form when the top of the entrainment zone reaches the saturation level for mixed layer air. When the clouds become positively buoyant due to the latent heat release, they start to vent mixed layer air into the free atmosphere, and thus can be expected to alter the characteristics of the entrainment zone and mixed layer considerable. This aspect will not be dealt with in this paper.

The scheme adopted here for the entrainment zone, a so-called first-order model (Deardorff, 1979), is portrayed in Fig. 1. The turbulence is assumed to be sufficiently intense to maintain a uniform distribution of potential temperature within the mixed layer up to the height h_0. The air immediately above constitutes an entrainment zone, of thickness Δh, consisting of very stably stratified but turbulent air with a potential temperature change of $\Delta \theta_t$. Above the entrainment zone in the free atmosphere the air is also stable but assumed free of turbulence and with generally less stratification. h constitutes the height of the mixed layer when the temperature jump is assumed to be infinitesimally thin. It is introduced in the entrainment zone in such a way that the heat deficit is conserved and is the height of the mixed layer that can be derived from zero-order jump models such as Batchvarova and Gryning (1991).

The rate of consumption of turbulent kinetic energy associated with the entrainment of free atmosphere air is related to the net rate of turbulent kinetic energy production in the mixed layer. An expression for the balance between the consumption of energy by the entrainment process and the production of turbulent kinetic energy is derived by integrating the turbulent kinetic energy equation over the depth of the mixed layer, and then parameterizing the result. A considerable literature exists on the parameterization of the terms in this equation. In simplified form, the turbulent kinetic energy equation can be written in terms of top-down bottom-up scaling (Gryning and Batchvarova, 1994)

$$- W_*^3 = A w_*^3 + B u_*^3 \quad , \tag{6}$$

where

$$w_* \equiv (\frac{g}{T}(\overline{\theta' w'})_s h)^{1/3} \tag{7}$$

is the usual convective or bottom-up velocity scale (Deardorff, 1970) and

$$W_* \equiv (\frac{g}{T}(\overline{\theta' w'})_h h)^{1/3} \tag{8}$$

is the analogous top-down velocity scale. As can be seen from Eq. (6) the effect of both convective and mechanical turbulence is contained in W_*. Its use is therefore not limited to convective conditions; it can also serve as a scaling parameter when mechanical turbulence is important for the growth of the mixed layer.

The vertical distance an air parcel penetrates into the stable entrainment zone is found by equating the initial kinetic energy of the air parcel at the top of the mixed layer with its maximum potential energy (Stull, 1976b; Deardorff et al., 1980). Setting the overshoot distance proportional to Δh, and with the top-down velocity scale, W_*, as the natural choice for the characteristic velocity of the air parcel leads to

$$W_*^2 \propto \frac{g}{T} \frac{\Delta \theta_t}{\Delta h} \Delta h^2 \quad , \tag{9}$$

which is applicable in neutral as well as convective mixed layer conditions. By use of Eq. (8) this can be expressed as

$$\frac{gh}{T} (\overline{\theta' w'})_h \propto (g/T \; \Delta \theta_t \; \Delta h)^{3/2} \quad . \tag{10}$$

For the left hand side of Eq. (10), representing buoyancy destruction in the entrainment zone, we adopt the usual parameterization (Mahrt, 1979)

$$- (\overline{\theta' w'})_h = C_\theta \; \Delta \theta_t \; w_e \quad , \tag{11}$$

where C_θ is the constant of proportionality and w_e is the entrainment rate

$$w_e = \frac{dh}{dt} - w_s \quad . \tag{12}$$

Since $\Delta \theta_t$ is the potential temperature difference across a layer of finite thickness, C_θ is probably less than unity owing to the heat storage effect (Mahrt, 1979; Deardorff, 1979), but not much less according to Kamada (1988). Inserting Eq. (11) into Eq. (10) gives

$$\frac{gh}{T} \Delta \theta_t \; w_e \propto (g/T \; \Delta \theta_t \; \Delta h)^{3/2} \quad . \tag{13}$$

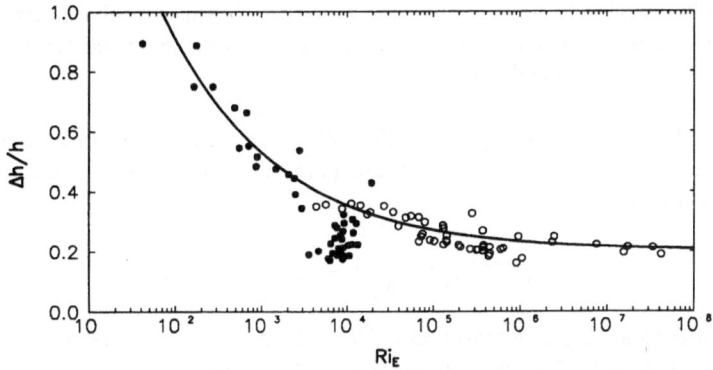

Figure 2. Normalized depth of the entrainment zone $\Delta h/h$ plotted as function of the entrainment Richardson number Ri_E (Gryning and Batchvarova, 1994). Data from Deardorff *et al.* (1980) are indicated by (o) and those of Boers and Eloranta (1986) of which only clear sky conditions are considered by (•). The solid line represents $\Delta h/h = 3.3 Ri_E^{-1/3} + 0.2$.

By defining an entrainment Richardson number,

$$Ri_E = \frac{(g/T)\,\Delta\theta_t\,h}{w_e^2} \quad, \tag{14}$$

this can be written as

$$\frac{\Delta h}{h} \propto (Ri_E)^{-1/3} \quad. \tag{15}$$

This constitutes the basis for the parameterization of $\Delta h/h$.

The data–sets of Boers and Eloranta (1986) and Deardorff *et al.* (1980) were used to determine the parameterization constants. From Figure 2 it can be seen that when the normalized entrainment zone depth, $\Delta h/h$, is plotted as a function of the entrainment Richardson number, Ri_E, the data orders rather nicely. The data–sets of Boers and Eloranta (1986) and Deardorff *et al.* (1980) supplement each other to cover a range of Ri_E from 10^2 to 10^8 with overlap among the data–sets around $Ri_E \sim 10^4$. The data of Deardorff *et al.* (1980) suggest an asymptotic limit for $\Delta h/h$ of the order of 0.2. In Figure 2 the solid line

$$\frac{\Delta h}{h} = \frac{3.3}{Ri_E^{1/3}} + 0.2 \tag{16}$$

combines the empirical asymptotic limit for $\Delta h/h$ of 0.2 with the -1/3 power law of Ri_E suggested in Eq. (15).

Combined model

Applying geometrical considerations, Figure 1, a relation between $\Delta\theta$ and $\Delta\theta_t$ can be obtained

$$\frac{\Delta\theta}{\gamma\,\Delta h} - \sqrt{\left(\frac{\Delta\theta}{\gamma\,\Delta h}\right)^2 + \frac{1}{4}} = \frac{1}{2} - \frac{\Delta\theta_t}{\gamma\,\Delta h} + \sqrt{\frac{\Delta\theta_t}{\gamma\,\Delta h}\left(\frac{\Delta\theta_t}{\gamma\,\Delta h} - 1\right)} \tag{17}$$

and the models described above can be combined to obtain the mixed layer height and the entrainment zone depth with the meteorological input required for the zero-order model of the mixed layer height.

To begin, the entrainment rate, w_e, and mixed layer height, h are derived from Eqs. (12) and (1). $\Delta\theta$ is calculated from Eq. (2). Then an iterative procedure is used to obtain Δh and $\Delta\theta_t$ from Eqs. (14), (16) and (17). The iterations can be avoided using an approximation to Eq. (16)

$$\frac{\Delta h}{h}(1 + 0.25 Ri_s^{-1/3}\frac{h}{\Delta\theta}\gamma) = 3.3 \; Ri_s^{-1/3} + 0.2 \quad , \tag{18}$$

where Ri_s is identical to the entrainment Richardson number but is composed with $\Delta\theta$ instead of $\Delta\theta_t$. Knowing Δh and $\Delta\theta$ we obtain Δh_1 from

$$\frac{\Delta h_1}{\Delta h} = \frac{1}{2} + \frac{\Delta\theta}{\gamma \, \Delta h} - \sqrt{(\frac{\Delta\theta}{\gamma \, \Delta h})^2 + \frac{1}{4}} \quad . \tag{19}$$

Then the height to which the air is well mixed is $h_0 = h - \Delta h_1$ and the upper limit of the entrainment zone is $h_2 = h_0 + \Delta h$.

DATA COMPARISON

Boers et al. (1984) and Boers and Eloranta (1986) report data on the atmospheric entrainment zone, collected during the Central Illinois Rainfall Chemistry Experiment, so–called CIRCE experiment. The structure of the entrainment zone was obtained by a lidar, scanning from the surface to an elevation of 60° in about one minute. From the back scatter intensity the outer edge of the entrainment zone was identified out to a typical horizontal distance of 6 km. With this near-instantaneous picture, they assumed the height of the entrainment zone to be the point at which 90% of the horizontal area was occupied by clean air from the free atmosphere, the mixed layer height h with 50%, and the lower edge height with only 10% clean air. Although the experiments were carried out under rather convective conditions, an influence on the structure of the entrainment zone owing to wind-shear in the mixed layer cannot be excluded. Typically, $w_*=1.5 \; ms^{-1}$ and $u_*=0.3 \; ms^{-1}$. The entrainment velocity was not included in the data–set but was derived from the growth rate of the mixed layer, dh/dt, taking into account the effect of subsidence.

Measurements of the mixed layer height and entrainment zone thickness over the sea were carried out during an experimental campaign at the island Anholt (56° 43′ N, 11° 31′ E) in Kattegat, the sea between Denmark and Sweden. The entrainment zone was identified from temperature, humidity, wind-speed and direction profiles obtained from special launches of radiosondes (type Vaisala RS-80) every three hours. The measurements were performed with a frequency of 0.5 Hz and a radiosonde ascent velocity of ∼2.5 ms^{-1}, which corresponds to a vertical resolution of ∼5 m. The time constants for the temperature and humidity sensors were 2 and 1 s., respectively. Measurements were only taken for those wind-directions for which the water fetch exceeded 200 km. The radio soundings were supplemented by conventional ground based measurements of heat flux and friction velocity. Typically, $w_*=0.5 \; ms^{-1}$ and $u_*=0.2 \; ms^{-1}$.

A total of 6 experiments were available for the model validation - 4 from CIRCE and 2 from Anholt. The agreement between measurements and model simulations is good, Figure 3.

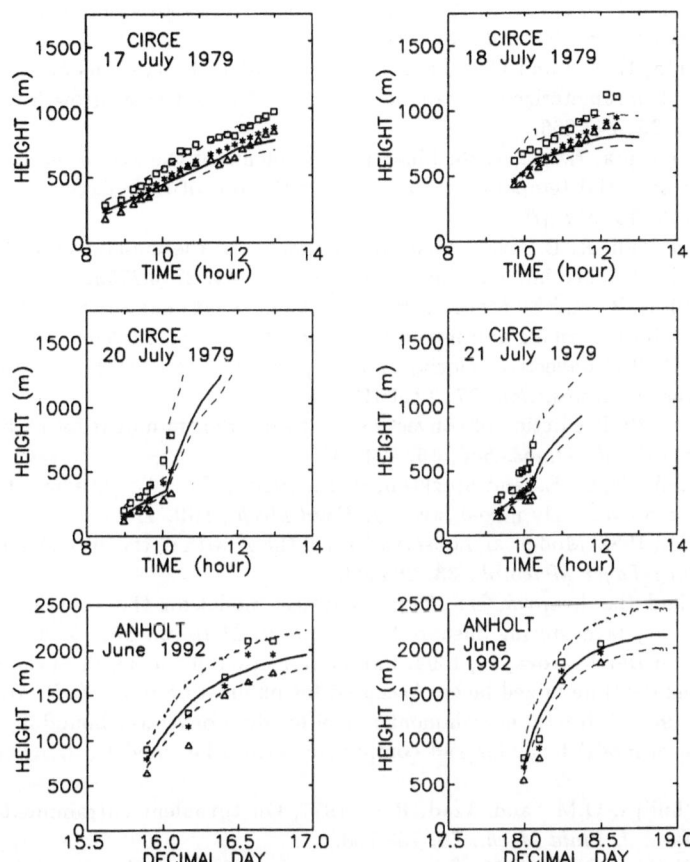

Figure 3. Measurements and simulation of the mixed layer height and the entrainment zone. The full line represents the simulated mixed layer height and the dashed lines show the predicted extend of the entrainment zone. Measurements of the mixed layer height (∗) and the upper (□) and lower (△) limits of the entrainment zone are also shown.

Acknowledgments

The study was supported by the the Danish Environmental Research Program 1992-1996 and the CEC mobility action under the 1992 Program for Cooperation in Science and Technology with Central and Eastern European Countries.

REFERENCES

Batchvarova, E., and Gryning, S. E., 1991, Applied model for the growth of the daytime mixed layer, *Boundary-Layer Meteorol.*, **56**, 261:274.

Boers, R., Eloranta, E. W., and Coulter, R. L., 1984, Lidar observations of mixed layer dynamics: Test of parameterized entrainment models of mixed layer growth rate, *J. Clim. Appl. Meteor.*, **23**, 247:266.

Boers, R., and Eloranta, E. W., 1986, Lidar measurements of the atmospheric entrainment zone and the potential temperature jump across the top of the mixed layer, *Boundary-Layer Meteorol.*, **34**, 357:375.

Crum, T. D., and Stull, R. B., 1987, Field measurements of the amount of surface layer air versus height in the entrainment zone, *J. Atmos. Sci.*, **44**, 2743:2753.

Crum, T. D., Stull, R. B., and Eloranta, E. W., 1987, Coincident lidar and aircraft observations of entrainment into thermals and mixed layers, *J. Clim. Appl. Meteor.*, **26**, 774:788.

Deardorff, J. W., 1970, Convective velocity and temperature scales for the unstable planetary boundary layer, *J. Atmos. Sci.*, **27**, 1211:1213.

Deardorff, J. W., 1979, Prediction of convective mixed layer entrainment for realistic capping inversion structure, *J. Atmos. Sci.*, **36**, 424:485.

Deardorff, J. W., Willis, G. E., and Stockton, B. H., 1980, Laboratory studies of the entrainment zone of a convectively mixed layer, *J. Fluid Mech.*, **100**, 41:64.

Driedonks, A.G.M., 1982, Models and observations of the growth of the atmospheric boundary layer, *Boundary–Layer Meteorol.*, **23**, 283:306.

Gryning, S. E., and Batchvarova, E., 1990, Analytical model for the growth of the coastal internal boundary layer during onshore flow, *Q. J. R. Meteorol. Soc.*, **116**, 187:203.

Gryning, S. E., and Batchvarova, E., 1994, Parameterization of the depth of the entrainment zone above the daytime mixed layer, Accepted for publication in *Q. J. R. Meteorol. Soc.*

Kamada, R. F., 1988, A fractal entrainment model for dry convective boundary layers. Part II: Discussion of model behavior and comparison with other models, *J. Atmos. Sci.*, **45**, 2375:2383.

Kantha, L.H., Phillips, O.M., and Azad, R.S., 1977, On turbulent entrainment at a stable density interface, *J. Fluid Mech.*, **79**, 753:768.

Kato, H., and Phillips, O.M., 1969, On the penetration of a turbulent layer into stratified fluid, *J. Fluid Mech.*, **37**, 643:655.

Mahrt, L., 1979, Penetrative convection at the top of a growing boundary layer *Q. J. R. Meteorol. Soc.*, **105**, 469:485.

Rayment, R., and Readings, C. J., 1974, A case study of the structure and energetics of an inversion, *Q. J. R. Meteorol. Soc.*, **100**, 221:223.

Stull, R.B., 1976a, The energetics of entrainment across a density interface, *J. Atmos. Sci.*, **33**, 1260:1267.

Stull, R. B., 1976b, Internal gravity waves generated by penetrative convection. *J. Atmos. Sci.*, **33**, 1279:1286.

Tennekes, H., 1973, A slab model for the dynamics of the inversion above a convective boundary layer, *J. Atmos. Sci.*, **30**, 558:567.

Tennekes, H., and Driedonks, A.G.M., 1981, Basic entrainment equations for the atmospheric boundary layer, *Boundary-Layer Meteorol.*, **20**, 515:531.

Zilitinkevich, S. S., 1975, Comment on a paper by H. Tennekes, *J. Atmos. Sci.*, **32**, 991:992.

DISCUSSION

S. J. ROBERTO: Have you made any comparison between entrainment heat flux and the sensible heat flux?

E. BATCHVAROVA: The model works very well with an entrainment heat flux of 20 % of the sensible heat flux at the ground, which corresponds to the value typically found in experiments (see table 3 in Zilitinkevich, 1991).

APPLICATIONS OF THE MIXED SPECTRAL FINITE-DIFFERENCE (MSFD) MODEL AND ITS NONLINEAR EXTENSION (NLMSFD) TO WIND FLOW OVER BLASHAVAL HILL

John L. Walmsley,[1] Wensong Weng,[1] Stephen R. Karpik,[2] Dapeng Xu,[3] and Peter A. Taylor[3]

[1]Atmospheric Environment Service
4905 Dufferin Street
Downsview, Ontario M3H 5T4 Canada

[2]Department of Mechanical Engineering
University of Toronto
5 King's College Road
Toronto, Ontario M5S 1A4 Canada

[3]Department of Earth and Atmospheric Science
York University
4700 Keele Street
North York, Ontario M3J 1P3 Canada

INTRODUCTION

In Walmsley et al. (1990), a comparison was made among four models for surface-layer flow in complex terrain. The models were applied to Blashaval Hill, Scotland, the site of a 1982 field experiment. In the present study we compare the results of two additional models for the same terrain.

The first model, MSFD-PC, is a recent implementation of the three-dimensional MSFD model (Beljaars et al., 1987; Karpik, 1988) on a personal computer (PC). The second is a two-dimensional nonlinear extension of MSFD that was first introduced by Xu and Taylor (1992) and called NLMSFD.

In this study both MSFD and NLMSFD are applied to Blashaval and results are compared with those of the MS-Micro model in Walmsley et al. (1990). In the following sections, the specific features of each model are described, information on terrain data is given, and the model results are presented.

Air Pollution Modeling and Its Application X, Edited by S-V. Gryning
and M. M. Millán, Plenum Press, New York, 1994

MODEL DESCRIPTIONS

MS-Micro

This model is the PC version of MS3DJH/3R, the latest in a series of models developed by scientists at AES, Canada (Walmsley et al., 1982; Taylor et al., 1983; Walmsley et al., 1986). The model is based on Mason and Sykes (1979) modifications and 3D extension of the original Jackson and Hunt (1975) 2D theory. In this model, wavenumber-dependent scaling is introduced and a blending method is employed between the inner- and outer-layer solutions. The model can also treat variable surface roughness. Fourier transformation in the horizontal directions, when combined with simple mixing-length turbulence closure and other assumptions, permits an analytic solution in the vertical.

MSFD-PC

MSFD is a linear Mixed Spectral Finite Difference model for the study of neutral surface-layer flow over topography and surface roughness changes. The PC version of MSFD is based on a mainframe model developed by Beljaars et al. (1987) and subsequently improved by Karpik (1988). Like MS-Micro, the model uses the spectral approach in the two horizontal directions. Unlike MS-Micro, however, finite-differencing in the vertical combines the simplicity and computational efficiency of the linear method with a choice of mixing-length, E-ϵ or algebraic-stress closures. To ensure a rational distribution of grid points in the vertical direction, a wavenumber-dependent, log-linear coordinate transformation is used. The set of difference equations is solved by using a block LU factorized algorithm that is robust, accurate and an absolutely stable method.

NLMSFD

This model is a nonlinear extension of MSFD, initially developed by Xu and Taylor (1992). Subsequent modifications by the original authors have improved the model convergence. Although linear models (e.g., MS-Micro, MSFD and others) provide good estimates of mean flow quantities on the upstream side of hills and on hill crests even for moderately steep complex terrain, prediction on the lee side of the hill is often rather poor. Therefore, it is of considerable importance to understand the nonlinear effects on airflow over inhomogeneous topography and desirable to improve the model by inclusion of nonlinear terms.

In this model, all the nonlinear terms, neglected in MSFD-PC model, were retained and treated as additional source terms. Horizontal diffusion terms are also incorporated. As in MSFD-PC, the governing equations are transformed from physical to spectral space using the Fast Fourier Transform in the horizontal coordinates and a finite-difference solution method is used in the vertical direction. Numerical solutions are computed iteratively in spectral variables; solutions in physical space are then obtained by using the inverse Fourier transform. At each iteration the nonlinear terms are evaluated in physical space and then transformed back to the spectral space. A wavenumber-independent logarithmic coordinate transformation is used (in contrast to MSFD-PC) to minimized the number of grid points in the vertical direction, while avoiding unwanted truncation error. NLMSFD is presently implemented only for 2D topography.

The first iteration of NLMSFD is the linear solution of MSFD-PC, which gives a reasonably accurate prediction for topography of gentle slope. As expected, convergence is reached very rapidly when the terrain perturbation is small, but more iterations and an under-relaxation factor are needed as terrain slope increases.

TERRAIN DATA

Blashaval Hill is located on North Uist, Scotland. For details, see Mason and King (1985). Since our main interest is the difference among models, a single source of terrain data was used by all models (see Figure 1). For all 2D runs, the input terrain data is the cross-section along 210°-30° of the original topographic data (for position, see Figure 1; for cross-section, see Figure 2b, where heights above sea level have been reduced by 10 m, the approximate height of the upstream terrain). The maximum slope is rather large (about 0.38) and is located about 190 m upstream of the summit. The terrain is quite smooth and, as in Walmsley et al. (1990), a uniform surface roughness length of 1 cm is used here in all calculations.

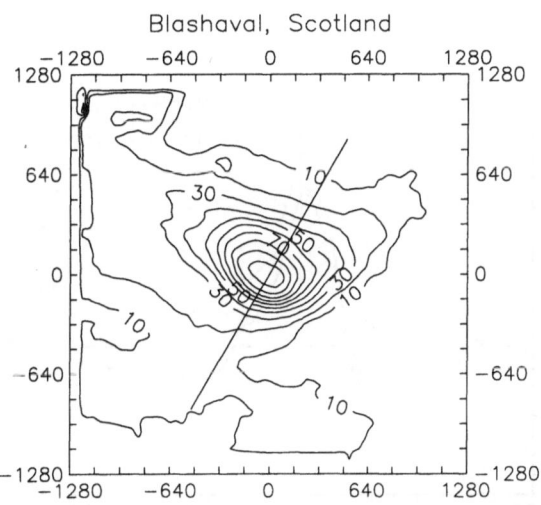

Figure 1. Blashaval Hill, heights above sea level. The x and y coordinates are distances (m) from the summit. Straight line indicates the position of cross-section results and the 2D topography used for some model calculations.

RESULTS AND DISCUSSION

Figure 2a shows normalized wind speed (i.e., wind speed divided by upstream speed at the same height above terrain) profiles between 2 and 64 m at the summit of Blashaval Hill for a wind direction of 210°. Figure 2b shows a cross-section at 8 m above the ground for the same experiment, with direction 210° oriented to the left. All three runs are obtained from MS-Micro. (The topographic cross-section is superimposed.) The results of two runs with different horizontal resolutions are almost identical; so for all subsequent comparisons we use 128 horizontal grid points. As expected, MS-Micro predicts larger wind speeds for flow over 2D terrain than 3D, since air can flow around the hill in the 3D case. The model value at the summit for the 3D case is 1.63, slightly lower than the mean observed value of 1.75, but within the range of variation of 1.55 to 1.95 (Mason and King, 1985; Walmsley et al., 1990).

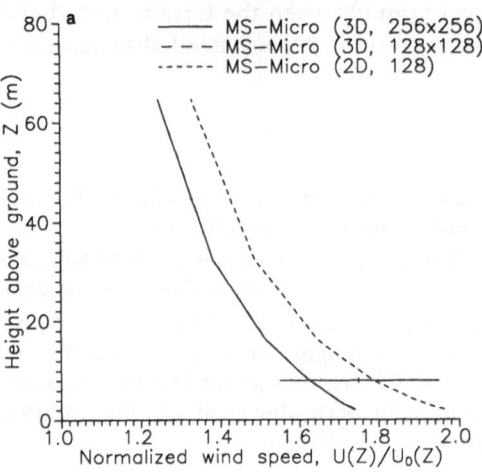

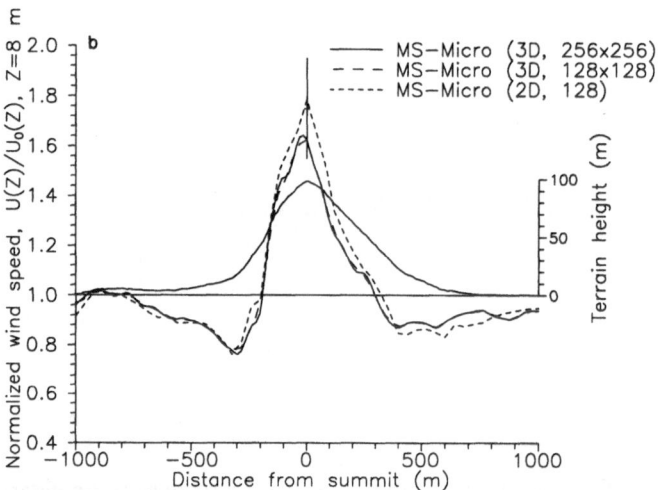

Figure 2. MS-Micro model results for Blashaval: 2D and 3D applications with two grid resolutions. Domain size: 5120 x 5120 m. Field data are from Mason and King (1985); mean value and variations in 1-h observations for a 20° wind-direction band are indicated. Only the 3D model results should be compared with the data. (a) Normalized wind speed profiles at the summit. (b) Cross-sections at 8 m above terrain along the line shown in Figure 1. Also shown is the topographic cross-section above the upstream height of 10 m.

Figures 3a and 3b show comparisons between the results of the MS-Micro and MSFD models. These are all 2D runs with 128 horizontal grid points. Figure 3a shows very good agreement above about 15 m among all three experiments. Close to the surface, however, E-ε closure of MSFD predicts slightly larger wind speeds than either MSFD with mixing-length closure or the MS-Micro model. This can also be seen in Figure 3b, which shows a cross-section of the normalized wind speed at 8 m above the ground. The E-ε closure of MSFD model gives slightly smaller wind-speed perturbations upstream, larger values near the summit and larger wind-speed reductions on the lee side of the hill.

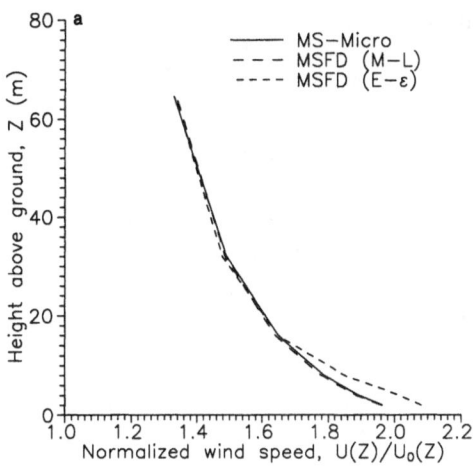

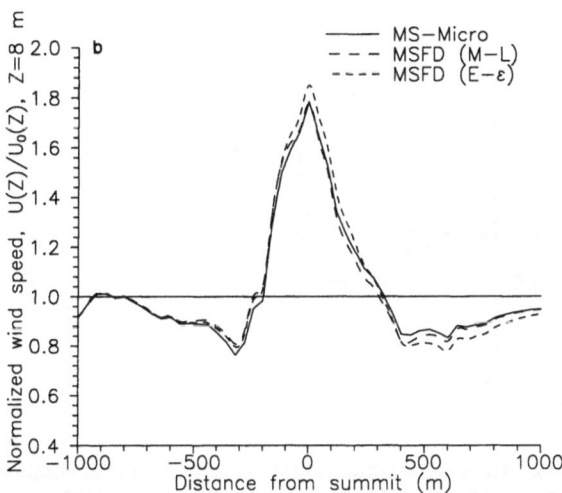

Figure 3. Same as Figure 2, except 2D MS-Micro and MSFD applications with 128 grid points and different turbulence closure schemes.

Results of the NLMSFD, MSFD and MS-Micro comparisons are shown in Figures 4a and 4b. The results of NLMSFD are quite different from those of MSFD and MS-Micro. At the summit of the hill, NLMSFD predicts smaller wind-speed values, compared with MSFD and MS-Micro (Figure 4a). Upstream and downstream of the summit, NLMSFD predicts smaller values than the other two models, especially on the lee side of the hill, confirming the importance of nonlinear effects.

Sensitivity of the NLMSFD results to horizontal domain size was investigated. Results with a domain of 5120 m, used for the other models, were poor. Increasing the domain to 8000 m gave some improvement. It was found, however, that the domain needed to be increased to 12000 m to obtain reliable results, shown here; tests with a domain of 16000 m were almost identical.

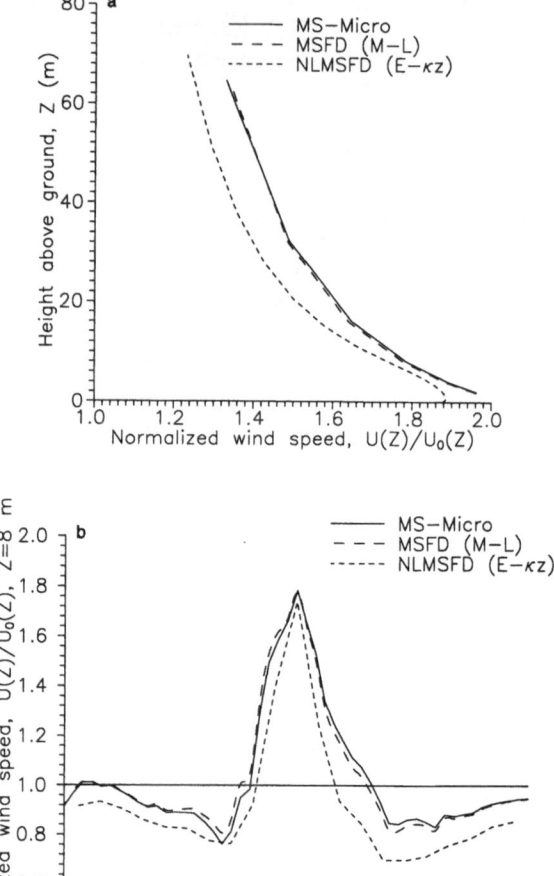

Figure 4. Same as Figure 2, except 2D MS-Micro, MSFD and NLMSFD applications with 128 grid points and different closure schemes. The NLMSFD run had a domain size of 12000 m. See text for explanation.

Results for E-ϵ closure are shown in Figures 5a and 5b, together with the NLMSFD E-κz results from Figure 4. The effects of nonlinearity are evident. The NLMSFD model predicts smaller summit wind-speed perturbations than MSFD, with the E-ϵ closure producing lower values than E-κz at all levels except in a thin layer near 3 m above terrain (Figure 5a). NLMSFD also gives lower speeds than MSFD at 8 m above terrain along the length of the cross-section, with E-ϵ yielding lower values than E-κz, especially on the lee side of the hill. Field data from the Askervein experiment (Beljaars et al., 1987, Figure 10) would suggest that, of the models tested here, NLMSFD E-ϵ has the best potential to simulate 3D flow over hills, including the lee side.

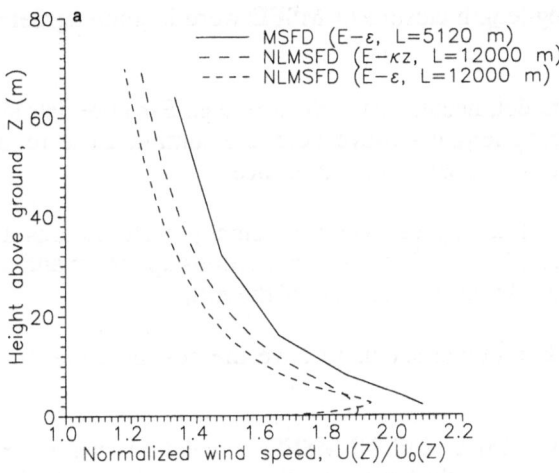

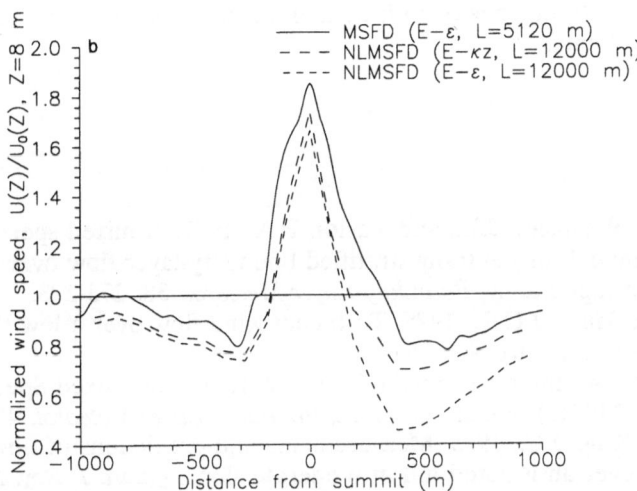

Figure 5. Same as Figure 2, except 2D MSFD and NLMSFD applications with different closure schemes and domain sizes.

SUMMARY

In this preliminary study comparing the MS-Micro, MSFD and NLMSFD models for surface-layer flow over complex terrain, we found that:

i) summit wind speeds over a two-dimensional cross-section of terrain were higher than those over the corresponding three-dimensional terrain;

ii) results of mixing-length closure of MSFD were in good agreement with those of MS-Micro;

iii) for the linear model, higher-order closure (e.g., E-ϵ) predicts slightly larger wind speed than mixing-length closure near the summit close to the surface, while higher up there is no significant difference;

iv) the effect of nonlinearity reduces the velocity perturbations at the hill summit area and produces larger wind-speed reductions upstream and downstream of the summit, particularly on the lee side of the hill;

v) the NLMSFD E-ϵ model seems to have the best potential to simulate 3D flow over hills.

Since all of the MSFD and NLMSFD calculations were for flow over a 2D topographic cross-section of Blashaval, no direct comparison with field observations is valid. We are, however, planning more detailed comparisons between MSFD and NLMSFD, including turbulence quantities, and intend to evaluate 3D flow results with the Askervein field data.

REFERENCES

Beljaars, A.C.M., Walmsley, J.L., and Taylor, P.A., 1987, A mixed spectral finite-difference model for neutrally stratified boundary-layer flow over roughness changes and topography, *Boundary-Layer Meteorol.* 38: 273-303.

Jackson, P.S., and Hunt, J.C.R., 1975, Turbulent wind flow over a low hill, *Quart. J. Roy. Meteorol. Soc.* 101: 929-955.

Karpik, S.R., 1988, An improved method for integrating the Mixed Spectral Finite Difference (MSFD) model equations, *Boundary-Layer Meteorol.* 43: 273-286.

Mason, P.J., and King, J.C., 1985, Measurements and predictions of flow and turbulence over an isolated hill of moderate slope, *Quart. J. Roy. Meteorol. Soc.* 111: 617-640.

Mason, P.J., and Sykes, R.I., 1979, Flow over an isolated hill of moderate slope, *Quart. J. Roy. Meteorol. Soc.* 105: 383-395.

Taylor, P.A., Walmsley, J.L., and Salmon, J.R., 1983, A simple model of neutrally stratified boundary-layer flow over real terrain incorporating wavenumber-dependent scaling, *Boundary-Layer Meteorol.* 26: 169-189.

Walmsley, J.L., Salmon, J.R., and Taylor, P.A., 1982, On the application of a model of boundary-layer flow over low hills to real terrain, *Boundary-Layer Meteorol.* 23: 17-46.

Walmsley, J.L., Taylor, P.A., and Keith, T., 1986, A simple model of neutrally stratified boundary-layer flow over complex terrain with surface roughness modulations (MS3DJH/3R), *Boundary-Layer Meteorol.* 36: 157-186.

Walmsley, J.L., Troen, I., Lalas, D.P., and Mason, P.J., 1990, Surface-layer flow in complex terrain: comparison of models and full-scale observations, *Boundary-Layer Meteorol.* 52: 259-281.

Xu, D., and Taylor, P.A., 1992, A non-linear extension of the Mixed Spectral Finite Difference model for neutrally stratified boundary-layer flow over topography, *Boundary-Layer Meteorol.* 59: 177-186.

DISCUSSION

M. MILLAN   Are these results for neutral stratification?

J.L. WALMSLEY   Yes, they are. Conditions during the collection of the Blashaval data were near neutral. We have begun work on extending the model to stable stratification (see Karpik and Walmsley, 1992 in Air Pollution Modeling and Its Application, IX) and plan to continue that work when we have completed improvements to the neutral version.

M. MILLAN   Are your models applicable to non-isolated terrain?

J.L. WALMSLEY   Yes, they are. Because of the Fourier transformation in the horizontal coordinates, we have periodic lateral boundary conditions. This gives us two options: (1) we could handle a series of two-dimensional ridges or a region of three- dimensional "rolling hills" or (2) we could embed the feature of interest within a larger area where the terrain is smoothed down to a uniform height at the edges. The latter option is similar to what would be done on a physical terrain model in a wind tunnel. Nevertheless, there is no requirement that the terrain be an isolated feature. It just happened that Blashaval was an isolated hill.

IMPACT OF A FULLY SPECTRAL MICROPHYSICAL SCHEME UPON GAS SCAVENGING IN A MESOSCALE METEOROLOGICAL MODEL

N. Huret, N. Chaumerliac and S. Cautenet

LAMP URA CNRS 267, Université Blaise Pascal
24, avenue des Landais, 63177 AUBIERE Cedex, France

1- INTRODUCTION

The process of wet deposition is the main mechanism for removing trace gases and aerosol particles from the atmosphere (Iribarne and Cho (1989)). They have shown that modeling cloud chemistry and wet deposition is a very complex task because dynamical and microphysical processes are linked together and interact strongly with chemical processes. Dynamical processes are responsible for the transport of water, gas and aerosol particles, but the chemical concentrations in aqueous phase are determined by the microphysical history of the droplets, including condensational and coalescence events. Modeling studies such as those of Hales (1989) and Ferretti et al. (1992) use a bulk microphysical parameterization like the Kessler one (1969), which treats droplet spectra as a single well mixed ideal solution of droplets that maintain vapor liquid equilibrium with surrounding trace gases. However, experimental results of Noone et al. (1988) and theoretical results obtained by Flossmann et al. (1987) show a dependency between drop size and the chemical concentration inside the drops.

In this paper we examine the impact of three different microphysical parameterizations on gas scavenging and wet deposition in a mesoscale model. The parameterizations range from the very simple to the very complex, and include the bulk approach developed by Kessler (1969), a semi-spectral approach developed by Berry and Reinhardt (1974a, b) and a complete spectral approach developed by Le Cam and Isaka (1989).In order to simplify the explanations in the rest of the paper we designate results obtained by the Kessler scheme as "K", Berry and Reinhardt as "BR" and spectral as "SP". Since we are focussing on gas scavenging and wet deposition, the condensation stage on aerosol particles is neglected as a first step in all three parameterizations.

The mesoscale framework of this study enables us to take into account interactions between microphysical and dynamical processes. We will examine the propensity of the three parameterizations to partition gases between air, cloud water and rainwater, and for the spectral scheme, the distribution of chemical species inside the raindrops. Two different gaseous species, one very soluble and one less soluble are transported, absorbed and desorbed in both cloud droplets and raindrops. Gas scavenging by the three microphysical schemes is then discussed in the framework of a two-dimensional mountain wave scenario that couples dynamics, microphysics and aqueous phase chemistry.

2 - RESULTS ON THE MOUNTAIN WAVE SCENARIO

A complete description of the model and of the microphysical schemes used in this paper can be found in Huret et al.(1994) and are just summarized here. A mountain wave scenario has been simulated with the mesoscale model developed by Nickerson et al.(1986) . The microphysical schemes, that have been coupled with this meteorological model are the SP scheme which explicitly resolves the stochastic coalescence equation and is based on the work of Le Cam and Isaka (1989) and the semi-spectral BR scheme (Richard and Chaumerliac, 1989).

The treatment of rainwater in the two parameterizations differs with respect to the shape of the droplet spectra, the number of prognostic variables and the representation of the microphysical processes. Then, mass transfer limitations are considered following Schwartz (1986) for cloud and raindrops considering the mean diameter of cloud droplets for BR and SP schemes, the mean diameter of raindrops for BR, and the predicted diameter of raindrops range from 60 mm to 5 mm for SP.

A series of tests on the chemical module has been carried out to determine the efficiency of drops, as a function of their diameter, in reaching equilibrium between gas and aqueous phase at 25 °C for both a highly soluble gas and a less soluble one. The results obtained have been compared to the theoretical results obtained by Iribarne and Cho (1989). The evolution of the concentration in aqueous phase by absorption of gas are presented for three different diameters of drops $D = 20$ mm, $D = 0.2$ mm and $D = 1$ mm. Characteristic times for the attainment of equilibrium between gas phase and aqueous phase obtained by Iribarne are summarized in Table 1. We can consider that the characteristic time to reach solubility equilibrium can be neglected when it is less than the microphysical time step $\Delta t = 10$ s.

Table 1. Relaxation time for acquiring solubility equilibrium

H_{eff} (M atm^{-1})	τ (s) D = 20 mm	τ(s) D = 0.2 mm	τ(s) D = 1 mm
10^5	4	400	4000
10^2	10^{-3}	10^{-1}	1

For a highly soluble gas, the limitation by mass transfer is not negligible for raindrops ($t = 400$ s for raindrops of diameter 0.2 mm). Differences between two size categories of raindrops are significant when one diameter is five times greater than the other, because then the characteristic time is higher by a factor of ten. As, previously mentioned, the limitation of gas phase transfer is inversely proportional to the surface area of the drops. In the case of a less soluble gas we can consider that all drops are in equilibrium with their gas phase concentration because for large raindrops the characteristic time is equal to 1 s.

These results serve to confirm our chemical module. Variations in characteristic times for reaching equilibrium are significant in term of drop diameter and the chemical species considered. The spectral approach allows us to take into account these significant variations.

In the calculations described above the spectral distribution of raindrops and cloud drops and microphysical events have not been considered. However, Perdue and Beck (1988) show that the simultaneous occurrence of drops with different pollution levels may result in a situation where each individual drop is in chemical equilibrium with the gas phase, but the bulk chemical composition of the drops is not. Consequently we also treat the scavenging of gas by clouds drops as a function of their different diameter.

A complete description of the mountain wave scenario can be found in Richard and Chaumerliac (1989) and Chaumerliac et al. (1990). It consists of a two-layer atmosphere over an idealized bell-shaped mountain. The horizontal homogeneous initial wind speed is 20 ms^{-1}, and a relative humidity of 80 % is assumed below 3 km. Simulations are performed using a two-dimensional version of the mesoscale model of Nickerson et al. (1986) over a horizontal domain of 300 km and a horizontal grid length of 10 km. The time step is 10 s and the simulation time is

6 hours. The initial gas profile is equal to the constant value of 1 ppb over the entire domain. In Fig. 1 results are presented for a highly soluble gas, where $H_{eff} = 10^5$ M atm^{-1}. For the three microphysical parameterizations, vertical cross sections of concentration of chemical species are shown in the gas phase and in the aqueous phase for both cloud and rain.

Maxima of concentration in the gaseous phase, Figs. 1a, for the three parameterizations are of the same order of magnitude (60 nanomole per cm^3 of air). We can see that the presence of the cloud generates a depletion of the concentration in the gas phase, but that the spatial distribution and the intensity of this depletion is very different for the three parameterizations. With the K and BR schemes the concentration of gas decreases to 40 nanomole per cm^3 of air in the presence of cloud. In contrast, with the SP scheme the minimum value of the concentration in gas phase is much lower (10 nanomole per cm^3 of air) and the spatial extent of the depletion stretches downwind. As we have seen in the previous section, the spectral scheme results in the uptake of more gas because as Pandis and Seinfeld (1991) mentioned, when the spectral distribution of cloud drops or raindrops is considered instead of a bulk mixing ratio, more gas is scavenged. With the SP scheme absorption of gas is not linearly dependant on the variation of cloud water mixing ratio, and the balance between gas phase and aqueous phase chemistry at each time step is made in terms of the totality of cloud water and rainwater mixing ratios. The spatial extent of the depletion on the downwind side of the mountain can be attributed to the presence of small raindrops reaching equilibrium in this area and scavenge more gas than the larger drops present in the BR and K schemes. At low levels, on the downwind side of the mountain the concentration in gas phase increases in the BR and SP schemes. Total evaporation of small drops permits the release of gas, and partial evaporation on larger drops concentrates the gas in aqueous phase and results in the transfer of the excess of chemical species to the gas phase. Considering the K scheme, this process does not occur because relatively few small drops are present and partial evaporation on large drops is not taken into account. With the SP scheme, the increase of the concentration in gas phase is more significant than with the BR scheme. More small drops which completely evaporate are present in the SP scheme, and in addition, partial evaporation of large drops significantly increases the aqueous concentration.

The aqueous concentrations in cloud water are comparable for the BR and K schemes and correspond exactly with the spatial extent of the cloud. However, with the SP scheme, the intensity of the cloud concentration is twice as large and explains the greater depletion in the gas phase. The scavenging of gas by cloud drops is not linearly dependent on the evolution of the cloud water content and at each time step the balance between gas phase and aqueous phase is made. The underestimation of cloud concentration obtained by the K and BR schemes originates from this linear dependance which does not take into account the advection of gas during two time steps and because bulk representation of cloud water is less capable of scavenging the gas.

This underestimation can be see also in rainwater, Figs. 1c, because the chemical species is transferred from cloud water to rainwater by coalescence. But in the SP scheme this behavior of high concentration in water is emphasized. With this scheme more small drops are present in the rainwater and partial evaporation on the downwind side of the mountain concentrates chemical species in raindrops. In the formulation of BR we consider the mean diameter of raindrops and an average value for the coefficient correction factor, f, to scavenge gas and these approximations increases the effects of mass transfer limitations. We can also notice a greater absorption of gas by rainwater upwind from the mountain top with BR and SP scheme because rainwater is carried by smaller drops more efficient to uptake gas, than greater raindrops in K scheme.

In addition to the comparisons made with the BR and K schemes, the SP scheme allows us to obtain the distribution of the concentration in mol per liter of water as a function of raindrop diameter, which is plotted in Fig. 2 for three different points of the domain at three different altitudes. The results obtained for a highly soluble gas (solid line) and a less soluble gas (dashed line) are presented. For the two gases considered, the concentration increases during the formation of rain. On the upwind side of the mountain, corresponding to the area where autoconversion occurs, the two gases react similarly, all drops present are at the same concentration, because as we have previously seen small drops reach chemical equilibrium quickly. Differences between the two gases appears when large drops are formed on the top of the mountain. For a highly soluble gas, the aqueous concentration in raindrops decreases when their diameter increases. Chemical concentration increases in all drops but mass transfer limitations prevent large drops from reaching equilibrium. The differences are smaller for a less soluble gas since the time to reach equilibrium is smaller, and microphysical processes act to homogenize the concentration for all diameters of

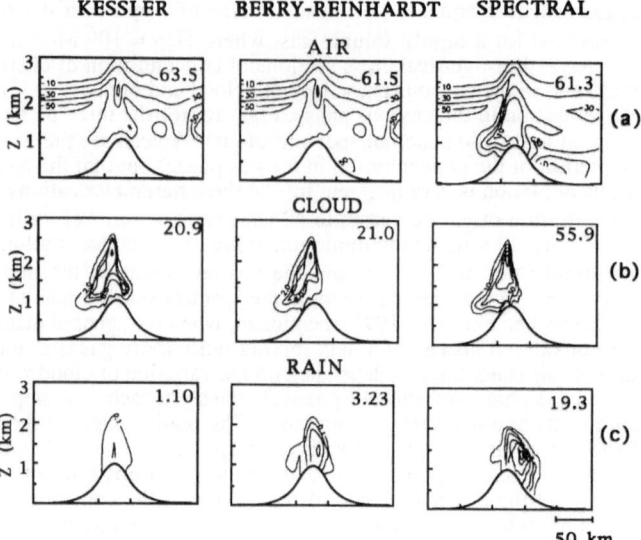

Figure 1 Vertical cross section of chemical concentrations in air, in cloud and in rain environments for an highly soluble gas in nanomole per liter of air, for the Kessler scheme (1969), the Berry and Reinhardt one (1974 a, b) and the spectral parameterizations. Maximum values are given in the upper right corner of each figure.

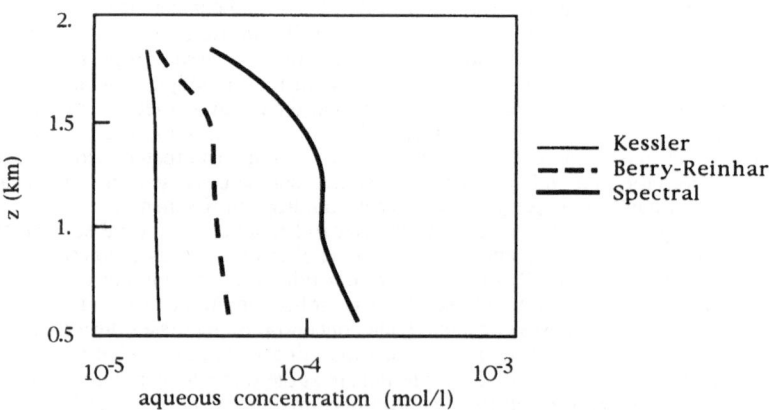

Figure 2 Aqueous concentrations in mole per liter of water for an highly soluble and a less soluble gas considered, as a function of raindrops diameter obtained with the spectral scheme, for three grid points located at the top of the mountain and on upwind and downwind sides from the mountain top for three vertical levels of the model.

raindrops (Flossmann et al.,1987). On the downwind side of the mountain large drops are created by the self-collection process. For the two chemical species considered the aqueous concentration decreases when the drop diameter increases. This feature is emphasized for highly soluble gas, where, for instance, at 360 m in altitude the concentration for a drop with a diameter of 0.1 mm is equal to 2.10^{-4} mol per liter of water, and is equal to 2.10^{-5} mol per liter of water for a diameter of 2 mm. Mass transfer limitations are responsible for the decrease of the aqueous concentration as a function of the raindrop diameter, but another process acts because this behavior also occurs with a less soluble gas. In this part of the domain evaporation processes play an important role. The detailed treatment of partial evaporation increases the aqueous concentration in drops and acts to concentrate chemical species inside drops. Drops are then enable to release gas. This behavior is consistent with the increase in gas concentration. Because the SP scheme considers in detail the evolution of the chemical concentration as a function of the raindrop diameter, partial evaporation is more important in the SP scheme than in the BR scheme. Individual small drops quickly reach chemical equilibrium and release more gas than the well-mixed solution with the same raindrop diameter obtained from the BR scheme. This explains the greater desorption area in gas phase on the downwind side of the mountain observed with the SP scheme than with BR scheme. When considering the scavenging of gas it is important to consider the distribution of raindrop diameter because significant differences appear when partial evaporation process concentrate the chemical species inside the drops and subsequently release gas.

3 - CONCLUSION

A comparative study of three microphysical schemes in a mesoscale model was made to examine the impact of each scheme on gas scavenging and wet deposition. The coupling of scavenging process with a bulk microphysical approach appears insufficient because intricate chemical processes are not linearly dependent on the evolution of water content as the spectral scheme shows. As bulk parameterizations assumed spectral scheme does not consider rainwater as a well mixed solution and this assumption conducts to a larger efficiency of rainwater to scavenge gas. The spectral scheme provides additional information on raindrop chemical concentration spectra which is important when considering evaporation processes. Partial evaporation concentrates gas in the liquid phase and allows chemical equilibrium to be reached more quickly for large drops with the subsequent release of gas. This behavior also shows up in the Berry and Reinhardt scheme. The same behavior is observed in the spatial and temporal evolutions of gases in BR and SP schemes but the efficiency of releasing gases through evaporation is lower in BR than in SP. This phenomenon is not even observed in K scheme. For the study of gas scavenging and chemical reactions in aqueous phase it is necessary to have a microphysical approach which considers diameter of drops and also to have a chemical module completely coupled with microphysical and dynamical processes because wet deposition is not linearly dependant on cloud processes. The parameterization of Berry and Reinhardt appears as a good compromise for studying chemical processes but absorption should be reviewed in terms of the linear dependance between scavenging of gas and the evolution of water mixing ratio.

REFERENCES

Berry, E.X., and R. L. Reinhardt, An analysis of cloud drops growth by collection : Part II. Single initial distributions, *J. Atmos. Sci.*, **31**, 1825-1831, 1974a.
——, and R. L. Reinhardt, An analysis of cloud drops growth by collection : Part III. Accretion and self-collection, *J. Atmos. Sci.*, **31**, 2118-2126, 1974b.
Chaumerliac, N., E. Richard, R. Rosset, and E. C. Nickerson, Impact of two microphysical schemes upon gas scavenging and deposition in a mesoscale meteorological model, *J. Applied Meteor.*, **30**, 88-97, 1990.
Flossmann, A. I., H. R. Pruppacher, and J. R. Topalian, A theoretical study of the wet removal of atmospheric pollutants. Part II : The uptake and redistribution of $(NH_4)_2SO_4$ particles and SO_2 gas simultaneously scavenged by growing cloud drops, *J. Atmos. Sci.*, **44**, 2912-2923, 1987.
Ferretti, R. , G. Visconti, and F. Giorgi, Chemistry and transport of sulfur compounds from large oil fires studied with 2D mesoscale model, *Ann. Geophysicae.*, **11**, 68-77, 1993.
Hales, J. M., A generalized multidimensional model for precipitation scavenging and atmospheric chemistry, *Atmos. Environ.*, **23**, 2017-2031, 1989.

Huret, N., N. Chaumerliac and E.C. Nickerson, Impact of different microphysical schemes on the dissolution of highly and less soluble non-reactive gases by cloud droplets and raindrops, to appear in *J. Atmos. Meteorol.*, 1994

Iribarne, J. V., and H. R. Cho, Models of cloud chemistry, *Tellus*, **41B**, 2-23, 1989.

Kessler, E., On the distribution and continuity of water substance in atmospheric circulations, *Meteor. Monogr.*, **10**, N° 32, 84pp, 1969.

Le Cam, M. N. et H. Isaka, Retrieval of microphysical variables by diagnostic modeling study : comparison between parameterized and detailed warm microphysics. *Tellus*, **41A**, 338-356, 1989.

Nickerson, E.C., E. Richard, R. Rosset, and D.R., Smith, The numerical simulation of clouds, rain and airflow over the Vosges and Black Forest Mountain : a meso-b model with parameterized microphysics. *Mon. Wea. Rev.*, **23**, 477-487, 1986.

Noone, K. J., R. J. Charlson, D. S. Covert, J. A. Ogren, and J. Heintzenberg, Cloud droplets: solute concentration is size dependent. *J. Geophys. Res.*, **93**, 9477-9482, 1988.

Pandis, S. N., and J. H. Seinfeld, Should bulk cloud water or fogwater samples obey Henry's law ? , *J. Geophys. Res.*, **96**, 10791-10798, 1991.

Perdue, E. M., and K. C. Beck, Chemical consequences of mixing aerosol droplets of varied pH, *J. Geophys. Res.*, **93**, 691-698, 1988.

Richard, E., and N. Chaumerliac, Effects of different parameterizations on the simulation of mesoscale orographic precipitation. *J. Appl. Meteor.*, **28**, 1197 - 1212, 1989.

Schwartz, S. E., Mass-transport considerations pertinent to aqueous phase reactions of gases in liquid water clouds. *Chemistry of multi-phase atmospheric systems (ed. Jaeschke)*, NATO ASI Series, Springer-Verlag, **G6**, 415-471, 1986.

A NUMERICAL STUDY OF DMS-OXIDATION

IN THE MARINE BOUNDARY LAYER

Karsten Suhre and Robert Rosset

Laboratoire d'Aérologie (UA CNRS 354)
Université Paul Sabatier
118, Route de Narbonne
31062 Toulouse, France
e-mail: suhk@aero.ups-tlse.fr

INTRODUCTION

During the last decade, dimethyl sulfide (DMS) has been invoked as an important source of non-anthropogenic sulfur (Andreae et al., 1983; Nguyen at al., 1983). Produced by marine phytoplankton, DMS is transferred to the marine boundary layer (MBL) where it is oxidized to sulfur dioxide (SO_2), methanesulfonic acid (MSA) and to sulfuric acid (H_2SO_4). The relationship between DMS emission, cloud condensation nuclei (CCN) and cloud albedo is a problem of climatic interest (Charlson et al., 1987) to be treated on the mesoscale due to the fact that the processes involved in DMS-oxidation in the MBL have timescales typically of the order of some hours or less.

A comprehensive mesoscale meteorological model has been coupled to a chemical module with complex gas-phase and simple SO_2 in-cloud chemistry. In this study, the interaction of vertical turbulent transport with DMS-chemistry and its oxidation product SO_2 is studied in a one-dimensional, marine boundary layer (MBL) case, with reference to a zero-dimensional box-model.

0D BOX MODEL

In the box-model approach, the marine boundary layer is treated as a well mixed chemical reactor. The kinematic equations defined by the reaction scheme are solved including the specified sources and sinks due to sea-air fluxes and dry deposition pro-

Air Pollution Modeling and Its Application X, Edited by S-V. Gryning
and M. M. Millán, Plenum Press, New York, 1994

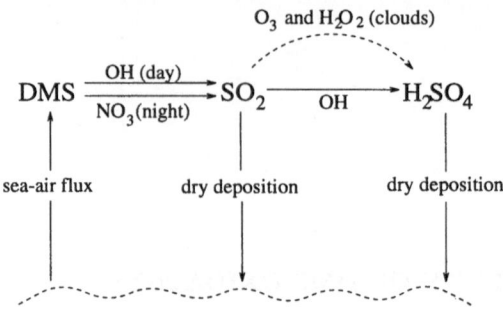

Figure 1. DMS oxidation pathways in the MBL simulated in this study.

cesses. This model is used first to obtain balanced initial values for the full 1D-model, by starting the computation with all prognostic species initialized to zero and running it for about 50 days. A quasi-equilibrium between the chemical conversion terms and the sea-air source and loss terms is established by this process. Then, the box model is run for a second time, initialized with these equilibrated concentrations to calculate the diurnal cycle of the 27 prognostic chemical tracers. Figure 2 displays the 24-hour cycles of DMS, SO_2 and their oxidants OH and NO_3, table 1 gives 24 hour averaged values and the imposed boundary conditions. They are in good agreement with measurements and simulations made by other authors (Andreae at al., 1985; Warneck, 1988 and references therein): the DMS diurnal cycle is marked by net DMS-destruction during the day, when oxidation by OH cannot be balanced by the sea-air flux, and by DMS-production at night, when DMS-oxidation by NO_3 is over-compensated by the DMS-flux. Let F be the DMS-flux and z_i the boundary layer height, then the 0D differential equation for DMS reads as

$$\partial_t[DMS] = \frac{F}{z_i} - (k_{70}[OH] + k_{73}[NO_3])[DMS]. \qquad (1)$$

SO_2, being produced by DMS-oxidation and lost by dry deposition, displays the opposite behaviour (production during the day, destruction at night)

$$\partial_t[SO_2] = -\frac{v_{depot}}{z_i} + (k_{70}[OH] + k_{73}[NO_3])[DMS] - k_{71}[SO_2][OH]. \qquad (2)$$

Here we ignore the effects of cloud-chemistry, but they will be taken into account in the full 1D model.

1D MODEL

The full one-dimensional model consists of a comprehensive meteorological model (SALSA), coupled with a chemical module for complex gas-phase chemistry and optionally a simple cloud chemistry part for in-cloud SO_2-oxidation. Keywords for the model SALSA are: prognostic equation for turbulent kinetic energy, mixing length scheme for turbulent convective boundary layers (Bougeault and Lacarrère, 1989), detailed cloud microphysics, partial cloudiness, radiation scheme (from ECMWF). A description may be found in Bechtold et al. (1991). The chemical reaction mechanism is an updated version of Kreidenweis et al. (1991). The basic equations to be solved here, considering

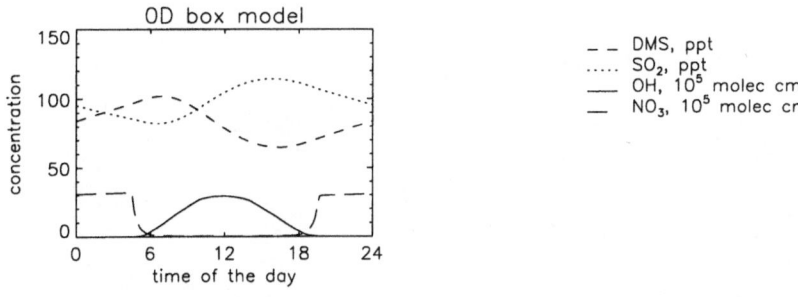

Figure 2. Diurnal cycles of DMS, SO_2, OH and NO_3 as calculated by the box model.

N chemical species $C_i(z)$, $i = 1..N$ as a function of height z are

$$\partial_t C_i(z) = \partial_z K(z)\partial_z C_i(z) + Q_i(C_1(z)..C_N(z)),\qquad (3)$$

where $K(z)$ is the turbulent mixing coefficient and Q_i the chemical transformation terms. Surface fluxes and dry deposition losses are now included as boundary conditions. Thus, the differential equation for DMS in the 1D model to be soved is

$$\partial_t[DMS](z) = \partial_z K(z)\partial_z[DMS](z) - (k_{70}[OH](z) + k_{73}[NO_3](z))\,[DMS](z).\qquad (4)$$

The meteorological scenario simulated consists of a convective boundary layer in the trade wind regime, covered by a deck of stratocumulus clouds. A key feature in this situation is the turbulent decoupling between the cloud layer and the subcloud layer between 12 LST and 17 LST, due to interaction between radiation and clouds. The impact of this decoupling process on DMS-chemistry is discussed in this study as an example for how complex vertical structures in the MBL can modify atmospheric chemistry. This is a point that should be taken into account when interpreting experimental data sampled e.g. on ships. Figure 3 displays the diurnal cycle of cloud water, water vapor and the turbulent mixing coefficient calculated by the model SALSA. Note that this scenario is of cyclic nature, so that the results presented here result from an equilibrium between the different forcings like radiation, evaporation and dynamics for the meteorological part and chemical transformations, turbulent diffusion and source-sink terms for the chemical part.

DISUSSION

Figure 4 shows the diurnal cycles for DMS and SO_2 as obtained with the full 1D model, but without cloud-chemistry, and compares them with the box-model; figure 5 displays timeseries for these sulfur species together with OH and NO_3 diurnal cycles and two vertical profiles of the turbulent mixing coefficient, in order to highlight the two competing forcings in DMS and SO_2 atmospheric chemistry: turbulent vertical transport in a complex turbulence field, and oxidation by the radicals OH and NO_3.

The following points may be retained from the simulations: Superposed on the principal features of the dirnual cycles of the chemical tracers, the effect of differentiate vertical transport can be observed: DMS concentration decrease with height, due to the fact that DMS has its source at the surface and is oxidized on its way up through the atmosphere, while SO_2 concentrations increase, since it is produced in situ from

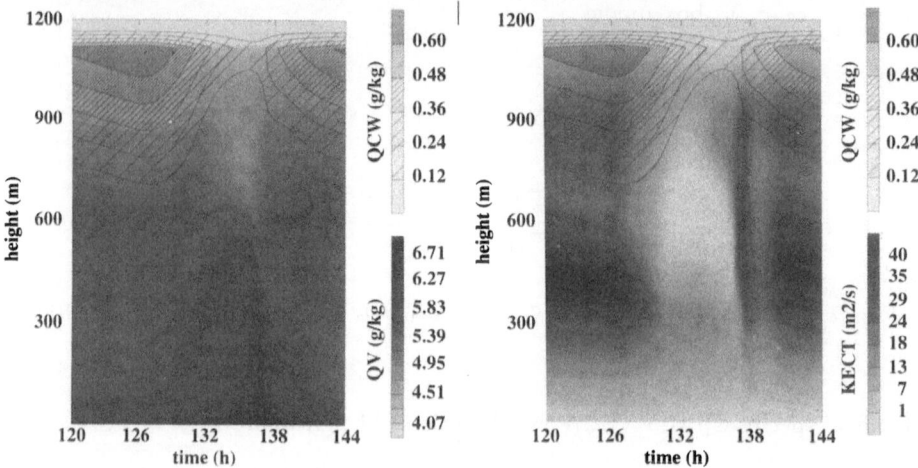

Figure 3. Timeseries for the vertical profiles of the turbulent mixing coefficient (KECT, shaded), the water vapor mixing ratio (QV, shaded) and the cloud water mixing ratio (QCW, hatched) as calculated by the mesoscale meteorological model SALSA: sixth day of the simulation, starting at midnight (t=120h), displayed up to 1200 meters ($\approx$ boundary layer height).

DMS and lost by dry deposition processes at the surface. Between 12 and 17 LST, when the vertical turbulent structure of the MBL confines turbulent transport to either the upper layer (above 600m) or the lower layer (below 600m), chemistry of these two layer evolutes independantly. In the upper layer, being insulating from surface exchanges, DMS destruction is enhanced since there is no source term at that state, whereas in the lower layer, DMS loss by oxidation is partially compensated by the DMS flux due to the reduced effective mixing height (600m instead of 1200m). Similary, SO_2 is accumulated in the upper layer, but lost more effectively in the lower layer during the decoupling period. At about 17 LST, layer recoupling takes place in a burst of turbulence, due to liberation of latent heat from raising humid air (cf. figure 3). The concentration gradient of DMS and SO_2 that has been build up by the decoupling process is then rapidly dissolved (cf. figure 5).

The effect of vertical turbulent transport and layer decoupling is of particular interest when more complex processes like SO_2 oxidation in cloud-droplets is considered. Figure 6 compares vertical SO_2 profiles for simulations performed without and including cloud-chemistry. Note first that in the cloud simulation SO_2 concentrations are much lower than in the simulation without cloud chemistry, where the values are too high when compared to experimental data (e.g. Andreae et al., 1988; Berresheim et al., 1990), indicating that heterogeneous processes are important for SO_2 chemistry. Moreover, the effect of turbulent decoupling now leads to SO_2 accumulation in the lower layer, since oxidation in cloud droplets is much more effective than gas-phase oxidation and dry deposition. The small hook at 13 LST is a consequence of a small subcloud layer, too small to be displayed in figure 3.

CONCLUSIONS

The following points may be retained from this study:

- Given the chemical reaction mechanism and the source and sink terms, the box-model reproduces a chemical "environment" consistent with experimental data

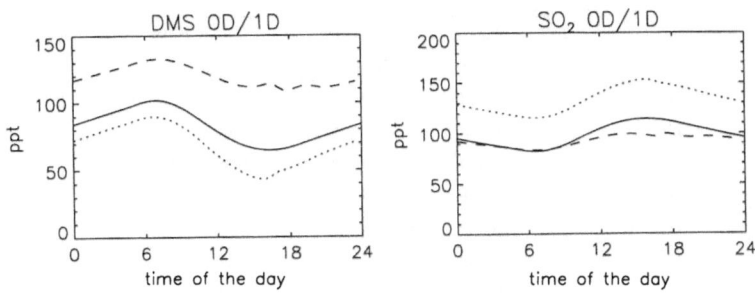

Figure 4. *DMS* and *SO₂* diurnal cycles as obtained by the box model (solid) compared to those obtained with the full 1D model at 8m (dashed) and 1000m (dotted).

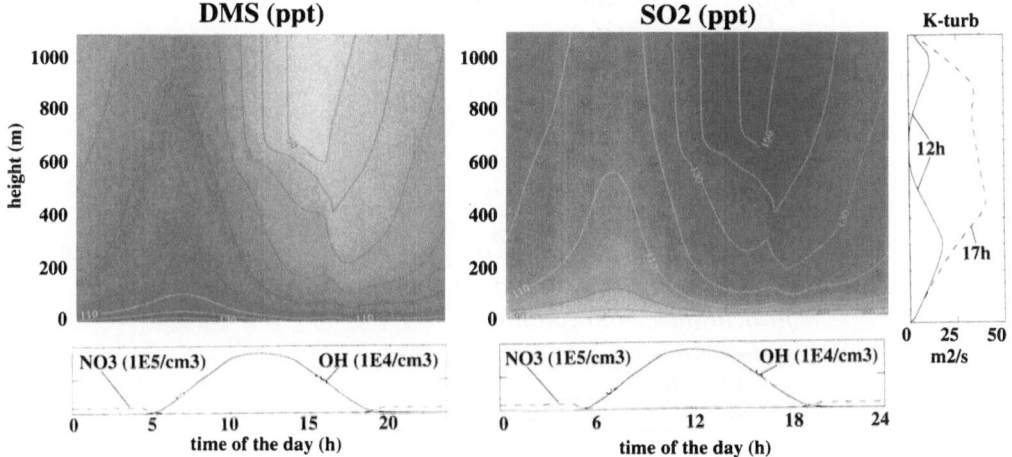

Figure 5. Timeseries of *DMS* and *SO₂*, starting at midnight; profiles below: vertically averaged diurnal cycles of the oxidants *OH* and *NO₃*; profiles to the right: turbulent mixing coefficient at 12 and 17 LST.

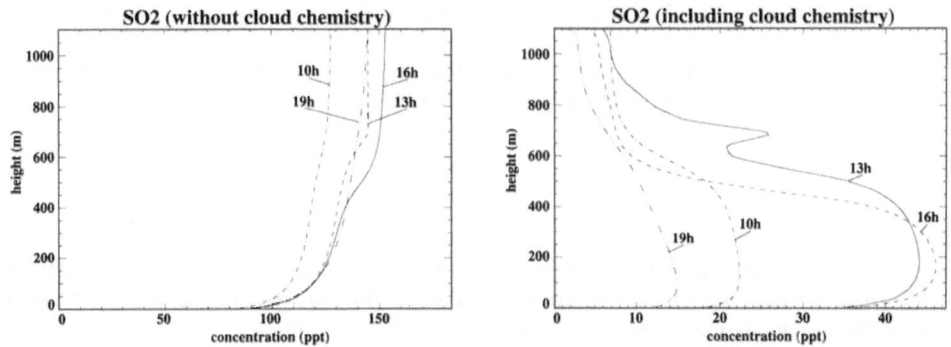

Figure 6. *SO₂* vertical profiles at 10, 13, 16 and 19 LST without (left) and including (right) cloud-chemistry

Table 1. 24 hour averaged tracer concentration as obtained by the box-model; imposed deposition velocities and source fluxes.

Species	24h Averaged Mixingratio	Deposition Velocity [cm/s]	Sea-Air-Flux [mol./$cm^2 s$]
O_3	$15ppb$	0.05	
NO	$6.6ppt$		$6.5 \cdot 10^7$
NO_2	$30ppt$	0.0012	$5 \cdot 10^3$mol./$cm^3 s$ *
$HONO_2$	$123ppt$	0.2	
$HCHO$	$300ppt$	0.3	
DMS	$82ppt$		$3.0 \cdot 10^9$
SO_2	$97ppt$	1.0	
H_2SO_4	$13ppt$	1.0	
OH	$0.037ppt$	0.83	
HO_2	$5.9ppt$	0.63	
H_2O_2	$610ppt$	1.0	
NO_3	$0.047ppt$	0.5	
CH_3O_2	$2.0ppt$	0.24	
CH_3OH	$5.1ppt$	1.0	
CH_3OOH	$180ppt$	1.0	
$HCOOH$	$83ppt$	1.0	
HO_2NO_2	$0.061ppt$	1.0	
N_2O_5	$0.0010ppt$		

other reactants (prognostic variables):

$O(^3P), O(^1D), CH_3, CH_3O, CH_3O_2NO_2, CH_3ONO,$

$HNO, HO_2NO_2, HONO, O_2CH_2OH$

species held constant:

H_2O, CO, CH_4, H_2

* in situ production of NO_x (lightning/advection)

(Warneck, 1988 and references therein).

- Timeseries for the vertical profiles of all prognostic tracers chemically reacting in a "realistic" MBL are obtained.

- Vertical turbulent transport clearly modifies DMS- and SO_2-chemistry, a point of interest for the evaluation of experimental data at sea-level.

- The structure of vertical turbulent mixing is of major importance when simulating heterogeneous processes like SO_2-cloud chemistry.

REFERENCES

Andreae, M. O., Berresheim, H., Andreae, T. W., Kritz, M. A., Bates, T. S., and Merrill, J. T., 1988, Vertical distribution of dimethylsulfide, sulfur dioxide, aerosol ions, and radon over the northeast pacific ocean, *J. Atmos. Chemistry*, 6:149-173.

Andreae, M. O., Ferek, R. J., Bermond, F., Byrd, K. P., Engstrom, R. T., Hardin, S., Houmere, P. D., LeMarrec, F., Raemdonck, H., and Chatfield, R. B., 1985, Dimethyl sulfide in the marine atmosphere, *J. Geophys. Res.*, 90:12891-12900.

Andreae, M. O. and Raemdonck, H., 1983, Dimethyl sulfide in the surface ocean and the marine atmosphere, *Science*, 221:774-747.

Bechtold, P., Pinty, J. P., and Mascart, P., 1991, A numerical investigation of the influence of large scale winds on sea breeze an inland breeze type circulations, *J. Appl. Meteor.*, 30:1268-1279.

Berresheim, H., Andreae, M. O., Ayers, G. P., Gillett, R. W., Merrill, J. T., Davis, V. J., and Chameides, W. L., 1990, Airborne measurements of dimethylsulfide, sulfur dioxide, and aerosol ions over the southern ocean south of australia, *J. Atmos. Chemistry*, 10:341-370.

Bougeault, P., and Lacarrère, P., 1989, Parameterization of orography-induced turbulence in a meso-beta model, *Mon. Wea. Rev.*, 117:1872-1890.

Charlson, R. J., Lovelock, J. E., Andreae, M. O., and Warren, S. G., 1987, Oceanic phytoplankton, atmospheric sulphur, cloud albedo and climate, *Nature*, 326:655-661.

Kreidenweis, S. M., Penner, J. E., Yin, F., and Seinfeld, J.H., 1991, The effects of dimethylsulfide upon marine aerosol concentrations, *Atmos. Env.*, 25A:2501-2511.

Nguyen, B. C., Bonsang, B., and Gaudry, A., 1983, The role of the ocean in the global atmospheric sulfur cycle, *J. Geophys. Res.*, 88:10903-10914.

Warneck, P., 1988, "Chemistry of the Natural Atmosphere", Academic Press, San Diego.

DISCUSSION

J. LANGNER: Could you comment on the relative contribution to the oxidation of DMS from reaction with OH and NO_3?

K. SUHRE: OH concentrations in this simulation are of the same order than NO_3 concentrations (ref table 1), whereas the reaction constant for OH oxidation is about ten times higher than that for NO_3. Hence, DMS oxidation by OH is ten times more effective than by NO_3

J. LANGNER: The relative contribution of the oxidation of DMS from gasphase and liquidphase reactions

K. SUHRE: Looking at figure 6 one sees that the ratio between average SO_2 without and including cloud chemistry is about 4. Assuming in average [PROD] - ([LOSS] + [CLOUDS]) * $[SO_2] = 0$, where [PROD] and [LOSS] are the production and loss terms for SO_2 by homogeneous chemistry and deposition, and [CLOUDS] is the loss term due to liquidphase chemistry, one can deduce [CLOUDS] = 3*[LOSS]. Hence, cloud chemistry is in THIS case three times more effective than homogeneous chemistry.

O. HERTEL: It seems to me that you consider SO_2 as the only oxidation product of DMS.

K. SUHRE: This is right. But as this is a purely numerical study of processes, we decided for simplicity to consider only SO_2 as a product of DMS oxidation, but we could easily include other products like MAS and DMSO into our model.

M. KROL:

Question about emissions: Emission of DMS depends on wind speed and biological processes. You took your emissions constant. Could you comment on that?

K. SUHRE:

We wanted to study the impact of vertical turbulent transport on DMS chemistry. Introduction of a variable DMS emission would have introduced an extra perturbation in the light of this aim. It is true that we have to take this dependence into account when modelling experimental data.

H. VAN DOP:

Are the predicted DMS concentrations and diurnal cycles in agreement with over sea/ocean observations?

K. SUHRE:

Yes, they agree well with measurements made by different investigators (cf. references in the paper by Andreae et al, Berresheim et al and Nguyen et al)

H. VAN DOP:

Is cloud formation a result from your 1-D model? If so, would you be able to study the effect of DMS emissions on cloud formation and cloud characteristics (e.g. more small droplets)?

K. SUHRE:

Cloud formation is in fact calculated by the meteorological model (cf reference Bechtold et al). In order to study the effect of DMS emissions on cloud formation, an aerosol module is needed. Presently, there is a PhD student working in our group on this project, aiming exactly on this problem.

A COMPARISON OF FAST CHEMICAL KINETIC SOLVERS IN A SIMPLE VERTICAL DIFFUSION MODEL

Oswald Knoth and Ralf Wolke

Institut für Troposphärenforschung
Permoserstr. 15
D–04303 Leipzig
Germany

1. INTRODUCTION

The photochemical reaction mechanisms used in regional air quality models usually consider 20 to 100 pollutant species. The equations resulting from these chemical mechanisms are nonlinear, highly coupled and extremely stiff depending on the time of the day. Therefore, the simulation time of the models is determined to a large degree by the computational burden associated with the solution of the chemistry equations.

When solving these stiff systems by classical explicit methods, the time steps must be kept very small to avoid numerical instabilities. The implicit techniques such as the Backward Differentiation Formulas (BDF) methods with automatic step size control can provide solutions with high accuracy. However, these methods are very expensive since they require the solution of large systems of linear equations. In recent years, the Quasi Steady State Analysis (QSSA) method is favoured for solving the chemistry equations. In QSSA the solution of large linear systems is not necessary. Therefore, this method is faster than Gear's method.

Here we present a new integration method and compare this method with respect to its computational efficiency as well as to the accuracy with QSSA and a Gear solver (the LSODE Code of Hindmarsh (1980)). For testing we use the chemical mechanism CHEMSAN developed in the chemistry department of our institute. It has 67 species and includes more than 150 reactions. Beside the chemical box model we investigate a simple one–dimensional vertical diffusion model (including deposition and sources) and a 2D model with an upwind scheme of Roe (1982) for the advection. This model can be considered as an ingredient in a 3D transport model which handles chemical kinetics and vertical diffusion together in every horizontal grid point.

Our new approach is based on a BDF method of maximal order 2 where the Jacobian is computed explicitly and not by finite differences. The part of the Jacobian which originates from the chemistry is approximated by its upper triangular part. Only the nonzero entries of this part have to be stored. It turned out that our approach reaches the QSSA (with a step size of 30 s) with respect to the computational speed. But the new method is more accurate than QSSA in many situations.

We consider the 2D transport model

$$\frac{\partial c_i}{\partial t} = \frac{\partial}{\partial z}\left(K(z)\frac{\partial c_i}{\partial z}\right) - u(z)\frac{\partial c_i}{\partial x} + R_i(t, c_1, \ldots, c_N) + Q_i(t, z, x),$$

$$(t, z, x) \in [t_0, T] \times [0, z_H] \times [0, X], \quad i = 1, \ldots, N , \qquad (1)$$

with the initial condition

$$c_i(t_0, z, x) = c_{0,i}(z, x)$$

and the boundary conditions

$$K(z_H)\frac{\partial c_i}{\partial z}(t, z_H, x) = 0, \quad c_i(t, z, 0) = c_{in,i}(t, z), \quad v_{D,i}(t)c_i(t, 0, x) + K(0)\frac{\partial c_i}{\partial z} = 0.$$

Here $c_i = c_i(t, z, x), i = 1, \ldots, N$, are the concentrations of several species. $u(z)$ denotes the horizontal wind. We assume that it is independent of x and t. Furthermore, the nonlinear function $R_i(t, c)$ describes the chemical reaction part, $Q_i(t, z, x)$ represents sources and sinks, and $K(z)$ is the vertical diffusion coefficient.

2. THE BOX–MODEL

The dynamics of the chemical kinetics in one computational cell can be described by a system of ordinary differential equations

$$\frac{d c_i}{d t} = R_i(c, t) + Q_i(t), \qquad i = 1, \ldots, N, \qquad t \in [t_0, t_1], \qquad c_i(t_0) = c_0. \qquad (2)$$

Systems describing chemical kinetics are often very stiff which is due to the large differences in the reaction rates of different species. To integrate stiff systems in an effective way one has to resort to implicit integration methods which in general involve the computation of the first derivative $\partial R/\partial c$ of R and the solution of linear systems. Theses linear systems have normally a coefficient matrix of the form $I - h(\partial R/\partial c)$. Let us illustrate this by the implicit Euler method. For a given time step Δt_n and an approximate value c^n of $c(t^n)$ the new value c^{n+1} is the solution of the nonlinear equation

$$g(c^{n+1}) = c^{n+1} - c^n - \Delta t_n(R(c^{n+1}, t_{n+1}) + Q(t_{n+1})) = 0 \qquad (3)$$

where $t_{n+1} = t_n + \Delta t_n$. The solution of this nonlinear equation is normally found by a variant of the Newton method where one or two iterations are sufficient. With $c^{n,0}$ as the starting value (often equal to c^n or some explicit predictor) this iteration has the form

$$(I - \Delta t_n J_n)\Delta c^{n,k} = -g(c^{n,k}) \qquad (4)$$
$$c^{n,k+1} = c^{n,k} + \Delta c^{n,k}, \qquad k = 0, 1, \ldots , \qquad (5)$$

with $J_n = \partial R/\partial c^n$.

Integration algorithms which are often used in environmental modelling are based on a representation of the right hand side

$$R_i(c) + Q_i = P_i(c) - L_i(c)c_i \qquad (6)$$

where P_i and $L_i c_i$ are its production and loss rate, respectively. In the absence of special chemical reactions $-L_i$ is i-th diagonal element of the Jacobian $\partial R / \partial c$. Under the assumption that P_i and L_i are constant in the time interval $[t_n, t_n + \Delta t_n]$ equation (2) has the solution

$$c_i(t) = \frac{P_i}{L_i} + (c_i(t_n) - \frac{P_i}{L_i})e^{-(t-t_n)L_i}, \qquad t \in [t_n, t_n + \Delta t_n]. \tag{7}$$

2.1. The Quasi–Steady State Approximation (QSSA) Scheme

In the QSSA method the exponential $e^{-\Delta t L_i}$ of equation (7) is approximated by different formulas depending on the ratio of the time step Δt_n to the characteristic time $1/L_i$. Such an approximation is necessary because the right hand side of (7) could not be computed stable on a computer for all values of $-\Delta t_n L_i$ by a straightforward implementation of the formula. This results in the following computational scheme

$$c_i^{n+1} = \begin{cases} \frac{P_i^n}{L_i^n}, & \Delta t_n L_i^n > 10 \\ \frac{P_i^n}{L_i^n} + (c_i^n - \frac{P_i^n}{L_i^n})e^{-\Delta t_n L_i^n}, & .01 \leq \Delta t_n L_i^n \leq 10 \\ c_i^n + \Delta t_n(P_i^n - L_i^n c_i^n), & \Delta t_n L_i^n < .01 \end{cases} \tag{8}$$

where c_i^n denotes an approximated value of $c_i(t_n)$, $P_i^n = P_i(c^n)$ and so on. One of the main drawbacks of the above scheme is that in each time step a decision has to be made by which formula the new value should be computed. To avoid this overhead a better way is to approximate the exponential function with the help of a Pade–formula (Hairer and Wanner, 1991). In the CTM2 code the (1,1) Pade–approximation

$$e^{-\Delta t L_i} = \frac{1 - \frac{1}{2}\Delta t L_i}{1 + \frac{1}{2}\Delta t L_i} \tag{9}$$

is used which leads to the scheme (Hass, 1991)

$$c_i^{n+1} = \frac{\Delta t_n P_i^n + (1 - \frac{1}{2}\Delta t_n L_i)c_i^n}{1 + \frac{1}{2}\Delta t_n L_i^n} \tag{10}$$

The integration form (10) does not preserve positivity for the case $P_i \geq 0$ and $c_i^n \geq 0$ in contrast to the QSSA method. The (0,2) Pade–approximation

$$e^{-\Delta t L_i} = \frac{1}{1 + \Delta t L_i + \frac{1}{2}(\Delta t L_i)^2} \tag{11}$$

which has the same approximation order as the (1,1) yields the integration form

$$c_i^{n+1} = \frac{\Delta t_n P_i^n(1 + \frac{1}{2}\Delta t_n L_i^n) + c_i^n}{1 + \Delta t_n L_i^n + \frac{1}{2}(\Delta t_n L_i^n)^2} \tag{12}$$

and has the mentioned property of positivity.

2.2. The BDF–method

The implicit Euler method (3) can be generalized to

$$c^{n+1} = \sum_{l=0}^{q-1} \alpha_l c^{n-l} + \Delta t_n \beta_0 \frac{dc^{n+1}}{dt},$$ (13)

q is the order of the method and α_l and β_0 are constants for a particular order. The name BDF comes from the fact that (13) can be written in a form that gives dc^{n+1}/dt as a combination of the c^{n-l}. In the atmospheric literature this method is also well known under the name Gear's method. To initialize a BDF method the integration is started with $q = 1$ and then the order is increased successively. There are a lot of high quality implementations of the BDF method under which the family of LSODE (Livermore Solver for Ordinary Differential Equations) integrators developed by Hindmarsh (1980) is mostly used. It is often argued that the LSODE code are not suitable in general atmospheric chemistry codes. In the following we will see that changing the linear algebra subroutines, the usage of a user–supplied Jacobian and a qualified choice of the input parameters lead to a speedup which makes the modified LSODE code faster than the QSSA method. Note that we have not changed any other part of the code. Especially the same procedures are used for error control, step size selection and so on. The LSODE code offers the possibility to supply only the diagonal part of the Jacobian instead of the whole Jacobian. Choosing this which is comparable to QSSA the program proposes small step sizes. On the other hand with a full Jacobian the integration is performed with very large step sizes which requires only a small number of right hand side evaluations. Inspired by the work of Sillman (1991) we fed the LSODE code with the upper triangular part of the Jacobian. The main advantage of this approach is that no matrix decomposition is necessary. Only one backward sweep have to be carried out for solving the linear systems. We implemented special routines for doing this (only for the nonzero elements) in an effective way.

The number of nonzeroes in the Jacobian is approximately four times the number of chemical species involved. Obviously, the performance of our approach depends on the ordering of the species, but a good one is from the fastest reacting species to the lowest reacting species.

3. THE VERTICAL DIFFUSION MODEL

We extend the model by vertical diffusion which results in

$$\frac{\partial c_i}{\partial t} = \frac{\partial}{\partial z}\left(K(z)\frac{\partial c_i}{\partial z}\right) + R_i(c, t) + Q_i(t, z), \; i = 1, \ldots, N \;,$$ (14)

and the boundary conditions

$$v_{D,i}(t)c_i(t, 0) + K(0)\frac{\partial c_i}{\partial z} = 0, \quad K(z_H)\frac{\partial c_i}{\partial z}(t, z_H) = 0,$$

In contrast to the traditional splitting approach vertical diffusion and kinetics will be treated simultaneously. At first the spatial coordinate is discretized. Then the resulting large system of ordinary differential equations is solved by a BDF method taking into account all the ideas described above. For doing so we introduce a non–equidistant grid $[z_0 = 0, z_1, \ldots, z_{nz} = z_H]$ and $h_j = z_j - z_{j-1}$, $j = 1, \ldots, nz$. For the spatial discretization we will use a simple cell centered scheme. Let $c_i^j(t)$ denote the cell value of $c(z,t)$ in the cell $[z_{j-1}, z_j]$ at time t, $K_j = K(z_j)$. Then the scheme can be written as

$$\frac{dc_i^1}{dt} = 2K_1 \frac{c_i^2 - c_i^1}{(h_1 + h_2)h_1} + v_{D,i}\frac{c_i^1}{h_1} + R_i(t,c^1) + Q_i^1(t),$$

$$\frac{dc_i^j}{dt} = 2K_{j+1}\frac{c_i^{j+1} - c_i^j}{(h_{j+1} + h_j)h_j} - 2K_j \frac{c_i^j - c_i^{j-1}}{(h_j + h_{j-1})h_j} + R_i(t,c^j) + Q_i^j(t),$$

$$j = 2, \ldots, nz - 1 \qquad (15)$$

$$\frac{dc_i^{nz}}{dt} = -2K_{nz}\frac{c_i^{nz} - c_i^{nz-1}}{(h_{nz} + h_{nz-1})h_{nz}} + R_i(t,c^{nz}) + Q_i^{nz}(t),$$

or in matrix form

$$\frac{dc_i}{dt} = T_i c_i + R_i(c) + Q_i, \qquad i = 1, \ldots, N. \qquad (16)$$

where $c_i = (c_i^1, \ldots, c_i^{nz})^T$, $c^j = (c_1^j, \ldots, c_N^j)^T$, and $R_i(c) = (R_i(c^1), \ldots, R_i(c^{nz}))^T$. The matrix T_i is tridiagonal. Two matrices T_i and T_l differ only in the element $(1,1)$ which is due to different deposition velocities $v_{i,D}$. Equation (16) has the same structure as (2). The differences are that c_i is now a vector and that we have the additional part $T_i c_i$. If we apply a BDF method to (16) and approximate again the Jacobian of R with respect to c by its upper triangular part the linear algebra consists in the solution of an upper block triangular system. The diagonal blocks are the tridiagonal matrices $I - \Delta t \beta_0 (T_i + diag(\partial R/\partial c))$ where $diag(\partial R/\partial c)$ denotes the diagonal part of $\partial R/\partial c$. Since the diagonal part is nonpositive the diagonal blocks can be decomposed by Gaussian elimination without pivoting. Therefore the solution of the whole system can be realized in a time which is nz times the time for one box plus the solution of N tridiagonal systems.

4. THE ADVECTION DIFFUSION MODEL

To integrate the full model (1) a usual approach is the splitting scheme

$$c_i^{**}(0) = c_{0,i}$$

$$\textbf{for} \quad n = 0, 1, \ldots \quad \textbf{do} \qquad (17)$$

$$\frac{\partial c_i^*}{\partial t} = -u\frac{\partial c_i^*}{\partial x}, \qquad\qquad c_i^*(t_n) = c_i^{**}(t_n), \qquad t \in [t_n, t_{n+1}]$$

$$\frac{\partial c_i^{**}}{\partial t} = \frac{\partial}{\partial z}(K\frac{\partial c_i^{**}}{\partial z}) + R_i(c^{**}) + Q_i, \quad c_i^{**}(t_n) = c_i^*(t_{n+1}), \quad t \in [t_n, t_{n+1}]$$

where the length of the time interval $[t_n, t_{n+1}]$ is equal to the time step for one integration step of the advection equation. The second equation is than integrated with a smaller time step inside the interval $[t_n, t_{n+1}]$. Cell centered explicit advection schemes have the general form

$$c_i^{jl}(t_{n+1}) = c_i^{jl}(t_n) + (t_{n+1} - t_n)\Delta c_i^{jl} \qquad (18)$$

where the index jl indicates one cell in a two dimensional cartesian grid. The corrections Δc_i^{jl} are computed from the fluxes between adjacent cells. If the values $c_i^*(t)$ and $c_i^{**}(t)$ are plotted against time you can see that the two curves look like two smooth parallel trajectories. This means that during the integration of the chemistry–diffusion part in the interval $[t_n, t_{n+1}]$ the value $c_i^{**}(t)$ has to be moved from one trajectory to the other one. So we have a transient phase for every time interval. This fact was also observed by using the code LSODE for following the intermediate curve. But our real goal is to follow the trajectory $c_i^{**}(t)$. This can be achieved by modifying the scheme (17) in the following way.

$$\textbf{for} \quad n = 0, 1, \ldots \quad \textbf{do}$$

$$\text{Compute} \quad \Delta c_i^{jl} \tag{19}$$

$$\text{Solve} \quad \frac{\partial c_i}{\partial t} = \frac{\partial}{\partial z}(K\frac{\partial c_i}{\partial z}) + R_i(c) + Q_i + \Delta c_i, \quad t \in [t_n, t_{n+1}]$$

This scheme is also mass conserving. But instead of changing the initial values the correction from the advection is brought in smoothly as an additional source/sink term.

5. TEST RESULTS

All algorithms described in the previous sections are tested with different scenarios. Because of the restricted space it is not possible to give a complete report about the used scenarios and all results in this paper. About this we refer to Knoth and Wolke (1993). However, here we want to try to summarize and discuss the main results. A detailed description of the chemical mechanism CHEMSAN (Renner et al., 1993) you can also find in Knoth and Wolke (1993).

As mentioned before the BDF code used here is the Livermore Solver LSODE with an automatic error control (Hindmarsh, 1980). The estimated local error in a concentration c_i is controlled by

$$error \leq RTOL \cdot c_i + ATOL$$

where $RTOL$ and $ATOL$ are input parameters given by the user. We tested several variants for the choice of tolerance $RTOL$. $ATOL$ is always set to 10^{-16}.

The accuracy of the methods is estimated by calculating the difference between results predicted by the method and results from LSODE with $RTOL = 10^{-5}$ (Hertel et al., 1993). As the measure for the error we use the largest absolute deviation $DCMAX$ and the relative deviation $RERR$ for the whole integration period

$$DCMAX_i = max_{t_n}(|c_i^{Code}(t_n) - c_i^{LSODE(10^{-5})}(t_n)|) \; ; \quad RERR_i = \frac{DCMAX_i}{c_i^{LSODE}(t_{nmax})} * 100$$

for $i = 1, \ldots, N$ and t_{nmax} refers to the time in which the largest deviation occur. The deviations are calculated for each half hour.

In the presentation of the test results we use the following abbreviations

LSODE(tol)	LSODE with $RTOL = tol$ and $MAXORD = 2$
LSODE_O5(tol)	LSODE with $RTOL = tol$ and $MAXORD = 5$
LSODE_FD(tol)	LSODE with $RTOL = tol$ and $MAXORD = 2$ where the Jacobian is calculated by finite differences
TRI(tol)	LSODE with a triangular approximation of the Jacobian
QSSA($time$)	QSSA with step size $time$ and PADE–approximation (12)
QSSA_P1($time$)	QSSA with step size $time$ and PADE–approximation (10)
QSSA_EX($time$)	QSSA with step size $time$ and approximation (8)

Box Model. For testing we use five different scenarios, two without sources. Three of them are the scenarios introduced by Hertel et al. (1993). The results for their scenario A are given in Table 1. The integration period was always 5 days.

The results of all tests show that it is very important for the efficiency to compute the Jacobian explicitly. LSODE with a finite–difference approximation is not a practicable way for solving such models. A tolerance of 10^{-2} for TRI and LSODE is sufficient for the integration. QSSA(30 s) differs sometimes markedly from the "exact" solution. It

Table 1. Numerical efficiency and accuracy for a box model

algorithm	O_3 DCMAX RERR	NO DCMAX RERR	OH DCMAX RERR	PAN DCMAX RERR	steps funct	CPU
LSODE(10^{-2})	4.14E-08 1.12	9.02E-09 0.74	3.85E-13 3.04	5.68E-09 0.94	843 1502	8.79 s
LSODE_O5(10^{-2})	1.50E-08 0.51	2.37E-09 0.20	8.46E-14 0.57	1.24E-09 0.36	832 1548	8.94 s
LSODE_FD(10^{-2})	4.14E-08 1.12	9.02E-09 0.74	3.91E-13 2.63	5.68E-09 0.94	813 21423	37.43 s
TRI(10^{-3})	1.25E-08 0.58	2.18E-08 1.21	1.60E-13 0.82	1.65E-08 0.73	1782 3656	8.88 s
TRI(10^{-2})	1.99E-07 7.66	1.04E-07 4.84	2.22E-12 8.98	3.73E-08 7.27	1099 2462	6.28 s
TRI(10^{-1})	1.17E-06 27.73	8.45E-07 69.56	1.03E-11 39.99	2.36E-07 36.94	580 1400	3.93 s
QSSA($10\,s$)	3.63E-07 10.63	1.18E-07 3.54	2.76E-12 14.67	4.07E-08 8.35	43200 43200	114.66 s
QSSA($30\,s$)	1.10E-06 32.24	3.62E-07 10.84	6.48E-12 38.57	1.61E-07 25.81	14400 14400	38.40 s
QSSA($60\,s$)	2.10E-06 58.49	6.39E-07 19.16	2.06E-11 71.99	3.22E-07 50.43	7200 7200	19.30 s
QSSA_P1($30\,s$)	1.58E-06 43.93	4.44E-07 13.31	1.30E-11 47.74	2.49E-07 39.05	14400 14400	38.29 s
QSSA_EX($30\,s$)	1.03E-06 30.16	3.18E-07 9.52	5.82E-12 34.61	1.47E-07 24.35	14400 14400	39.29 s

Table 2. Numerical efficiency and accuracy for a 1D model

integration code	splitting time	O_3 DCMAX RERR	NO DCMAX RERR	OH DCMAX RERR	PAN DCMAX RERR	steps funct	CPU
TRI(10^{-2})	30 s	6.05E-08 0.55	1.00E-08 1.22	1.27E-13 7.25	6.90E-09 0.32	17864 28041	208.92 s
TRI(10^{-2})	60 s	9.66E-08 0.87	1.93E-08 2.36	1.47E-13 7.73	1.11E-08 0.55	16521 28165	194.02 s
TRI(10^{-3})	no	2.61E-08 0.28	3.65E-09 0.41	5.92E-14 0.92	3.68E-09 0.18	2403 5204	46.23 s
TRI(10^{-2})	no	1.73E-07 1.64	9.82E-09 4.00	6.45E-13 13.62	2.81E-08 1.79	1235 2904	26.03 s
TRI(10^{-1})	no	1.53E-06 31.11	2.91E-07 76.17	5.76E-12 54.35	3.65E-07 17.83	609 1406	13.40 s
QSSA($10\,s$)	–	4.16E-07 10.18	1.09E-07 27.97	8.67E-13 12.45	5.28E-08 4.25	25920 25920	116.38 s
QSSA($30\,s$)	–	1.44E-06 15.77	2.06E-07 52.87	2.47E-12 12.06	2.41E-07 13.05	8640 8640	39.44 s

Table 3. Numerical efficiency for a simple 2D model

integration code	advection according to	steps	funct	CPU
TRI(10^{-2})	(17)	121 802	171 774	1957 s
TRI(10^{-2})	(19)	91 440	104 941	1132 s
TRI(10^{-1})	(19)	85 005	89 944	1054 s
QSSA($10\,s$)	(19)	803 520	803 520	3650 s
QSSA($30\,s$)	(19)	267 840	267 840	1182 s

seems that a step size of 60 s leads to non realistic simulation results. Furthermore, LSODE is too expensive for the integration. The TRI(10^{-2}) is favourable. In most cases it needs a smaller execution time than QSSA(30 s) but it is much more accurate.

Simple Vertical Model. We test two scenarios with sources, vertical diffusion and dry deposition. The choice of initial concentrations, emission rates, and diffusion coefficients depends from the heigth z and is similar to the scenario described in Renner and Rolle (1989). For the discretization in the z–direction we use a non–equidistant grid with 10 and 30 layers, respectively. In all runs the simulation time was 72 hours.

The results for one simulation with 10 layers are presented in Table 2. The main conclusions of our tests are that the simultaneous integration techniques work robust. Especially TRI(10^{-2}) is very efficient and accurate. Splitting reduces the numerical efficiency dramatically when methods with step size control are used. QSSA(30 s) needs about the same CPU-time as TRI(10^{-2}). But the results are not so accurate. Therefore we recommend a simultaneous integration with TRI for the numerical solution of such 1D models.

2D Advection Model. In two scenarios we extend the 1D vertical model to a 2D model. Additional this includes advective transport. For the advection the x–axis is aligned along the wind vector. The horizontal modelling domain is divided into 31 equidistant intervals of length 1 km. The emission is placed only in the second column. The rate of emission in this column is multiplied by a factor 10 compared to the 1D model. The wind $u(z)$ depends only from the height z. The integration period is 78 hours. During the first 6 hours the integration is done without emission and advection.

A comparison of the numerical efficiency for one scenario is given in Table 3. In all cases we integrate chemistry and vertical diffusion together. For the advection step we use always 180 s. It turns out that our approach for handling the advective fluxes works very well. The CPU-time for this method connected with TRI(10^{-2}) is smaller than by using QSSA(30 s). However, the simulation results are much more accurate.

6. REFERENCES

Hairer, E., and Wanner, G., 1991, "Solving Ordinary Differential Equations II," Springer Verlag, Berlin.

Hass, H., 1991, "Description of the EURAD Chemistry–Transport–Model Version 2 (CTM2)," Mitteilungen aus dem Institut für Geophysik und Meteorologie, Universität Köln.

Hertel, O., Bercowicz, R., Christensen, J., and Hov, Ø., 1993, Test of two numerical schemes for use in atmospheric transport–chemistry model, to be published in Atmos. Environ.

Hindmarsh, A. C., 1980, LSODE and LSODI, two new initial value ordinary differential equation solver, ACM–SIGNUM Newsl., 15, 4, 10–11.

Knoth, O., and Wolke, R., 1993, "A Comparison of Numerical Schemes for Use in Atmospheric Transport–Chemistry Models," Working Paper, Institute for Tropospheric Research, Leipzig.

Renner, E., and Rolle, W., 1989, Modelling of the formation of photooxidants by a Lagrangian grid cell model under characteristic conditions of Central Europe, Atmos. Environ., 23, 1841–1847.

Renner, E., Rolle, W., and Helmig, D., 1993, Comparison of computed and measured photooxidant concentrations at a forest side, Chemosphere, 27, 881–898.

Roe, P. L., 1981, "Numerical Algorithms for the Linear Wave Equation," Royal Aircraft Establishment, Technical Report 81047.

Sillman, S., 1991, A numerical solution for the equations of tropospheric chemistry based on an analysis of sources and sinks of odd hydrogen, J. Geophys. Res., 96, 20.735-20.744.

MODELLING FLUX-GRADIENT RELATIONSHIPS FOR CHEMICALLY REACTIVE SPECIES IN THE ATMOSPHERIC SURFACE LAYER

Jordi Vilà-Guerau de Arellano, Peter G. Duynkerke and Karl F. Zeller[*]

IMAU, Utrecht University, Princetonplein 5, 3584 CC Utrecht,
The Netherlands
* USDA Forest Service, Rocky Mountain Forest and Range Experimental
Station, Fort Collins, Colorado 80526, USA

SYMBOLS

Capital letters are average quantities
Small letters are fluctuating quantities
Horizontal bar denotes time average

C_1:	closure constant for "return-to-isotropy" in the temperature eqs.
C_2:	closure constant for "return-to-isotropy" in the momentum (co-)variances eqs.
C_4:	closure constant for dissipation in the momentum (co-)variances eqs.
C_5:	closure constant for dissipation in the temperature variance eq.
C_6:	closure constant for "return-to-isotropy" in the concentration flux eqs.
C_7:	losure constant for dissipation in the concentration variance eq.
C_8:	closure constant for dissipation in the heat-concentration covariance eq.
C_B:	constant for buoyancy length scale
C:	average concentration
c:	fluctuating concentration
c_*:	concentration scale ($-\overline{wc}/u_*$)
D:	Damköhler number
e:	turbulent kinetic energy
g:	acceleration of gravity
J:	first-order chemical reaction rate
K_m:	exchange coefficient for momentum
K_h:	exchange coefficient for heat
K_c:	exchange coefficient for concentration
L:	Monin-Obukhov length
l_t:	turbulence length scale
l_B:	buoyancy length scale
p:	pressure

Air Pollution Modeling and Its Application X, Edited by S-V. Gryning
and M. M. Millán, Plenum Press, New York, 1994

R:	the depletion rate in the concentration flux eq. due to chemical reactions
$r_{ij}, r_{i\theta}, r_{ic}$	dimensionless (co-)variances momentum, temperature and concentration
$r_{\theta\theta}, r_{cc}, r_{c\theta}$:	dimensionless (co-)variances temperature and concentration
U:	large-scale wind in x direction
u_*:	friction velocity $\left(\left(-\overline{wu} \right)^{1/2} \right)$
u_i:	wind component
w:	vertical wind component
x:	horizontal distance
z:	vertical distance from surface, height
$\alpha_1, \alpha_2, \alpha_3$:	closure constants shear-stress part of the pressure covariance term (momentum eq.)
β:	buoyancy parameter (g/Θ_0)
$\epsilon, \epsilon_\theta, \epsilon_c, \epsilon_{\theta c}$:	dissipation rate of turbulent kinetic energy, temperature fluctuations, concentration fluctuations and covariance between temperature and concentration
ζ:	dimensionless stability parameter
η:	dimensionless turbulence length scale
κ:	von Karman constant (= 0.4)
Θ:	average potential temperature
θ:	fluctuating potential temperature
θ_*:	temperature scale ($-\overline{w\theta}/u_*$)
ρ_0:	reference air density.
ϕ_m, ϕ_h, ϕ_c:	flux-gradient relationships
$\phi_\epsilon, \phi_{\epsilon\theta}, \phi_{\epsilon c}, \phi_{\epsilon\theta c}$:	dimensionless dissipation rates

INTRODUCTION

In atmospheric surface layer studies, the turbulent flux of variable A is approximated by the formula (Stull, 1988; Garratt, 1992):

$$\overline{wa} = - \frac{\kappa z u_*}{\phi_a(\zeta)} \frac{\partial A}{\partial z} . \tag{1}$$

The flux-gradient profile ϕ_a is a dimensionless quantity introduced to take the atmospheric stability (ζ) into account. In the absence of direct flux measurements, it is often used to calculate the flux from mean-gradient measurements. It is generally assumed that the same ϕ_a should be applied to heat and concentration fluxes. However, several experimental (Duyzer et al., 1983; Delany et al., 1986; Wesley et al., 1989; Zeller, 1993) and modelling studies (Fitzjarrald and Lenschow, 1983; Gao et al., 1991; Hamba, 1993 and Vilà-Guerau de Arellano et al., 1993) have already pointed out the influence of chemical reactions on the relationship between turbulent fluxes and mean-gradient concentrations as described by (1). It is the purpose of this paper to investigate to what extent the flux-gradient profile ϕ_a of a reactive scalar is altered by a first-order chemical reaction. Typical reactions of this kind are photodissociation reactions, which play an important role in atmospheric chemistry. Important chemical species which photodissociate in the troposphere are NO_2, H_2O_2, HCHO and O_3.

The use of second-order closure models have proved to provide insight into flux-gradient relationships dependence upon stability (Zeman, 1981; Mellor and Yamada, 1982; Launder, 1989 and Wyngaard, 1992). In this study, a second-order model is developed to calculate explicitly the flux-gradient profiles of momentum, heat and concentration as a function of stability. The model used is similar to the one developed by Yamada (1985,

1987). Moreover, chemical reaction terms are taken into account on the concentration-governing equations. As a consequence, the flux-gradient profile ϕ_a is now a function of the stability (ζ) and the Damköhler number (D) defined as the ratio of the turbulence time scale and the chemical reaction time scale. Moreover, new second-order closure constants are determined from the micrometeorological measurements of Zeller (1993).

MODEL DESCRIPTION

The main features of the second-order model will be given below; for a more detailed and complete description see Yamada (1985, 1987). The flux-gradient profiles of momentum, heat and concentration are studied in the atmospheric surface layer using second-order moment equations. A derivation of the second-order equations can be found elsewhere (Tennekes and Lumley, 1972; Nieuwstadt and van Dop, 1982; Stull,1988 and Garratt, 1992). In the atmospheric surface layer, under quasi steady-state and horizontal homogeneous conditions, one can write the following equations for $\overline{wu}$, $\overline{w\theta}$ and $\overline{wc}$:

$\overline{wu}$
$$0 \approx -\overline{w^2}\frac{\partial U}{\partial z} - \frac{1}{\rho_o}\left(\overline{w\frac{\partial p}{\partial x}} + \overline{u\frac{\partial p}{\partial z}}\right) + \beta\,\overline{u\theta} \qquad (2)$$

$\overline{w\theta}$
$$0 \approx -\overline{w^2}\frac{\partial \Theta}{\partial z} - \frac{1}{\rho_o}\left(\overline{\theta\frac{\partial p}{\partial z}}\right) + \beta\,\overline{\theta^2} \qquad (3)$$

$\overline{wc}$
$$0 \approx -\overline{w^2}\frac{\partial C}{\partial z} - \frac{1}{\rho_o}\left(\overline{c\frac{\partial p}{\partial z}}\right) + \beta\,\overline{c\theta} + R. \qquad (4)$$

The right-hand terms in Equations (2), (3) and (4) can be interpreted as a balance between production and destruction of the respective fluxes. From left to right, one can distinguish the gradient production, the pressure gradient covariance and the buoyancy effect. Third-order covariance terms (turbulence transport) have been neglected (Wyngaard et al., 1971). Note that in equation (4) an additional term has been added (R) , i.e. a sink term, to take into account the depletion of chemical species by chemical reactions. This term is often omitted, assuming that the chemical reaction is slow compared to turbulence. However, as obtained in several experimental and modelling studies, and mentioned in the Introduction, when the time scale of turbulence is similar to the time scale of chemistry, R has to be taken into account.

For a simple chemical decay, R takes the following form (Fitzjarrald and Lenschow, 1983; Gao et al., 1991; Vilà-Guerau de Arellano and Duynkerke, 1992 and Hamba, 1993):

$$R = -J\,\overline{w\,c}. \qquad (5)$$

The equations (2)-(4) can be solved if the pressure covariance and dissipation terms are approximated. The flux equations must be solved simultaneously with the variances equations for momentum ($\overline{u^2}$, $\overline{v^2}$, $\overline{w^2}$), temperature ($\overline{\theta^2}$) and mass ($\overline{c^2}$) and the horizontal heat flux equation and the horizontal mass flux equation ($\overline{u\theta}$, $\overline{uc}$). In addition, André et al. (1978) introduced the equation of covariance for temperature and moisture. Similarly, the covariance equation between temperature and concentration ($\overline{c\theta}$) is added to the above-mentioned governing equations.

A parameterization for the pressure-gradient covariance term can be deduced from the Poisson equation (Launder, 1975). The three main contributions to this term can be recognised: (1) the "return-to-isotropy" (Rotta, 1951), (2) the shear-stress generation and (3) buoyancy generation (for the last two terms see the papers of Launder et al. (1975) and Zeman and Lumley (1976)). The near-wall effect is not included in the parameterization of the pressure covariance term (Gibson and Launder, 1978). The dissipation terms are typically approximated (see Stull, 1988 and Garratt, 1992); as a function of the turbulent kinetic energy, the variances for the heat and concentration and a turbulence length scale. This turbulence length scale also requires a closure equation. For unstable conditions, it is assumed that the length scale is proportional to z. In stable conditions the turbulence length scale is proportional to z as well, except in very stable conditions where z loses its significance and drops out of the turbulence length scale (Brost and Wyngaard, 1978).

For the atmospheric surface layer, the governing equations can be written in a dimensionless form using similarity theory. The Appendix summarises the dimensionless quantities and the final set of equations. The final governing equations of the second-order closure model consists of 11 diagnostic equations (as a function of the stability parameter z/L) and 6 additional closure equations.

MICROMETEOROGICAL MEASUREMENTS

The micrometeorological data used were collected at the US Forest Service's Pawnee eddy correlation research site which is located on a flat shortgrass prairie of north central Colorado within the US Agricultural Research Service's Central Plains Experimental Range (CPER) for Long-term Ecological Research (LTER) area. A complete description of the site is given by Zeller et al. (1989). Data were collected at 14 Hz by means of the eddy correlation micrometeorological technique and were averaged every 30 min. Special sensors included: sonic anemometers, fast response chemiluminescence ozone analysers, platinum resistance thermometers and a krypton hygrometer. A complete description of the instrument configuration and data analysis is given by Zeller (1993).

The data were collected over a seven-day period (13-20 June 1989) of typical warm summer days separated (on the 17th June 1989) by a cold front passage and high winds. The measurements were collected at 3 m and 8 m. Daily maximum temperatures ranged from 20 to 34.5 C. Global radiation peaked daily at a little less than 1000 W m^{-2}. Ozone concentration and flux values ranged from 16 to 76 ppb and from -0.01 to -0.6 ppb m s^{-1}, respectively.

All second-order empirical constants were obtained from observations in the neutral limit ($\zeta=0$). Table 1 summarises the values of the closure constants obtained for this study from the above-mentioned data, except C_B which is taken from Brost and Wyngaard (1978).

Table 1. Second-order closure constants based on measurements taken by Zeller (1993)

C_1	C_2	C_4	C_5	C_6	C_7	C_8	C_B	α_1	α_2	α_3
5.44	4.	0.15	0.12	5.44	0.12	0.25	1.69	0.6	0.71	0.6

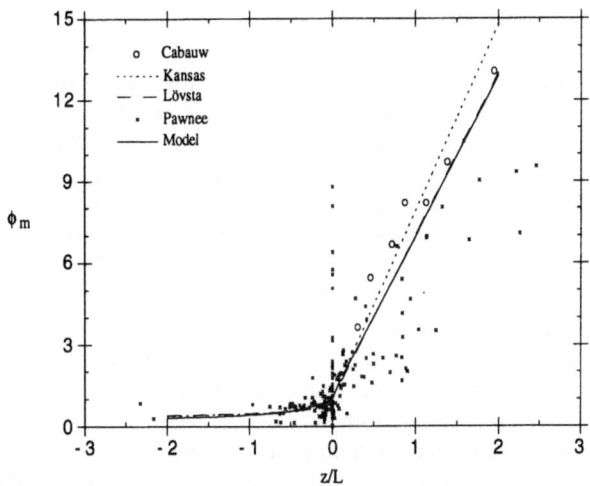

Figure 1. Measured and modelled flux-gradient function for momentum as a function of stability.

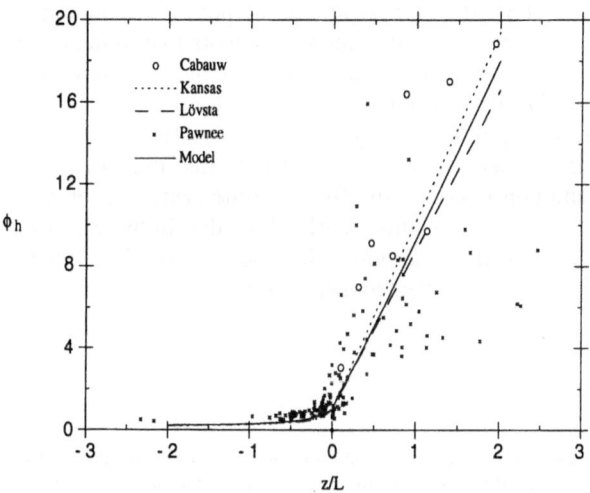

Figure 2. Measured and modelled flux-gradient function for heat as a function of stability.

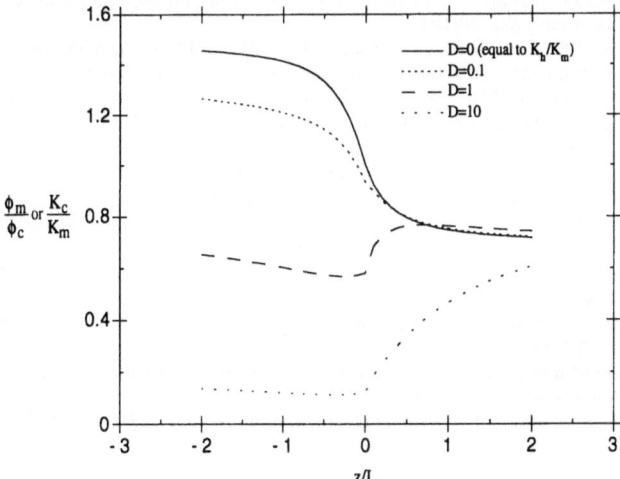

Figure 3. The modelled ratio of the flux-gradient functions for momentum and concentration as a function of stability.

RESULTS AND DISCUSSION

The flux-gradient profiles for momentum and heat are evaluated against micrometeorological measurements and are shown in Figures 1 and 2. The data used were collected in Kansas (USA) (Businger et al., 1971 and Wieringa, 1980), in Lövsta (Sweden) (Högström, 1988) and Pawnee (Zeller, 1993). For the stable case, the data measured at Cabauw (The Netherlands) (Nieuwstadt, 1984) are also included. As the equations in the Appendix shows, the quantities for momentum and heat are not affected by the chemical terms. For the unstable case, the model results agree very well with all the observations. In the stable case, the modelled ϕ_m perfectly agrees with the Lövsta data and slightly underestimates the Kansas and Cabauw data. For ϕ_h, the model is in between the values measured in Lövsta (overestimates) and Cabauw and Kansas (underestimates). The other (co-)variances (not plotted) show similar agreement which indicates that the model give reasonable results for heat and momentum. Therefore, we will now study the influence of chemical reactions terms on the flux-gradient profile of the concentration.

Figure 3 shows the ratio ϕ_m/ϕ_c (= K_c/K_m) as a function of the Damköhler number ($D=\kappa\ z\ J\ /u_*$; the ratio of the turbulence time scale to the chemical time scale). In the absence of chemistry (D=0) the flux-gradients of heat and concentration are equal. If chemistry is introduced, ϕ_c increases if D increases. Note that in the neutral limit (ζ =0), ϕ_c is larger than 1. The differences are larger in the unstable case than in the stable case and already noticeable for D=0.1, a typical transition number between slow and moderate chemistry (Vilà-Guerau de Arellano and Duynkerke, 1993). For the stable case, the model results only show differences for the case D=10. Notice that for D -> ∞, the exchange coefficient K_c -> 0 and consequently the flux becomes zero, i.e. only chemical terms are important in equation (4). These results clearly show that in the case of a simple chemical decay an error is introduced if one assumes that the flux-gradient functions (or exchange coefficients) are both the same for heat and concentration.

REFERENCES

André J.C., De Moor G., Lacarrère P., Therry G. and du Vachat R., 1978, Modeling the 24-hour evolution of the mean and turbulent structures of the planetary boundary layer, *J. Atmos. Sci.* 35:1861.

Brost R.A. and Wyngaard J.C., 1978, A model study of the stably stratified planetary boundary layer, *J. Atmos. Sci.* 35:1427.

Businger J.A., Wyngaard J.C., Izumi Y and Bradley E.F., 1971, Flux-profile relationships in the atmsopheric surface layer, *J. Atmos. Sci.* 28:181.

Delany A.C., Fitzjarrald D.R., Lenschow D.H., Pearson Jr.R., Wendel G.J. and Woodruff B., 1986, Direct measurements of nitrogen oxides and ozone fluxes over grassland, *J. Atmos. Chem.* 4:429.

Duyzer J.H., Meyer G.M. and van Alst R.M., Measurement of dry deposition velocities of NO, NO_2, O_3 and the influence of chemical reactions, 1983, *Atmos.Environ.* 17:2117.

Fitzjarrald D.R. and Lenschow D.H., 1983, Mean concentration and flux profiles for chemically reactive species in the atmospheric surface layer, Atmos. Environ. 17:2505.

Garratt J.R., 1992, "The Atmospheric Boundary Layer", Cambridge University Press.

Gao W., Wesley M.L. and Lee I.Y., 1991, A numerical study of the effects of air chemistry on fluxes of NO, NO_2, O_3 near the surface, *J. Geophys. Res.* 96 :18,761.

Gibson M.M. and Launder B.E., 1978, Ground effects on pressure fluctuations in the atmospheric boundary layer, *J. Fluid. Mech.* 86:491.

Hamba F., 1993, A modified K model for chemically reactive species in the planetary boundary layer, *J. Geophys. Res.* 98:5,173.

Högström U., 1988, Non-dimensional wind and temperature profiles in the atmospheric surface layer; a re-evaluation, *Bound.-Layer Meteor.* 42:55.

Launder B. E., 1989, Second-moment closure: present... and future?, *Int. J. Heat and Fluid Flow* 10:282.

Launder B.E., 1975, On the effects of a gravitational field on the turbulent transport of heat and momentum, *J. Fluid. Mech.* 67:569.

Launder B.E., Reece G.J. and Rodi W., 1975, Progress in the development of a Reynolds-stress turbulence closure, *J. Fluid Mech.* 68:569.

Mellor G. L. and Yamada T., 1982, Development of a turbulence closure model for geophysical fluid problems, *Rev. Geophys. Space Physics.* 20:851.

Nieuwstadt F.T.M. and van Dop H. (Eds.) , 1982, "Atmospheric Turbulence and Air Pollution Modelling", Reidel, Dordrecht.

Rotta J.C., 1951, Statistische Theorie nichthomogener Turbulenz, *Z. Phys.* 129:547.

Stull R.B., 1988, "An Introduction to Boundary Layer Meteorology", Kluwer Academic Publishers,.

Tennekes H. and Lumley J. L., 1972, "A First Course in Turbulence". MIT Press.

Vilà-Guerau de Arellano J. and Duynkerke P.G, Second-order closure study of the covariance between chemically reactive species in the surface layer, *J. Atmos. Chem.* 16:145(1993).

Vilà-Guerau de Arellano J. and Duynkerke P.G., 1992, Influence of chemistry on the flux-gradient relationships for the $NO-O_3-NO_2$ system, *Bound.-Layer Meteor.* 61:375.

Vilà-Guerau de Arellano J., Duynkerke P.G. and Builtjes P.J.H., 1993, The divergence of the turbulent diffusion flux due to chemical reactions in the surface layer: the $NO-O_3-NO_2$ system, *Tellus* 45B:23.

Wesley M.L., Sisterson D.L., Hart R.L., Drapcho D.L. and Lee Y., 1989, Observations of nitric oxide fluxes over grass, *J. Atmos. Chem.* 9:447.

Wieringa J., 1980, A revaluation of the Kansas mast influence on measurements of stress and cup anemometer overspeeding, *Bound.-Layer Meteor.* 18:411.

Wyngaard J.C., 1992, Atmospheric Turbulence, *Annu. Rev. Fluid Mech.* 24:205.

Wyngaard J.C., Coté O.R. and Izumi Y., 1971, Local free convection, similarity, and the budgets of shear stress and heat flux, *J. Atmos. Sci.* 28:1171.

Yamada N., 1985, Model for the pressure terms in the equations for second-order turbulence moment, and its application to the atmospheric surface layer, *J. Met. Soc. Japan* 63:695.

Yamada N., 1987, A trial prediction of the values of some surface-layer turbulence constants using Schumann's method for promoting realizability, *Bound.-Layer Meteor.* 38:1.

Zeller K, Massman W., Stocker D., Fox D.G., Stedman D., Hazlett D., 1989, Initial results from the Pawnee eddy correlation system for dry deposition research. Research paper RM-282, Rocky Mountain Forest and Range Experimental Station, Ft. Collins, CO, USA.

Zeller K.F., Eddy diffusivities for sensible heat, ozone and momentum from eddy correlation and gradient measurements, 1993, Research paper RM-313, Rocky Mountain Forest and Range Experiment Station, Ft. Collins, CO, USA.

Zeman O., 1981, Progress in the modeling of planetary boundary layer, *Ann. Rev. Fluid Mech.* 13:253.

Zeman O. and Lumley J.L., 1976, Modeling buoyancy driven mixed layer, *J. Atmos. Sci.* 33:1974.

APPENDIX

The definition and nomenclature of the dimensionless quantities and second-order constants have been kept similar to those proposed by Yamada (1985).

Dimensionless quantities

Flux-gradient relationships

$$\phi_m = \frac{\kappa z}{u_*} \frac{\partial U}{\partial z}, \quad \phi_h = \frac{\kappa z}{\theta_*} \frac{\partial \Theta}{\partial z}, \quad \phi_c = \frac{\kappa z}{c_*} \frac{\partial C}{\partial z},$$

dissipation rates

$$\phi_\varepsilon = \frac{\kappa z}{u_*^3} \varepsilon, \quad \phi_{\varepsilon\theta} = \frac{\kappa z}{u_* \theta_*^2} \varepsilon_\theta, \quad \phi_{\varepsilon c} = \frac{\kappa z}{u_* c_*^2} \varepsilon_c,$$

(co-)variances

$$r_{jk} = \frac{\overline{u_j u_k}}{u_*^2}, \text{ except when } (j,k)=(1,3)$$

$$r_{j\theta} = \frac{\overline{u_j \theta}}{u_* \theta_*}, \quad r_{jc} = \frac{\overline{u_j c}}{u_* c_*} \text{ except when } j = 3$$

$$r_{\theta\theta} = \frac{\overline{\theta^2}}{\theta_*^2}, \quad r_{cc} = \frac{\overline{c^2}}{c_*^2}, \quad r_{c\theta} = \frac{\overline{c\,\theta}}{c_*\,\theta_*},$$

turbulent kinetic energy, length and time scales

$$e_t = \frac{e}{u_*^2}, \quad \zeta = \frac{z}{L}, \quad D = \frac{\kappa\,z\,J}{u_*}, \quad \eta = \frac{l_t}{z}, \quad \text{where}$$

$$\frac{1}{l_t} = \frac{1}{\kappa z} + \frac{1}{\kappa l_B}\left(\frac{1}{\kappa l_B} = 0 \text{ for } \zeta < 0 \text{ and } l_B = \frac{C_B\,\kappa\,(L\,z\,r_{33})^{1/2}}{\phi_h^{1/2}} \text{ for } \zeta \geq 0\right)$$

Flux equations

$\overline{wu}$
$$(\alpha_1 - 1)\,r_{33}\,\phi_m + \frac{7}{10}\,\zeta\,r_{1\theta} + C_2\,\frac{\phi_\varepsilon}{e_t} + \alpha_3\,r_{11}\,\phi_m - \alpha_2\,e_t\,\phi_m = 0 \qquad (A1)$$

$\overline{w\theta}$
$$-\,r_{33}\,\phi_h - \frac{1}{5}\,r_{1\theta}\,\phi_m + \frac{2}{3}\,\zeta\,r_{\theta\theta} + C_1\,\frac{\phi_\varepsilon}{e_t} = 0 \qquad (A2)$$

$\overline{wc}$
$$-\,r_{33}\,\phi_c - \frac{1}{5}\,r_{1c}\,\phi_m + \frac{2}{3}\,\zeta\,r_{c\theta} + C_6\,\frac{\phi_\varepsilon}{e_t} + D = 0 \qquad (A3)$$

Variance equations

$\overline{u^2}$
$$\left(2 - \frac{4}{3}\,\alpha_1 + \frac{2}{3}\,\alpha_3\right)\phi_m - \frac{1}{5}\,\zeta - C_2\,\frac{\phi_\varepsilon}{e_t}\left(r_{11} - \frac{2}{3}\,e_t\right) - \frac{2}{3}\,\phi_\varepsilon = 0 \qquad (A4)$$

$\overline{v^2}$
$$\frac{2}{3}\,(\alpha_1 + \alpha_3)\,\phi_m - \frac{1}{5}\,\zeta - C_2\,\frac{\phi_\varepsilon}{e_t}\left(r_{22} - \frac{2}{3}\,e_t\right) - \frac{2}{3}\,\phi_\varepsilon = 0 \qquad (A5)$$

$\overline{w^2}$
$$\frac{2}{3}\,(\alpha_1 - 2\,\alpha_3)\,\phi_m - \frac{8}{5}\,\zeta - C_2\,\frac{\phi_\varepsilon}{e_t}\left(r_{33} - \frac{2}{3}\,e_t\right) - \frac{2}{3}\,\phi_\varepsilon = 0 \qquad (A6)$$

$\overline{\theta^2}$
$$\phi_h - \phi_{\varepsilon\theta} = 0 \qquad (A7)$$

$\overline{c^2}$
$$\phi_c - \phi_{\varepsilon c}\,D = 0 \qquad (A8)$$

Covariance equations

$\overline{u\theta}$
$$\phi_h + \frac{1}{5}\,\phi_m - C_1\,\frac{\phi_\varepsilon}{e_t}\,r_{1\theta} = 0 \qquad (A9)$$

$\overline{uc}$
$$\phi_c + \frac{1}{5}\,\phi_m - C_6\,\frac{\phi_\varepsilon}{e_t}\,r_{1c} - D\,r_{1c} = 0 \qquad (A10)$$

$\overline{c\theta}$
$$\phi_h + \phi_c - \phi_{\varepsilon\theta c} - D\,r_{c\theta} = 0 \qquad (A11)$$

Turbulent kinetic energy

$$e_t = \frac{1}{2}\,(r_{11} + r_{22} + r_{33}) \qquad (A12)$$

Closure assumptions
Turbulence length scale

$$\begin{aligned}
\eta &= \kappa && \zeta < 0 \\
\eta &= \frac{\kappa}{1 + \dfrac{1}{\kappa\,C_B}\left(\dfrac{\zeta\,\phi_h}{r_{33}}\right)^{1/2}} && \zeta \geq 0
\end{aligned} \qquad (A13)$$

Dissipation rate of turbulent kinetic energy

$$\eta\,\phi_\varepsilon = \kappa\,C_4\,e_t^{3/2} \qquad (A14)$$

Dissipation rate of temperature fluctuations, concentration fluctuations

$$\eta\,\phi_{\varepsilon\theta} = \kappa\,C_5\,e_t^{1/2}\,r_{\theta\theta} \qquad (A15)$$

$$\eta\,\phi_{\varepsilon c} = \kappa\,C_7\,e_t^{1/2}\,r_{cc} \qquad (A16)$$

$$\eta\,\phi_{\varepsilon c\theta} = \kappa\,C_8\,e_t^{1/2}\,r_{c\theta} \qquad (A17)$$

DISCUSSION

J. MATTHIJSEN: Can you explain why the deviation of the flux-
gradient functions is so large for fast chemistry (D=10)?

J. VILA-GUERAU DE   For large Damkohler numbers (fast chemistry) the
ARELLANO: only important terms in the flux equations are the chemical
                           terms. Consequently, the flux-gradient relationships tend to
                           infinite and become meaningless. A possible explanation is
                           that the Reynolds decomposition applied to the velocity and
                           concentration variables is not longer valid if the chemical re-
                           action time scale is much faster compared to the turbulence
                           time scale.

NEURAL NETWORK TECHNIQUES FOR SO₂
EPISODE PREDICTION

S. J. Perantonis, N. Vassilas, G. T. Amanatidis,
S. J. Varoufakis and J. G. Bartzis

National Center for Scientific Research "Demokritos"
153 10 Aghia Paraskevi, Athens, Greece

INTRODUCTION

Air quality has become a major issue during the last decades for a great number of cities and a matter of concern for both citizens and scientists. The quality of life has deteriorated in urban regions due to high pollution levels and a large number of people suffer from pollution effects on health. The prediction of air pollution episodes is very useful in that it enables the local authorities to give a warning against high pollutant concentrations or to take limitation measures on the emission sources. In the past, efforts for pollution level forecasting have been reported using mathematical models (simulating the physical process) or statistical methods (Zannetti, 1990). The mathematical model operation requires high computational capacity, meteorological data and emission inventory. The required meteorological data are very difficult to collect and process, while the creation of an emission inventory is not an easy task.

A statistical approach has the advantage that no knowledge of the mechanism of pollution dispersion is needed. Most statistical forecasting work dealing with urban air-pollution utilizes traditional well known techniques such as linear regression models, time-series analysis etc. (Zannetti, 1990; Finzi and Tebaldi, 1982; Inoue et al., 1986). However, conventional statistical techniques suffer from several drawbacks. The statistical distribution of the pollution data is not *a priori* known and therefore parametric regression techniques that assume a given distribution, such as multivariate Gaussian, may not work optimally. Linear regression techniques in general are not capable of capturing the inherently non-linear nature of the problem, while non-linear regression techniques work for low-dimensional input spaces (up to approximately 5-D). Moreover, the number of weather and pollution parameters implies high dimensional input spaces with the well known problem of the *curse of dimensionality* (Duda and Hart, 1973).

For the above reasons, it is believed that artificial neural network (ANN) techniques, which have recently been successfully applied to various problems involving forecasting (e.g. stock price (Siriopoulos et al., 1992), energy consumption, sunspot

Air Pollution Modeling and Its Application X, Edited by S-V. Gryning
and M. M. Millán, Plenum Press, New York, 1994

levels (Weigend et al., 1991) etc.) may provide a good alternative to statistical techniques, mainly because of their speed, robustness, non-linear characteristics, non parametric regression capabilities, generalization properties and easiness of working with high-dimensional data.

In this paper, we apply two ANN techniques to the prediction of SO_2 episodes in the Athens basin. The ANNs used are trained using the back propagation (BP) and learning vector quantization (LVQ) algorithms. These two techniques are described and applied to meteorological and air-pollution data sets. The performance of the ANNs is examined as the percentage of correctly predicted episodes or non-episodes.

ARTIFICIAL NEURAL NETWORK BASICS

ANNs are computational systems whose architecture and operation are based on our present-day knowledge about biological nervous systems. Such systems are composed of cells called *neurons*, which are interconnected via *synapses*, i.e. connections that transmit chemical signals from one neuron to another. By analogy, ANNs consist of a set of suitably positioned simple processing elements (nodes) representing the neurons. Each node receives signals X_i from a fixed number of other nodes and determines its response (activation) Y as a function of these signals and of the strength of the synapses, expressed in terms of *synaptic coefficients* or *weights* W_i. Thus $Y = g\left(\sum_i W_i X_i\right)$, where g is a monotonic increasing function. In turn, this response serves as input signal to other nodes.

Different ANN models can be constructed by suggesting different ways of connecting processing elements. An example is the multi-layered feedforward network (Rumelhart et al., 1986) which consists of an input layer, intermediate or "hidden" layers of nodes and a layer of output nodes, with each node receiving inputs only from nodes in the previous layer. Depending on the number of layers and nodes in each layer, as well as on the values of the synaptic weights, an ANN can realize any arbitrarily complicated, generically non-linear functional relationship between its inputs and its outputs (Funahashi, 1989) by superposition of the elementary node functions. This relationship can be continuous or discrete (e.g. in a classification problem with two classes: All inputs belonging to one class correspond to output equal to 1, while those belonging to the second class correspond to output equal to 0).

The solution of a problem by an ANN is achieved in two stages. In the supervised *training* stage, the network is provided with a "training set" of examples (input plus desired output) of the relationship to be learned, and by implementing specific algorithms, usually iterative in nature, which change the values of the synaptic weights, until the network becomes able to reproduce these examples. Once the training stage has been completed, the values of the synaptic weights are fixed and the *testing* stage can begin. A "test set" of new examples (not contained in the training set) are presented to the ANN. In this way, we can test whether the ANN can *generalize*, i.e. realize the correct associations using data not previously encountered.

FORECASTING USING NEURAL NETWORKS

Forecasting using ANNs bears resemblance to statistical non-parametric forecasting methods. In the simplest case, consider a time-series $z(t_1)$, $z(t_2), \ldots, z(t_n), \ldots$ where sampling is performed at equal times $\Delta t = t_{i-1} - t_i$. An optimal functional relationship is sought associating each value $z(t)$ with the values of z in N previous times

$$z(t) = f\left(z(t - N_1 \Delta t), \ldots, z(t - N_2 \Delta t), v_1, v_2, \ldots, v_R\right) \qquad (1)$$

The integer $N = N_1 - N_2$ represents the width of a "window" of time points. The R parameters v_i are determined using known examples of the behavior of z in the past. Once v_i have been determined, the function f can be used for the estimation-forecasting of z in the future.

An ANN to be used for forecasting has one node in the output layer, whose activation represents $z(t)$ when the values $z(t - N_1 \Delta t), \ldots, z(t - N_2 \Delta t)$ are presented as input to the N nodes of the input layer. In this sense, the ANN implements the functional relationship of each value on N previous values. The role of the parameters v_i is played by the synaptic weights, which are evaluated in the training phase.

Apart from the simple model described by equation (1), where it is supposed that the value of $z(T)$ depends exclusively on previous values of the same quantity, ANNs can be used for simulating much more complex models, whereby

- $z(T)$ depends on previous values of other quantities $y_1, y_2, \ldots y_n$. In this case, it is sufficient to increase the number of input nodes of the ANN and to present to each extra input node the value of one of these quantities.
- rather than producing $z(T)$ itself, the network is called upon to learn how to classify $z(T)$ in one of a number of categories. In this case, binary activation is used in the ANN output nodes.

In this latter case falls the problem of SO_2 episode prediction. We shall use ANNs to decide whether or not an episode will occur in the morning maximum of SO_2 concentration, using as input the values of SO_2 concentration and meteorological parameters at times prior this morning (e.g. by 8 pm of the previous day). We shall use two well known ANN prototypes, namely a multilayered feedforward network trained by the BP algorithm and networks trained by a cascade of LVQ algorithms. These prototypes are explained in detail in the next sections.

THE BACK PROPAGATION ALGORITHM

Consider a multilayered feedforward neural network with one layer of input, M layers of hidden and one layer of output units. Unit outputs are denoted by $Y_{ip}^{(m)}$, where the superscript (m) labels a layer within the structure of the ANN ($m = 0$ for the input layer, $m = 1, 2, \ldots, M$ for the hidden layers, $m = M + 1$ for the output layer), i labels a unit within a layer and p labels the input patterns. Similarly, synaptic weights emanating from unit i and directed towards unit j in layer m are denoted by $w_{ij}^{(m)}$. We also include thresholds (biases), to be considered as weights emanating from units of constant, pattern-independent output equal to one. Thus, node outputs are given by

$$Y_{jp}^{(m)} = g\left(\sum_i W_{ij}^{(m)} Y_{ip}^{(m-1)}\right) \qquad (2)$$

where g is taken to be the logistic function $g(x) = 1/(1 + \exp(-x))$.

The BP algorithm (Rumelhart et al., 1986) attempts to match the network outputs with the desired targets. To this end, the cost function

$$E = \frac{1}{2}\sum_{jp} \varepsilon_{jp}, \quad \varepsilon_{jp} = \left(T_{jp} - Y_{jp}^{(M+1)}\right)^2 \qquad (3)$$

307

is minimized via a search in the weight space using gradient descent. The algorithm is initialized using random initial weights, which are updated iteratively by applying the following equation:

$$\delta W_{ij}^{(m)} = -\eta \frac{\partial E}{\partial W_{ij}^{(m)}} + \alpha \Delta W_{ij}^{(m)} \qquad (4)$$

where the second term is proportional to the difference of the current and previous epoch weights (momentum acceleration term). At each iteration (epoch), the derivatives of E with respect to the weights and thresholds can be calculated in a systematic way, starting from evaluations involving the top layer and working backwards. For weights involved in the top layer we obtain:

$$\delta W_{ij}^{(M+1)} = \eta \sum_p \delta_{jp}^{(M+1)} Y_{ip}^{(M)} + \alpha \Delta W_{ij}^{(m)}, \quad \delta_{jp}^{(M+1)} = (T_{jp} - Y_{jp}^{(M+1)}) Y_{jp}^{(M+1)} (1 - Y_{jp}^{(M+1)}) \quad (5)$$

while in subsequent layers

$$\delta W_{ij}^{(m)} = \eta \sum_p \delta_{jp}^{(m)} Y_{ip}^{(m-1)} + \alpha \Delta W_{ij}^{(m)}, \quad \delta_{jp}^{(m)} = Y_{jp}^{(m)} (1 - Y_{jp}^{(m)}) \sum_k W_{jk}^{(m+1)} \delta_{kp}^{(m+1)} \quad (6)$$

Note that $\delta_{jp}^{(M+1)}$ is proportional to the error $T_{jp} - Y_{jp}^{(M+1)}$ corresponding to pattern p and that effort in the computation of $\delta_{jp}^{(m)}$ is saved, for each j, m by using the already calculated value of $\delta_{jp}^{(m+1)}$. Thus, starting at the top layer, we "propagate the errors" backwards to lower layers, hence the name "back propagation algorithm".

There has been criticism that the BP method is slow and has rather poor convergence properties in large-scale problems (Fahlman, 1988), e.g. problems with a relatively large number of training patterns such as the pollution level forecasting problem which is of interest in this paper. However, a modified version of the BP algorithm, known as on-line BP, has been found to exhibit very good convergence and learning speed properties in large scale problems (Fogelman Soulie, 1991; Karras and Perantonis, 1993) and will be used in this paper. While in equations 5 and 6 updating of the synaptic weights takes place after performing the appropriate sum over all pattern labels, weight updating in on-line BP takes place each time a pattern is presented.

THE LEARNING VECTOR QUANTIZATION ALGORITHMS

The LVQ Algorithms have been proposed by Kohonen (Kohonen, 1989;1990; Kohonen et al., 1992) and are described next. Let $x \in R^n$ denote an input vector to be classified in one of k categories. We select k class-labelled reference vectors $W_i, i = 1, \ldots, k$, also known as *codebook vectors*, which represent the synaptic weights of our ANN. The objective of the learning procedure is to place the codebook vectors in the input space in such a way, as to capture the distribution of the data. Thus, the LVQ algorithm variants attempt the optimal placement of codebook vectors in the input space – so as to optimally describe class boundaries – through an adaptive iterative process. Class boundaries are segments of hyperplanes placed at the mid-distance of two neighboring codebook vectors that belong to different classes. Once learning is complete, new patterns x in the test set are assigned to the class C whose codebook vector W_c is the nearest to x. Three LVQ variants proposed by Kohonen are briefly described in table 1.

Before an LVQ algorithm is implemented, the total and relative numbers of codebook vectors for each class are picked and initial values are assigned to the codebook vectors. Too few codebook vectors may not be enough for good class separation, while too many lead to prohibitive learning times and bad generalization results. Thus, a compromise has to be achieved through experimentation for best results. In the absence of *a priori* knowledge about the class probability densities, the codebook vectors are split equally among the classes. Codebook vector initialization is usually done with the values of real training data that carry the necessary class labels to correspond with the numbers of codebook vectors in each class. Moreover, these data should be chosen carefully to ensure that the codebook vectors are always placed in the correct side of the class boundaries. This can be achieved by accepting training samples whose class labels agree with those given by a tentative classification scheme, such as the k-nearest neighbors (k-NN) (Kohonen et al., 1992).

Table 1. Learning Vector Quantization algorithm variants.

The LVQ1 Algorithm	
$W_c(p+1) = W_c(p) + a(p)[x(p) - W_c(p)]$	for x, W_c in the same class
$W_c(p+1) = W_c(p) - a(p)[x(p) - W_c(p)]$	for x, W_c in different classes
$W_k(p+1) = W_k(p)$	for $k \neq c$
The LVQ2 Algorithm	
$W_i(p+1) = W_i(p) - a(p)[x(p) - W_i(p)]$	where W_i, W_j are the two closest codebook vectors, with x and W_i in different classes,
$W_j(p+1) = W_j(p) + a(p)[x(p) - W_j(p)]$	while x and W_j in same class. x must fall in a window described in the text.
$W_k(p+1) = W_k(p)$	for $k \neq i$ or j
The LVQ3 Algorithm	
$W_i(p+1) = W_i(p) - a(p)[x(p) - W_i(p)]$	where W_i, W_j are the two closest codebook vectors, with x and W_i in different classes,
$W_j(p+1) = W_j(p) + a(p)[x(p) - W_j(p)]$	while x and W_j in same class. x must fall in a window described in the text.
$W_l(p+1) = W_l(p) - \epsilon a(p)[x(p) - W_l(p)]$	for $l = i$ or j and x, W_i, W_j in same class. ϵ is a parameter set between 0.1 and 0.5.
$W_k(p+1) = W_k(p)$	for $k \neq i$ or j

In table 1, p denotes a learning iteration (epoch). The learning rate $a(p)$ is given a small initial value (less than 0.05) and decreases monotonically as learning proceeds (e.g. linearly with p or as $1/p$). A variation of LVQ1, known as OLVQ1 (Optimized LVQ1), provides for faster convergence than LVQ1 by associating different learning rates $a_i(p)$ to each $W_i(p)$ (for details see Kohonen et al.,1992). The window referred to in LVQ2 and LVQ3 is defined as follows: Let d_i and d_j be the distances from x to W_i and W_j. x is defined to lie in the window if $\min(d_i/d_j, d_j/d_i) > 1 - w$ with w depending on the number of available samples (a typical value for w is 0.35).

EXPERIMENTAL DATA AND EVALUATION OF ALGORITHMS

Continuous records of SO_2 concentration hourly values measured at the Patission station in Athens, as well as hourly values of temperature, humidity, wind direction and wind speed and daily values of precipitation measured at the National Observatory in Athens for the period 1985-1991 were used for evaluating the performance of the algorithms described above. It is noticed that the SO_2 levels remain relatively high during the examined period, after a decreasing trend in the 70s (Zerefos et al., 1989).

The purpose of all simulations is to forecast whether or not an episode will occur concerning the morning maximum of SO_2 concentration which is defined as the maximum value of this concentration between 8 am and 2 pm. Episodes are considered to occur when the morning maximum of SO_2 concentration exceeds 250 $\mu g/m^3$. Forecasting is based on a number of parameters (previous hourly pollutant concentration, temperature, wind speed and direction, relative humidity, daily rainfall, day of the week) which are given as input to ANNs. Naturally, it is assumed that a decision has to be reached sometime before 8 am, and thus collection of information stops accordingly. To test the prediction accuracy of the ANNs, three different sets of simulations were conducted with decision times at 8 pm of the day before the event (episode or non-episode), at 12 midnight of this day and finally, at 8 am of the day of the event.

The parameters on which forecasting is based are the following:

- Two four-hour windows with values of SO_2 concentration (8 values in total). The windows are selected using the statistical Student-t test criterion: For each hourly SO_2 concentration value, it is tested whether the training set distributions corresponding to episodes and non-episodes are sufficiently separated. Windows are chosen which correspond to large t-values, indicating adequately separated distributions.

- One four-hour window for temperature, one for humidity and one for wind speed values using the t-test criterion as above.

- Daily precipitation value of the day prior to forecasting. The determination of the actual removal of SO_2 by rainfall is described in Amanatidis (1992).

- An index designating the predominant direction of the wind. In the Athens basin, northerly winds tend to reduce pollutant concentrations, while southerly winds related with the sea-breeze mechanism generally increase pollution levels (Paliatsos and Amanatidis, 1993). The average wind direction is determined for a period of 8 hours prior to the "decision time", and a binary number is given as input to the ANNs indicating whether this direction is above or below the East-West axis.

- An indicator related to the day of the week (to account for the fact that pollutant emission is usually reduced on weekends).

In all, 23 parameters are given to the ANNs as input. Two ANN prototypes are used for forecasting, namely a feedforward network with 23 input nodes, one layer of hidden nodes and one output node trained using on-line BP, and a network with 23 input nodes trained using three LVQ variants in succession (OLVQ1, LVQ2 and LVQ3). The use of a cascade of LVQ variants, rather than just one variant, is an innovation of this paper and led to much better results than those obtained using each LVQ variant on its own.

Since practically all SO_2 concentration episodes occur during autumn and winter months, we have prepared training and test sets with information in the period October to February included. Thus our training set consists of 643 training patterns in the period 1985-1989, of which 88 correspond to episodes and 555 to non-episodes, while our test set consists of 294 patterns in the period 1990-1991. It is well known that neural networks used for classification give optimal results when the numbers of training samples corresponding to each category are approximately equal. To achieve this balance, we enhance the "episode" patterns by presenting them to the ANN 5 times for each time that non-episode patterns are presented. Forecasting results are presented in table 2. ANN parameters used are quoted (N_c denotes the number of codebook vectors for the LVQ cascade, N_h denotes the number of hidden nodes for the feedforward network) and the percentage of correct predictions for episodes and non-episodes is shown. For reasons of comparison, we also show results obtained using a powerful conventional statistical method (k-NN classification). In this method, given an input vector x, its k nearest neighbors in the training set are determined, and x is assigned to the category to which the relative majority of neighbors belong (Kohonen et al., 1992).

Table 2. Forecasting accuracy of neural networks and k-NN classifier.

LVQ Algorithm					
Decision time	Parameters			Prediction Rate	
	no of epochs (OLVQ1/LVQ2/LVQ3)	N_c	Episodes(%)	Non-Episodes(%)	
8 am	1500/5000/5000	90	90.6	90.3	
12 midn.	200/800/4800	30	81.3	83.2	
8 pm	300/1200/2400	30	75.0	84.0	
BP Algorithm					
Decision time	Parameters			Prediction Rate	
	η	α	N_h	Episodes(%)	Non-Episodes(%)
8 am	0.7	0.3	25	87.5	86.3
12 midn.	0.7	0.3	25	84.4	82.1
8 pm	0.7	0.3	25	78.1	81.3
k-NN Algorithm					
Decision time	Parameters			Prediction Rate	
	k			Episodes(%)	Non-Episodes(%)
8 am	11			87.5	86.6
12 midn.	37			81.3	81.3
8 pm	9			68.8	76.5

It is evident from the results in table 2 that both neural network prototypes are quite efficient in predicting SO$_2$ episodes. The results obtained using ANNs are better than (in just one case comparable to) those of the k-NN classifier. It is important that the superiority of the ANN results over those of the k-NN classifier becomes more evident with increasing separation between decision time and actual time at which the event (episode or non-episode) takes place. It is therefore worthwhile to study the application of ANNs in air pollution level forecasting further, as improvements in prediction accuracy can lead to effective early warning against high pollution concentrations.

CONCLUSION AND PROSPECTS

In this paper, two well known ANN prototypes were applied to the prediction of SO$_2$ episodes, with encouraging results. Further improvement of the results can originate from more sophisticated ANN techniques with better generalization capabilities (Karras and Perantonis, 1993; Gatos et al., 1993), which we intend to apply to air pollution forecasting in the near future. Collection of pollution and meteorological data from a spatial grid of stations (Finzi and Tebaldi, 1982) and usage of weather predictions for the date of the events – rather than weather data values measured hours before the events – can also lead to increased prediction rates. It is also possible to predict actual pollution levels using ANNs, rather than just forecasting episodes. Finally, we plan to extend our work to forecasting of photochemical pollutants (O_3, NO_x etc.) which are responsible for acute pollution episodes in Athens and other Mediterranean cities.

REFERENCES

Amanatidis, G. T., 1992, Sulphate, sulphur dioxide and rainfall relationships in the Athens basin, Greece, *Fresenius Environmental Bulletin* 1:462.

Duda, R. O., and Hart, P. E., 1973, "Pattern Classification and Scene Analysis," Wiley, New York.

Fahlman, S. E., 1988, Faster learning variations on back propagation: an empirical study, in "Proceedings of the 1988 Connectionist Models Summer School," San Mateo, 38.

Finzi, G., and Tebaldi, G., 1982, A mathematical model for air pollution forecast and alarm in an urban area, *Atmospheric Environment* 16(9):2055.

Fogelman Soulie, F., (1991), Neural network architectures and algorithms: a perspective, in "Artificial Neural Networks," T. Kohonen, K. Makisara, O. Simula, and J. Kangas, eds., Elsevier.

Funahashi, K-I., 1989, On the approximate realization of continuous mappings by neural networks, *Neural Networks* 2:183.

Gatos, V., Karras, D. A., and Perantonis, S. J., Optical character recognition using novel feature extraction and neural networks classification techniques, in "Proceedings of the Workshop on Neural Networks: Techniques and Applications" Liverpool, UK.

Inoue, T., Tagauri, M., and Hoshi, M., 1986, Prediction of nitrogen oxide concentration by a regression model, *Atmospheric Environment* 20:2325.

Karras, D. A., and Perantonis, S. J., 1993, Efficient constrained training algorithms for feedforward networks, *IEEE Trans. on Neural Networks*, under review.

Karras, D. A. and Perantonis, S. J., 1993, Comparison of learning algorithms for feedforward networks in large scale networks and problems, in "Proceedings of the International Joint Conference on Neural Networks," Nagoya, Japan.

Kohonen, T., 1989, "Self-Organization and Associative Memory," Springer-Verlag, Berlin-Heidelberg-New York-Tokyo.

Kohonen, T., 1990, Improved versions of learning vector quantization, in "Proceedings of the International Joint Conference on Neural Networks," San Diego, 1:545.

Kohonen, T., Kangas, J., Laaksonen, J. and Torkkola, K., 1992, LVQ-PAK: the learning vector quantization program package, Version 2.0.

Paliatsos, A. G. and Amanatidis, G. T., 1993, Smoke concentrations in Athens, Greece: trends and strong episodes, *The Science of the Total Environment*, in press.

Rumelhart, D. E., Hinton, G. E., and Williams, R. J., 1986, Learning internal representations by error propagation, in "Parallel Distributed Processing: Explorations in the Microstructure of Cognition," D. E. Rumelhart and J. L. McClelland, eds., MIT Press, Cambridge, MA.

Siriopoulos, C., Karakoulas, G., Doukidis, G., Perantonis, S. and Varoufakis, S., 1992, Applications of neural networks and knowledge-based systems in stock investment management: a comparison of performances, *Neural Network World* 2(6):785.

Weigend, A. S., Rumelhart, D. E. and Huberman, B. A., 1991, Generalization by weight elimination with application to forecasting, in "Advances in Neural Information Processing Systems," 875.

Zannetti, P., 1990, "Air Pollution Modeling," Computational Mechanics Publ., New York.

Zerefos, C. S., Bais, A., Ziomas, I., Paliatsos, A. G., Amanatidis, G. T., and Tourpali, K., 1989, Review of air pollution trends in Greece, in "Proceedings of the Technical Conference on the Monitoring and Assessment of Changing Composition of the Troposphere," Sofia, Bulgaria; WMO 724: 47.

DISCUSSION

P. SEIBERT: The first equation presented in the paper ($Y = g \sum_j (W_i X_i)$) is a linear combination of input parameters. In the introduction of the paper it is mentioned that ANN include non linear interactions. How is the nonlinearity brought into the model?

G.T. AMANATIDIS: In the equation mentioned above, g is a non-linear function of one variable, which acts on the linear combination $\sum_j (W_i X_i)$ of the input parameters. This represents the action of one processing element (elementary node), whose output in turn serves as input to other elements etc. In effect, the overall model, which consists of many processing elements, achieves the approximation of arbitrarily complex non-linear functions by a superposition of the functions corresponding to each elementary node.

R. SAN JOSE GARCIA: Similar results were observed in hourly O_3 data in Spain.

G.T. AMANATIDIS: It is encouraging that ANN techniques are beginning to be applied with success to air pollution problems, as is evident from the paper by Mlakar et al. (in this volume), from our work and, as we are now glad to hear, from work on O_3 data in Spain. We believe that these results can be further improved by combining more sophisticated ANN techniques and meteorological information, as analysed in our conclusions.

SOURCE FOOTPRINT ANALYSIS FOR SCALAR FLUXES MEASURED IN FLOWS OVER AN INHOMOGENEOUS SURFACE

Ashok K. Luhar[†] and K. Shankar Rao

Atmospheric Turbulence and Diffusion Division, NOAA/ARL
P. O. Box 2456, Oak Ridge, TN 37831, USA

INTRODUCTION

Any source near the ground could potentially contribute to the vertical scalar flux measured downwind, depending on the distance from the source and the height of the measurement, as well as the flow and turbulence characteristics of the atmospheric boundary layer. The 'footprint' is a measure of the relative importance of upwind sources to the vertical flux measurements at a given point.

Footprint estimation is important for the measurements of atmospheric fluxes of greenhouse gases such as water vapor, methane, and ozone, and for assessing the relative importance of their sources in climate change studies. For many problems of horizontal inhomogeneities, such as changes in surface roughness or moisture, footprint analysis can provide an estimate of the key 'height-to-fetch' ratio which determines the optimum siting of instruments for flux measurements. It could also facilitate the interpretation of aircraft flux measurements in relation to tower data.

THE FOOTPRINT

We consider a two-dimensional flow in which surface emissions are represented by an area source of infinite extent in the crosswind direction. The area source may be infinite or finite in size along the mean wind direction depending on the type of emissions (or depositions) considered. We divide the surface area source into several elements, which may or may not be equal in size depending on the resolution required in footprint calculations. The term footprint, f, is defined as the contribution, per unit emission, of each element of the upwind surface area to the measured vertical flux (Schuepp et al., 1990; Horst and Weil, 1992). This implies that the measured flux is the integral of contributions from all elements of upwind surface emissions, whereas the footprint is the relative weight given to each elemental emission flux. The above can be written mathematically as:

$$F_a(0, z_m) = \int_0^\infty Q_o(x') f(x', z_m) \, dx'. \tag{1}$$

[†] NOAA/NRC Resident Research Associate.

Air Pollution Modeling and Its Application X, Edited by S-V. Gryning
and M. M. Millán, Plenum Press, New York, 1994

Here, $F_a(0, z_m)$ is the vertical scalar flux (with units of g m^{-2} s^{-1}, for example) measured at a point $(x' = 0, z = z_m)$, and $Q_o(x')$ is the surface emission rate (g m^{-2} s^{-1}) of an area source element located at a distance x' upwind of the flux instrument.

It is clear from Equation (1) that the footprint, $f(x', z_m)$ (m^{-1}), is the vertical flux F (g m^{-2} s^{-1}) at $(0, z_m)$ due to a continuous line source of unit strength located upwind at x' (Wilson and Swaters, 1991), i.e.:

$$f = (1/Q_o)\, dF_a/dx' = F/q, \tag{2}$$

where q is the line source strength (g m^{-1} s^{-1}) which may vary with downwind distance. Equation (1) suggests that if the surface emission rates and footprints are known, we can make a quantitative identification of the upwind sources that make dominant contributions to the flux F_a measured at a given point.

There have been a few analytical as well as Lagrangian stochastic dispersion model studies of footprints (e.g., Leclerc and Thurtell, 1990; Schuepp et al., 1990; Wilson and Swaters, 1991; Horst and Weil, 1992). However, the existing analytical models of footprints are applicable only in horizontally homogeneous surface-layers, and the stochastic modeling technique has been used only in one-dimensional boundary layers. The assumption of horizontal homogeneity is rarely realized for natural surface conditions. For measurement of fluxes from surfaces of limited extent (fetch), such as small lakes or irrigated fields within arid lands, the effects of the surface inhomogeneities can be important. Also, aircraft data are sometimes obtained above the surface layer. Under such circumstances stochastic models are necessary, because the existing analytical models are applicable only in the surface-layer.

Guided by the need for footprint analysis for flux data over inhomogeneous surfaces, Rao and Luhar (1993) considered a simple step-change in surface roughness and used the second-order closure atmospheric boundary layer (ABL) model of Rao et al. (1974a) and Rao (1975) to compute the flow and turbulence fields downwind of the surface discontinuity under different stability conditions. These fields were then utilized in LSDM2D, a 2-D Lagrangian stochastic dispersion model described by Luhar and Rao (1993), to estimate the footprints for scalar fluxes measured at locations downwind of the surface discontinuity. The effects of measurement height, measurement location, and atmospheric stability on the estimated footprints were discussed. Though the model results provided insight into the effects of a simple surface inhomogeneity, this study did not consider any change in the surface moisture conditions. Such change is typical of heterogeneous land surfaces.

In this study, we consider the flow over a surface inhomogeneity involving a sudden change of both surface roughness and moisture conditions, for example, the flow from a relatively smooth arid area to an irrigated crop field. An internal boundary layer forms downwind of the discontinuity. The presence of moisture over the downwind surface leads to changes in the partitioning of solar insolation into sensible and latent heat fluxes near the ground, and influences the flow and turbulence properties over that surface. We modified Rao et al.'s (1974b) second-order closure ABL model to account for the effects of vegetation on surface energy and moisture balances through a 'big-leaf' approach as described in the following section. The modified ABL model is used to compute the flow and turbulence fields which are utilized in the LSDM2D to estimate the footprints for moisture fluxes measured at various locations and heights near the surface.

THE HIGHER-ORDER TURBULENCE CLOSURE MODEL

Rao et al. (1974a) investigated the structure of the internal boundary layer over a sudden change of surface roughness using a two-dimensional higher-order closure ABL model. They concluded that flux-profile relations based on the "local similarity" assumption are valid only in a thin equilibrium layer above the surface. Rao et al. (1974b) also modeled the flow from a smooth dry area to a well-watered grassy surface, and successfully simulated the observed mean wind, temperature, and humidity profiles. This model also predicted the advection effects on turbulent flux distributions, surface evaporation rate, and Bowen ratio.

The original model of Rao et al. (1974b) did not include the effects of vegetation on surface energy and moisture budgets. Using an approach similar to that described by Kroon and De Bruin (1993), we have recently modified the surface boundary condition in this higher-order-closure ABL model by incorporating Penman-Monteith's 'big-leaf' equation (Monteith, 1965) for estimating energy fluxes over vegetation canopies. In this equation, the influence of vegetation on energy fluxes is taken into account by including the stomatal resistance (r_s) of the vegetation layer. The model of Rao et al. (1974b) carries transport equations for turbulent fluxes, and the unknown third moment terms in these equations are closed approximately in terms of the second moments, the mean gradients, and a turbulence time scale, $\overline{u_i' u_i'}/\epsilon$, where ϵ is the dissipation rate of turbulent kinetic energy. This time scale is determined by the model itself, not specified a priori, since the model includes a dynamical equation for ϵ.

In this study, we consider the atmospheric flow from an extensive smooth dry area to a grassy wet field. As in Rao and Luhar (1993), $z_{01} = 0.001$ m is the roughness of the surface upwind of the discontinuity (Surface 1) and $z_{02} = 0.1$ m is that of the downwind surface (Surface 2). The model of Rao et al. (1974b), modified as described above, was utilized to simulate the flow and turbulence fields. The first part of the numerical model code, which calculates the flow over Surface 1, was run for the diurnal cycle and one output flow field for daytime unstable conditions corresponding to a Monin-Obukhov length (L) of -30 m, was selected. The parameters used for Surface 1 are: friction velocity $u_{*1} = 0.35$ m s^{-1}, sensible heat flux $H_{01} = 0.21$ m K s^{-1}, moisture flux $F_{01} = 0.23 \times 10^{-4}$ g g^{-1} m s^{-1}. The second part of the model code simulates the flow over Surface 2 using the flow field over Surface 1 as initial conditions. In this run, r_s was assumed to be constant over Surface 2 with a value of 90 s m^{-1} (e.g., Kroon and De Bruin, 1993).

An output of the model is presented in Fig.1 which shows advection effects on surface fluxes over Surface 2. Most of the variation in the sensible heat and moisture fluxes occurs within a few hundred meters from the discontinuity. The surface sensible heat flux abruptly changes sign from positive (unstable stratification) to negative (stable stratification) at the discontinuity due to the presence of moisture at the surface and the resulting demand on energy for evaporation, which exceeds the available energy. This leads to a loss of sensible heat by the air near the surface, and the formation of an advective inversion. The surface shear stress (hence the friction velocity) also reacts abruptly to the roughness transition as shown in Fig.1. The variations of other mean and turbulent quantities are qualitatively similar to those shown in Rao et al. (1974b).

THE LAGRANGIAN STOCHASTIC DISPERSION MODEL

A Lagrangian stochastic model simulates turbulent dispersion by calculating the trajectories of individual fluid elements (or particles). In this work, we used the 2-D Lagrangian Stochastic Dispersion Model (LSDM2D) developed at ATDD (Luhar and Rao, 1993). In this model, the position (X, Z) at time t of a particle released from a line source in an inhomogeneous Gaussian turbulent flow, neglecting the streamwise

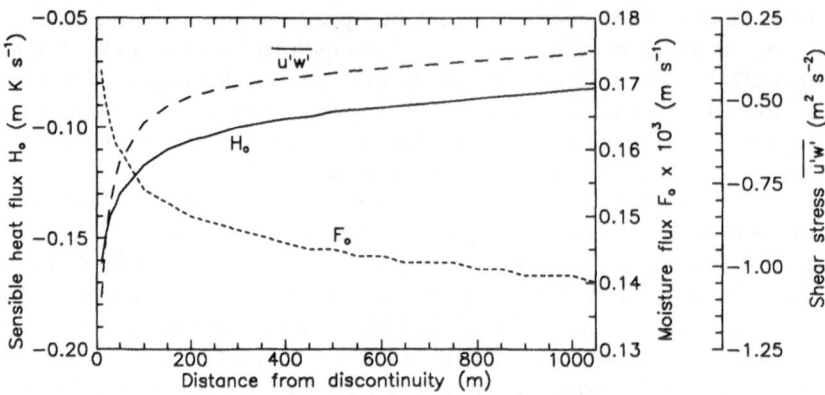

Figure 1. Variation of surface fluxes of momentum $(\overline{u'w'})$, sensible heat (H_o), and moisture (F_o) over the wet grassy surface (Surface 2).

diffusion, can be determined by the following equations (Thomson, 1987; Luhar and Britter, 1989):

$$\left.\begin{aligned} dW' &= \{-[C_o\epsilon/(2\,\sigma_w^2)]W' + (1/2)[1 + (W'^2/\sigma_w^2)](\partial\sigma_w^2/\partial z)\}\,dt + \sqrt{C_o\epsilon}\,d\xi(t), \\ dZ &= (W' + \overline{w})\,dt, \\ dX &= \overline{u}\,dt, \end{aligned}\right\}$$

(3)

where X, Z, and W' are the downwind location, height, and fluctuating vertical velocity, respectively, of the particle in Lagrangian coordinates. C_o is a constant, σ_w^2 is the vertical turbulent velocity variance, $\overline{u}$ and $\overline{w}$ are the mean Eulerian velocities in the horizontal and vertical directions, respectively, and $d\xi(t)$ is a Gaussian random forcing with mean zero and variance dt. The above stochastic model does not include the effects of skewness of the vertical turbulent velocities. A proper incorporation of skewness is important in unstable conditions, especially when sources are elevated (e.g., Luhar and Britter, 1989). In our footprint predictions, the sources are at the ground where the atmospheric conditions are stable. The skewness close to the ground is usually small compared to that in the bulk of the boundary layer. We have therefore neglected skewness effects.

When the turbulent flow is horizontally homogeneous, the footprint depends only on the distance between the locations of the flux measurement point and the emission element. The footprint depends on their actual locations, however, when the flow is horizontally inhomogeneous as in the present study. In this case, the footprints are calculated by placing a number of crosswind line sources upwind of the measurement point. Each line source correspond to an element of the upwind area source. If $F(0, z_m)$ is the vertical flux at point $(x' = 0, z_m)$, contributed by a crosswind surface line source of *unit strength* located at a distance x' upwind, then the flux footprint of this source at point $(0, z_m)$ is given from Equation (2) by $f(x', z_m) = F(0, z_m)$. $F(0, z_m)$ is determined by releasing thousands of particles from the line source and calculating their trajectories using Equations (3), following the procedure described in Rao and Luhar (1993). These calculations are repeated for other line sources.

Footprints were calculated for the measurement points: distances $x_m = 250, 500,$ and 1000 m downwind of the discontinuity, and heights $z_m = 3, 6, 9,$ and 15 m. A total of 30 line sources were placed at log-linear intervals at the ground-level for $x_m = 1000$ m, with the finer logarithmic spacing closer to the measurement location. The number of sources for $x_m = 500$ m was 21 and that for $x_m = 250$ m was 17. The number of particles released at each source was 10,000.

RESULTS AND DISCUSSION

Figure 2 shows footprint plots for moisture flux calculated from the stochastic model for several measurement heights and locations downwind of the surface discontinuity. These footprint plots are also valid for any other emissions that are confined to Surface 2 (*e.g.*, CO_2 flux from vegetation). The occurrence of the footprint peak value for a measurement point corresponds to the upwind location of the area source element to which the flux measurement at that point is most sensitive. For example, the footprint peak for $x_m = 1000$ m and $z_m = 6$ m occurs at ~ 50 m upwind of the measurement point (Fig.2a). This indicates that the area source element located 50 m upwind primarily affects the flux measurement at this point.

The footprint peaks decrease as the measurement height increases, as shown in Fig.2. This can be explained as follows: the higher the measurement point, the more diffused is the plume from the surface source that is contributing the most to the flux at that point. As the measurement height increases, sources farther away from the measurement point will contribute more to the flux because plumes from the nearer sources have not spread wide enough to reach the measurement point; consequently, footprint peaks occur farther from the measurement point. As the measurement height increases, the footprint curves have longer tails to the right hand side and therefore the widths of the curves increase. The footprint curves terminate at an upwind distance equal to x_m, because there are no sources of moisture flux upwind of the discontinuity. For $x_m = 1000$ m, the sources contributing to the flux measurements at $z_m = 6$ m are all located within the fetch from the surface discontinuity to the measurement location (Fig.2a); this is also true for $z_m = 9$, and 15 m. However, for $x_m = 250$ m (Fig.2c), the fluxes measured at $z_m = 15$ m may not be representative of the surface flux because of the inadequate fetch. The measurement heights of 15 m and above are relevant to airborne flux observations, while the lower heights are of interest to tower measurements. We have shown that footprint peaks and locations are different at various heights. Footprint analysis is useful in establishing correspondence between aircraft data and ground-based tower measurements, and in determining an optimum flight altitude when the condition of horizontal homogeneity is not met.

Figure 3a presents the footprint variation for the three downwind measurement locations at a height of 6 m. The footprint curves for this measurement height do not show any noticeable variation with x_m. However, when flux measurements are made at higher elevations, the variation of the footprint with x_m becomes significant (Figs. 3b and 3c). A general conclusion based on these plots is that, as the measurement location is moved away from the surface discontinuity, the footprint peak value decreases and occurs farther from the measurement point, and that such behavior is more pronounced for higher measurement levels. Amongst other factors, the footprint is a function of wind speed (U) and the vertical turbulence intensity (σ_w^2). A larger σ_w^2 causes higher values of footprint peak, while a larger U leads to smaller peak values. Near the surface, both U and σ_w^2 decrease with increasing downwind distance from the discontinuity. The higher footprint peak values for smaller x_m values in the

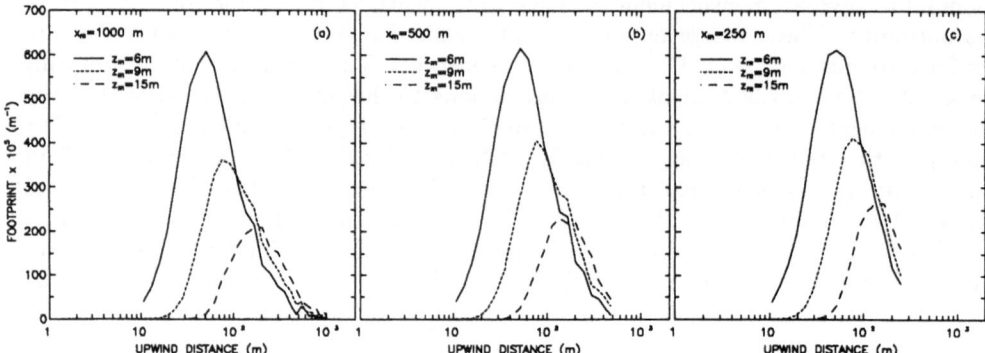

Figure 2. Plots of moisture flux footprint predictions for the measurement locations (x_m) of (a) 1000 m, (b) 500 m, and (c) 250 m.

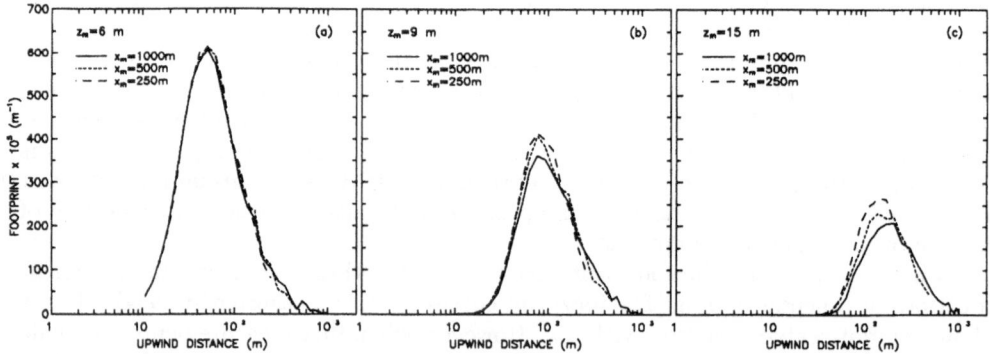

Figure 3. Plots of moisture flux footprint predictions for the measurement heights (z_m) of (a) 6 m, (b) 9 m, and (c) 15 m.

plots shown in Figs 3 suggest that σ_w^2 effects dominate over those due to change in wind speed.

For moisture flux, Equation (1) can be written as:

$$F_a(0, z_m) = \int_0^{x_m} F_o(x') f(x', z_m) \, dx'. \tag{4}$$

Because we know the variations of F_o (Fig.1) and f as functions of upwind distance, we can calculate the moisture flux at a given position (x_m, z_m) by numerically evaluating the integral in this equation, and compare it to that computed directly by the higher-order closure ABL model. Fig.4 shows this comparison. In this plot, the four points for each x_m value correspond to the four heights: 3, 6, 9, and 15 m. The agreement is quite good. For $x_m = 1000$ m, the moisture fluxes calculated using Equation (4) at $z_m =$ 3, 6, 9, and 15 m are 90, 87, 82, and 74 %, respectively, of the surface moisture flux computed by the ABL model at that x_m; these figures are 86, 83, 76, and 63% for $x_m = 500$ m, and 84, 72, 60, and 44% for $x_m =$ 250 m. Clearly, the fetch is not long enough for some moisture flux measurement points.

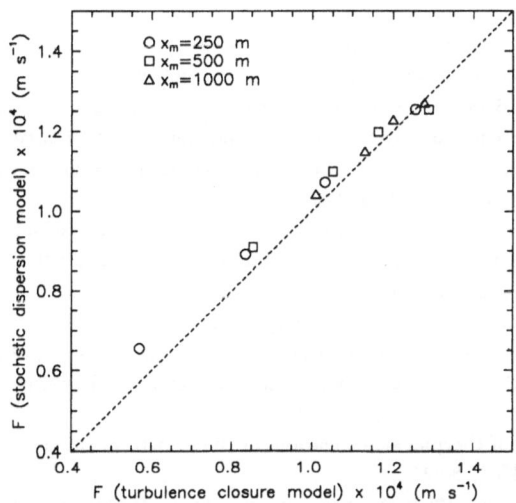

Figure 4. Comparison of the moisture fluxes predicted by the higher-order closure model and computed using the Lagrangian stochastic dispersion model.

CONCLUSIONS

The relative contribution of each element of the upwind surface area source to the measured vertical scalar flux has been referred to as the footprint. In this work, we have applied a Lagrangian stochastic dispersion model to calculate footprints at a number of measurement locations and heights for a simple inhomogeneity involving a step-change in surface roughness and surface moisture conditions (encountered by the flow from an arid region to an irrigated crop field, for example, which changes partitioning of insolation into sensible and latent heat fluxes and hence leads to a change in stability close to the surface). The flow fields driving the stochastic model are calculated from a second-order closure ABL model which included the Penman-Monteith approach with a constant stomatal resistance to account for the effects of vegetation on surface energy and moisture balances. Only the daytime unstable conditions are considered. The results show that the peak footprint value decreases and occurs farther from the measurement location as the measurement height increases. The peak footprint values decrease as the location of the measurement point moves away from the surface discontinuity. This behavior is more pronounced for measurements at higher elevations. The moisture fluxes calculated by the Lagrangian stochastic dispersion model and the higher-order closure model are in good agreement. Footprint analysis is useful for assessing the relative importance of the sources of greenhouse gases in climate change studies, for estimating the 'height-to-fetch' ratio for the siting of flux measurement instruments over inhomogeneous surfaces, and for interpreting aircraft flux measurements in relation to tower data.

Acknowledgements–The research presented in this paper was conducted while the first author held a Research Associateship of the National Research Council at ATDD during 1991-93.

REFERENCES

Horst, T. W., and Weil, J. C., 1992, Footprint estimation for scalar flux measurements in the atmospheric surface layer, *Boundary-Layer Met.* 59:279.

Kroon, L. J. M., and De Bruin, H. A. R., 1993, Atmosphere-vegetation interaction in local advection conditions: effect of lower boundary conditions, *Agricultural Forest Met.* 64:1.

Leclerc, M. Y., and Thurtell, G. W., 1990, Footprint prediction of scalar fluxes using a Markovian analysis, *Boundary-Layer Met.* 52:247.

Luhar, A. K., and Britter, R. E., 1989, A random walk model for dispersion in inhomogeneous turbulence in a convective boundary layer, *Atmos. Environ.* 23:1911.

Luhar, A. K., and Rao, K. S., 1993, Random-walk model studies of the transport and diffusion of pollutants in katabatic flows, *Boundary-Layer Met.* (in press).

Monteith, J. L., 1965, Evaporation and environment, *in* "The State and Movement of Water in Living Organisms," G. E. Fogg, ed., Symposium of Society for Experimental Biology 19:205, Academic Press, N.Y.

Rao, K. S., 1975, Effects of thermal stratification on the growth of the internal boundary-layer, *Boundary-Layer Met.* 8:227.

Rao, K. S., and Luhar, A. K., 1993, Footprint analysis for scalar fluxes measured in the internal boundary layer, *Atmos. Environ.* (submitted).

Rao, K. S., Wyngaard, J. C., and Coté, O. R., 1974a, The structure of a two dimensional internal boundary layer over a sudden change of surface roughness, *J. Atmos. Sci.* 31:738.

Rao, K. S., Wyngaard, J. C., and Coté, O. R., 1974b, Local advection of momentum, heat, and moisture in micrometeorology, *Boundary-Layer Met.* 7:331.

Schuepp, P. H., Leclerc, M. Y., MacPherson, J. I., and Desjardins, R. L., 1990, Footprint prediction of scalar fluxes from analytical solutions of the diffusion equation, *Boundary-Layer Met.* 50:355.

Thomson, D. J., 1987, Criteria for the selection of stochastic models of particle trajectories in turbulent flows, *J. Fluid Mech.* 180:529.

Wilson, J. D., and Swaters, G. E., 1991, The source area influencing a measurement in the planetary boundary layer: the 'footprint' and the 'distribution of contact distance', *Boundary-Layer Met.* 55:25.

DISCUSSION

B. SMITH: Is it possible to replace the second-order closure scheme solution, to determine the wind turbulence structure downwind of the discontinuity, by a structure determined by properties carried by the (Langrangian) particles? If not, why not.

C. NAPPO: No. The wind and turbulence quantities from the second-order closure model are inputs to the particle model. We are not aware of any way of determining these quantities from the properties carried by the particles.

B. PHYSICK: Can you tell me what is the averaging period over which the fluxes are calculated.

C. NAPPO: We considered a steady state internal boundary layer (IBL), and ensemble-averaging is implied in the model results. Since atmospheric conditions can be considered to be in steady state for a period of roughly one hour during daytime, this period may be considered as the averaging period.

MULTIPLE MASTER LENGTH SCALES DERIVED FROM A STATISTICAL

DIFFUSION THEORY

G.A. Degrazia,[1]A.P. de Oliveira,[2]and O.L.L. Moraes[1]

[1]Departamento de Física,Universidade Federal de Santa Maria-
97.119.900-Santa Maria,RS,Brasil and
Radar Meteorológico, UFPel, Pelotas,RS, Brasil
[2]Departamento de Ciências Atmosféricas,Universidade de São Paulo-
01.065.970-São Paulo,SP,Brasil

INTRODUCTION

The second-order closure models have been claimed to be one of the most appropriated tools to simulate numerically the planetary boundary layer. They reconcile the numerical amenability with a more basic physical description of the turbulent process. Several second-order closure models simulate appropriately the convective boundary layer(CBL). Simulating the stable planetary boundary layer(SBL) on the basis of a second-order closure model is a more difficult task. This certainly explains why SBL has been numerically simulated less frequently than the corresponding CBL(Moeng and Wyngaard,1989). The worst difficulty has to do with the proper choice of a turbulent master length scale(Mellor and Yamada ,1982) that leads to an adequate parameterization of the undetermined terms in the equations for evolution of the second-order moments. In the SBL turbulence length scales are also relatively small and limited ultimately by the local Obukhov length rather than SBL Depth H. Recently Lacser and Arya(1986) have shown that the schemes which incorporate the local Monin-Obukhov length as a stability limit on the turbulent master length scale predict shallower and more stable boundary layers. Detailed comparisons between model predictions and data from the Cabauw mast have shown that Delage's(1974) mixing-length formulation performed better than the other schemes in describing vertical profiles and temporal behavior of the mean and turbulent variables.

In this work, we use the statistical diffusion theory to derive an turbulent master length scale for the SBL. In relation to Degrazia et al.(1992) work the present approach represents a new method to obtain characteristic turbulence length scales. In its general formulation, these multiple turbulent master length scales show a form

Air Pollution Modeling and Its Application X, Edited by S-V. Gryning
and M. M. Millán, Plenum Press, New York, 1994

similar to earlier proposals by Delage (1974) with the energy containig eddies expressed in terms of the local Monin-Obukhov length.

THE MODEL

An expression for the time dependent diffusivity coefficients $K\alpha\alpha$ was derived by Batchelor(1949) and represented by Degrazia and Moraes(1992) in the form

$$K\alpha\alpha = \frac{\sigma_i^2 \beta_i}{2\pi} \int_0^\infty \frac{S_i(n)sin(2\pi nt/\beta_i)}{n} dn$$

(1)

where $\alpha=x,y,z$, $i=u,v,w$, $S_i(n)$ is a local energy spectrum, β_i is defined as ratio of the Lagrangian to the Eulerian timescales and $\sigma_i{}^2$ is the turbulent velocities variances. Wandel and Kofoed-Hansen(1962) have shown that

$$\beta_i = \sqrt{\frac{\pi}{4}} \frac{u}{\sigma_i}$$

(2)

where u is the mean wind speed.

The asymptotic behavior of equation(1) for large diffusion travel times (lim t → ∞) when $K\alpha\alpha$ is just a function of the turbulence can be expressed by(Degrazia and Moraes, 1992)

$$K\alpha\alpha = \sigma_i^2 \beta_i S_i(0)/4$$

(3)

Now, $K\alpha\alpha$ is the eddy diffusivity defined by

$$K\alpha\alpha = \sigma_i l_i$$

(4)

where l_i represents the multiple mixing lengths.

By comparision (3) and (4) we obtain a general formulation for the multiple master length scales

$$l_i = \frac{\sigma_i \beta_i S_i(0)}{4}$$

(5)

The value of the spectrum at the origin as given by Degrazia and Moraes (1992) is of the form

$$S_i(0) = \frac{0,64}{(f_m)_{n,i}} \frac{Z}{uq}$$

(6)

with $(f_m)_{n,i}$ being the frequency of the spectral peak in neutral stratification and q is a stability function given by (Sorbjan ,1986)

$$q = 1 + 3,7 \frac{z}{\Lambda}$$

(7)

Finally, the substitution of (2) and (6) in (5) yields the following expression for the multiple turbulent master length scales

$$\frac{1}{l_i} = \frac{1}{\left[0,072 \Big/ (f_m)_{n,i} \right] z} + \frac{1}{\left[0,019 \Big/ (f_m)_{n,i} \right] \Lambda}$$

(8)

The eq. (8) is a characteristic length for turbulent transport processes that has Blackadar (1962) and Delage (1974) form. It is important to stress at this point that this derivation of the multiple master length scale leads simultaneously to the determination of the characteristic size of the energy containing eddies which in the present case has different values for each spacial direction (anisotropic transports) and is determined in terms of local Monin-Obukhov length Λ.

As a particular case we select in (8) the vertical length scale by taking the value of the frequency of the maximum energy containing eddy in the neutral case $(f_m)_{n,w}$ from the experiment as analysed by Sorbjan (1986). In this case study, it was shown that $(f_m)_{n,w} = 0,33$, so that from (8) result.

$$\frac{1}{l_w} = \frac{1}{0,22 z} + \frac{16,5}{\Lambda}$$

(9)

It is here interesting to notice that the numerical factor of 16.5, which is associated to the characteristic size of the energy containing eddies is in good agreement with the value of 14,3 when $\beta \sim 5$ and $k \sim 0,35$ is substituted in eq.(6) of Delage paper (1974).

We stress here that the main effect of buoyancy in the SBL is to decrease the turbulence length scale (Grant ,1992). This decrease in the turbulence length scale increases the dissipation rate and leads to realistic values of turbulence characteristics in regions where the structure of turbulence does not respond to the ground conditions.

CONCLUSIONS

In this paper we presented a method of derivation of characteristic turbulence length scales. This was achieved by first presenting what is proposed as more adequate way of determining a stable boundary layer master length scale. The novel feature of this derivation is the provision for multiple master length scales, one for each different spacial direction. In its general formulation, the multiple master length scales l_i show a form similar to that proposed earlier by Blackadar. An interesting feature of these multiple characteristic length scales l_i is that they become, like Delage length scale, proportional to the local Monin-Obukhov length on the upper part of the SBL.

ACKNOWLEDGEMENTS

We acknowledge financial support provided by Fundação de Amparo a Pesquisa do Estado do Rio Grande do Sul (FAPERGS), Conselho Nacional de Desenvolvimento Científico e Tecnológico (CNPq) and Fundação de Amparo a Pesquisa do Estado de São Paulo (FAPESP).

REFERENCES

Batchelor,G.K.,1949,Diffusion in a field of homogeneous turbulence, I.Eulerian analysis,Aust.J.Sci.Res.2:437.

Blackadar,A.K.,1962,The vertical distribution of wind an turbulent structures of the planetary boundary layer,J.Geophy.Res.67:3095.

Degrazia,G.A., Oliveira,A.P.and Moraes,O.L.L.,1992,Multiple master length scales for stable atmospheric boundary layer,II Nuovo Cim.15c:409.

Degrazia,G.A.and Moraes,O.L.L.,1992,A model for eddy diffusivity in a stable boundary layer,Bound.-Layer Metor.58:205.

Delage,Y.,1974,A numerical study of the nocturnal atmospheric boundary layer,Quart.J.Roy.Meteorol.Soc.100:351.

Grant,A.L.M.,1992,The structure of turbulence in the near-neutral atmospheric boundary layer,J.Atmos.Sci.49:226.

Lacser,A.and Arya,S.P.S.,1986,A comparative assessment of mixing-length parameterizations in the stably stratified nocturnal boundary layer(NBL),Bound.-Layer Meteor.36:53.

Mellor,G.L.and Yamada,T.,1982,Development of a turbulence closure model for geophysical fluid problems,Rev.of Geophys.and Space Phys.20:851.

Moeng,C.H.and Wyngaard,J.C.,1980,Evaluation of turbulent transport and dissipation closures in second-order modeling,J.Atmos.Sci.46:2311.

Sorbjan,Z.,1986,Local similarity of spectral and cospectral characteristics in the stable-continuous boundary layer,Bound.-Layer Meteor.35:257.

Wandel,C.F. and Kofoed-Hansen,O.,1962,On the Eulerian-Lagrangian transform in the statistical theory of turbulence,J.Geophys.Res.67:3089.

DISCUSSION

C. NAPPO: How, if at all, your work relate to the sporadic nature of turbulence in the stable PBL?

A.P. DE OLIVEIRA: The scales are derived in this work with the assumption that the turbulence is continous in space. Therefore intermittency can not be inclued here.

DEVELOPMENT OF A LAGRANGIAN STOCHASTIC MODEL
FOR DISPERSION IN COMPLEX TERRAIN *

G. Brusasca,[1] G. Tinarelli,[1] D. Anfossi,[2]
E. Ferrero,[3] F. Tampieri,[4] F. Trombetti [4]

[1] ENEL/CRAM Unita' studi e monitoraggi ambientali, Milano, Italy
[2] Istituto di Cosmogeofisica, C.N.R., Torino, Italy
[3] Universita' di Alessandria, Italy
[4] Istituto FISBAT, C.N.R., Bologna, Italy

INTRODUCTION

A particle model SPRAY suitable for dealing with the atmospheric dispersion of buoyant emission in complex terrain has been developed. It is an extension of the model LAMBDA (Anfossi et al. 1991, Brusasca et al. 1989 and 1992) based on the Langevin equation, designed to simulate the dispersion on flat terrain, which was validated   in various atmospheric conditions. In this last model the vertical profiles of wind and turbulence, defined at the source locations, were kept constant in all the computational domain.

In order to apply our random walk model in complex terrain the 3-D inhomogeneous structure of the mean wind and turbulence fields have been accounted for. In particular the crosscorrelation terms $\overline{u'w'}$, $\overline{u'v'}$, $\overline{v'w'}$ on the random forcing has been explicitly introduced.

WIND AND TURBULENCE FIELDS

A peculiar feature of lagrangian stochastic models is their capability of using in the best way all the available meteorological measurements and, in particular, the information

---

* The paper was presented during the XIXth ITM in Crete and erroneously was not published in the earlier proceedings

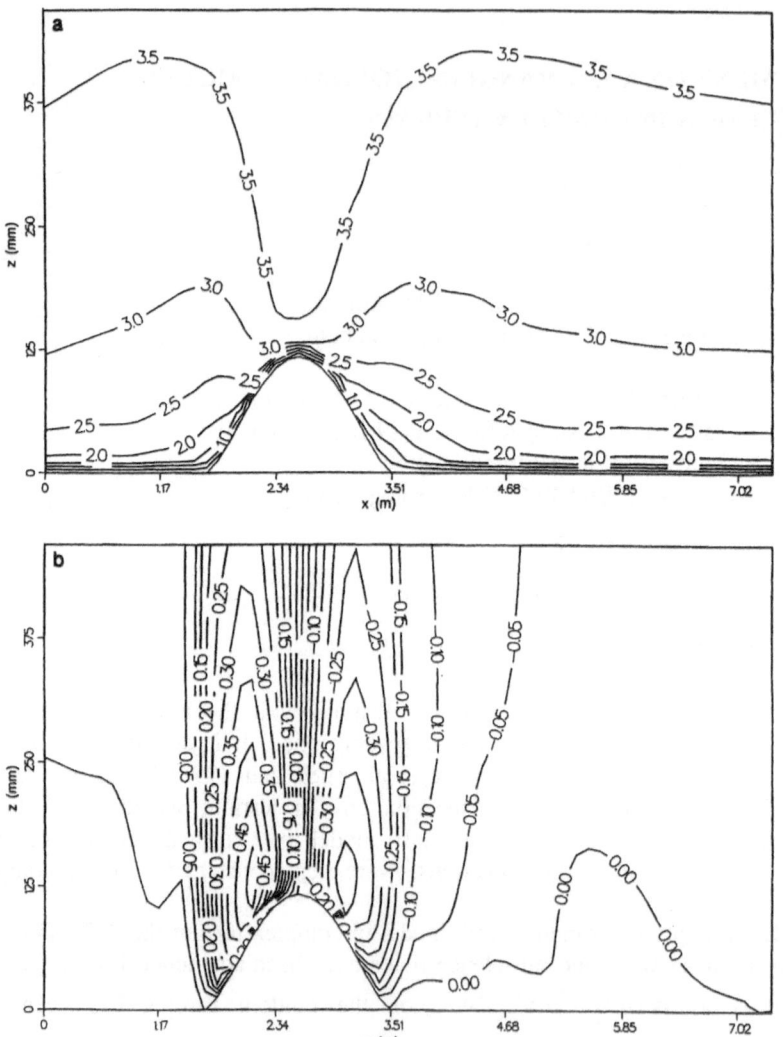

Fig. 1. Vertical cross section of the horizontal (a) and vertical (b) wind component for a 2-D hill in the EPA wind tunnel experiments. Units in X-axis are in metres and in Y-axis are in mm.

supplied by advanced instruments (like Doppler Sodars, Rass, sonic anemometers and so on).

However, before applying our new model to field conditions, we thought it appropriate to test it against data sets obtained in controlled conditions. As a consequence we utilized the EPA wind tunnel data set (Khurshudyan et al., 1981). This experiment simulates the boundary layer flow and the dispersion from point sources over flat terrain and schematic 2-D hills. Vertical profiles of mean horizontal velocity $\overline{u(z)}$, mean flow angle. Reynold stresses (σ_u', σ_v', σ_w', $\overline{u'w'}$) were measured. The tracer concentration field (C(x,y,z)) was measured too.

In order to apply our model, a proper inizialization of these fields was needed. The mean wind field was adjusted by means of a mass consistent diagnostic model (Geai, 1987) whose input was the complete set of $\overline{u(z)}$ profiles. The vertical velocities generated by this model were found to be in excellent agreement with those derived from the measurements. Fig.s 1 show an example of the resulting 2-D vector wind field (Finardi et al. 1993).

As regards the turbulent quantities, a smoothing of data has been performed to reduce the strong oscillations of the spatial gradients due to the data scatter. Lacking a theoretical foundation, a smoother based on splines (routine CSSMH: IMSL, 1989) was used. This routine calculates a smoothing spline by using a "weight" parameter that coincides with the experimental standard deviation of the data as deduced from the observations of Khurshudyan et al. (1981) and from the error analysis reported by Gong and Ibbetson (1989). Fig.s 2 show some Reynolds stresses obtained by a suitable interpolation of the smoothed vertical profiles.

It should be noted that in the hill cases $\sigma_v'(z)$ was not measured and we assumed $\sigma_v'(z) = 0.75 \, \sigma_u'(z)$ on the basis of the value found on the flat terrain case (Trombetti and Tampieri, 1991). This choice is quite similar to that one used in a valley study in the same EPA wind tunnel by Snyder et al. (1991).

LANGRANGIAN TIME SCALES

The lagrangian time scales T_{Li} (with i=x,y,z) necessary to run the model have been estimated using the available observations of dispersion of plumes released at two different heights over flat floor. The resulting formula for T_{Li} have been used also for the simulation over the hills. To evaluate T_{Li} the measured standard deviation σ_y of the lateral concentration distribution of the above mentioned plumes have been fitted to the Taylor formula (see Hanna, 1982). In terms of lagrangian quantities, the length scale l_y is proportional to $T_{Ly} \, \sigma_v'$; by assuming $l_y^{-1} = l_{0y}^{-1} + (cz)^{-1}$, we obtain

$$T_{Ly}(z) = \frac{c \, l_{0y} \, z}{(cz + l_{0y}) \, \sigma_v(z)} \tag{1}$$

(with $l_{0y} = 0.11$ m and c = 3.05 to fit the previously obtained $T_{Ly}(z)$ values). $T_{Lx}(z)$ was assumed equal to $T_{Ly}(z)$.

As far as $T_{Lz}(z)$ are concerned, it was necessary to take into account the wind shear effects on the vertical dispersion (Hunt, 1985). According to Trombetti and Tampieri (1991), we defined

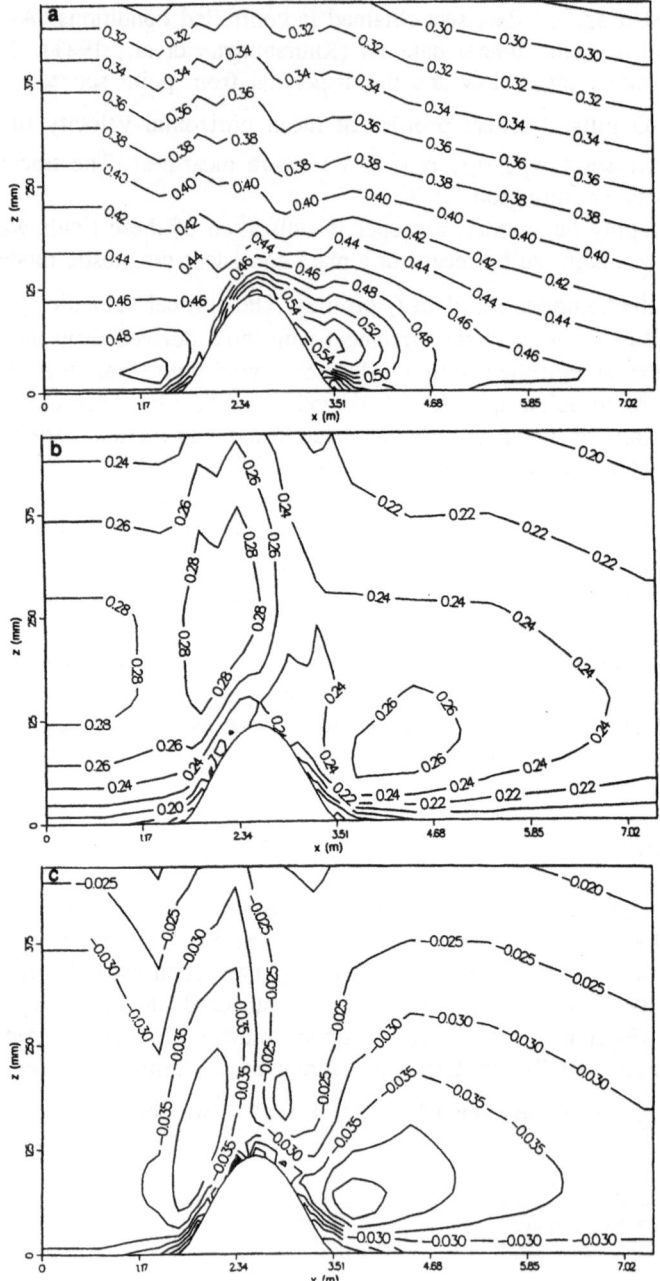

Fig. 2. As in fig. 1 but for $\sigma_{u'}$ (a), $\sigma_{w'}$ (b) (a), and $\overline{u'w'}$ (c).

$$T_{Lz}(z) = \frac{1}{1.2}\frac{l_z}{\sigma_{w'}(z)} \tag{2}$$

where, in this case, the vertical lenght scale l_z can be given by:

$$l_z^{-1} = l_{0z}^{-1} + (kz)^{-1} \tag{3}$$

as in the T_{Ly} case (where k is the Von Karman constant and having estimated $l_{0z} = 0.2$ m), or by

$$l_z^{-1} = l_{0z}^{-1} + \frac{\partial u/\partial z}{\sigma_{w'}(z)} + (2.4z)^{-1} \tag{4}$$

in order to account for the vertical wind shear when this last is different from the logarithmic one (Hunt et al., 1989).

NUMERICAL SCHEME

The particle positions X_i (i=x,y,z) are computed at each time step Δt as follows:

$$X_i(t+\Delta t) = X_i(t) + \left(\overline{U_i}(t) + U_i'(t)\right)\Delta t \tag{5}$$

where $\overline{U_i}$ represent the transport which are provided by the mass consistent model output, and U'_i refer to the turbulent terms (random forcing) which are computed from the Langevin equations (De Baas et al., 1986)

$$U'(t+\Delta t) = U'(t)\left(1 - \frac{\Delta t}{2T_{Li}}\right)\left(1 + \frac{\Delta t}{2T_{Li}}\right)^{-1} + \mu\left(1 + \frac{\Delta t}{2T_{Li}}\right)^{-1} \tag{6}$$

To obtain the numerical scheme for evaluating the random forcing μ it is necessary to derive the analytical expression of the moments of the random forcing distribution, to solve an algebraic system to evaluate the parameters used to pick up at random from that distribution and to designe its relative computer technique. The complete set of moment equations is developed according to Thomson (1984) for the first and second moments $\overline{\mu}$, $\overline{\mu^2}$ and $\overline{\mu_i \mu_j}$ along the three directions x, y and z (Ferrero and Anfossi, 1991). We took into account the third order moment $\overline{\mu_z^3}$ too because in the boundary layer here studied large vertical gradients were found. These moments are calculated using the measured eulerian values of velocity statistics and the lagrangian time scale profiles.

A skew distribution for the vertical random velocity can be obtained by a linear combination of two Gaussian distributions (G.d.) as suggested by Baerensten and Berkowicz (1984) and De Baas et al. (1986). Extending such method to the 3-D case we construct it out of the combination of bidimensional G.d.. Therefore we consider two couples of joint G.d. P (u,w_1), P(u, w_2), P(v, w_1), P(v, w_2) and one joint G.d. P(u,v). Hence we can calculate the following equations:

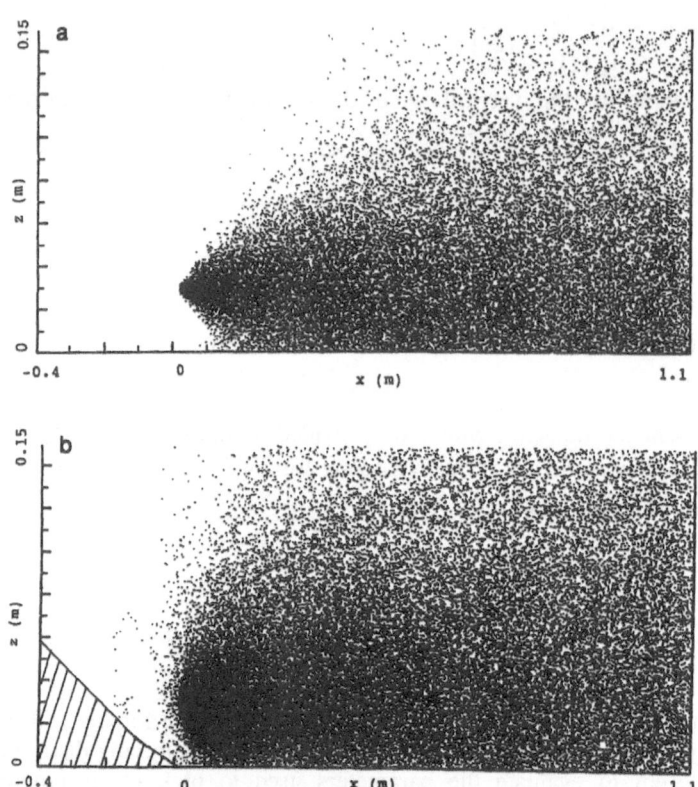

Fig. 3. Vertical cross section of particle simulations emitted from a source at 29 mm: (a) flat terrain case; (b) hill case (the source is located at the downwind hill base).

$$\iint u\,v\,P(u,v)\,du\,dv = \overline{\mu_x \mu_y}$$

$$\sum_{j=1}^{2} a_j \iint u_i w_j\, P(u_i, w_j)\,du_i\,dw_j = \overline{\mu_i \mu_z} \qquad i = x, y; \quad u_x = u; \quad u_y = v$$

$$\iint u_i^n\, P(u_i, w)\,du\,dw = \overline{\mu_i^n} \qquad i = x, y; \quad n = 1, 2$$

$$\sum_{j=1}^{2} a_j \iint w_j^n\, P_j(u, w_j)\,du\,dw_j = \overline{\mu_z^n} \qquad n = 0, 1, 2, 3$$

These eleven equations give the values of the means (m_u, m_v, m_{wj}), standard deviations (s_u, s_v, s_{wj}) and cross correlations (ρ_{uv}, ρ_{uw_j}, ρ_{vw_j}) of the distributions (having set $m_{u_i}^2 = s_{u_i}^2$ and $\rho_{u_i w_1} = \rho_{u_i w_2}$. From these distributions we get three random cross correlated variables (μ_x and μ_y normally distributed and μ_z skew distributed) as follows:

$$\mu_x = b_1 + a_{11} R_1$$
$$\mu_y = b_2 + a_{21} R_1 + a_{22} R_2$$
$$\mu_z = b_3 + a_{31} R_1 + a_{32} R_2 + a_{33} R_3$$

where R_k ($k = 1, 2, 3$) are three normally (0,1) distributed numbers and:

$$b_k = m_1 \quad (k = 1, 2, 3 \text{ and } 1 = u, v, w); \quad a_{11} = s_u; \quad a_{21} = s_v\,\rho_{uv}; \quad a_{31} = s_w \rho_{uw}$$

$$a_{22} = s_v \sqrt{1 - \rho_{uv}^2}; \qquad a_{32} = s_w \frac{\rho_{vw} - \rho_{uw}\rho_{uv}}{\sqrt{1 - \rho_{uv}^2}};$$

$$a_{33} = s_w \sqrt{1 - \rho_{uv}^2 - \frac{(\rho_{vw} - \rho_{uw}\rho_{uv})^2}{1 - \rho_{uv}^2}}$$

being $w = w_1$ or w_2

CONCLUSIONS

An example of the preliminary results obtained by SPRAY simulations is shown in Fig.s 3. It illustrates the different behaviour of the same tracer released from an elevated ($z = 29$ mm) source for flat floor (a) and hill case in which the source was located at the downwind hill base. In both runs the upwind turbulence and wind fields were the same. These drawings evidentiate the distortions induced to the plume trajectory and width by the presence of the hill upwind the source. In particular it clearly appears that the model is able to simulate the main expected features, i.e.: in the hill case the maximum ground level concentrations is higher than in the flat floor case and its position is nearer to the source.

ACKNOWLEDGEMENTS

Some authors (Anfossi, Ferrero, Tampieri and Trombetti) were partially supported by the CNR-ENEL Project "Interactions of energy systems with human health and environment", Roma, Italy.

The numerical code of the mass consistent model MINERVE was developed by EdF, Directions des Etudes et Recherches.

REFERENCES

Anfossi D., Ferrero E., Brusasca G. Tinarelli G., Giostra U., Tampieri F., and Trombetti F., 1992, Dispersion simulation of a wind tunnel experiment with lagrangian particle models. Il Nuovo Cimento 15C-2 139-158

Baerentsen J.H. and Berkowicz R., 1984, Monte Carlo simulation of plume dispersion in convective boundary layer. Atmospheric Environment, 18, 701-712

Brusasca G., Tinarelli G. and Anfossi D., 1989, Comparison between the results of a Monte Carlo atmospheric diffusion model and tracer experiment, Atmospheric Environment, 23, 1263-1280

Brusasca G., Tinarelli G. and Anfossi D., 1992, Particle model simulation of diffusion in low windspeed stable conditions, Atmospheric Environment, 26, 707-723.

De Baas H.F., Van Dop H. and Nieuwstadt F.T.M., 1986, An application of the Langevin equation for inhomogeneous conditions to dispersion in a convective boundary layer. Q. Jl. R. Met. Soc., 112, 165-180

Finardi S., Brusasca G., Morselli M.G., Trombetti F. and Tampieri F., 1993, Boundary-Layer flow over analytical two-dimensional hills: a systematic comparison of different models with wind tunnel data, Boundary Layer Met, 63, 259-291

Ferrero E. and Anfossi D., 1991, Numerical schemes for determining the stochastic component in the random walk atmospheric dispersion models, ICG/CNR Report No 241/91

Geai P., 1987, Method d'interpolation et de reconstitution tri-dimensionnelle d'un champ de vent: le code d'analysis objective MINERVE, EdF Rpt HE/34-87 .03

Gong W. and Ibbetson A, 1989, A wind tunnel study of turbulent flow over model hills, Boundary Layer Met. 49, 113-148

Hanna S.R., 1982, Turbulent diffusion: chimneys and cooling towers, in: Engineering Meteorology, Plate E. Ed., Elsevier, Amsterdam, 429-479

Hunt J.C.R., 1985, Turbulent diffusion from sources in complex flows, Ann. Rev. Fluid. Mech. 17, 447-485

Hunt J.C.R., Moin P., Lee M., Moser R.D., Spalart P., Mansour N.N., Kaimal J.C. and Gaynor E., 1989, Cross correlation and length scales in turbulent flows near surfaces, in: Advances in Turbulence 2, Fernholz H.H. and Fiedler H.E. Eds., Springer-Verlag, Berlin, Heildelberg, 128-134

IMSL Math/Library, 1989, FORTRAN Subroutines for mathematical applications, MALB-USM-PERFECT-EN8912-1.1, User Manual, 550-553

Khurshudyan L.H., Snyder W.H. and Nekrasov I.V., 1981, Flow and dispersion of pollutants over two-dimensional hills. U.S. Envir. Prot. Agcy Rpt. No. EPA-600/4-81-067, Res. Tri. Pk., N.C, U.S.A.

Snyder W.H., Khurshudyan L.H., Nekrasov I.V., Lawson R.E. and Thompson R.S.,1991, Flow and dispersion of pollutants within two-dimensional valleys, Atmospheric Environment 25, 1347-1375

Thomson D.J., 1984, Random walk modelling of diffusion in inhomogeneous turbulence. Q. Jl. R. Met. Soc., 110, 1107-1120

Trombetti F. and Tampieri F., 1991, Analysis of wind tunnel dispersion data over two-dimensional obstacles, Boundary Layer Met. 59, 209-226

DISCUSSION

R. SALERNO: What were the stability conditions you have used in your simulations?

G. BRUSASCA: The EPA wind tunnel experiment we have simulated was performed in neutral conditions.

R. SALERNO: Do you think that the Lagrangian time scale you have found is underestimated?

G. BRUSASCA: In the lee region the vertical Lagrangian time scales used in our simulations (eqs. (2)-(4)) are probably underestimated, in fact in the continuation of this work, we found that Tlw parametrization proposed by Berlyand and Genikhovich (1971) and quoted by Khushudyan et al. (1981), generates higher values of these parameters in the lee region, producing better agreement between observed and predicted ground level concentrations.

ACCIDENTAL RELEASE

chairman: H. ApSimon

rapporteur :   M. Coutinho

RECURRENCE OF EXTREME CONCENTRATIONS

Leif Kristensen

Risø National Laboratory
4000 Roskilde, Denmark

INTRODUCTION

We have known for many years that atmospheric dispersion modeling should include not only predictions of mean concentrations of pollutants, but also information about possible extreme events, i.e. probabilistic statements about deviations from the predicted mean concentrations. One way of complying with this requirement is to predict, as a function of time and space, the variance together with the mean of the concentration. With an assumption about the form of the probability density it will then be possible to calculate the probability that, at any given time, the concentration will exceed a particular value. Relevant discussions can be found in Chatwin and Sullivan (1993) and references therein.

However, the concentration probability density only tells how long a fraction of the time the concentration, on average, is larger than a certain value and not, as is often of more interest from a regulatory point of view, how often a certain value is exceeded. The problem has been stated very pointedly by Deardorff and Willis (1988): "How long must one wait in exposure to a pollutant before there is an even chance that a fluctuation in the concentration exceeding a particular value will actually occur?". Equivalently, one can ask the question: "On average, how often will a particular threshold be exceeded?". To answer this question it is necessary to know more than the concentration probability density.

Before the question about "how often?", i.e. the *recurrence*, one should realize that we are talking about gases that are volatile, toxic or just inconvenient because of their smell. This implies that we are not considering the instantaneous concentration at a particular point in space, but rather the concentration integrated over a certain time and/or volume, because a certain "exposure" is required to trigger a spontaneous explosion or for an odor to be detected. In other words, we need to specify the volume- and/or time averaging, i.e. the *filtering*, of the concentration observation of the gas in question. Many other factors are decisive in recurrence considerations. Here I will

Air Pollution Modeling and Its Application X, Edited by S-V. Gryning
and M.M. Millán, Plenum Press, New York, 1994

discuss, in simple terms, extreme value statistics and the possibility for prediction of extreme events and emphasize some of these factors:

- Reference time interval.

- Filtering in time.

- Type of source.

The following is to a large extent based on the discussions in Kristensen *et al.* (1989) and Kristensen *et al.* (1991).

BASIC CONSIDERATIONS

I have already mentioned that the gas concentrations we are considering are averaged with a filter. I will limit our discussion to temporal filters which can be characterized by a time constant τ_0.

There is another time or rather, time interval T, we need to specify when we are trying to determine the statistics of the largest values. To understand this we start by realizing that there is no well-defined, absolutely largest value of a concentration. Instead we could ask for the probability that the largest value of a gas concentration $c(t)$, we are going to observe during an experiment of duration T, does not exceed a certain value C. Therefore, we imagine that there exists a large number, M, of statistically equivalent, continuous records or time series of $c(t)$, all of duration T. From each of this large number of records, which is also what we call an ensemble of realizations, we pick the largest value. Let C_i be that pertaining to i's realization. Then we can find the probability $P(<C)$ that the largest value is smaller than C by introducing the indicator function

$$
B_i(C) = \begin{cases} 1 \text{ for } & C_i \leq C \\ \\ 0 \text{ for } & C_i > C \end{cases}
\tag{1}
$$

and use that to obtain

$$
P(<C) = \lim_{M \to \infty} \frac{1}{M} \sum_{i=1}^{M} B_i(C).
\tag{2}
$$

Theoretically it can be shown (Gumbel, 1958) that $P(<C)$ is often well approximated by

$$
P(<C) = \exp\left(-\exp\left(-\gamma \frac{C - [C]}{\langle C \rangle - [C]}\right)\right),
\tag{3}
$$

where

$$
\gamma = 0.5772156649\ldots
\tag{4}
$$

is the Euler constant, $\langle C \rangle$ the mean of the largest values, viz.,

$$\langle \mathcal{C} \rangle = \lim_{M \to \infty} \frac{1}{M} \sum_{i=1}^{M} \mathcal{C}_i, \tag{5}$$

and $[\mathcal{C}]$ the mode, i.e. the position of the maximum of the probability density function corresponding to (3):

$$p(\mathcal{C}) = \frac{\gamma}{\langle \mathcal{C} \rangle - [\mathcal{C}]} \exp\left(-\gamma \frac{\mathcal{C} - [\mathcal{C}]}{\langle \mathcal{C} \rangle - [\mathcal{C}]}\right) \exp\left(- \exp\left(-\gamma \frac{\mathcal{C} - [\mathcal{C}]}{\langle \mathcal{C} \rangle - [\mathcal{C}]}\right)\right). \tag{6}$$

Note that, because $[\mathcal{C}] \neq \langle \mathcal{C} \rangle$, the Gumbel probability density function is always skew.

An alternative way of discussing large concentrations is to ask how many times during the *reference time* T the concentration $\mathcal{C}$ is exceeded. This number will of course in general be different from realization to realization, so we must ask for an average number. If $N_i(\mathcal{C})$ is the actual number from the i'th realization, the average number of excursions becomes

$$N(\mathcal{C}) = \lim_{M \to \infty} \frac{1}{M} \sum_{i=1}^{M} N_i(\mathcal{C}). \tag{7}$$

It is possible to establish a connection between the two ways of discussing extreme events. If we assume that in each realization the individual excursions beyond the level $\mathcal{C}$ are so large that they can be considered statistically independent, then the probability $P_{\mathcal{C}}[n]$ for a particular number n of excursions will have a Poisson distribution, i.e.

$$P_{\mathcal{C}}[n] = \frac{e^{-N(\mathcal{C})}}{n!} N^n(\mathcal{C}). \tag{8}$$

In particular, the probability for no excursions beyond $\mathcal{C}$ ($n=0$) becomes

$$P_{\mathcal{C}}[0] = e^{-N(\mathcal{C})}. \tag{9}$$

This probability is the same as the probability, given by (2), that in a particular realization the maximum value is less than $\mathcal{C}$. In other words,

$$P_{\mathcal{C}}[0] = P(<\mathcal{C}). \tag{10}$$

Using (3) and (10), we obtain the relation

$$N(\mathcal{C}) = \exp\left(-\gamma \frac{\mathcal{C} - [\mathcal{C}]}{\langle \mathcal{C} \rangle - [\mathcal{C}]}\right). \tag{11}$$

This equation implies that if the average number of excursions beyond $\mathcal{C}$ is known for a range of $\mathcal{C}$ we can determine $\langle \mathcal{C} \rangle$ and $[\mathcal{C}]$, and, consequently, the probability $P(< \mathcal{C})$ that the largest value of the time series $c(t)$ of duration T does not exceed $\mathcal{C}$. And vice versa, if $P(< \mathcal{C})$ is known, then we can determine the average number of excursions $c(t)$ beyond $\mathcal{C}$ in a time interval of duration T. Interesting enough, if we decide to let this number be one, i.e. we want to find the value of $\mathcal{C}$ which on average is exceeded once, (11) shows that this value is $[\mathcal{C}]$. Stated differently: *The value exceeded on average once is the most probable largest value.* We see, by setting $N(\mathcal{C})$ equal to one in (8), that the probability that $[\mathcal{C}]$ is not exceeded is 37% and that the probability that there is at least one excursion beyond $[\mathcal{C}]$ is 63%.

In the next section we will show how it is often possible to determine $N(\mathcal{C})$ or rather the rate of excursions $\eta_{\mathcal{C}} = N(\mathcal{C})/T$ beyond $\mathcal{C}$ by determining the mean and the variance of $c(t)$. Following Panofsky and Dutton (1984), we call the theory we use *exceedance statistics*.

EXCEEDANCE STATISTICS

We consider a concentration time series $c(t)$ being discretely sampled with the temporal resolution Δt. Suppose $c_0 = c(n\Delta t) < \mathcal{C}$ and $c_1 = c((n+1)\Delta t) > \mathcal{C}$. Then we know that the level $\mathcal{C}$ has been up-crossed at least once in the period from $n\Delta t$ to $(n+1)\Delta t$. If we call the probability for this event $\text{Prob}(c_0 < \mathcal{C}, c_1 > \mathcal{C})$ then the lower limit of the average number of times the level $\mathcal{C}$ has been exceeded in the period of time $T = N\Delta t$ is $N\text{Prob}(c_0 < \mathcal{C}, c_1 > \mathcal{C})$ and the corresponding average rate becomes

$$\eta_{\mathcal{C}}' = \frac{N}{T}\text{Prob}(c_0 < \mathcal{C}, c_1 > \mathcal{C}) = \text{Prob}(c_0 < \mathcal{C}, c_1 > \mathcal{C})/\Delta t. \tag{12}$$

When the temporal resolution is extremely good, i.e. $\Delta t \to 0$, $\eta_{\mathcal{C}}'$ becomes the true excursion rate beyond $\mathcal{C}$:

$$\lim_{\Delta t \to 0} \eta_{\mathcal{C}}' = \eta_{\mathcal{C}}. \tag{13}$$

Let $\tilde{p}(c_0, c_1)$ the joint probability density of c_0 and c_1. Then

$$\text{Prob}(c_0 < \mathcal{C}, c_1 > \mathcal{C}) = \int_0^{\mathcal{C}} dc_0 \int_{\mathcal{C}}^{\infty} \tilde{p}(c_0, c_1)dc_1. \tag{14}$$

Introducing the transformation

$$\left.\begin{array}{rcl} c_m &=& \frac{1}{2}(c_0 + c_1) \\ \Delta c &=& c_1 - c_0 \end{array}\right\} \Longleftrightarrow \left\{\begin{array}{rcl} c_0 &=& c_m - \Delta c/2 \\ c_1 &=& c_m + \Delta c/2 \end{array}\right. , \tag{15}$$

we can reformulate (14) in the following way

$$
\begin{aligned}
\text{Prob}(c_0 < \mathcal{C}, c_1 > \mathcal{C}) &= \int_0^{\mathcal{C}} d\Delta c \int_{\mathcal{C}-\Delta c/2}^{\mathcal{C}+\Delta c/2} p(c_m, \Delta c)dc_m \\
&+ \int_{\mathcal{C}}^{\infty} d\Delta c \int_{\Delta c/2}^{\mathcal{C}+\Delta c/2} p(c_m, \Delta c)dc_m \\
&\approx \int_0^{\infty} d\Delta c \int_{\mathcal{C}-\Delta c/2}^{\mathcal{C}+\Delta c/2} p(c_m, \Delta c)dc_m \\
&\approx \int_0^{\infty} \Delta c\, p(\mathcal{C}, \Delta c)d\Delta c
\end{aligned} \tag{16}
$$

where

$$p(c_m, \Delta c) \equiv \tilde{p}(c_m - \Delta c/2, c_m + \Delta c/2). \tag{17}$$

The last expression in (16) is the conditional average $\langle \Delta c \rangle_C$ of Δc and the assumption behind the validity of this last expression is that C is chosen such that

$$\langle \Delta c \rangle_C \ll C. \tag{18}$$

This, of course is accomplished by letting Δt be short and/or C large.

Inserting (16) in (12), we get

$$\eta'_C = \int_0^\infty \frac{\Delta c}{\Delta t} p(C, \Delta c) d\Delta c. \tag{19}$$

In the limit $\Delta t \to 0$ we have

$$\lim_{\Delta t \to 0} \frac{\Delta c}{\Delta t} = \frac{dc}{dt} \equiv \dot{c} \tag{20}$$

so that

$$\eta_C = \lim_{\Delta t \to 0} \eta'_C = \int_0^\infty \dot{c} P(C, \dot{c}) d\dot{c}, \tag{21}$$

where

$$P(c, \dot{c}) = \lim_{\Delta t \to 0} \{p(c, \Delta c) \Delta t\} \tag{22}$$

is the joint probability density of c and its time derivative $\dot{c}$. The derivation of (21) is slightly different from that presented by Panofsky and Dutton (1984).

We note that $P(c, \dot{c})$ is independent of the reference time T. This implies that η_C is independent of T and that consequently $N(C) = T\eta_C$ is proportional to T. The total time $c(t)$ is larger than C during T is also proportional to T and is given by

$$\Theta_C = T \int_C^\infty dc \int_{-\infty}^\infty P(c, \dot{c}) d\dot{c}. \tag{23}$$

Therefore it is also possible to estimate the average duration θ_C of an excursion beyond C by

$$\theta_C = \frac{\Theta_C}{\eta_C T} = \frac{\int_C^\infty dc \int_{-\infty}^\infty P(c, \dot{c}) d\dot{c}}{\int_0^\infty \dot{c} P(C, \dot{c}) d\dot{c}}, \tag{24}$$

which, of course, is also independent of the chosen reference time T.

From these considerations we realize that useful information can be derived from the joint probability density $P(c, \dot{c})$. If we assume that $c(t)$ is statistically stationary, implying that the ensemble mean of any function of $c(t)$ is constant, it is easily seen that $c(t)$ and $\dot{c}$ are uncorrelated. First of all, the mean itself $C = \langle c \rangle$ is independent of time. This implies that the mean of $\dot{c}$ is zero:

$$\langle \dot{c} \rangle = \frac{dC}{dt} = 0. \tag{25}$$

The covariance of $c(t)$ and $\dot{c}(t)$, which is the mean of the product of $c'(t) \equiv c(t) - C$ and $\dot{c}(t) - \langle \dot{c} \rangle = \dot{c}(t)$, becomes

$$\langle c'(t) \dot{c}(t) \rangle = \langle c(t) \dot{c}(t) \rangle = \frac{d}{dt} \left(\frac{\langle c^2(t) \rangle}{2} \right) = 0, \tag{26}$$

345

which shows that $c(t)$ and $\dot{c}(t)$ are uncorrelated. In general (26) does not imply that $c(t)$ and $\dot{c}(t)$ are statistically independent, i.e. the joint probability can we written as a product of the form

$$P(c,\dot{c}) = Q_0(c)Q_1(\dot{c}). \tag{27}$$

On the other hand, if $c(t)$ and $\dot{c}(t)$ are statistically independent and obey an equation of the form (27) one can easily convince oneself that then $c(t)$ and $\dot{c}(t)$ are also uncorrelated. There is one case for which (26) implies (27), namely if $P(c,\dot{c})$ is Gaussian. Then

$$P(c,\dot{c}) = \frac{1}{2\pi\sigma_c\sigma_{\dot{c}}\sqrt{1-\rho^2}} \exp\left(-\frac{1}{1-\rho^2}\left\{\frac{c'^2}{2\sigma_c^2} - \rho\frac{c'\dot{c}}{\sigma_c\sigma_{\dot{c}}} + \frac{\dot{c}^2}{2\sigma_{\dot{c}}^2}\right\}\right), \tag{28}$$

where $\sigma_c^2 = \langle c'^2\rangle$, $\sigma_{\dot{c}}^2 = \langle \dot{c}^2\rangle$ and $\rho = \langle c'\dot{c}\rangle/(\sigma_c\sigma_{\dot{c}})$. Since (26) implies that $\rho = 0$, (28) is of the form (27).

Unfortunately, there are at the moment no observations to guide a choice of $P(c,\dot{c})$ and, since (26) can be considered circumstantial evidence for statistical independence, we will assume that $P(c,\dot{c})$ is of the form (27) and further that $Q_1(\dot{c})$ is Gaussian:

$$Q_1(\dot{c}) = \frac{1}{\sqrt{2\pi}\sigma_{\dot{c}}} \exp\left(-\frac{\dot{c}^2}{2\sigma_{\dot{c}}^2}\right). \tag{29}$$

Substituting (29) in (21) and (24), we get

$$\eta_C = \frac{\sigma_{\dot{c}}Q_0(C)}{\sqrt{2\pi}} \tag{30}$$

and

$$\theta_C = \frac{\sqrt{2\pi}}{\sigma_{\dot{c}}Q_0(C)} \int_C^\infty Q_0(c)dc. \tag{31}$$

Kristensen *et al.* (1989) derived the expressions for η_C and θ_C under the assumption the $Q_1(\dot{c})$ is symmetric and that $\dot{c}^2/\sigma_{\dot{c}}^2$ has a log-normal probability density. In this case the expressions for η_C and θ_C are identical to (30) and (31), except that $\sqrt{2\pi}$ is replaced by $2K^{1/8}$ where $K = \langle\dot{c}^4\rangle/\langle\dot{c}^2\rangle^2$ is the kurtosis, or flatness factor, of $\dot{c}$. According to Gibson *et al.* (1970), the temperature derivative, which we assume exhibits the same properties as $\dot{c}(t)$, has values of K which typically lies in the interval from 25 to 40. This means that $2K^{1/8}/\sqrt{2\pi}$ lies in the interval from 1.2 to 1.3. I take this as an indication that the results (30) and (31) are not particularly sensitive to the choice of $Q_1(\dot{c})$.

THE ATMOSPHERIC SURFACE LAYER

The considerations I have presented so far apply to any stationary time series and have been presented in more or less the same form by Kristensen *et al.* (1991) and by Kristensen (1993) in connection with wind gusts.

We are interested in the statistics of a diffusing gas with concentration $c(\boldsymbol{x},t)$[†] and assume the atmospheric transport equation

---

[†]The concentration is in general a function of both time t and space coordinates $\boldsymbol{x}$. However, since I am considering the statistics of the temporal variation of c at a particular location, the argument $\boldsymbol{x}$ is occasionally suppressed for convenience.

$$\frac{\partial c}{\partial t} + u_i \frac{\partial c}{\partial x_i} = \gamma_c \frac{\partial^2 c}{\partial x_i \partial x_i}. \tag{32}$$

Here u_i ($i = 1, 2$ and 3) is the velocity field and γ_c the molecular diffusivity. Summation over repeated indices is implied.

Decomposing the velocity and the concentration in means and fluctuations as

$$\left\{ \begin{array}{c} u_i \\ c \end{array} \right\} = \left\{ \begin{array}{c} U_i \\ C \end{array} \right\} + \left\{ \begin{array}{c} u_i' \\ c' \end{array} \right\}, \tag{33}$$

we obtain the turbulent budget equation

$$\begin{aligned}
\frac{\partial}{\partial t} \langle c'^2 \rangle = & -2\langle u_i' c' \rangle \frac{\partial C}{\partial x_i} - U_i \frac{\partial}{\partial x_i} \langle c'^2 \rangle - \chi_c \\
& - \frac{\partial}{\partial x_i} \langle u_i' c'^2 \rangle + \gamma_c \frac{\partial^2}{\partial x_i \partial x_i} \langle c'^2 \rangle,
\end{aligned} \tag{34}$$

where

$$\chi_c = 2\gamma_c \left\langle \frac{\partial c'}{\partial x_i} \frac{\partial c'}{\partial x_i} \right\rangle, \tag{35}$$

is the molecular rate destruction of the variance of the concentration fluctuations. The other terms on the right-hand side of (34) represent, from left to right, gradient production, mean advection, turbulent transport and molecular diffusion (Panofsky and Dutton, 1984).

Assuming steady-state conditions and a horizontally homogeneous velocity field with the mean-wind velocity $U = U_1$ in the direction of the x-axis (x_1-axis), (35) becomes

$$\chi_c = -2\langle u_i' c' \rangle \frac{\partial C}{\partial x_i} - U \frac{\partial}{\partial x} \langle c'^2 \rangle - \frac{\partial}{\partial x_i} \langle u_i' c'^2 \rangle, \tag{36}$$

where the molecular diffusion term has been neglected because it is negligible compared to the turbulent transport term.

Since the molecular destruction takes place mainly at so small length scales that the turbulence can be considered isotropic, (35) can be written

$$\chi_c = 6\gamma_c \left\langle \left(\frac{\partial c'}{\partial x} \right)^2 \right\rangle = \frac{6\gamma_c}{U^2} \left\langle \left(\frac{\partial c'}{\partial t} \right)^2 \right\rangle, \tag{37}$$

where we have used Taylor's 'frozen-turbulence' hypothesis to replace differentiation with respect to x with differentiation with respect to t.

Looking at (37) one could feel tempted to jump to the conclusion that we have found $\sigma_{\dot{c}} = \sqrt{\langle (\partial c'/\partial t)^2 \rangle}$. This would be true if we were able to detect concentration fluctuations down to the smallest scales, the scales of molecular destruction. However, observation or detection of c nearly always implies filtering away the small scales through averaging as already mentioned in the introduction. We can determine the 'practical' value of $\sigma_{\dot{c}}$ when we know the filter characteristics. Here I will just assume that we have a linear, first-order filter in time.

First we write the temporal autocovariance

$$R(\tau) = \langle c'(t)c'(t + \tau) \rangle \tag{38}$$

in terms of the power spectrum

$$S(\omega) = \frac{1}{2\pi} \int_{-\infty}^{\infty} R(\tau)e^{-i\omega\tau} d\tau \tag{39}$$

as

$$R(\tau) = \int_{-\infty}^{\infty} S(\omega)e^{i\omega\tau} d\omega. \tag{40}$$

Here $S(\omega)$ is the filtered spectrum which is given as the product of the unfiltered spectrum $S_0(\omega)$ and, in this case, the *transfer function* pertaining to a first-order linear filter with the time constant τ_0.

$$S(\omega) = \frac{S_0(\omega)}{1 + \omega^2\tau_0^2}. \tag{41}$$

By differentiating (38) twice with respect to τ and utilizing (26) we find

$$\langle \dot{c}(t)\dot{c}(t + \tau) \rangle = -\ddot{R}(\tau) \tag{42}$$

so that

$$\sigma_{\dot{c}}^2 = -\ddot{R}(0) = 2 \int_0^{\infty} \omega^2 S(\omega) d\omega = 2 \int_0^{\infty} \frac{\omega^2 S_0(\omega)}{1 + \omega^2\tau_0^2} d\omega. \tag{43}$$

The integral (43) over frequency ω gets relative little weight from the lower frequencies and, consequently, we assume that only the high-frequency part of the spectrum $S_0(\omega)$ is important. Following Kristensen *et al.* (1989), we use

$$S_0(\omega) = b\chi_c\epsilon^{-1/3}U^{2/3}|\omega|^{-5/3}, \tag{44}$$

where $b \approx 0.2$ is a dimensionless constant and ϵ the rate of dissipation of turbulent kinetic energy.

If the temporal resolution were infinitely good, i.e. $\tau_0 = 0$, the integral (43) would be divergent if we used the expression (44) for $S_0(\omega)$ in the limit $|\omega| \to \infty$. In that case we cannot use (44) for values of $|\omega|$ much larger than the frequency $\omega_c = U(\epsilon/\gamma_c^3)^{1/4}$, corresponding to the microscale $\eta_c = (\gamma_c^3/\epsilon)^{1/4}$ of the scalar $c(\boldsymbol{x}, t)$ (see e.g., Tennekes and Lumley, 1972). We assume that the Schmidt number $Sc = \nu/\gamma_c$, where ν is the kinematic viscosity of air, is about the same as the Prandtl number, i.e. of the order one, and then ω_c is typically of the order 10^4 s^{-1}. Since τ_0 is presumably at least of the order 0.01 s, we can from a practical point of view safely use (43) together with (44) and we get

$$\sigma_{\dot{c}}^2 = 2b\chi_c\epsilon^{-1/3}U^{2/3} \int_0^{\infty} \frac{\omega^{1/3}}{1 + \omega^2\tau_0^2} d\omega = \frac{2\pi}{\sqrt{3}}b\chi_c\epsilon^{-1/3}U^{2/3}\tau_0^{-4/3}. \tag{45}$$

In this expression we can determine ϵ when the atmospheric surface layer can be considered horizontally homogeneous, whereas χ_c depends on the type of source of the pollutant.

The rate of dissipation ϵ can be written in terms of the height $z = x_3$, the friction velocity

$$u_* = \lim_{z \to 0} \sqrt{-\langle u_3' u_1' \rangle} \tag{46}$$

and the Monin-Obukhov length

$$L = -\frac{u_*^2}{\kappa g} \frac{T}{T_*}. \tag{47}$$

Here κ is the von Kármán constant, equal to about 0.4 according to Zhang et al. (1988), g the acceleration of gravity, T the mean temperature of the surface layer and

$$T_* = \frac{\langle u_3' \vartheta' \rangle}{u_*} \tag{48}$$

the upward turbulent flux of potential temperature ϑ' divided by the friction velocity (46). The ratio z/L is proportional to the ratio between the production of specific turbulent kinetic energy by buoyancy and shear and thus a quantization of the atmospheric stability.

We get

$$\epsilon = \frac{u_*^3}{\kappa z} \varphi_\epsilon \left(\frac{z}{L} \right), \tag{49}$$

where (Panofsky and Dutton, 1984)

$$\varphi_\epsilon \left(\frac{z}{L} \right) = \varphi_m \left(\frac{z}{L} \right) - \frac{z}{L}. \tag{50}$$

The function φ_m is the dimensionless wind profile gradient, defined by

$$\varphi_m \left(\frac{z}{L} \right) = \frac{\kappa z}{u_*} \frac{dU}{dz}, \tag{51}$$

which, again according to (Panofsky and Dutton, 1984), has been determined experimentally to be

$$\varphi_m \left(\frac{z}{L} \right) = \begin{cases} \left(1 - 16\frac{z}{L} \right)^{-1/3} & \text{for} \quad \frac{z}{L} < 0 \\ 1 + 5\frac{z}{L} & \text{for} \quad \frac{z}{L} \geq 0 \end{cases} . \tag{52}$$

The last two equations can be used to determine $U(z)$ as a function of height. Straightforward operations lead to

$$U(z) = \frac{u_*}{\kappa} \left\{ \ln \left(\frac{z}{z_0} \right) - \psi_m \left(\frac{z}{L} \right) \right\}, \tag{53}$$

where z_0 is the surface roughness length and where, for $z/L < 0$,

$$\psi_m \left(\frac{z}{L} \right) = \frac{3}{2} \ln \left(\frac{1 + \xi + \xi^2}{3} \right) - \sqrt{3} \arctan \left(\frac{1}{\sqrt{3}} \frac{\xi - 1}{\xi + 1} \right) \tag{54}$$

with

$$\xi = \left(1 - 16\frac{z}{L} \right)^{1/3} = \left(1 + 16 \left| \frac{z}{L} \right| \right)^{1/3}. \tag{55}$$

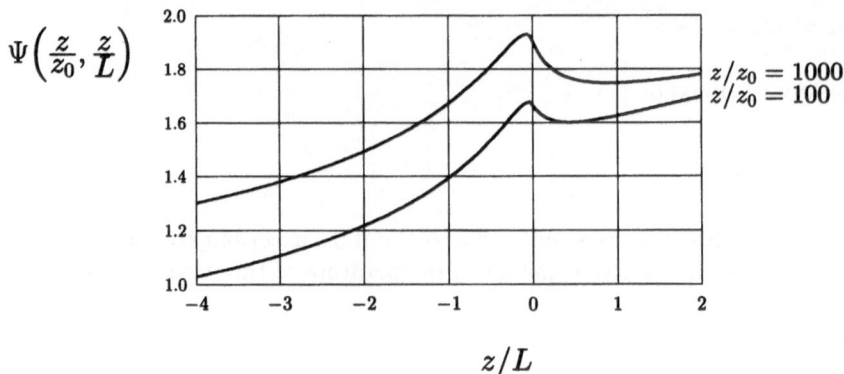

Figure 1. The function $\Psi(z/z_0, z/L)$ (58) with z/L as the independent variable and for two different values of the parameter z/z_0.

When $z/L \geq 0$ we simply have

$$\psi_m\left(\frac{z}{L}\right) = -5\frac{z}{L}. \tag{56}$$

Inserting (49) and (53) in (45), we obtain from (30) the following expression for the excursion rate:

$$\eta_c = \sqrt{\frac{b}{\sqrt{3}}}\,\Psi\left(\frac{z}{z_0}, \frac{z}{L}\right)\chi_c^{1/2}\sqrt{\frac{\kappa u_*}{z}}\left(\frac{z}{\kappa u_* \tau_0}\right)^{2/3} Q_0(\mathcal{C}), \tag{57}$$

where

$$\Psi\left(\frac{z}{z_0}, \frac{z}{L}\right) = \left\{\ln\left(\frac{z}{z_0}\right) - \psi_m\left(\frac{z}{L}\right)\right\}^{1/3}\varphi_\epsilon^{-1/6}\left(\frac{z}{L}\right). \tag{58}$$

The function $\Psi(z/z_0, z/L)$ is shown in Figure 1.

In general, χ_c and $Q_0(c)$ depend on the type of source. In the two next subsections I shall illustrate this by discussing two idealized source types, the infinite surface source and the elevated point source, both in a horizontally homogeneous atmospheric boundary-layer flow.

Infinite Area Source

The concentration fluctuations inside a large area with a homogeneous surface-source distribution is considered horizontally homogeneous.

To determine χ_c we return to (36). The assumption about horizontal homogeneity implies that the second term, the mean advection, is identically zero. We shall assume that the third term, the turbulent transport, is small compared to the first term, the gradient production. This last assumption is supported to some extent by Wyngaard et al. (1978) who showed that this term in the unstable surface layer is at least about five times larger than the turbulent transport term. Thus, we have

$$\chi_c \approx -2\langle u_3' c' \rangle \frac{\partial C}{\partial z} \approx 2u_* c_* \frac{dC}{dz}, \tag{59}$$

where we have introduces the surface-concentration scale

$$c_* = -\frac{\lim_{z \to 0}\langle u_3' c' \rangle}{u_*}. \tag{60}$$

Following a suggestion by Panofsky and Dutton (1984) that scalars in general follow the same flux-gradient relations as heat, we write

$$\frac{dC}{dz} = \frac{c_*}{\kappa z}\varphi_h\left(\frac{z}{L}\right) \tag{61}$$

and

$$c_* = 0.5\sigma_c \varphi_h^{-1}\left(\frac{z}{L}\right) \tag{62}$$

so that

$$\chi_c = 0.5\frac{u_*}{\kappa z}\varphi_h^{-1}\left(\frac{z}{L}\right)\sigma_c^2. \tag{63}$$

The diabatic correction $\varphi_h(z/L)$ is according to general considerations (Panofsky and Dutton, 1984) equal to $\varphi_m^2(z/L)$ when $z/L < 0$, else to $\varphi_m(z/L)$.

I have no good reference concerning the form of $Q_0(c)$ for a homogeneous area source. If we, for the sake of argument, assume that $Q_0(c)$ is Gaussian, we get from (56)

$$\eta_c = \sqrt{\frac{b}{4\pi\sqrt{3}}}\Psi_s\left(\frac{z}{z_0}, \frac{z}{L}\right)\frac{u_*}{z}\left(\frac{z}{\kappa u_* \tau_0}\right)^{2/3}\exp\left(-\frac{(C-C)^2}{2\sigma_c^2}\right), \tag{64}$$

where

$$\Psi_s\left(\frac{z}{z_0}, \frac{z}{L}\right) = \Psi\left(\frac{z}{z_0}, \frac{z}{L}\right)\varphi_h^{-1/2}\left(\frac{z}{L}\right) = \left\{\ln\left(\frac{z}{z_0}\right) - \psi_m\right\}^{1/3}\varphi_\epsilon^{-1/6}\varphi_h^{-1/2}. \tag{65}$$

This function is shown in Figure 2.

Although the assumption that $Q_0(c)$ is Gaussian is not well founded experimentally or theoretically, it is worthwhile pointing out that when $C - C \gg \sigma_c$, the estimate θ_c of the excursion duration (31) becomes proportional to $1/(C - C)$—a result which seems intuitively appealing.

Elevated Point Source

In this example, we consider concentration fluctuations of effluents from an elevated point source in a neutrally stratified boundary layer. It was argued by Kristensen *et al.* (1989) that, on basis of wind-tunnel measurements by Fackrell and Robins (1982) and Netterville (1979) and the assumption that $\sigma_c^2(x)$ develops downstream in the $x = x_1$ direction in a self-similar way (Csanady, 1967), the budget equation (36) near the surface reduces to

$$\chi_c \approx -U\frac{\partial\langle c'\rangle^2}{\partial x} \approx U\frac{\sigma_c^2}{x}. \tag{66}$$

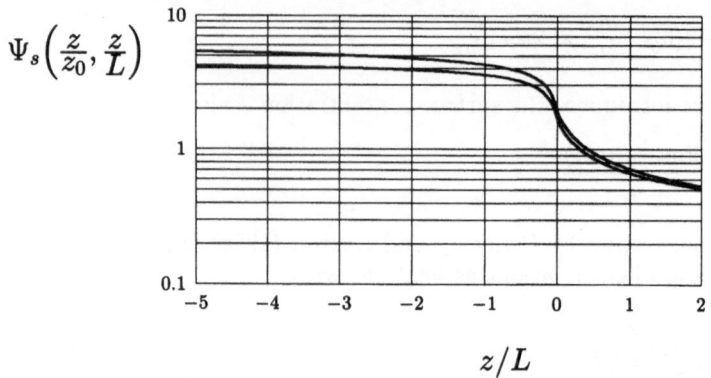

Figure 2. The function $\Psi_s(z/z_0, z/L)$ (65) with z/L as the independent variable and for two different values of z/z_0. The upper curve corresponds to $z/z_0 = 1000$ and the lower to $z/z_0 = 100$.

This approximation is based on the assumptions that the source height is less than one fifth of the boundary layer thickness h and that x is greater than $3h$.

Deardorff and Willis (1988) found, albeit in an unstable boundary layer, that the probability density for c at particular point inside the plume is well-described by the Gamma distribution

$$Q_0(c) = \frac{\alpha}{\Gamma(\alpha)} \frac{1}{C} \left(\alpha \frac{c}{C}\right)^{\alpha-1} \exp\left(-\alpha \frac{c}{C}\right),$$

(67)

where

$$\alpha = \frac{C^2}{\sigma_c^2}.$$

(68)

We assume without further justification that this probability can be used in a neutrally stratified boundary layer and get, by substituting in (57) with $z/L = 0$,

$$\eta_C = \sqrt{\frac{b}{\sqrt{3}}} \ln^{5/6}\left(\frac{z}{z_0}\right) \frac{u_*}{\sqrt{xz}} \left(\frac{z}{\kappa u_* \tau_0}\right)^{2/3} \frac{\alpha^{1/2}}{\Gamma(\alpha)} \left(\alpha \frac{c}{C}\right)^{\alpha-1} \exp\left(-\alpha \frac{c}{C}\right).$$

(69)

In this case the assumption that $Q_0(c)$ is a Gamma distribution implies that the estimate θ_C in the limit where $C/C \gg 1/\alpha = \sigma_c^2/C^2$ goes to a constant which is proportional to $1/\sqrt{\alpha}$.

CONCLUSION

I have shown examples of the possibility of determining the recurrence of events of large concentrations of a gas, being dispersed in the turbulent atmosphere, by just applying the parameters describing the state of the atmospheric flow and the mean and variance of the gas concentration. Two examples of sources have been discussed, the infinite, horizontally homogeneous surface source and the elevated point source. Admittedly, these idealized cases with the corresponding, somewhat speculative the surface-layer parameterizations, cannot be directly applied from a regulatory point of view. There is a great need for experimental verification. However, they could be useful when asking relevant questions to be settled experimentally. If we could obtain a more precise information about, in particular,

1. the probability density function $Q_0(c)$ of the concentration for different types of sources, including surface point sources,

2. the variation of functional form of $Q_0(c)$ with the averaging of $c(t)$ in time, and

3. the relevant surface-layer scaling for different types of scalar fields, including possible chemical and photo-chemical reactions,

it would be possible, on the assumption that the concentration $c(t)$ and its time derivative are statistically independent, to make predictions about the probabilities of extreme concentrations by just measuring the mean and the variance of $c(t)$. This approach would then allow us, as I have shown, not only to determine how often a particular concentration C is exceeded on average, but also to estimate the average duration of an excursion beyond C and the probability for zero or any number excursions beyond C during a given duration of time T.

Behind these considerations it has been tacitly assumed that the time scales are no larger than one day, say. This implies that when T is larger that one day, the approach will have to be modified accordingly. The surface-layer parameterization will certainly not be correct since they are based on small-scale turbulence spectra. However, the discussions in the second and the third section are still relevant.

There is another limitation. I have only considered the extreme statistics of $c(t)$ in one point. This is probably relevant when we are concerned with gases which are toxic or just smell bad. In the case where we are dealing with a volatile gas which may explode if the concentration exceeds a certain value *at any place* within a cloud of linear dimensions ℓ, the approach must be generalized accordingly to four dimensions, one temporal and three spatial, where any one of the spatial dimensions will have to be treated equivalently to the temporal dimension with ℓ replacing T.

ACKNOWLEDGEMENTS

This work was supported by the Danish Environmental Programme Project: Odour and Concentration Fluctuations.

REFERENCES

Chatwin, P.C., and Sullivan, P.J., 1993, The structure and magnitude of concentration fluctuations, *Boundary-Layer Meteorol.* 62:269.

Csanady, G.T., 1967, Concentration fluctuations in turbulent diffusion, *J. Atmos. Sci.* 24:21.

Deardorff, J.W., and Willis, G.E., Concentration fluctuations within a laboratory convectively mixed layer, *in:* "Lectures on Air Pollution Modeling", A. Venkatram and J.C. Wyngaard, eds., American Meteorological Society, Boston, MA.

Fackrell. J.E. and Robins, A.G., 1982, Concentration fluctuations and fluxes in plumes from point sources in a turbulent boundary layer, *J Fluid Mech.* 117:1.

Gibson, C.H., Stegun, G.R., and Williams, R.B., 1970, Statistics of the fine structure of turbulent velocity and temperature fields measured at hight Reynolds number, *J. Fluid Mech.* 41:153.

Gumbel, E.J., 1958, " Statistics of Extremes," Columbia University Press, New York, N.Y.

Kristensen, L., Weil, J.C., and Wyngaard, J.C., 1989, Recurrence of high concentration values in a diffusing, fluctuating scalar field, *Boundary-Layer Meteorol.* 47:263.

Kristensen, L., Casanova, M., Courtney, M.S., and Troen, I., 1991, In search of a gust definition, *Boundary-Layer Meteorol.* 55:91.

Kristensen, L., 1993, "The cup anemometer and other exciting instruments", Risø-R-615(EN), Risø National Laboratory, Roskilde, Denmark.

Netterville, D.D.J., 1979, Concentration fluctuations in plumes, *in:* "Syncrude Environmental Research Monograph 11979-4," Syncrude Canada Ltd., Edmonton, Alberta.

Panofsky, H.A., and Dutton, J.A., 1984, "Atmospheric Turbulence," John Wiley & Sons, New York, NY.

Tennekes, H., and Lumley, J.L., 1972, "A First Course in Turbulence", MIT Press, Cambridge, MA.

Wyngaard, J.C., Pennell, W.T., Lenschow, D.H., and LeMone, M.A., 1978, The temperature-humidity covariance budget in the convective boundary Layer, *J. Atmos. Sci.* 35:47.

Zhang, S.F, Oncley, S.P. and Businger, J.A., 1988, A critical evaluation of the von Kármán constant from a new atmospheric surface layer experiment, *in:* "Proc. Eighth Symp. on Turbulence and Diffusion", April 25-29, San Diego, CA, American Meteorological Society, Boston, MA.

DISCUSSION

H.M APSIMON

Can your work be used to estimate:

a) the volume of a cloud above a flammable limit

b) interaction between chemically reacting species

L. KRISTENSEN   a) It seems to me that this part of the question relates to the last paragraph in my paper. As I understand it,
a necessary condition that a flammable gas actually bursts
into flames is that a minimum concentration is sustained
for a minimum of time in a minimum volume; there simply must be molecules enough in close contact long enough
for a spontaneous chain reaction to occur with a high probability (~ 1). Generalizing the model from one temporal
dimension to one temporal and three spatial dimensions, it
should be possible to estimate the average volume - as well as
the average duration - of a flammable gas with a concentration higher than the threshold than that for bursting if that
concentration is known from other sources, e.g. chemistry
information.

b) Yes, from the same point of view the model could
probably be used to estimate the average rate of a chemical
transformation.

MAJOR INDUSTRIAL HAZARDS: THE SEVEX PROJECT - SOURCE TERMS AND DISPERSION CALCULATION IN COMPLEX TERRAIN

C. Delvosalle, J-M. Levert and F. Benjelloun[1]
G. Schayes and B. Moyaux[2]
F. Ronday, E. Everberg, T. Bourouag and J. P. Dzisiak[3]

[1]Faculté Polytechnique de Mons
[2]Université Catholique de Louvain
[3]Université de l'Etat à Liège, Belgium

In 1982, the European Economic Community has adopted a Directive on the prevention of major accidents of industrial activities ("SEVESO Directive"). In this frame, Public Authorities have namely to delineate "risk areas" and organise "external emergency planning". For this purpose, the Southern Region of Belgium (Walloon Region) supports a multidisciplinary approach known as the **"SEVEX Project"** (<u>SEV</u>eso <u>EX</u>pert). The project results from the close co-operation of three Belgian university departments, each being involved in a specific part of the computation of data needed to assess the effects of a major industrial hazard. The determination of the source has been done by the Faculté Polytechnique de Mons, the mesoscale wind field has been developped by the Université Catholique de Louvain, and the dispersion calculation was performed by the Université de l'Etat à Liège.

This paper will focus on the source and dispersion parts. The mesoscale model has been also used in the APSIS intercomparison experiment and will be presented in this session[12].

1. SOURCE TERM AND SHORT DISTANCE CONSEQUENCES

To predict the environmental impact of an accidental release of a flammable or toxic material, one must determine the Source Term with a fair degree of precision. This involves estimating all the source term related phenomena such as the release rate, the data associated with the jet depressurisation, its mixing with the atmosphere, the characteristics of the vapour cloud generated and its dispersion.

For this aim, we developed an original powerful and user-friendly software (named SOURCE) which is written in Turbo-Pascal and runs on IBM-compatible PC's.

Air Pollution Modeling and Its Application X, Edited by S-V. Gryning
and M.M. Millán, Plenum Press, New York, 1994

Starting from the safety analysis performed on a chemical plant, the most important potential accidental scenarios are retained. For each primary scenario, the main data describing the potential accident are introduced in the computer through various screens.

Namely, the characteristics of the storage are defined: the chemicals under consideration, geometry and capacity of the vessel, temperature, eventual over pressure of an inert gas. The data concerning the accidental release are given (instantaneous or continuous release, size of a hole in the storage wall, size and length of a broken pipe, duration of the release, the size of an eventual retaining dike). Lastly, the meteorological data (stability classes, wind speed, air temperature and humidity, ground temperature, solar radiation), the surface roughness of the land surrounding the plant, the parameters of the jet and the cloud dispersion models are provided by the user.

With these data, the software automatically links the various models required to describe the source term and the scenario consequences (see fig. 1)

1.1. Instantaneous releases

In the case of instantaneous releases, two secondary scenarios are taken into account: the occurrence of a BLEVE and a fireball on one hand, the formation of a dense cloud and its dispersion with toxic or explosive consequences on the other hand.

BLEVE and fireball. A BLEVE results from the catastrophic failure of a vessel containing a liquefied gas stored at a temperature well above its normal boiling point at atmospheric pressure. A BLEVE is an explosion involving both the rapid vaporisation of the liquid and the rapid expansion of the vapour out of the vessel. This expansion creates a shock wave which is first calculated. But, if the liquid is flammable, the principal risk factor is the ignition of the cloud and the formation of a fireball. The radiation effects of this fireball are also computed[1].

Dense cloud dispersion. If the liquid is not enough superheated to produce a BLEVE or if the cloud does not form a fireball, the software considers the atmospheric dispersion of a gas cloud denser than the air. This dispersion is described in the near field by a "box model" and by a passive (gaussian) model in the medium field.

The dense gas cloud is treated as a cylinder with a specified size and uniform concentration. The density difference causes the edges of the cloud to slump on the ground. The increase of the cloud radius R is calculated using:

$$\frac{dR}{dt} = \sqrt{gh(\frac{\rho_c - \rho_a}{\rho_a})}$$

where ρ_c and ρ_a are respectively the cloud and the air density and h is the cloud height. Thus the cloud radius increases and, in the initial stages, its height is reduced. But the cloud will also incorporate air both through its edge and through its top. The rate at which air is entrained into the cloud is given by (m_a is the mass of air in the cloud):

$$\frac{dm_a}{dt} = \rho_a(\pi R^2)U_T + \rho_a(2\pi Rh)U_E$$

where U_T is the top entrainment velocity and U_E is the edge entrainment velocity which are calculated using:

$$U_E = \alpha_1 \frac{dR}{dt} \quad ; \quad U_T = \alpha_2 U_l Ri^{-1} \text{ with } Ri = (gl_s/U_l^2)(\rho_c - \rho_a)/\rho_a.$$

U_l is the longitudinal turbulence velocity proportional to the friction velocity u^*(Monin[2]) and l_s is the turbulence length scale which vary with the stability class and the cloud height (Taylor et al.[3]). The best values for the constants α_1 and α_2 seem to be respectively 0.5 and 0.05.

The cloud is advected downwind at the wind velocity at mid height of the cloud. A rigorous enthalpy balance taking into account liquid droplets vaporisation, heat transfer with the ground and humid air incorporation in the cloud is performed to compute the evolution with the downwind distance of the cloud temperature, vapour fraction and composition. The cloud composition is supposed to be uniform at any moment and is computed as the mass (or volume) of chemicals released divided by the cloud volume.

Transition between the dense and the passive phases is based on two alternative tests. The first test is based on the density difference between the cloud and the air. The second test compares the rate of increase of the radius due to the gravitational slumping to the one expected from the atmospheric turbulence alone.

In the passive phase, the cloud concentrations are given at any point and any time t by the classical gaussian equations[1] (the radial and vertical dispersion coefficients are computed according Briggs[4]). To assure a smooth transition between the dense and the passive phase, virtual sources are used to compute the radial and vertical dispersion coefficients[7].

In case of a flammable release, the shock wave effects of the cloud explosion (UVCE) are computed[1]. All the mass of the release is supposed to participate to the explosion.

1.2. Continuous releases

In the case of a continuous release, the SOURCE software computes the release rate and its evolution in time. All cases are taken into account : liquid release with or without frictional losses through ducts, gaseous releases (choked or not) with or without frictional losses through ducts, flashing liquid flow through ducts (choked). The discharge rate and the physical conditions of the release at the orifice plate are determined (pressure, temperature, vapour fraction, velocity, ...). The accuracy of these data is essential for the following computations.

The modelling of the atmospheric dispersion of a continuous release is divided in three regions: a jet region, a dense cloud region and finally a neutral cloud region.

Jet Region. In the jet region, the fraction of unflashed liquid entrained as an aerosol and the rainout liquid forming a pool are directly computed. The jet region description is based on a model proposed by Iannello et al.[5]. The jet region is subdivided in two modules: an expansion module and an entrainement module.

In the expansion module, the jet atomises to droplets and flashes. A 1D momentum balance is used to determine the jet velocity after expansion while a 1D energy balance is used to determine the vapour fraction of the jet (the pressure is supposed to be the atmospheric pressure and the temperature is the normal boiling point of the chemicals released). A thermal or a mechanical atomisation model is used to determine the mean diameter of droplets and the probability distribution of these diameters. The fraction of the unflashed liquid that leaves the jet and falls to the ground (rain out) is also determined[5].

In the entrainment module, the jet is diluted by turbulent mixing with the atmosphere. This mixing is described by a model based on the work of Ricou and Spalding[6]. A rigorous energy balance is performed on the entrainement region and the evolution of the jet temperature, composition and vapour fraction is computed with the distance from the source. The jet model is considered as long as its momentum dominates over buoyancy and atmospheric turbulence effects. A criterion based on a Richardson number is used.

An accurate modelling of the jet region is of prime importance for the computation of near field dispersion and namely explosion (UVCE) consequences.

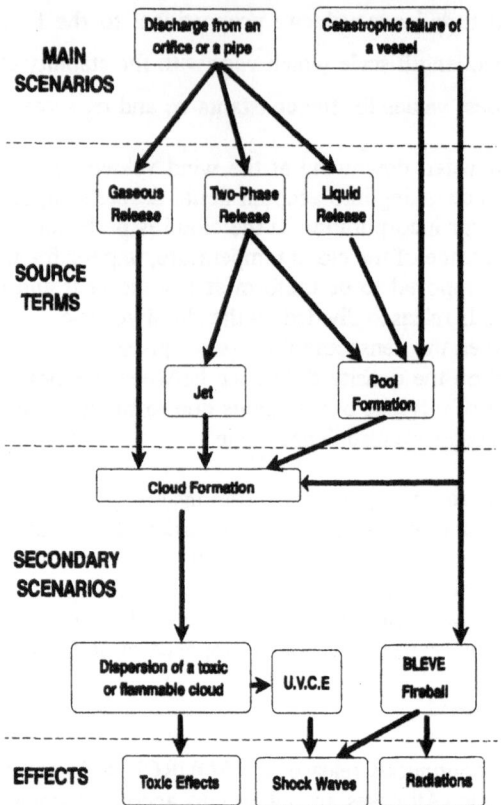

Fig. 1. Various models linked in SOURCE software.

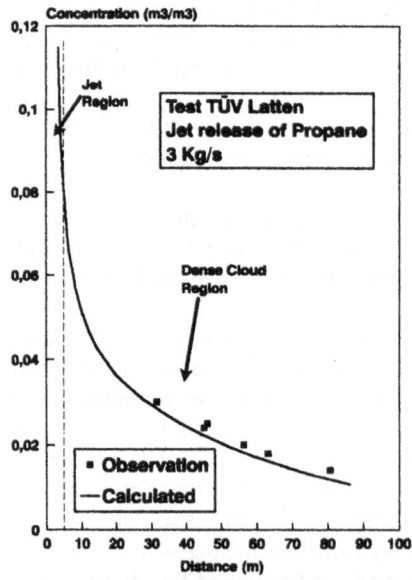

Fig. 2. Comparison of SOURCE results with TÜV test EEC55 [9].

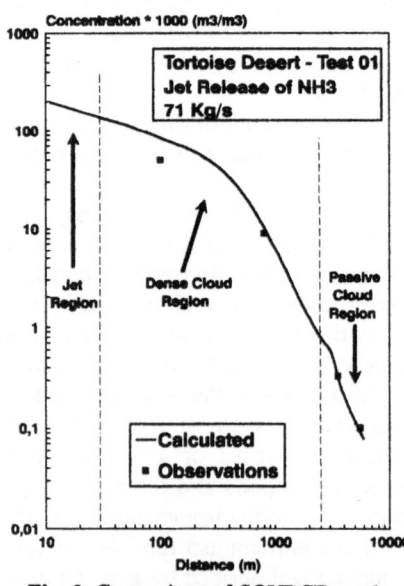

Fig. 3. Comparison of SOURCE results with Tortoise test 01 [10].

Dense Cloud Region. At the end of the jet region, the gas cloud is supposed to form a steady plume with a rectangular cross-section. As for the instantaneous releases, the cloud slumps on the ground and incorporates air through its edges and its top. The equations are similar to those for an instantaneous release (with L representing the cloud half-height).

$$\frac{dL}{dt} = \sqrt{gh(\frac{\rho_c - \rho_a}{\rho_a})} \quad \text{and} \quad \frac{dQ_a}{dt} = 2\rho_d hU_T + 2\rho_d LU_E$$

where Q_a is the mass flow rate of air through a given cross-section of the plume, U_T is the top entrainment velocity and U_E is the edge entrainment velocity which are calculated using:

$$U_E = \alpha_1 \frac{dL}{dt} \quad ; \quad U_T = \alpha_2 U_I Ri^{-1} \quad \text{with} \quad Ri = (gl_s/U_I^2)(\rho_c - \rho_a)/\rho_a$$

A rigorous enthalpy balance is performed as in the instantaneous case to compute the evolution with the downwind distance of the cloud temperature, vapour fraction and composition. The concentration is computed as the mass (volume) flow rate of the chemicals release divided by the volume flow rate of the cloud.

Transition between the dense and passive phases is based on the tests used for the instantaneous releases. In the passive phase, the cloud concentrations are given at any point by the classical gaussian equations[1]. To assure a smooth transition between the dense and the passive phase, virtual sources are again used to compute the dispersion coefficients[8].

In case a flammable release, the mass of the chemicals included in the cloud between the upper and lower flammability limits is determined and the shock wave effects of the explosion (UVCE) of this mass are computed[1].

Computed results compare very well with known experimental data for distances ranging from some ten meters to some kilometres (see fig. 2 and 3).

When the passive dispersion is reached, the characteristics of the cloud are passed to the lagrangian particle dispersion model (see section 3).

2. MESOSCALE MODEL FOR THE WIND AND TURBULENCE FIELD

The meteorological module, calculates the 3D wind and turbulence fields around the industry in complex terrain. The model in use is a adapted version of the TVM model which is hydrostatic, boussinesq, incompressible and solves the dynamical equations in vorticity form[11]. The turbulent diffusivities are obtained from a turbulent energy equation asociated with the mixing lengths of Therry and Lacarrère. This model is presented in the APSIS session[12].

For each site, the calculations are run for a set of typical meteorological situations, namely geostrophic winds of 4, 8, 12 m/s, for 16 directions (each 22.5°) and 2 stability classes (simulating covered sky during day, and clear sky during night), thus in total 96 cases.

3. DISPERSION MODEL

An efficient dispersion model must fulfil several conditions: (i) the calculation must be accurate (no mass loss); (ii) the actual topography and the terrain characteristics must be taken into account.

In order to satisfy these conditions, a lagrangian particule diffusion technique (Monte-Carlo scheme) is used to compute the concentration of the effluent in the air, following the scheme of Yamada and Bunker (1988)[14]. In order to minimise the number N of particles needed, a gaussian kernel is associated to each particle.

Each puff particule is advected by the velocity of the air $\mathbf{u}_p$ composed of two parts:

$$\mathbf{u}_p = \mathbf{u}_m + \mathbf{u}_t.$$

The mean velocity $\mathbf{u}_m$ is deduced and interpolated from the mesoscale model wind field, and a characteristic turbulent velocity $\mathbf{u}_t = \left(u_{t,x}, u_{t,y}, u_{t,z}\right)$ is deduced from the turbulent characteristics of the air also given by the meso-scale model. The three components of this velocity take into account the standard deviations of the velocity $(\sigma_u, \sigma_v, \sigma_w)$ and the lagrangian times (T_{Lu}, T_{Lv}, T_{Lw}) by formulae similar to the following:

$$u_{t,x}(t+\Delta t) = a_x u_{t,x}(t) + b_x \sigma_u \xi$$

where ξ is a random number from a gaussian distribution with a zero mean and a unit variance and where $a_x = \exp\left(\dfrac{-\Delta t}{T_{Lu}}\right)$ and $b_x = (1-a_x^2)^{0.5}$.

The lagrangian times scales and the σ values used are those of Hanna[16]. They are functions of the parameters u^* and w^*, representing the friction velocity and the convective velocity respectively, obtained from the mesoscale wind field model.

The spatial extensions of the air particules are function of the time and of the standard deviations of the velocity field $(\sigma_u, \sigma_v, \sigma_w)$. Evolution equations for the spatial standard deviations are:

$$\sigma_{x,p}(t+\Delta t) = \sigma_{x,p}(t) + \sigma_u \Delta t \quad \text{if } t \le 2T_{Lu}$$

$$\sigma_{x,p}^2(t+\Delta t) = \sigma_{x,p}^2(t) + 2\sigma_u \Delta t \quad \text{if } t > 2T_{Lu}$$

and similar relations exist for $\sigma_{y,p}$ and $\sigma_{z,p}$.

Finally the concentration field is given by :

$$C(x,y,z,t) = \frac{1}{(2\pi)^{1.5}} \sum_{p=1}^{N} \frac{M_p}{\sigma_{x,p}\sigma_{y,p}\sigma_{z,p}} \exp\left[\frac{-(x_p-x)^2}{2\sigma_{x,p}^2}\right] \exp\left[\frac{-(y_p-y)^2}{2\sigma_{y,p}^2}\right]$$

$$\left\{+\exp\left[\frac{-(z_p-z)^2}{2\sigma_{z,p}^2}\right] + \exp\left[\frac{-(z_p+z-2z_g)^2}{2\sigma_{z,p}^2}\right]\right\}$$

where: x_p, y_p, z_p are the coordinates of the particule p,

$\sigma_{x,p}, \sigma_{y,p}, \sigma_{z,p}$ the spatial extensions of the particule p,

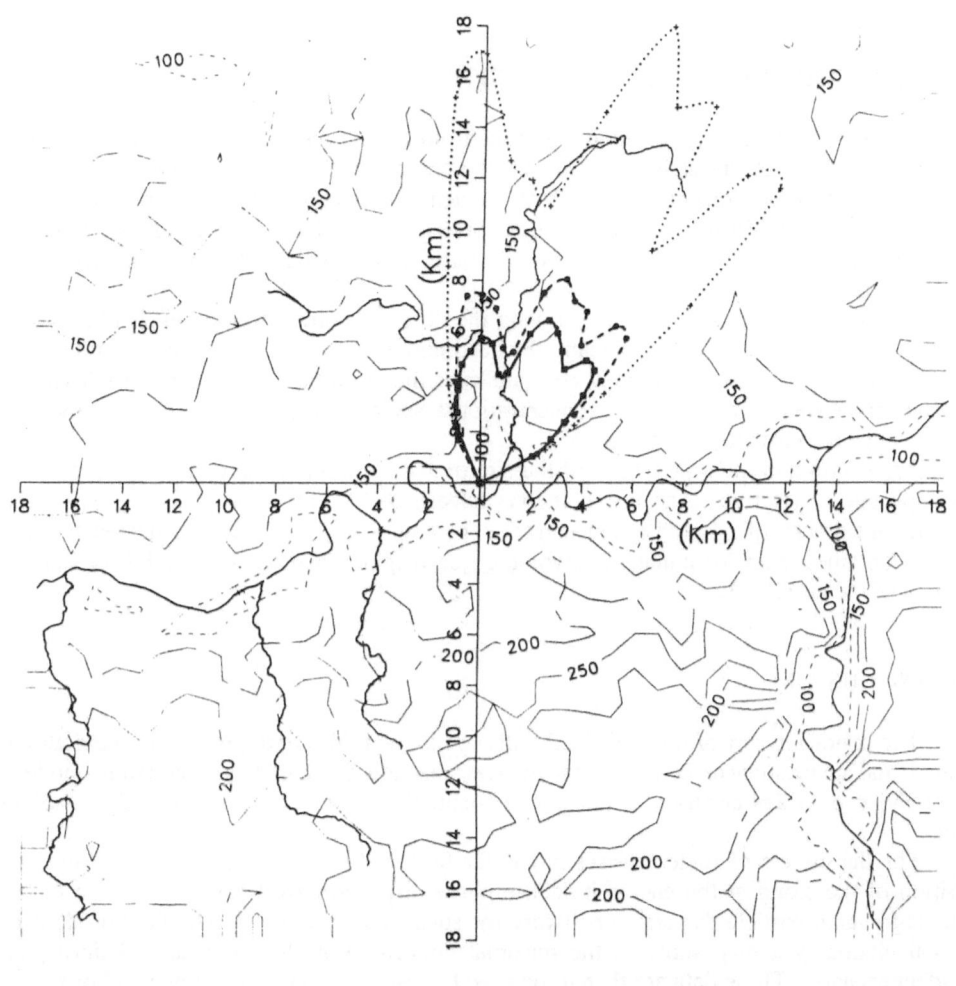

Site de Jemeppe
Cas de JOUR
3—DISPERSION INST 55 T CHLOR
+   CONCENTRATION : 3.8 ppm
●   CONCENTRATION : 15.0 ppm
■   CONCENTRATION : 25.0 ppm

Vent a 10.0 metres

Fig. 4. Topography of Jemeppe and resulting
envelope curves at the concentrations shown
in the legend, corresponding to a chlorine
release of 55T.

M_p the mass of particule p,

Z_g the altitude of the ground at (x,y).

The model has been calibrated for various sites and weather situations.

4. APPLICATION TO AN INDUSTRIAL SITE

In the case of a passive cloud resulting from a continuous release, after the initial gravitational phase of dispersion (section 1), the characteristics of the cloud are passed to the particle dispersion scheme. The mass of accidental pollutant is divided in the number of particles (we use 4000) in space and time according to the characteristics of the cloud. In this part, the dispersion is calculated on the base of the wind fields resulting from the mesoscale model.

As a result, a map is produced showing the different concentration levels obtained in function of time. We retain as significant, the maximum levels occurring at any time in space.

As the wind field is not exactly constant over the duration of the pollutant release we calculate the envelope curves of isoconcentration associated to a direction of the wind and the two adjacent directions (+ or - 22.5°).

Fig. 4 presents the topography of the site of Jemeppe-sur-Sambre in Belgium and the envelope curves relative to an instantaneous release of 55T of chlorine. For getteing the absolute maximum of possible concentration in any place, the values indicated are the maxima obtained with 3 simulations involving geostrophic winds of 4, 8 and 12m/s and a stability of the air class D.

5. CONCLUSIONS

The major interest of the SOURCE software is twofold. Firstly, it can compute all the aspects and consequences of accidental releases through a set of linked coherent models. Secondly, it provides accurate results for atmospheric dispersion in the near and medium field.

For the effects of toxic releases in the far field, data describing the state, shape and position of the cloud at the end of its dense phase dispersion are transferred to the meso-scale lagrangian particle dispersion software for subsequent treatment. The final product of the calculation is a map showing the maximal concentrations never exceeded during an accident scenario. These data are then to be used by rescue planners and public authorities.

The authors express their acknowledgements to the Walloon Region which provides financial support to the project.

REFERENCES

[1] Levert J-M., Delvosalle C. and Benjelloun F., (1992), Rapport de Synthèse, Projet Sevex, Vol. 1, Région Wallonne.
[2] Monin A.S. (1962), J. Geophys. Res., 67, 3103.
[3] Taylor R.J. et al. (1970), Quart. J. Roy. Met. Soc., 96, 750.
[4] Briggs G.A. (1973), Diffusion Estimation for Small Emissions, ATDL Rep. No. 79, ATDL, P.O. Box E, Oak Ridge, TN 37830.
[5] Iannello V. et al., (1989), 6th Symp. on Loss Prevention and Safety Promotion in the Process Industries, 58-1, Oslo, Norway.
[6] Ricou F.P. and Spalding D.B., (1961), J. Fluid Mechanics, 11(1), 21

[7] Fryer L.S. and Kaiser G.D., (1979), DENZ - a computer program for the calculation of the dispersion of dense toxic or explosive gases in the atmosphere, United Kingdom Atomic Energy Authority.

[8] Jagger S.F., (1983), Development of CRUNCH : a dispersion model for continuous releases of a denser-than-air vapour into the atmosphere,United Kingdom Atomic Energy Authority.

[9] Jones S.J., Martin D., Webber D.M. and Wren T., (1991), The Effects of Natural and Man-Made Obstacles on Heavy Gas Dispersion - Part II: Dense Gas Dispersion over Complex Terrain, SRD - Commission of European Communities Program "Major Technical Hazard".

[10] Havens J., (1992), J. Loss Prev. Process Ind., vol 5(1), 28.

[11] Schayes G., Moyaux B., (1992), Projet Sevex, Rapport de synthèse, Vol 3, Région Wallonne.

[12] Gallée H., Giorgi F., Graziani G., Schayes G., Thunis Ph., (1993), Flow simulations over the Attic peninsula: description and evaluation of the TVM, MAR and MM4 models. Preprint 20th ITM on Air Pollution Modelling, Valencia, Spain.

[13] Bourouag T., J.F. Deliege, J.P. Dzisiak, E. Everbecq and F. Ronday (1992), Les industries à risques majeurs en région Wallonne. Rap. N°4. Centre Environnement, Université de Liège, 215pp.

[14] Yamada T. and Bunker (1988), Development of a nested grid second moment turbulence closure model and application to the 1982 Ascot Brush Creek data simulation. J. Atmos. Sci. 27, 562-578.

[15] Zannetti P. (1990), Lagrangian dispersion models. In Air Pollution Modelling. Comp. Mech. Publ., Van Nostrand Reinhold. New-York, 185-222.

[16] Hanna S. (1982), Application in air pollution modelling. In Atmospheric Turbulence and Air Pollution Modelling. Ed. Nieuwstadt F. and Van Dop H. Reidel, 275-310.

DISCUSSION

M. FERNANDO: Concerning to the dense gas dispersion models, have you compared the results of your model against other models, for example, HEGADAS, SLAB or DEGADIS?

G. SCHAYES: Until now, we do not have the opportunity to compare directly our model results with other dense gas models (we plan to do it with HEGADAS in the next few months). Nevertheless, we have compared our model predictions with many experimental data (such as Tortoise runs, for instance) and we have observed that our results fit these data at least as well as other models. Moreover, our approach consists in modelling all the release circumstances, including the source term (the leak) and the jet region as well as the dense phase dispersion. Lastly, you know that comparison between models is always difficult because the inputs of different models are often not perfectly consistent.

R. KUNZ: What was the horizontal resolution of the mesoscale wind simulation in the SEVEX?

G. SCHAYES: 1 km - Near the limit of the hydrostatic assumption.

A MESOSCALE BOUNDARY LAYER METEOROLOGICAL
MODEL FOR INHOMOGENEOUS TERRAIN

S. M. Daggupaty[1], R. S. Tangirala[1], and H. Sahota[2]

[1] Atmospheric Environment Service
4905 Dufferin street
Downsview, Ontario, Canada M3H 5T4

[2] Air Resources Branch
Ontario Ministry of Environment
125 Resources Road
Rexdale, Ontario, Canada M9W 5L1

INTRODUCTION

Limited area mesoscale meteorological models are very useful tools for regional weather forecast on short temporal and spatial scales, and for air pollution studies. With particular interest in air pollution transport and dispersion in the boundary layer and for responding to air pollution emergencies in coastal regions, a mesoscale model restricted to the lower troposphere and capable of providing boundary layer meteorology simulations within a short time has been developed. This model is programmed for a microcomputer and is in research mode. Daggupaty and Sahota (1991) report results with an earlier version of this mesoscale model by assuming the land area of the model domain as one homogeneous land type. In the present, paper we discuss the model results by considering land heterogeneities.

Air Pollution Modeling and Its Application X, Edited by S-V. Gryning
and M.M. Millán, Plenum Press, New York, 1994

MESOSCALE MODEL

This boundary layer mesoscale model (named BLFMESO) is a prognostic three dimensional model based on primitive equations with hydrostatic assumption, and is formulated in terrain following coordinates. The thermal forcing in the model, due to the differential heating is through diurnally varying ground temperature and constant lake temperature. The ground temperature is estimated by the force-restore method (Deardorff, 1978) based on surface energy budget. Only dry thermodynamics are considered in the present version.

Turbulence is parameterized by first order K-theory closure. The vertical eddy diffusivity coefficients are determined by surface similarity theory and O'Brien (1970) profile method. The planetary boundary layer depth and its temporal variation over land points is parameterized with Deardorff (1974) model as modified by Benoit (1976). Whereas, the boundary layer depth over water is invariant throughout the simulation period and is assigned a value of 50 meters.

Surface layer similarity relations (Arya 1988) are used in the constant flux layer (10m deep) to solve for wind and temperature, whereas, full set of primitive equations are solved above the surface layer. The model has 10 irregularly spaced levels in the vertical (nine levels are respectively at 0, 1.5, 3.9, 10, 100, 350, 700, 1200, 2000 meters above ground; model top is the 700mb pressure level above ground). The vertical levels near the ground are closely spaced so as to simulate stronger vertical gradients in the lower levels. In the horizontal x and y directions, the mesoscale integration domain has 37 grid points with 5 km grid space. But model results are analyzed on the inner 29x29 domain leaving four grids of sponge area on each side.

Land surface heterogeneities are accounted through roughness changes and albedo changes over the model domain. Topographic maps are used to find the different land-use types in each 5 km grid square of the model domain. Effective surface roughness length ($z_{0\text{eff}}$) at each grid is estimated by taking the area weighted logarithmic average of z_0 of different land types inside the grid area (Taylor, 1987). Likewise, effective albedo at each grid is specified as area weighted average value.

Model Initialization

The BLFMESO is initialized with the Canadian Meteorological Centre's (CMC) objectively analyzed numerical weather prediction data of the nearest time analysis. The temperature and pressure heights data at the available four levels (1000, 850, 700, and 500mb) from the 11x11 grid CMC window are interpolated horizontally to the mesoscale integration domain. Temperature is interpolated linearly in the vertical to 10 meso-model levels. Whereas, hydrostatic relation is integrated downward from mesoscale model top (700mb level) to obtain pressure fields. Geostrophic winds are then computed at each mesoscale vertical level utilizing the pressure and the interpolated potential temperature fields. One dimensional momentum equations are solved at each model grid for about four inertial periods with geostrophic wind components as initial guess values for wind components. The resulting winds form the initial conditions for the mesoscale model.

Boundary Conditions and Numerics

The model equations are integrated by implicit finite difference schemes with a 300sec timestep. Upstream space differencing scheme is used for advection terms and semi-implicit central space differencing scheme is adopted for diffusion terms.

At the lower boundary, wind components are zero. Lake temperature, given from the satellite data, is invariant during the model integration. The ground temperature is predicted by the force-restore method. At the top boundary, meteorology data are supplied by initialization procedure and are assumed time independent. At the lateral boundaries, normal gradients of dependent variables are assumed to be zero. The model simulations are conducted for a 15 hour period starting at 0500 hours local time. BLFMESO requires about 20min of computing time for a 12 hour simulation on an IBM-PC i486 microcomputer.

RESULTS

The BLFMESO is implemented for a 150km x 150km domain covering Toronto city (Canada) and surroundings on the Lake Ontario. Figure 1 is a simplified map of the study area. Several case studies are being conducted to evaluate the model performance. Here we discuss one particular case of 12 hour simulation for May 2, 1988. On this day a quasi-stationary synoptic scale high pressure system was situated over the study area and northeasterly winds were prevailing. A high pressure system is usually characterised by weak winds and clear skies. The dynamic responses of differential frictional and heating effects are predominant under these conditions over lakeshore environments. The model was initialized with 1200 GMT CMC data. As the surface temperature increased due to solar insolation a southerly component developed over Lake Ontario by 1000 hours and persisted until 1800 hours. Figure 2 shows the modeled winds at 100m (above ground) over the study area at 1400 hrs local time. The lake-breeze front has penetrated to about 15-20km inland with a vertical extent of approximately 1000m. Modeled planetary boundary layer depth development/decay are in general agreement with theory.

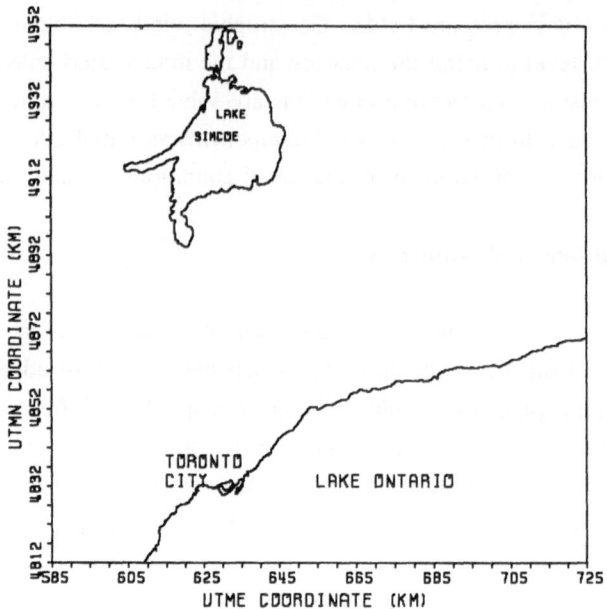

Figure 1 Simplified map of the study area

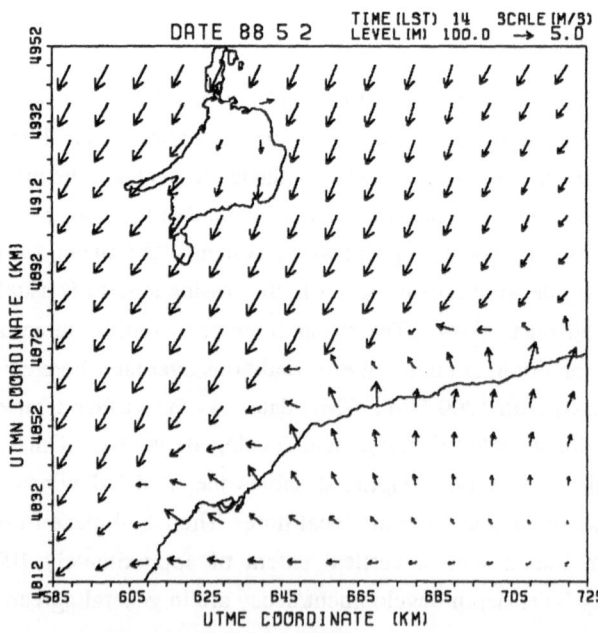

Figure 2 Modeled 100m winds at 1400 hours local time on May 2, 1988

Boundary layer depth increased with time to reach a maximum at about noon and dropped off suddenly in the late afternoon. Eddy diffusivity also evolved systematically with time during the simulation period.

A comparison of the model simulations with and without land heterogeneities indicates:

  i) a reduction of wind speed over roughness elements,

 ii) backing of winds near large horizontal gradients of surface roughness

iii) overall improvement when land inhomogeneities are included.

Model simulated wind and temperatures are compared with the observed data from a mesometeorological observational network around the Pickering nuclear power station (near Toronto). Winds and temperatures are generally overpredicted for this particular case. The root mean squared errors (rmse) are 1.4m/s and 60^0 for wind speed and direction respectively. These overestimates may partly be attributed to inadequacy of the model to account for the unresolved scales of motion of less than 10km. Also that the present version does not include the effects of vegetation and diabatic heating processes.

CONCLUSIONS

A mesoscale boundary layer model is applied for Toronto city and vicinity on the lake Ontario, Canada. Effects of inhomogeneous land cover are included in the present version of the model. The model has simulated the mesoscale circulations due to effects of topography and differential frictional and thermal forces at a lakeshore environment. The model simulated the onset and gradual development of lake breeze. Comparison of the model results with observations are quite encouraging. The computational time of 20 minutes for a 12 hour forecast on a microcomputer makes the model very useful for air pollution emergencies and related applications.

ACKNOWLEDGEMENTS

Authors would like to thank Prof. Peter Taylor for useful discussions at different stages of this model development project. Dr. John Walmsely is acknowledged for his suggestions.

REFERENCES

Arya, S.P., 1988, Introduction to Micrometeorology. *Intl. Geophysics Series* Vol. 42, Academic Press, p307.

Benoit, R., 1976, A comprehensive parameterization of the atmospheric boundary layer for general circulation models. Ph.D. Thesis Department of Meteorology, *McGill University*, Montreal, Canada.

Daggupaty, S.M., and H. Sahota, 1991, A mesoscale boundary layer forecast model and its use for air pollution emergencies. *Proc. OECD Specialists' meeting on Advanced Modelling and Computer Codes for Calculating Local Scale and Mesoscale Atmospheric Dispersion of Radionuclides and their Application*, March 6-8, 1991, Saclay, France.

O'Brien J.J. (1970): A note on the vertical eddy exchange coefficient in the planetary boundary layer. *J. Atmos. Sci.* 27:1213-1214.

Deardorff, J.W., 1974, Three dimensional numerical study of the height and mean structure of a heated planetary boundary layer. *Boundary-Layer Meteorol.* 7:81-106.

Deardorff, J.W., 1978, Efficient prediction of ground surface temperature and moisture, with inclusion of a layer of vegetation. *J. Geophys. Res.* 83(C4):1889-1903.

Taylor, P.A., 1987, Comments and further analysis on effective roughness lengths for use in numerical three-dimensional models. *Boundary-Layer Meteorol.* 39:403-418.

USE OF DMI-HIRLAM FOR
OPERATIONAL DISPERSION CALCULATIONS

Jens Havskov Sørensen, Leif Laursen, and Alix Rasmussen

Danish Meteorological Institute (DMI)
Meteorological and Oceanographic Research Division
Lyngbyvej 100, DK-2100 Copenhagen Ø, Denmark

INTRODUCTION

During the latest decades there has been a rapid progress in the ability to forecast weather. This progress has been possible because of the development of advanced numerical weather-prediction (NWP) models running on the most powerful computers available. Especially, there has been progress in the capability to make numerical forecasts in the range from one day to about a week ahead in the Northern Hemisphere which is dominated by the travelling weather systems. On longer time scales, there has been some progress, but severe difficulties are met due to the intrinsic chaotic nature of the atmosphere.

There has been a considerable increase in available computer resources in recent years enabling the development of numerical models of the atmosphere with high spatial resolution. This favourable development has made computers in the gigaflop range with memories of a gigabyte or more reachable for many meteorological institutes. This has enabled a number of national weather services in Europe to run and utilise products from forecasting systems capable of resolving meso-scale phenomena. These systems consist of elements taking care of quality control of observations, analysis, initialisation, forecast, and some post-processing including archiving and graphics.

In order to achieve a horizontal resolution of the order 10 to 50 kilometers, it is necessary to apply either a Limited Area Model (LAM) with the associated problem of obtaining boundary values, or an approach with variable resolution with the possibility to move the lateral boundaries far away from the area of interest. An example of the first approach is shown in Fig. 1. Simultaneously with the development in computer technology, substantial progress in the theoretical understanding of the physical phenomena has been achieved. These phenomena have a tendency to interact strongly with surface conditions like mountains and land/sea contrasts. Such interactions demand more complicated surface-parameterisation schemes which in turn demand detailed physiographic databases.

In order to provide the duty forecasters with information on fine-mesh/short-term developments, there are obvious constraints on the scheduling of this type of models.

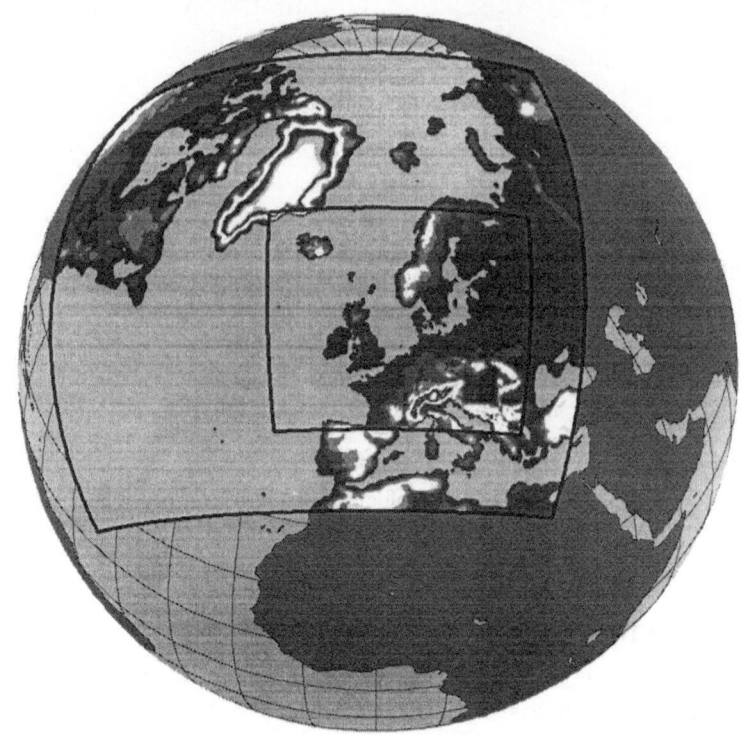

Figure 1. Operational HIRLAM at the Danish Meteorological Institute (DMI). The version with the large area (GRV) is used for forecast in Greenland, and supply the higher-resolution version (DKV) with boundary values. Both versions have 31 levels in the vertical. GRV has a horizontal resolution of about 46 km with 194 × 163 points in each level, and for DKV the numbers are 23 km and 162 × 136 points.

In practical applications, the products should be available within three to four hours after the initial synoptic time, and, preferably, updated forecasts should be available every six hours. The direct model results are used not only by the duty forecaster, but they are also applied in a number of different models, e.g. designed to handle certain environmental issues.

The main advantages of an increased resolution are obviously to be found in the better representation of local geographical effects. As an example, Fig. 2 shows the kind of details one may expect from a high-resolution surface-wind forecast. In Fig. 3, a wind forecast is compared with observations from a weather-recording station.

There is a strong need for a coordinated effort between several countries in order to advance the meso-scale NWP modeling research. An example of such effort is the rather successful HIRLAM (HIgh-Resolution Limited-Area Model) project (Kållberg, 1990; Machenhauer, 1988; Hansen Sass and Sørensen, 1992) initially started by the Nordic countries and the Netherlands. The HIRLAM project has later been joined by Ireland, and partly by France and Spain.

There is little doubt that further increases in computing power and memory, as well as in (mass-)storage devices, will lead to more operational applications from results produced by models having still finer resolutions, a more accurate description of atmospheric processes (including a generalisation of the step from hydrostatic to non-hydrostatic modeling), and the capacity to assimilate more kinds of observations.

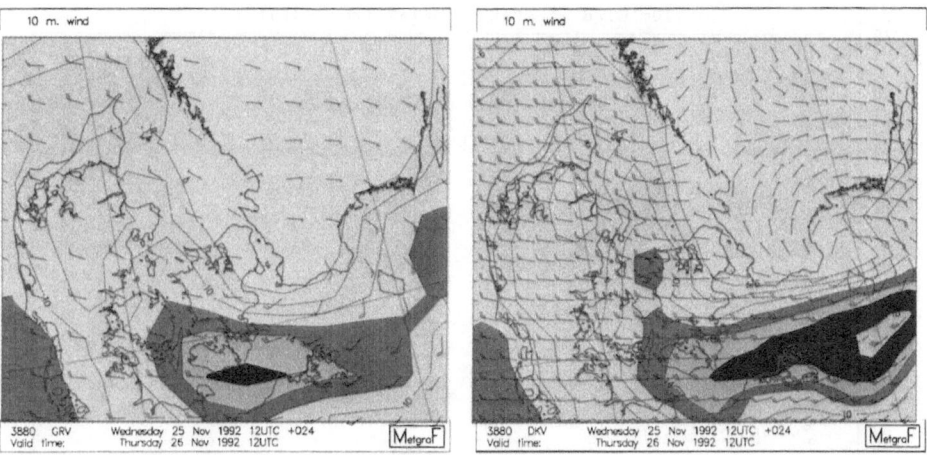

Figure 2. 24-hour forecast of 10-m wind valid at 12 UTC, November 26, 1992. The wind arrows are drawn at each grid point for the two HIRLAM models GRV and DKV. Both position and strength of the wind maxima are better described by the higher-resolution model DKV. The areas with wind speeds above 16 m/s are hatched. The contour interval is 2 m/s.

THE OPERATIONAL HIRLAM FORECASTING SYSTEM AT DMI

At the Danish Meteorological Institute (DMI), a complete three-dimensional regional atmospheric forecasting system is operational. The system is based on the HIRLAM forecast model, level 2.

HIRLAM is a primitive-equation NWP model using a grid-point representation with second-order difference approximations for the spatial derivatives. The horizontal grid is a regular spatially-staggered latitude/longitude grid (the Arakawa C grid) in a rotated spherical coordinate system with "North pole" at 180° E, 25° N. The vertical coordinate is a terrain-following hybrid coordinate which at the surface is identical with the σ coordinate ($\sigma = p/p_\mathrm{s}$), and approaches the pressure p with increasing height.

The forecasting system is run on two different limited areas, cf. Fig. 1. The boundary fields for the limited-area model GRV are obtained from the global model run by the European Center for Medium-range Weather Forecast (ECMWF). The small-area version (DKV) is nested in the large-area version (GRV) which provides the boundary values for DKV. Both models are run with the same vertical resolution (31 hybrid levels), but the horizontal resolution is 0.42° for GRV and 0.21° for DKV. The time step is 4 minutes for GRV and 3 minutes for DKV. Currently, a model nested in DKV with an even higher horizontal resolution of about 0.04° and 38 vertical layers is studied.

The HIRLAM forecasting system consists of pre-processing, analysis, initialisation, forecast, post-processing, and verification. Both models (GRV and DKV) are run with their own 6-hourly data-assimilation cycle.

ATMOSPHERIC BOUNDARY-LAYER HEIGHT

For atmospheric dispersion models, it is in general of great significance to estimate the height of the atmospheric boundary layer (ABL) well. In many dispersion models, empirical techniques are adopted. But for long-range transport models, which are based upon output from NWP models, it may well be beneficial (and dynamically consistent) to compute the ABL height from the NWP model profiles. Therefore, it is contemplated

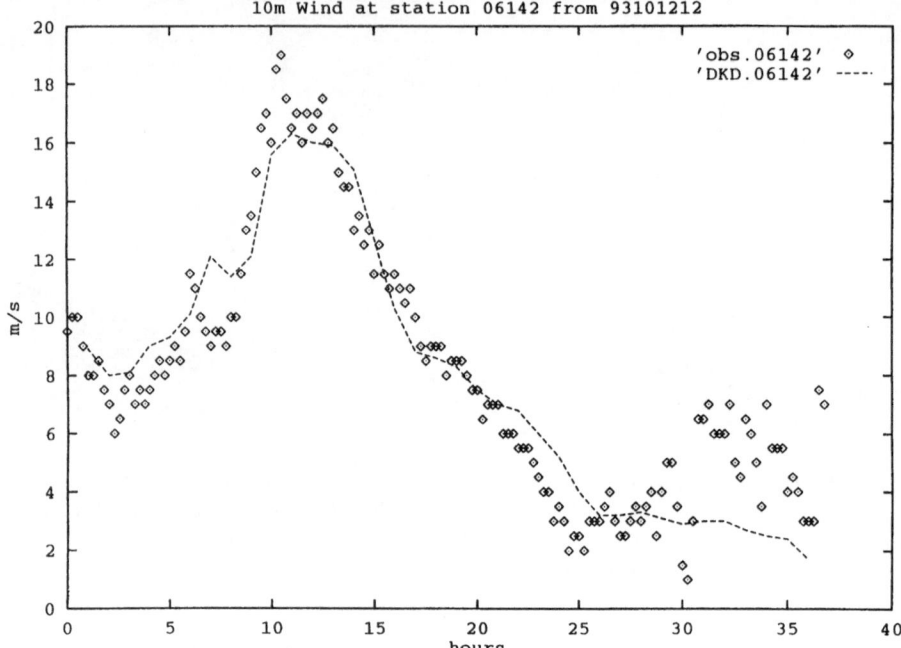

Figure 3. Comparison of observed values of the 10-m wind with the corresponding DKV forecast at the site of a weather-recording station.

to incorporate a calculation of the height of the atmospheric boundary layer directly in the post-processing of DMI-HIRLAM.

The model proposed (Holzlag and Moeng, 1991; Troen and Mahrt, 1983) is suited for use in NWP models where some resolution is possible within the boundary layer, but where the resolution is still insufficient for resolving the detailed boundary-layer structure. DMI-HIRLAM has nine model levels inside the boundary layer for a typical day-time ABL height of about 1500 m. The ABL height is represented in terms of a modified bulk Richardson number. The boundary-layer top is thus determined as the height where the bulk Richardson number over the whole boundary layer becomes critical.

LAGRANGIAN DISPERSION MODEL BASED ON HIRLAM

A simple Lagrangian long-range atmospheric transport model (Havskov Sørensen, 1993) is developed at DMI. The model simulates release of material in the atmosphere from one or several simultaneous emission points which may be moving. The calculations are based on analysed and forecast meteorological fields from the HIRLAM models in operation at DMI or from the global model at ECMWF, as read from the operational database at DMI.

In order to describe well an emission of material in the atmosphere from a source extending a large land area, it may be important to use more than one emission point. Such situations may occur e.g. in case of burning oil wells as the ones in Kuwait. In case of emergency situations such as a radioactive release following a shipwreck involving a nuclear-power driven ship or submarine, it is relevant to describe moving emission points. Furthermore, the dispersion program may be used in attempting to locate potential emission sites in situations where no information about accidental releases of hazardous material in the atmosphere has been given. Such use of the program involves integration backwards in time.

The program is capable of calculating trajectories in either model levels or pressure levels. In order to simulate the emission, air parcels are released in the levels above the emission point(s) up to a given initial plume height in (hourly) batches subsequently being advected passively by the model wind. The trajectories for these parcels are calculated by successive approximations with a prescribed tolerance, cf. (Kållberg, 1984; Kållberg; Petterssen, 1940). The meteorological fields used in this calculation are bilinearly interpolated in the horizontal and linearly interpolated in time.

The effect of sub-grid scale diffusion is indicated by drawing a circle, centered at the position of each released air parcel, with radius e.g. σ_y where $\sigma_y(t)$ is the standard deviation in a Gaussian distribution from the centerline of the plume at travel time t since the parcel was released. The following analytical result from Gifford's random-force theory (Gifford, 1982; Gifford, 1984) is calculated by the dispersion program,

$$\sigma_y^2 = 2\,K_E\,t_L \left\{ \tau - \left(1 - e^{-\tau}\right) - \tfrac{1}{2}\left(1 - e^{-\tau}\right)^2 \right\}, \tag{1}$$

where the large-scale eddy diffusivity $K_E = 5 \times 10^4$ m^2 s^{-1} = 200 km^2 hour^{-1}, the Lagrangian time scale $t_L = 10^4$ s = 3 hours, and the quantity τ is the travel time t in units of the Lagrangian time scale, $\tau = t/t_L$. The formula (1) has the "right" asymptotics: the well-known 3/2-power law for short-time diffusion and the 1/2-power law for long-range transport,

$$\sigma_y^2 = \begin{cases} \tfrac{1}{3}\,K_E\,t_L^{-2}\,t^3 & \text{for } t \ll t_L \\ 2\,K_E\,t & \text{for } t \gg t_L \end{cases}. \tag{2}$$

As an example, a simulation of a *hypothetical* release of radioactive material in the atmosphere from an accident on the nuclear power plant Ignalina in Lithuania is presented. Due to heat production, the initial plume height was assumed to be 1500 m above Ignalina. Fig. 4 is a 'snapshot' showing the positions of the released parcels at a given time. The figure is taken from a time series which may be animated on a work station. From such a series of snapshots, the time of arrival (and departure) of the radioactive cloud over a given area may be directly read from the verification time written on the figures. The positions of the released air parcels are marked by letters indicating the geopotential height according to the tables on the figures. In order to indicate the effect of sub-grid scale diffusion, the particle positions are surrounded by shaded circles. The radius of these circles is here twice the standard deviation (1) in a Gaussian-distributed plume from the centerline. In Fig. 5, trajectories corresponding to one batch of released air parcels are shown. The effect of the vertical wind shear deserves notice.

The dispersion model described above has its virtues: it is simple and computationally fast. In an emergency situation, it may be more important to obtain a result very quickly from a simple dispersion program than to obtain a more accurate result from a time-consuming elaborate model. Besides, the initial conditions of the emission, such as the emission profile, will probably not be known very accurately at the time the first informations concerning an accidental emission of harmful material in the atmosphere are received.

The shortcomings of the model are the following: two-dimensional trajectories are not reliable e.g. for particles traversing frontal systems, the effect of sub-grid scale diffusion is merely treated schematically, we have not taken into account the wet and dry deposition of the released material, in some cases the square-box emission profile is not very realistic, and we have made no attempts of taking into account vertical diffusion. Thus, there are no atmospheric boundary-layer considerations in the model, and the model is not capable of calculating concentrations. It should be noted that e.g. the advanced stochastic Lagrangian dispersion model NAME (Maryon et al., 1992) complies with these requests.

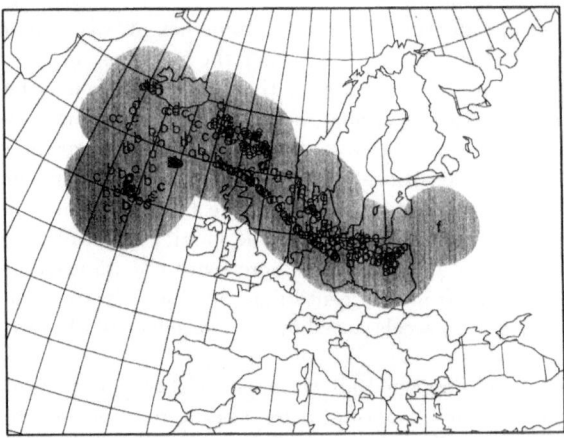

Atmospheric Dispersion based on GR-HIRLAM   DMI

Run time	: 12.09.93, 15:33 UTC
Emission:	
Emission point	: (26.17, 55.35)
Initial time	: 08.09.93, 00:00 UTC
Final time	: 12.09.93, 00:00 UTC
Initial plume height	: 0-1500 m
Verification time	: 12.09.93, 18:00 UTC

Height above msl:
a : < 250 m
b : 0250-0500 m
c : 0500-0750 m
d : 0750-1000 m
e : 1000-1250 m
f : 1250-1500 m
g : 1500-1750 m
h : 1750-2000 m
i : 2000-2250 m
j : 2250-2500 m
k : 2500-2750 m
l : 2750-3000 m
m: > 3000 m

Figure 4. Snapshot corresponding to a *hypothetical* release of radioactive material in the atmosphere from the nuclear power plant Ignalina in Lithuania.

Atmospheric Dispersion based on GR-HIRLAM DMI

Run time : 12.09.93, 15:33 UTC

Trajectories:
Emission point : (26.17, 55.35)
Emission time : 10.09.93, 03:00 UTC
Period : 5 days 0 hours 0 minutes
Marks each 6 hours
Initial plume height : 0-1500 m

Height above msl:
a : < 250 m
b : 0250-0500 m
c : 0500-0750 m
d : 0750-1000 m
e : 1000-1250 m
f : 1250-1500 m
g : 1500-1750 m
h : 1750-2000 m
i : 2000-2250 m
j : 2250-2500 m
k : 2500-2750 m
l : 2750-3000 m
m: > 3000 m

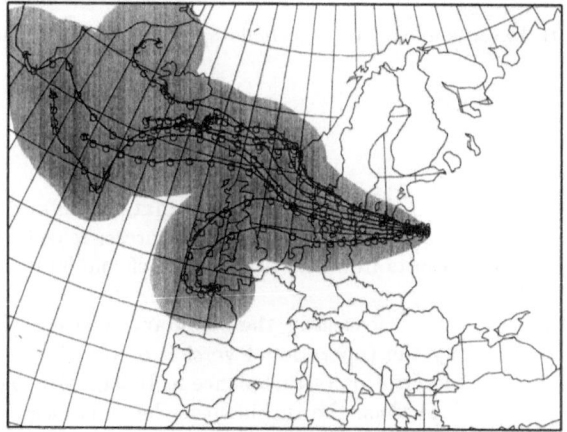

Figure 5. Trajectories corresponding to a *hypothetical* release of radioactive material in the atmosphere from the nuclear power plant Ignalina in Lithuania.

379

Gifford, F.A., Horizontal diffusion in the atmosphere: A Lagrangian-dynamical theory. *Atmos. Environ.* **16** (1982) 505-512

Gifford, F.A., The random-force theory: Application to meso- and large-scale atmospheric diffusion. *Boundary-Layer Met.* **30** (1984) 159-175

Hansen Sass, B., and Sørensen, L.S., The DMI operational HIRLAM forecasting system. *DMI Technical Report* **92-12** (1992)

Havskov Sørensen, J., Operational dispersion program, version 1, documentation manual. *DMI Technical Report* **93-5** (1993)

Holzlag, A.A.M., and Moeng, C.H., Eddy diffusivity and countergradient transport in the convective atmospheric boundary layer. *J. Atmos. Sci.* **48** (1991) 1690-1698

Kållberg, P., Air parcel trajectories from analyzed or forecast windfields. *In:* Workshop on Simplified Models for Short-Range Forecasting on the Mesoscale. SMHI R & D notes **37** (1984)

Kållberg, P., Kinematic air parcel trajectories. *SMHI Memorandum*

Kållberg, P., HIRLAM forecast model, level 1. *Documentation Manual, SMHI* (1990)

Machenhauer, B., (Ed.), HIRLAM final report. *HIRLAM Technical Report* **5** (1988)

Maryon, R.H., Smith, F.B., Conway, B.J., and Goddard, D.M., The U.K. nuclear accident model. *Progress in Nuclear Energy* **26** (1992) 85-104

Petterssen, S., *in:* Weather Analysis and Forecasting, Vol. I, Chapter IV (New York, McGraw-Hill, 1940)

Troen, I., and Mahrt, L., A boundary layer formulation for atmospheric models. (1984)

DISCUSSION

P. SEIBERT: Why do you believe that the bulk Richardson number is a better indicator for the presence of turbulence than the results from the turbulence scheme built in HIRLAM?

J.H. SØRENSEN: Presently, the boundary-layer height is not estimated by the (operational version of the) HIRLAM model. The description of turbulence in HIRLAM is based on K theory, and thus, the turbulence scheme is local. As a method for estimating the boundary-layer height, the bulk Richardson number approach (as described in the presentation) was chosen for the following reasons. The method is robust, simple, and to some extent non-local. In a future implementation of the HIRLAM model, the different contributions to the TKE equation are likely to be calculated implying a better description of the turbulence and, thus, the boundary-layer height.

J. WALMSLEY: You plotted the bulk Richardson number as a function of height, yet it did not seem to be height-dependent in the formula. Would you explain?

J.H. SØRENSEN: The bulk Richardson numbers shown at the presentation as a function of height are calculated for a discrete set of heights only, i.e. the bulk Richardson numbers are calculated for the layers between the surface and the HIRLAM model levels. In order to estimate the boundary-layer height, these bulk Richardson numbers are compared with a critical value. By interpolation, the boundary-layer top is thus determined as the height where the bulk Richardson number over the whole boundary layer becomes critical.

S. ZILITINKEVICH: The use of the bulk Richardson number approach can be physically grounded for stable stratification only. My question is: How do you calculate the boundary-layer height in unstable stratification?

J.H. SØRENSEN: In the case of unstable stratification, the bulk Richardson number approach is also utilised. Obviously, the method does not take into account non-local effects in any detailed manner. However, contrary to methods based on calculations of the gradient Richardson number, which are purely local approaches, the bulk Richardson number method does include non-local effects to some extent.

T. IVERSEN: If you have internal boundary layers (several layers of turbulence separated by stable layers), which height would your model choose as the boundary-layer height?

J.H. SØRENSEN: Assuming that the vertical resolution of the HIRLAM model is sufficient for resolving such detailed structure, the boundary-layer height estimated by the model is the lowest height in which the bulk Richardson number becomes critical. Unfortunately, it may happen that this does not occur at the top of the boundary layer in such a complicated situation. However, the bulk Richardson method is much less sensitive to internal stable layers than purely local approaches such as the ones based on calculations of gradient Richardson numbers.

EXPERIMENTAL EVALUATION OF A PC-BASED REAL-TIME DISPERSION MODELING SYSTEM FOR ACCIDENTAL RELEASES IN COMPLEX TERRAIN

Søren Thykier-Nielsen, Torben Mikkelsen and Josep M. Santabàrbara [1]

[1] Sponsored from the Catalan Research Council - Spain
Risø National Laboratory
DK-4000 Roskilde, Denmark

INTRODUCTION

The local-scale real-time dispersion modelling system LINCOM/RIMPUFF is evaluated using data from the complex terrain Guardo experiments carried out in Northern Spain November 1990.

Releases of tracer gas (SF_6) were simulated from a 185 m tall buoyant power plant stack, and from a valley floor release (10 m). The observed wind flow and plume diffusion is modelled and compared with tracer gas measurements. Studies reported in this paper include one daytime up-valley breeze situation and one nighttime down-valley drainage flow situation.

Wind and turbulence data were obtained from a network of 9 meteorological towers over the 40 kmx40 km experimental domain, accompanied by balloon and acoustic soundings.

The study shows that a linearized dynamic flow models, when properly initialized (assimilated) by observations, is an attractive alternative to the diagnostic mass consistent models for real-time applications.

BACKGROUND

Dispersion from sources located in mountainous regions is significantly influenced by terrain-induced flow and turbulence. Some important local flow-driving forces include differential heating/cooling on the slopes that in addition to the orography modify the wind and temperature profiles otherwise well-known over homogeneous terrain.

Atmospheric dispersion over hills is consequently influenced, not only by the local turbulence, but also by topographically induced winds, including channelling, speed-up effects, wind-shear, local up-valley breezes and down-valley drainage flows etc.

Over homogeneous terrain, atmospheric dispersion is often parameterized in terms of form-invariant and formula-based distribution functions, for instance specified by a single sigma-parameter as a function of travel time (Olesen and Mikkelsen, 1992).

Air Pollution Modeling and Its Application X, Edited by S-V. Gryning
and M.M. Millán, Plenum Press, New York, 1994

For hilly and topographically influenced regions however, dispersion modelling in addition requires a detailed flow-modelling capability for the terrain in question.

For accidental type real-time applications, the time constraints imposes flow modelling to be balanced as a compromise between what is most reliable within the available short time.

THE REAL-TIME DOSE ASSESSMENT SYSTEM RODOS

Within the Radiation Protection Research Programme of the European Communities, eighteen laboratories and institutes presently contribute with research and development to a joint real-time decision support system for nuclear emergencies named RODOS (Real-time On-line DecisiOn Support).

The aim of the project is to develop a system which is able to support decision makers in the events of future nuclear accidents in determining how best and most effectively mitigate their consequences.

RODOS involves multiple radiological disciplines, including real-time dose assessments and decision support for early emergency actions. Its ability to estimate a specific atmospheric dispersion scenarios in real-time is therefore identified as a first-priority.

A first prototype of RODOS was presented in 1992 at an International Workshop on Real Time Systems for Off-site Emergency Management (Mikkelsen and Desiato (1993), Thykier-Nielsen et al., 1993)

Atmospheric dispersion models are for RODOS being adopted and nested for the three different scales of atmospheric motion: 1) local scale (out to a few tens of kilometers), 2) the mesoscale (out to a few hundreds of kilometers), and 3) the continental (regional) scale (out to several thousands kilometers). In each case, modelling approach and sophistication level represent a compromise between making the most reliable estimate possible within the available time. For the local scale predictions need to be made very quickly (within a few minutes) to enable a rapid response. The modelling approach must therefore, for the local scale, be made much simpler compared to the grater distances.

For RODOS, a suite rather than a single atmospheric dispersion model have been identified and ranked in accordance with their suitability for real-time calculation from different emissions, meteorological condition, terrain type and scale of interest within the RODOS framework (Mikkelsen and Desiato, 1993).

Our combined flow and diffusion model system LINCOM/RIMPUFF is included in this suite for real time flow and dispersion calculations on the local or near site scales (0 - 20 (30 km).

THE LINCOM/RIMPUFF DISPERSION MODEL

Flow model LINCOM

LINCOM (Troen and de Baas, 1986; Walmsley et al. 1990; Santabàrbara et al. 1993) is an extremely fast diagnostic, non-hydrostatic dynamic flow model based on the solution of linearized versions of the three momentum equations, with a first order spectral turbulent diffusion closure in addition to the continuity equation.

Its truncated physics (linearization, neutral stratification) of course restrict it from application to severe non-uniform terrain, but considerable realism in the resulting wind fields is achieved by use of assimilation techniques to match the models resulting windfield to locally measured tower or forecast winds, as the paper will show.

Stratification effects is in LINCOM accounts for only by matching (based on least square error) the resulting flow field to the available observations. For near-neutral stability conditions, however, LINCOM accounts for many terrain features with only modest mathematical and computational effort. LINCOM runs extremely fast in "look-up" mode (100 by 100 grid points are processed in less than 10 seconds on a 50 MHz 486 DX PC).

A thermodynamic energy balance, based on an estimated or measured temperature field) is presently under development (version: LINCOM-T) whereby stratified flow effects, such as valley and local cold air drainage jets, in principal can be modelled. But even so equipped, this model is not an alternative to a full prognostic, primitive-equation based non-hydro-static flow model that accounts for differential heating (sea-breeze, valley slope and drainage winds etc).

However, by responding in seconds on a standard PC, LINCOM is attractive as a "driver" for fast real-time atmospheric dispersion models like RIMPUFF in assessment of emergency response and decision support.

Diffusion model RIMPUFF

RIMPUFF (Mikkelsen et al., 1984; Thykier-Nielsen et al., 1989; Thykier-Nielsen and Mikkelsen, 1993; Santabàrbara (1992)) is a fast and operational puff diffusion code which is suitable for real-time simulation of puff and plume dispersion during time and space changing meteorology. Also optimized for fast response on a PC this model is provided with a puff splitting feature to deal with plume bifurcation and flow divergence due to channelling, slope flow and inversion effects in non-uniform terrain.

For real-time applications, RIMPUFF can be driven by wind data from a combination of:

1) A permanent network of meteorological towers,
2) The flow model LINCOM (or similar), and
3) Numerical Weather Forecast data.

The puff or plume diffusion process is (averaging-time dependent) controlled by local turbulence levels, either provided directly from on-site measurements, or provided via pre-processor calculations (Mikkelsen and Desiato, 1993). Rimpuff is further equipped with plume rise formulas, inversion and ground level reflection capabilities, gamma dose algorithms and wet/dry depletion.

Prognostic capability of LINCOM/RIMPUFF is obtained via data link to regional atmospheric forecasting centers operation at several european meteorological institutes. The danish GRV-HIRLAM forecast model (Machenhauer et al. 1991) covers presently all of the CEC countries on a 0.42 deg horizontal resolution polar grid, with a 31 level vertical resolution. Running in a 6 hour updated data-assimilation cycle, this model forecast wind, temperature, humidity and surface pressure up to + 48 hours ahead. A small area version (DK-HIRLAM) covers the Nordic countries and central Europe and runs on a resolution of 0.21 deg (23 km) and forecast at 06 hours intervals up to + 36 Hours. These models are at present operational at several European meteorological services (Denmark, Holland, Ireland, Sweden, Finland), from where local wind and precipitation forecasts are envisioned to be transferred on-line to RODOS users via fast digital telephone network (ISDN), or via computer networks (Internet).

385

Evaluation record

Concurrently with the improvements in code, the Lincom/Rimpuff modelling system is continuously being evaluated using full scale experimental data. For non-homogeneous terrain applications, recent references include: Thykier-Nielsen et al. (1989); Massmeyer et al. (1990); Thykier-Nielsen et al. (1990a); Thykier-Nielsen et al. (1990b); Kamada et al. (1991); Thykier-Nielsen et al. (1991); Kamada et al. (1992); Thykier-Nielsen et al. (1993).

THE GUARDO FIELD STUDY

The Guardo experiment

An intensive field experiment organized by the spanish utility company Iberdrola II (Project PIE-134.036) took place in the fall of 1990 in order to quantify atmospheric dispersion processes over non-uniform terrain: Ibarra (1991), Ibarra (1992), Ibarra 1993).

Diffusion experiments based on SF_6 tracer gas and SO_2 emissions were conducted around the Guardo coal fired power plant, which is located in the river Carrion valley of the southern foothill of "Picos de Europe" near the plains of the Palencia region, see Fig. 1.

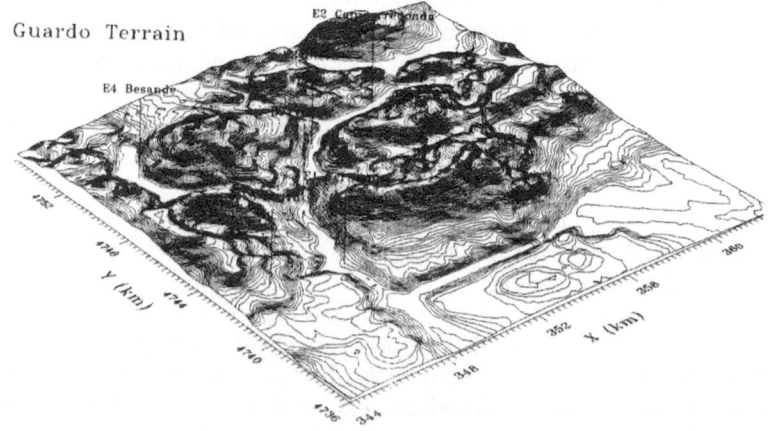

Figure 1. The Guardo experimental area with positions of the met-towers used in the modelling is shown. The River Carrion valley extends north-south in the northern part of the Palencia Province. A large plateau extends to the south from the outlet of the valley where the town of Guardo is located. A smaller north-west oriented valley merges into the Carrion valley some two kilometers north of Guardo.

The highest part of the valley extends to the north-east and merges into the *"Picos de Europa"* mountain chain that parallels the Cantabrico sea. The width of the valley is 1-2 km near the power plant, hoverer the narrower part close to the Compuerto Embalsa dam is only some 300 m wide, to the south the valley broadens further into the town of Guardo.

Atmospheric transport and diffusion processes studied over a 40kmx40km domain centered around the Guardo power plant (TERMINOR) shown in Fig. 1.

The purpose was to investigate the dispersion characteristics from the 185 meter buoyant power plant chimney, and also to study the dispersion characteristics from ground-level down-valley SF_6 releases during drainage conditions. The local meteorology of the valley was significantly influenced by a diurnal flow reversal between daytime up-valley advection and night time down-valley drainage flow.

A large number of micro- and mesoscale meteorological instruments were deployed along the valley to study winds, pressure and thermal gradients, and also outside the valley for more general wind patterns to be used in the mesoscale models, Ibarra (1991, 1992,1993).

Meteorological measurements

Fig.1 shows the locations where the meteorological measurements were carried out.

Horizontal measurements of 10-min mean wind speed and direction were recorded at 8 locations using 10 m masts. Turbulence measurements were in addition carried out from a 25 m met-mast instrumented with cups, vanes and a sonic anemometer. 10-min averaged mean and turbulence quantities (L, u*, heat flux etc) were obtained every 10-min (Nielsen and Mikkelsen, 1992).

A near by Doppler sodar monitored the mean wind speed and direction components and mixing heights in the lowest 600 m of the atmosphere.

A total of 14 tracer experiments of approximately 2-4 hours duration each were conducted during the campaign. Seven of the experiments were performed for ground level releases and another seven experiments for stack releases.

Some experiments were on purpose conducted during non-stationary transition periods between valley breeze and drainage. During daytime, the elevated stack releases were typically dispersed in an up-valley S-SW convective flow that developed over noon and continued until late afternoon, at what point a south-running valley drainage layer started to extend vertically. Sometime during the night this would reach the upper ridges of the valley (300-400m) and could exceed 12 m/s at the Vivero micro-met tower site. Ground level valley releases were captured by this NE drainage flow.

Case analysis

In this paper we have chosen to simulate one daytime up-valley breeze situation and one nighttime down-valley drainage flow situation:

Case #1: Valley breeze - Experiment No. 5 (Nov. 14, 1990)

Experiment No. 5 was launced during mid-afternoon when a local up-valley breeze was well developed. No significant wind shear was observed within the local valley flow. However, a strong wind direction shear occurred at the upper boundary of the mixing layer (500 m) where the mean wind vector veered to NW after passing a minimum in speed. A slight upward movement within the valley was observed while a much stronger downward flow occured over the hills associated with the synoptic flow aloft. A strong shear layer was observed at the interface between the local southerly breeze and the above synoptic N-NW flow at approximately 500 meters height.

Otherwise a clear sunny day with good visibility after early morning frost. Global radiation evolves from nearly 500 W/m^2 at the beginning of the experiment to some 100 W/m^2 near the end.

Initially, the plume was advected in the valley breeze towards the N-NNW impacting on an inversion or subsidence observed at approximately 400-500 meters height. The top of the plume was observed to veer towards ESE-SE influenced by a weak upper level NE flow. Vertical plume diffusion was observed to be suppressed by the presence of the subsidence. Otherwise, a "typical" up-valley" breeze case.

During clear sky and a moderate SSW breeze (3-4 m/s), SF_6 was released from the chimney during the period 15:00 to 18:00. The plume was observed to rise to approx. 400 meters total height with a strong horizontal diffusion. Sampling took place in 4 consecutive 1/2 hour periods between 16:00 and 18:00.

Case #2: Drainage flows - Experiment No. 14 (Nov. 30, 1990)

Starting at 7:30 (am), SF_6 was released from the 10-m level of the micro-met tower at the Vivero site in the upper part of the Carrion Valley.
Cold, and with clear skies during the nocturnal freeze, a strong mountain valley drainage jet (peaking at 8-10 m/s near the ground) has established over night. It extendeds vertically to approx. 200-300 m befor vanishing into very weak aloft winds.

Tracer gas sampling took place between 8:00 and 10:00 am.

MODELLING

LINCOM data assimilation procedure

Wind fields produced by LINCOM reflects the terrain-perturbed response to a single up-wind "free" or un-perturbed input wind. That is, the input wind is the mean speed and direction that would apply uniformly over the entire model domain in the absence of hills or mountains.

Tower observations made in valleys, ridges or on mountain tops are not "free winds", they are to the contrary often heavily influenced by their surrounding topography.

The problem therefore arises on how to utilize terrain-influenced observations for driving a diagnostic Navier-Stoke Eqs. based flow model:

Based on the fact that LINCOM is a linearized model, and that the response from two orthogonal inputs (East and North, say) results in two independent perturbation fields, the following procedure have been adapted to obtain model windfields that assimilate observations obtained within the terrain best possible:

1   Calculate the terrain-induced perturbations over the entire modelling domain from two unit-amplitude "East" and "North" free winds.
    This (most time consuming part) is only done once.

2. Backfit, by a linear-combination of the perturbations obtained in step 1, the free wind vector that best possible fits (by least-square-error method) the available met-tower observations. (For cases where a single met-tower is used, this can be done exactly).

3. By use of the "best-fitted" free wind vector obtained during step 2 now calculate the wind field at the actual positions and heights where puffs are being advected.

Further details are listed in Thykier-Nielsen et al.(1993a).

Case 1: Simulation of experiment no. 5 (Valley Breeze)

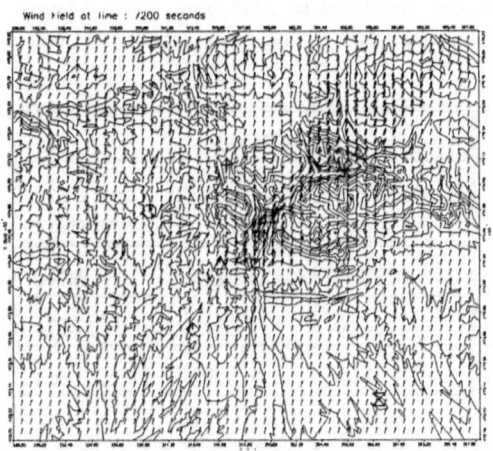

Wind Field at time : /200 seconds   Z-plan 0.00 / MaxD: 100 = 1.0000E-03 gs/c / Time 10800

Figure 2. LINCOM: Wind field at 17:00   **Figure 3.** RIMPUFF: Mean plume at 18:00

Data for wind speed and direction were available from meteorological masts at the following 9 stations as 10 or 15 min averages : TERMINOR, CAMPORREDON-DO,OTERO, BESANDE, MARIU, MOHECILLAS, MORGOVEJO, SAN PEDRO and VIVERO.

The VIVERO 25-meter micromet tower furthermore provided on-line turbulence data for plume parameterization. The turbulence data used were 10 min averages of σ_ϕ and σ_θ calculated from wind variances measured.

The effective plume rise was assumed to be limited   the subsidence at 400 m, ie. the inversion height.

From the meteorological observations, the above described fitting procedure was used to obtain LINCOM wind fields. We used the mid-valley height (200 m a.g.l.) as a representative mean height for puff advection in the valley for this case.

The winddata assimilation procedure was repeated for each 10 minute averaging period and feed to RIMPUFF. The on-line available turbulence measurements from the micromet-tower was also provided to Rimpuff at 10 minutes consecutive intervals.

We studied 4 different "modes" for supplying input winds:

I: The 25-m micromet-tower at VIVERO (one met-tower only).
II: The 5 most local met-towers: TERMINOR (source), VIVERO, BESANTE, OTERO, CAMPORREDONDO.
III: All available 9 stations.
IV: Interpolation (no LINCOM) All 9 stations.

As an example the resulting wind field at 17:00 (in the middle of the sampling period) from mode-I is shown on Fig. 2.

Source terms, release rates and many other relevant parameters were obtained from Ibarra (1991) and provided to RIMPUFF. The resulting mean concentrations for the entire sampling period (16:00-18:00) is shown in Fig. 3.

Table 1. Modelling performance based on measured wind and SF₆ observations: Performance statistics for LINCOM vs. tower observations are listed as RMS error [m/s], whereas RIMPUFF mean concentration assessment is based on: Fractional bias; Correlation coefficient and Index of agreement N. (Cf. Ibarra (1993)).

Mode	LINCOM	DOSE		
	RMS	Correlation Coefficient	Fractional Bias	Index of Agreement
I: Risø Micromet Tower	0.00	0.3709	0.0475	0.6090
II: 5 local stations	1.60	0.7419	-0.2645	0.7874
III: All 9 masts	1.49	0.0865	0.3583	0.3864
IV: Interpolation		0.0955	0.4895	0.3918

LINCOM calculations have alternatively been based on data from the 200 meter level of the acoustic sounders co-located near the VIVERO site, but with no significant different results compared to "mode 1" above. Also a "one over distance square" interpolation scheme is inherent in RIMPUFF, cf mode-IV.

Best results are obtained using representative and well exposed valley located met-towers in combination with LINCOM.

Total CPU-time on a 50 Mhz 486 PC: 10-(15) min. LINCOM: 60 sec for initial 4 fields (two at the 10 meter level + two at the 200 meter level) plus approx. 10 sec in look-up mode for each additional 10-min period.

Case 2: Simulation of experiment no. 14 (Drainage)

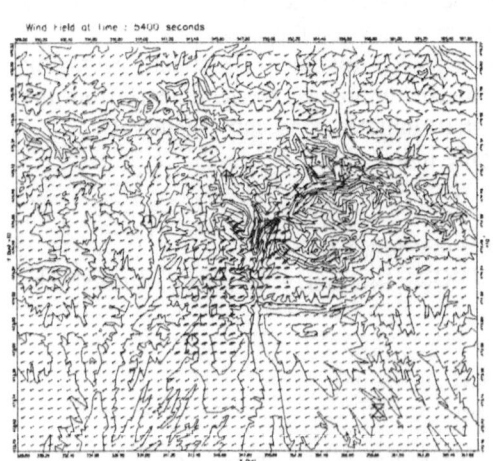

Figure. 4. LINCOM: Wind field at 09:00

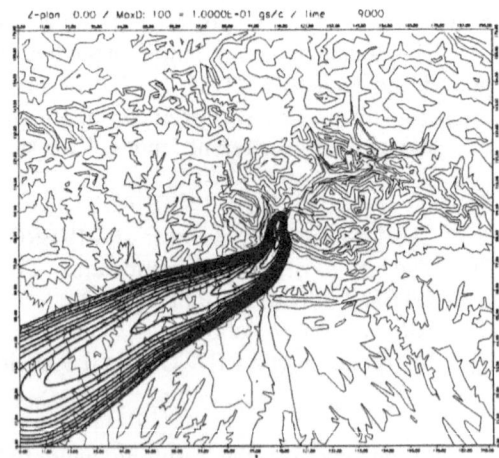

Figure 5. RIMPUFF: Mean plume at 10:00

Again, data for wind speed and direction were available as in case 1.

The non-buoyant tracer gas release took this time place from the 10 meter level of the VIVERO micromet tower.

This time we used the 10 meter wind field of LINCOM for the down-valley puff advection.

Otherwise the winddata assimilation procedure was as in Case 1. Again, the on-line available turbulence measurements from the micromet-tower was provided to Rimpuff at 10 minutes consecutive intervals.

For wind "mode-II", we this time used the following 4 most local met-towers: VIVERO (source), TERMINOR, MARIU and SAN PEDRO.

The wind field at 9:00, again from mode-I is shown on Fig. 4. whereas the corresponding mean concentrations for the entire sampling period (8:00-10:00) is shown in Fig. 5.

Table 2. Presents as above the corresponding statistical modelling performance.

Total CPU-time for this case: 5 min. LINCOM: 30 sec for initially 2 fields at the 10 meter level, plus again approx. 10 sec for each additional 10-min period.

Mode	LINCOM	DOSE		
	RMS	Correlation Coefficient	Fractional Bias	Index of Agreement
I: Risø Micromet Tower	0.00	0.4511	0.3357	0.5279
II: 4 local stations	2.11	0.4086	0.0508	0.6121
III: All 9 masts	2.11	0.4075	-0.0028	0.6005

DISCUSSION

We have at this point modelled 2 out of at least 7 well-defined Guardo experiments with our PC base real-time modelling system LINCOM/RIMPUFF.

Apparently this simplified and comparatively fast concentration assessment system compares well regarding statistical performance with other and more advanced modelling system including the Mathew/Adpic modelling system of Lawrence Livermore, cf. Ibarra (1993) and Ibarra (Ibid).

Although preliminary (there are at least still 5 experiments to modelled), our preliminary study have shown that a linearized dynamic flow models, when properly initialized (assimilated) to observations, has potential as an alternative to diagnostic mass consistent models for real-time applications. In particular so, when 1: Only a single or a few met-tower measurements are available, and 2: For near-neutral, relatively strong wind situations.

Work is now in progress in order to modify LINCOM to implicit include thermal stratification effects by inclusion of buoyancy in its vertical momentum equation. By prescribing mean temperature field either from measurements, modelling, or a combination of both, the model can be extended to encompass, to first order still, stratification effects such as the up-valley convection and cold-air drainage flows as considered in this paper. Data from the Guardo field experiments are excellent for this to be evaluated.

ACKNOWLEDGEMENTS
George Lai, NASA Goddard Space Flight Center, USA is gracefully acknowledged for invention on the tower assimilation method used in LINCOM.

Our participation in the Guardo experiment was sponsored by the Spanish utility company Iberdrola II, S.A. Madrid, as part of Project PIE-134.036. We are sincerely graceful to José I. Ibarra for organizing and coordinating these experiments very successfully, and for distribution of the data. The modelling part of this work has been sponsored by the CEC DGXII Radiation Protection Research Programme, under contracts No.FI3P-CT92-0044.

REFERENCES

Ibarra, J.I. (1991) Atmospheric dispersion experiments over complex terrain in a Spanish valley site (Guardo-1990). In Proceedings of the: Advanced Modelling and Computer Codes for Calculating Local Scale and Meso-scale Atmospheric Dispersion of Radionuclides and their Applications. (AD-LMS'91). Ed. E. Sartori. OECD NEA Data Bank, Saclay,91191 Gif-sur Yvette, France, pp 74-85.

Ibarra, J.I. (1992) Observations of valley winds in Spains Guardo valley.Proceedings of the Forth International Conference ENVIROSOFT92, Portsmouth, 7-9 September 1992, Computational mechanics publications, ISBN 1-85312-177-0, Southhampton,pp 652-6672.

Ibarra, J.I. (1993) Atmospheric dispersion under local valley breeze conditions during the Guardo experiment. In: International Symposium on Air Pollution '93, Monterrey, Mexico 23-25 February 1993. Available from: Wessex Institute of Technology, Ashurst, Southhampton, UK.Kamada, R.F., S.A. Drake, T. Mikkelsen and S. Thykier-Nielsen (1992). LINCOM/RIMPUFF vs. Mt. Iron, a data/modelling comparison. Proceedings of the 7th JANNEF Envir. Safety & Protection Subcomm. Meet. Aug 10-14, 1992, NPS, Monterey, CA.

Kamada, R.F., S.A. Drake, T. Mikkelsen and S. Thykier-Nielsen (1991). A Comparison of eight cases from the Vandenberg AFB MT. Iron tracer study with results from the LINCOM/RIMPUFF Dispersion Model.Naval Postgraduate School, Monterey, California, NPS report No.: NPS-PH-92-006.Dec 1991.

Kamada, R.F., S.A. Drake, T. Mikkelsen and S. Thykier-Nielsen (1992). LINCOM/RIMPUFF vs. Mt. Iron, a data/modelling comparison. Proceedings of the 7th JANNEF Envir. Safety & Protection Subcomm. Meet. Aug 10-14, 1992, NPS, Monterey, CA.

Machenhauer B., U.B. Nielsen and A. Rasmussen (1991): Evaluation of the Quality Especially of Wind-forecasts by the Operational Meteorological HIRLAM-system. Proceedings of the OECD/NEADB Specialists' Meeting on Advanced Modelling and Computer Codes for Calculating Local Scale and Meso-Scale Atmospheric Dispersion of Radionuclides and their Applications (AD-LMS'91), Saclay, France. OECD Publication No. 75907. pp 152-163.

Massmeyer, K., K. Born, B. Erbshäusser, C. Hauk, Th. Flassak (1990). Regional flow fields in North Rhine Westfalia - A Case study comparing flow models of different complexity. 18 th ITM on Air Pollution Modelling and its Applications, NATO-CCMS, Vancouver, Canada.

Mikkelsen, T., S.E. Larsen and S. Thykier-Nielsen (1984). Description of the Risø Puff Diffusion Model. Nuclear Technology, Vol. 67, pp. 56-65.

Mikkelsen, T. (1992) Atmospheric dispersion models for real-time application in the decision support system being developed within the CEC. In: Olesen, H.R. and T. Mikkelsen (Eds.) 1992: Proceedings of the workshop "Objectives for Next Generation of Practical Short-Range Atmospheric Dispersion Models", Risø, Denmark, May 6-8 Danish Center for Atmospheric Research (DCAR), P.O.BOX 358, DK-4000 Roskilde. pp 109-130.

Mikkelsen, T. and F. Desiato(1993): Atmospheric Dispersion Models and Pre-processing of Meteorological Data for Real-time Application. In: Proceedings of the Third International Workshop on Real-time Computing of the Environmental Consequences of an Accidental Release to the Atmosphere from a Nuclear Installation, Schloss Elmau, Bavaria, October 25-30 1992. To appear in: Journal of Radiation Protection Dosimetry (Dec. 1993).

Nielsen, M. and T. Mikkelsen (1992): Micro-meteorological data report from the Guardo dispersion experiments in complex terrain. Risø-R-634(EN), 148 pp. ISBN 87-550-18165-5. Available on request from: Risø Library, Risø National Laboratory, P.O. Box 49, DK-4000 Roskilde, Denmark.

Olesen, H.R. and T. Mikkelsen (Eds.) 1992: Proceedings of the workshop "Objectives for Next Generation of Practical Short-Range Atmospheric Dispersion Models", Risø, Denmark, May 6-8 Danish Center for Atmospheric Research (DCAR), P.O.BOX 358, DK-4000 Roskilde. 262 pp.

Santabàrbara, J.M. (1992): S. User'& Reference Manual. Interim Risø report, October 1992. Available on request from: Department of Meteorology and Wind Energy, Risø National Laboratory, P.O. Box 49, DK-4000 Roskilde, Denmark.

Santabàrbara, J.S., T. Mikkelsen, R. Kamada, G. Lai and A.M. Sempreviva (1993): LINCOM Wind Flow Model, Risø-R-report(EN), 37 pp. Available on request from: Department of Meteorology and Wind Energy, Risø National Laboratory, P.O. Box 49, DK-4000 Roskilde, Denmark.

Thykier-Nielsen, T. Mikkelsen, S.E. Larsen, I. Troen, A.F. de Baas, R. Kamada, C Skupniewicz and G. Schacher. (1989) A model for accidental releases in complex terrain. In: Air Pollution Modelling and its Application VII (Ed. H. van Dop) Plenum Publ. Corporation) pp 65-76.

Thykier-Nielsen, S., T. Mikkelsen, R. Kamada and S.A. Drake(1990a): Wind Flow Model Evaluation Study for Complex Terrain. In: Proceedings from the ninth Symposium on Turbulence and Diffusion. American Meteorological Society. Risø, Denmark, April 30 - May 3. 1990. pp 421-424.

Thykier-Nielsen, S., T. Mikkelsen, F. Gassmann and V. Herrnberger(1990b): Comparison of Wind field Dispersion Models with Tracer Experiments in Weak Neutral Flow Conditions of Complex Terrain. In Proceedings of: Jahrestagung Kerntechnik '90. Deutches Atomforum e.V. Nürnberg, 15-17th May 1990.

Thykier-Nielsen, S., T. Mikkelsen and V. Herrnberger (1991): Real-time Wind and Dispersion Simulation of tracer experiments conducted over complex terrain during weak and neutral flow conditions. In proceedings of:OECD/NEADB Specialists' meeting on Advanced Modelling and Computer Codes for Calculating Local Scale and Meso-Scale Atmospheric Dispersion of Radionuclides and their Applications (AD-LMS'91), 6-8 March 1991, OECD NEA Data Bank, Saclay, France. pp. 86-119 (OECD PUBLICATION No. 75907 1991).

Thykier-Nielsen, S., J.M. Santabarbara and T. Mikkelsen (1993a). A real-time dispersion scenario over complex terrain. In: Proceedings of the 3. International Workshop on Real-time Computing of the Environmental Consequences of an Accidental Release to the Atmosphere from a Nuclear Installation. Schloss Elmau, Bavaria, Oct 25-30, 1992. Accepted for publication in: Radiation Protection Dosimetry, Nov. 1993.

Thykier-Nielsen, S. and T. Mikkelsen (1993b): **RIMPUFF USER'S GUIDE** (Version 33 - PC version). Available on request from: Department of Meteorology and Wind Energy, Risø National Laboratory, P.O. BOX 49, DK 4000 Roslilde.

Troen I. and de Baas, A.F., 1986: A spectral diagnostic model for wind flow simulation in complex terrain, Proc. Euro. Wind Energy Assoc. Conf., Rome, Oct. 7-9, 1986.

Walmsley, J.L, I. Troen, D.P. Lalas and P. J. Mason (1990): Surface-layer Flow in Complex terrain: Comparison of Models and Full-scale Observations. Boundary layer Meteorology Vol. 52, pp 259-281.

DISCUSSION

G. ADRIAN: A model only for neutral stratification should not be used in mountainous terrain with strong termal effects.

S. THYKIER-NIELSEN: Yes, you are right, - a neutral flow model applies in this case only approximately, and best during relatively strong (externally forced) winds.

LINCON is used in this study merely as an "orography-influenced" interpolation scheme, assuming that stratification effects are already represented in the met-tower observations.

A version is under development that explicitly takes the stratification effects into account by inclusion of the (linearized) temperature equation.

RADIOACTIVE DISPERSION MODELLING AND

EMERGENCY RESPONSE SYSTEM AT THE GERMAN WEATHER SERVICE

Barbara Fay, Hubert Glaab, Ingo Jacobsen, Reinhold Schrodin

Deutscher Wetterdienst
Frankfurter Str. 135
63067 Offenbach, Germany

INTRODUCTION

Following the Chernobyl accident in 1986, the German Weather Service (DWD) was called upon by law to perform emergency response calculations in case of nuclear incidents. It was already charged with supervising atmosphere and precipitation for radioactive substances and their transport, and a monitoring network comprising measurements and trajectory calculations was operational.

In the framework of a national emergency scheme a radioactivity emergency response system is being completed at the DWD consisting of integrated measurements, a trajectory model, and a Lagrangian particle dispersion model.

A Eulerian transport model was also developed which was validated against measurements of the North America Tracer Experiment ANATEX together with the Lagrangian particle dispersion model.

RADIOACTIVITY EMERGENCY SYSTEM

In 1987, the Integrated Measurement and Information System (IMIS) was initiated to co-ordinate and unify all routine and emergency action like measurements of radioactivity, modelling, and information of government bodies on a national and federal basis.

In the framework of IMIS a radioactivity emergency system is under completion at the DWD. It includes measurements and dispersion modelling and consists of a routine precautional and an actual emergency part.

The radioactivity monitoring network is installed at meteorological observation stations. It was extended from 12 to 36 stations and is concentrating on nuclide-selective gamma-spectrometric measurements of radioactivity carried by aerosol particles. Artificial α- and β-activity are also determined. Daily results are transmitted from the stations to the

Air Pollution Modeling and Its Application X, Edited by S-V. Gryning
and M.M. Millán, Plenum Press, New York, 1994

Central Office at Offenbach, they are verified and transferred into the IMIS network ready to be used in dispersion models. In case of an emergency the measurement frequency is intensified.

The IMIS emergency model system consists of a trajectory model and a Lagrangian particle dispersion model. These models mainly use the meteorological fields calculated by the 'Europa-Modell' of the DWD (see below) but can also be run on a meso-β or a global data base. Precautional daily trajectories are calculated for immediate information. The emergency system proper is menu-operated and designed to be run by the Meteorological Operations Department. It automatically produces tables and plots, and feeds trajectory positions and radioactivity concentration fields directly into the IMIS emergency network. The results are stored in the IMIS data bank. From there they are retrieved together with the radioactivity measurements as input to the program package that evaluates radiological consequences and serves for decision making by governmental bodies. All components of the system have been tested repeatedly, with the final test performed in autumn 1993.

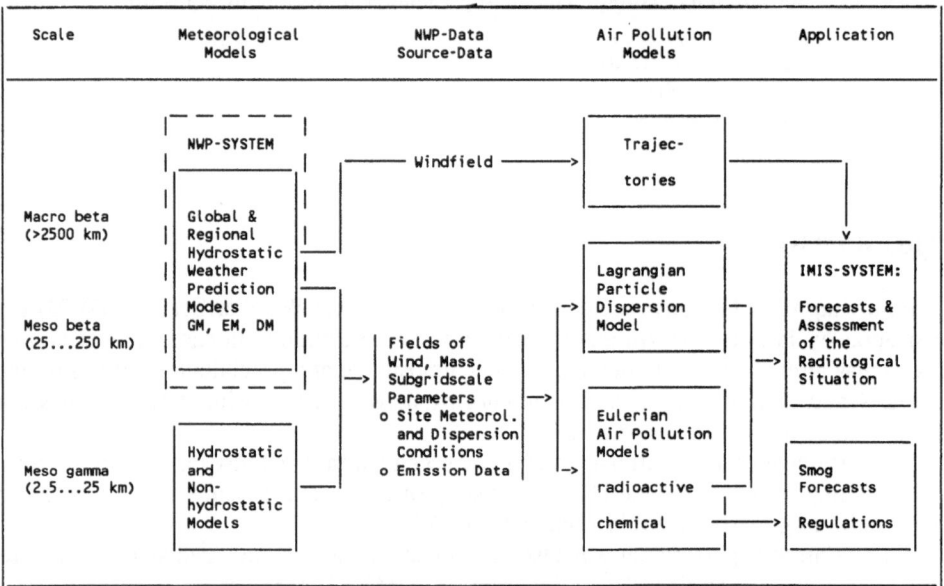

Figure 1. Scheme of NWP and air pollution models at the DWD.

NUMERICAL WEATHER PREDICTION MODELS

The quality of any dispersion model is crucially dependent on the quality of the employed meteorological fields. Since January 1991, the model suite of global model (Global-Modell, GM) and meso-α model (Europa-Modell, EM) is running operationally supplemented by a meso-β model (Deutschland-Modell, DM) since July 1993 (NWP Progress Report,1992).

The GM is a global spectral model with T106L19 resolution based on the ECMWF model but with different physical and numerical treatment. The EM covers Europe and the North Atlantic with a 55 km horizontal grid and 20 vertical layers (Majewski,1991) while the DM focuses on Germany and its surroundings with a 14 km horizontal resolution and 20 vertical layers.

Verification of the GM of DWD resulted in a performance comparable to the ECMWF-model in the northern hemisphere. Among the DWD models the EM shows a better performance than the GM while the DM provides more detailed surface meteorological fields and performs better compared to measurements than the EM. Thus a good meteorological basis for dispersion calculations is guaranteed. In Figure 1 the present scheme of NWP and air pollution models at the DWD is displayed.

AIR POLLUTION MODELS

Trajectory Model

Based on the GM-EM-DM model suite at the DWD, a new trajectory model system was introduced. The 3D-trajectory calculations are performed off-line, with the wind updated from the NWP-model most suited to the application. At the boundaries of a nested model the data supply can be switched to the next coarser model. Trajectories are calcula-

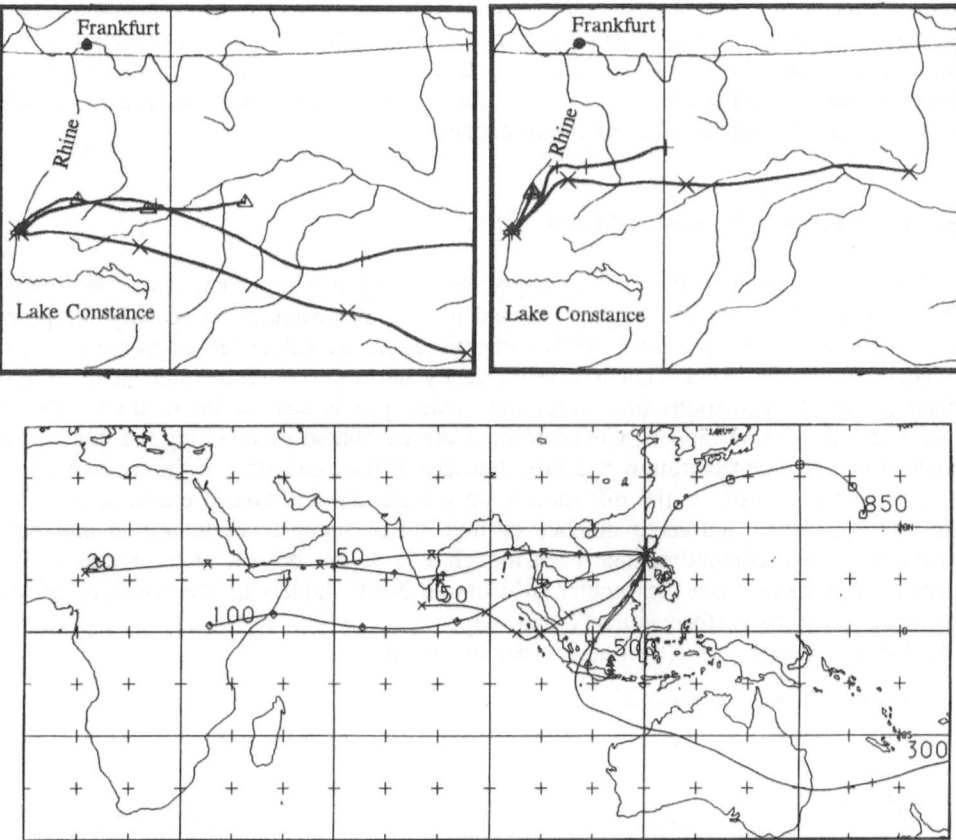

Figure 2. Trajectory calculations. Top: comparison of 36h EM(left) and DM(right) forward trajectories starting 23 Jul 1993, 00z, in Rhine valley at 50m △ , 250m + , 500m x above model orography showing chanelling of DM trajectories in Rhine valley due to higher model resolution. Symbols at 00z and 12z. Bottom: 120h GM forward trajectories from erupting Pinatubo, Philippines, starting 16 Jun 1991, 00z, in 850hPa ⊙ , 500hPa △ , 300hPa + , 150hPa x , 100hPa ◇ , 50hPa ↑ , 20 hPa x . Symbols at 00z.

ted using an iterative first order Euler-Cauchy scheme with third order spatial and first order temporal interpolation of meteorological fields.

On the precautional side, trajectories for the domain of the EM are calculated operationally. Twice a day 48h forward resp. backward trajectories are computed at three different height levels for about 30 European nuclear power installations resp. 35 German radioactivity measuring sites, and for the German regional forecast centers. Their results can be fed into the IMIS-network and serve as an instant source of information in case of a nuclear accident. A selection of trajectories are displayed graphically on workstations in the forecast center and transmitted to users via weather fax. DM-based forward trajectories are run twice daily, too, for all German, Swiss, Czech and a few French nuclear power installations, producing IMIS-conform results.

The menu-operated emergency trajectories were put to the test successfully during various instances (reactor incidents in Sosnovy Bor and Tomsk, bush fires near Chernobyl, German alert exercises etc.).

The emergency system was also designed for calculating trajectories for scientific investigations which are in constant extensive demand. Backward trajectories to 5 African cities are distributed daily via METEOSAT. Kuwait trajectories were provided to the Federal Environmental Agency for planning aircraft measurements in the Gulf war oil fire plumes. Trajectories for several German and European research projects concern(ed) among others chemical measurements on various ships and stations in the Alps and the polar regions, trajectories for TRANSALP, EASOE, the long-term preparation of the European Tracer Experiment (ETEX), and for the estimation and measurement of the transport of aircraft exhausts across the tropopause.

Lagrangian Particle Dispersion Model

For use in the radioactivity emergency system a Lagrangian particle dispersion model (LPDM) was chosen because of physical and numerical advantages especially for point sources. From the source a large number of trajectories are calculated representing contaminated air parcels. Their path is determined by the mean wind and stochastically described turbulent fluctuations (therefore, the model type is also called random walk or Monte-Carlo dispersion model). Concentrations are calculated by counting the number of particles in a given concentration grid box. Because of the exact start of the trajectories at the source no unrealistic initial diffusion as in a Eulerian grid box is encountered. The mutual independence and large number of trajectories supports vectorization and parallelization to save computing time even though a large number of interpolations of the meteorological fields from the regular grid of the NWP models to the positions of the trajectories has to be performed every time step.

In the LPDM the position $x_i(t)$ of each trajectory is given as

$$x_i(t+\Delta t) = x_i(t) + v_i(t)\Delta t.$$

The velocity

$$v_i = \bar{v}_i + v_i'$$

is composed of a mean velocity $\bar{v}_i$ and the turbulent fluctuation v_i' computed by

$$v_i'(t+\Delta t) = R_i v_i'(t) + \sqrt{1-R_i^2}\, v_i'' + T_{Li}(1-R_i)\frac{\partial \sigma_i^2}{\partial x_i}$$

(Legg and Raupach,1982;de Baas,1988,etc.). Terms 1 and 2 on the right-hand side comprise a Markov chain. In the time-correlated term 1 the correlation function R_i is usually approximated exponentially using the Lagrangian time scale T_{Li} which describes how long turbulent fluctuations are remembered by the particle. This concept serves to simulate the nature of atmospheric turbulence more closely than is done in a Eulerian transport model. In the stochastic term 2 the random turbulent fluctuation v_i'' is calculated using the velocity variance σ_i and a normally distributed random number. The numerical correction term 3 (drift velocity) is necessary to prevent an artificial accumulation of particles in regions of low turbulence.

The appropriate parameterization of σ_i and T_{Li} poses one of the theoretical problems in constructing a LPDM . It should be consistent with the scale of the NWP model supplying the meteorological fields. σ_i in the LPDM is parameterized as the square root of turbulent kinetic energy (derived from a second order closure scheme, a modified version of Mellor and Yamada(1974);Müller(1981)) multiplied by stability dependent partitioning factors for the three spatial directions. The Lagrangian time scale is the relation of the horizontal (vertical) diffusion coefficient K_H (K_V) to σ_i^2 (Taylor's theorem).

There still remains uncertainty about the horizontal diffusion coefficient because it is chosen in NWPs partly to serve not only physical but also numerical demands. Differences in horizontal and vertical scales have to be considered as well. Therefore, the LPDM now uses a time-split integration scheme with a time step of 5 sec in the vertical compared to

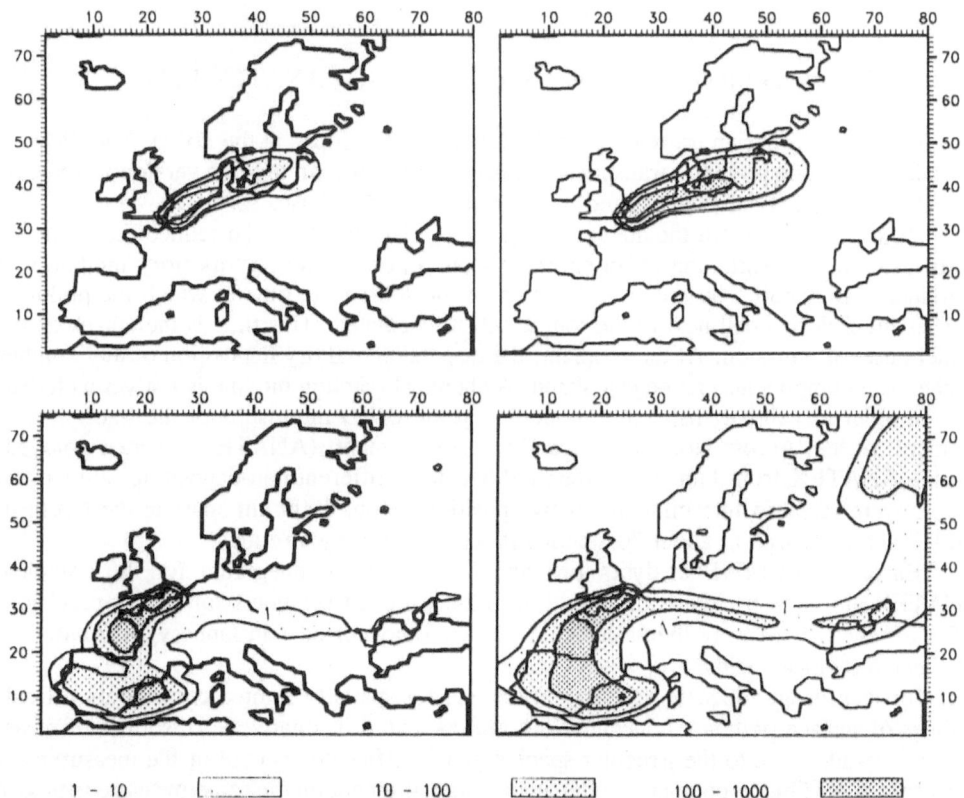

1 – 10   10 – 100   100 – 1000

Figure 3. LPDM forecast for 30h(top) and 78h(bottom) following an assumed emission from the nuclear generating site of Gravelines, Belgium, on 8 Jan 1992, 06z, based on EM meteorological fields.
Emission height 30m, source strength 1GBq/sec.
Left: surface concentration (mBq/m³), right: vertically integrated air concentration (Bq/m²).

900sec in the horizontal in order to make the simulation of vertical diffusion processes in the LPDM more realistic.

The mass m(t) of each radioactive particle is reduced by radioactive decay which is a linear process with constant coefficients (decay rates). It is also diminished by dry and wet deposition which are modelled in the LPDM using a probability concept (Axenfeld,1984). Due to these probabilites a portion of each particle is deposited by dry or wet deposition reducing its original mass $m(t_0)$.

Dry deposition probability is parameterized by a deposition velocity depending on a resistance approach including the atmospheric and surface resistances. This Eulerian deposition velocity is used to define a deposition probability w_d implying that a portion w_d of each particle touching the ground is deposited while the portion $(1-w_d)$ of the same particle is reflected back to the atmosphere.

At present a particle's portion to be lost due to wet deposition is evaluated using different scavenging coefficients for rain and snow. As the EM contains a complex parameterization of cloud microphysical processes like condensation, freezing, melting, accretion, sedimentation and evaporation (Müller et al.,1986) the basis for a more comprehensive description of wet deposition exists.

The LPDM has been fully vectorized and needs about 30 min. for a 72-h forecast on one CPU of a CRAY Y-MP including deposition and calculating up to 100 000 trajectories. The model can be run with several nuclides simultaneously and parallel on all 4 CPUs and then performs at a peak rate of more than 400 Mflops. An example of a dispersion calculation with the LPDM is given in Figure 3.

Eulerian Transport Model and Model Verification using ANATEX Data

Recently, a Eulerian transport model (ETM) was developed at the DWD. The EM was extended to include pollutant transport equations (ETM-on). An off-line version (ETM-off) uses the meteorological fields stored in the EM output data base.

Both models contain the advection algorithm of Bott(1992). To reduce the influence of diffusion at the beginning of an emission from a point source a trajectory module was introduced. Depending on the exact position of the source the time at which the pollutant reaches the Eulerian cell next to the source cell is calculated. This time is then taken as the model onset of emission. At the moment, the dispersion and dry deposition of any number of chemical components can be calculated. A chemical reaction module is not yet included.

In order to validate dispersion modelling at the DWD the results of the tracer measurements of the Across North America Tracer Experiment (ANATEX) were employed. During ANATEX from January to March 1987, three different non-depositing fluorcarbon tracers were released intermittently at two positions about 1000 km apart in the North of the US and measured at about 80 stations in the area and by aircraft.

Based on ECMWF analyses EM and ETM-on were integrated for the complete ANATEX emission period using only 12h to 24h forecast segments for higher precision. ETM-off and LPDM were then calculated for two 5-day episodes in January 1987 based on EM meteorological fields.

For lack of an adequate verification scheme for spatially limited concentration fields isolines of surface pollutant concentrations are compared qualitatively to isolines of measurement results. Due to the irregular spacing and insufficient coverage of the measurement stations it is sensible to compare model against measurement on the same measurement grid constituted by the measurement points.

The qualitative agreement of measurements and model results of all three models during the first episode (example in Figure 4) is the best of all cases. ETM and LPDM dispersion results in general are similar, and both model types seem to be adequate in simulating the tracer measurements considered. This was also shown in another dispersion modelling exercise of ETM and LPDM.

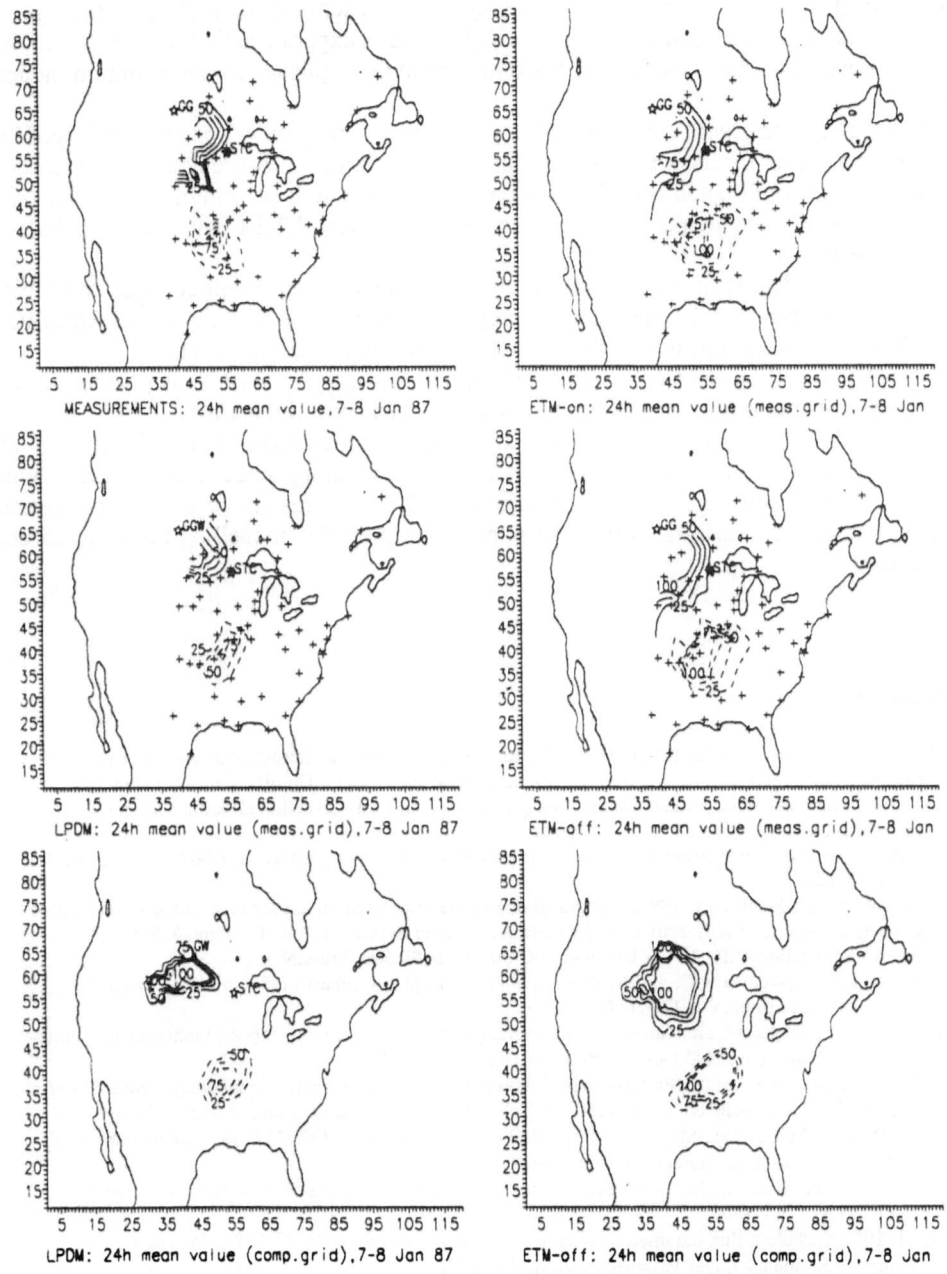

Figure 4. Comparison of ANATEX tracer measurements and transport calculations with ETM-on, ETM-off, and LPDM. Tracer emission from two sources: (40,65) ——— and (55,56) - - - on 5 Jan 1987, 17 to 20z. Displayed is a 24h mean finishing at 14z on 8 Jan, thus covering the 3rd day after emission .
Top and middle: concentration isolines relating to grid of measurement stations (meas.grid),
bottom   : concentration isolines relating to EM computational grid (comp.grid).
Input of meteorological fields to ETM-off and LPDM at hourly intervals.

FUTURE DISPERSION MODELLING AT THE DWD

In future the physics and chemistry of the atmosphere will increasingly be dealt with together under the heading of protection of man and environment. Requirements are manifest or arise in the fields of emergency systems , chemical monitoring of the atmosphere, and air quality assessments. At the DWD there is experience with and direct access to NWP models and their results which are important prerequisites for air pollution modelling.

Concerning emergency models the radioactivity emergency system IMIS will be enhanced to provide for the assimilation of ground and aircraft measurements. Co-operation with government bodies and organisations on the federal state level will be intensified. The DWD will participate in the European Tracer Experiment ETEX with the LPDM and possibly the ETM.

A model to deal with chemical emergencies on the local scale will be taken over from the German military weather service. Already, the winter smog forecast model SMOROP of the Kernforschungszentrum Karlsruhe is run operationally at the DWD.

The ETM will be developed and fitted with a chemistry module. It is planned to use it for continual monitoring of the atmophere and for episode simulations.

These developments would best be concentrated in a national center for applied atmospheric environment simulation which is in its first planning stage in co-operation with the Federal Environmental Agency. It would build on the results of already performed research and could concentrate efforts to improve air pollution modelling for research and applications.

REFERENCES

Axenfeld,F.,Janicke,L.,and Münch,J.,1984,Entwicklung eines Modells zur Berechnung des Staubniederschlages,Environm.Res.Plan,Min.Interior,No.104 02 562,Dornier System GmbH,Friedrichshafen.

Bott,A.,1992,Monotone flux limitation in the area-preserving flux-form advection algorithm,Mon.Wea.Rev. 120:2592-2602.

de Baas,A.F.,1988,Some properties of the Langevin model for dispersion,Report Risø-M-2627,Risø Nat.Lab., Risø,Roskilde.

Fay,B.,Glaab,H.,and Jacobsen,I.,1991,A Lagrangian particle dispersion model for long-range simulation of accidental releases of radioactivity at the German Weather Service,in:"Air Pollution Modeling and Its Application VIII",H.van Dop and D.G.Steyn,ed.,Plenum Press,N.Y..

Gifford,F.A.,Barr,S.,Malone,R.C.,and Mroz,E.J.,1988,Tropospheric relative diffusion to hemispheric scales, Atm.Environm.Vol.22,9:1871-1879.

Glaab,H.,1986,Lagrangesche Simulation der Ausbreitung passiver Luftbeimengungen in inhomogener atmosphärischer Turbulenz,PhD thesis,Technische Univ.,Darmstadt.

Legg,B.J.,and Raupach,M.,1982,Markov-chain simulation of particle dispersion in inhomogeneous flows: the mean drift velocity induced by a gradient in Eulerian velocity variance,Bound.Lay.Met.24:3-13.

Majewski,D.,1991,The Europa-Modell of the Deutscher Wetterdienst,in:"ECMWF Seminar on numerical methods in atmospheric models",Vol.2:147-191.

Mellor,G.L.,and Yamada,T.,1974,A hierarchy of turbulence closure models for planetary boundary layers, J.Atm.Sci.31:1791-1806.

Müller,E.,1981,Turbulent flux parameterization in a regional-scale model,in:"ECMWF Workshop on Planetary Boundary Layer Parameterization",193-220.

Müller,E.,Frühwald,D.,Jacobsen,I.,Link,A.,Majewski,D.,Schwirner,J.-U.,and Wacker,U.,1986,Results and prospects of mesoscale modeling at the Deutscher Wetterdienst,in:"Collection of papers presented at the WMO/IUGG NWP Symposium",Tokyo.

NWP Progress Report for 1992,1993,NWPP Report Series No.19,Tech.Doc.WMO/TD No.548:73-86.

DISCUSSION

K. NODOP: Can you explain to me why Lagrangian Particle Dispersion Models give lower dispersion results than Eulerian Transport Models?

B. FAY: Due to numerical diffusion, the dispersion especially near a point source in an Eulerian Transport Model is larger than in a Lagrangian Particle Dispersion Model. However, the pollution concentration is sensitive to the parameterization of the vertical turbulent velocities which in turn influence the horizontal spread because of the vertical wind shear. This leads to different concentrations in the atmospheric levels and accordingly to differing surface concentrations. The numerical solution scheme of the dispersion process is also important.

R. SAN JOSE: Have you made some sensitivity analysis of the emission inventory?

B. FAY: We have not yet become involved with emission inventories.

THE EMBEDDING OF THE LAGRANGIAN DISPERSION
MODEL LASAT INTO A MONITORING SYSTEM FOR
NUCLEAR POWER PLANTS

Lutz Janicke

Ingenieur-Büro
88662 Überlingen, Germany

INTRODUCTION

The Lagrangian dispersion model LASAT (LAgrangian Simulation of Aerosol Transport) has its origin in a research model developed in 1980. One key problem then was the correct treatment of the dispersion in inhomogeneous turbulence fields. When the problem of the drift velocity in the case of inhomogeneous turbulence was solved[1,2], the model was used as the reference for calibrating simpler dispersion models of Gaussian type[3]. In 1989, the model was implemented on a PC and became available for other users.

At that time the authority in charge of the nuclear power plants in the state Niedersachsen in Germany decided to improve the dispersion modelling in their monitoring system. Two power plants were equipped with cooling towers taller than the stack of the plant. The Gaussian plume model in use was to be replaced by a Lagrangian model better suited to describe the interaction of the plume with the turbulence field generated by the tower(s). This interaction has been studied in extensive wind tunnel experiments for both sites whereby wind field, turbulence field and plume dispersion have been measured[4]. Thus, the requirements for LASAT are:

- Model the boundary layer on the basis of meteorological data measured online, including a 3–component SODAR profile.

- Take into account the influence of cooling towers.

- Calculate concentration and gamma cloud radiation for 10 tracer species up to a source distance of 25 km.

- Run on a PC within a computer network.

- Perform in real time with a cycle time of 10 minutes.

The following sections discuss the necessary steps to fulfill these requirements.

Air Pollution Modeling and Its Application X, Edited by S-V. Gryning
and M.M. Millán, Plenum Press, New York, 1994

CALCULATION OF THE WINDFIELD

The wind tunnel experiments provide data on the wind velocity and fluctuations over cross sections perpendicular to the wind direction, at different distances from the cooling tower. They have been used to adjust the coefficients of empirical wind models for the near and far wake behind obstacles[5]. These models modify the plane wind and turbulence field derived from the SODAR measurements and a plane boundary layer model. The result is used as initialisation field of a diagnostic wind model to get a divergence free wind flow.

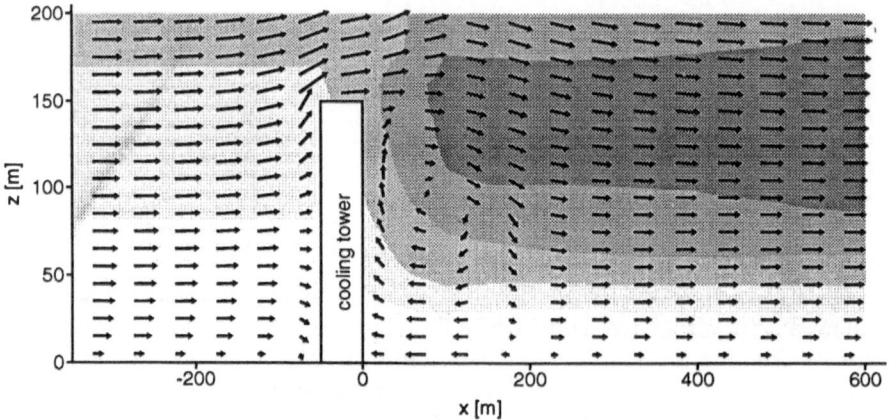

Figure 1. Wind vectors in a vertical cross section through the cooling tower (note the anisotropic scaling). The shading sketches the size of the vertical diffusion coefficient which increases by $5\,\mathrm{m^2/s}$ with each gray level.

The wind field resulting from the diagnostic wind model is shown in Figure 1. The zone of recirculation just behind the obstacle and the reduction of wind speed in the wake can be seen. The vertical diffusion is increased considerably. Therefore, the maximum ground level concentration of the plume increases by the factor 2, if the stack is on the lee-side of the tower (for a stack height of 132 m, and a distance from the tower of 350 m).

OPTIMIZATION OF THE TIME STEP

Lagrangian dispersion models usually apply a time step that is small with respect to the smallest Lagrangian correlation time because the drift velocity term is derived using this assumption. Thus, the time step often becomes unnecessarily small as for the required resolution, less particle trajectories can be computed within a given time span, and the sampling error in the resulting concentration field increases. A first step to improve the situation is to choose a spacially variing time step $\tau(x)$. Furthermore, it is possible to derive a drift velocity term without restricting the time step to smaller values than the Lagrangian correlation time. Only the result will be presented here.

Let us assume that the boundary layer is described by the vector of the mean wind field $V(x)$, the tensor of the turbulent velocity fluctuations $\Sigma(x)$ and the diffusion tensor $\mathbf{K}(\boldsymbol{x})$. It is practical to define a tensor $\boldsymbol{\Phi}$ by

$$\mathbf{K} = \boldsymbol{\Phi}^{-1} \cdot \boldsymbol{\Sigma} \,. \tag{1}$$

In one-dimensional simulations $\boldsymbol{\Phi}^{-1}$ is the Lagrangian correlation time. Then, in LASAT the algorithm to calculate the new position $\hat{\boldsymbol{x}}$ and the new velocity $\hat{\boldsymbol{u}}$ at time $\hat{t} = t + \tau$ from the old values $\boldsymbol{x}$ and $\boldsymbol{u}$ at time t is

$$\hat{\boldsymbol{u}} = \boldsymbol{\Psi}(\boldsymbol{x}) \cdot \boldsymbol{u} + \boldsymbol{w}, \tag{2}$$

$$\hat{\boldsymbol{x}} = \boldsymbol{x} + \tau [\boldsymbol{V}(\boldsymbol{x}) + \hat{\boldsymbol{u}}], \tag{3}$$

where $\boldsymbol{\Psi}$ is related to $\boldsymbol{\Phi}$ by

$$\boldsymbol{\Psi} = \frac{\mathbf{I} - \frac{1}{2}\tau\boldsymbol{\Phi}}{\mathbf{I} + \frac{1}{2}\tau\boldsymbol{\Phi}} \tag{4}$$

$$\approx \mathbf{I} - \tau\boldsymbol{\Phi} \quad \text{if } \|\tau\boldsymbol{\Phi}\| \ll 1 \,. \tag{5}$$

The stochastic velocity $\boldsymbol{w}$ of the particle algorithm (2) is generated from a random vector $\boldsymbol{R}$ (the components are independent, have mean 0 and variance 1) by

$$\boldsymbol{w} = \boldsymbol{W}(\boldsymbol{x}) + \boldsymbol{\Lambda}(\boldsymbol{x}) \cdot \boldsymbol{R}, \tag{6}$$

$$\text{with} \quad \boldsymbol{\Lambda} \cdot \boldsymbol{\Lambda}^{\mathrm{T}} = \boldsymbol{\Sigma} - \boldsymbol{\Psi} \cdot \boldsymbol{\Sigma} \cdot \boldsymbol{\Psi}^{\mathrm{T}} \,. \tag{7}$$

The correct form of the drift velocity $\boldsymbol{W}(\boldsymbol{x})$ can be derived from the entropy condition: In a closed system with uniform particle density, the particle density must remain uniform, otherwise the entropy would decrease. An ensemble of noninteracting particles is described by the distribution density with respect to position and velocity, $f(\boldsymbol{x}, \boldsymbol{u}, t)$. Within one time step this distribution density changes according to

$$f(\hat{\boldsymbol{x}}, \hat{\boldsymbol{u}}, \hat{t}) = \int\!\!\int f(\boldsymbol{x}, \boldsymbol{u}, t) \, \delta(\hat{\boldsymbol{x}} - \boldsymbol{x} - \tau\boldsymbol{a}) \, \delta(\hat{\boldsymbol{u}} + \boldsymbol{V}(\boldsymbol{x}) - \boldsymbol{a}) \, p(\boldsymbol{R}) \, \mathrm{d}^3\boldsymbol{x} \, \mathrm{d}^3\boldsymbol{u} \, \mathrm{d}^3\boldsymbol{R}, \tag{8}$$

$$\text{with} \quad \boldsymbol{a} = \boldsymbol{V}(\boldsymbol{x}) + \boldsymbol{\Psi}(\boldsymbol{x}) \cdot \boldsymbol{u} + \boldsymbol{W}(\boldsymbol{x}) + \boldsymbol{\Lambda}(\boldsymbol{x}) \cdot \boldsymbol{R} \,. \tag{9}$$

$p(\boldsymbol{R})$ is the probability density of the random vector $\boldsymbol{R}$. The evaluation of this relation for constant time step and uniform particle density $n(\boldsymbol{x}, t) = \int f(\boldsymbol{x}, \boldsymbol{u}, t) \, \mathrm{d}^3\boldsymbol{u}$ yields

$$\boldsymbol{W}(\boldsymbol{x}) = \frac{1}{2}\tau[\mathbf{I} + \boldsymbol{\Psi}(\boldsymbol{x})] \cdot [\nabla \cdot \boldsymbol{\Sigma}(\boldsymbol{x})] \,. \tag{10}$$

A corresponding derivation for the case of spatially variing time step could not be performed. However, empirically the following relation was found:

$$\boldsymbol{W}(\boldsymbol{x}) = \frac{1}{2}[\mathbf{I} - \boldsymbol{\Psi}(\boldsymbol{x})] \cdot \{\nabla \cdot [\tau(\boldsymbol{x})\boldsymbol{\Sigma}(\boldsymbol{x})]\} + \tau(\boldsymbol{x})\boldsymbol{\Psi}(\boldsymbol{x}) \cdot [\nabla \cdot \boldsymbol{\Sigma}(\boldsymbol{x})], \tag{11}$$

The only condition on the size of the time step is that the system should be only weakly inhomogeneous as seen from the particle during one time step. In practice the time step is limited locally by twice the value of the smallest Lagrangian correlation time. If all Lagrangian correlation times have the same value, $\boldsymbol{\Psi}$ becomes zero at this point and the algorithm changes from the simulation of turbulent diffusion to the simulation of classical diffusion.

OPTIMIZATION OF THE PARTICLE NUMBER

In order to reduce the sampling error a large particle number should be used in the simulation. On the other hand, one computational cycle including calculation of the wind field, particle trajectories, concentration and cloud radiation, communication and data transfer within the computer network, compressing and archiving the results must not take over 10 minutes. Certain meteorological situations can lead to an accumulation of particles within the computational region. To avoid a violation of the time constraint the program

- estimates the time available for calculating the particle trajectories,

- dynamically changes the emission rate of particles,

- reduces if necessary the population of already existing particles.

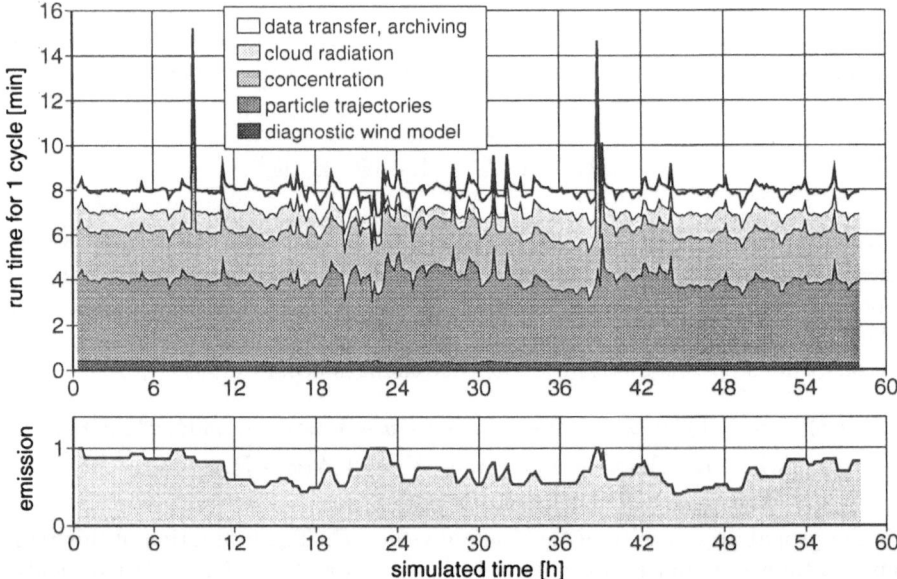

Figure 2. The lower part shows the reduction in particle emission (and population) for a test case, where accumulation of particles occured. The upper part shows the time spent by the program for different tasks. The program was scheduled to complete a cycle of 10 minutes within a run time of 8 minutes.

A demonstration of the real time behaviour of the program is shown in Figure 2. A weather situation with weak winds causes an increase in particle number within the computational region. Accordingly the emission rate and total particle population is reduced by the program and the time spent by the program for the calculation of particle trajectories remains roughly constant. The spikes that appear in the diagram are due to a shortcoming of the special hardware used.

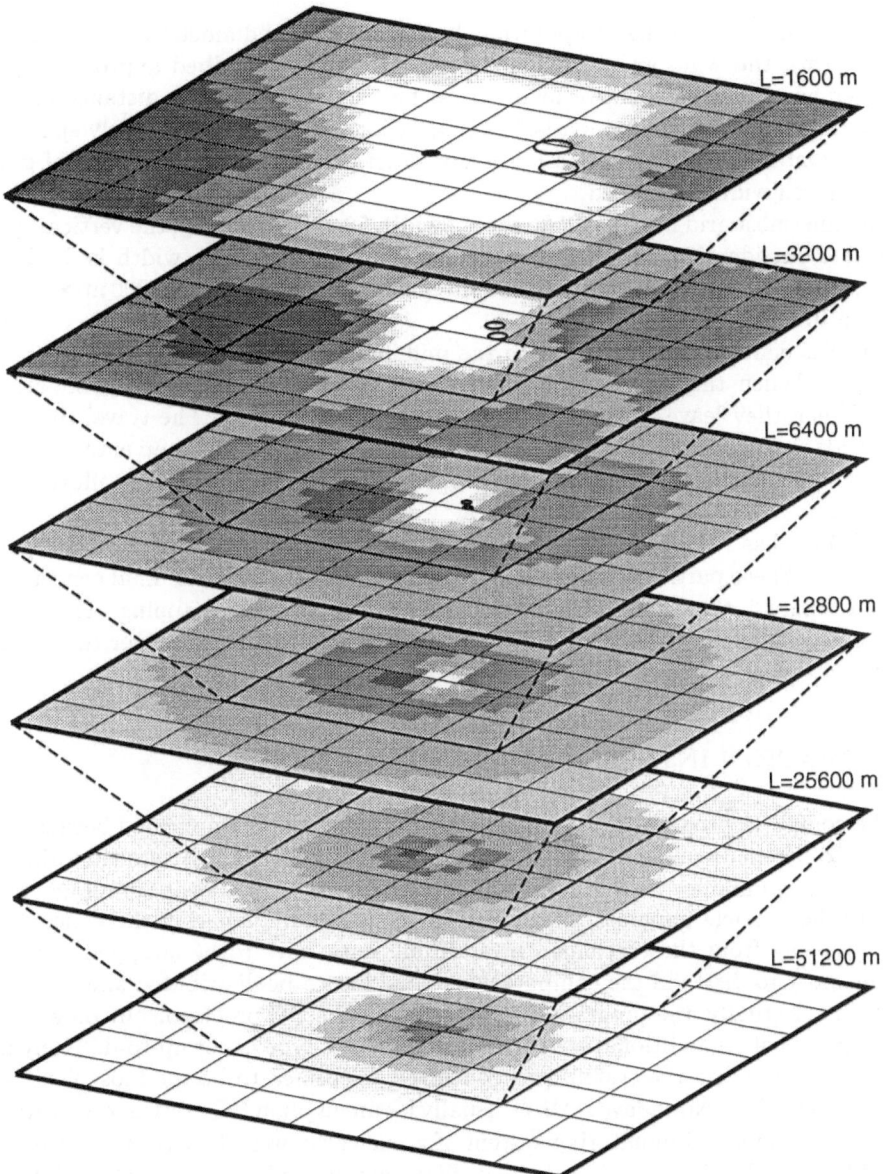

Figure 3. Nesting of grids with six levels. With each level the mesh width changes by a factor of 2. Therefore, the innermost grid has a mesh width of 50 m, the outermost grid has 1600 m. For each grid the horizontal extension is given. The shading shows the ground level concentration in the case of an isotropic wind windrose with neutral stability. With each gray level the concentration value changes by a factor of 2 or 2.5. The source is in the middle of the grids, two cooling towers can be seen in the finer grids at a distance of about 350 m from the source.

NESTING OF GRIDS

If the mesh width of the computational grid exceeds the diameter of a cooling tower (about 50 m), the wind and turbulence field can not be described appropriately. On the other hand, the radioactive load has to be computed up to a distance of 25 km, so that the computational region is 50 km × 50 km. Clearly, an uniformly spaced grid with horizontally 1000 × 1000 meshes is not practical. Therefore, a nesting of grids of different mesh width is applied.

The innermost grid has 32 × 32 meshes of width 50 m horizontally, the vertical spacing is variable but identical for all grids. In the next grid the mesh width is double and with six levels of grids the computational region has increased to 51 200 m × 51 200 m as shown in Figure 3.

Particles emitted from the stack in the middle of the region first move in the innermost grid. When they cross the boundary they are intercepted by the next grid and so on. When they leave the outermost grid they are discarded. The travel time in the outer grids is longer than in the inner ones. Therefore, a particle group entering a wider grid has to be diminished to obtain the proper balance in computational effort between the grids.

Each grid has its own concentration field built up from a part of the total particle population. These partial fields are mapped on each other to get the final concentration field. In Figure 3, these concentration fields are shown after the mapping. The mapping of the cloud radiation is more involved because the radiation reaches over the boundary of a grid.

INTEGRATION INTO THE MONITORING SYSTEM

The computer program of the dispersion model is rather complicated because of the usage of different boundary layer models, nesting of grids and dynamical adjustment of the particle number. Furthermore, the program can estimate the sampling error by dividing the particle population into several groups and comparing the concentration fields resulting from these groups. In addition, data have to be swapped temporally from memory to disk and the computational load has to be distributed among different processors. Actually, the program does not run on the PC processor but on a network of transputers of an additional board. A system of this type is not up-to-date any more, but at the time when the system was conceived, it seemed to be the most flexible and extensible solution. Nowadays LASAT usually is run on plain PC's or Unix workstations.

The environmental monitoring system of a nuclear power plant (in this case the system TIS mase by Siemens AG) is in itself a complicated program. To facilitate the system integration, both programs are not combined into one large program. Rather, the dispersion model is running on a separate computer (PC) where it can also be used offline (actually two computers are running in parallel for safety reasons). It is controlled by the monitoring system through the exchange of message and data files over a local area network. The data computed during an alarm case are stored by the PC on an optical disk.

SUMMARY

The Lagrangian dispersion model LASAT has evolved from a research model in 1980 to a practical tool suited to perform real time dispersion modelling in an environmental monitoring system for nuclear power plants. The influence of large buildings (cooling towers) on the dispersion of the plume is modelled using a diagnostic wind model calibrated by specific wind tunnel experiments. The performance has been improved by generalizing the particle algorithm to allow even larger time steps than the Lagrangian correlation time. Real time performance is warranted by a dynamical adjustment of the total particle number. A good spatial resolution in the vicinity of the source together with a large computational region is achieved by using a set of nested grids. The system is in operation since June 1993 in the state Niedersachsen, Germany.

REFERENCES

1. L. Janicke, Particle simulation of inhomogeneous turbulent diffusion, *in*: "Air Pollution Modeling And Its Application II", C. De Wispelaere, ed., Plenum Press, New York (1983).

2. D. Wilson, G.W. Turtell, G.E. Kidd, Numerical simulation of particle trajectories in inhomogeneous turbulence, II: systems with variable turbulent velocity scale, *Boundary-Layer Meteorol.* **21**, 423 (1981).

3. L. Janicke, Particle simulation of dust transport and deposition and comparison with conventional models, *in*: "Air Pollution Modeling And Its Application IV", C. De Wispelaere, ed., Plenum Press, New York (1985).

4. M. Schatzmann, A. Lohmeyer. "Kernkraftwerk Grohnde: Veränderung der Abluftausbreitung durch Kühlturm, Gebäude und Topografie — Windkanalversuche —", im Auftrag des Niedersächsischen Landesamtes für Immissionsschutz, Hannover (1991).

5. H.G.C. Woo, J.A. Peterka, J.E. Cermak. "Wind-Tunnel Measurements in the Wakes of Structures", Report CER75-76HGCW-JAP-JEC40 Colorado State University (1976).

DISCUSSION

K. NESTER:　　　　　　　I think that the cooling towers are never in operation, because the flow field around the cooling towers is completely different from that you have shown. Can you comment on this.

L. JANICKE:　　　　　　　Your observation is correct, the windfield shown is modeled for a cooling tower not in operation. The wind tunnel experiments show that near the ground the highest concentration values occur in this case.

A TRANSPORT AND DISPERSION MODEL PERFORMANCE EVALUATION USING THE RESULTS OF A TRACER EXPERIMENT IN COMPLEX TERRAIN

R. Lamprecht and D. Berlowitz

Paul Scherrer Institute
CH-5232 Villigen PSI, Switzerland

INTRODUCTION

Within the framework of the TRANSALP project, a short range tracer experiment was carried out on October 3, 1991 under calm, purely thermic wind conditions over a terrain that is representative in irregularity for the northern prealpine region of Switzerland. TRANSALP is a project in the framework of the environmental research program EUROTRAC of the European Community and is designed to investigate the transport and diffusion of air pollutants in the planetary boundary layer over complex terrain.

The experiment of October 3, 1991 was designed to imitate a hazardous gas emergency situation in the agglomeration of Lucerne. A harmless atmospheric tracer (C_8F_{16}) was released in order to test the performance of the Monte Carlo dispersion code PARTRAC (Lamprecht, 1989) which calculates the spatio-temporal concentration fields of passive air constituents over complex terrain. This code has previously been validated for simplified asymptotic conditions against an analytical solution of the diffusion-advection equation (Lamprecht, 1989).

The required three-dimensional flow fields were generated diagnostically by the mass consistent wind model CONDOR (Moussiopoulos et al., 1988) on a half hourly basis using the wind data which was continuously recorded during the experiment at five different weather stations and with two Doppler acoustic sounders. The sensitivity of the wind fields to varying input data was evaluated by comparing the calculated concentration fields to observational data. The purpose of this sensitivity analysis was to evaluate how much of the overall uncertainty of the model output was associated with the lack of spatial representativeness of the wind measurements at the different monitoring sites.

Furthermore, the performance of the concentration predictions was evaluated using prognostic flow fields generated by the non-hydrostatic mesoscale meteorological model MEMO (Flassak, 1990, Moussiopoulos et al., 1993) . For this prognostic model performance evaluation, MEMO was initialized by the output of the High Resolution regional weather forecast Model (HRM) which has been in a preoperational state since spring 1992, both at the weather services in Germany (DWD) and Switzerland (SMA) (Schubiger, 1993). The grid resolution of the HRM model is approximately 14 km and is operational in Switzerland with a grid size of 109 x 101 points. This model again is initialized by the European weather forecast Model (EM) which is operated by the DWD in Offenbach a.M., Germany.

As a fair-weather phenomenon thermo-topographically induced diurnal flow circulations between the prealpine part of the Swiss Plateau and the Alps may occur (Burger and Ekhart, 1937). The occurrence of this wind regime was suggested by the model results relying on the diagnostic wind fields. The prognostic numerical representation of this large-scale flow is superposed on the computed wind fields in the study area by applying a model nesting technique.

The statistical analysis for evaluating model reliability is based on observed and predicted concentration fields paired at equal locations and times - one of the most stringent test of an air quality model (Hanna, 1988). The evaluation of diagnostic wind fields for which the best performance measures (FB

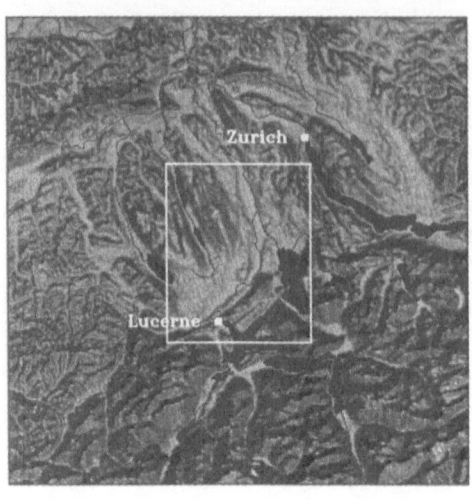

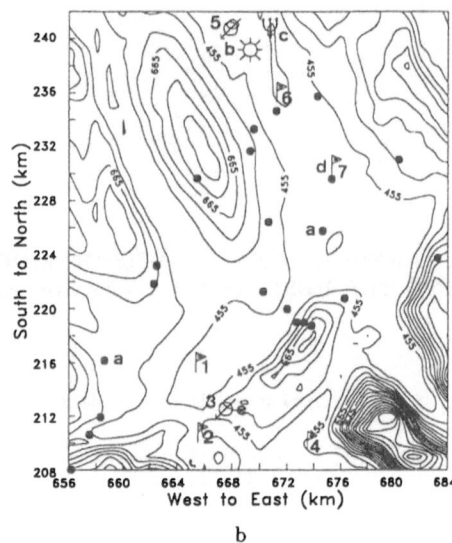

a

b

Figure 1. (a) The study area on which prognostic flow fields were calculated and embedded the smaller experiment area (b) showing the different meteorological instruments as well as the 22 ground-level air monitoring locations. **Legend:** a: tracer sampling unit, b: tracer release point, c: Sonic anemometer, d: meteorological mast, e: SODAR.

and $NMSE$) for the concentration fields were achieved resulted in a benchmark against which the purely prognostic results from the models MEMO and PARTRAC could be judged.

SITE DESCRIPTION, METEOROLOGICAL CHARACTERIZATION AND INSTRUMENTATION

The experiment took place in the lower Reuss Valley on an area of about 950 km^2 in the prealpine region of Switzerland, north of Lake Lucerne. Figures 1(a) and 1(b) show the test region and the locations of the different meteorological instruments and ground-level air monitoring stations.

The meteorological condition during the experiment was characterized by a flat pressure distribution over the Central Alps which is known to produce thermally induced wind regimes with a prevailing flow towards the Alps and the prealpine valleys in the south during daytime and a reversed drainage flow at night. The collected wind data indeed revealed an oscillation of the wind direction of 180° during the course of the day. The wind velocities over the study area were generally low and did not exceed 2 m/s. During the morning hours, up to about 11:30 am (UTC+1), the bottom of the lower Reuss Valley was still covered with ground fog.

In order to characterize the air flow 10 m above ground level, 5 meteorological masts were mounted which recorded wind direction and velocity with a time resolution of 10 minutes. Two Doppler acoustic sounders (SODAR) characterized the wind field at upper air levels up to approximately 500 m above ground level. At the tracer release point, 10 m above ground level, a Kaijo-Denki sonic anemometer-thermometer measured the sensible heat flux and the turbulent velocity fluctuations and hence delivered the basic information for a subsequent determination of the friction velocity u_* and the Monin-Obukhov length L. Both of these meteorological parameters are required by the model PARTRAC for an estimation of the turbulent diffusivity σ_u, σ_v and σ_w.

The tracer was released during 3.5 hours between 11:00 am and 2:30 pm (UTC+1) with a mean source strength of 4 g/s on a hill in the middle of the lower Reuss Valley approximately 90 m above the bottom of the valley (see Figure 1(b)). Due to the low detection limit of perfluorocarbons using ECD technique and the very low background of these species in the atmosphere, perfluorodimethyl-cyclohexane ($PP3$) was released. The Environmental Institute of the Joint Reserach Center at Ispra, Italy, which participated in the experiment, analyzed the air samples by gas chromatography and contributed also the experimental equipments for tracer release and tracer sampling.

The tracer was registered with 32 sampling units which were distributed in the area south of the release point at receptor locations expected to be representative for the respective local surroundings. At the monitoring locations more distant from the release point, two sampling units were installed in each case. Each sampling device can consecutively fill 8 bags with air. The integration time for taking an air sample was set to 30 minutes in synchronization with the data collection periods of the two Doppler SODAR systems for measuring the vertical wind profiles. Except for the sampling devices closest to the tracer release point, the sampling procedure was usually interrupted for 30 minutes each time before taking a new sample. Hence, concentrations were obtained maximally over a period of up to 16 h.

Moreover, three motorgliders recorded the concentration fields at upper air levels. A motorglider sampled air with increasing distance around the release point. The other aircraft flew either on pre-scribed transects or measured vertical profiles. One of the motorgliders was additionally instrumented to enable the measurement of NO_2, O_3 as well as temperature and humidity. NO_2 and humidity are highly useful for the determination of the height of the inversion layer. These parameters exhibit large concentrations immediately below the mixing height and then drop off sharply. For that purpose, vertical profiles of these parameters were recorded periodically.

MODELING CONCEPTS

Using Monte Carlo methods in air pollution applications, the emitted polluting material is characterized by fictitious computer particles which consist of different moving position vectors in the computational domain. Each particle is moved at each time step by pseudo velocities, which take into account both the average wind transport and the random turbulent fluctuations of the wind components.

The average wind transport was calculated using the non-divergent 3D-wind-fields as delivered either from the diagnostic or prognostic mesoscale boundary layer models CONDOR or MEMO, respectively. The diagnostic wind model CONDOR interpolates only the measured wind data with respect to a topography following coordinate system under the constraint of satisfying mass consistency, i.e. no physics is involved. Since no physical criterion supports any reasonable choice for the interpolation weight, the distance-weighting factor r^{-2} was chosen for all the flow field generations. Goodin et al. (1979) and Porch et al. (1987) heuristically determined that the r^{-2} weighting scheme interpolates the measured wind data adequately to phyisically realistic three-dimensional flow fields. The dynamic mesoscale model MEMO is based on the coupled numerical integration of the conservation equations of momentum, mass, energy, turbulent kinetic energy and water vapor. The conservation of mass is written in terms of the non-hydrostatic pressure thus yielding an elliptic partial differential equation which is numerically solved applying Fourier analysis. The conservation equation system is solved with respect to the same terrain following coordinates as proposed by Clark (1977). Additional information about the meteorological models CONDOR and MEMO is provided by Flassak (1990) and Moussiopoulos et al. (1993).

Primary influence on the tracer transport and hence on the concentration fields are due to changes in the flow over time. Therefore 30 minute averages of the wind fields were calculated and supplied to the dispersion model PARTRAC which subsequently calculated a sequence of 24 concentration fields covering the whole period of the experiment from 11:00 am to 11:00 pm (UTC+1).

The physical background of the dispersion model PARTRAC is similar to that of the model LASAT developed by Janicke which is described in a report by Axenfeld et al. (1984). The transport equations as derived by Janicke for inertia-less particles in inhomogeneous turbulent flows is based on the Fokker-Plank equation and on the theory of Markovian stochastic processes according to Kolmogorov (1933).

PARTRAC is a single particle model, since only one particle is released at a time. An ensemble of trajectories is build up by repeating the random walk process with a large number of particles in order to improve the dispersion statistics for the subsequent concentration calculations.

The model assumes that each component of the particle velocity can be split into two terms:

$$U_{i,n} = < U_{i,n} > (X_{i,n}) + \sum_j \Delta_{i,j} U'_{j,n} \tag{1}$$

where the subscripts i and n refer to the three cartesian components and the time step, respectively. $X_{i,n}$ denotes the component i of the position vector at the n-th time step. The first term of Equation (1) represents the transport due to the average Eulerian flow conditions as calculated e.g. by one of the above mentioned boundary layer models. The second term is computed according to the Lagrangian stochastic movements of the particles.

The fluctuations $U'_{j,n}$ of the wind components are given in a Lagrangian wind coordinate system in which the x-axis is parallel to the mean wind direction and the two other components are perpendicular to it. Since the Eulerian wind components are given with respect to a fixed coordinate system in which the x-axis points towards the east and the y-axis towards the north, the components of the wind fluctuations have to be transformed after each time step by an orthogonal rotation $\Delta_{i,j}(n)$.

The fluctuation variances σ_i^2 and the Lagrangian time-scales T_{Li} which are both height dependent functions, were assigned according to the scheme proposed by Hanna (1982). The subscript i refers to the three wind components u, v, w. The wind variances σ_i^2 are measures for the intensity of turbulence and depend sensitively on the Monin Obukhov length L. The components of the Lagrangian time-scale T_{Li} indicate for all coordinate directions the correlation times during which an inertia-less airborne particle remains suspended within a particular eddy. These parameters constitute along with the friction velocity u_*, the roughness length z_0 and the mixing height h the main input for PARTRAC. Although the application of the profiles for σ_i^2 and T_{Li} according to Hanna is restricted to boundary layers over flat terrain, one is tempted to apply these parameterizations to diffusion problems over irregular terrain.

By discrete integration of Equation (1) one immediately obtains the position vector of a particle along a trajectory

$$\vec{X}_{n+1} = \vec{X}_n + \Delta t \cdot \vec{U}_n. \tag{2}$$

At any time step, the actual position of a particle has to be related to the corresponding receptor cell of the grid which resolves concentration. In the limit of large numbers of particle trajectories the total time all particles spend in a cell is proportional to the concentration in that cell.

Perfect reflection of the particles is assumed at the top of the mixing height. At the ground each incoming particle is reflected or deposited with prescribed respective probabilities.

MODEL PERFORMANCE ANALYSIS ON THE BASIS OF DIAGNOSTIC FLOW FIELDS

The objective of the study in this section is to determine the sensitivity of concentration predictions to the amount of meteorological data used for generating diagnostic flow fields. The diagnostic wind fields are generated by spatial interpolation of the ground-level and upper-air wind measurements using the wind model CONDOR. This model is designed to utilize local surface and upper-air observations, where available, while at the same time providing some information on terrain-generated airflows in regions where local observations are unavailable. Since half hourly averaged observations of vertical winds were measured by the Doppler acoustic sounders, the concentration fields were computed with respect to the same time resolution. The model performance measures were thus evaluated by comparing predicted half hourly concentrations $C_{p,i}$ with observed values $C_{o,i}$ paired at equal locations and times. The comparison is between station measurements and grid-averaged predictions where the lateral resolution of the calculated concentration fields is $1~km^2$. For the model performance evaluations, we used the two simple performance measures: fractional bias FB and normalized mean square error $NMSE$, as suggested by Hanna (1988).

$$FB := 2\frac{\bar{C}_p - \bar{C}_o}{\bar{C}_p + \bar{C}_o} \tag{3}$$

where $\bar{C}_p$ denote $\bar{C}_o$ the mean predicted and observed concentrations, respectively. Values for FB range between -2 and +2 and indicate the tendency of the model to underestimate or overestimate the observed concentrations. Good model performance is indicated by FB values close to 0.

$NMSE$ describes the scatter in the entire data set of calculated concentrations with respect to measured ones:

$$NMSE := \frac{\overline{(C_p - C_o)^2}}{\bar{C}_p \bar{C}_o} \tag{4}$$

The uncertainty in the wind fields can be traced to the density and representativeness of observations used to construct the fields. The sensitivity of the wind fields to varying input data was assessed by comparing the calculated concentration fields to observational data. No sensitivity check, however, has been done with parameters which are assumed to have inferior influence on the concentration predictions, as for instance the depth of the mixing layer or the intensity of turbulence according to atmospheric stability and friction velocity. Table 1 summarizes the performance data of the calculated concentration fields as a function of different combinations of wind stations #1 to #4 used to generate the 3D-diagnostic flow fields. The locations of all the wind stations #1 to #7 are indicated on the map of Figure 1(b). The meteorological stations #1 to #4 were placed in the area of Lucerne whereas the other three wind stations were installed either at elevated topographical locations (station #5 and #7) in the northern part of the area under investigation or at the bottom of the lower Reuss Valley (station #6). In view of the small values achieved for the performance measures FB and $NMSE$ at receptor sites near the wind stations #5, #6 and #7 these stations were assumed to measure representative data for the regarded region and have thus not been subjected to a sensitivity analysis. The influence of Lake Lucerne on the wind system along with the irregularity of the terrain at the northern border of the Lake was expected to increase the complexity of the regional wind field. As a consequence the density of the meteorological network in this area was chosen to be higher than anywhere else.

Table 1. Performance data of the calculated concentration fields for various combinations of the wind stations #1 to #4 with stations #5, #6 and #7, used for the generation of the diagnostic flow fields.

comb. of wind stations		samplers near stations # 1 to 4		all ground based samplers		comb. of wind stations			samplers near stations # 1 to 4		all ground based samplers	
#		FB	$NMSE$	FB	$NMSE$		#		FB	$NMSE$	FB	$NMSE$
	1	1.39	10.91	0.64	5.27		2	4	0.83	2.93	0.16	2.56
	2	1.20	20.41	0.44	8.44		3	4	0.38	1.90	-0.08	2.52
	3	0.87	3.20	0.18	2.51	1	2	3	1.55	23.44	0.88	10.78
	4	0.58	2.23	0.02	2.46	1	2	4	1.02	4.10	0.30	2.81
1	2	1.41	14.28	0.68	6.62	1	3	4	0.73	3.78	0.10	2.90
1	3	1.35	12.29	0.61	5.72	2	3	4	0.78	2.31	0.14	2.38
1	4	0.89	3.14	0.19	2.56	1	2	3	4	1.03	4.67	0.28	2.93
2	3	1.44	13.51	0.72	6.32							

Table 1 shows the results for the performance measures FB and $NMSE$ for an ensemble of 49 and 101 concentration data which were collected either by sampling units at receptor locations in the vicinity of the wind stations #1 to #4 or by all ground based sampling devices, respectively. The first four lines of Table 1 contain the performance data if each time only one of the wind stations #1 to #4 were added to the bulk of wind stations #5 to #7 for generating the flow fields. The wind station #4 in Table 1 was installed at the eastern border of Lake Lucerne and according to both, the FB and the $NMSE$ values, this wind station seems to be the most significant one for representing the flow field in the area around Lucerne. Moreover, the addition of mast #4 to any combination of wind stations is superior with respect to the FB and $NMSE$ values than any other combination of wind stations without mast #4. Conversely, the inclusion of mast #1 or mast #2 to any combination of wind stations results in a deterioration of the performance statistics so that the conclusion can be drawn that these wind stations are less representative and impair the generated wind fields.

Since the overall FB and $NMSE$ values for the concentration fields based on the wind stations #4 to #7 are the smallest ones for all ground based samplers these results agree best with the observations.

A few time sequences of these calculated ground level concentration fields are depicted in Figures 2(a) to (c). Monte Carlo dispersion simulations on diagnostic flow fields may involve relatively simple estimation of complex terrain effects, such as the deflection and blocking of air constituents by complex terrain. Such effects are shown in Figure 2(b) on a hill east of Lucerne in the evening between 7:30 pm and 8:00 pm (UTC+1) under the additional influence of a decreased mixing depth of 200 m and a corresponding stably stratified atmosphere.

The preceding sensitivity study, which aimed at providing PARTRAC with the most representative diagnostic flow fields to numerically reproduce the observations as reliable as possible, established a benchmark to which the prognostic results to be attained in the following section can be compared.

ANALYSIS OF THE TRACER EXPERIMENT USING DYNAMIC FLOW FIELDS

The scope of this section is to investigate the performance of prognostic flow fields to numerically reproduce the results of the experiment. This enables one to check the forecasting capabilities of a mesoscale meteorological model in case of a hazardous gas release over complex topography.

In the case of thermally driven flows the influence of the topography is highly relevant both on a local and in particular for the current study also on a supra-regional scale where larger circulations may occur. A pressure gradient arising from differential heating between the air above the Alps and the air at the corresponding height level in the free troposphere over the prealpine Swiss Plateau is assumed to result in a prevailing plain-mountain flow (Burger and Ekhart, 1937) during day time. At night, the mechanism and the circulation are reversed. The assumption that such a large scale

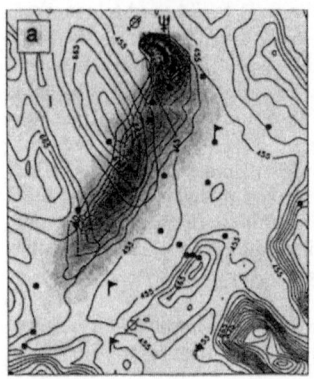 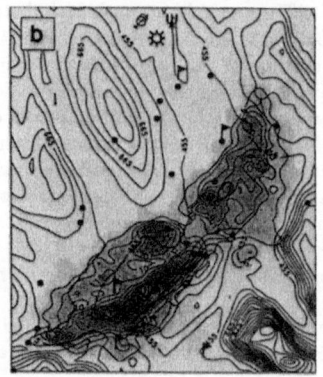 

Figure 2. A few selected time sequences of the diagnostically calculated ground level concentrations for the following time intervals: (a) 14.00-14.30, (b) 19.30-20.00, (c) 22.30-23.00 (UTC+1).

circulation for the present study might be relevant is due to the observed nocturnal flow reversal over the experiment area that was documented by the dispersion scenario based on the diagnostic flow fields in the previous section.

On the background of the physical rationale given above and the many observations confirming the existence of large-scale thermal circulations between mountains and plateaus the assumption of attributing the observed southerly nocturnal winds to the cold drainage flows from the Alps is plausible.

Since the different thermal flows are induced by topographical features of various scales, each computational domain of a certain scale requires an adequate grid resolution in order to numerically generate the scale-specific flow fields. Hence the present study makes use of a nested simulation. The computer code provides a 'one-way' or 'passive' nesting of an outer model domain into an inner one.

All the prognostic flow fields were generated by the non-hydrostatic mesoscale meteorological model MEMO. MEMO was initialized by the output of the prognostic High Resolution Model (HRM), which is operated by the Swiss Meteorological Institute in Zurich to provide additional information for making the short term weather forecast (Schubiger, 1993).

On the basis of the radiation budget calculated by a preprocessor routine within the model MEMO the recurrent mountain-plain flow circulation was generated by MEMO in a domain with a horizontal grid resolution of 5 km, enclosing the relevant parts of the Alps and the Swiss Plateau. The resulting air mass exchanges between the Alps and the Swiss Plateau along a vertical section from north to south two hours after sunset at 20 (UTC+1) is shown in Figure 3.

This large scale circulation could only be simulated in this larger model domain with the reduced grid resolution of 5 km. Attempts in simulating the mountain-plain flow on the smaller area with a grid resolution of 1 km (see Fig 1(a)), however, failed. Therefore it was necessary to superpose

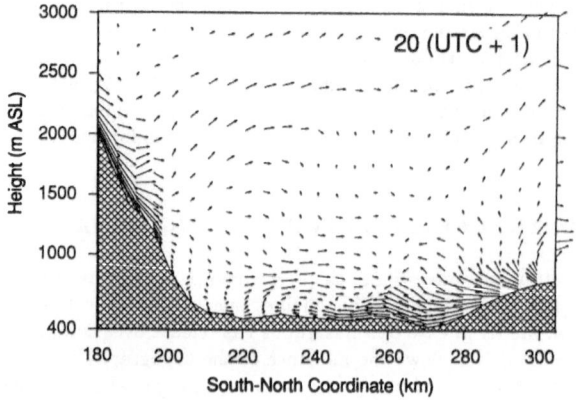

Figure 3. Vertical section from south to north in the middle of the experiment area showing the air mass exchange at 20 (UTC+1).

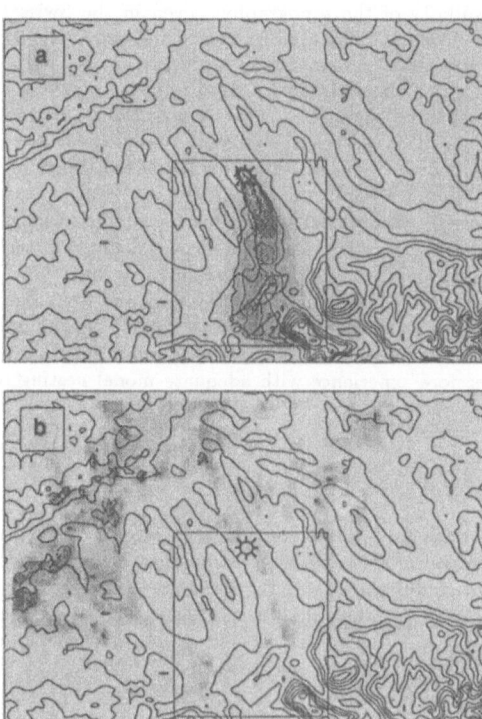

Figure 4. Two selected time sequences of the prognostically calculated ground level concentrations for the time intervals: (a) 14.00-14.30, (b) 22.30-23.00 (UTC+1).

the supra-regional plain-mountain flow and its nocturnal counterpart on the wind fields of the higher resolved study area by an adequate grid-nesting procedure.

While the initialization process of MEMO consists of an interpolation of wind and temperature profiles over the entire 3D-modeling domain, other meteorological data which represents revised flow conditions at a later stage of the simulation is extrapolated to the domain boundaries. Hence the grid-nesting procedure used here consists of providing MEMO with the additional data at the boundaries of the higher resolved modeling domain. This data represents the mountain-plain flow circulation from the larger domain.

The assumption of nocturnal mountain winds and the approach of a nested prognostic simulation to incorporate these flows made it possible to simulate the observed flow reversal after sunset.

The concentration fields as obtained from this simulation are shown in Figures 4(a) and (b). The time intervals are in accordance with those of the diagnostic modeling results in Figures 2(a) and (c). Since a model is always an abstraction of the real world the results of any simulation over complex topography can only partially be in quantitative agreement with the observations.

CONCLUSIONS

In this paper the simulation of an air pollution dispersion scenario was presented with reference to a tracer experiment performed under fair-weather conditions over complex terrain on the Swiss Plateau. Since the knowledge of accurate wind fields is of primary importance for reliable air quality modeling, a model performance evaluation was carried out with respect to the sensitivity of the dispersion model output to changes in the flow field input.

In the case of a real accidental release of toxic air constituents in Switzerland, only the permanently operating network of weather stations (ANETZ and ENET) provide wind information for the diagnostically based prediction of concentration patterns. This network of automatically working weather stations is operated by the Swiss Meteorological Institute and consists of 115 (1991) meteorological masts that are distributed all over Switzerland.

In view of the high irregularity of the topography - even in the Swiss Plateau - the density of the ANETZ-network appears to be too small to be considered as a sufficient data base for reliable concentration predictions in the case of an emergency situation.

Therefore a prognostic mesoscale model may offer a promising alternative in providing a characterization of air flow in the absence of representative observations. However, apart from the high computational costs, the prognostic mesoscale modeling of flow fields in complex terrain suffers from the same deficiency as the diagnostic approach inasmuch as the reliability of the results is difficult to be estimated quantitatively.

Additionally, applying a prognostic model to an area of complex topography often requires the nesting of simulations of different scales. The data selection for such a grid nesting procedure, however, can be accomplished in different ways which do not necessarily produce identical results.

The application of a model such as MEMO to any given dispersal scenario requires a phenomenological knowledge of the different meteorological processes influencing the flow field generation in the respective modeling area. These include thermally induced wind fields by topographical features of the various mesoscales. Therefore experience with adequate model nesting techniques is essential to enhance the reliability of the results.

ACKNOWLEDGEMENT

We greatfully acknowledge the participation of the CEC tracer group at Ispra / Italy in the experiments. Moreover, we are indepted to SMA-Zurich for providing the HRM-model output for October 3, 1991 and SMA-Payerne for the radiosonde data. Furthermore we are grateful to Prof. N. Moussiopoulos and Dr. Th. Flassak for receiving their models CONDOR and MEMO. Moreover, we would like to thank Dr. J. Keller and Mr. M. Tinguely for providing some of their plot software. Finally thanks are addressed to the Bundesamt für Bildung und Wissenschaft (BBW) for funding the TRANSALP project.

REFERENCES

Axenfeld, F., Janicke, L., Münch, J., 1984, Entwicklung eines Modells zur Berechnung des Staubniederschlages, Forschungsbericht 104 02 562, Dornier System GmbH, Friedrichshafen

Burger, A., Ekhart, E., 1937, Über die tägliche Zirkulation der Atmosphäre im Bereich der Alpen. Beitr. Geophys. 49:341

Clark, T.L., 1977, A small-scale dynamic model using a terrain-following coordinate transformation, J. Comput. Phys. 24:186

Flassak, Th., 1990, Ein nicht-hydrostatisches mesoskaliges Modell zur Beschreibung der Dynamik der planetaren Grenzschicht, Fortschritt-Bericht Nr. 74, Reihe 15: Umwelttechnik, VDI Verlag

Goodin, W.R., McRae, G.J., Seinfeld, J.H., 1979, A comparison of interpolation methods for sparse data: Application to wind and concentration fields, J. Appl. Meteor. 18:761

Hanna, S.R., 1982, Applications in air pollution modelling, in: Atmospheric Turbulance and Air Pollution Modelling, chap. 7 (ed. F.T.M. Nieuwstadt and H. van Dop), Reidel, Dordrecht

Hanna, S.R., 1988, Air quality model evaluation and uncertainty, JAPCA 38:406

Kolmogorov, A.N., 1933, Zur Theorie der stetigen zufälligen Prozesse, Math. Ann. 108:149

Lamprecht, R., 1989, Modeling of air pollution dispersion with a Monte Carlo diffusion model, in: PSI Annual Report 1989, Annex V, General Energy Technology

Moussiopoulos, N., Flassak, Th., and Knittel, G., 1988, A refined diagnostic wind model, Environmental Software 3(2):85

Moussiopoulos, N. et al., 1993, Simulations of the wind fields in Athens with the nonhydrostatic mesoscale model MEMO, Environmental Software, 8(1):29-42

Porch, W., Rodriguez, D., 1987, Spatial interpolation of meteorological data in complex terrain using temporal statistics, J. Climate Appl. Meteor. 26:1696

Schubiger, F., 1993, Verifikation des hochauflösenden Regionalmodells während der präoperationellen Phase. Arbeitsbericht der Schweizerischen Meteorologischen Anstalt

DISCUSSION

G. KALLOS: What do you mean by nesting in the model you used? Is it a one-way or two-way interactive nesting? In the case of one nesting how do you control the overdiffusion at the boundaries? The same for the case you are doing a simple interpolation which you call it nesting. It is better to avoid simple interpolation at the boundaries for cases with significant physiographic variations at the model boundaries.

D. BERLOWITZ: The nesting applied here consists of a passive one way nesting, where results from a calculation on the larger domain were used to determine the boundary values for temperature and wind velocities during the calculations on the smaller domain.

K. JUDA-REZLER: Which tracer gas has been used for the experiment?

D. BERLOWITZ: Perfluorodimethylcyclokexane ($C_8 F_{16}$)

AN EMERGENCY RESPONSE AND LOCAL WEATHER

FORECASTING SOFTWARE SYSTEM

Craig J. Tremback, Walter A. Lyons, William P. Thorson
and Robert L. Walko

ASTeR, Inc.
P.O. Box 466
Fort Collins, CO USA 80522

INTRODUCTION

Recent advances in computer technology have now placed supercomputer power on the desktop for a small fraction of the price. Many traditional supercomputer applications have benefited greatly in the move from the realm of the supercomputer center to more direct local control of the end user. Two of the atmospheric applications that have and will continue to benefit greatly from these advances in computer technology is in the arenas of local weather forecasting and emergency response systems.

ASTeR has continued the development of the Emergency Response Dose Assessment System (ERDAS) (funded by the United States Air Force Space Systems Division) which has been designed to run on the high-performance RISC (Reduced Instruction Set Computers) workstations (Lyons and Tremback, 1993). The purpose of ERDAS is to provide emergency response guidance to operations at Cape Canaveral Air Force Station/Kennedy Space Center (CCAFS/KSC) in case of an accidental hazardous material release or an aborted vehicle launch. Although the system is being designed for use at CCAFS/KSC, it can and will be applied to a wide range of other emergency response and/or local weather forecasting applications.

ERDAS uses several key software components to accomplish its tasks: 1) the Regional Atmospheric Modeling System (RAMS) software (developed at Colorado State University and ASTeR) to provide for local meteorological forecasting and analysis, 2) HYPACT (HYbrid Particle And Concentration Transport package), a model under development by ASTeR, will be used to simulate the dispersion of the hazardous materials using the RAMS meteorological output, 3) visualization software, currently NCAR Graphics, AVS (from AVS, Inc.), and savi3d (from SSESCO) to display and analyze the results from RAMS and HYPACT, and 4) a customized Graphical User Interface (GUI) to provide a consistent and easily-used interface for all of the software components. These components provide the control and visual depiction of the results of the local weather forecasts and resultant dispersion estimates which will account for three-dimensional, time-dependent, complex mesoscale phenomena including topographic effects, wind discontinuities (land and sea breezes), thermal internal boundary layers (TIBLs), wind shears, and mesoscale vertical motions. This should greatly improve the estimates of mesoscale dispersion of pollutants within the complex coastal flow regimes at CCAFS/KSC or other geographic locations. Further general design details and the specific requirements for CCAFS/KSC will be described in the following sections.

THE REGIONAL ATMOSPHERIC MODELING SYSTEM - RAMS

At the core of ERDAS is the Regional Atmospheric Modeling System (RAMS) which was developed at Colorado State University and ASTeR, Inc. RAMS is a multipurpose, numerical prediction model that is designed to simulate weather systems spanning in scale from the hemisphere down to large eddy simulations (LES) of the planetary boundary layer. Its primary use is to simulate atmospheric phenomena on the mesoscale (horizontal scales from 2 km to 2500 km) for purposes ranging from support of basic research to air quality regulatory applications. RAMS was developed initially to perform research into the areas of modeling physiographically-driven weather systems (i.e. sea breeze circulations, thermally-driven mountain circulations, and circulations driven by contrasts in land-use) (Pielke, 1974; Mahrer and Pielke, 1977) and simulating convective clouds, mesoscale convective systems, cirrus clouds, and precipitating weather systems in general (Tripoli and Cotton, 1982; Tremback, 1990). Since then, the use of RAMS has increased greatly, with a variety of applications from regional climate simulation to large eddy simulations to operational forecasting applications.

RAMS contains a number of options which makes it amenable for use in an emergency response and local weather forecasting system. It is designed so that the code contains a variety of structures and features ranging from hydrostatic to non-hydrostatic codes, resolution ranging from a few meters to the order of a hundred kilometers, domains from a few kilometers to an entire hemisphere, and a suite of physical options. This allows for an easy selection of the appropriate options for a different spatial scale or different locations. A particular feature of RAMS that makes it attractive for mesoscale and microscale applications is the interactive (or two-way) nesting procedure that allows the user to specify any numer of telescoping grids or even moving grids that could float through a mesoscale grid while calculating, for example, transport and dispersion of a pollutant or the propagation of a thunderstorm. This also allows a high resolution forecast for a target area while simultaneously providing coarser resolution forecasts (but still higher resolution than NMC or ECMWF) for a much larger surrounding area.

THE HYBRID PARTICLE AND CONCENTRATION TRANSPORT PACKAGE -HYPACT

ERDAS will employ HYPACT (developed by ASTeR, Inc.) for computing the dispersion estimates of the contaminants. HYPACT is a combination of a LaGrangian particle model and an Eulerian concentration transport model, combining the main advantages of both. The Lagrangian model attempts to deal with the subgrid-scale aspects of a pollutant release while the Eulerian model takes over when the contaminant "cloud" becomes adequately resolved on the computational grid, thus reducing the number of particles needed and increasing the computational efficiency. With the velocity and turbulence fields simulated by RAMS, the LaGrangian model in HYPACT will advect the particles with mean and random turbulent wind components. HYPACT uses a level 2.5 turbulent closure scheme based on the prognostic turbulent kinetic energy from RAMS. When the pollutant is adequately resolved on the grid, the particles will be converted to a concentration field and handled in an Eulerian manner.

The source configurations for HYPACT are very flexible. Any number of sources may be specified anywhere in the domain and configured as a point, line, area, or volume source. The emissions from these sources can be instantaneous, intermittent, or continuous. The pollutants can be treated as gases or aerosols and a radioactive half-life or a settling velocity can be specified. Plume rise parameterizations and a dry deposition scheme have recently been added to HYPACT.

COMPUTATIONAL ASPECTS

Recent developments in the computer industry have now made it possible to consider employing numerical weather forecasting and dispersion models, such as RAMS and HYPACT, for a wide range of applications and spatial scales. Because of the increase of computer throughput that has been achieved and the additional advances that are expected in the next several years, emergency response and weather forecasting systems will not be limited to simple types of models that have been used in the past, but will benefit from the more sophisticated models that have been developed for atmospheric research and are now starting to be used for operational purposes. Atmospheric simulation and forecast models had operated almost exclusively on large, multi-user mainframes (such as CRAY's and CDC CYBER's) until the late 1980's. In the last several years (since about 1987), RISC-based, superworkstations have come on the market which have sufficient computational power to run large forecasting codes at speeds within an order of magnitude of supercomputers. Aside from the cost benefits of using the smaller workstations, the conveniences of local control of the computer system, local data storage, and sophisticated visualization tools on the same computer system are just a few of the additional advantages.

The RAMS and HYPACT codes are quite portable as they currently execute on CRAY supercomputers (X-MPs, Y-MPs and IIs), superworkstations (Sun, SGI, IBM, HP), DEC VAX's and even PC's. RAMS primarily runs on the UNIX operating system, although porting to a different system is not difficult. The performance level we have achieved on RISC workstations computational efficiencies of one-third to one-fifth the speed of a CRAY Y-MP.

Even more recently (within the last 2-3 years), software and hardware have become available which allow RISC workstations to be clustered to provide a very cost-effective parallel, high-performance computing system. Such clusters of workstations are very cost-effective when they are dedicated to a task such as running these forecast models. The capabilities of the ERDAS system would be greatly enhanced if efficient use could be made of these parallel clusters.

And, indeed, such an effort is underway. ASTeR has obtained funding from NASA to extend ERDAS to include the forecasting of sea-breeze-induced deep convection. This project will require additional computing resources because of the added complexity of the physics and the necessary increase in spatial resolution needed for the numerical forecast. We are targeting the workstation cluster as the parallel platform because of the cost-effectiveness and the flexibility of such a parallel system. RAMS is well suited for parallelization since it does not use physical/numerical routines that are global. Pressure, for example, is solved locally either using the hydrostatic approximation or non-hydrostatically using a time-split compressible approximation. Advection is calculated using local finite difference operators rather than using non-local spectral methods. For the same truncation scale, spectral methods have been shown to be more accurate than finite difference approximations. However, finite difference approximations have been refined enormously in recent years allowing higher-ordered accuracy (i.e., Tremback et al., 1987), at much less computational expense than spectral methods. The advantage of the non-global character of finite-difference schemes can thus be used to great advantage for parallelization with good computational accuracy.

ERDAS CONFIGURATION FOR CCAFS/KSC

As an example of the use of ERDAS for a specific location, we will detail the configuration of the system for the CCAFS/KSC application. The system will be configured for use on one IBM RS/6000 Model 550 workstation with 64 Mb of central memory and 2.8 Gb of disk storage.

Meteorological Forecast

Even with their relatively impressive performance, workstations have limitations. As the mesh size decreases, run times increase dramatically. Certain RAMS options, such as explicit cloud microphysics, also require substantial computational resources. For these reasons, the RAMS configuration in the initial ERDAS system at CCAFS/KSC will be limited to a 3 km inner mesh size and will not explicitly treat convective cloud formation. These restrictions will disappear as faster processors and clustered workstation systems become available at CCAFS/KSC. The initial goal is to provide forecasters with new 24 hour forecast model output within six hours of initialization. The grid configuration for RAMS, therefore, has been chosen to be a 60 km mesh covering the southeastern United States, a 15 km grid covering most of the Florida peninsula, and a 3 km mesh covering a 110 x 110 km region around KSC/CCAFS. The 3 km grid represents the coarsest mesh that we feel will resolve adequately both the sea breeze and island/estuary perturbations. As workstation processors become even more powerful, spawning a finer mesh grid (approximately 1000 meters) directly over KSC/CCAFS is readily accomplished.

For the local meteorological forecasts, RAMS will be initialized and run twice daily from the standard data times of 0000 UTC and 1200 UTC. Thus, there will always be a forecast available from which to run the dispersion estimates in case of an accidental or planned contaminant release. Fixed data inputs into RAMS include topography, USGS-provided land use/land characteristics, and climatological sea surface temperature patterns. RAMS will have a non-homogeneous initialization with non-stationary boundary conditions since temporal variability in the outer boundary conditions is needed. Data sets such as NMC grid point analysis and NGM forecast, NWS rawinsonde and SAO data, as well as local mesonetwork and boundary layer wind profiler data will be utilized within the RAMS isentropic analysis package used as part of the model initialization procedure.

As mentioned, the initial ERDAS configuration will not activate the explicit cloud microphysics modules. Therefore, while the local flow features which trigger sea-breeze thunderstorms will be represented, the resultant convective response of the atmosphere will not. RAMS is capable of simulating convective storms explicitly. Considerable success was achieved in modeling Merritt Island Thunderstorm and Atlantic sea breeze convection in a recently completed project. Since the inclusion of the microphysical module is more computational demanding, work is proceeding on a faster microphysics package. With this faster code, along with workstation performance doubling every 12 to 18 months, it is simply a matter of time before ERDAS will treat convective cloud impacts upon local dispersion at CCAFS/KSC. We note, however, that previous work has suggested that a non-convective configuration of RAMS can successfully diagnose the initiation and development of thunderstorms generated by local thermal forcing. Since a 6 to 24 hour dispersion forecast not accounting in some way for the potential for convective disturbances in the boundary layer is potentially misleading, we propose an interim solution. Previous thunderstorm forecasting experiments at KSC showed that relatively simple diagnostics applied to a "dry" prognostic model demonstrated skill at predicting the initiation of sea breeze storms during the upcoming day. Various candidate storm diagnostics will be examined, including the K index, the KLIW index, convective available potential energy, and the output of 1-D diagnostic cloud models. As part of the ERDAS display, the spatial and temporal evolution of the convective storm potential diagnostic will be available to the forecaster. This will help flag those upcoming periods in which the dispersion estimates may be disrupted by deep convective clouds. Even with these initial limitations due to the computer power at CCAFS/KSC, we feel that the meteorological forecasts will be suitable for about 75% of the hours in a typical year.

Dispersion Estimates

RAMS, then will produce a 24-hour meteorological forecast continuously on call. The design for the dispersion modeling portion of ERDAS involves: 1) developing a generalized treatment of key dispersion processes using HYPACT, and 2) specifying the source terms and accident scenarios relevant to KSC/CCAFS. In order to maintain continuity with systems currently in use within the Air Force, we will provide emulations of both the OB/DG and AFTOX dispersion models . In addition to running with current data, these codes will then have the additional feature of providing estimates using forecasted weather. This will allow predictions of both long-lasting releases and analysis of potential impacts for planned activities later in the day. The types of accidental or planned release scenarios that we will attempt to handle initially include:

- LAUNCH VEHICLE ABORT CLOUDS; use of HYPACT to specify the dispersion of both gases and aerosols (over a spectrum of sizes) for debris clouds from a launch vehicle abort.

- SPILLS OF TOXIC CHEMICALS AT LAUNCH PADS AND STORAGE FACILITIES; the "cold spill" scenario, in which evaporation takes place from pools; also treatment of momentum and buoyant jets if adequately specified; a suite of products will be provided ranging from single trajectories to emulations of OB/DG and AFTOX using RAMS forecast data, and HYPACT, ranging from plume visualization using a small sample of particles to detailed concentrations and dosage estimates using a large number of particles.

- PLUMES FROM SRM GROUND FIRES AT LAUNCH PADS, PROCESSING, AND STORAGE FACILITIES; use HYPACT with either specified or computed plume rise for one or more burn locations.

- METEOROLOGICAL INPUTS TO REEDM, BLAST AND MARSS, AND SIMPLE THUNDERSTORM PROBABILITY GUIDANCE; files will be created which will allow other systems to access forecasted data fields in the same format as current observations; the thunderstorm potential display will alert users to possible degradation of system performance due to convective storm development during the forecast period.

- VENTING OF TOXIC CHEMICALS FROM STORAGE FACILITIES; either momentum or buoyant releases, using OB/DG, AFTOX and HYPACT.

- EXHAUST GROUND CLOUDS FROM NOMINAL LAUNCHES OF TITAN, ATLAS, DELTA AND STS VEHICLES; using HYPACT to simulate dispersion of the ground cloud and exhaust plume gaseous and aerosol species.

The resultant dispersion fields produced by the component models will consist of contaminant concentration estimates in either a three-dimensional field (for HYPACT and AFTOX) or a surface concentration field (for OB/DG). The ERDAS system will display either the instantaneous concentrations or a time-averaged dosage estimate.

User Interface and Visualization

The forecasting staff at the Cape Canaveral Forecast Facility is already heavily tasked. Therefore, the level of effort required to initialize and run the models of the ERDAS system needs to be kept to a minimum. ERDAS will have two modes: 1) an *operational mode* in which a limited subset of the features are available to the operator, and 2) a *planning mode* in which the full range of features are available to be exercised by the user.

The *operational mode* will be primarily driven from a single graphical user interface (GUI) menu. From this mouse-driven menu, the operator can view the results of the

RAMS forecasts or the current weather from the local mesonet. Horizontal plots of winds, temperature, moisture, etc. will be drawn using NCAR Graphics at the ground surface or at elevated levels for all of the available forecast times. The displayed domain may be chosen as the southeast US, the Florida peninsula, or the CCAFS/KSC area. A mouse-driven arbitrary zoom feature will allow the operator to closely examine any local area. On the smaller scales, a high resolution map will be displayed showing details such as roads, rails, and even buildings.

At the forecast times of 0000 UTC and 1200 UTC, the operator will have the ability to display, check, and edit the meteorological input data to the RAMS forecast. The same maps will be displayed, as mentioned above, with icons designating observation type (rawindsonde, profiler, tower, etc.). By clicking on the icon, an appropriate display of the observation will be made (e.g., a skew-T plot for the rawindsonde). The operator at this point can decide if the data is valid and optionally remove or edit the data. The RAMS forecast will then run automatically as soon as all data is received.

If an accidental release occurs, the operator will click on a button and a menu will be displayed which will control the dispersion estimates. The operator will have a choice of choosing a pre-programmed accident scenario which will already have defined a chemical type, release rate, spill amount, location, etc. and then have the ability to override any of the parameters if needed. The operator will choose which dispersion model to run (HYPACT, AFTOX, OB/DG) and whether to use the forecast meteorology, just the observed meteorology, or an objective combination of the two. The dispersion estimates will be computed and the operator will then be able to display the results.

In contrast, in the *planning mode*, the user will have access to many more of the system's features. The meteorology will be able to be displayed not only in with horizontal plots but also with vertical cross-sections. More RAMS output parameters will be available. The user can also view the meteorology or dispersion results with the 3-dimensional graphics packages of AVS and savi3d, if desired. If operational execution time constraints can be relaxed, RAMS can be set up in higher resolution than the operational mode. Sensitivity runs can be made with either RAMS or HYPACT. The user can pose "what-if" scenarios to the ERDAS system and determine the system's response.

SUMMARY

One of the atmospheric applications that will benefit greatly from recent advances in computer technology is the area of local weather forecasting and its use in emergency response systems. ASTeR has continued the development of the Emergency Response Dose Assessment System (ERDAS) (funded by the United States Air Force Space Systems Division) which will run on UNIX workstations. The purpose of ERDAS is to provide emergency response guidance to operations at Cape Canaveral Air Force Station/Kennedy Space Center (CCAFS/KSC) in case of an accidental hazardous material release or an aborted vehicle launch. The system can and will be applied to a wide range of other emergency response and/or local weather forecasting applications. ERDAS uses the Regional Atmospheric Modeling System (RAMS) software (developed at Colorado State University and ASTeR) to provide for local meteorological forecasting and analysis. A new model under development by ASTeR, HYPACT (HYbrid Particle And Concentration Transport package) will be used to simulate the dispersion of the hazardous materials using the RAMS meteorological output. The resultant dispersion estimates account for three-dimensional, time-dependent, complex mesoscale phenomena including topographic effects, wind discontinuities (land and sea breezes), thermal internal boundary layers (TIBLs), wind shears, and mesoscale vertical motions. This should greatly improve the estimates of mesoscale dispersion of pollutants within the complex coastal flow regimes at CCAFS/KSC. The models and visualization software will be controlled by a customized Graphics User Interface (GUI) tailored to allow for efficient operation of the system and rapid interpretation of the dispersion model results. Future work will involve the inclusion of RAMS' explicit cloud parameterizations and the porting of the system to a parallel RISC workstation cluster.

ACKNOWLEDGEMENTS

This research is supported primarily by the U.S. Air Force (Space and Missle Systems Center) under contract No. F04701-91-C-0058 with contributions from NASA/KSC. Dr. Marek Uliasz is assisting in the development of HYPACT. The supporting efforts of Bart Lundblad (The Aerospace Corporation), William Boyd (Patrick Air Force Base), and David Stuck, Matt Willis, and Chris Knear (USAF Space and Missile Systems Center) are also acknowledged.

REFERENCES

Lyons, W.A. and C.J. Tremback, 1993: A prototype operational mesoscale air dispersion forecasting system using RAMS and HYPACT. Preprints. Air and Waste Management Association 86th Annual Meeting. Denver, Colorado, 13-18 June 1993.

Mahrer, Y. and R.A. Pielke, 1977: A numerical study of the airflow over irregular terrain. *Beitrage zur Physik der Atmosphare,* **50,** 98-113.

Pielke, R.A., 1974: A three-dimensional numerical model of the sea breezes over south Florida. *Mon. Wea. Rev.,* **102,** 115-139.

Tremback, C.J., J. Powell, W.R. Cotton, and R.A. Pielke, 1987: The forward-in-time upstream advection scheme: Extension to higher orders. *Mon. Wea. Rev.,* **115,** 540-555.

Tremback, C. J., 1990: Numerical simulation of a mesoscale convective complex: model development and numerical results. Ph.D. dissertation, Atmos. Sci. Paper No. 465, Colorado State University, Dept. of Atmospheric Science, Fort Collins, CO 80523.

Tripoli, G.J., and W.R. Cotton, 1982: The Colorado State University three-dimensional cloud/mesoscale model – 1982. Part I: General theoretical framework and sensitivity experiments. *J. de Rech. Atmos.,* **16,** 185-220.

COMPARISON OF MODELS FOR AEROSOL VAPORISATION
IN THE DISPERSION OF HEAVY CLOUDS

J. Kukkonen[1], M. Kulmala[2], J. Nikmo[1], T. Vesala[2], D. M. Webber[3] and T. Wren[3]

[1] Finnish Meteorological Institute, Air Quality Department
   Sahaajankatu 22 E, SF-00810, Helsinki, Finland
[2] University of Helsinki, Department of Physics
   Po. Box 9, SF-00014, Helsinki, Finland
[3] AEA Technology, Consultancy Services (SRD), Thomson House
   Warrington Rd, Risley, Warrington, Cheshire WA3 6AT, UK

INTRODUCTION

The dispersion of heavier-than-air aerosol clouds is of great relevance to the analysis and assessment of hazards in the chemical and petrochemical industries. There are many simple mathematical models for heavy gas dispersion, but two-phase dynamics are either ignored or treated in a very simple way, without any convincing scientific justification. In fact essentially all of those models which do treat two-phase phenomena rely on the "homogeneous equilibrium" assumption, whether they be for jet dispersion, for gravity dominated dispersion, or for any other aspect of a two-phase release.

Here we test the homogeneous equilibrium model for two-phase ammonia clouds released in dry and moist air, by comparing its predictions with those of a more sophisticated aerosol approach. It is shown that the simpler model does indeed provide a good description for some envisaged release situations, and guidance is given on where the homogeneous equilibrium model is not likely to be adequate.

The presence of liquid drops in a cloud can be crucial to determining its overall behaviour. Ammonia is a particular case in point. Ammonia gas at its boiling point is lighter than ambient temperature air. Ammonia clouds, however, are now well known to behave as a heavy gas. This can happen, for example because ammonia which is released from pressurised liquefied storage at ambient temperature flashes to form an aerosol cloud.

The simplest possible model of a two phase cloud which can account for the overall behaviour described above is the "homogeneous equilibrium" (HE) model. This has been adopted in various integral models for jets and clouds, including for instance the two-phase jet model TRAUMA (Wheatley, 1987) and the dense gas dispersion model DRIFT (Webber et al., 1992).

The HE model follows the spirit of integral models of dispersion by assuming that the liquid drops are statistically distributed, both spatially throughout the cloud and in size, either uniformly or with more general self-similar profiles, and that there is one temperature characteristic of the cloud. The liquid and gas phases are taken to be in thermodynamic equilibrium at that temperature. In the simplest case the cloud will contain contaminant vapour, contaminant liquid and air. In cases one stage more complex, there will also be liquid water and water vapour in the cloud contributing to a five-component equilibrium.

The more sophisticated (aerosol) approach considers the actual phase composition of the cloud (Kukkonen et al. 1989, Vesala et al., 1989, Kukkonen 1990, Vesala 1991, Vesala and Kukkonen 1992, Nikmo et al., 1993). Mathematical models for droplet vaporisation and condensation have recently been developed also by Hewitt and Pattison (1992), with the aim of incorporating the aerosol model into a heavy gas dispersion program. Woodward and Papadourakis (1991) have presented a two-phase jet model, including a description of droplet vaporisation.

We assume that the cloud contains binary (two-component) droplets together with the surrounding gas. The droplets contain contaminant liquid and liquid water, and the gas phase consists of inert gas (dry air) and the vapours of species forming the droplet. In the following we refer to the gaseous phases of the two condensing or evaporating species simply as "vapours", and to the mixture of air and the two vapours as "the surrounding gas", or "the gas".

The mass and heat transfer processes from the droplets into the surrounding gas are then modelled in detail. The rates of these processes are strongly dependent on the rates of diffusion and thermal conduction in the gas. For large droplets the ventilation of droplets due to their free fall is also an important factor, as it causes enhancement of mass and heat transfer by forced convection. The droplets and the gas are not generally in thermal equilibrium, and the deviation from equilibrium is determined by the rates at which mass and heat can be transported in the mixture. Two temperatures are needed to characterise the cloud, the droplet temperature and the gas temperature.

The HE model is in fact a limiting case of the more complex aerosol model: the limit is that where heat and mass transfer processes are very rapid compared to all other time scales, most importantly the time scale over which the cloud dilutes. The validity of such a limiting approximation is therefore strongly dependent on the rates of the competing processes.

The principal reason for adopting the HE model in dispersion models is that it is simple enough, even in fairly complex cases, to couple into the overall model. The homogeneous equilibrium model has not been shown to be manifestly wrong in its predictions for the overall behaviour of the cloud, and is probably the simplest model with any chance of being realistic.

On the other hand, as far as we are aware, no theoretical justification has been presented for the validity of the HE model in any dispersion context. It is therefore the purpose of this work to investigate the validity of the HE model, by comparing its predictions with those of the more detailed aerosol model. To do this we shall use the homogeneous equilibrium model as embodied in the dense gas dispersion code **DRIFT** (**D**ense **R**eleases **I**nvolving **F**lammables or **T**oxics) given by Webber et al. (1992) and the aerosol model **AERCLOUD** (**AER**osol **CLOUD**), the latest version of which has been reported by Nikmo et al. (1993).

This paper is an abbreviated version of the complete report, which is in press (Kukkonen et al., 1993).

THE MODELS

The dispersion model DRIFT models an instantaneously released cloud in the form of a now-standard slumping cylinder. The cloud is considered to contain different numbers of moles of contaminant vapour, contaminant liquid, dry air, water vapour and water liquid, which evolve in time. For instantaneous releases the total number of moles of contaminant is assumed to be constant, and the total number of moles of dry air and water is controlled by the entrainment rate.

The balance between the phases, and the evolution of the temperature depends on the homogeneous equilibrium model which defines

$$x_i = \chi_i (X_i, T) \tag{1}$$

where x is the mole fraction in the vapour phase, χ is the vapour pressure over the binary liquid mixture (normalised by atmospheric pressure), X is the mole fraction in the liquid phase, T is the temperature and the equation holds for i = contaminant and i = water independently. For ammonia, DRIFT uses the correlations for the χ functions derived by Wheatley (1987). The vapour pressure functions are often written in the form

$$\chi_i(X_i, T) = X_i \Gamma_i(X_i, T) \chi_i^0(T) \tag{2}$$

where χ^0 is the normalised vapour pressure over the pure liquid and Γ is an activity coefficient. Substances which form ideal solutions have unit activity coefficients; this is far from the case for ammonia where strong hygroscopic behaviour is accompanied by a significant heat of solution.

In earlier papers, a model was presented for estimating the evolution of an aerosol containing contaminant droplets, contaminant vapour and dry air (Kukkonen et al. 1989, Vesala et al., 1989), and the model was validated against experimental results on the laboratory scale (Vesala et al. 1989). Later, the model was generalised to allow for the effects caused by water vapour in the entrained air (Vesala, 1991, Vesala and Kukkonen, 1992), the experimental validation on the laboratory scale being reported by Vesala (1991). The model was originally designed for constant mass of air, but it has recently been generalised to admit a time-varying ambient dilution (Nikmo et al. 1993).

The aerosol model takes account of the detailed temperature and concentration gradients near the surface of a binary droplet. The essential problem is the modelling of the mass and heat transfer processes from the droplets into the surrounding gas. The droplets are assumed to be in the continuum regime and the droplet growth and evaporation are assumed to be quasistationary. That is to say the droplets are large enough (>1μm) to see the surrounding gas as a continuum, and the concentration and temperature profiles around the droplets are at any time essentially the same as in the steady state, with changes following changes in the boundary conditions directly. The mass fluxes are calculated using the well-known formula taking into account ordinary diffusion; the heat flux is governed by Fourier's law of heat conduction and by the enthalpy carried by diffusing species.

If the droplets are moving with respect to the surrounding gas, the rates of mass and heat transfer are increased due to forced convection. If the droplet ventilation is allowed for, the diffusive mass fluxes are multiplied by the droplet Sherwood number, and the thermal conductivity is multiplied by the droplet Nusselt number.

RESULTS AND DISCUSSION

It is important at the outset to distinguish between vaporisation and deposition phenomena. Our prime objective is to examine the vaporisation model and so we shall largely ignore deposition for most purposes.

First we shall obtain predictions of the code DRIFT for instantaneous releases of ammonia into moist air. Following this, we shall model a gas cloud with the same initial content using the code AERCLOUD, importing air into it at the time-dependent rate given by DRIFT. In this way we can study the differences of model predictions for the thermodynamic behaviour, excluding possible differences in modelling, for instance entrainment, spreading or transport processes. We also need not couple the aerosol model explicitly to a heavy gas dispersion model.

We have made predictions for instantaneous releases of a pure ammonia cloud into dry and humid air. Liquid and vapour deposition were neglected, and we assumed that there is no heat transfer from the ground into the cloud. The ambient temperature was taken to be + 15 °C, the Pasquill-class was D, the average wind velocity at a height of 10 m was 2 m/s, and the roughness length was 6 mm.

The total mass of contaminant was taken to be 10 000 kg, which is roughly from two to three times the mass of the initial gas cloud in the Thorney Island field experiments (McQuaid and Roebuck, 1985). The flashing is assumed to have taken place, and therefore the pressure is atmospheric and the initial contaminant temperature is equal to the boiling point of ammonia (- 33 °C). The initial contaminant liquid fraction by mass was assumed to be 85%, 60% or 30%. The value of 85% corresponds approximately to the largest possible liquid fraction after flashing has taken place. But ammonia is stored with different degrees of overpressure and refrigeration, and so the values of 30 and 60% have been taken to correspond to different possible storage conditions.

In order to illustrate the effects of atmospheric moisture, we have included the case of dry air and that of 99.99% relative humidity; this is effectively the maximum humidity short of introducing fog. Two options were used for the interactions of ammonia and water in the liquid phase: assuming an ideal solution, and allowing for the actual interactions. Ammonia and water behave attractively in the liquid phase, and therefore the activity coefficients are smaller than unity; correspondingly, the partial mixing enthalpies are negative (Vesala and Kukkonen, 1992, Wheatley, 1987). For an ideal solution, the activity coefficients are equal to unity and the partial mixing enthalpies vanish.

We are considering here only monodisperse aerosol populations. The initial droplet size is required by the aerosol model as input information; we have chosen the radii of 1000 µm, 100 µm and 10 µm. The droplet radius values in the laboratory scale measurements in this context range typically from a few to a few hundred micrometers (for instance, Nolan et al., 1990, Moodie and Ewan, 1990 and Schmidli et al., 1990). We considered it appropriate to select a fairly broad range of droplet sizes for the computations, in order to obtain more general results particularly in view of the possibly occurring non-equilibrium effects. The droplets are assumed to consist initially of pure ammonia, and they are at the assumed source term temperature (- 33 °C).

For the sake of a clear comparison, we disregard depletion of liquid due to deposition, which would surely be an important process at large droplet sizes. The gravitational settling velocities of 1000 µm, 100 µm and 10 µm ammonia droplets are about 5 m/s, 0.5 m/s and 0.01 m/s, respectively (Vesala et al., 1989).

The AERCLOUD results have been computed using three model options: including and excluding droplet ventilation, and in the HE limit. The values at the HE limit are expected to duplicate the respective results computed with DRIFT.

If we multiply the heat and mass transfer coefficients of AERCLOUD by a sufficiently large factor, we obtain the HE model results as limit values. The numerical simulations showed that a multiplicative factor of 10^4 was needed to achieve a complete equilibrium; an equilibrium state can always be recognised by comparing the droplet and gas temperatures, as these become equal at the HE limit. Clearly, the molecular rates of diffusion and conduction (or convection) have certain theoretical maximum values, which may be exceeded in this process. Nevertheless the derivation of HE limit values is very useful for model comparison purposes.

Just some example results are illustrated in the following, for complete results the reader is referred to Kukkonen et al. (1993). Figure 1 shows the numbers of moles of ammonia, vapour and liquid phases. The headlines of figures show also the selected initial droplet sizes (from left to right 1000 µm, 100 µm and 10 µm). The curves marked with "no ventilation" and "ventilation" are the non-equilibrium model predictions, in which ventilation of droplets due to forced convection has been excluded or included, respectively. The curves marked with "HE-limit" show the model predictions in the homogeneous equilibrium limit.

We compared the HE limit predictions of the model AERCLOUD with the corresponding results computed with the model DRIFT. The maximum differences for the numbers of moles of contaminant vapour and liquid, and for the temperature were smaller than 4%.

The differences of various modelling options can be seen clearly in Figure 1. The rates of mass and heat transfer are most slow for the non-equilibrium computation, where droplet ventilation has been excluded ("no ventilation"). Including droplet ventilation in the non-equilibrium computations enhances the mass and heat transfer. In the HE-limit these transfer rates are the largest possible, and ammonia is therefore vaporising more quickly.

The effect of ambient air moisture is again to increase the rate of vaporisation at small times, and suppress it at larger times. The latent heat released by the condensing water vapour at the droplet surface tends to raise the droplet temperature and therefore enhances the evaporation of ammonia. At larger times the concentration of ammonia in the droplet decreases with increasing water content. For a large dilution, the mole fraction of ammonia and the activity coefficient are small, reducing the evaporation rate of ammonia (Vesala and Kukkonen, 1992).

For the initial droplet radii of 100 µm, the differences of model predictions of the three model options is less than one per cent for both the cases considered. For a droplet radius of 10 µm, the curves are practically identical, implying that thermodynamic equilibrium prevails at all times. We can conclude that the HE model gives good results concerning contaminant vaporisation for the selected cases, if the droplet radius is not larger than about 100 µm.

CONCLUSIONS

The homogeneous equilibrium model is often adopted in dispersion models of two-phase jets and clouds, but has never, to our knowledge, undergone a critical test. We have therefore now tested the homogeneous equilibrium model, in the context of the dispersion of a dense gas cloud, against a more sophisticated aerosol model.

The first stage in performing the comparison between the models was to demonstrate agreement if we force the aerosol model to its homogeneous equilibrium limit by increasing

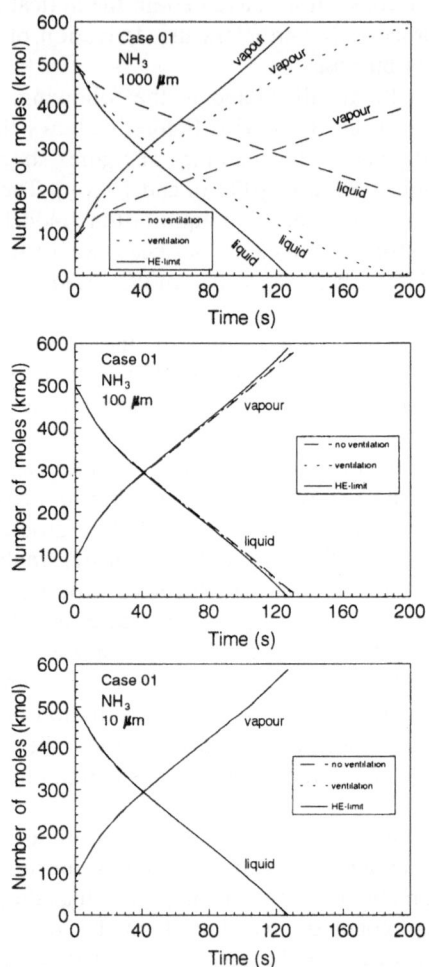

Figure 1. Number of moles of ammonia vapour and liquid vs. time predicted by AERCLOUD. The initial droplet radius ranges from 1000 µm to 10 µm. The curves have been computed using three model options: including and excluding droplet ventilation and in the homogeneous equilibrium limit.

heat and mass transfer coefficients from their physical values. This test was remarkably successful considering that the derivation and solution of equations is not trivial in either model, and in view of the fact that different correlations for the thermodynamics of the ammonia-water mixture are used. We are therefore confident that both models and computer programs are performing accurately, and that the results are not overly sensitive to the small differences in the thermodynamic correlations.

Having first checked for agreement of the models in the appropriate limit, we then compared the predictions of the models. The results of the comparison show that the homogeneous equilibrium model is perfectly adequate in dense gas dispersion models in cases where droplets are small. In the cases illustrated, we have seen that it is manifestly adequate for 100 µm droplets, and as droplets become smaller it gets progressively better. In the aerosol model, for droplets larger than 1 µm, the total heat and mass transfer rates are effectively controlled by the inverse square of the droplet radius. Smaller droplets in this régime therefore tend to give results closer to the homogeneous equilibrium limit which corresponds to very rapid transfer.

In practice, we may expect that an accidental breach of high pressure storage will result in the production of small droplets in the flashing process, and in this case we believe that the homogeneous model is adequate for dense aerosol cloud dispersion. However there are cases of lower pressure storage, for example semi-refrigerated ammonia, where a less violent flashing process may result in formation of larger drops. In these cases the homogeneous model may not be appropriate.

Physically we understand this result as implying that equilibration processes are fast compared with the entrainment time scales in the dispersion. The validity of the homogeneous equilibrium model has not therefore been demonstrated for case where entrainment may be faster - for example in the high momentum region of a jet. This is an interesting open question to which we hope to return in the future.

In order to compare the models we have neglected deposition. This is quite appropriate for the comparison, but may not be in modelling realistic situations. Predictions for deposition rates are readily obtained from aerosol models as they already contain the crucial information about droplet size, allowing a deposition model to be coupled to the vaporisation model relatively straightforwardly.

The main conclusion of this study was that if the droplets are small enough, then the heat and mass transfer processes to the droplet are fast enough (compared to the overall dispersion rate) to make homogeneous equilibrium a good approximation. In the heavy gas dispersion tests done here, 100 µm droplets are seen to be small enough. The question of the adequacy of the homogeneous equilibrium model is thus reduced to that of estimating droplet sizes. It is clear that in many processes involving violent flashing of pressure-liquefied gases the droplets, the drop sizes will be expected to be small enough to admit homogeneous equilibrium as a good approximation in the subsequent advection and dispersion of a heavy cloud.

Acknowledgements

The work presented here has been done as part of a contribution to the STEP programme of the Commission of the European Communities. The contribution of the Finnish Meteorological Institute and the University of Helsinki to this work has also been supported by the Ministry of Environment of Finland, the Academy of Finland and the Neste Co. The contribution of AEA Technology to this work has also been supported by the UK Health and Safety Executive and the Corporate Research Programme of AEA Technology. The support of all these bodies is gratefully acknowledged.

REFERENCES

Hewitt, G.F. and Pattison, M.J., 1992. Modelling of release and flow of two-phase jets. The Safe Handling of Pressure Liquefied Gases (IBC), London, 26-27 November 1992. London.

Kukkonen, J., 1990. Modelling source terms for the atmospheric dispersion of hazardous substances. Ph.D. thesis, University of Helsinki; Commentationes Physico-Mathematicae, Vol. 115, The Finnish Society of Sciences and Letters, Helsinki.

Kukkonen, J., Vesala, T. and Kulmala, M., 1989. The interdependence of evaporation and settling for airborne freely falling droplets. *J. Aerosol Sci.* 20:7.

Kukkonen, J., Kulmala, M., Nikmo, J., Vesala, T., Webber, D.M. and Wren, T., 1993. Aerosol Cloud Dispersion and the Suitability of the Homogeneous Equilibrium Approximation, Atomic Energy Authority AEA/CS/HSE/R 1003 /R (in press).

McQuaid, J. and Roebuck, B., 1985. Large scale field trials on dense vapour dispersion, Final Report on the Heavy Gas Dispersion Trials at Thorney Island 1982-1984, Report EUR 10029, Commission of the European Communities, Brussels.

Moodie, K. and Ewan, B.C.R., 1990. Jets discharging to the atmosphere. *J. Loss Prev. Process Ind.* 3.

Nikmo, J., Kukkonen, J., Vesala, T. and Kulmala, M., 1993. A model for mass and heat transfer in an aerosol cloud (submitted to *J. Hazard. Materials*).

Nolan, P.F., Pettitt, G.N., Hardy, N.R. and Bettis, R.J., 1990. Release conditions following loss of containment. *J. Loss Prev. Process Ind.* 3.

Schmidli, J., Banerjee, S. and Yadigaroglu, G., 1990. Effects of vapour/aerosol and pool formation on rupture of vessels containing superheated liquid. *J. Loss Prev. Process Ind,* 3.

Vesala, T., 1991. Binary droplet evaporation and condensation as phenomenological processes. Ph.D. thesis, University of Helsinki, Commentationes Physico-Mathematicae, Vol. 127, The Finnish Society of Sciences and Letters, Helsinki.

Vesala, T., and Kukkonen, J., 1992. A model for binary droplet evaporation and condensation, and its application for ammonia droplets in humid air. *Atmos. Environ.*, 26A.

Vesala, T., Kukkonen, J. and Kulmala, M., 1989. A model for evaporation of freely falling droplets. Finnish Meteorological Institute, Publications on Air Quality 6, Helsinki.

Webber, D.M., Jones, S.J., Tickle, G.A. and Wren, T., 1992. A model of a dispersing dense gas cloud and the computer implementation DRIFT: I. Near-instantaneous releases. AEA Report SRD/HSE R586.

Wheatley, C.J., 1987. Discharge of liquid ammonia to moist atmospheres - Survey of experimental data and model for estimating initial conditions for dispersion calculations. UKAEA Report SRD/HSE/R410, London.

Woodward, J. and Papadourakis, A., 1991. Modeling of droplet entrainment and evaporation in a dispersing jet. International Conference and Workshop on Modelling and Mitigating the Consequences of Accidental Releases of Hazardous Materials, May 6-10, 1991, New Orleans, Louisiana. American Institute of Chemical Engineers, New York

MDGP: A NEW NUMERICAL MODEL FOR DENSE GAS DISPERSION. SENSITIVITY ANALYSIS AND FIRST VALIDATION TRIALS.

Roberto Bellasio[1] and Matteo Tamponi[2]

[1]Private Consultant
 Rho, MI 20017 (Italy)
[2]National Health Service
 PMIP USSL 16
 Lecco, CO (Italy)

INTRODUCTION

Modern industries produce and use different substances that are dangerous for the human health because they can be toxic, flammable or explosive. Some of these substances, if accidentally released in the atmosphere, have a density greater than that of the air. So the result is often the formation of a cloud at ground level that can travel over the populated area near the industrial plant.

The accidents caused by releases of denser than air gases sometimes are very tragic. For example we can remember the Bhopal accident in which a release of methylisocynate killed over 2000 people. There is consequently the necessity to understand and to model the different physical processes involved in this kind of releases.

There are different causes of heavy gas behaviour[1], for example materials can have a molecular weight greater than that of air or a low temperature of release; also they can be stored at particular conditions and chemical reactions can play an important role.

It is common practice to divide the evolution of dense gas releases in four phases, obviously this is only a convenience that can help to model the phenomena. The initial phase is characterized by the release conditions and by the atmospheric flow. In fact, under some particular conditions[2], a potentially denser than air gas release can behaves like a passive release. The gravity slumping phase is governed by the cloud negative buoyancy and by the atmospheric flow, so at the same time there is the collapsing of the cloud and its displacement along the wind direction. Successively there is the density stratification phase, this is a phase of transition from the dominance of the internal buoyancy to the dominance of the ambient turbulence. In all this three stages the cloud is denser than air, some experiments have shown that, in this conditions, the vertical turbulence inside the cloud is attenuated. The final phase is the passive one in which the gas emitted has a density less than or equal to that of air. This stage is controlled by the atmospheric wind and by the ambient turbulence.

Due to the permanence of the cloud near the ground it is important the heat exchange between ground and cloud because it determines a variation in temperature and so a variation in density. For the same reason the heat exchange between cloud and air is also important.

The physical processes involved in dense gas releases are different and complex. A mathematical model that want to describe all these processes will require many calculation resources and large calculation time. In this paper will be shown a three dimensional time dependent numerical model in which some work hypotheses are done in order to reduce the calculation time. Sensitivity analysis and first comparisons with experimental results are also included.

THE MDGP MODEL

All the proposed dense gas models are based on the same set of equations: the equation of continuity, the equation of balance of mass of contaminant, the equation of balance of energy and the equation of balance of momentum. Also is needed one equation of state of gases, for example the ideal gas equation. All these equations are in turbulent form because we are considering an atmospheric flow.

The MDGP model make use of the above mentioned equations but, in order to work with a model that can run on computers of modest dimensions and in relatively short calculation time, some work hypotheses are assumed. Some turbulent terms are neglected respect to others, the remaining are parameterized with the k-theory. MDGP assumes that the turbulent dispersion tensors are diagonal and identical for the dispersion of enthalpy and mass[3]. Also in this first version of the model we suppose to work in flat orography and in absence of obstacles. The flow field is supposed to be two dimensional with the x component imperturbed by the presence of the cloud and the vertical component perturbed by the cloud. The MDGP model solves the following equations:

$$u(z) = f(z) \qquad \qquad (1)$$

$$\rho \, \frac{\partial w}{\partial t} + \rho \, u \, \frac{\partial w}{\partial x} = \alpha \, (\rho_a - \rho) \, g \qquad \qquad (2)$$

$$\frac{\partial C}{\partial t} + \frac{\partial Cu}{\partial x} + \frac{\partial Cw}{\partial z} = \frac{\partial}{\partial x}\left(K_x \frac{\partial C}{\partial x}\right) + \frac{\partial}{\partial y}\left(K_y \frac{\partial C}{\partial y}\right) + \frac{\partial}{\partial z}\left(K_z \frac{\partial C}{\partial z}\right) + S_C \qquad (3)$$

$$\rho\,\frac{\partial h}{\partial t} + \rho\, u \frac{\partial h}{\partial x} + \rho\, w \frac{\partial h}{\partial z} = \frac{\partial}{\partial x}\left(\rho\, K_x \frac{\partial h}{\partial x}\right) + \frac{\partial}{\partial y}\left(\rho\, K_y \frac{\partial h}{\partial y}\right) + \frac{\partial}{\partial z}\left(\rho\, K_z \frac{\partial h}{\partial z}\right) + S_h \qquad (4)$$

$$P = \frac{\rho\, RT}{M} \qquad (5)$$

$$dh = (C_{pa}\,(1-\omega) + C_{pg}\omega)\, dT \qquad (6)$$

$$C = \rho\,\omega \qquad (7)$$

where u and w are the x component and the z component of the wind, ρ is the density of the mixture dense gas-air, h is the specific enthalpy, C is the gas concentration, T is the temperature, C_p is the specific heat at constant pressure, g is the gravity acceleration, R is the universal constant of the gases, M is the molar weight of the mixture, ω is the mass fraction and α is an empirical parameter for the buoyancy effect. The quantity S_c and S_h are respectively the source of mass contaminant and of enthalpy. The K_x, K_y and K_z represent the turbulent dispersion coefficients.

The MDGP model considers also other relations as, for example, the relation that gives the molar weight of the mixture as a function of the molar weight of the air and of the gas released, or the relation that express the specific heat at constant pressure as a function of the temperature[4].

The turbulent dispersion coefficients K are described with a turbulence submodel. The horizontal coefficients are assumed to be equal and proportional to the vertical coefficient[4]. This last coefficient is described by means of an internal Richardson number and it recovers the turbulent dispersion coefficient of the surface layer as the density of the mixture tends to that of the air.

The equations are discretized with the control volume method[5] which guarantees the conservation of the quantities transported. The temporal derivative is discretized with an explicit method.

SENSITIVITY ANALYSIS

Sensitivity analysis is conducted in order to verify the

correct physical answer of the model to the variation of one single input parameter.

In particular the MDGP model was tested with respect to the variation of the geometry of the calculation domain, of the meteorological parameters, of the characteristics of the gas released and of the characteristics of the source.

The dimension of the domain in the reference situation is of 440 m in the downwind direction and 70 m in the upwind direction, 250 m in the crosswind direction and 30 m in the vertical direction. The length of the grids is of 10 m in downwind and crosswind direction and 3 m in vertical direction. The source is supposed a cube at ground level with sides of 9 m, the rate of release is of 2 $Kgm^{-3}s^{-1}$ and the duration of release is of 1 s. The gas released is a mixture of Freon12 and air at the temperature of 280 K resulting in an initial density of about 2.2 Kgm^{-3}. The ground temperature and the ambient temperature are of 293 K, the Monin Obukhov length is of 5 m and the wind is of 4 ms^{-1} at the height of 10 m with a logarithmic profile.

Figure 1 shows the concentration as a function of the time with two different wind velocities of 2 ms^{-1} and 4 ms^{-1}. The positions at which the concentration is calculated are of 55 m and 135 m downwind with y=0 m and z=1.5 m.

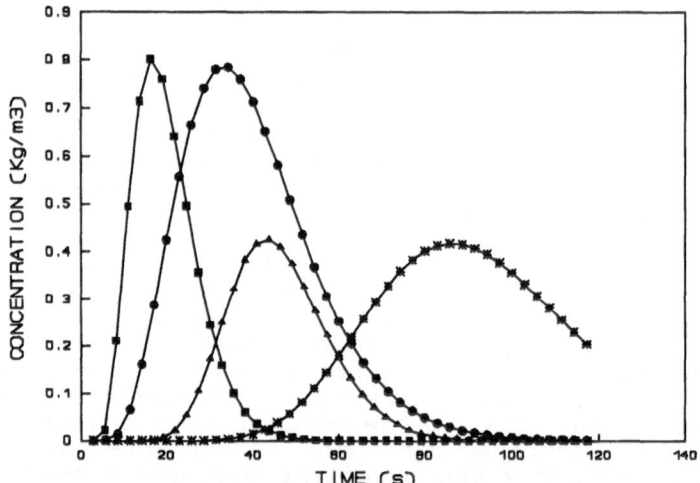

Figure 1. Effects of the wind velocity. Concentration against time at two different downwind distances. ■ x=55 m, v=4 ms^{-1}; ● x=55 m v=2 ms^{-1}; ▲ x=135 m v=4 ms^{-1}; * x=135 m v=2 ms^{-1}. The other coordinates are y=0 m and z=1.5 m.

It is possible to see that at 55 m downwind, with a wind of 4 ms^{-1} the maximum concentration is reached at a time of about 20 s. At the same distance but with a wind of 2 ms^{-1} the maximum concentration is reached at a time of about 40 s. So if the velocity is half the initial velocity, the time at which the maximum concentration is reached is about twice the time at which the maximum concentration is reached with the initial velocity. The same happens at 135 m downwind from the source.

Shown in figure 2 are the temporal evolutions of the concentrations at the downwind distances from the source of 55 m and 135 m. It is possible to note that in the near neutral situation (L=50 m) the concentration is lower than that in a more stable situation (L=5 m). This is due to the fact that in the second case the heavy gas cloud is more compact and the concentration is calculated along the y=0 m axis.

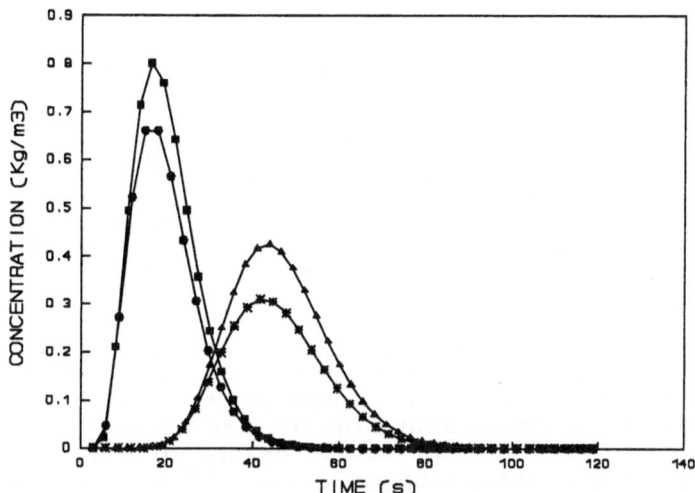

Figure 2. Effects of the Monin Obukhov length. Concentration against time at two different downwind distances. ■ x=55 m, L=5 m; ● x=55 m L=50 m; ▲ x=135 m L=5 m; * x=135 m L=50 m. The other coordinates are y=0 m and z=1.5 m.

Other results shown that in the crosswind direction the situation become the reverse. At some distance from the y=0 m axis the concentration is greater for the near neutral situation than for the stable situation.

It is interesting to note the ability of the model to simulate the release of gases with density variable with the time. It was simulated the continuous release of acetylene, from a source placed at the height of 3 m, at the temperature of about 189 K. At this temperature the density of the gas is about 1.7 Kgm^{-3} while at the ambient temperature of 293 K the density is about 1.1 Kgm^{-3}. Figure 3 shows the concentration as a function of the downwind distance from the source at two different temporal instants from the release. At 10 s from the release the concentration at the height of 1.5 m is higher than that at 4.5 m only before about 50 m from the source. The contact with the ground and the entrainment of air cause an increase in temperature and so a decrease in density. At the time of 30 s the concentration at 1.5 m of height is higher than that at 4.5 m before 170 m from the source.

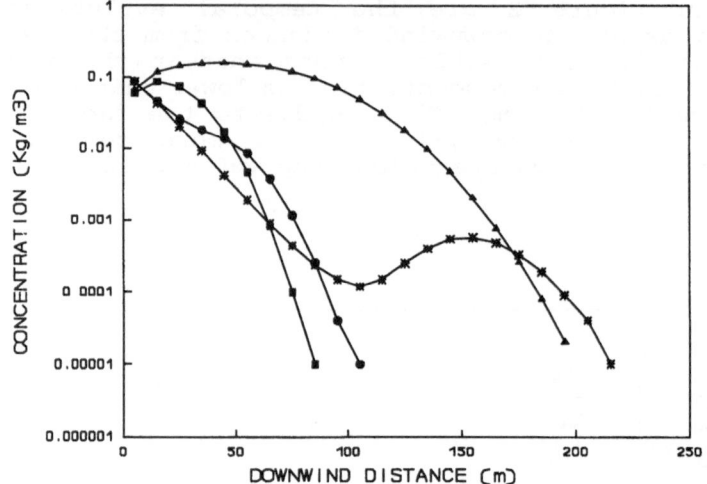

Figure 3. Release of acetylene at the temperature of 189 K, the source is posed at z=3 m, the duration of release is of 30 s. Concentration as a function of the downwind distance at the times of 10 s and 30 s. ■ z=1.5 m, t=10 s; ● z=4.5 m t=10 s; ▲ z=1.5 m t=30 s; * z=4.5 m t=30 s.

At about 100 m downwind there is a minimum in the concentration at z=4.5 m, this is due to the initial negative buoyancy of the acetylene. As a consequence of the heat exchange with the ground the temperature of the gas increase, so it gains a positive buoyancy, returns at the height of 4.5 m where there is a maximum in concentration at about 150 m, and goes upward. This example shows the ability of the MDGP model to simulate the release of gases with variable buoyancy.

Shown in figure 4 are the concentrations at two different heights, symmetrical respect the vertical position of the source, for a passive gas and for a dense gas. For the passive gas there is the same concentration in z=1.5 m and in z=7.5 m, the only difference is in the downwind distance at which the maximum concentration is reached, this difference is due to the logarithmic profile of the wind. For the dense gas the situation is very different, due to the negative buoyancy the concentration at ground level is higher than that at z=7.5 m.

FIRST COMPARISON WITH EXPERIMENTAL RESULTS

The MDGP model is a young model, so a satisfactory validation remains to do.

However a first comparison with experimental data was done with the Thorney Island experiments, in particular trials 8 and 18 were used.

The experimental maximum concentration and the time at which it was reached at fixed positions was compared with the calculated results. Table 1 shows a satisfactory agreement between the calculated and experimental results.

It is possible to note that the calculated time is always lower than the observed time. This is due to the hypothesis that the wind is not influenced by the dense gas cloud in the horizontal direction.

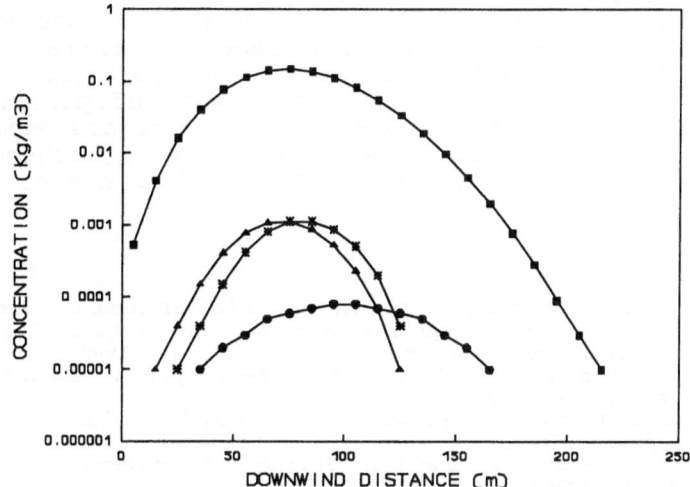

Figure 4. Release of a passive gas PG (Nitrogen) and of a dense gas DG (Freon12-air), the source is posed at z=3 m and has a vertical extension of 3 m, the duration of release is of 10 s. Concentration as a function of the downwind distance at the time 30 s. ■ z=1.5 m DG; ● z=7.5 m DG; ▲ z=1.5 m PG; * z=7.5 m PG.

Table 1. First comparisons with experimental results. The letter C indicates calculated results, the letter O indicates observed results.

Position	$Cmax_c$%	$Cmax_o$%	$t_c(s)$	$t_o(s)$
Trial 8				
1	0.08	0.2	73.80	90
2	0.86	0.7	184.97	220
3	1.28	0.9	140.13	240
4	1.21	0.96	161.62	250
5	0.13	0.7	102.76	140
Trial 18				
6	4.73	7.4	7.59	10
7	0.31	0.14	82.77	100
8	1.59	0.5	48.60	70
9	0.39	1.1	35.69	50
10	0.37	0.28	65.30	70

CONCLUSIONS

This paper contains a three dimensional time dependent numerical model to describe the heavy gas dispersion in the atmosphere. This model describes in a simple manner the flow field, so it doesn't require long calculation time. It is implemented with the control volume method which assure the mass conservation. Sensitivity analysis show the correct physical answers of the model and its ability to simulate the release of gases with variable density or the release of passive gases. Preliminary comparisons with experimental data are encouraging, they have also shown that one of the work hypotheses is to reformulate. A good validation of the model is needed.

REFERENCES

1. R.P. Koopman, D.L. Ermak and S.T. Chan "A review of recent field tests and mathematical modelling of atmospheric dispersion of large spills of denser than air gases." Atmospheric Environment (23) 731-745 (1989).
2. R.E. Britter "Atmospheric dispersion of dense gases." Annual review of fluid mechanics, 317-344 (1989).
3. S.T. Chan, D.L. Ermak and L.K. Morris "FEM3 model simulations of selected Thorney Island phase I trials." J. Hazardous Materials (16) 267-292 (1987).
4. R.C. Reid, J.M. Prausnitz and B.E. Poling "The properties of gases and liquids." 4^ed., McGraw Hill, New York (1987)
5. S.V. Patankar "Numerical heat transfer and heat flow." Mc Graw Hill (1980).

EVALUATION OF THE ATMOSPHERIC RELEASE ADVISORY CAPABILITY EMERGENCY RESPONSE MODEL FOR EXPLOSIVE SOURCES

Ronald L. Baskett, Robert P. Freis, and John S. Nasstrom

EG&G Energy Measurements, Inc.
P. O. Box 8051
Pleasanton, CA 94588 USA

INTRODUCTION

The Atmospheric Release Advisory Capability (ARAC) at the Lawrence Livermore National Laboratory (LLNL) uses a modeling system to calculate the impact of accidental radiological or toxic releases to the atmosphere anywhere in the world (Sullivan et al., 1993). Operated for the U. S. Departments of Energy and Defense, ARAC has responded to over 60 incidents in the past 18 years, and conducts over 100 exercises each year. Explosions are one of the most common mechanisms by which toxic particulates are injected into the atmosphere during accidents. Automated algorithms with default assumptions have been developed to estimate the source geometry and the amount of toxic material aerosolized. The paper examines the sensitivity of ARAC's dispersion model to the range of input values for explosive sources, and analyzes the model's accuracy using two field measurement programs.

MATHEW/ADPIC MODELING SYSTEM

ARAC's emergency response computer system is built around the MATHEW (mass-consistent three-dimensional wind field) and ADPIC (atmospheric dispersion particle-in-cell) diagnostic models (Sherman, 1978, Lange, 1978). These models are run on a three-dimensional Eulerian grid typically with 40 x 40 x 14 uniform rectangular cells scaled to encompass the desired domain. Wind speed and direction from up to 50 surface stations and 15 upper-air profiles within and surrounding the domain are used to initialize MATHEW. Surface data are first interpolated using inverse-distance-squared weighting of the input data. MATHEW then minimizes the divergence in the initialized field based on the continuity equation. Vertical motions are created and horizontal motions are adjusted according to the effects of atmospheric stability and terrain.

ADPIC simulates atmospheric releases by partitioning the mass or radioactivity of the source material into thousands of Lagrangian "marker" particles. A diffusivity velocity is determined for each cell from the concentration gradient based on K-theory and similarity scaling relationships for the boundary layer. Assuming incompressibility, ADPIC then computes three-dimensional "pseudo-velocities" from the sum of the advection and diffusivity velocities at each cell corner. A particle's total vertical velocity is the sum of its pseudo-velocity interpolated to its position in the cell, its gravitational settling velocity, and its dry deposition velocity. Deposition velocities are applied in the lowest model layer (surface layer). For each time step, the Lagrangian marker particles are moved in three dimensions on the Eulerian grid according to their total local vertical velocities and their horizontal pseudo-velocities. Four inner nested grids, each with half the previous grid dimension, provide detailed resolution near the source.

EXPLOSIVE SOURCE SIMULATION

Aerosolized Fraction and Particle Size

If surrounded by explosive charge, the toxic source material is assumed to be completely aerosolized. State changes or chemical reactions in the high-temperature, high-pressure explosive environment are not modeled. Instead, we begin with a stabilized particle size distribution (PSD) of the source material after the condensation of detonation products. PSDs depend on the chemical and physical structure of the source material and can vary dramatically. Each source is parameterized with single or multiple log-normal PSDs. For heavy elements, the median aerodynamic diameter (MAD) is typically 20-40 µm, while the MAD for lighter elements is usually 1 µm or less. As a rule of thumb when calculating the inhalation of the toxic material, 20% of heavy metal or transuranic element source mass and 90 to 100% of light metal or powder sources are in the respirable size range (<10 µm MAD).

Static-Source Geometry

Two methods are used to initialize the size and shape of the explosive cloud in ADPIC. The simplest of these is an empirical relationship fit to data from measured high-explosive (HE) detonations. Figure 1 illustrates the mushroom-shaped geometry characteristic of a cloud that stabilized a few minutes after a ground-level detonation. Particulate clouds were tracked by flash photography and theodolites for 22 evening and nighttime explosions during the May to June, 1963 Project Roller Coaster at the Tonopah Test Range, Nevada (Church, 1969). The measurement technique was considered accurate to within 20%. The cloud top heights began to stabilize about 2 to 3 min after detonation during stable lapse rate conditions but did not stabilize within the 5-min tracking period when the temperature soundings were mostly adiabatic. Since cloud heights were only available at 2 min after detonation for all 22 experiments, only measurements at 2 min were used to establish the relationship to explosive amount. Based on a theoretical 1/4 power-law form, regression analysis produced the following equation using TNT-equivalent masses ranging from 54 to 1020 kg:

$$H_t = 92.6 \ M_{TNT}^{1/4} \tag{1}$$

where H_t = cloud top height (m) at 2 min after detonation, and
M_{TNT} = mass (kg) of TNT-equivalent explosive.

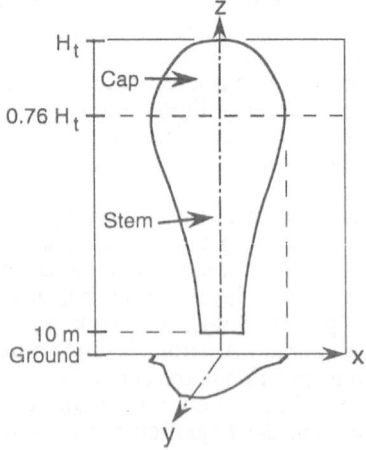

Figure 1. Static-source geometry for explosive cloud.

The three-dimensional static-cloud geometry is specified by assuming symmetry and employing truncated Gaussian particle cross-section distributions on the vertical and horizontal axes. Two truncated half-Gaussians, one for the top or cap cloud and another for the bottom or stem portion, are used to initialize the distribution of particles. Based on visual observations at 2 min after detonation, a 10-m "lift-off" height is applied to the stem.

The fraction of source mass in the stem and cap parts of the initial cloud is also based on limited Roller Coaster vertical particle measurements. For transuranic elements with a median aerodynamic diameter of 40 µm, 76% of the mass was in the stem and 24% in the cap cloud. For lighter elements or powders, the majority of the mass is in the cap cloud 2 min after detonation. While determining the source distribution values for the static-source geometry may be somewhat subjective, an advantage of this method is that it uses a minimal amount of computer time to initialize the source particle distribution in ADPIC.

Time-Dependent Explosive Cloud Rise

A more sophisticated method of simulating explosive cloud rise in a numerical dispersion model involves calculating the time-dependent evolution of both physical and thermodynamic properties of the high-explosive detonation. An explosive cloud rise computer code developed by Boughton and DeLaurentis (1987) was implemented in ADPIC. Table 1 summarizes the submodel inputs, physics and outputs. Boughton and DeLaurentis integrated the three-dimensional conservation equations of mass, momentum and energy over the explosive cloud cross section. Included are entrainment and the crossflow-induced pressure drag on the cloud via empirical relationships. The assumption of spherical symmetry reduces the integral equations to a set of ordinary differential equations that produce the radius, centerline height, and velocity, temperature and density of the cloud each time step.

Table 1. Summary of time-dependent explosive cloud rise submodel.

INPUTS	PHYSICS	OUTPUTS
Source amount	Conservation Equations: 1. Mass/water vapor	For the cloud:
Source PSD	2. Momentum 3. Thermal energy	1. Radius
Sounding (temperature, wind)	Assumptions:	2. Height
Relative humidity	1. Cloud symmetry 2. Uniform cloud	3. Vertical velocity
Surface roughness	cross-section	
Surface pressure	Parameterizations:	4. Temperature
Particle-cloud coupling coefficient	1. Entrainment 2. Pressure drag 3. Particle-cloud coupling	5. Density

The explosive cloud is a sphere of hot gas that rises, expands and cools. Initially the sphere contains dry air and water, either as vapor or liquid. For a surface detonation, the radius of the initial sphere of ambient air is equal to the distance at which the blast overpressure has decayed to one atmosphere. Particles are be distributed uniformly within the initial sphere with an initial radius, R_0 (m), based on the empirical relationship:

$$R_o = 3.2 \, M_{TNT}^{1/3} \qquad\qquad (2)$$

Figure 2 shows the velocity components affecting the cloud and particles. Buoyancy causes the cloud to simultaneously expand radially and rise vertically. However, the cap cloud rises vertically within the first few seconds at a faster rate than predicted by buoyancy alone. To account for the initial "bounce" after surface detonations, an initial cloud liftoff height was added as an input parameter to the model.

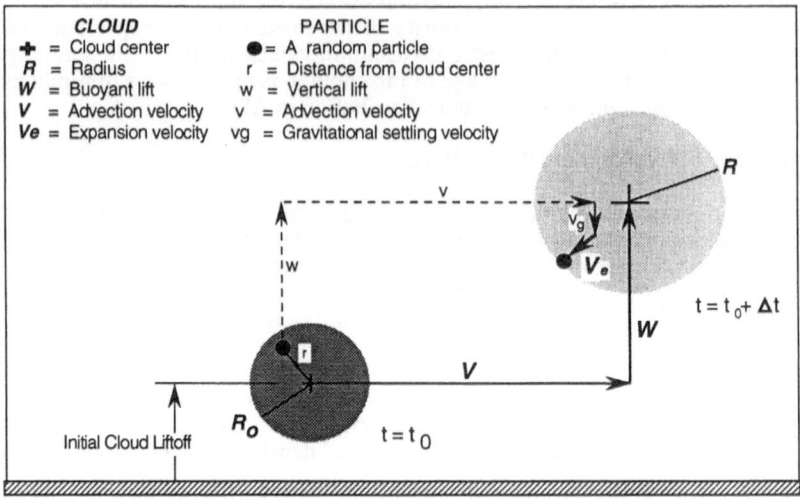

Figure 2. Time-dependent explosive cloud rise submodel.

Particles in the explosive cloud are treated differently from those outside the cloud in two ways. First, particles in the cloud experience the radial expansion velocity of the cloud. Second, each cloud particle experiences a fraction of the cloud's vertical velocity due to its buoyant lift. Boughton and DeLaurentis (1987) parameterized the fraction of the cloud's vertical velocity that is transferred to each particle by the following relationship:

$$w = We^{-c\left(\frac{r}{R}\right)^2}$$

(3)

where w = the particle's vertical velocity (m/sec) due to the cloud's buoyancy,
W = the cloud's vertical buoyant velocity (m/sec),
c = the particle-cloud velocity coupling coefficient,
r = the particle's distance (m) from the cloud center, and
R = the cloud's radius (m).

Expression (3) does not simulate the details of the toroidal evolution inside the cap cloud or entrainment of air around the cloud's edges. Instead, it provides a simple mechanism to approximate the overall size of the cloud and remove particles from its lower side. Particles at the center of the sphere can be lifted as much as the sphere, while those near the edge only receive a fraction of the cloud's vertical velocity. Each particle's vertical velocity depends on its position in the sphere and the coupling coefficient, a non-dimensional empirically-derived model input parameter.

Figure 3 illustrates the difference between strong coupling ($c = 0.1$) and weak coupling ($c = 1.0$) for large particles (100 μm MAD) one min after a 23-kg detonation. The explosive cloud rises over sloping terrain under neutral, 11 m/sec winds moving from left to right. With strong coupling over half the particles are still within the confines of the spherical explosive puff. Weak coupling results in most of the particles leaving the explosive puff and many depositing on the ground.

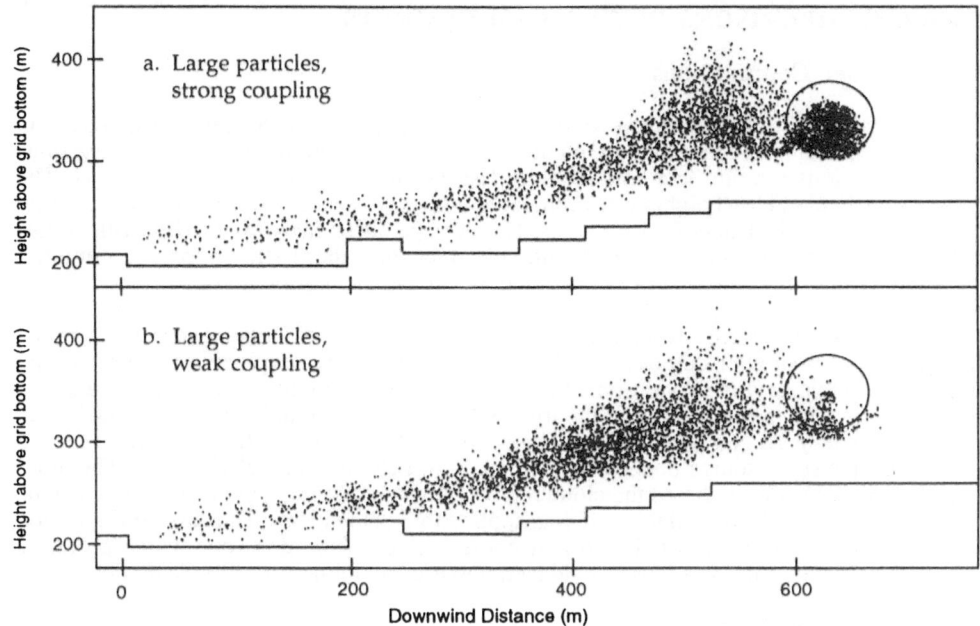

Figure 3. Effect of coupling coefficient on particle distribution one mintue after detonation.

MODEL SENSITIVITY

Because model inputs are not always well known in an accident, it is useful to know how sensitive the model results are to the inputs. Baskett and Cederwall (1991) tested the sensitivity of the static and time-dependent puff source methods to typical ranges of explosive mass, particle size and atmospheric stability. An example result is shown in Figure 4. For the typical range in explosive mass, the time-dependent cloud gives up to 10 times higher centerline concentrations near the source. This is because the time-dependent model places more particles closer to the ground at small times compared to the vertical distributed particles in the static source. Beyond 2.5 km, the time-dependent cloud produces about 35% less air concentration than the fixed cloud. This difference is caused by greater depletion and deposition of particles near the source in the time-dependent method.

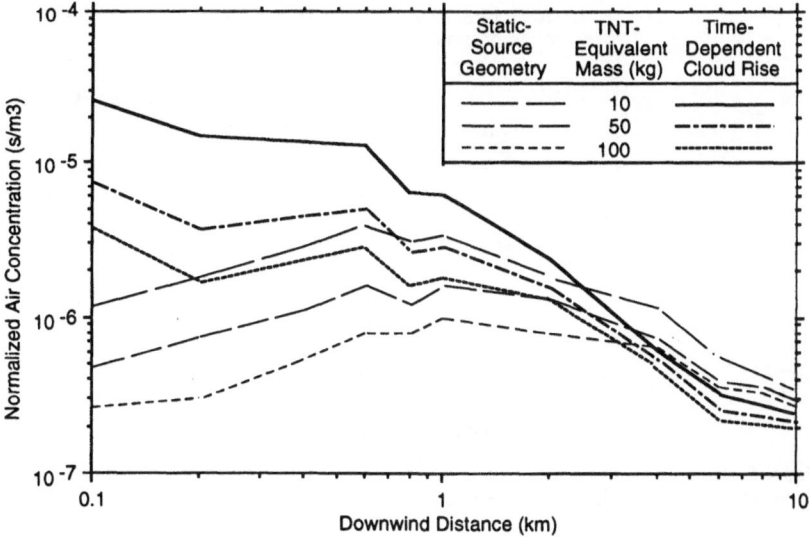

Figure 4. Sensitivity of static-source and time-dependent cloud rise to typical ranges of explosive mass.

MODEL COMPARISONS WITH MEASUREMENTS

Source Model Comparisons

Both static and time-dependent sources were tested against the Roller Coaster data. On the average, the static source produced cloud heights at 2 min within 20% of the observations for 22 detonations ranging from 54-1020 kg HE during stable nighttime conditions. The time-dependent cloud top heights were compared with instantaneous observations 1 to 5 min after detonation for 15 Roller Coaster cases. The average difference of only 10% with almost no bias indicated a better agreement with the data than the static source. A coupling coefficient of 0.35 gave the best fit between the model and the observations.

The cloud top heights from the two source methods were also compared against observations of 6 smaller detonations (6-31 kg HE) at LLNL's explosive test site during neutral to unstable daytime conditions. Figure 5 shows an example comparison for a 23-kg HE shot during 11 m/sec winds. Comparisons between photographs and the model were made about every 10 sec for the first 2 min after detonation. On the average for the 6 detonations, the static source underestimated the 2-min cloud top height by 12%. The time-dependent model underpredicted the cloud top by 8% on the average, but nearly matched the observations when a 20-m initial liftoff was applied to 3 detonations. The shot configuration determined if there was an initial liftoff. A conservative assumption (one which produces higher ground-level concentrations) would be not to use any initial liftoff in the model.

Dispersion Model Comparisons

A limited number of air concentration and deposition measurements from the Roller Coaster experiments were also compared with the output from ADPIC. Figure 6 compares the static and dynamic source simulations with air concentration measurements of a 482-kg TNT experiment. The puff traveled along a relatively flat valley floor for the first 10 km, and then it interacted with a ridge. The test occurred during 3- to 5-m/sec stable nightime conditions. Measurements of 40-μm-median diameter transuranic particles were made downwind. The plot shows a comparison along the computed plume centerline. Using the static source, ADPIC underestimates the centerline concentration for the first kilometer, does well for a few kilometers, but overestimates the measurement past 5 km. Overall, the model better matches the observed data better using explosive cloud rise with c = 0.35.

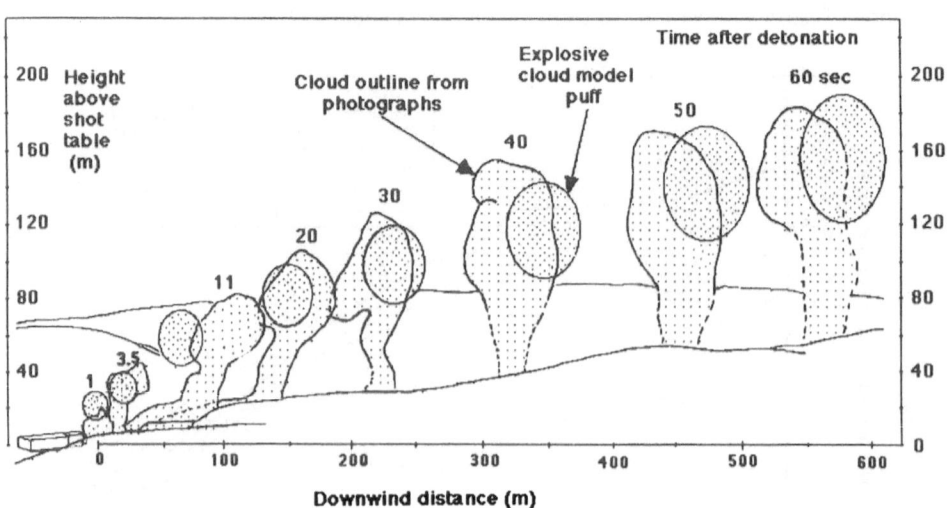

Figure 5. Comparison between cloud photography and time-dependent puff locations for a 23-kg detonation at LLNL's explosive test site.

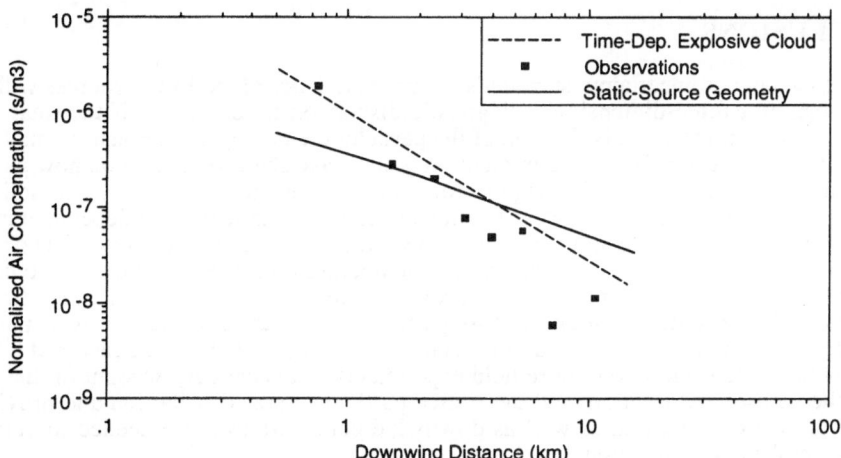

Figure 6. Comparison of downwind centerline particulate air concentrations with observations from a 482-kg detonation during Project Roller Coaster.

Figure 7 compares the observed deposition from the 23-kg detonation containing light powder particles at LLNL's explosive test site. With 5% of the source material as 100-µm particles, a coupling coefficient of 1.0 produced values that were within a factor of 2 for the first 150 m. Measurements further downwind distances were not available.

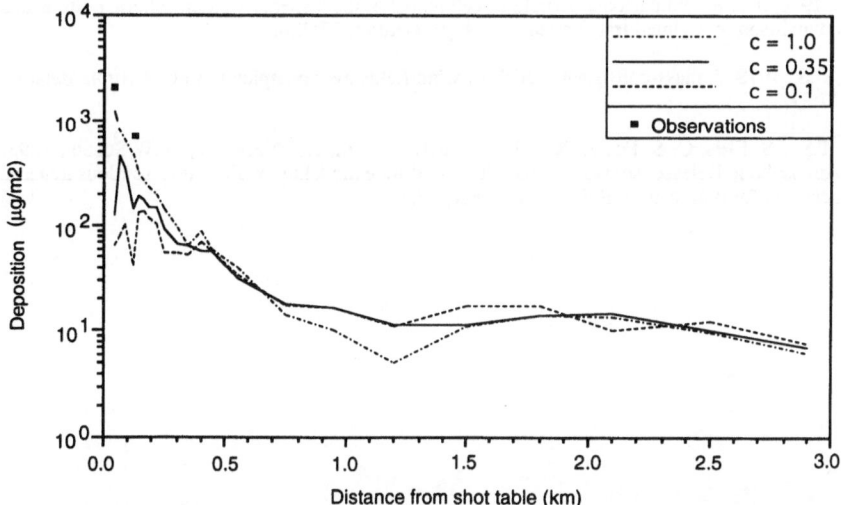

Figure 7. Comparison of downwind centerline deposition from the time-dependent cloud rise model with particulate observations from a 23-kg detonation at LLNL's explosive test site.

CONCLUSIONS

A great deal of uncertainty surrounds the characteristics of explosive sources within the framework of a three-dimensional diagnostic dispersion model. Even if the total source amount is known, the size distribution of the particles created by the detonation may not be well understood. A few field measurement programs have shed some light on how to reduce uncertainties when determining inputs for modeling the toxic material aerosolized in explosions. The cloud geometry can be simulated either by static or time-dependent models. The explosive cloud can be simulated by both source models to within 20% of its observed height if the explosive mass and the ambient temperature profile are known. The detailed time-dependent explosive puff model provides greater accuracy within the first few kilometers. If the source amount and its particle size distribution are known, the time-dependent cloud rise simulation can reproduce concentrations to within a factor of 2 for the first few kilometers. However, more field experiments that accurately account for the source mass, the aerosolized fraction, and the source particulate size distribution and provide the time evolution of the cloud as well as downwind concentrations are needed to verify the accuracy of the dispersion model.

This work was performed under the auspices of the U.S. Department of Energy at Lawrence Livermore National Laboratory under contract number W-7405-Eng-48 and EG&G contract number DE-AC08-93NV11265. By acceptance of this article, the publisher and/or recipient acknowledges the U.S. Government's right to retain a nonexclusive, royalty-free license to any copyright covering this paper.

REFERENCES

Baskett, R.L. and R.T. Cederwall, 1991, Sensitivity of numerical dispersion modeling to explosive source parameters, Air & Waste Management Association 84th Annual Meeting, Vancouver, BC. Paper 91-85.6.

Boughton, B.A. and J.M. DeLaurentis, 1987, An integral model of plume rise from high explosive detonations, 1987 ASME/AIChE National Heat Transfer Conference, Pittsburgh.

Church, H.W., 1969, Cloud rise from high-explosive detonations, SC-RR-68-903, Sandia National Laboratory, Health and Safety Report UC-41, Albuquerque.

Lange, R., 1978, A three-dimensional particle-in-cell model of the dispersal of atmospheric pollutants and its comparison to regional tracer studies, *J. Appl. Meteor.*, 17:320.

Sherman, C.S., 1978, A mass-consistent model for wind fields over complex terrain, *J. Appl. Meteor.*, 17:312.

Sullivan, T.S, J.S. Ellis, C. S. Foster, K.T. Foster, R.L. Baskett, J.S. Nasstrom, W.W.Schalk, 1993, Atmospheric Release Advisory Capability: Real-time modeling of airborne hazardous materials, accepted for publication, *Bull. Amer. Meteor. Soc.*

MODEL ASSESSMENT AND VERIFICATION

chairmen: R.D. Bornstein
S.E. Gryning

rapporteurs: K. Nodop
E. Batchvarova

IMPROVING THE SCIENCE OF REGULATORY DISPERSION MODELS

FOR SHORT-RANGE APPLICATIONS

J.C. Weil[1]

National Center for Atmospheric Research [2]
P.O. Box 3000
Boulder, Colorado 80307, USA

INTRODUCTION

Atmospheric dispersion models used in assessing environmental impact and regulatory issues are often labeled "regulatory." Such models are typically of a simple form due to their repeated use and the fast turnaround required. In the United States for example, models often are run for a year of hourly meteorological inputs to determine the highest or "worst-case" concentration for a source. Despite the simplicity requirement, however, regulatory models *must* be based on sound physical principles to generate confidence in their results.

The most common regulatory approach is the Gaussian plume model with lateral (σ_y) and vertical (σ_z) dispersion parameters given by the Pasquill-Gifford (PG) curves (Gifford, 1961). The dispersion curve or "class" is estimated from Turner's (1964) criteria, which include near-surface winds, cloud cover, ceiling height, etc. The limitations of this combined approach—the Pasquill-Gifford-Turner (PGT) technique—are well-known (Weil, 1985). For example, the PG curves are based on tracer releases from a ground-level source and on concentration measurements out to only ~ 800 m from the source. They do not apply to an elevated plume which can have quite different dispersion characteristics (Lamb, 1982; Hunt, 1982).

In most applications, one is concerned with dispersion in the planetary boundary layer (PBL), the turbulent air layer next to the earth's surface that is controlled by the surface heat and momentum fluxes. The PBL typically ranges from a few 100 m in depth at night to 1 - 2 km during the day. Major developments in understanding the PBL began in the 1970s through numerical modeling, field observations, and laboratory simulations (Wyngaard, 1988). For the convective boundary layer (CBL),

---

[1] Permanent Affiliation: Cooperative Institute for Research in Environmental Sciences, University of Colorado, Boulder, Colorado 80309, USA

[2] The National Center for Atmospheric Research is sponsored by the National Science Foundation.

Air Pollution Modeling and Its Application X, Edited by S-V. Gryning
and M.M. Millán, Plenum Press, New York, 1994

a milestone was Deardorff's (1972) numerical simulations which revealed the CBL's vertical structure and the important turbulence scales. Major insights into dispersion followed from laboratory experiments (Willis and Deardorff, 1976, 1978), numerical simulations (Lamb, 1978), and later field observations (Eberhard et al., 1988). For the stable boundary layer (SBL), advancements occurred more slowly. Dispersion near the surface was put into a solid framework by theoretical and experimental work (e.g., Horst, 1979; van Ulden, 1978). However, dispersion in the upper part of the SBL, while having a theoretical base (Hunt, 1985; Venkatram et al., 1984), was more variable than its CBL counterpart. Nevertheless, for both boundary layers, our understanding of turbulence and dispersion was fairly mature by the early-to-mid 1980s.

During the early 1980s, researchers began to apply this information to simple dispersion models for applications. This consisted of eddy-diffusion techniques for surface releases, statistical theory and PBL scaling for estimating σ_y and σ_z, a new probability density function (p.d.f.) approach for dispersion in the CBL, etc. Much of this work was reviewed and promoted in workshops (Weil, 1985), revised texts (Pasquill and Smith, 1983), and in short courses and monographs (Nieuwstadt and van Dop, 1982; Venkatram and Wyngaard, 1988). Thus, a substantial scientific base existed in the early-to-mid 1980s for overhauling regulatory dispersion models. However, this did not occur. One exception was the Danish OML model (Berkowicz et al., 1986).

In the past three years, however, government groups have supported the development of scientifically-improved regulatory models for short-range dispersion. Three examples can be cited: 1) the United Kingdom Advanced Dispersion Modeling System (UK - ADMS) (Carruthers et al., 1992), 2) a series of workshops within the Commission of European Communities (CEC) to promote improved dispersion models (Olesen and Mikkelsen, 1992), and 3) the establishment of the AMS/EPA Regulatory Model Improvement Committee (AERMIC) by the American Meteorological Society (AMS) and the US Environmental Protection Agency (EPA). AERMIC's objective is to include state-of-the-art science in regulatory models (Weil, 1992).

This paper briefly reviews our understanding of the PBL and dispersion, the developments in applied dispersion models, and the use of this information in building better regulatory models with focus on the AERMIC activity. Attention is on point-source plumes, downwind distances less than about 30 km, and concentration averaging times of ~ 1 hr.

OVERVIEW OF THE PBL AND DISPERSION

Convective Boundary Layer

CBL Properties. The vertical structure of the CBL is controlled primarily by the surface heating and the large-scale eddies that extend from the ground to the capping inversion. Deardorff's (1972) simulations showed that for the bulk of the CBL, the turbulent length scale was h, the CBL depth, and the velocity scale was w_*, the convective velocity scale. $w_* = (g\overline{w\theta_o}h/\Theta_o)^{1/3}$, where g is the gravitational acceleration, $\overline{w\theta_o}$ is the surface kinematic heat flux, and Θ_o is the mean potential temperature. Turbulence quantities such as the vertical and lateral velocity variances, σ_w^2 and σ_v^2, when scaled by w_*^2 could be expressed as simple functions of z/h, where z is the height above ground; see Kaimal et al. (1976), Hicks (1985), and Deardorff and Willis (1985).

Mechanical turbulence due to flow over the rough surface scales with the friction velocity u_* and is confined mostly to shallow heights, $-z < L$, where L is the Monin-Obukhov length; $|L|$ is typically 10 to 100 m. The "stability" parameter characterizing the importance of convective and mechanical turbulence throughout the CBL is $-h/L$, with useful alternatives being u_*/w_* or w_*/U, where U is the mean wind speed in the CBL (Weil, 1988). One of these parameters replaces the PG class during daytime.

In the "mixed layer" ($0.1h \lesssim z \lesssim h$), the mean wind, σ_v, and σ_w vary little with z; in dispersion applications, they are often taken as constant in this layer with $\sigma_v, \sigma_w \simeq 0.6w_*$. For dispersion, an important feature for elevated plumes is the non-Gaussian p.d.f. of vertical velocity w; the p.d.f. is positively skewed with a skewness that averages about 0.6 across the CBL (Wyngaard, 1988). This is consistent with downdrafts occupying a greater fraction (0.6) of the horizontal surface area than updrafts (Lamb, 1982).

Dispersion Characteristics. The important aspects of dispersion in the CBL were first demonstrated through laboratory experiments (Willis and Deardorff, 1976, 1978, 1981) and numerical simulations (Lamb, 1978, 1982). The investigations showed the importance of source height z_s on the dispersion patterns and the utility of the convective scales, w_* and h, in organizing the data. Figure 1 shows contours of the crosswind-integrated concentration (CWIC) C^y in an x - z plane from the laboratory experiments, where x is the downwind distance. The C^y is made dimensionless using Uh/Q, where Q is the source strength. The dimensionless CWIC is shown as a function of z/h and the dimensionless distance X,

$$X = \frac{w_* x}{Uh} ,\tag{1}$$

which is the ratio of travel time x/U to the eddy-turnover time h/w_*.

Figure 1 shows that for the near-surface source, the average plume centerline as defined by the locus of maximum $C^y(X)$, ascends after a short distance downwind whereas the centerlines from the elevated sources descend until they reach the ground. The elevated plume descent is explained by the greater areal coverage by downdrafts, and hence, the higher probability of material being released into them. The surface plume ascent results from the sweep out of material into convergence zones near the surface before material carried aloft recirculates down (Lamb, 1982); from the viewpoint of a Lagrangian statistical model, this is caused by the vertical inhomogeneity of σ_w and the long turbulence time scale (Weil, 1990). The behavior in Fig. 1 has been confirmed by field measurements of tracers released from near-surface and elevated sources on a 300-m tower (Briggs, 1988; Eberhard et al., 1988). The behavior differs from the Gaussian plume model which predicts an elevated plume centerline to remain horizontal for a long distance downwind. The difference is caused by the skewness of the w p.d.f., which also leads to C^y values near the surface exceeding those predicted by the Gaussian plume model (see Lamb, 1982; Weil, 1988).

Numerical simulations and laboratory experiments also have given information on the behavior of σ_z and σ_y. Figure 2 shows Lamb's (1982) numerical results for σ_z/h as a function of X with a division into elevated ($z_s/h \gtrsim 0.1$) and surface layer ($z_s/h \lesssim 0.1$) sources. The elevated source results collapse to a nearly single curve, $\sigma_z/h = 0.5X$, for $X \lesssim 0.7$. This is consistent with the short-time limit of statistical theory, $\sigma_z = \sigma_w x/U$ (Taylor, 1921) assuming $\sigma_w = 0.5w_*$, and is explained by the

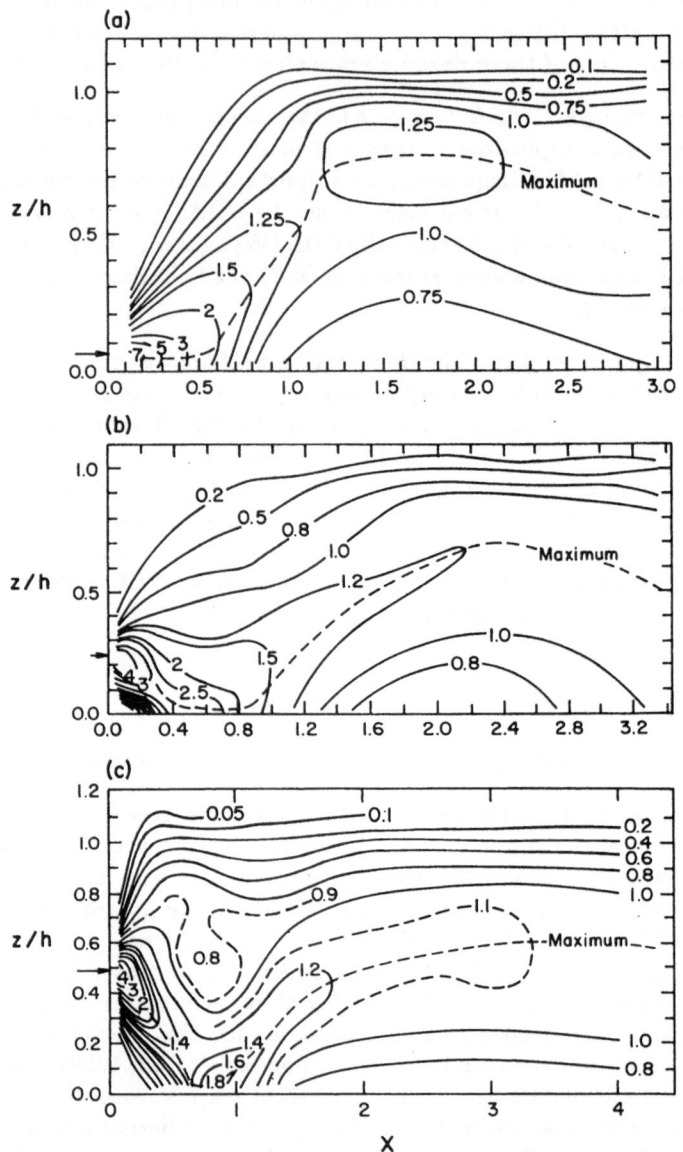

Figure 1. Laboratory convection tank results showing contours of dimensionless crosswind-integrated concentration as a function of dimensionless height and downwind distance for sources at three release heights in a convective boundary layer. Horizontal arrow denotes release height. (a) $z_s/h = 0.067$ from Willis and Deardorff (1976); (b) $z_s/h = 0.24$ from Willis and Deardorff (1978); (c) $z_s/h = 0.49$ from Willis and Deardorff (1981).

quasi-homogeneity of σ_w in the upper part of the CBL and the long turbulence time scale, $\sim h/w_*$. For the lowest source, σ_z/h initially increases more rapidly than X (i.e., $X^{6/5}$) in qualitative agreement with Yaglom's (1972) similarity theory, which predicts $\sigma_z/h \propto X^{3/2}$. For $X \gtrsim 1$, the σ_z/h tends to a constant, dependent on z_s, due to plume trapping below $z = h$. The results in Fig. 2 have been supported by field measurements—Nieuwstadt (1980) for a surface release and Eberhard et al. (1988) and Briggs (1988) for surface and elevated sources. For σ_y, a useful expression that matches the short- and long-time limits of statistical theory is

$$\frac{\sigma_y}{h} = \frac{0.6X}{(1 + \alpha X)^{1/2}} \qquad (2)$$

(e.g., Weil, 1985), where $\alpha = 0.5h/T_{Ly}$ and T_{Ly} is a Lagrangian time scale. A typical value of α is 0.7 (Weil, 1988) while an upper bound is 2 (Briggs, 1988).

For a surface source, the diffusion equation with an appropriate eddy diffusivity $K(z)$ can describe the concentration field near the source. This occurs because the characteristic eddy size ℓ is small and grows with z near the surface, i.e., $\ell \propto z$. van Ulden (1978) adopted Roberts' analytical solution to the diffusion equation:

$$C^y(x,z) = \frac{AQ}{\overline{u}\,\overline{z}}\exp\left[-\left(\frac{B\overline{z}}{\overline{z}}\right)^s\right], \qquad (3)$$

where A and B are constants depending on the exponent or shape factor s ($1 \lesssim s \lesssim 2$), $\overline{u}$ is the mean wind speed at $\overline{z}$, and $\overline{z}(x)$ is the mean plume height at x. van Ulden found analytical expressions for $x(\overline{z})$ and $\overline{u}(\overline{z})$ as a function of u_*, L, and the surface roughness length z_o. The predicted C^y along the surface agreed well with measurements from the Prairie Grass experiments.

Recently, Venkatram (1992) derived the limiting asymptotic behavior for σ_z and $C^y(x,0)$. He combined the long-time limit of statistical theory in the form $d\sigma_z^2/dt = 2K(t)$ with K given by the semi-empirical result for heat, $K = K_h(z/L)$, and K_h given by Businger et al. (1971). Evaluating K_h at $z = \sigma_z$, he obtained

$$\sigma_z \propto \frac{x^2}{|L|} \qquad \text{and} \qquad C^y(x,0) \propto \frac{Q|L|}{u_* x^2}, \qquad (4)$$

where it is implicit that $\overline{z} \propto \sigma_z$. Figure 3 shows van Ulden's results (solid lines) for $\overline{z}/z_o$ and $C^y u_* z_o/Q$ as a function of $(x + x_o)/z_o$, where x_o is the distance to a virtual source for an elevated release in the surface layer; results for convective conditions are given by negative values of z_o/L. As can be seen, the $\overline{z}$ and C^y approach an x^2 and x^{-2} dependence, respectively, in the limit of large instability ($z_o/L = -10^{-2}$) and large distances (see dashed lines).

Stable Boundary Layer

SBL Properties. Turbulence in the SBL is generated by wind shear and destroyed by the stable buoyancy forces and viscous dissipation. It is much weaker than its CBL counterpart and is characterized by small velocity fluctuations and length scales and is vertically inhomogeneous. The turbulence velocity scale is u_* which is typically of order 0.1 m/s in stable conditions. The SBL height typically ranges from 10s of meters to a few hundred meters. The characteristic eddies are proportional to z near

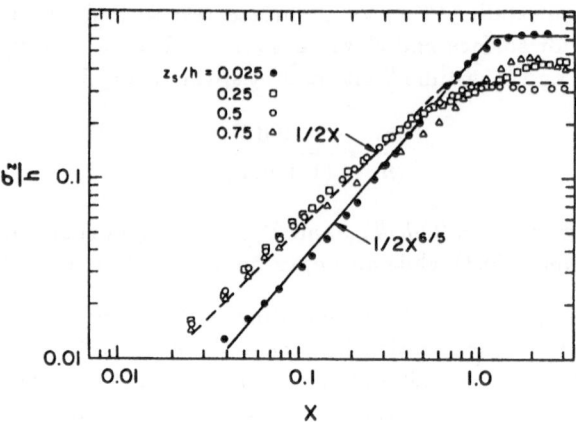

Figure 2. Numerically computed σ_z/h as a function of the dimensionless distance for a point source in a convective boundary layer. From Lamb (1979).

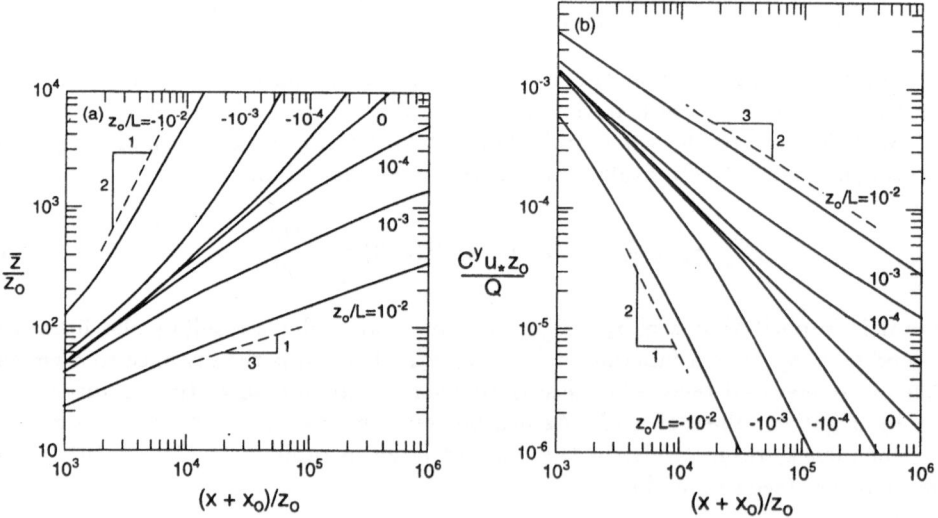

Figure 3. Dimensionless mean plume height and dimensionless crosswind-integrated concentration as a function of dimensionless distance for a surface source in the atmospheric surface layer. From van Ulden (1978) with addition.

the surface and attain a limiting value of σ_w/N far from the surface, where N is the Brunt-Vaissala frequency, $N = [(g/\Theta)(\partial\Theta/\partial z)]^{1/2}$; a typical value of σ_w/N is 10 m. See Wyngaard (1988) for further discussion.

Dispersion Characteristics. For the stable surface layer, van Ulden (1978) obtained analytical expressions for $x(\overline{z})$ and $\overline{u}(\overline{z})$ (Eq. 3) based on the appropriate similarity forms of $U(z)$ and K_h. Venkatram (1982, 1992) combined statistical theory with $K = K_h(z/L)$ using the appropriate forms for $U(z)$ and K_h. In the limit of small σ_z/L, he found $\sigma_z \propto u_* x/U$ and $C^y \propto Q/u_* x$, consistent with the neutral limit of eddy-diffusion theory (see Fig. 3, $z_o/L = 0$). For $\sigma_z/L \gg 1$, he obtained

$$\sigma_z \propto L^{2/3} x^{1/3} \qquad \text{and} \qquad C^y \propto \frac{Q}{u_* L^{1/3} x^{2/3}} . \tag{5}$$

Figure 3 gives van Ulden's results for $\overline{z}$ and $C^y(x,0)$ for stable conditions, $z_o/L > 0$. The limiting forms for $\overline{z} \propto \sigma_z$ and $C^y(x,0)$ given by (5) are shown as dashed lines and can be seen to give the appropriate slopes for $z_o/L = 10^{-2}$.

For elevated plumes, Venkatram et al. (1984) assumed that statistical theory described σ_z and adopted an interpolation expression that matched the short- and long-time limits of the theory:

$$\sigma_z = \frac{\sigma_w t}{(1 + 0.5t/T_{Lz})^{1/2}} \tag{6}$$

with σ_w evaluated at z_s. The Lagrangian time scale was taken as $T_{Lz} = \ell/\sigma_w$, with ℓ given by an interpolation expression, $\ell^{-1} = \ell_n^{-1} + \ell_s^{-1}$, similar to that of Brost and Wyngaard (1978). Venkatram et al. found ℓ in the neutral limit to be $\ell_n = 0.36z$ and in the stable limit (large z/L) to be $\ell_s = 0.27\sigma_w/N$.

A theory by Pearson et al. (1983) also predicted that $\sigma_z = \sigma_w t$ for short times, $t \ll N^{-1}$, but for large times ($t \gg N^{-1}$), it gave

$$\sigma_z = \frac{\sigma_w}{N}(\zeta_z^2 + 2N^2 t T_{Lz})^{1/2} . \tag{7}$$

Here, $\zeta_z \simeq 1.3$, $T_{Lz} \sim \gamma^2 N^{-1}$, and γ is a dimensionless parameter measuring the degree of mixing between fluid elements. If $\gamma \sim 0.1$, σ_z approaches a constant ($\sim \sigma_w/N$) over a considerable range of time but if $\gamma \gtrsim 0.3$, σ_z exhibits a $t^{1/2}$ dependence as in statistical theory. Both a constant σ_z and a $\sigma_z \propto t^{1/2}$ have been deduced for elevated plume dispersion using the Venkatram et al. (1984) data. The causes for the differences and a better understanding of γ are necessary.

IMPROVEMENTS IN APPLIED DISPERSION MODELS

Convective Boundary Layer: Surface-Layer Sources

The main problem in modeling dispersion in the surface layer is the vertical inhomogeneity in σ_w and wind speed. In strong convection ($h/|L| > 10$), the mean wind is approximately uniform except close to the ground. In this case, a useful approach is the Gaussian plume model with an appropriate σ_z. Based on laboratory experiments, Deardorff and Willis (1975) suggested an interpolation expression for σ_z:

$$\frac{\sigma_z}{h} = \left[1.8(z_s/h)^{2/3}X^2 + 0.25X^3 + (\mu u_* X/w_*)^2\right]^{1/2} , \tag{8}$$

where μ is a coefficient dependent on $u_* x / U z_s$. For appropriate ranges of X and z_s/h, Eq. (8) satisfies the short-time limit of statistical theory, Yaglom's (1972) similarity model, and the Lagrangian similarity theory for the neutral surface layer. For a σ_z matching Venkatram's (1992) limit instead Yaglom's, the $0.25X^3$ term would be replaced by $a_c k^2 (w_*/u_*)^2 X^4$, where the coefficient a_c must be determined.

Short-range vertical dispersion also can be estimated from van Ulden's (1978) model. Gryning et al. (1987) successfully applied the model to surface layer measurements from several experiments in the US and Europe and gave operational methods for implementing it. However, this approach is limited to distances such that $\bar{z}$ is still within the surface layer, i.e., $\bar{z} \lesssim 0.1h$.

Convective Boundary Layer: Elevated Sources

Gaussian Plume Model. Weil and Brower (1984) and Berkowicz et al. (1986) improved the Gaussian model by including dispersion parameters based on knowledge of CBL turbulence. Weil and Brower adopted Briggs empirical dispersion curves (see Gifford, 1975), which were intended for elevated releases and were consistent with the short-range limit of statistical theory, i.e., all curves have $\sigma_y \propto x$ and $\sigma_z \propto x$ as $x \to 0$. They defined the applicability of the very unstable (A) to neutral (D) curves in terms of U/w_* using the short-range limits of the curves and interpolation expressions for σ_v and σ_w. The expressions were of the form $\sigma_w = (\sigma_{wn}^2 + \sigma_{wc}^2)^{1/2}$, where $\sigma_{wn} \propto U$ and $\sigma_{wc} \propto w_*$ are the neutral and strongly convective limits, respectively.

In developing the OML Model, Berkowicz et al. (1986) gave dispersion parameters for both surface $(z_s/h \lesssim 0.1)$ and elevated sources with σ_y and σ_z expressed explicitly in terms of w_*, u_*, h, z_s, and x. They used statistical theory for elevated sources and an adaptation of it for surface-layer sources. Their model as well as Weil and Brower's was applicable to buoyant plumes and included Briggs' (1975, 1984) formulations for final rise in the CBL and plume penetration of elevated stable layers.

P.d.f. Model: Passive Releases. The laboratory experiments discussed earlier motivated the development of a new dispersion model—the p.d.f. approach—based on the w p.d.f., p_w, which is not Gaussian (Misra, 1982; Venkatram, 1983; Weil, 1988). For a passive or nonbuoyant source, particles are assumed to be emitted into an updraft or downdraft having a horizontal speed U and a random vertical velocity w specified by p_w. A key assumption is that the T_{Lz} which measures the velocity "memory" is effectively infinite. Thus, a particle's velocity at any x is uniquely determined by its initial velocity. Assuming that w is uniform in an updraft or downdraft, a particle follows a straightline path: $z_c = z_s + wx/U$, where z_c is the particle height.

The C^y is found from the p.d.f. of z_c which in turn is obtained from p_w:

$$C^y(x,z) = Q \frac{p_w[U(z_c - z_s)/x]}{x} . \tag{9}$$

The random velocity w in p_w is found from the particle trajectory: $w = U(z_c - z_s)/x$. For simplicity, p_w can be assumed to be a superposition of two Gaussian distributions, which is a good fit to observations and simulations (Misra, 1982; Weil, 1988). Assuming perfect reflection at $z = 0$, the dimensionless CWIC at the surface is

$$\frac{C^y U h}{Q} = \frac{\Gamma_1}{X} \exp\left(-\frac{R_1}{2}\left[\frac{Z_s}{a_1 X} - 1\right]^2\right) + \frac{\Gamma_2}{X} \exp\left(-\frac{R_2}{2}\left[\frac{Z_s}{a_2 X} - 1\right]^2\right) , \tag{10}$$

where $\Gamma_j = 2P_j/\sqrt{2\pi}b_j$, $a_j = \overline{w}_j/w_*$, $b_j = \sigma_j/w_*$, $R_j = (a_j/b_j)^2$, $Z_s = z_s/h$, and $j = 1,2$ (see Weil, 1988). P_j, $\overline{w}_j$, and σ_j are the total probability, mean vertical velocity, and standard deviation of distribution j. Equation (10) applies to the C^y for particles coming directly from the source; additional image source terms are added to account for particle reflections at $z = h$.

The p.d.f. model has been shown to give good agreement with the surface C^y values from the laboratory experiments and numerical simulations discussed earlier. In addition, the model can produce the entire $C^y(x,z)$ field as demonstrated by Li and Briggs (1988) for $z_s/h = 0.5$. Here, Fig. 4 shows the dimensionless C^y contours given by the Weil et al. (1986) model for a nonbuoyant release and $z_s/h = 0.067, 0.24$ and 0.5. As can be seen, the simple model captures the essential behavior of the contours near the source ($X \lesssim 1.5$) as observed in the laboratory experiments (Fig. 1). For the elevated sources, this behavior is highly dependent on the skewness of the w p.d.f. In contrast to Fig. 1, the contours in Fig. 4 are nearly normal to the boundaries due to the assumed particle reflection and the assumed vertical homogeneity of the turbulence. Nevertheless, the magnitudes of the contours near the surface are in approximate agreement with values from the experiments; this is considered sufficiently good for many applications.

P.d.f. Model: Buoyant Releases. The p.d.f. approach was extended to buoyant plumes but was divided into a "low" ($F_* < 0.1$) and "high" ($F_* > 0.1$) buoyancy regime as defined by the dimensionless buoyancy flux F_* (Weil et al., 1986). Here, $F_* = F_b/Uw_*^2h$, where F_b is the stack buoyancy flux. This somewhat artificial division was motivated by the laboratory experiments of Willis and Deardorff (1983).

For $F_* < 0.1$, Weil et al. superposed the vertical displacements due to source buoyancy and convection to find the "instantaneous" plume centerline height as

$$z_c = z_s + 1.6\frac{F_b^{1/3}}{U}x^{2/3} + \frac{wx}{U} \, . \tag{11}$$

The second term on the right is the rise due to buoyancy and is given by the "two-thirds" law (Briggs, 1975). The CWIC in the instantaneous plume was assumed to have a Gaussian distribution about z_c, and the ensemble-averaged C^y was found by summing or integrating the CWIC contributions due to all possible plume heights.

The dimensionless CWIC at the ground due to plume segments arriving directly from the source is given by

$$\frac{C^yUh}{Q} = \frac{1.2}{\sqrt{2\pi}\sigma_{z1}^*}\exp\left(-\frac{h_1^{*2}}{2\sigma_{z1}^{*2}}\right) + \frac{0.8}{\sqrt{2\pi}\sigma_{z2}^*}\exp\left(-\frac{h_2^{*2}}{2\sigma_{z2}^{*2}}\right) \, , \tag{12}$$

where

$$\sigma_{zj}^* = (1.3\beta'^2 F_*^{2/3} X^{4/3} + b_j^2 X^2)^{1/2} \tag{13a}$$

$$h_j^* = Z_s + 1.6F_*^{1/3}X^{2/3} + a_jX \, . \tag{13b}$$

In (13), a_j and b_j have the same values as in Eq. (10), $j = 1,2$ and β' ($= 0.4$) is the ratio of the buoyant plume radius to the rise. The terms involving buoyancy appear separately so that as $F_* \to 0$, the solution for a nonbuoyant source is recovered. Additional terms are included to satisfy the no-flux condition at $z = h$.

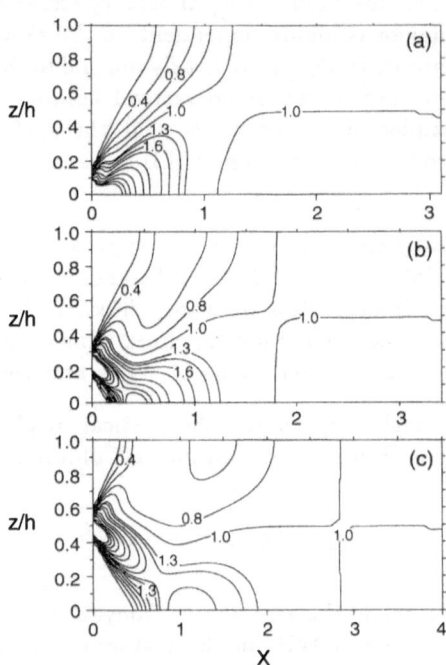

Figure 4. Contours of dimensionless crosswind-integrated concentration as a function of dimensionless height and downwind distance for sources at three release heights in a convective boundary layer. Contours computed from p.d.f. model of Weil et al. (1986) for a passive release at $z_s/h = 0.067, 0.24$ and 0.5 in panels (a) to (c).

For $F_* > 0.1$, the plume is assumed to rise to the top of the CBL, where it "lofts" or lingers until individual cross sections are mixed to the surface by downdrafts. Two key assumptions are made. First, a plume cross section remains within the same downdraft until it is uniformly mixed in the vertical. Second, plume segments become uniformly mixed only when a downdraft has sufficient kinetic energy $\rho w^2/2$ to overcome the potential energy difference between the plume and the environment. These assumptions together with an assumed Gaussian p_w at the CBL top and $\sigma_w = 0.56w_*$ resulted in the following expression for the surface CWIC:

$$\frac{C^y U h}{Q} = 1 - \text{erf}\left(\frac{1.8 A_f}{\sqrt{2}(X/F_*)^{1/3}}\right) , \tag{14}$$

where A_f ($= 2.5$) is an empirical parameter.

Evaluation of the p.d.f. model with laboratory measurements of the near-surface CWIC showed good agreement for both nonbuoyant and buoyant plumes (Fig. 5). Evaluation with SF_6 ground-level concentrations (GLCs) downwind of the Kincaid power plant (Fig. 6) showed fair-to-good agreement on average. The geometric mean (GM) and geometric standard deviation (GSD) of C_p/C_o were 1.1 and 2.1, respectively, where C_p and C_o are the predicted and observed GLCs. In addition, 68% of the predictions were within a factor of 2 of the observations. For reference, the CRSTER model was compared with the Kincaid data (Fig. 6). As can be seen, the results are poorer, as demonstrated by the greater scatter (GSD $=$ 4.6) and fewer predictions within a factor of 2 of the observations (33%).

Other Models. Venkatram's (1980) impingement model is similar to the p.d.f. approach in that plume rise is considered relative to updrafts and downdrafts, but the focus is on plume segments caught in downdrafts. The GLC is found from an assumed p.d.f. of the impingement distance, which is where plume segments caught in downdrafts first touch the ground. The mean impingement distance is obtained from Briggs' (1975) touchdown model. Evaluation of Venkatram's model around tall stacks showed good performance. For highly buoyant plumes ($F_* > 0.1$), Briggs' (1985) semi-empirical expression is also a method for estimating the CWIC and GLC.

Another approach is the Hybrid Plume Dispersion Model (HPDM; Hanna and Paine, 1989) which combines several submodels (e.g., low F_* p.d.f and Briggs' approaches, a high wind model) to deal with dispersion from tall stacks; different submodels apply depending on the values of F_*, L, etc. HPDM was extended to treat point sources with $z_s \gtrsim 50$ m, but it is limited to terrain heights $\lesssim 0.5z_s$ (Hanna and Chang, 1993). In general, HPDM performed better than the EPA Industrial Source Complex (ISC) Model in estimating the high concentrations of regulatory interest.

Stable Boundary Layer

By comparison to the CBL, there have been fewer attempts to develop improved dispersion models for the SBL. This is due mainly to two factors. First, our knowledge of the SBL vertical structure, while reasonable for moderate-to-strong winds and weak-to-moderate stability (e.g., Nieuwstadt, 1984), is more variable and incomplete in general; this is particularly true for light wind, strongly stable conditions (e.g., $h/L \gtrsim 10$) when waves and drainage flow effects are present. Second, there are fewer measurements of dispersion and concentration fields over a wide range of conditions (source, meteorology, distance) in the SBL for guiding and testing models. In spite

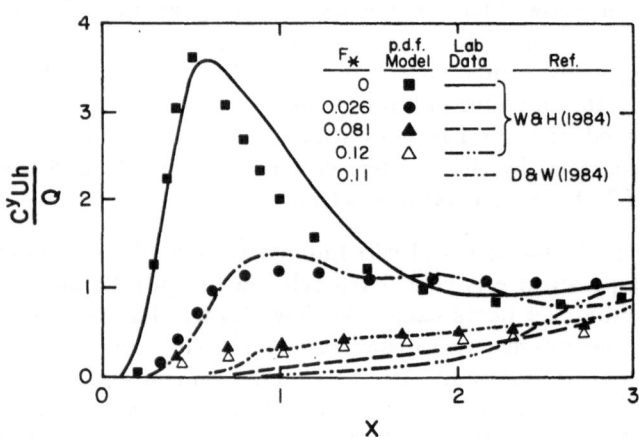

Figure 5. Nondimensional crosswind-integrated concentration at the surface versus dimensionless downwind distance for passive and buoyant plumes in the convective boundary layer. From Weil et al. (1986).

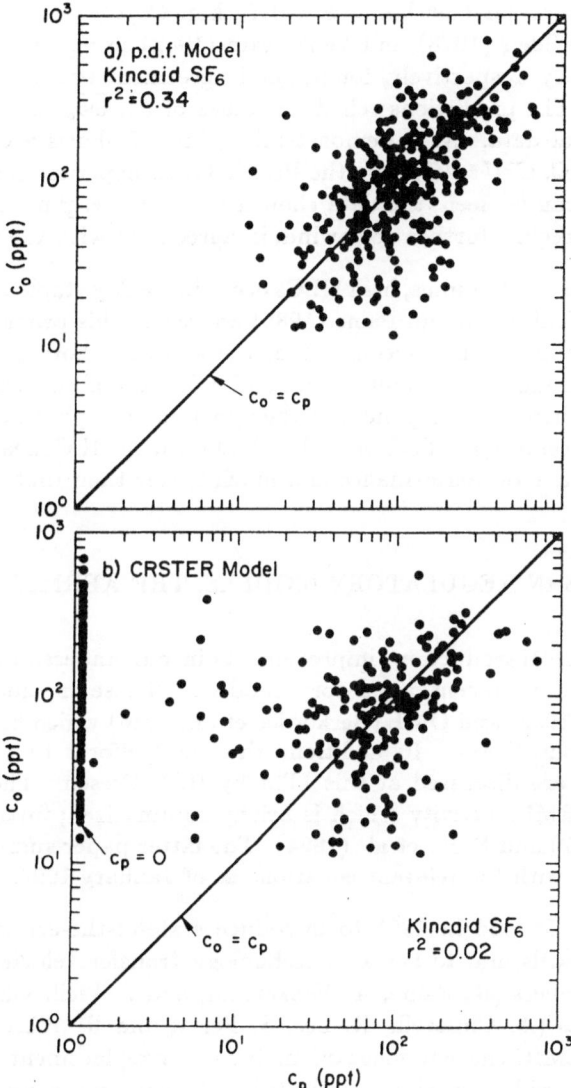

Figure 6. Observed versus predicted ground-level SF$_6$ concentrations for the p.d.f. and CRSTER models at the Kincaid power plant. From Weil et al. (1986).

of these limitations, there are two problems in which improved knowledge of the SBL and dispersion has given better concentration estimates for applications: 1) dispersion from surface sources, and 2) dispersion of buoyant plumes in a high wind SBL.

For surface sources, Irwin (1983) compared Gaussian model estimates of centerline GLCs with observations from ten surface-release experiments. Predictions based on the PG curves tended to underestimate the measurements with a C_p/C_o ratio that decreased with x; e.g., for 0.55 km $< x <$ 1.75 km, C_p/C_o was $\sim$ 0.5. However, as noted earlier, van Ulden (1978) and Venkatram (1982) using the diffusion equation and statistical theory, respectively, found good agreement between their predictions and observations. The predictions relied on values of u_*, $\overline{w\theta_o}$, L, etc. inferred from other meteorological data. As a demonstration, Fig. 7 shows a comparison of the dimensionless CWIC, $C^y U z_o/Q$, from the Prairie Grass experiment with Venkatram's model (lines). As can be seen, the data show a $C^y \propto x^{-1}$ regime close to the source and a $C^y \propto x^{-2/3}$ regime further downwind, in agreement with the model.

For elevated buoyant plumes, high GLCs occur in weakly stable conditions coupled with high winds. Venkatram and Paine (1985) addressed this problem with a form of the Gaussian plume model but accounted for the vertical inhomogeneity in σ_w. This was done by computing σ_z separately above and below the plume centerline assuming $d\sigma_z/dt = \sigma_w(z')$, where z' corresponded to the upper or lower plume edge for dispersion above or below the centerline. Their model evaluation with GLC measurements around tall stacks showed a good performance and much better than that obtained with the CRSTER Model.

IMPROVEMENTS IN REGULATORY MODELS: THE AERMIC ACTIVITY

As noted in the Introduction, improvements in our understanding of dispersion have been included in recent regulatory models. These include the UK-ADMS (Carruthers et al., 1992) and OML (Berkowicz et al., 1986) which have been discussed at this and previous ITMs. In addition, the CEC efforts to promote improved regulatory models are discussed at this ITM by H.R. Olesen. Therefore, the focus here is on the AERMIC activity which is briefly summarized; further details can be found in Weil (1992) and Perry et al. (1994). The latter paper summarizes the model development along with the relevant equations as of January 1994.

AERMIC was formed in 1991 to introduce state-of-the-art modeling concepts into regulatory models and to act as a technology transfer vehicle. It is comprised of three AMS members (R. Paine, A. Venkatram, and J. Weil, Chairman) and four EPA meteorologists (A. Cimorelli, R. Lee, S. Perry, and R. Wilson). In it's initial deliberations, AERMIC chose to focus on an update or replacement for the ISC Model because ISC: 1) is widely used in applications, 2) contains a number of outdated concepts and practices, and 3) has a new highly modular structure (ISC2), which facilitates the incorporation of improvements/modifications.

The model being developed is referred to as the AMS/EPA Regulatory Model or AERMOD and uses the ISC2 framework. AERMIC felt that a contemporary regulatory model should have the following attributes: 1) include state-of-the-art science, 2) be simple but capture the essential physical processes, 3) be robust in estimating regulatory design concentrations (i.e., provide reasonable concentration estimates over a wide range of meteorological and source conditions with a minimum of

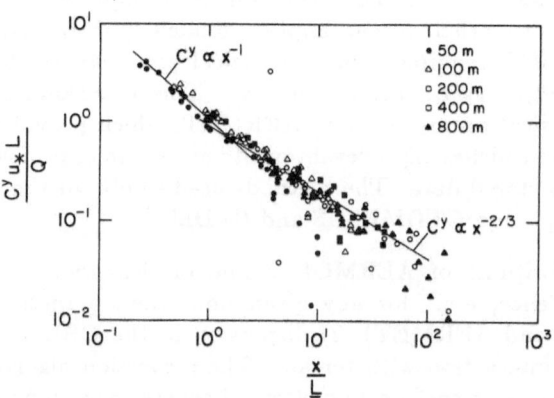

Figure 7. Dimensionless crosswind-integrated concentration at the surface for surface releases in a stable boundary layer as a function of the dimensionless distance; data are from the Prairie Grass experiment. From Venkatram (1982).

discontinuities), 4) be easily implemented, user friendly, and require simple inputs and resources, and 5) evolve as the science does and therefore accomodate modifications with ease. These attributes have served as guidelines in the model development.

From a scientific viewpoint, the following are considered to be the main deficiencies of the existing ISC Model: 1) dispersion based on the PGT scheme which does not account for the PBL's mean and turbulence structure, 2) a mixed-layer height found from an unphysical day-to-day interpolation method, 3) plume rise based on Briggs' (1971) model, which does not address convection, and plume penetration of elevated inversions using a simple "all-or-none" approach, 4) a complex or hilly terrain treatment that does not consider terrain greater than the stack height nor the concept of a dividing streamline height (Snyder et al., 1985) in stably-stratified flow, and 5) a building downwash treatment that has a discontinuity in the stack height dependence and an inadequate handling of plume buoyancy.

The first phase of the AERMOD development is due for completion in early 1995 and is aimed at resolving the first four deficiencies in the previous paragraph. An improved treatment of building downwash is a high priority item for future development. As with other recent applied models (OML; HPDM; CTMPLUS, Perry, 1992), AERMOD requires a meteorological processor to determine the PBL variables such as $\overline{w\theta_o}$, u_*, L, w_*, h, and z_o. This is accomplished through the AERMIC meteorological preprocessor or AERMET, which provides hourly values of the variables. It also furnishes any relevant onsite measurements of wind, temperature, and turbulence from stored data. The methods used to obtain the PBL variables are similar to those adopted in CTDMPLUS and HPDM.

The major algorithms of AERMOD define or describe: 1) profiles of wind, temperature, turbulence, etc. for any given hour (i.e., a meteorological interface between AERMOD and AERMET), 2) dispersion in the CBL, 3) dispersion in the SBL, and 4) plume interaction with terrain. The dispersion algorithms are designed to account for surface, near-surface, and elevated sources as well as source momentum and buoyancy. A number of ideas discussed in earlier sections of this paper (p.d.f. approach, convective scaling, etc.) are adopted but are integrated in a way to minimize discontinuities. The vertical inhomogeneity of σ_w, U, etc. near the ground is handled by determining "effective" values of these variables, which are an interpolation between the variable at source height and an average value over the PBL. The effective variables are a function of x (see Perry et al., 1994).

Dispersion in the CBL is based on the p.d.f. approach—a skewed p.d.f. for the vertical concentration distribution but a Gaussian p.d.f. for the lateral distribution. However, the approach differs from the Weil et al. (1986) model in three ways. First, the a_j, b_j, and P_j in p_w (see Eqs. 10, 13 and the subsequent discussion) are selected from an objective method using σ_w^2 and $\overline{w^3}$, the third moment of w. σ_w^2 is parameterized by a linear combination of u_*^2 and w_*^2 and $\overline{w^3}$ is taken as $\propto w_*^3$; thus, the a_j, b_j, and P_j vary continuously with u_* and w_*. Second, $\overline{w^3}$ is assumed to be constant above $0.1h$ but is z-dependent below that height, whereas σ_w^2 has a z-dependence throughout the CBL, albeit a weak one for $z > 0.1h$. The σ_w^2 and $\overline{w^3}$ expressions are based on observations and earlier parameterizations (e.g., Hicks, 1985; Hunt et al., 1988). Third, the dispersion and C^y vary continuously with F_* so that there are not different treatments depending on the magnitude of F_*.

472

The C^y and C are assumed to arise from three contributions: 1) a "direct" plume which emanates from the stack and accounts for plume segments caught in sufficiently strong downdrafts to bring them to the surface (this is the low F_* model discussed earlier); 2) an "indirect" plume originating (conceptually) above the CBL and accounting for plume segments that rise to the CBL top, remain there for some time due to buoyancy, and then disperse downwards; and 3) a penetrated plume that accounts for material with sufficient buoyancy to rise into the elevated inversion. The indirect plume is a mathematical device to satisfy the no-flux boundary condition at the CBL top. It plays the same role as the "image" plume in the Gaussian model.

Dispersion in the SBL is modeled by the Gaussian plume with a plume height that accounts for both the source momentum and buoyancy. The dispersion parameters are given by $\sigma_y = \text{Max}[\sigma_v x/U, \, 0.1x]$ and $\sigma_z = (\sigma_w x/U)/(1 + t/2T_{Lz})^{1/2}$ (Venkatram, 1988), with σ_v, σ_w, U, and T_{Lz} being effective values as discussed earlier.

Dispersion in complex terrain accounts for the dividing streamline (H_c) concept in stably-stratified flow. In such flow, ambient air tends to move around the terrain in a horizontal plane for $z < H_c$, whereas it travels over the terrain for $z > H_c$; H_c depends on the obstacle height and the terrain Froude number. AERMOD handles dispersion in this "two-layer" flow by assuming that the plume can exist in two states: one in which the plume is horizontal and one in which the plume travels over the terrain. The probability f of the plume being in the horizontal layer is related to the fraction of the plume below H_c. The total concentration C_T at receptor location x, y, z is given by

$$C_T(x, y, z) = fC(x, y, z) + (1 - f)C(x, y, z_{eff}) , \qquad (15)$$

where $C(x, y, z)$ is the flat-terrain concentration field, and z_{eff} is an "effective" receptor height for plume material traveling over the terrain. The formulation is simpler than that in CTDMPLUS; details are given in Perry et al. (1994).

In addition to model formulation, AERMIC is conducting two types of evaluations with field data: 1) a model physics evaluation to assess AERMOD's scientific components, and 2) an operational performance evaluation to determine how well the model predicts the high concentrations important in air quality regulations. Studies are also planned to assess the model sensitivity to inputs and the difference between AERMOD and ISC predictions over a wide range of inputs. Finally, it is recognized that the scientific background required for AERMOD may be substantially greater than for ISC. In anticipation of this, AERMIC is developing a Model Formulation Document that includes a complete description of AERMOD as well as a primer on boundary layer meteorology as it relates to the new model. This document also includes an extensive discussion making the links between the older technology in the PGT scheme and the approach taken in AERMOD.

ACKNOWLEDGMENTS

I am grateful to Seth White for producing Figure 4.

473

REFERENCES

Berkowicz, R., Olesen, H.R. and Torp, U., 1986, The Danish Gaussian air pollution model (OML): Description, test and sensitivity analysis in view of regulatory applications, in: "Air Pollution Modeling and Its Application V,", C. De Wispelaere, F.A. Schiermeier, and N.V. Gillani, Eds., Plenum, New York.

Briggs, G.A., 1971, Some recent analyses of plume rise observations, in: "Proceedings of the Second International Clean Air Congress," H.M. Englund and W.T. Berry, Eds., Academic Press, New York.

Briggs, G.A., 1975, Plume rise predictions, in: "Lectures on Air Pollution and Environmental Impact Analyses," D.A. Haugen, Ed., Amer. Meteor. Soc., Boston.

Briggs, G.A., 1984, Plume rise and buoyancy effects, in: "Atmospheric Science and Power Production," D. Randerson, Ed., U.S. Department of Energy, DOE/TIC-27601.

Briggs, G.A., 1985, Analytical parameterizations of diffusion: the convective boundary layer, J. Climate Appl. Meteor., 24:1167.

Briggs, G.A., 1988, Analysis of diffusion field experiments, in: "Lectures on Air Pollution Modeling," A. Venkatram and J.C. Wyngaard, Eds., Amer. Meteor. Soc., Boston.

Brost, R.A., and Wyngaard, J.C., 1978, A model study of the stably stratified planetary boundary layer, J. Atmos. Sci., 35:1427.

Businger, J.A., Wyngaard, J.C., Izumi, Y., and Bradley, E.F., 1971, Flux-profile relationships in the atmospheric surface layer, J. Atmos. Sci., 28:181.

Carruthers, D.J., Holroyd, R.J., Hunt, J.C.R., Weng, W.-S., Robins, A.G., Apsley, D.D., Smith, F.B., Thomson, D.J., and Hudson, B., 1992, UK atmospheric dispersion modelling system, in: "Air Pollution Modeling and Its Application IX," H. van Dop and G. Kallos, Eds., Plenum Press, New York.

Deardorff, J.W., 1972, Numerical investigation of neutral and unstable planetary boundary layers, J. Atmos. Sci., 29:91.

Deardorff, J.W., and Willis, G.E., 1975, A parameterization of diffusion into the mixed layer, J. Appl. Meteor., 14:1451.

Deardorff, J.W., and Willis, G.E., 1985, Further results from a laboratory model of the convective planetary boundary layer, Bound.-Layer Meteor., 32:205.

Eberhard, W.L., Moninger, W.R., Briggs, G.A., 1988, Plume dispersion in the convective boundary layer. Part I: CONDORS field experiment and example measurements, J. Appl. Meteor., 27:599.

Gifford, F.A., 1961, Uses of routine meteorological observations for estimating atmospheric dispersion, Nuclear Safety, 2:47.

Gifford, F.A., 1975, Atmospheric dispersion models for environmental pollution applications, in: "Lectures on Air Pollution and Environmental Impact Analyses," D.A. Haugen, Ed., Amer. Meteor. Soc., Boston.

Gryning, S.E, Holtslag, A.A.M., Irwin, J.S., and Sivertsen, B., 1987, Applied dispersion modelling based on meteorological scaling parameters, Atmos. Environ., 21:79.

Hanna, S.R., and Chang, J.C., 1993, Hybrid plume dispersion model (HPDM) improvements and testing at three field sites, Atmos. Environ., 27A:1491.

Hanna, S.R., and Paine, R.J., 1989, Hybrid plume dispersion (HPDM) development and evaluation, J. Appl. Meteor., 28:206.

Hicks, B.B., 1985, Behavior of turbulent statistics in the convective boundary layer, J. Climate Appl. Meteor., 24:607.

Horst, T.W., 1979, Lagrangian similarity modeling of vertical diffusion from a ground-level source, J. Appl. Meteor., 18:733.

Hunt, J.C.R., 1982, Diffusion in the stable boundary layer, in: "Atmospheric Turbulence and Air Pollution Modelling," F.T.M. Nieuwstadt and H. van Dop, Eds., Reidel, Dordrecht.

Hunt, J.C.R., 1985, Diffusion in the stably stratified atmospheric boundary layer, J. Climate Appl. Meteor., 24:1187.

Hunt, J.C.R., Kaimal, J.C., and Gaynor, J.E., 1988, Eddy structure in the convective boundary layer—new measurements and new concepts, Quart. J. Roy. Meteor. Soc., 114:827.

Irwin, J.S., 1983, Estimating plume dispersion—A comparison of several sigma schemes, J. Climate Appl. Meteor., 22:92.

Kaimal, J.C., Wyngaard, J.C., Haugen, D.A., Cote, O.R., Izumi, Y., Caughey, S.J., and Readings, C.J., 1976, Turbulence structure in the convective boundary layer, J. Atmos. Sci., 33:2152.

Lamb, R.G., 1978, A numerical simulation of dispersion from an elevated point source in the convective planetary boundary layer, Atmos. Environ., 12:1297.

Lamb, R.G., 1982, Diffusion in the convective boundary layer, in: "Atmospheric Turbulence and Air Pollution Modelling,", F.T.M. Nieuwstadt and H. van Dop, Eds., Reidel, Dordrecht.

Li, Z.-K., and Briggs, G.A., 1988, Simple pdf models for convectively driven vertical diffusion, Atmos. Environ., 22:54.

Misra, P.K., 1982, Dispersion of nonbuoyant particles inside a convective boundary layer, Atmos. Environ., 16:239.

Nieuwstadt, F.T.M., 1980, Application of mixed-layer similarity to the observed dispersion from a ground-level source, J. Appl. Meteor., 19:157.

Nieuwstadt, F.T.M., 1984, The turbulent structure of the stable nocturnal boundary layer, J. Atmos. Sci., 41:2202.

Nieuwstadt, F.T.M., and van Dop, H, Eds., 1982, "Atmospheric Turbulence and Air Pollution Modelling," Reidel, Dordrecht.

Olesen, H.R., and Mikkelsen, T., 1992, "Proceedings of the Workshop on Objectives for Next Generation of Practical Short-Range Atmospheric Dispersion Models," National Environmental Research Institute, Roskilde, Denmark.

Pasquill, F.A., and Smith, F.B., 1983, "Atmospheric Diffusion," Wiley, New York.

Pearson, H.J., Puttock, J.S., and Hunt, J.C.R., 1983, A statistical model of fluid-element motions and vertical diffusion in a homogeneous stratified turbulent flow, J. Fluid Mech., 129:219.

Perry, S.G., 1992, CTDMPLUS: A dispersion model for sources near complex topography. Part I: Technical formulations. J. Appl. Meteor., 31:633.

Perry, S.G., Cimorelli, A.J., Lee, R.F., Paine, R.J., Venkatram, A., Weil, J.C., and Wilson, R.B., 1994, AERMOD: A dispersion model for industrial source applications, To be presented at 87th Annual Meeting of Air and Waste Management Association, Air and Waste Management Association, Pittsburgh.

Snyder, W.H., Thompson, R.S., Eskridge, R.E., Lawson, R.E., Castro, I.P., Lee, J.T., Hunt, J.C.R., and Ogawa, Y., 1985, The structure of strongly stratified flow over hills: Dividing-streamline concept, J. Fluid Mech., 152:249.

Taylor, G.I., 1921, Diffusion by continuous movements, Proc. London Math. Soc. Ser. 2, 20:196.

Turner, D.B., 1964, A diffusion model for an urban area, J. Appl. Meteor., 3:83.

van Ulden, A.P., 1978, Simple estimates for vertical dispersion from sources near the ground, Atmos. Environ., 12:2125.

Venkatram, A., 1980, Dispersion from an elevated source in the convective boundary layer, Atmos. Environ., 14:1.

Venkatram, A., 1982, A semi-empirical method to compute concentrations associated with surface releases in the stable boundary layer, Atmos. Environ., 16:245.

Venkatram, A., 1983, On dispersion in the convective boundary layer, Atmos. Environ., 17: 529.

Venkatram, A., 1988, Dispersion in the stable boundary layer, in: "Lectures on Air Pollution Modeling," A. Venkatram and J.C. Wyngaard, Eds., Amer. Meteor. Soc., Boston.

Venkatram, A., 1992, Vertical dispersion of ground-level releases in the surface boundary layer, Atmos. Environ., 26A:947.

Venkatram, A., and Paine, R., 1985, A model to estimate dispersion of elevated releases into a shear-dominated boundary layer, Atmos. Environ., 19:1797.

Venkatram, A., Strimaitis, D., and Dicristofaro, D., 1984, A semiempirical model to estimate vertical dispersion of elevated releases in the stable boundary layer, Atmos. Environ., 18:923.

Venkatram, A., and Wyngaard, J.C., Eds., 1988, "Lectures on Air Pollution Modeling," Amer. Meteor. Soc., Boston.

Weil, J.C., 1985, Updating applied diffusion models, J. Climate Appl. Meteor., 24:1111.

Weil, J.C., 1988, Dispersion in the convective boundary layer, in: "Lectures on Air Pollution Modeling," A. Venkatram and J.C. Wyngaard, Eds., Amer. Meteor. Soc., Boston.

Weil, J.C., 1990, A diagnosis of the asymmetry in top-down and bottom-up diffusion using a Lagrangian stochastic model, J. Atmos. Sci., 47:501.

Weil, J.C., 1992, Updating the ISC model through AERMIC, in: "Proceedings 85th Annual Meeting of Air and Waste Management Association," Air and Waste Management Association, Pittsburgh.

Weil, J.C., and Brower, R.P., 1984, An updated Gaussian plume model for tall stacks, J. Air Pollution Control Assoc., 34:818.

Weil, J.C., Corio, L.A., and Brower, R.P., 1986, Dispersion of buoyant plumes in the convective boundary layer, in: "Fifth Joint Conference on Applications of Air Pollution Meteorology," Amer. Meteor. Soc., Boston.

Willis, G.E., and Deardorff, J.W., 1976, A laboratory model of diffusion into the convective planetary boundary layer, Quart. J. Roy. Meteor. Soc., 102:427.

Willis, G.E., and Deardorff, J.W., 1978, A laboratory study of dispersion from an elevated source within a modeled convective planetary boundary layer, Atmos. Environ., 12:1305.

Willis, G.E., and Deardorff, J.W., 1981, A laboratory study of dispersion from a source in the middle of the convectively mixed layer, Atmos. Environ., 15:109.

Willis, G.E., and Deardorff, J.W., 1983, Buoyant plume dispersion and inversion entrapment in and above a laboratory mixed layer, Atmos. Environ., 21: 1725.

Wyngaard, J.C., 1988, Structure of the PBL, in: "Lectures on Air Pollution Modeling," A. Venkatram and J.C. Wyngaard, Eds., Amer. Meteor. Soc., Boston.

Yaglom, A.M., 1972, Turbulent diffusion in the surface layer of the atmosphere, Izv. Akad. Nauk USSR, Atmos. Ocean. Phys., 8:333.

DISCUSSION

K. JUDA-REZLER: Do you have experience with modeling dispersion from very high stacks ($h \sim 300$ m), and can you comment on problems arising from high buoyancy usually connected with such high point sources?

J.C. WEIL. For the convective boundary layer (CBL), the dispersion of a highly buoyant plume, defined roughly as $F_* \sim 0.1$, is a difficult problem to model. Field observations and laboratory data show that a plume with $F_* \sim 0.1$ and completely trapped in the CBL remains near the CBL top, due to its buoyancy, and disperses downwards slowly over distance. As F_* increases, part of the plume penetrates the elevated inversion while the rest remains in the CBL, undergoing a slow vertical dispersion. For a sufficiently large F_*, the plume may completely penetrate the inversion. The high F_* p.d.f. approach discussed in the paper is a simple way of dealing with this problem, but further experimental and modeling work is necessary. The problem is being addressed in the new model AERMOD discussed in the paper.

A. GERHARD: Where do you get the information about the dividing streamline or the flow deformation over complex terrain you mentioned?

J.C. WEIL: In stably-stratified flow, the terrain or hill height h_t, U, and N are needed to calculate the dividing streamline height; here, U is the mean wind speed and N is the Brunt-Vaisala frequency, which depends on the potential temperature gradient as given in the paper. h_t is obtained from terrain maps while the wind speed and temperature gradient are either measured or estimated from a meteorological pre-processor. The flow deformation information used for example in the CTDMPLUS model is based on two-dimensional, horizontally-layered potential flow for $z < h_c$ and on the linearized equations of motion for $z > h_c$, i.e., for flow over the hill.

B. SMITH. How would you extend your techniques to cope with very light wind situations, when $w_* > U$, which are of particular importance in tropical countries?

J.C. WEIL. I believe that the best way to address very light winds is with a simple puff model that can be integrated analytically over time to yield a "plume" description. My colleague, Roger Brower, and I did this for a plume in the CBL based on the approach of Frenkiel (Frenkiel, F.N., 1953, Turbulent diffusion: mean concentration distribution in a flow field of homogeneous turbulence, Adv. Appl. Mech., 3:61) with dispersion about an "effective" stack height. The concentration field varies continuously with w_*/U, remaining well behaved as $U \to 0$. This approach has not been adapted yet to the p.d.f. model with a distance-dependent plume height. As for a rough interim approach using the p.d.f. model, one might replace U by αw_*, where α is an empirical coefficient of order 1.

M.M. MILLAN. Real world experience with tall stacks (Sudbury) shows that the dispersion is a fully three-dimensional process with the plume becoming sheared to the side. The impact area can be displaced by as much as 20 km at 30 km from the stack and from the upper plume.

J.C. WEIL: I agree with your description when strong wind direction shear is present. Wind shear can be included and is in some simple models. However, we do not have routine measurements of the shear. Furthermore, while the shear may laterally displace the "touchdown" zone of the ground-level concentration from the plume aloft, it would not increase the concentration over the value that would exist in the absence of shear. That is, the simple models which ignore the shear should yield conservative estimates (i.e., too high) of the ground-level concentration.

D. CARRUTHERS: What type of model do you use to describe flow and dispersion over complex terrain in AERMOD, and what input data are required?

J.C. WEIL: The dispersion model for complex terrain is a simple approach in which the plume is considered to be in two states: the plume either flows horizontally around the terrain with a probability f of occurrence or flows over the terrain with a probability of $1 - f$. The ground-level concentration is the sum of the concentrations in these two states, each weighted by the appropriate probability as given in Eq. (15). Further details are given in the paper by Perry et al. (1994).

D. CARRUTHERS: Why does σ_z reach a larger far-field value for ground-level sources than for elevated sources in Lamb's simulations?

J.C. WEIL: The σ_z from Lamb's simulations was computed from second moments of neutrally-buoyant "particle" displacements about the source height z_s rather than the mean plume height $\langle Z \rangle$, i.e., $\sigma_z = \langle (Z - z_s)^2 \rangle^{1/2}$, where Z is the particle height. Thus, this σ_z includes the plume mean vertical displacement which is more significant for surface than for elevated releases; for $z_s = 0$, $\langle Z \rangle$ varies from 0 at $x = 0$ to $0.5h$ far downwind.

EUROPEAN COORDINATING ACTIVITIES CONCERNING
LOCAL-SCALE REGULATORY MODELS

H.R. Olesen

National Environmental Research Institute (NERI)
P. O. Box 358
DK-4000 Roskilde
Denmark

1. INTRODUCTION

During the 19th ITM in Ierapetra, a round-table discussion on harmonization within atmospheric dispersion modelling took place. Since then, there has been a number of activities within Europe related to this issue. In particular, two workshops dealing with topics discussed at the Ierapetra meeting have been held - one of them in May 1992 at Risø, Denmark, and the other in August 1993 in Manno, Switzerland. The workshop at Risø was entitled *"Objectives for Next Generation of Practical Short-Range Atmospheric Dispersion Models"*, and that in Manno *"Intercomparison of Advanced Practical Short-Range Atmospheric Dispersion Models"*.

Here, an overview of this series of activities is given, with special emphasis on the results of the Manno workshop. A major issue there was the testing of model evaluation procedures.

The main focus in this paper will be on dispersion models for *regulatory purposes*. These models are used - especially during the planning stage - to determine whether a polluter complies with certain rules and regulations with regard to air quality management.

2. BACKGROUND

Several initiatives have been taken within Europe during the past few years concerning the management of dispersion models for regulatory purposes. This is due to two main reasons:
- There is a great number of such models in use within Europe.
- Generally, the models applied are not scientifically up-to-date.

As to the first reason, the existence of a multitude of models poses several problems. Throughout Europe, a great number of models is being used for regulatory purposes. Typically, a model is used only in the country where it was developed. The fact that different models may produce differing results for the same scenario is inconvenient from an administrative point of view. Further, it is difficult to compare the merits of the various models.

A model should be fit for the purpose for which it is applied. Thus, a key question to be answered by the modelling community is: *How well is a given model fit for a specific purpose?* There seems, however, to be a lack of certain basic standards and tools which would make it feasible to make statements on model merits in a satisfactory way.

As to the second reason stated above, a lot of knowledge and experience is documented in the scientific literature. The role of models is to generalize this information into sets of rules and procedures which can be used in practice. Thus, it can be regarded as the primary purpose of modelling to ensure that *the greatest possible amount of available knowledge is incorporated into the decision-*

making process. However, at present new scientific developments do not seem to be transferred to the user community as efficiently as they could be. This was expressed in a resolution agreed upon during the workshop at Risø:

> *The practical atmospheric dispersion models which are presently used for regulatory purposes are based on 25-year old research (Pasquill-Gifford type stability classes, simple Gaussian dispersion schemes, i.a.). During the past 25 years, research in atmospheric dispersion has progressed substantially, thus rendering the presently used regulatory models outdated. Consequently, there is now a need for development and practical implementation of computer codes based on our present knowledge of atmospheric dispersion.*

The state of affairs as described above was the background for a European initiative started in June 1991. The purpose was to achieve increased cooperation on and standardization within regulatory models. A steering committee was formed with the task of arranging a series of workshops.

3. WORKSHOP ON OBJECTIVES FOR MODELS, MAY '92

The first in the series of workshops was entitled *"Objectives for Next Generation of Practical Short-Range Atmospheric Dispersion Models"*. It was held in May 1992 at Risø, Denmark, and it was hosted by DCAR (Danish Centre for Atmospheric Research, an umbrella organization for several Danish research institutions). The focus of the workshop was on the *management of model development* and the *definition of model objectives*, rather than on detailed model contents. It was the intention to point out actions to be taken in order to improve the development and use of atmospheric dispersion models.

The papers presented at the workshop are assembled in a volume of proceedings (Olesen and Mikkelsen, 1992). In the proceedings, various topics within the broad spectrum of matters related to up-to-date practical dispersion models are dealt with, such as the scientific basis for such models, requirements for model input and output, meteorological preprocessing, standardization within modelling, electronic information exchange as a potentially useful tool, model evaluation, and data bases for model evaluation.

A number of recommendations for the improvement of "modelling culture" were suggested by the workshop. The most important of these are:

- *There should be systematic comparisons of model predictions versus existing data sets from experiments. Further, future experiments needed to fill out knowledge gaps should be pointed out.*
- *Model users should be aware of the uncertainties inherent in model calculations. Work should be undertaken on how to determine model uncertainties and how to present them to the model users.*
- *There should be an action for setting up guidelines for model development and documentation. The aim is to promote more correct use of models.*
- *There should be an action for harmonization of meteorological input for "next-generation models".*
- *One or more electronic Bulletin Board Services (BBS) or similar services should be established to help the coordination of model development and management.*

The organizers of the workshop were not in a position to back up these recommendations with economic support. Nevertheless, after the workshop, some of the proposed activities have been initiated, as described in the following sections.

4. WORKSHOP ON INTERCOMPARISON OF MODELS, AUGUST '93

One of the recommendations of the workshop at Risø was to stimulate work on model evaluation. Therefore, a subsequent workshop was held in Manno, Switzerland, on *"Intercomparison of Advanced Practical Short-Range Atmospheric Dispersion Models"*. The organizing body was ERCOFTAC (European Research Community On Flow Turbulence And Combustion).

A major concern of the workshop was to investigate model evaluation procedures and - on the basis of practical experience - discuss the problems involved in the process. The workshop was not an attempt to perform an in-depth evaluation of models, but can rather be characterized as a demonstration exercise. It was the intention to seek agreement on methodologies which can later be used for studies involving more data sets and more models. It was sought in practice to start establishing a toolbox of recommended methods for model evaluation.

The type of problem primarily considered was relatively simple, namely a case where a single source emits a non-reactive gas in homogeneous terrain. Prior to the workshop, data sets from three atmospheric field experiments (Kincaid, USA; Copenhagen, Denmark; Lillestrøm, Norway) had been distributed to the participants. They were accompanied by a package of model evaluation software (developed by Sigma Research Inc.). The software package includes the options of producing an *operational evaluation* as well as pursuing *a diagnostic approach* to model evaluation. It further includes graphical software suited to the purpose of model evaluation.

The *operational (statistical) evaluation* yields a number of statistical performance measures such as mean fractional bias, normalized mean square error, fraction of observations within a factor of two etc. Further, the software package offers the possibility of computing confidence limits on these measures, making use of a bootstrap resampling procedure.

The *diagnostic (scientific) approach* aims at obtaining an understanding of cause-and-effect relationships by analysis of model residuals, stratified by primary parameters.

These two approaches complement each other. A regulator may be primarily interested in an operational evaluation, but the diagnostic approach is indispensable. In order to have confidence in a model, it should be verified that it gives the right answer *for the right reason*, which can only be accomplished by a diagnostic approach.

A central question posed at the workshop was: *Why is model evaluation so difficult - and what can be done to overcome the difficulties?*

Table 1 briefly sums up some basic problems in model evaluation to be discussed in the following.

Table 1. Model evaluation - difficulties and reactions.

Difficulty	Reaction	Implied problems
Data sets reflect only few of the possible scenarios	a) Extrapolate model behaviour to other conditions	a) Does the model give the right result for the right reason? We must understand model behaviour!
	b) Use many data sets	b) Hard work!
The appropriate evaluation method depends on the context of the application	An array of various evaluation methods must be developed	
Processing of input data is not trivial	Take care! When comparing models, they should be run on the same data	Numerous problems!
Inherent uncertainties exist	Accept them! Quantify them!	

Data sets reflect only few of the possible scenarios

One fundamental difficulty is that experimental data sets are limited in several respects. Usually a data set is from a campaign with only *one* source configuration, *one* set of terrain conditions, and a limited number of meteorological scenarios. Further, the number of points where the concentration has been measured is small.

On the other hand, users want models that can be used for a broad range of source configurations, for a large number of meteorological scenarios, and that can compute concentrations at all points in space. As a reaction to this difficulty, one is inevitably forced to *extrapolate model behaviour* to conditions where the model has not been validated. This can only be done in a credible manner if the model has a sound description of physical processes. Therefore, model evaluations should be used to develop an understanding of model behaviour and diagnose the causes for model bias. This is the reason why the diagnostic approach to model evaluation is indispensable.

Another reaction to the difficulty is simply to use many data sets. It sounds simpler than it is, because processing data requires a substantial amount of work. One has to become acquainted with data and recognize potential pitfalls. For example, it may require careful analysis to recognize that a certain monitor is placed in unrepresentative surroundings.

As a matter of fact, one should try not only to assess the information contained in a data set, but also be aware of the information *not contained* in it. If e.g. nighttime conditions are rare, this affects the validity of extrapolations.

Experience shows that there is an obvious danger in using data from only one or two sites to draw conclusions regarding model performance. As stated by Hanna (1993), it is not unusual for the same model to overpredict by 40% at one site and underpredict by 40% at another site.

There is no easy way to get around the difficulties imposed by data set limitations. But in order to have models validated on many data sets, and in order to have the data sets used properly, it is recommended that data sets be prepared for ease-of-use, with their peculiarities and pitfalls well documented.

Still, model evaluation does require hard work.

The appropriate evaluation method depends on on the context of the application

Another basic difficulty in model evaluation is that there will never be just *one* recommended method for validating models. Just as models should be fit for purpose, so should evaluation methods.

In an excellent paper discussing the intricacies of model evaluation, Robin Dennis (1986) comments:

It should be clear from the above examples that one must do much more than simply calculate statistics to evaluate the performance of a model. One must first be clear about the relevant questions and then design a set of quantified measures to evaluate those questions. Otherwise, one does not know if the emperor is clothed or not. Experience indicates likely not.

Basically, model validation consists of comparing observed concentrations with modelled. But there is an infinite variety of ways to perform this task. Which numbers should be compared and what should be done to the result?

Much experience has been gained during the past two decades, and we are now in a position to point to some fruitful methods while others can be considered less useful. No single universal method can be pointed out, but a process can be initiated leading to the selection of recommended methods.

One lesson that has been learned is that when comparing values paired in both space and time, i.e. model predictions with observations at the same point and time, the correlation is virtually zero (see e.g. Reynolds et al., 1984). One of the reasons for this is an understandable, legitimate source of uncertainty: we cannot specify wind direction precisely, and the concentration at a point is totally dependent on whether a plume hits it centrally or misses it by a few degrees. In relation to many applications, however, this error is not important. This is an essential observation: one should avoid mixing up large - *but unimportant* - errors with the really important ones.

Accordingly, one must choose other ways of comparing data than a straightforward pairing in space and time. One way is to focus on the highest concentrations. This simplifies the problem to one where models have a greater chance for success. There are several different concentration variables that can be considered for the purpose; however, space does not allow a thorough discussion here (cf. Olesen, 1993)

An example of an evaluation procedure developed for a particular application is the concept of a *composite performance index* which has been adopted by the US EPA. This number represents a weighted mean of performance indices based on one-hour averages, 3-hour averages and 24-hour averages. Such a composite index is relevant in context with the US regulations, but has less interest in Europe.

A useful reaction to the difficulty of choosing a relevant evaluation method is to develop an array of methods to be used in various contexts. However, as pointed out by Dennis, before making any final judgement about a model, one should be convinced that the test undertaken really corresponds to the questions asked.

Processing of input data is not trivial

Beneath this modest statement a host of problems is concealed. Some of them can be exemplified with reference to the Kincaid data set discussed at the Manno workshop.

In the first place, a model needs *meteorological input data*. For the Kincaid data set, one has a choice between using *routine measurements* or the more specialized *research-grade measurements* that are also available from the experimental campaign. After choosing one of these - for instance the research-grade data - one is faced anew with a number of choices. Should an *observed* boundary layer depth or a *predicted* boundary layer depth be used? Should an observed or predicted value of σ_v be used? And on the basis of which data should heat flux be calculated? Etc. ...

According to some preliminary results for one specific model shown at the workshop, the effect of choosing observed values of σ_v - as opposed to computed values - results in predictions of the maximum concentration for the entire data set which are a factor of three larger than otherwise! This illustrates the magnitude of the problem.

Processing of *tracer concentration data* is also far from simple. All three data sets selected for the Manno workshop were based on field tracer experiments where the monitoring networks were dense. Therefore, it was possible to determine maximum concentrations for crosswind arcs. Such maximum arc-wise concentrations were compared to the computed plume centre line concentration at the same distance. By following this procedure instead of comparing concentrations paired in space and time, one eliminates a serious, but unimportant source of error, namely uncertainty in wind direction. Valuable information such as the distance to the observed concentration is still retained in the data set.

However, it was clearly demonstrated at the workshop that determining an arc-wise maximum cannot always be done in a hands-off, objective manner. The Kincaid experiment was characterized by the fact that very often the concentration pattern was highly irregular. Determining whether an observed arc maximum is representative enough to be used requires a manual, subjective analysis.

An example can serve to illustrate the importance of the subjective quality control of tracer data. Computed plume centre line concentrations were compared to measured maximum arc-wise concentrations as described above. One model had 36% of its predictions within a factor of two when a first version of the Kincaid data set was used, in which arc-wise maxima were indicated for *all* arcs, no matter their quality. After exerting a quality control on data, resulting in a more reliable data set which was also distributed to workshop participants, 51% of the predictions were within a factor of two. Thus, quality control is of utmost importance to the apparent performance of a model.

Quality control of tracer data was also performed independently by one of the participants. The result was another data set, for which also 51% of predictions were within a factor of two. Interestingly, there was a substantial difference in terms of another performance measure, namely the mean fractional bias FB (FB = $2 (C_{obs}-C_{pred}) / (C_{obs}+C_{pred})$). For the two quality controlled data sets, FB was respectively -0.38 and -0.03 (where -0.67 corresponds to an overprediction by a factor of two). Evidently, quality control is not straightforward. During the workshop, opinions converged on how quality control should be performed, but a subjective element unavoidably remains.

To conclude the discussion on processing of input data it can be stated once again that careful work should be done and pitfalls identified. For a diagnostic evaluation it can be of value to consider several different levels of input data (e.g. routine data and research-grade data). However, when it comes to *comparing models*, the models should all be run on the same data according to a common protocol.

Inherent uncertainties exist

When validating models against an experimental data set, a large scatter between observations and predictions is to be expected. Several contributions to the scatter can be identified:
- data errors
- inadequacy of the model formulation
- stochastic nature of the problem.

The two first reasons are relatively straightforward. The third - the stochastic nature of the problem - implies that there is an inherent uncertainty in air quality modelling. A model prediction is meant to represent an ensemble average, whereas any given observation reflects a specific realization that will almost always differ from the prediction - no matter how perfect the model and the input data.

How should this problem be dealt with?

It is a rule of the game and as such has to be accepted. A useful reaction is to try to quantify the uncertainties. One way of doing this is to establish a reference data base of model performance.

Workshop conclusions

In the discussion presented above, emphasis has been on model evaluation *methods* as opposed to *results*. This is a so because the results at the workshop were of a preliminary nature, and great care should be exercised in drawing conclusions based on the limited amount of investigations reported. Readers interested in details are referred to the workshop proceedings which are in preparation[1]. A few remarks on model tests are appropriate, however.

For the workshop, a few modern Gaussian plume models based on boundary layer parameterization were tested on the workshop data sets (the Danish OML, the American HPDM, the British UK-ADMS). Also, a more traditional Pasquill-Turner-based model, the American ISC2, was considered.

To be very brief it can be stated that in many respects the newer models performed similarly, tending to have their highest values closer to the source than traditional Pasquill-Turner based models. Results from the ISC2 model which showed severe underprediction were presented at the workshop; these results were, however, later shown to be erroneous. Work will be continued to resolve peculiarities concerning model performance on the data sets.

Despite the difficulties with objective quality control of concentration measurements, the procedure of considering arc-wise maxima was regarded as valuable. Use of this method is restricted to data sets for single isolated sources and with a good coverage of monitors.

The approaches which had been proposed for comparing models were generally accepted as appropriate, and were regarded more or less as a *de facto* standard for model intercomparison. The software package used for model evaluation was considered to be a tool of great practical value, though a few additions to it were suggested.

---

[1] Editor: C. Cuvelier, JRC Ispra, TP 690, 21020 Ispra, Italy

5 RELATED ACTIVITIES

A number of activities are presently in progress, fitting into the list of recommendations from the workshop at Risø.

Forthcoming workshop

The Manno workshop dealt with model evaluation which was one of the subject areas pointed to at Risø. In Manno, a number of questions requiring further exploration emerged. Therefore, a subsequent workshop to be held in Mol, Belgium in November 1994 will continue the work of the Manno workshop. Its title is *"Operational Short-Range Atmospheric Dispersion Models for Environmental Impact Assessments in Europe"*. Besides following up on the work done in Manno, a main theme will be the demonstration of administrative implications when various practical models are applied to realistic sources in Europe[2].

Meteorological input

In order to satisfy the requirements of the "Pasquill-Gifford-Turner type" regulatory dispersion models widely in use at present, meteorological data processing need not be very sophisticated. Besides determining the Pasquill stability classes, not much else need be done. However, up-to-date dispersion models with a better representation of atmospheric physics require as meteorological input more fundamental atmospheric boundary-layer parameters. These parameters include surface heat flux, friction velocity and boundary-layer height.

During the past decade, a number of different schemes and methods for deriving such parameters from standard operational meteorological data have been developed. It is now recognized that there is a need to understand how well the various meteorological preprocessors perform.

Therefore, a project in the framework of COST - a European collaboration within science - has just been started. The project will include a study of methodologies used in current meteorological preprocessors. One major undertaking within the project to test the results from these methods against parameter values derived from reliable boundary-layer experiments and from radiosonde data.

Presently, there is an official call for European countries to indicate whether they wish to take part in the project labelled COST-710, *"Harmonization in the pre-processing of meteorological data for dispersion models"*. About 20 research groups in Europe have unofficially announced their interest in the project[3].

Electronic information exchange

At the first workshop it was recommended to start an electronic Bulletin Board Service (BBS) or other similar services in order to help in the coordination of model development and management in Europe. No such BBS has yet been set up, but a somewhat simpler service has been established, namely an anonymous FTP service. It is operated by the National Environmental Research Institute, Denmark, and provides information on activities in relation to the Risø and Manno workshops[4]. Another anonymous FTP service of relevance is that of ERCOFTAC, i.a. containing information on a Special Interest Group on Turbulence and Dispersion in the Urban Atmosphere[5]

---

[2] Information from G. Cosemans, VITO, Boeretang 200, 2400 Mol, Belgium.

[3] Temporary project coordinator is Dr. R. Maryon, British Meteorological Office, London Road, Bracknell, Berkshire RG12 2SZ, U.K.; fax +44 344 854493.

[4] The service can be reached on the Internet at the node 130.226.32.4. The user ID is *anonymous* and there is no password (just press Carriage Return).

[5] Internet node 128.178.163.130 (dmehpb-f.epfl.ch); username: *anonymous*, password: *your e-mail address*, directory: ercoftac.

Also of relevance to the modelling community is an electronic mail list within the Global Research Network on Sustainable Development (GRNSD). A thematic group on *Atmospheric Dispersion of Chemicals* has been established which will be used, i.a., for discussions on intercomparison of models[6].

The above mentioned services are directed mostly towards an audience of model developers. However, it seems relevant also to mention the SCRAM BBS operated by the US EPA which is directed towards model users (SCRAM: Support Centre for Regulatory Air Models). SCRAM is available via telephone as well as through the Internet[7].

Risk assessment

Within a research program dealing with major industrial hazards supported by the CEC DG XII (Commission of the European Communities Directorate-General for Science, Research and Development), efforts are in progress to set up a more systematic framework for model evaluation. This activity relates not only to atmospheric dispersion models, but to all types of technical models used in risk assessment.

AERMIC

Finally it should be mentioned that in the USA and in Canada there are efforts to modernize regulatory dispersion models. In the USA, the so-called AERMIC committee - the AMS/EPA Regulatory Model Improvement Committee - was established about two years ago. The role of the committee is to introduce state-of-the-art modelling concepts into regulatory dispersion models by providing recommendations for changes in the various model components. The work of AERMIC is the subject for another presentation (by J. Weil) at this conference.

REFERENCES

Dennis, R.L., 1986, Issues, design and interpretation of performance evaluations: Ensuring the emperor has clothes, *in:* "Air Pollution Modeling and Its Application V", pp. 411- 424. Plenum Press, New York.

Hanna, S.R., 1993, Uncertainties in air quality model predictions, *Boundary-Layer Meteorology*, 62:3.

Olesen, H.R., and Mikkelsen, T., 1992, "Proceedings of the workshop on Objectives for Next Generation of Practical Short-Range Atmospheric Dispersion Models", Risø, Denmark. Available at National Environmental Research Institute, P.O. Box 358, DK-4000 Roskilde, Denmark.

Olesen, H.R., 1993: Review of earlier model evaluation work. Proceedings of the workshop on *"Intercomparison of Advanced Practical Short-Range Atmospheric Dispersion Models"*, Manno, Switzerland. Available from C.Cuvelier, JRC Ispra, TP 690, 21020 Ispra, Italy.

Reynolds, S.D., Seigneur, C., Stoeckenius, T.E., Moore, G.E., Johnson, R.G., 1984, "Operational Validation of Gaussian Plume Models at a Plains Site", Electric Power Research Institute, Palo Alto, California.

---

[6] Further information from Dr. Ivo Bouwmans, e-mail ivo@dutw239.tudelft.nl

[7] Telephone +1 (919) 541 5742; Internet address: ttnbbs.rtpnc.epa.gov

DISCUSSION

C. NAPPO: Have you considered the use of mesoscale/Lagrangian Particle Diffusion models to produce data sets for model evaluation?

H.R. OLESEN: The work described concerning model evaluation can be regarded as a demonstration exercise which has until now been of limited extent. In this first phase of model evaluation, field experiments have had a high priority. However, in the next phases of model evaluation work, other types of data sets can also be included. It is the plan that before the next workshop in Mol, large eddy simulations (LES) will be undertaken for a few experiments selected from the datasets used in Manno.

UK ATMOSPHERIC DISPERSION MODELLING SYSTEM

VALIDATION STUDIES

D.J. Carruthers[1], C.A. McHugh[1], A.G. Robins[2], D.J. Thomson[3],
B. Davies[3], M. Montgomery[3]

1. Cambridge Environmental Research Consultants Ltd., Cambridge
2. University of Surrey, Surrey and National Power, Swindon
3. Meteorological Office, Bracknell

1. INTRODUCTION

The UK Atmospheric Dispersion Modelling System (UK-ADMS) is a new PC based dispersion model for continuous releases, or releases of finite duration, based on an up to date understanding of the boundary layer and dispersion and, including the complex effects of underlying complex terrain, buildings and coastlines. The scientific approach used has been described at previous ITM's (Hunt et al 1990, Carruthers et al 1993). After extensive validation and testing the system has now been released. The purpose of this paper is to summarise the main features of the system and present important aspects of the validation.

2. SUMMARY OF PRINCIPAL FEATURES OF THE MODEL

UK-ADMS runs under Microsoft Windows on a PC. It typically takes only a few seconds to run, except where the effects of complex terrain are included or, multiple runs for sequences of meteorological data, or, statistical meteorological data are required. The principal modules of UK-ADMS are:

Meteorological Data Input Takes logged data which may be raw or statistically processed and uses them to produce appropriate quantities required by the boundary layer structure module. These include the wind speed and direction (U, ϕ) the boundary layer height h and and the Monin Obukhov length (L_{MO}).

Boundary Layer Structure Uses the processed met. data to calculate required profile of viscous flow $U(z)$, turbulence quantities (r.m.s. velocities σ_w, σ_v, σ_u and length scales L_w, L_v) and temperature and buoyancy frequency ($T(z)$, $N(z)$).

Mean Concentration Calculates the dispersion parameters σ_z, σ_y, σ_x (finite release only) and then calculates the mean concentration. The vertical profiles of concentration are Gaussian for neutral and stable flows, but skewed in unstable flows.

Plume Rise This is based on an integral model using an entrainment parameterisation and includes the effects of initial buoyancy and momentum of the plume.

Air Pollution Modeling and Its Application X, Edited by S-V. Gryning
and M.M. Millán, Plenum Press, New York, 1994

491

Concentration Fluctuations Calculates the concentration fluctuations occurring on time scales less than one hour. Both the variance in the concentration and the probability that a given concentration is exceeded are output as a function of the averaging time, which can be as small as is required.

Complex Effects

Complex Terrain The effect of underlying complex terrain on dispersion, including changes in both elevation and surface roughness, are accounted for by making use of the airflow model FLOWSTAR (Carruthers et al, 1990). This calculates changes in both mean and turbulent flow due to variation in surface elevation and surface roughness.

Buildings A model of dispersion influenced by a main site building is included using a model for flow round a building which includes the near wake (recirculation region), main wake where there is strong downflow, and the far wake (Puttock 1978).

Coastline This considers the case where a release into an external layer flowing off the sea, passes through a density interface into a growing thermal boundary layer. Material is then mixed rapidly within this layer.

Removal Processes The model includes dry and wet deposition and gravitational sedimentation. These effects are parameterised using deposition velocities, which may be specified or calculated within the system, and a washout coefficient.

Radioactivity This module can be used to calculate the decay of radioactive releases and the γ-dose rate at the ground, under the plume centreline.

3. VALIDATION STUDIES

The emphasis in this paper is on studies using the flat terrain version of the model. Comparisons are made with a coded version of the R-91 algorithms (Clarke, 1979), the ISC model (US E.P.A.) and with field and laboratory data.

Before describing details of the comparisons it is important to first of all describe the salient features of UK-ADMS and how these are different from the R-91 and ISCST models[*]. In UK-ADMS the calculation of plume spread and concentration is a two stage process,

(i) the mean profiles of wind speed, turbulence (transverse component σ_v, vertical component σ_w) and buoyancy frequency (temperature gradient) are expressed in terms of the surface roughness z_0, and three meteorological parameters: the measured (given) mean wind speed at a specified height (z), the boundary layer height (h) and the Monin Obukhov length L_{MO}[†] which depends on the surface heat flux and surface shear stress. h and L_{MO} may be input directly into the model or may be calculated from a range of other related variables such as time of day, cloud cover and the surface heat flux. The vertical profiles are functions of both z/L_{MO} and z/h.

---

[*] R-91 and ISCST are based on the same principle. For conciseness we will henceforth refer only to the R-91 model in discussions of model details.

[†] Vertical motions on length scales larger than $|L_{MO}|$ are dominated by buoyancy induced effects (unstable) or are inhibited (stable). Length scales smaller than $|L_{MO}|$ are dominated by shear induced turbulence.

(ii) the variances σ_y, σ_z and thence concentration (C) are calculated from the calculated values of the mean wind speed $U(\bar{z})$, $\sigma_v(\bar{z})$, $\sigma_w(\bar{z})$ and buoyancy frequency $N(\bar{z})$, and an averaging time or mean wind meandering term. $\bar{z}$ is the mean height of the plume (calculated) which is initially equal to z_S, the source height. Thus the concentration at a given distance downstream depends principally on U, z_s/h, z_s/L_{MO} and h/L_{MO}.

By contrast, in R-91 σ_y, σ_z are determined solely by $U(10)$, the mean wind speed at 10m, and the Pasquill stability category P, surface roughness z_0 and averaging time; they do not depend on z_s or h (some dependence of C on h and z_s occurs by virtue of reflection at the ground and at the boundary layer top). This is clearly an approximation which, since P is determined by surface meteorology, is likely to be most appropriate for low level sources. The R-91 method does not consider, for example, the large variation of σ_w with height which occurs in unstable conditions. Another disadvantage of R-91 is that it cannot be adapted to include measured boundary layer parameters such as σ_v, σ_w and N; these are central to UK-ADMS.

Further significant differences between the models occur in convective flows, where UK-ADMS includes the effect of the skewed vertical velocity fluctuation ($w^3 \neq 0$) on the concentration distribution by using a non-Gaussian distribution (the R-91 distribution is Gaussian in all cases), and also in the treatment of plume penetration of the overlying inversion. UK-ADMS calculates the amount of material lost above the boundary layer by considering the buoyancy of the plume and the temperature step at the boundary layer top, if any, and the temperature gradient above. Computer codes based on R-91 reflect all material back into the boundary layer.

3.1 Unstable Flows - Laboratory Experiments

The basis of this study is the laboratory experiment of Willis and Deardorff (1978). The convective boundary layer is simulated using a laboratory model in which there is penetrative convection (with zero mean wind) of a deep fluid layer heated from below into an overlying stable layer. Although the experiment had zero mean wind, the atmospheric flow in UK-ADMS which models this convective dominated turbulence, has a non-zero wind speed and a relatively flat velocity profile, so that it may be said to represent the tank translated at a uniform speed. For all but very low-level sources the effect of wind shear and hence the surface friction velocity (u_*) is negligible and the results can be expressed in terms of the non-dimensional parameter

$$X = \frac{x}{h}\frac{w_*}{U}$$

where x is the distance downstream of the source and w_* is the mixed layer velocity scale which is represented in terms of the UK-ADMS parameters h and L_{MO} by

$$w_*^3 = \frac{h}{|L_{MO}|} \cdot \frac{u_*^3}{\kappa}$$

where κ is von Karman's constant (0.4). Thus w_* (and hence σ_w) increases with h/L_{MO} which increases as the surface heat flux (F_{θ_0}) increases. The parameters used in the experiments are shown in Table I.

Table I. Parameters for the Willis and Deardorff Experiments.

Experiment	Case	h(cm)	w_* (cms^{-1})	F_{θ_0} (cms^{-1} °C)	z_s/h
2	1	31.0	1.06	0.165	0.25

Figures 1-4 (σ_y, σ_z, C, $\bar{z}$) show data from experiment 2 (elevated source). In this case all the models agree well with the experimental data for σ_y as does the numerical simulation of Lamb (1978). However, the surface layer based models (R-91 and ISCST) markedly under-predict σ_z. Note that the calculated values of σ_z do not include the effects of reflection and are thus expected to be larger than the 'effective σ_z' for $X > 0.6$. The contours of C show that UK-ADMS brings the position of maximum concentration down towards the surface (contrast R-91) and also predicts the increase of $\bar{z}$.

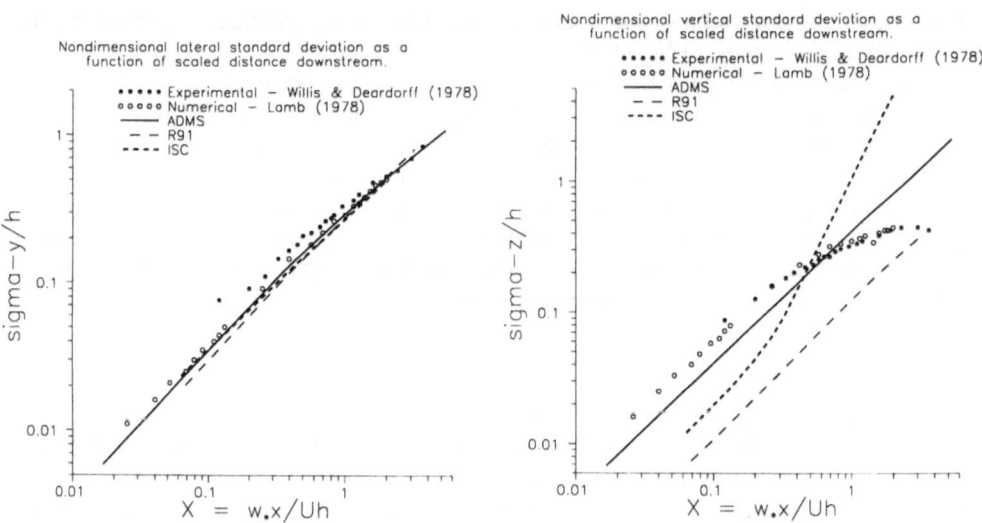

Figure 1. Comparison of ADMS with experimental and numerical results for σ_y.

Figure 2. Comparison of ADMS with experimental and numerical results for σ_z.

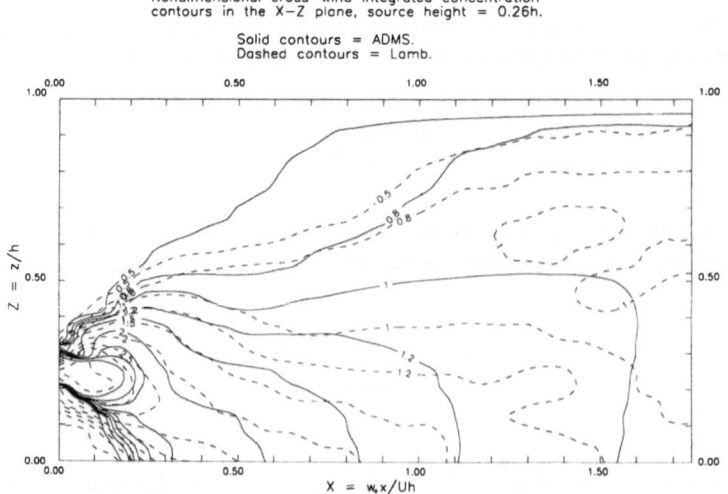

Figure 3. Comparison of ADMS with numerical results, contours of non-dimensional cross-wind integration concentration.

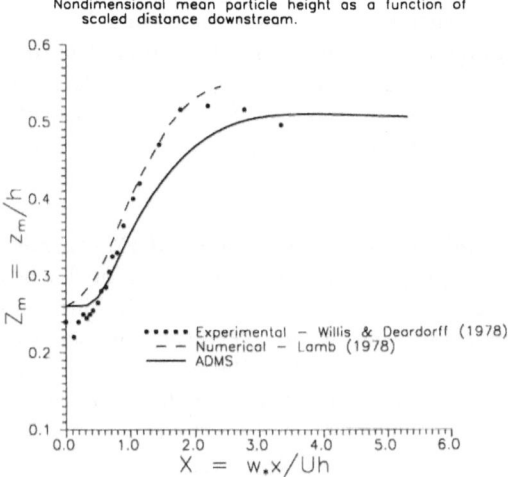

Figure 4. Comparison of ADMS with experimental and numerical results for $\bar{z}$.

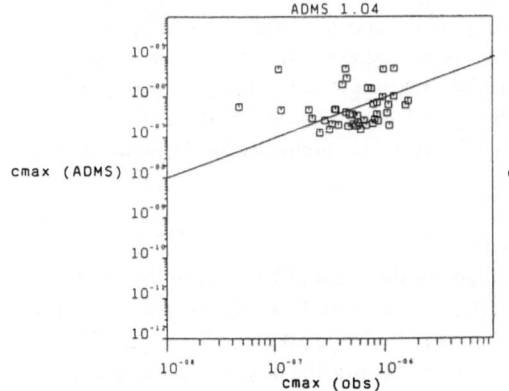

Figure 5. Scatter plot of ADMS maximum ground level concentration and measured values at Kincaid.

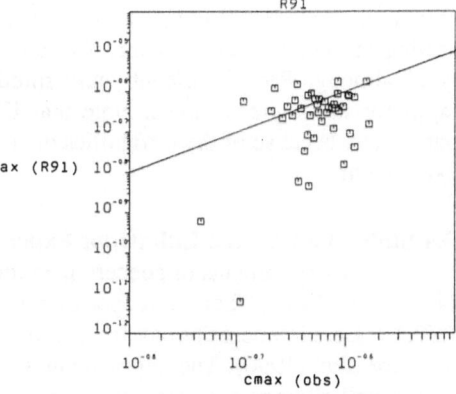

Figure 6. Scatter plot of R91 maximum ground level concentration and measured values at Kincaid.

3.2 Unstable Flows - Kincaid Field Data

Extensive field data have been obtained for sulphur hexaflouride releases from the Kincaid power plant in Illinois, USA (Hanna and Paine, 1989). The stack at the plant is 187m high with diameter 9m. Approximately 350 hours of data were collected mainly in convective conditions. For each hour ground level concentrations were measured along a series of arcs downstream of the stack. These measurements could be used to estimate the maximum ground level concentrations a given distance downstream of the stack. Figures 5 and 6 show scatter plots of maximum ground level concentrations for UK-ADMS and R91 for a high quality subset of data, that is, data for which the cross-wind profile shows a relatively well-defined maximum. The scatter in the plot using the UK-ADMS is seen to be much smaller than that of R91. The few overestimates predicted by UK-ADMS occurred near to the stack. R91 sometimes failed to predict significant concentration in these cases.

3.3 Neutral Flow -Wind Tunnel Measurements of Robins and Fackrell

Extensive laboratory experiments of the concentration downstream of point sources at different height in the boundary layer were made by Robins and Fackrell (Robins and Fackrell 1979). In this study we compare the results of their experiment B, with the predictions of UK-ADMS, R-91 and ISCST. The source height z_s was varied between $z_s/h = 0$ and $z_s/h = 0.5$. The flow field properties are shown in Table II. U_e is the external flow velocity .

Table II. Flow field properties for the Robins and Fackrell Experiment.

Flow	h(m)	z_0/h	u_*/U_e	z_s/h
B	0.6	10^{-4}	0.045	0.0 -0.5

The maximum ground level concentration (glc) and distance downstream of the maximum concentration (x'_{max}) are plotted in Figures 7 and 8. Given the uncertainties in scaling the wind tunnel to atmospheric conditions the agreement with all the models is good, although R91 significantly underpredicts the maxim glc while ISCST underpredicts (x_{max}) for the higher sources. Note that ISCST could not be used for the full range of parameters because of the difficulties of scaling the surface roughness for the wind tunnel experiment.

3.4 Stable Flows - the Lillestrøm Experiment

The experiments of concern here took place in the town of Lillestrøm (near Oslo), Norway in 1987. They were performed by the Norwegian Institute of Air Research (NILU). Detailed descriptions are given in the papers by Haugbakk and Tønnesen (1989) and Grønskei (1990). The experiments were carried out in a flat residential area with 6-10m high buildings and trees. A tracer system was used in which bromotriflouromethane ($CBrF_3$) was released from ground level (1m). Each experiment consisted of two sequential 15-min periods. Meteorological measurements were carried out along the 36m high mast. Sonic anemometer measurements were processed to give 10 min average values for wind speed and wind directions at the 10m level. Furthermore, covariances were determined between velocity components, and between velocity components and temperature fluctuations. The temperature during the tracer experiments was low ($\approx$ -20° Celsius), and the ground was snow-covered. The surface roughness was about 0.5m.

For all runs during the experimental campaign, the crosswind profiles of tracer concentrations were well determined, thus making a relatively accurate estimate of crosswind integrated concentration (C_y) possible.

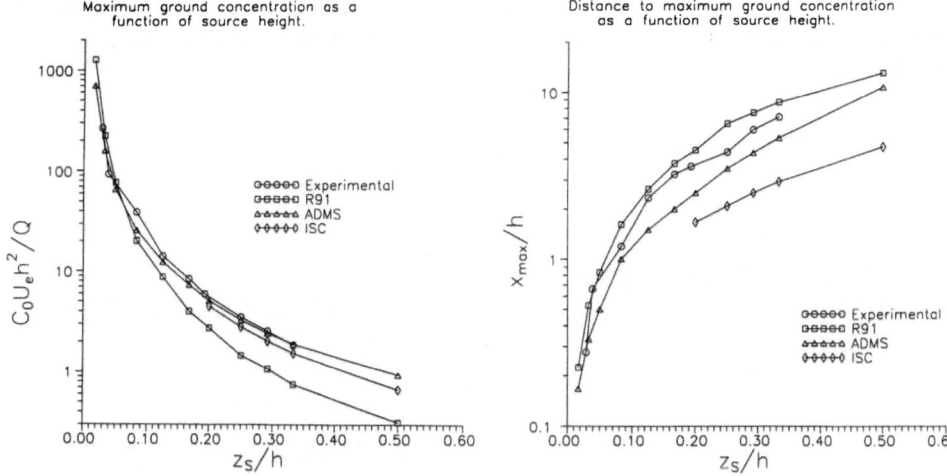

Figure 7. Comparison of ADMS maximum ground level concentration with wind tunnel experiments and other models.

Figure 8. Comparison of ADMS distance to ground level maximum concentration with wind tunnel experiments and other models.

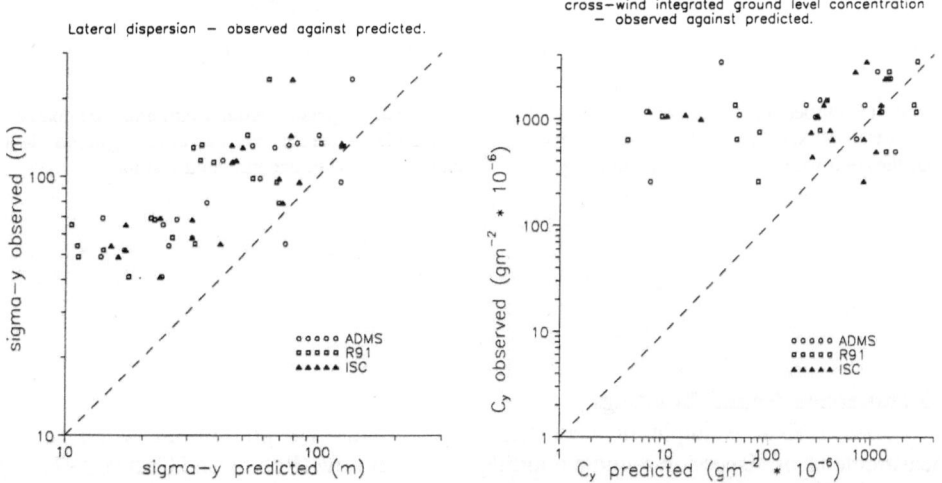

Figure 9. Comparison of ADMS σ_y with observed values at Lillestrøm.

Figure 10. Comparison of ADMS cross-wind integrated ground level concentration with observed values at at Lillestrøm.

Figures 9 and 10 show scatter plots of observed values versus predicted values for σ_y and C_y for an elevated release. None of the models perform well. Concentrations are much higher than expected; this appears to be due to the fact that the centre of the plume is brought rapidly down to the surface either by internal gravity waves or by increased turbulence levels over the built up area either because of local convection or increased surface drag. The experiment highlights the great difficulty of making good predictions of dispersion in stable flows since such features as gravity waves cannot be predicted with certainty.

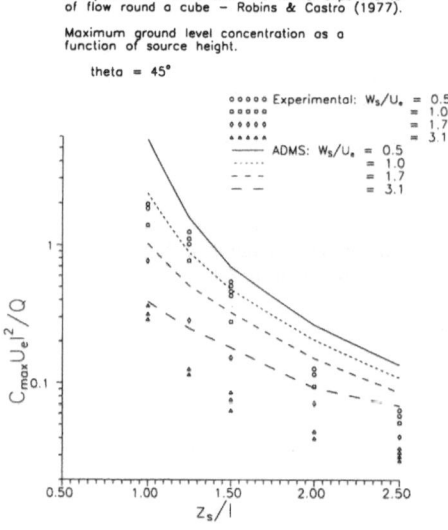

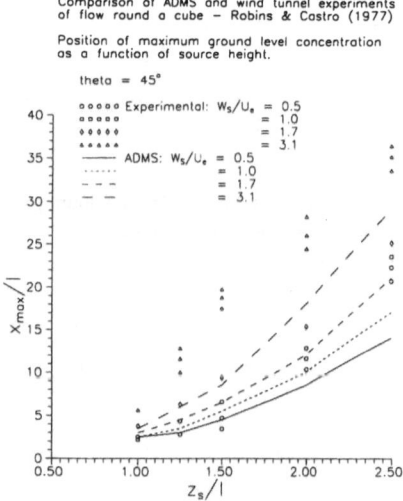

Figure 11. Dispersion around buildings, comparison of ADMS maximum ground level concentration with experimental results.

Figure 12. Dispersion around buildings, comparison of ADMS distance to maximum ground level concentration with experimental results.

3.5 Dispersion Round Buildings

As a final example of comparisons with data, we present the wind tunnel measurements of dispersion round a building in neutral flow (Robins and Castro, 1977a,b) and compare them with the predictions of the building module of UK-ADMS. Figures 11 and 12 show the maximum surface concentration for one set of experiments when the block was angled at 45° to the flow and the distance downstream of the maximum. The stack is situated at the centre of the cubic building, at (0,0), at a height equal to 1.5 times the cube length.

Figure 13 shows vertical and horizontal cross-sections of the calculated concentration with, and without, the presence of the buildings. Notice how the material is brought down into the building wake. Similar results were obtained with the block angled at 0° to the wind.

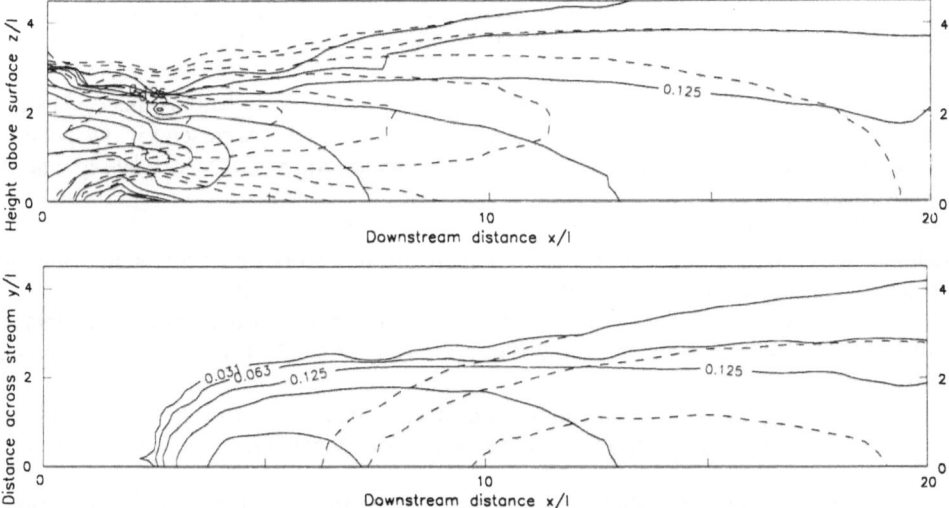

Figure 13. Dispersion around buildings, comparison of ADMS concentration field, with and without building.

4. DISCUSSION

This paper presents some comparisons betweem UK-ADMS and laboratory and field data. We have presented both case by case comparisons and scatter plots. The conclusions from the studies are that UK-ADMS is a model which has great flexibility, firstly in terms of the input meteorological parameters, and hence the physical situations which may be successfully modelled. Secondly, it is flexible in its range of application, here illustrated by its use with buildings, but it may also be used with complex terrain (hills and variable roughness) and near a coastline. It has been shown that there are benefits from the improved physics in UK-ADMS, in particular, it avoids the serious underpredictions of the other 2 models for elevated sources in convective conditions. The cases shown form a small subset of our extensive validation studies which include further case by case and statistical comparison studies, using mean concentration calculations over flat terrain. It also includes studies with other modules of the system, such as the meteorological input (e.g. estimates of boundary layer height), boundary layer structure (comparison between predicted vertical profiles of the boundary layer variables and field and large eddy simulation data), complex terrain (comparison with field data of both airflow and concentration distribution) and concentration fluctuations.

5. REFERENCES

Carruthers, D.J., Holroyd, R.J., Hunt, J.C.R., Weng, W.S., Robins, A.G., Apsley, D.D., Smith, F.B., Thomson, D.J., Hudson, B. 1991. UK Atmospheric Dispersion Modelling System. Proc. 19th International Technical Meeting on Air Pollution Modelling and its Applications. Crete.

Carruthers, D.J., Holroyd, R.J., Hunt, J.C.R., Weng, W.S., Robins, A.G., Apsley, D.D., Smith, F.B., Thomson, D.J., Hudson, B. 19921. UK Atmospheric Dispersion Modelling System. *Proc. 20th NATO-CCMS Int. Conf. on Air Pollution Modelling and its Applications, Greece.*

Clarke, R.H. (1979) A model for short and medium range dispersion of radionuclides released to the atmosphere. (First report of a working group on atmospheric dispersion.) National Radiological Protection Board Report NRPB-R91.

Haugbakk, I. and Tønnesen, D.A. (1989): Atmospheric Dispersion Experiments in Lillestrøm. 1986-1987 Data Report. Lillestrøm, Norwegian Institute for Air Research (NILU OR 41/89).

Grønskei, K.E. (1990): Variation in dispersion conditions with height over urban areas - results of dual tracer experiments. 9th AMS Symposium on Turbulence and Diffusion, 1990.

Hanna, S. and Paine, 1989. Hybird Plume Dispersion Model (HPDM) development and evaluation. *J. Appl. Met.* **28**, 206-224.

ISC Industrial Source Complex Model, US E.P.A., Office of Air Quality Planning and Standards.

Lamb, R.G. (1979): A numerical simulation of dispersion from an elevated point source in the convective planetary boundary layer. *Atmospheric Environment* **12** 1297-1304

Robins, A.G. and Castro, I. 1977a,b. *Atmospheric Environment* **11** 291-297, 299-311.

Robins, A.G. & Fackrell, J.E. (1979): Continuous plumes - their structure and prediction in Mathematical Modelling of Turbulent Diffusion. ed. C.J. Harris, Academic Press.

Willis, G.E. & Deardorff, J.W. (1978): A laboratory study of dispersion from an elevated source within a modelled convective planetary boundary layer. *Atmospheric Environment* **12** 1305-1311.

DISCUSSION

J. WEIL: 1) How are plume rise and/or buoyancy effects included in the non Gaussian model for unstable conditions?

2) How is plume rise due to source momentum and buoyancy included in the building wake model and do you account for possible "fumigation" of a plume that initially rises above the wake (and intersects the wake further downstream)?

D. CARRUTHERS: 1) The plume rise model (momentum and buoyancy) is included within our non Gaussian model for unstable conditions in a manner similar to that used in the Gaussian model (stable and neutral conditions). The vertical (σ_z)plume spread is the sum of contributions from plume rise spread (σ_{zpr}), the spread due to mechanically driven turbulence σ_{zN} (the u_* contribution) and the spread due to convectively driven turbulence σ_{zC} (the w_* contribution), ie $\sigma_z^2 = \sigma_{zpr}^2 + \sigma_{zN}^2 + \sigma_{zC}^2$. Only the convective part causes a non Gaussian contribution to the vertical profile of concentration σ_{zC}. The height of the maximum concentration $z_p = z_s + z_{pr}$ where z_{pr} is the plume rise.

2) Plume rise is treated in the wake in a manner similar to other regions of the flow using the local (calculated) mean velocity in the plume rise calculation. "Fumigation" of a plume which initially rises above the wake is treated by calculating the vertical spread of the part of the plume within wake using the calculated local mean velocity and turbulence within the wake.

C. NAPPO: Please explain how an atmospheric gravity wave can transport an elevated plume to the ground surface?

D. CARRUTHERS: In the exceptionally stable conditions prevailing at Lillestrøm stationary gravity waves (like mountain waves) can be caused by buildings. These can bring material down towards the surface (or indeed move material away from the surface).

THE USE OF SIMULTANEOUS CONFIDENCE INTERVALS TO EVALUATE

CARBON MONOXIDE (CO) INTERSECTION MODELS

Donald C. DiCristofaro[1], David G. Strimaitis[1],
Thomas N. Braverman[2], and William M. Cox[2]

[1]Sigma Research Corporation
196 Baker Avenue
Concord, Massachusetts   01742

[2]U.S. Environmental Protection Agency
Office of Air Quality Planning and Standards
Technical Support Division
Research Triangle Park, North Carolina   27711

INTRODUCTION

The United States (U.S.) Environmental Protection Agency (EPA) has recently evaluated the performance of eight modeling techniques in simulating carbon monoxide (CO) concentrations at six intersections in New York City (EPA, 1992). The eight intersection modeling techniques evaluated include: CAL3QHC (1985 Highway Capacity Manual Modified CAL3Q Model), FHWAINT (Federal Highway Administration (FHWA) Intersection Model), GIM (Georgia Intersection Model), EPAINT (EPA Intersection Model), CALINE4 (California Line Source Model), VOL9MOB4 (MOBILE4 Modified Volume 9 Technique), TEXIN2 (Texas Intersection Model), and IMM (Intersection Midblock Model). The New York City database includes hourly meteorological, carbon monoxide, and traffic observations for six intersections in the city. This paper describes a method developed by EPA for aggregating component results of model performance into a single performance measure which is then used to compare the overall performance of the modeling techniques.

PHASE I: SCREENING TEST

The EPA (Cox, 1988) has suggested the use of a screening test for model performance, which would normally be applied to reduce the number of models evaluated using refined methods. This screening test was applied to the results obtained during phase I of this study, in which the eight models using the MOBILE4 emissions factor model were evaluated at all six intersections. The performance measure used for the screening test is the absolute fractional bias (AFB) defined as

Air Pollution Modeling and Its Application X, Edited by S-V. Gryning
and M.M. Millán, Plenum Press, New York, 1994

$$AFB = 2 \left| \frac{OB - PR}{OB + PR} \right| \qquad (1)$$

where OB and PR refer to the averages of the observed and predicted highest 25 values matched only by rank. The absolute fractional bias of the standard deviation is also used where OB then refers to the standard deviation of the 25 highest observed values and PR refers to the standard deviation of the 25 highest predicted values. If AFB tends to exceed 0.67 (factor of two) for either the average or the standard deviation, consideration may be given to excluding that model from further evaluation due to its limited credibility for refined regulatory analysis. In this evaluation of intersection models for CO, the eight techniques were ranked by AFB in order to help indicate which of the models would be evaluated in phase II of the study, in which the MOBILE4.1 model is used to estimate emissions. Of the three EPA intersection models (EPAINT, VOL9MOB4, and CAL3QHC), CAL3QHC performed best using MOBILE4. Of the two models utilizing the FHWA advocated average speed approach rather than explicit queuing (FHWAINT and GIM), GIM performed better. Therefore, the Phase II MOBILE4.1 analysis was performed for the following five models: CAL3QHC, GIM, IMM, TEXIN2, and CALINE4. The three intersections with the best quality assurance procedures and unhindered approach wind flows and wind field uniformity were also used for the Phase II analysis.

PHASE II: REFINED EVALUATION

The U.S. EPA has developed a method for aggregating component results of model performance into a single performance measure that may be used to compare the overall performance of two or more models (Cox, 1988; Cox and Tikvart, 1990). The bootstrap resampling technique (Efron, 1982) is used to determine the significance of differences in composite performance between models. Results from different data bases are combined using a technique related to meta-analysis to produce an overall result.

The EPA's scoring system for refined evaluations is divided into two separate components. The "scientific or diagnostic component" refers to the evaluation of peak concentrations during specific meteorological conditions at each monitor and the "operational component" refers to the evaluation of peak averages independent of meteorological condition or spatial location. The averages evaluated in the operational component are those for which regulatory standards must be met (e.g., 3-hour and 24-hour averages). The capability of models to predict concentrations at specific locations and meteorological conditions subject to the limitations of the data base is tested using the scientific component. The New York City data base contains mostly non-consecutive, one-hour observations, thereby limiting the evaluation to one-hour averages. There is a U.S. regulatory standard for one-hour average CO concentrations, so the dataset allows both diagnostic and operational components to be evaluated. Typically, monitors are located adjacent to an intersection, so that they record near-field concentrations during varying meteorological and traffic conditions. A diagnostic evaluation could focus on aspects of the performance that are related to wind speed, wind direction, stability, and traffic counts, for example. However, the wind direction aspect will not be addressed in this evaluation. In essence, it is believed that uncertainties in the "true" wind direction, coupled with a sparse monitoring network and a distributed source (intersecting line sources), preclude any attempt to accurately delineate the ability of a model to reproduce spatial relationships contained in the measured concentrations. Instead, the performance will be evaluated only on the basis of the peak modeled and observed concentrations during each hour at each intersection. This choice eliminates any difference between datasets for a diagnostic and an operational evaluation.

Blocking

Several subsets of the dataset for each site are formed in order to block the data according to parameters related to significant modeling variables. Thus, differences in model performance under different model regimes may be assessed. In this case, the relevant parameters are wind speed, stability class, and traffic counts (a crude measure of emission rate). There is very little variation in traffic counts from one hour to the next because most of the traffic data are associated with rush-hour conditions. This is not surprising, since the hours were selected on the basis of the maximum observed CO concentrations. Since there is not much variation in the traffic data, this parameter does not appear to be useful when examining the scientific component.

Overall the wind speed/stability classification seems to be a good manner in which to classify the data. It is important to choose a classification that maintains an equitable distribution of the hours across subsets. When confidence intervals are estimated for each class, these should be based on as many data points as possible. The blocked bootstrap resampling method is used to estimate confidence limits, as described later in this paper. The following wind speed (u)/stability classification was used: u ≤ 6 mph, neutral/stable; u ≤ 6 mph, unstable; and u > 6 mph, all stabilities.

Primary Performance Measure

Both AFB (absolute fractional bias) and FB (fractional bias) are used in the comprehensive evaluation. FB is used in the diagnostic evaluations, so that the tendency of a model to underpredict or overpredict can be characterized. However, the AFB is used when combining results for various categories or sites so that cancellation of overpredictions or underpredictions do not occur.

When calculating either FB or AFB, the "robust highest concentration," RHC, is used rather than the mean of the highest 25 concentrations. As discussed by Cox and Tikvart (1990), the RHC is preferred in this type of statistical evaluation because of its stability. Also, the bootstrap distribution of the RHCs is not artificially bounded at the maximum predicted or observed concentration, which allows for a continuous range of concentrations. The RHC is based on a tail exponential fit to the upper end of the distribution and is calculated as follows

$$RHC = x(n) + (\bar{x} - x(n)) \ln \left[\frac{3n - 1}{2} \right] \qquad (2)$$

where

$\bar{x}$ = average of the n-1 largest values,
$x(n)$ = nth largest value,
n = number of values exceeding the threshold value (n=26 or less).

The size of the three intersection data sets requires the value of n to be less than 26. For this study, the value of n is nominally set to 11 so that the number of values averaged $(\bar{x})$ is 10. A threshold of 0.5 ppm is used.

Composite Performance Measure

A composite performance measure (CPM) is calculated for each model as a weighted linear combination of the individual absolute fractional bias components. The operational component is given a weight that is equal to the weight of the combined scientific components. The results from the different data bases (intersections) are given equal weight. The CPM is defined as

$$CPM = \frac{1}{2} \overline{AFB_d} + \frac{1}{2} AFB_1 \qquad (3)$$

where

AFB_d = Absolute fractional bias weighted for each diagnostic category d,

$$AFB_1 \quad = \quad \text{Absolute fractional bias for the operational one-hour averages.}$$

The wind speed (u) ≤ 6 mph and neutral/stable category is weighted more than the other two categories because of the importance of this category for regulatory modeling purposes. Thus, the average of AFB_d is

$$\overline{AFB_d} = 0.5 \; AFB(u \leq 6 \; mph, \; Neutral/Stable) + \\ 0.25 \; AFB(u \leq 6 \; mph, \; Unstable) + \\ 0.25 \; AFB(u > 6 \; mph, \; All \; stabilities) \tag{4}$$

Model Comparison Measures

Because the purpose of the analysis is to contrast the overall performance among the models, differences in model performance are characterized by calculating pairs of differences in the composite performance measure between the models. The difference between the composite performance of one model and another is the model comparison measure (MCM) defined simply as

$$MCM_{i,j} = CPM_i - CPM_j \tag{5}$$

where CPM_i = Composite performance measure for model i,
CPM_j = Composite performance measure for model j.
For the five models compared using the MOBILE4.1 emissions methodology, there are ten comparison measures computed. The MCM is used to judge the statistical significance of the apparent superiority of any one model over another.

Confidence Intervals

The bootstrap resampling technique is used to estimate confidence intervals on the various measures described above. In applying the bootstrap procedure, observed and predicted one-hour data are resampled for each intersection. Sampling is done with replacement, so some hours are represented more than once. This process is repeated 1000 times so that sufficient samples are available to calculate the standard error of each measure. At each site, the resampling recognizes the blocks selected for the diagnostic evaluation. This assures that each of the 1000 variants of the original dataset retained the same number of samples from each diagnostic category. Had the data not been blocked in this way, one of the 1000 variants might, for example, only consist of a few samples associated with the largest wind speed (repeated many times). The bootstrap resampling method allows the standard deviation, s_{ij}, of any performance measure to be estimated, from which confidence limits can be calculated:

$$95\% \; Confidence \; Limits = Measure \pm c \; s_{ij} \tag{6}$$

The standard error is simply the standard deviation of the measure over all of the bootstrap-generated outcomes. If the measure involves a single comparison, such as FB for a single model, then the value of c can be set equal to the student-t parameter.

Difference measures such as ΔFB or MCM require that simultaneous confidence intervals be found for each pair of models in order to ensure an adequate confidence level and to protect against falsely concluding that two models are different The method of Cleveland and McGill (1984) is used to calculate c. In this method, c is found such that for 95 percent of the 1,000 bootstrap i-tuples,

$$\frac{|\Delta_{ij} - \Delta_{ijk}|}{s_{ij}(\Delta_{ijk})} \leq c \tag{7}$$

where Δ_{ij} = model comparison difference measure for model pair i, j,

Δ_{ijk} = model comparison difference measure for model pair i, j and bootstrap replication k, and

s_{ij} = standard deviation of all the Δ_{ijk} values.

For this analysis, c is found for each of the three intersections (c_l),

$$\frac{|M_{ijl} - M_{ijkl}|}{s_{ij}(M_{ijkl})} \le c_l \qquad (8)$$

where M_{ijl} = model comparison difference measure (MCM) for the lth database, and ith and jth model,

    i and j = 1 to 5 for each model combination,

    k = 1 to 1000 for each bootstrap,

    l = intersection database, and

    $s_{ij}(M_{ijkl})$ = lth standard deviation of M_{ijkl} for bootstrap replications 1 to 1000.

The model comparison difference measure (M_{ijl}) is based on differences in CPM and FB in between models. Using CPM, for example, the difference in CPM values between models i and j is calculated as

$$M_{ijl} = CPM_{il} - CPM_{jl} \qquad (9)$$

for the primary data set, and

$$M_{ijkl} = CPM_{ikl} - CPM_{jkl} \qquad (10)$$

for each bootstrap replication of the dataset.

Composite Model Performance Measure

The foregoing sections have identified how performance is quantified, how specific performance measures are found, how these are combined into a single measure for each model at each intersection, how differences in these measures between models are calculated, and how confidence intervals are found for all of these. What remains to be done is to calculate composite measures across all sites (intersections) used in the evaluation. A composite model comparison measure (CMCM) is suggested by Cox and Tikvart (1990):

$$CMCM = \frac{\sum (W_l \, M_l)}{\sum (W_l)} \qquad (11)$$

where M_l = model comparison measure for the lth data base,

    W_l = $1.0/S_l^2$, and

    S_l = bootstrap estimated standard error for the lth data base,

    Σ = summation over l = 1, L (for L databases).

Using the model comparison difference measure of Equation 9, bootstrap outcomes for the composite measure can be written as

$$CMCM_{ijk} = \frac{\sum_{l=1}^{L} \left(\dfrac{M_{ijkl}}{s_{ij}(M_{ijkl})^2} \right)}{\sum_{l=1}^{L} \left(\dfrac{1}{s_{ij}(M_{ijkl})} \right)^2} \qquad (12)$$

With this definition, a confidence interval on CMCM follows that of Equation 8, so that

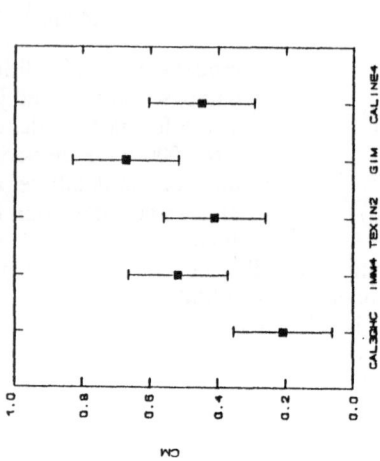

Figure 2. The composite model performance measure (CMCM) with 95% confidence limits using CPM statistics.

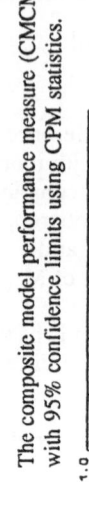

Figure 3. The CMCM with 95% confidence limits using the AFB of diagnostic category 1 (u ≤ 6 mph, neutral/stable) statistics.

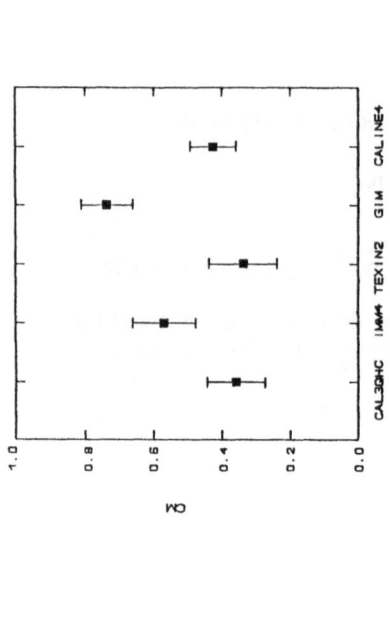

Figure 1. The composite performance measure (CPM) for each model as a function of site.

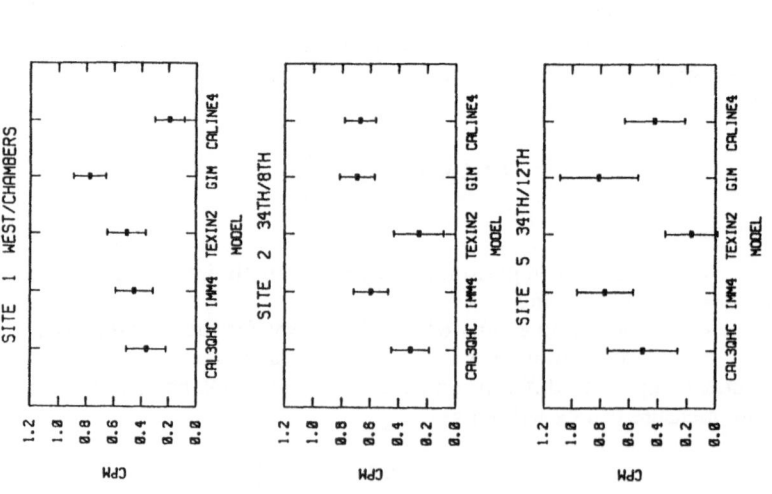

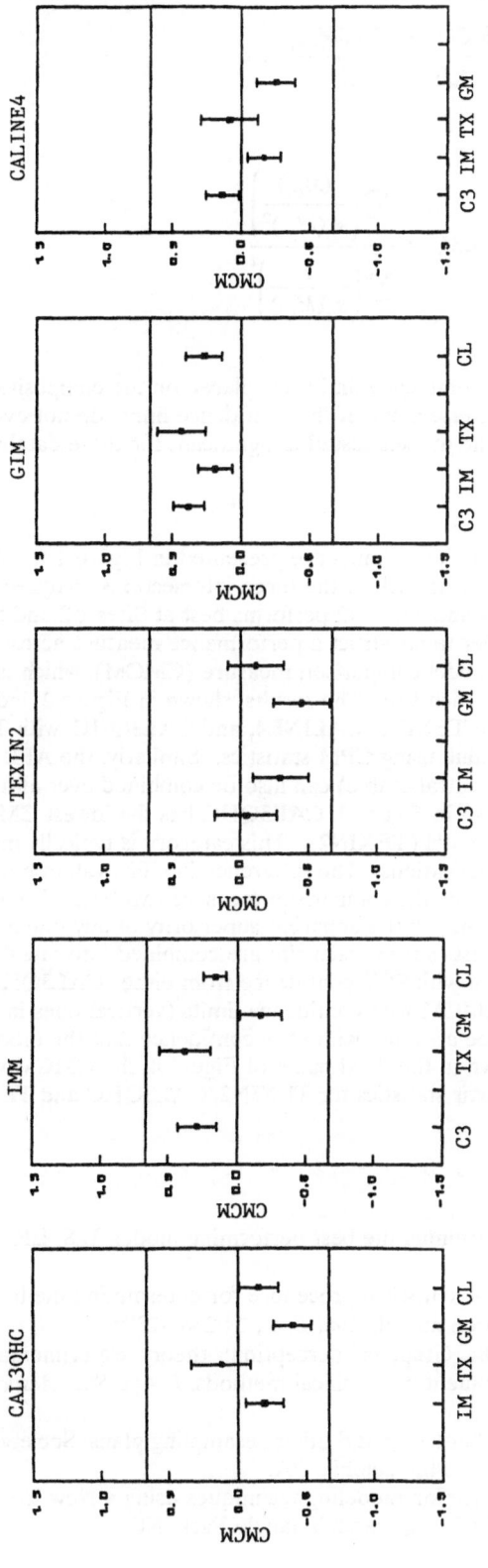

Legend: C3 = CAL3QHC
         IM = IMM
         TX = TEXIN2
         GM = GIM
         CL = CALINE4

Figure 4.   The composite model performance measure (CMCM) with 95% confidence limits
             for each model pair using model comparison measure (MCM) statistics.

the value of c for FB or CPM is the value that satisfies the 95 percent criterion for

$$\frac{|CMCM_{ij} - CMCM_{ijk}|}{s_{ij}(CMCM_{ijk})} \leq c \tag{13}$$

where

$$CMCM_{ij} = \frac{\sum_{1}^{l} \left(\frac{(M_{ijl})}{s_{ij}(M_{ijl})^2} \right)}{\sum_{1}^{l} \left(\frac{1}{s_{ij}(M_{ijl})} \right)^2} \tag{14}$$

For each model pair, simultaneous confidence limits are placed on the composite performance CM_{ij} as with the lth intersection. If the confidence limits do not overlap zero, then the difference between the models tested is significant for these databases.

RESULTS

The CPM values with 95% confidence limits are presented in Figure 1 for the five models tested in the Phase II analysis at each of the three intersections analyzed. CALINE4 performs best at Site #1 and TEXIN2 performs best at Sites #2 and 5. A further combination is made in order to construct a performance measure across all three sites. This is the composite model comparison measure (CMCM), which is made up of the CPM values calculated at each site. The results, shown in Figure 2, indicate that the best performing models are TEXIN2, CALINE4, and CAL3QHC, with TEXIN2 having the lowest overall CMCM value using CPM statistics. Similarly, the AFB from diagnostic category 1 (u ≤ 6 mph, neutral/stable) can also be combined over all three sites into a single CMCM. As shown in Figure 3, CAL3QHC has the lowest CMCM by a factor of two from the next best model (TEXIN2). This category is typically most important in terms of regulatory applications. The difference in CPM values between one model and another model is the model comparison measure (MCM). The MCM is used to judge the statistical significance of the apparent superiority of any one model over another. When the MCM statistics from each site are combined into one CMCM, TEXIN2 is not significantly different with 95% confidence from either CAL3QHC or CALINE4 (see Figure 4). If the CMCM with confidence limits (vertical lines in Figure 4) cross the zero line, then it may be assumed with 95% confidence that the models are not significantly different. As shown in the third panel of Figure 4, the CMCM with confidence intervals of the model pair statistics for TEXIN2/CAL3QHC and TEXIN2 /CALINE4 cross the zero line.

REFERENCES

Cox, W.M., 1988, Protocol for determining the best performing model, U.S. EPA, Research Triangle Park, NC.

Cox, W.M. and J.A. Tikvart, 1990, A statistical procedure for determining the best performing air quality simulation model, Atm. Env., 24:2387-2395.

Cleveland, W.S. and R. McGill, 1984, Graphical perception: theory, experimentation, and application to the development of graphical methods, J. Am. Stat. Assoc., 79:531-554.

Efron, B., 1982, The jackknife, the bootstrap and other resampling plans, Society for Industrial Applied Mathematics, Philadelphia, PA.

EPA, 1992, Evaluation of CO intersection modeling techniques using a New York City database, EPA-454/R-92-004, Research Triangle Park, NC.

DISCUSSION

S.T. RAO: Since MOBILE5 is the current version of the emission model, have you considered assessing the performance of the models with emissions derived from MOBILE5? We have found that the bootstrap method is very unreliable in providing the confidence bounds for small samples or for highly skewed distributions. Have you performed a simulation study to verify that 95% confidence bounds, estimated from the bootstrap method, do in fact contain the true mean 95% of the time? This is essential since the basis for your model performance evaluation is confidence intervals. Non-overlapping and overlapping confidence intervals need to be interpreted carefully in the context of a model's behavior assessment.

D. DICRISTOFARO: The model evaluation study used measured meteorological data with the MOBILE4.1 emissions factor model. Using MOBILE 4.1 emissions, the CAL3QHC model underpredicted by 25 to 30%. When worst-case meteorological conditions were used, the CAL3QHC model predictions were close to the observed concentrations. The MOBILE5A emissions factor model for the study year of 1989 gives approximately 25% higher emissions. Therefore, if the model evaluation study was redone using the measured meteorological data, the CAL3QHC model would perform better. Furthermore, the relative ranking of the models would not change. The U.S. EPA has agreed to a one year grace period (from the November 11, 1994 promulgation date) during which (1) users may use the MOBILE4.1 model and (2) the EPA will develop a refined modeling procedure for using the CAL3QHC model with measured hourly meteorological data.

The authors are aware of potential problems in application of the bootstrap to estimate confidence intervals with small samples for selected statistics which may have a highly skewed distribution. In this application, these potential problems are of little concern for several reasons. First, the evaluation study used 575 hours which we do not regard as being a small sample. Second, the fractional bias (which is the basic statistic we use for measuring model performance) is both symmetrical and bounded. Third, and perhaps more important, our model comparison evaluation statistic is based on a weighted composite of the fractional bias statistics among the various monitoring sites. By the central limit theorem, this composite statistic should tend towards normality which further supports our assertion that the 95% confidence bounds based on the bootstrap in this application are indeed good unbiased approximations to reality.

A STUDY OF THE DISPERSION OF AIR POLLUTANTS RELEASED FROM MAJOR ELEVATED SOURCES LOCATED NEAR ATHENS, GREECE

Pavlos Kassomenos, George Kallos, Maria Varinou and Anastasios Papadopoulos

University of Athens, Dept. of Applied Physics, Meteorology Lab., Ippocratous 33
Athens 10680, Greece

ABSTRACT

In the area around Athens is a number of elevated sources mainly power plants, refineries, cement plants etc. These industrial installations are using fuels of different types and characteristics. Almost all the examined installations are located near the coast and therefore the dispersion characteristics show significant spatial and temporal variations.

INTRODUCTION

The Attic peninsula is located at the SE part of the Greek one. To the E of the Attic peninsula is the southern part of the island of Evoia. The Attic peninsula is separated from the island of Evoia with a narrow strip of sea, the Evoic Gulf. To the W of the Attic peninsula is the Corinthian Gulf while to the S is the Saronic one. The Saronic Gulf is a closed Gulf with an opening to the SE. In the SW of the Attic peninsula is the NE Peloponnese. There is a narrow strip of land which separates the Saronic from the Corinthian Gulf (Fig. 1). There are three small plains in the Attic peninsula. In the middle of it, is the Athens basin while in the W and the E are the Thriassion and Mesogea plains respectively. The Attic peninsula and the surrounding areas have complicated physiographic characteristics and land water distribution. Almost the entire Athens basin could be considered as an urban area, while the other two plains are partly cultivated. The slopes of the mountains and hills are almost bald with patches of pine forests. Because of the complexity of the physiographic characteristics of this part of SE Greece, the flow fields and the Planetary Boundary Layer (PBL) structure are complicated. The same is true for the dispersion characteristics. The Athens basin with the Saronic Gulf, the NE Pelopennese, the Thriassion and Mesogea plains as well as the Evoic Gulf could be considered as an entire system where air masses from one could be transferred over the other. When and how this occurs are two questions directly related to the air quality of the region.

In the area around Athens there is a number of large industrial installations with tall stacks. Their height is varying between 30-150 m. Three power plants are located at Keratsini (at Piraeus harbour, marked by P_1), Lavrio (at the SE edge of Attic peninsula, marked by P_2) and Aliveri (marked by P_3) in southern Evoia at Evoic Gulf (see Fig. 1). The power plant of Keratsini is not working since 1980 but we have done the simulation because the Power production Company of Greece is planning to set it in operation again. The other significant elevated sources are located at Thriassion plain and W of it at the W coast of Saronic Gulf. From these sources the most important ones are three oil refineries (Agioi Theodoroi, marked by R_1, Elefsis, marked by R_2, ELDA, marked by R_3). Their positions are indicated in Fig. 1. There is a number of other minor sources like cement plants, steel smelters, etc. but most of them are not in operation now. A major cement plant is located

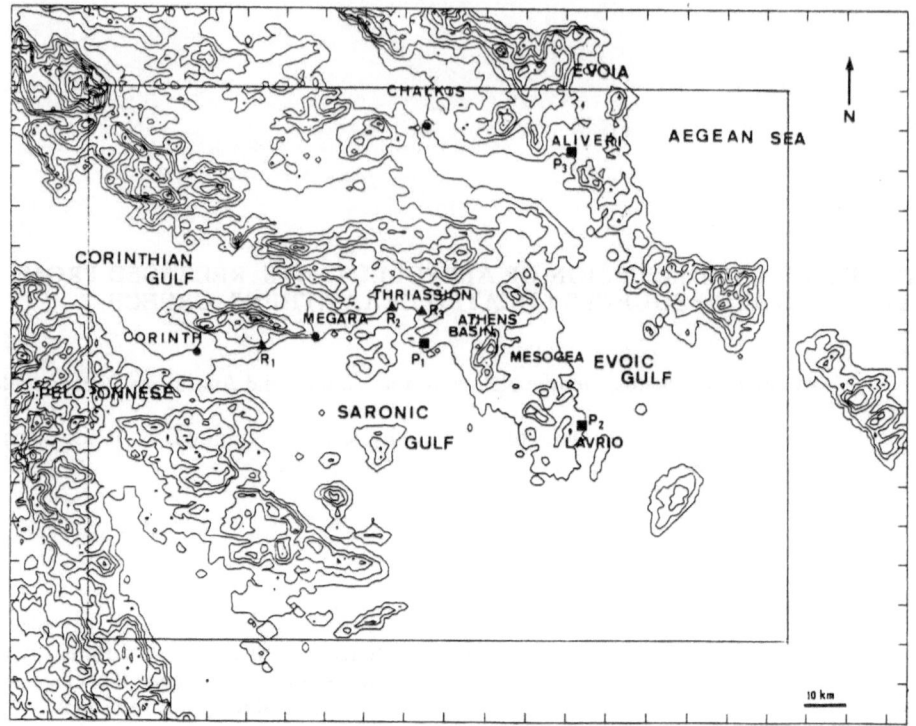

Fig. 1. The topography of SE Greece. Contours are every 200 m. The distance between two consecutive tick marks is 10 Km. The refineries (▲) and power plants (■) are also shown.

near Aliveri. One of these three refineries is located west of mount Aegaleo (ELDA), at the eastern part of Thriassion plain, the second is located at the western edge of it (Elefsis) and the third near the Isthmus of Corinth (Agioi Theodoroi). All these industrial installations are located at the coast of Saronic and Evoic Gulf. The location of these sources, the characteristics of their stacks, the complicated land-water distribution and in general the physiographic characteristics of the region make the dispersion of the plumes very complicated. Phenomena like plume fumigation, impingment, recirculation splitting etc. occur very often in this region. Because of the air pollution problem in the Greater Athens Area (GAA) the questions related to "if", "when", "which" and "under which circumstances" the plumes from these industrial installations contribute to this problem need clear answers. These answers cannot be provided in a simple manner. Simple techniques like Gaussian plume dispersion calculations cannot be considered as the adequate approach for obvious reasons. In the present study, a two-stage procedure was employed: In the first stage the 3-D atmospheric fields were resolved with the use of the Colorado State University Regional Atmospheric Modelling System (CSU-RAMS). In the second, a Langrangian Particle Dispersion Model (LPDM) was used in order to describe the dispersion characteristics of each plume. A brief description of these two models is given below.

MODEL DESCRIPTIONS

Atmospheric Model: The atmospheric model, the CSU-RAMS, has been developed at the Department of Atmospheric Science, Colorado State University from the groups of R. Pielke and W. Cotton (Walko and Tremback, 1991; Pielke et al., 1992). It is an advanced model with unique capabilities which make it one of the most appropriate models for simulations with complex flow interactions. One of the most significant features of CSU-RAMS is the two-way interactive nesting with any number of either telescoping or parallel fine nest grids. The model uses terrain following coordinate surfaces with cartesian or polar stereographic horizontal coordinates and cloud microphysics parameterization at various levels of complexity. Other significant capabilities of the model are the various schemes for turbulence or radiative transfer parameterizations (short and longwave) through clear and

514

cloudy atmospheres. For the boundary conditions the model uses various options for upper and lateral boundary conditions and various levels of complexity for the surface-layer parameterization (soil model, vegetation etc.). Finally, for initialization it uses either horizontally homogeneous or variable (isentropic analysis) data sets (e.g. NMC, ECMWF). It may also use outputs from other models.

Lagrangian Particle Dispersion Model (LPDM): It is a Langrangian-type dispersion model (McNider et al., 1988, Moran, 1992). The motions of discrete mass elements (which represent pollutants) are tracked inside the model domain as they move with the synoptic, mesoscale and microscale (turbulent) wind components. Synoptic and mesoscale wind components are these calculated directly from the mesoscale model while the turbulent ones are deduced from the atmospheric model closure scheme. This model has been modified accordingly for the use with CSU-RAMS.

MODEL SETUP-DATA USED

In order to provide an accurate description of the subregional-scale phenomena the nesting capabilities of the atmospheric model, were used. Two nested grids were used. The coarse grid, with horizontal grid increments of 16 Km, covers a portion of the Ionian Sea to the W and the western part of the Asia Minor to the E. To the S, it starts S of Crete and ends N to the Balkan peninsula. The fine grid, with horizontal grid increment of 4 km, covers the NE Peloponnese, the Saronic Gulf, the S Evoia with the Evoic Gulf and a large portion of SE Greece. The fine grid is presented in Fig. 1 (the area within the frame). The coarse grid was used in order to provide an accurate description of the subregional-scale flow in the area and mainly in the Aegean Sea.

For all the model simulations presented below the horizontally uniform initialization option of CSU-RAMS was used. For this kind of initialization the night radiosonde (02:00 LT) from the airport of Athens was used. This type of initialization was chosen because emphasis is given in simulations with no rapid changes in the synoptic scale and with no strong pressure gradients. All the simulations started at 02:00 LT and ended after 36 hours. Land-use data (e.g. vegetation index, soil type etc.) were derived from satellite images and cartographic charts. The roughness length over land was determined at each grid cell according to the land-use. Some tests were made with a third finer nest with 2 Km grid increment. Since this study aims at the description of the behaviour of these plumes over a larger area, the third grid was not used.

For the dispersion simulations, the resulted meteorological fields from the CSU-RAMS were used (stored every 10 minutes) while additional data such as stack parameters were also provided. The effective stack height was calculated every 10 minutes.

DISCUSSION

In order to present an accurate description of the dispersion characteristics in the above mentioned area of interest, several simulations were made. These simulations are typical cases chosen from the synoptic classes described in Kallos et al., (1993b). These classes are briefly described in Table. 1. Emphasis is given in cases characterized by relatively weak synoptic conditions because of the complicated transport mechanisms created from the local circulations. As it is known, the plumes from sources located near the shoreline exhibit some unique characteristics. Such characteristics are plume fumigation due to surface inversion break-up, plume fumigation due to Internal Boundary Layer (Misra, 1980), plume impignment, splitting, recirculation etc.

One category with the above characteristics is this which is in favour of the formation of thermal circulations, mainly sea breezes. During such a day a typical flow field for day and night-hours is presented in Kallos and Kassomenos, (1992). During these days, the plumes from sources, with either tall or short stacks, located in the coast of Thriassion plain are directly transported over the GAA during the day-hours (Fig. 2a). During the night, the air pollutants from these tall stacks are released above the mixing layer, they are passing over GAA and transported out of it through the gap between Partitha and Penteli (Fig. 2b). Pollutants released from relatively short stacks (30-50 m), which are located at the Thriassion Plain, are transported over the Athens basin through the gap between Aegaleo and Parnitha because the mixing height is lower than the height of the surrounding mountains during the night hours. During the early-morning hours the air pollutants released either from

Table 1. The main characteristics of the synoptic classes as they are described in Kallos et al. (1993b).

Synoptic Class	Main Characteristics
1	Behind a low pressure system which passed over Greece or South of it. Strong pressure gradient over the Aegean Sea and the area of Vosporus.
2	Behind a cold front which passed over Greece. Strong pressure gradient over the Aegean and/or Vosporus.
3	High pressure system over the Balkan area caused by the shifting of the circulation from the Central Europe.
4	Extension of the high pressure system located over Central Europe toward the Balkan area.
5	High pressure system over the Balkan area caused from the extension of the Siberian anticyclone toward W or SW.
6	Greece is within the warm sector of a deep low.
7	High pressure system over the Mediterranean Sea and the Balkan, associated with a thermal low over Asia Minor (Etesians).
8	A cold front passing over Greece.
9	Western flow over Greece caused by a deep low located over Europe.
10	Greece is behind a cold front or in the cold sector of a low with weak pressure gradients in the area around.
11	A trough or the edge of a cold front is passing over Northern and/or Central Greece.
12	A weak anticyclonic system covers the Mediterranean Sea and South Europe during the cold period of the year.
13	Greece is ahead of a cold front.
14	A weak anticyclonic system covers the Mediterranean Sea and Southern Europe during the warm period of the year.
15	Greece is within the warm sector of a weak low.
16	A warm anticyclone over the Central Mediterranean and South Europe which favours warm spells over Greece.

tall or short stacks are initially transported over the sea and after a few hours, with the aid of the sea breeze cell developed in the Saronic Gulf, over the GAA. The same is true for the pollutants released from the sources located near the Isthmus of Corinth in the northwestern coast of Saronic Gulf. These pollutants are initially transported over the Saronic Gulf and then, during the afternoon hours, with the aid of the sea breeze, over the GAA (Fig. 2c).

When the thermal circulations are weak (which mainly occurs during transient or winter seasons) the air pollutants from Keratsini, are released above the mixing height all the day and transported to the S. This is due to the high effective stack height. When the thermal circulations are strong enough (during the warm period of the year) the pollutants from this are released below the mixing height during the noon and early afternoon hours and therefore they affect the air quality within the Athens basin (Fig. 2d). Of course this power plant is not in operation during the last 13 years but there are plans to reactivate it. During the day, the air pollutants released from the sources located at the coast of Evoic Gulf are initially transported over the sea and then with the aid of the sea breeze, developed in the area, are moving toward the Mesogea plain. During the afternoon hours the pollutants released from Lavrio are transported to the S, while these from Aliveri they do not disperse and are moving around the point of release within the small valley. During the night-hours the pollutants released from Aliveri are moving over the island toward the Aegean Sea. The air pollutants released from Lavrio are moving, over the sea (South Evoic Gulf).

As it was found in previous studies (e.g Lalas et al., 1987; Kallos, 1987) recirculation of air pollutants may occur in the area of Athens and Saronic Gulf mainly during days with thermal circulations. Recirculation of air pollutants released from the elevated sources mentioned above may also occur. As it is seen in Fig. 3 the trajectories of air pollutants released from some of the sources mentioned above are moving initially over the sea and then with the aid of sea breezes cells developed in the area return to the land. The trajectories shown in this figure are the positions of particles released instantaneously and tracked in

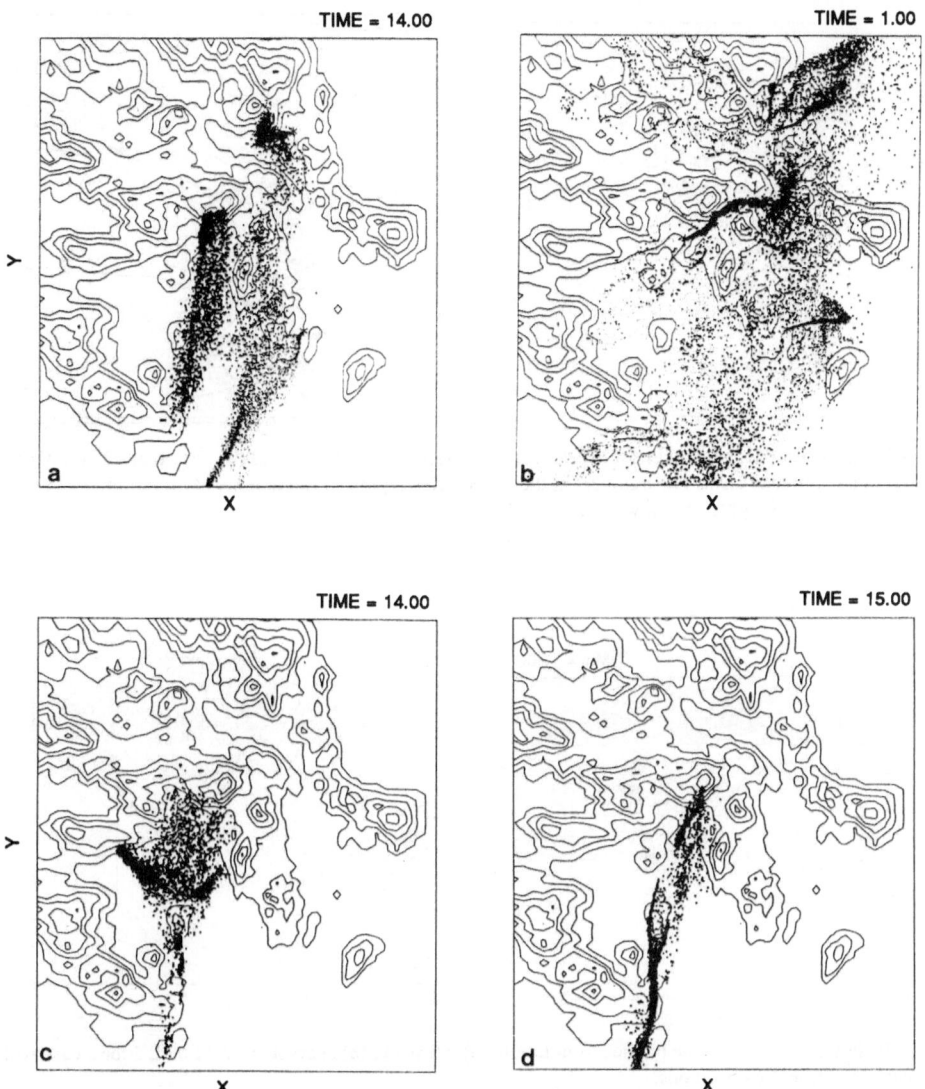

Fig. 2. Plain view of the particle plumes from Lavrio (P$_2$), Aliveri (P$_3$) and ELDA (R$_3$) (a) at 14.00 LT, (b) at 01:00 LT, (c) from Agioi Theodoroi (R$_1$) at 14:00 LT and (d) from Keratsini (P$_1$) at 15:00 LT for a summer sea breeze case.

time. A version of the LPDM was used for these simulations. The numbers at the trajectories represent the heights of air pollutants in tens of meters every two hours.

Days with very stable atmospheric conditions (e.g. classes 12,14,16) appear very often in the above mentioned area of interest, during all the seasons of the year, but they are more frequent during winter and transient seasons. The plume behaviour under very stable atmospheric conditions is quite interesting. As it was described in Kallos et al. (1993a) such atmospheric conditions are usually associated with warm advection in the lower troposphere. During these days, the mixing layer is relatively shallow and the thermal circulations are not strong. Air pollutants released from tall stacks are moving above the PBL and therefore they do not directly affect areas of great interest, such as urban areas. A classical example of such a case is shown in Fig. 4. In this figure the plumes from two different stacks located in Lavrio are shown. One is from a relatively short stack (40 m) and another one from a taller one (150 m). As it is seen in this figure the plume from the shorter stack is moving within the mixing layer, over South Evoic Gulf (Fig. 4a), while the other is moving above the

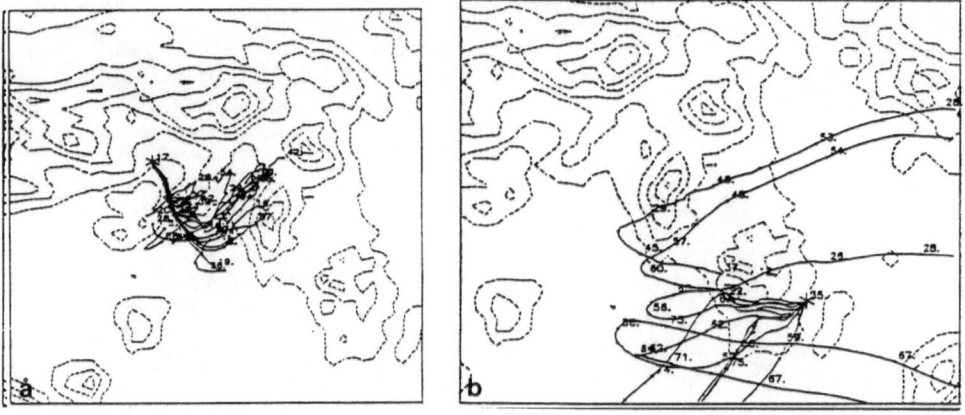

Fig. 3. Trajectories of air pollutants released from (a) Elefsis at 06:00 LT and (b) Lavrio at 12.00 LT for the summer sea breeze case. The numbers at the trajectories represent the heights of air pollutants in tens of meters. The heights are shown every two hours.

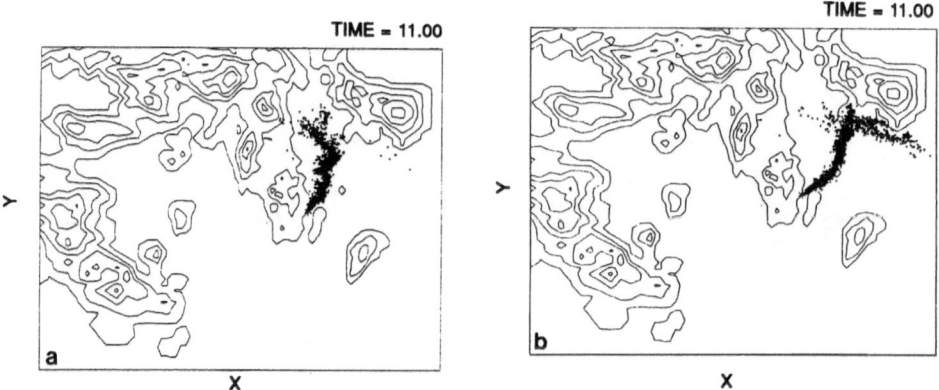

Fig. 4. Plain view of the particle plumes from Lavrio (a) tall stack, (b) short stack at 11.00 LT for a case with very stable atmospheric conditions.

mixing layer toward NE and sticks in the slopes of the mountainous area of southern Evoia (Fig. 4b). Similar behaviour of these plumes was found, during the transient and winter seasons when the synoptic flow and the thermal circulations are relatively weak.

Other interesting phenomena observed are the plume fumigation which occurs during morning hours because of the inversion break-up (Fig. 5a) and the splitting of the plume around a physical barrier, as a hill (e.g. a hill), when the Froude number is less than 1 (Fig. 5b). Because of the complicated topography of the area, impignment of the plume in the surrounding hills may happen especially in these cases when the plume is transported under stable conditions over the mixing layer in sort distances (Fig. 5c). Such dispersion phenomena do not affect the Athens basin, but the areas near the sources.

Days with strong northerly synoptic or sub-regional flow are very usual during summer or winter which are due to different reasons. The northerly flow across the Aegean during summer is due to the formation of a relatively strong pressure gradient which is due to the high pressure system over the Mediterranean region associated with the thermal low over the Asia Minor. It is also known as Etesian (Kassomenos, 1993 and references there in). This sub-regional circulation exhibits significant diurnal variation, since it is stronger during

TIME = 8.00

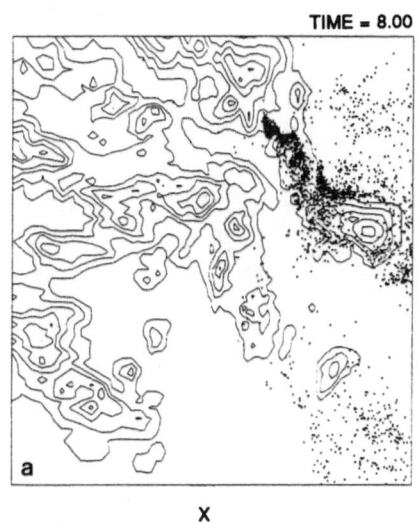

TIME = 2.00   TIME = 20.00

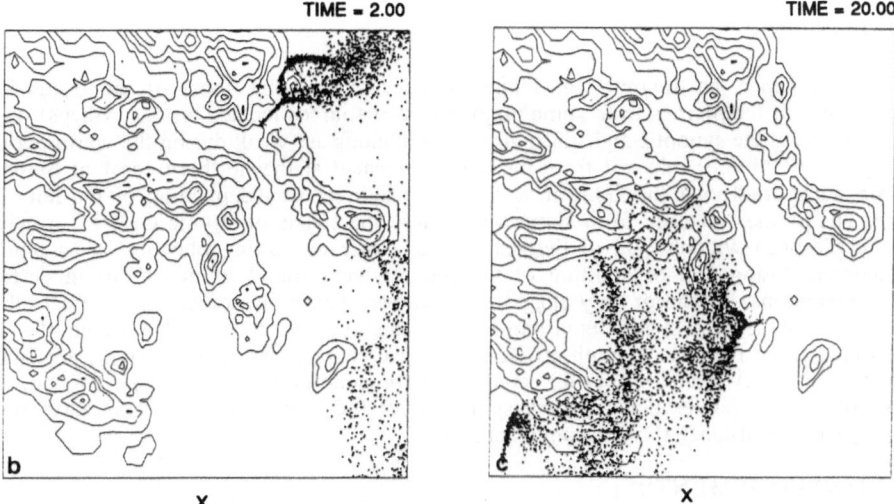

Fig. 5. Plain view of the particle plume from (a) Aliveri at 8:00 LT, (b) Aliveri at 02:00 LT and (c) Lavrio at 20:00 LT for a case with very stable atmospheric conditions.

the day-hours and weaker during night. During these days the plumes from the sources located at the coast of Saronic Gulf are directed to the S or SE. The same is true for the plumes released from the stacks of Lavrio. During most of the cases, the plumes from Aliveri are directed towards NE Attic and over the Athens basin for several hours (Fig. 6). During winter (e.g. classes 1,2,3,4,5), the strong northerly flow doesn't exhibit significant diurnal variations and therefore the plumes from the above mentioned sources may travel at large distances to the S or SW. These plumes can easily reach either the mountainous area of NE Peloponnese or even the island of Crete under the appropriate conditions. It is worth to mention also that these plumes are travelling to these large distances at relatively low distances from the ground because of the strong horizontal wind components.

Days with strong southerly flow (e.g. class 6) occur for a significant number of days annually. During these days the diurnal variation of the flow is negligible. The plumes from the sources located at the coast of Saronic Gulf are moving toward the mountainous area of the NW Attic and they not affect the Athens basin. The plumes from the sources located at the coast of South Evoic Gulf area moving to the N too. Figures from such simulations are not shown here because of the lack of space.

TIME = 20.00

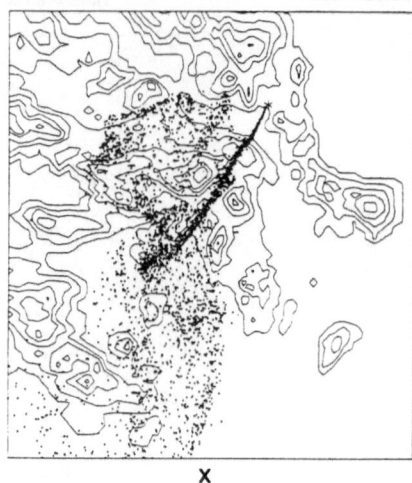

X

Fig. 6. Plain view of the particle plume from Aliveri at 20.00 LT for a summer case with strong northerly flow (Etesians).

CONCLUSIONS

In this paper, an attempt was made in order to describe some of the dispersion characteristics of the plumes from some major elevated sources located around Athens. As it was found, when the synoptic and sub-regional circulations are weak during the warm period of the year, the plumes released from the sources located at the northern coast of Saronic Gulf are transported over the GAA in a direct and indirect way depending of the hour of release. The plumes released from sources located at the coasts of Evoic Gulf are moving over the Mesogea with the aid of the sea breeze. Other interesting phenomena as plume impingement, fumigation or splitting of the plumes were found to occur during certain weather conditions, around or very close to the sources. These phenomena may affect other areas around Athens, probably environmentally sensitive. As it was also found the air pollutants released from stacks with different technical characteristics (e.g. height) presented completely different behaviour during certain synoptic conditions. As it was also found, the plumes of Aliveri power plant are transported over the Athens basin during days with relatively strong northerly flow, during all seasons.

ACKNOWLEDGEMENTS

Financial support for this work was provided by the Greek Public Power Corporation, the General Secretariat of Research and Technology, the IBM-Hellas and the Commission of the European Communities DG-XII (Contract #EV5 VCT 910050).

REFERENCES

Kallos, G., 1987: Simulation of the atmospheric circulation over Attic with the help of a mesoscale atmospheric model. Sea Breeze case. Final report prepared for The General Secretariat of Research and Technology, pp. 159 (In Greek).

Kallos, G., P. Kassomenos, and R.A. Pielke, 1993a: Synoptic and mesoscale weather conditions during air pollution episodes in Athens, Greece. *Bound. Layer. Meteorol.*, 62, 163-184.

Kallos, G., P. Kassomenos, M. Varinou and A. Papadopoulos, 1993b: Synoptic and mesoscale conditions in SE Greece: Their association with the air quality in the Greater Athens Area. Intermediate Report prepared for the Commission of the European Communities Directorate General XII, Contract # EV5 VCT 910050.

Kassomenos, P., 1993: Analysis of the weather conditions during air pollution episodes in the Greater Athens Area. Ph. D., Dissertation, University of Athens, Department of Applied Physics, pp. 362 (In Greek).

Lalas, D.P., Tombrou-Tsella M., Petrakis M., Asimakopoulos D. N. and C. G. Helmis, 1987: An experimental study of the vertical distribution of ozone over Athens. *Atmospheric Environment*, 12, 2681-2693.

McNider, R.T., M. Moran, and R.A. Pielke, 1988: Influence of diurnal and inertial boundary layer oscillations on long range dispersion. *Atmospheric Environment*, 22, 2445-2462.

Misra, P.K., 1980: Dispersion from tall stacks into a shoreline environment. *Atmospheric Environment*, 14, 397-400.

Moran, D. M., 1992: Numerical modelling of mesoscale atmospheric dispersion. Ph.D., Dissertation, Colorado State University, Fort Collins, Colorado, pp. 758.

Pielke, R.A., W.R. Cotton, R.L. Walko, C.J. Tremback, W.A. Lyons, L.D Grasso, M.E. Nicholls, M.D. Moran, A. Wesley, T.J. Lee, and J.H. Copeland, 1992: A comprehensive meteorological modelling system - RAMS. *Meteorol. Atmos. Phys.*, 49, 69-91.

Walko, R.L. and C.J. Tremback, 1991: RAMS- The Regional Atmospheric Modelling System Version 2c.: User's Guide. Published by ASTER, Inc., PO Box 466, Fort Collins, Colorado, 86 pp.

DISCUSSION

D. ANFOSSI   How did you compute plume rise in your Langrangian particle model and how did you threat the interaction of plumes with elevated inversion layer?

P. KASSOMENOS: We are using the Briggs formulation for the estimation of the effective stack height. We calculate it every 10 minutes from the meteorlogical fields provided by the meteorological model (RAMS). The plumes are moving with the 3D wind and turbulence fields. It does not matter if the plume is moving below or above the inversion height

EVALUATION OF ATMOSPHERIC DISPERSION MODELS DURING THE GUARDO EXPERIMENT

José I. Ibarra

Iberdrola,S.A.
Environmental Department
Goya 4, 3ª Planta
28001 Madrid (Spain)

INTRODUCTION

In November 1990, an intensive field tracer experiment was carried out in the Guardo valley (northern Spain). The main purpose of the tracer study was the atmospheric characterization of dispersion patterns over complex terrain. To do so, an evaluation exercise based on different modelling platforms was developed. The selected modelling tools embraced a set of different dispersion techniques: (1) a straight-line gaussian plume model, (2) a gaussian puff model driven by a simple interpolated wind field, (3) a particle-in-cell model driven by a mass-consistent wind model and (4) a lagrangian particle model driven by a prognostic mesoscale model.

The selected models, in the order mentioned above, were: (1) MPTER (E.P.A. model), (2) MESOI-2, (3) MATHEW-ADPIC and (4) RAMS-HYPACT.

A total set of 7 SF_6 tracer experiments were used in the validation exercise, each of them being representative of different meteorological conditions in the Guardo valley. SF_6 surface and elevated releases were performed; elevated releases from a 185m stack height at the Velilla power station and surface releases at different locations within the valley[1].

The main objectives of the validation exercise can be summarized in the following terms:

a. Evaluation of representative algorithms used to characterize plume-rise, wind shear and diffusion coefficients.

Air Pollution Modeling and Its Application X, Edited by S-V. Gryning
and M.M. Millán, Plenum Press, New York, 1994

b. Performance of selected models in reproducing ground level concentrations patterns and receptor concentrations.

c. Identification of most relevant properties and shortcomings of selected models and algorithms.

DESCRIPTION OF SF$_6$ TRACER EXPERIMENTS

The simulated cases that configured the validation process were drawn from a total set of 14 tracer experiments carried out during the Guardo field tracer campaign. Based on previous meteorological and turbulence analyses[5], the finally selected experiments were identified on the basis of being representative of the most relevant meteorological features observed during the field campaign[6]. **Table 1** depicts the final model runs.

Each of the 7 cases presented in **Table 1** took place under different stability and turbulence conditions making possible the evaluation of different numerical schemes. All of them were representative of an estimated 80% of present meteorological conditions at the Guardo site. SF$_6$ release conditions also varied from case to case.

EVALUATION OF CLASSICAL ALGORITHMS

A full set of the most widely used algorithms in dispersion applications are reviewed in Ibarra[3]. They were used to evaluate wind shear, plume-rise and diffusion coefficients.

Wind shear

Wind shear is traditionally used to characterize wind conditions at a height representative of plume transport. Most applications extrapolate ground based wind observations (typically at 10m) to the height where the plume is supposed to be dispersed in the environment. In complex terrain such practices become uncertain as terrain interactions with local flows distort the theoretical logarithmic wind profile.

Classical wind shear algorithms based on formulations by,

- **Busse & Zimmerman**
- **CRSTER model by the E.P.A.**
- **Ermak & Nyholm**
- **Irwin**
- **Novak & Turner**
- **XOQDOQ model by the U.S.N.R.C.**

were evaluated in order to check their usefulness in complex terrain applications. A detailed analysis of such set of algorithms can be found in Ibarra[3].

The main conclusions of this analysis can be identified by:

1. The application of the logarithmic wind profile overestimates plume transport a factor ranging from 1.4 at a mid-level to 2.0 at the final rise.

2. If surface wind speed readings were directly applied to the height of the estimated plume transport, advection would be underestimated a factor of ≈ 2.

3. Minor differences were observed in algorithms with a power exponent close to zero, which indicates that a constant profile was on the average the best approach.

Plume-rise (δh)

A realistic evaluation of the center of mass of an elevated plume becomes a crutial term in modelling applications. Errors in the right parameterization of δh may lead to incorrect predictions of maximum concentrations and their relative distances to the source. A good design of a monitoring network also depends on the precise evaluation of such parameter.

Plume-rise is frequently modelled based on Briggs's formulas although there is a set of algorithms to do such work. According to the review presented in Ibarra[2], the final set of formulas evaluated were:

- **Briggs (1969, 1970, 1975)**
- **Holland (1953)**
- **Moses & Carson (1969)**
- **Carpenter (1971)**
- **Montgomery (1972)**
- **CONCAWE (1966)**

The main conclusions drawn from this analysis can be summarized as follows:

1. The CONCAWE algorithm yielded the best overall agreement with the observations, regardless stability conditions.

2. Briggs's algorithms showed a tendency to overestimate the final rise in a factor of 2 under neutral-unstable conditions, whereas they showed a 20% underestimate under stable conditions.

3. The length scale of momentum (l_m) prevailed over the length scale of buoyancy (l_b) in a 74% of the 23 cases examined, yielding a mean ratio l_m/l_b equal to 3.54. As shown by Ibarra[11], a higher dominance of momentum would make the plume reach the final rise at a distance closer to the source than that predicted by the Briggs's formulas. In the Guardo case the final rise is attained at an average distance of ≈ 450m making necessary a "sub-grid" treatment of the plume-rise process in grid-based models[7,8].

4. The computed efflux Froude number was much higher than unity which means that buoyancy was not a significant factor and downwash effects were prevented. Large efflux velocities are detrimental to the rise of a buoyant plume because they are accompanied by a more rapid entrainment of ambient air into the plume. In fact, an entrainment coefficient $\beta = 0.71$ was evaluated compared with 0.6 assumed in Briggs's formulas.

5. If using wind data at an intermediate level between stack height and final rise, only in an 18% of the cases the wind speed would be higher than that observed at the stack height, with an average increment lower than 1m/s. Consequently, wind

profile considerations would only explain a minor fraction of the overestimation observed in those algorithms were δh is proportional to u^{-1}.

Figure 1 shows the mean ratios and the standard deviations of each formulation compared with observations during unstable conditions. An investigation of modifications to some of the tested plume-rise algorithms can be seen in Ibarra[10].

Diffusion Coefficients

The evaluation of gaussian diffusion coefficients σ_y and σ_z was carried out following general criteria gathered in Ibarra[3]. The following set of formulas were analyzed:

- **CRSTER model by the E.P.A.**
- **MPTER model by the E.P.A.**
- **Smith algorithm**
- **Pasquill algorithm**
- **Irwin algorithm**
- **Venkatram algorithm**
- **Hanna algorithm**

Venkatram and Hanna algorithms are based on the Taylor statistical theory making use of turbulence intensities, whereas the reimainder algorithms are based of the classical Pasquill-Gifford-Turner (P-G-T) stability classification.

The main clonclusions drawn from this analysis can be summarized as follows:

1. Observations of turbulence intensities gave rise to horizontal diffusion coefficients higher than those calculated according to the P-G-T stability scheme[9]. In neutral conditions a mean ratio of 1.5 was attained while in stable conditions it rised to 6. In unstable conditions both schemes provided similar results.

2. All algorithms based on the P-G-T stability scheme, except the Irwin algorithm, underestimated vertical diffusion in a factor ranging from 1.5 to 2.0. Irwin, Venkatram and Hanna provided the best agreement with observations, particularly the Hanna's algorithm.

PERFORMANCE OF NUMERICAL MODELS

The evaluation of selected models is performed taken up, as basic information, SF_6 ground level concentrations integrated over 2 hours at a mean number of 20 samplers for each experiment. Concentrations are given in $\mu g/m^3$ and a 95% confidence level is assumed in the entire statistical evaluation. The validation procedure was based on the following statistics[3,11]: normalized mean square error (NMSE), fractional bias (FB), index of agreement (IA), average gross error (AGE), critical success index (CSI), true skill statistic (TSS) and the correlation coefficient (r). Standard estimators (mean, median, variance, correlation and frequency distribution) were also tested against classical methods (t-test, Wilcoxon-Mann, F-test, Fisher-z and Kolmogorov-Smirnov) to check the overall performance of each model.

The "bias" was tested against zero value. Two estimates of bias were used: the average bias and the mean bias. Two statistics were calculated; the first is a paired comparison and the second is an unpaired comparison statistic. Values of the correlation

Table 1. Generic description of simulated cases in the validation exercise of numerical models.

Case	Exp.	SF6 release	sampling[1] time	meteorological conditions
1	5	chimney	16h-18h	SW up-valley breeze
2	10	"	14h-15h	Northerly synoptic wind
3	9	"	13h-15h	Westerly synoptic wind
4	12	"	14h-16h	N-NE synoptic wind
5	14	surface	8h-10h	NE drainage flow
6	4	"	10h-14h	transition local flows
7	8	"	8h-10h	low wind - near calm

[1]Local central time.

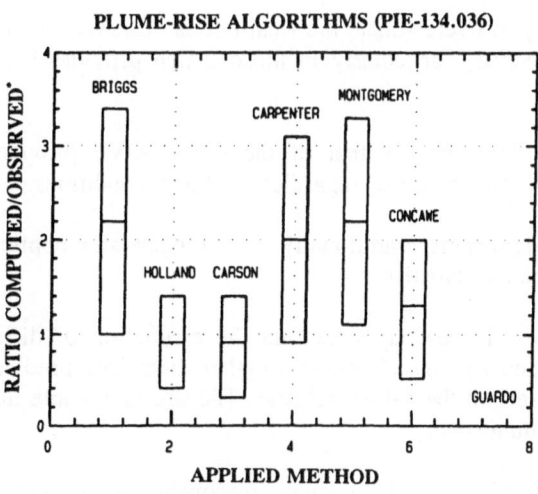

(*) UNSTABLE CONDITIONS.

Figure 1. Mean ratios and standard deviations obtained in the evaluation of plume-rise algorithms under unstable conditions.

coefficients and the frequency distribution comparison were tested against certain hypothesized values to deduce meaningful indicators (**Table 3**). Variability was tested using the estimated variance of the observed SF_6 concentrations compared with the estimated variance of the computed SF_6 concentrations. Ideally, the variance ratio should be equal.

All models were previously calibrated and adjusted to assimilate as much of the data gathered at field. This process was only limited by their own numerical capabilities.

Results of models' performance are depicted in **Tables 2 and 3**. **Figure 2** shows the percentage of computed air concentration values within a factor R of measured, for the entire tracer experiments. A complete analysis of the entire validation process can be found in Ibarra *et al.*[4]

CONCLUSION

The evaluation of the selected dispersion models in this paper bring us to the following findings:

1. Scarce representativity of the wind power law in complex terrain and in local valley winds. Application of such a rule would lead to a significant uncertainty in the estimation of ground concentrations. Design of environmental monitoring networks would also be negatively influenced.

2. Plume-rise evaluation presents various aspects of interest as follows:

2.1 A tendency to overestimate the final rise is observed in an ample ensemble of algorithms checked, particularly in those which provides better agreement with observations.

2.2 A simple formula like that of the CONCAWE provides the best overall agreement with observations, regardless stability conditions.

2.3 Briggs's algorithms, traditionally used in dispersion applications, produce the worst performance statistics.

3. Hanna gives de best agreement in the estimation of diffussion coefficients. Observations show plume dimensions higher than those predicted by conventional algorithms based on the P-G-T scheme. The use of reliable turbulence intensities is highly recommended.

4. MATHEW-ADPIC (M/A) gives reasonable agreement with observed air concentrations. Results indicate the model is unbiased, with variability, prediction of impaction areas and time-spatial dependent distribution concentrations similar to those observed at field. A certain tendency to underestimate average ground concentrations is detected.

5. MPTER and MESOI predictions produce important deviations when compared with observations. This model lacks of a significative bias but variability, prediction of impaction areas and distribution concentrations are, in general terms, well different from observations. Thus, the use of MPTER or models with similar characteristics for licensing and environmental applications pose a high degree of

Table 2. Summary statistics obtained in the validation exercise.

Statistic	M/A	MPTER	MESOI	RAMS[2]
mean bias	1.16	1.94	0.49	12.37
RMS error	4.58	5.95	3.75	26.42
fractional bias	0.42	1.34	0.39	-0.79
average gross error	2.85	30.36	9.31	6.39
NMS error	1.68	2.00	1.01	3.66
critical success index	0.744	0.506	0.554	0.545
true skill statistic	0.663	0.487	0.495	0.222
correlation coeff.	0.867	0.178	0.400	0.747
index of agreement	0.914	0.328	0.550	0.610
[1]Xc/Xo<2	42.1%	19.6%	26.9%	0.0%
Xc/Xo<5	73.2%	43.9%	52.7%	22.2%
Xc/Xo<10	82.0%	61.0%	70.0%	---
Xo/Xc (T-statistic)	1.69	3.08	3.56	0.45
No. observations	164	107	119	27

[1]Xc= computed ground level concentrations.
 Xo= observed ground level concentrations.
[2]Results for only one experiment (Exp.#14).

Table 3. Testing performance measures[1].

performance measure	test	tested hypothesis	critical value	test statistic M/A	MESOI	MPTER
mean bias	t-Student	average=0	<2.15	1.31	1.01	3.45
" "	Wilcoxon	median=0	>3.00	5.16	0.49	2.47
" "	Wilcox-Mann	-U	>15.0	14.64	0.22	12.66
variance	F	$\sigma_o^2 = \sigma_c^2$	<5.00	1.53	2.75	39.31
correlation	Fisher-z	r=0.0	0.738	0.66	0.59	0.16
frequency distribution	Kolmogorov -Smirnov	$F_o = F_c$	$K_D < 9.0$	6.27	7.00	10.90

[1]For all tests, the confidence level is 95%.

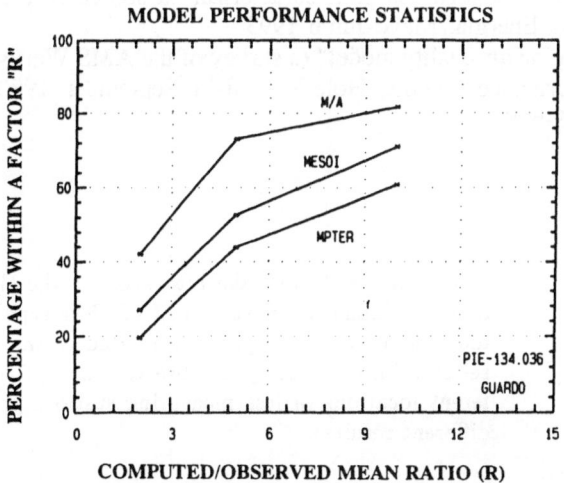

Figure 2. Percentage of computed samples agreeing to within a factor R with those measured (Guardo data set).

uncertainty. A tendency to underestimate average ground concentrations is also observed, particularly in MPTER.

REFERENCES

1. J.I. Ibarra,"Atmospheric dispersion experiments over complex terrain in a spanish valley site (Guardo-90)",Proceedings of the Specialists' Meeting on Advanced Modelling and Computer Codes for Calculating Local Scale and Meso-Scale Atmospheric Dispersion of Radionuclides and their Applications, 6-8 March 1991, OECD/NEA Data bank, Paris.
2. J.I. Ibarra,"Análisis de relaciones paramétricas aplicadas en modelos de dispersión atmosférica", Revista Energía, Septiembre-Octubre (1992).
3. J.I. Ibarra,"Especificación técnica para la validación numérica de modelos de dispersión atmosférica", Proyecto PIE-134.036, Documento PIE-013E, Iberdrola S.A., (1991).
4. J.I. Ibarra,M. Rojo and D. Olivé,"Validación de modelos de dispersión atmosférica en terreno complejo", Proyecto PIE-134.036, Documento PIE-022E, Iberdrola-Asinel-Ibersaic, (1992).
5. J.I. Ibarra,"Turbulence and boundary layer parameterization during valley wind flows", ENVIROSOFT-92 Conference, 2-5 September 1992, Portsmouth, U.K.
6. J.I. Ibarra,"Observations of valley winds in Spain's Guardo valley", Sixth Conference on Mountain Meteorology, AMS, 29 Sept.- 2 Oct. 1992, Portland, Oregon.
7. J.I. Ibarra,"Atmospheric dispersion under local valley breeze conditions during the Guardo experiment", International Symposium on Air Pollution'93, 16-18 February 1993, Monterrey, Mexico.
8. W. Lyons,J.I. Ibarra,"Evaluation of complex terrain dispersion predictions using a fine mesh prognostic mesoscale model", 86th Annual Meeting & Exhibition, Air & Waste Management Association, 13 June 1993, Denver, Colorado.
9. D.H. Slade,"Meteorology and Atomic Energy", TID-24190, Air Resources Laboratory, U.S.A.E.C., Oak Ridge, Tennessee (1968).
10. J.I. Ibarra,"Nuevos algoritmos para caracterizar la sobreelevación de penachos de SO_2", Revista Energía, Mayo-Junio 1993.
11. D.G. Fox, "Judging air quality model" (a survey of the AMS Workshop on Dispersion Model Performance, Woods Hole,MA, 8-11 September 1980), *Bulletin of the AMS*,**62**, 599-609.

DISCUSSION

G. KALLOS
How did you take into account the thermal effects in your simulation with the MATHEW-type codes? How did you suppress the problem related to the bias from the selected locations for your observations? (if you choose different locations of less measuring stations you will obtain different results).

J.I. IBARRA   MATHEW as a mass consistent model lacks in modelling thermal effects. The procedure followed involved the assimilation of 9 surface meteorological stations and 1 local sounding (either acoustic or rawinsonde profiles) deployed in an area of 40 x 40 km. The adjustment between vertical and horizontal components was based on the Strouhal number scheme for α^2 proposed by Moussiopoulos (1988). Such a scheme was applied in a spatially variable manner. Calibration of the model outputs with observations was finally carried out in order to guarantee the goodness of the diagnostic approach provided by the model. A different gridded domain was used in terms of the nature of the SF_6 releases: a 500 x 500 x 50 m grid was used for elevated releases, whereas a 200 x 200 x 20m grid was applied when dealing with surface releases. A maximum of 15 vertical levels was modelled in both release conditions.

G. KALLOS   How did you initialize RAMS? What kind of data sets did you use?

J.I. IBARRA   RAMS was initialized from a standard 2.5° gridded data provided by the US NMC at 0Z and 12Z, making use of a nested grid assimilation technique made up of 5 nested grids with a maximum local horizontal resolution of 200 x 200 m. An anelastic solution and a non-hydrostatic approach was used in the simulations. Local features such as albedo, vegetation cover, soil type, surface roughness were also took into account to parametrize ground fluxes. A Langrangian particle model drived by RAMS was used to calculate ground SF_6 concentrations.

DISPERSION MODELLING AND OBSERVATIONS
FROM ELEVATED SOURCES IN
COASTAL TERRAIN

J. A. Noonan[1], W. L. Physick[1], J. N. Carras[2] and D. J. Williams[2]

[1]CSIRO, Division of Atmospheric Research
Aspendale, Victoria, 3195, Australia
[2]CSIRO, Division of Coal and Energy Technology
North Ryde, New South Wales, 2113, Australia

INTRODUCTION

This study is focussed on the Central Coast region of New South Wales, on the east coast of Australia, where Pacific Power Inc. operates three power stations, situated on small lakes within a few kilometres of the coast. One of the objectives of the study was to assess the suitability and value of our modelling system to the air quality assessment needs of Pacific Power in the region, including the evaluation of green-fields sites for future power stations. The dispersion of plumes from all power stations throughout the year is not only influenced by the terrain blocking and channeling of the synoptic winds, but also by mesoscale wind systems such as sea breezes and drainage flows. Our Lagrangian Atmospheric Dispersion Model (LADM) predicts winds and turbulence and uses these to simulate the transport and diffusion of emissions from discrete sources, for impact distances ranging from hundreds of metres to a few hundred kilometres. In this paper we compare LADM results to available observations in the far field.

THE MODEL

There are two components to LADM (Physick et al., 1993). The first is a three-dimensional prognostic mesoscale model which predicts the windfield and turbulence characteristics on a grid. The model solves the primitive equations in an (x, y, σ) coordinate system where $\sigma = p/p_s$ (pressure normalized by surface pressure). The hydrostatic assumption is employed and the model is fully compressible. The momentum (τ), heat (H) and evaporation fluxes (E) in the equations are parameterized according to the scheme of Louis (1979). Short- and long-wave radiation parameterizations are used, sea-surface temperature is kept constant, and surface temperature over land is diagnosed from a surface heat balance condition at each timestep. A heat diffusion equation is solved at six levels in the soil to compute heat flux into the ground. The

Air Pollution Modeling and Its Application X, Edited by S-V. Gryning
and M.M. Millán, Plenum Press, New York, 1994

mesoscale model can be nested within itself, enabling finer resolution of topographic features, and a grid point can be representative of a grid square that consists of part land and part water. The horizontal advection and geostrophic adjustment processes in the model are evaluated separately and a semi-Lagrangian approach is applied to the computation of horizontal advection. An advantage of this scheme is that it allows the advective terms to be evaluated on a larger timestep than that determined by the Courant-Friedrichs-Lewy criterion for an Eulerian-based scheme, hence improving the speed of the model.

The second component of LADM is the Lagrangian particle dispersion model in which plume dispersion is simulated by transporting and diffusing neutrally-buoyant particles according to the wind and turbulence fields predicted by the first component. The particles are released from the power station locations at 20 second intervals. The model employs skewed homogeneous turbulence within the convective boundary layer (Hurley and Physick, 1993) and Gaussian inhomogeneous turbulence in stable conditions (McNider et al., 1988). Final-rise heights are determined by numerical solution of the Briggs (1975) equations for a bent-over plume, using diurnally varying emission characteristics for each station. Hourly-averaged ground-level concentrations (glcs) of SO_2 and NO_x are computed by counting particles in 'boxes' at the surface. Box sizes are 25 m deep and range horizontally from 250×250 to 1000×1000 m, according to the distance from the source. The number of particles released per timestep is such that one particle in a box represents a concentration of 20 μg m^{-3}.

RESULTS

Meteorology

We have chosen two case-study days, November 29 and 30, 1989. Both were clear-sky days with light synoptic winds and sea breezes. On the 29th the synoptic wind was from the northeast (onshore) quadrant and on the following day the synoptic wind had moved to the northwest (offshore) quadrant. Using synoptic wind profiles from surface and upper-air pressure charts, and initial temperature profiles from Williamtown radiosonde flights (for location see Fig. 1), we ran the model on both days for 48 hours. For each case-study day we used three levels of nesting, with grid resolutions of 10, 5 and 2 km, and 55×55 grid points in the horizontal. During the first 24 hours, the windfield adjusted to the diurnal heating cycle and the underlying terrain and at 0400 Local Solar Time (LST) on the second day we began releasing particles from the power stations on the 2 km grid. For each day, comparison of predictions from the second 24 hours with coastal wind data from anemometers, tethersonde and radiosonde flights and an acoustic sounder showed that the dominant features of the windfield were satisfactorily reproduced by the model.

During the early morning on November 29, in the coastal region, the modelled winds at plume height were from the northeast. These winds veered to become easterly with the onset of the sea breeze at about 0800 LST and by early evening they had backed to again be from the northeast. In contrast on November 30, at plume height, the sea breeze arrived at the power stations at about midday, switching early morning westerly flow to southerly flow. As the sea breeze deepened and moved inland the winds became south-easterlies. During the afternoon and evening the winds continued to back and gain a northerly component. Fig. 1 shows the modelled windfields at 600 m (about plume height) at 1200 LST for November 29 and 30. Clearly, very different plume directions will be predicted for each day at this time.

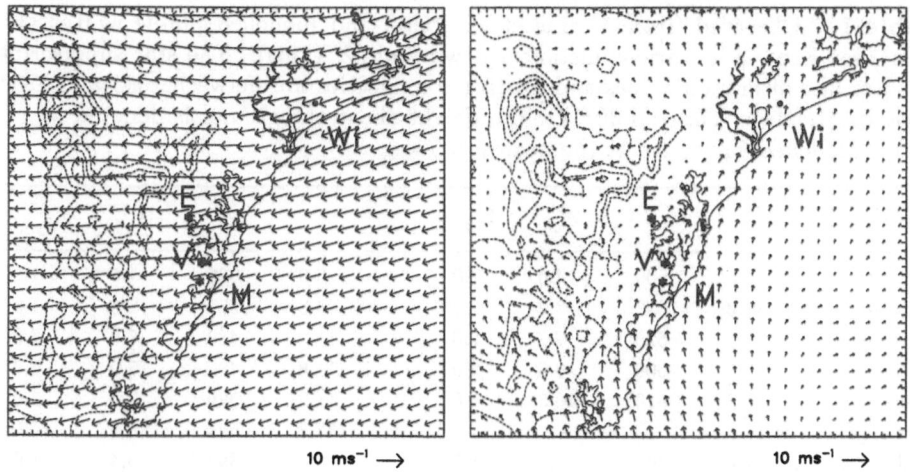

Figure 1. Modelled windfields at 1200 LST at a height of about 600 m above the ground for the 2 km grid, (left) November 29 and (right) November 30. Arrows denote strength and direction of horizontal winds. Orography contours (dashed lines) are shown at a spacing of 100 m, the solid line is the coastline and the tick mark interval on the axes represents a distance of 2 km. Labels: E, V, M - Eraring, Vales Point, Munmorah power stations. Wi - Williamtown.

Plume Widths

The CSIRO aircraft flew near the coastal power stations on both days and provided data for plume width comparisons (Carras et al., 1992) in stable and convective conditions, at distances up to 30 km from the respective sources.

We performed three experiments to compare different formulations for lateral dispersion and examined the effect of these formulations on far-field glcs. In the **first experiment** the particles were released at the downwind distance and placed at the effective stack height (*esh*) calculated from the Briggs equations. Above the convective boundary layer (CBL) $\sigma_{u,v}$ were set to the mixed layer value (= $0.6w_*$), and therefore overestimated lateral dispersion, and σ_w was set to a default value of 1 cm s^{-1}, assuming no turbulence.

In the **second experiment** non-zero turbulence was specified for $0.5z_i$ (z_i is CBL height) above the CBL (associated with entrainment and internal waves), with $\sigma_{u,v} = 0.1w_*$ and $\sigma_w = 0.3w_*$ (Caughey and Palmer, 1979). Above $1.5z_i$ $\sigma_{u,v}$ and σ_w were set to default values of 1 cm s^{-1}.

In the **third experiment** the particles at release time were distributed in a Gaussian manner about the *esh* in an attempt to parameterize the spread of the plume as it rises from the stack to its *esh*. The plume cross-section was assumed to have a ratio of 2:1 (width to height) - as is generally observed, with concentration distributed in a Gaussian manner laterally and vertically. If the plume cross-section is assumed to be *circular*, its radius at any time during plume rise has been shown by Briggs (1975) to be $R = \beta z_r$, where β is an entrainment constant usually taken to be 0.6 and z_r is the plume height above the stack height. Assuming that the mean radius of our *elliptical* plume is also R and that $\sigma = R/\sqrt{2}$, it can be shown that $\sigma_y = R$ and $\sigma_z = R/2$.

Table 1. Observed (from Carras et al., 1992) and modelled plume widths (W (km)) for Expt. 3 at various distances downwind (X (km)). The star superscript denotes late afternoon values and unstarred quantities are between 1045 and 1200 Eastern Standard Time on 30/11/89 . Height above the ground is denoted by (Z (m)). Modelled values are to the nearest 0.5 km.

Station	X	Z	W_{obs}	W_{mod}	X^*	Z^*	W^*_{obs}	W^*_{mod}
Eraring	10.8	500	4.0	2.0	5.0	300	1.5	1.5
Eraring	11.8	620	7.2	2.0	4.8	370	2.8	1.5
Eraring	8.5	620	7.2	2.0	11.9	480	7.0	5.0
Eraring	9.1	600	9.5	2.0	11.8	400	7.0	5.0
Eraring	18.5	510	10.5	3.0	17.2	520	7.0	7.0
Eraring					33.3	590	8.8	10.0
Vales Pt.	2.6	300	0.8	1.0	6.8	390	0.4	1.0
Vales Pt.	5.1	530	0.8	1.0	2.8	180	0.5	–
Vales Pt.	4.8	550	2.0	1.0	4.8	280	1.8	1.0
Vales Pt.					19.0	800	1.8	–
Vales Pt.					30.1	580	7.6	7.0
Munmorah	3.8	540	2.5	2.0	2.3	180	0.7	–
Munmorah	4.0	280	3.8	2.0	18.5	420	5.5	–
Erar-Vale	15.2	510	4.5	–	26.0	530	14.0	16.0
Vale-Mun	3.8	310	12.5	8.0	11.8	350	2.8	–

In the formulation of this plume spread in the model, each particle is simply positioned at the effective stack height (x_e, y_e, z_e) plus a random distance which is proportional to the plume dimension in each direction. The equations used are

$$x = x_e + 0.6 z_r r_u$$

$$y = y_e + 0.6 z_r r_v$$

$$z = z_e + 0.3 z_r r_w$$

where r_u, r_v and r_w are random numbers from a $(0,1)$ Gaussian distribution.

Comparison of plume widths from Expts. 1 and 2 with the aircraft data showed that these experiments seriously underestimated the lateral dispersion of the plume. In the steady afternoon winds, glcs of Expt. 2 were reduced by up to 20% over those of Expt. 1. Time of fumigation and location of plume footprints were virtually the same in each.

Observed and modelled plume widths for Expt. 3 are presented in Table 1. Except for the Eraring plume during the late morning, the distribution of particles about the effective stack height produces plume widths which compare favourably with aircraft observations. The improvement in Expt. 3 compared with Expt. 1 (Fig. 2) arises because wind shear at the *esh* is able to act on the plume over a reasonable depth. The comparison with afternoon observations when the plumes are in the mixed layer of the relatively steady sea breeze is especially good. There is still some under-estimation of the Eraring plume width in the morning when it lies above the CBL. This is probably due to the neglect of wind shear (through the top of the sea breeze) on the plume as it rises from the stack top to its final-rise height. Acoustic sounder winds show a backing of the wind by about 90° over this distance.

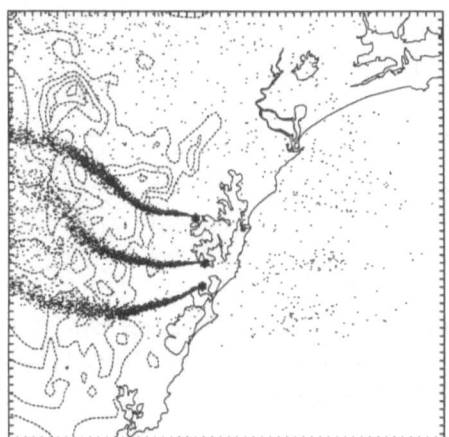

Figure 2. Plan view of particle distribution, below 1 km, at 1700 LST for November 30, (left) Expt. 1 and (right) Expt. 3.

The degree of reduction of far-field maximum glcs in Expt. 3 over those of Expt. 1 is variable, but large 'bullseye' values were decreased by up to 60%. In Expt. 3, some particles in the lower levels were advected by the sea breeze during the morning, an effect which was absent in Expt. 1. However, when results from both experiments are compared to the sparse monitor data, it is not possible to nominate which formulation is superior, but on the basis of plume width comparisons we conclude that the formulation that distributes the particles in a Gaussian manner about the *esh* is superior. This formulation is used in the following experiments.

Ground-level Concentrations

As a result of the different synoptic wind directions the two case study days differed in an important aspect - the thermal stability profile in the onshore sea-breeze flow. On the 29th when the synoptic wind was onshore, the stability was moderate and typical of an oceanic profile. The sea breeze thermal internal boundary layer (TIBL) later in the day was deep. However, the following day when the synoptic wind moved to become offshore, nocturnal and early-morning offshore flow led to a relatively strong stability profile over the sea, and subsequently in the sea breeze. Consequently the sea breeze TIBL later in the day was not as deep as for the 29th. The differing stabilities meant that the *final-rise heights* of plumes from the coastal power stations were within the sea breeze's TIBL on the 29th (Fig. 3a), but were above it on the 30th (Fig. 3b), leading to quite different dispersion patterns on the two days. For the onshore case predicted maximum glcs were significantly higher and occurred much closer to the stacks than for the offshore case. For example, maxima at 1600 LST were found 5 and 15 km from Vales Point power station for the two respective cases.

The prediction of hourly-averaged glcs of SO_2 and NO_x at the four monitor locations in the Central Coast region was not as successful as for the Phase I study in the Upper Hunter Valley (northwest of the region shown in Fig. 1, Physick et al.,

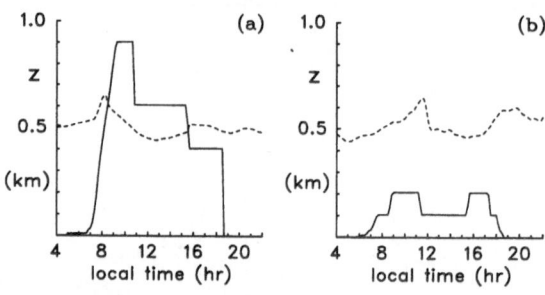

Figure 3. Modelled plume final-rise height (dashed line) and planetary boundary layer (PBL) height (solid line) at the Eraring power station for (a) November 29 and (b) November 30. In stable conditions, PBL height is set to 2 m, but is not used in any computations.

1991). The difference is probably due to the presence of the TIBL in the coastal region. Its variation in space and time strongly affects the locations at which plumes are fumigated to the ground, and the linking of it to discrete model levels in the current version of LADM further reduces accuracy. Given the inherent uncertainty in air quality modelling (including errors in model inputs and formulation, the stochastic nature of turbulence, and the fact that the winds are predicted in our model), it is unrealistic to expect accurate glc predictions at a specific point at a particular time. On both days, modelled plumes were in the vicinity of monitor locations at those times when non-zero readings were observed, with glc values which were similar to those registered at the monitors. As an example Fig. 4 shows the predicted glcs and a plan view of the particle positions at 1500 LST. The Wyee monitor registered 15 μg m^{-3} at this time and as Fig. 4 shows we predict glcs a few kilometres west of the monitor. If 'error bars' in space and time are specified with the model results, then it is possible to use the predictions in a meaningful way. We evaluate the model's performance by using it in a 'real-time' mode to predict events which can be verified to a great extent by data. Below we list our predictions and corresponding observations for various events. We present values for SO$_2$ only (NO$_x$ predictions give similar comparisons).

1. By 1300 EST on November 29 the plumes will have been dispersed a considerable distance to the southwest. By late afternoon the plumes will be in a similar position with less spread toward the south.

	Predicted	*Observed by aircraft*
	Yes	Yes

2. SO$_2$ glcs will be recorded at Wyee during the day on November 29.

	Predicted	*Observed*
Time	0800-1900 LST	1100-1300 EST

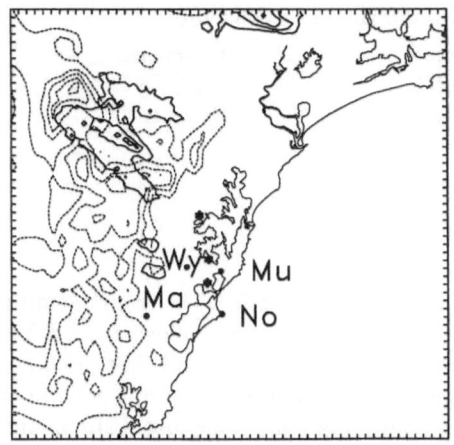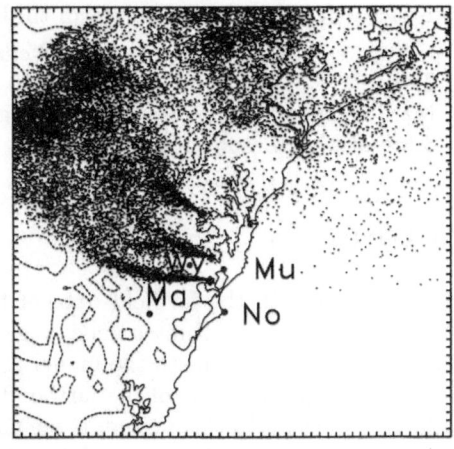

Figure 4. November 30, (left) the predicted glcs of SO$_2$ (thick solid lines) for the hourly period ending at 1500 LST, and (right) plan view of particle distribution at 1500 LST. Contour interval is 20 μg m^{-3}. Labels: Wy, Mu, Ma, No - Wyee, Munmorah, Mardi, Norah Head air quality monitors. Locations of Eraring, Vales Point and Munmorah power stations are indicated by asterisks.

3. SO$_2$ glcs will be recorded near Mardi during the late afternoon of November 29.

	Predicted	*Observed*
Time	1800, 1900 LST	1800, 1900 EST
Max. value within 3 km of Mardi monitor	19, 36 μg m^{-3}	45, 35 μg m^{-3}

4. On November 30, plumes will be advected offshore until about midday and by late afternoon the plumes will be over the hills west of the power stations.

	Predicted	*Observed by aircraft*
	Yes	Yes

5. No glcs will be registered at Norah Head on either day and none will be registered at Munmorah on November 29.

	Predicted	*Observed*
	Yes	Yes

6. On November 30, the Munmorah plume will come to ground between north and east of the stack.

	Predicted	*Observed*
Time	1000, 1100 LST	1000, 1100 EST
Max. value within 3 km of Munmorah monitor	16, 5 μg m^{-3}	15, 15 μg m^{-3}

7. Glcs will be recorded near Wyee on November 30.

	Predicted	*Observed*
Time	1100 - 1400 LST	1500, 1600 EST
Max. value within 3 km of Wyee monitor	32, 37, 45, 21 μg m^{-3}	15, 50 μg m^{-3}

8. No glcs will be registered at Mardi till late afternoon on November 30.

	Predicted	*Observed*
Time	1700 LST	1700 EST
Max. value within 3 km of Mardi monitor	43 μg m^{-3}	18 μg m^{-3}

CONCLUSION

Comparison with the far-field data of this study, and with those from a similar study of the neighbouring Hunter Valley region (Physick et al. 1991), allows us to assign error bars in space and time to the model predictions. Thus it appears that our point predictions are representative of the surrounding 3 to 8 km and that a timing error of up to 1.5 hours is likely. In this regard, it seems possible that the modelling system could be used in a real-time mode as a regulatory tool to control power station emission rates in regions where far-field glcs are a problem. However, a simpler procedure involving a convective scaling technique and real-time winds (and perhaps mixing depth) at the source would be more appropriate for near-source maxima.

REFERENCES

Briggs, G.A., 1975, Plume-rise predictions, *in:* "Lectures on air pollution and environmental impact", Workshop Proceedings, Boston, Mass. 29 Sep. - 3 Oct. 1975. American Meteorol. Soc. 59-111.

Carras, J.N., Lange, A.L., Thomson, C.J., and Williams, D.J., 1992, Behaviour of the power station plumes in the Hunter Valley/Central Coast region of New South Wales, Vol. 1: Results of the intensive field campaigns. CSIRO Division of Coal and Energy Technology Report CET/IR030.

Caughey, S.J., and Palmer, S.G., 1979, Some aspects of turbulence structure through the depth of the convective boundary layer, *Quart. J. Roy. Meteor. Soc.* 105:811.

Hurley, P.J., and Physick, W.L., 1993, A skewed homogeneous Lagrangian particle model for convective conditions, *Atmos. Environ.* 27A:619.

Louis, J.-F., 1979, A parametric model of vertical eddy fluxes in the atmosphere, *Bound. Layer Meteor.* 17:187.

McNider, R.T., Moran, M.D., and Pielke, R.A., 1988, Influence of diurnal and inertial boundary-layer oscillations on long-range dispersion, *Atmos. Environ.* 22:2445.

Physick, W.L., Noonan, J.A., Hurley, P.J., McGregor, J.L., and Abbs, D.J., 1993, The Lagrangian Atmospheric Dispersion Model. CSIRO Division of Atmospheric Research Technical Paper No. 23.

Physick, W.L., Noonan, J.A., Manins, P.C., Hurley, P.J., and Malfroy, H., 1991, Application of coupled prognostic windfield and Lagrangian dispersion models for air quality purposes in a region of coastal terrain, *in:* "Air Pollution Modelling and its Application IX", H.van Dop and G. Kallos, eds., Plenum, New York.

DISCUSSION

B. FISHER:	Could you use your modelling system to decide where to put monitoring instruments?
J. NOONAN:	Yes. Dr. W. Physick will present a paper on this topic in the video session.

SENSITIVITY ANALYSIS OF THE URBAN AIRSHED MODEL TO WIND FIELDS DERIVED FROM THE REGIONAL OXIDANT MODEL, DIAGNOSTIC WIND MODEL, AND THE URBMET/TVM MESOSCALE MODEL

G. Sistla and S. T. Rao
Division of Air Resources
New York State Department of Environmental Conservation
Albany, New York, USA

R. D. Bornstein and F. Freedman
Department of Meteorology
San Jose State University
San Jose, CA

P. Thunis
Environmental Institute
European Community Joint Research Center
Ispra, Italy

ABSTRACT

The United States Environmental Protection Agency has recommended the use of the Urban Airshed Model (UAM) with the Carbon Bond-IV chemical mechanism in evaluating the efficacy of emission control strategies and in determining controls necessary to reduce ambient ozone concentrations in urban areas to the level of the National Ambient Air Quality Standard (NAAQS) for ozone. UAM applications to urban areas in the northeastern United States utilize a one-way nesting of the UAM with the Regional Oxidant Model (ROM) for deriving the wind fields for the UAM. However, diagnostic and prognostic meteorological modeling techniques are being explored as alternatives for the development of wind fields for the UAM. In this study, we present the sensitivity of the UAM-predicted ozone concentrations to the wind field derived from three different approaches. The three techniques are, (1) interpolation of ROM wind fields to the UAM using the ROM-UAM Interface system, (2) interpolation of the observed surface and upper-air data using a diagnostic model (DWM), and (3) interpolation of winds from a mesoscale planetary boundary layer model (URBMET/TVM).

In this study, several UAM simulations have been performed for the greater New York Metropolitan area with these three wind fields for a high ozone episode in July 1988. Preliminary results indicates that the agreement between the predicted and measured ozone concentration pattern varies depending upon the wind field selected. The DWM-derived wind fields provide a better agreement between the predicted and observed ozone concentration fields than those derived from the ROM-UAM interface system. Examination of the impact of across-the-board VOC and NO_x emissions reduction on ozone levels indicates that the simulation with wind fields from the ROM-UAM interface yields a greater change in ozone concentrations for NO_x-focussed reductions than VOC-focussed reductions.

PHOTOCHEMICAL MODELS AND INPUT DATA

Briefly, the ROM[1] design consists of 3½ layers in the vertical with an approximate horizontal grid spacing of 18.5 km with the modeling domain covering the eastern half of the United States. The UAM[2] set-up in this study comprises of five layers in the vertical with 3-layers above and 2-layers below the diffusion break, and a horizontal grid spacing of 5 km. The UAM modeling domain (see Figure 1) extends from north of Philadelphia, PA as the southwest corner to the border of Massachusetts, Connecticut and Rhode Island as the northeast corner, covering an areal extent of 290 km east-west and 230 km north-south. Anthropogenic emissions used in the ROM simulations are based upon the 1985 National Acid Precipitation Assessment Program (NAPAP) inventory adjusted to 1988 together with biogenic emissions reflecting 1988 meteorological conditions. The anthropogenic and biogenic emissions for the urban area along with initial and boundary concentrations for the period of July 5 to 8, 1988 were derived from ROM[3]. The hourly diffusion break heights were estimated using the RAMMET-X algorthim[4] from the twice daily upper air sounding from Albany, NY involving a 3-point smoother technique. Assuming spatially invariant diffusion break, the other input data needed for the UAM were retrieved from the ROM-UAM[5] interface system.

WIND FIELDS

ROM-UAM Winds

The ROM-UAM interface system utilizes a height-weighted interpolation scheme to transer the ROM three layer, 18.5 km gridded wind fields to the 5-layer, 5 km grid structure of the UAM. Given the coarseness of the ROM grid size in the horizontal and the thickness of the vertical layers, the ROM-UAM wind field cannot resolve localized circulations such as land-sea breeze patterns. The ROM-UAM interface system provides a very smooth flow field over the domain. Figures 2a and 2b provide an example of the morning (0700-0800) and afternoon (1500-1600) wind fields from this method for July 8, 1988.

DWM Winds

The DWM[6] interpolates the surface and upper-air wind observations to each grid cell with an adjustment to account for the kinematic effects due to terrain. Vertical velocities are smoothed and the divergence/convergence is minimized to yield a three-dimensional wind field. A total of 60 surface and 6 upper-air stations were used in the

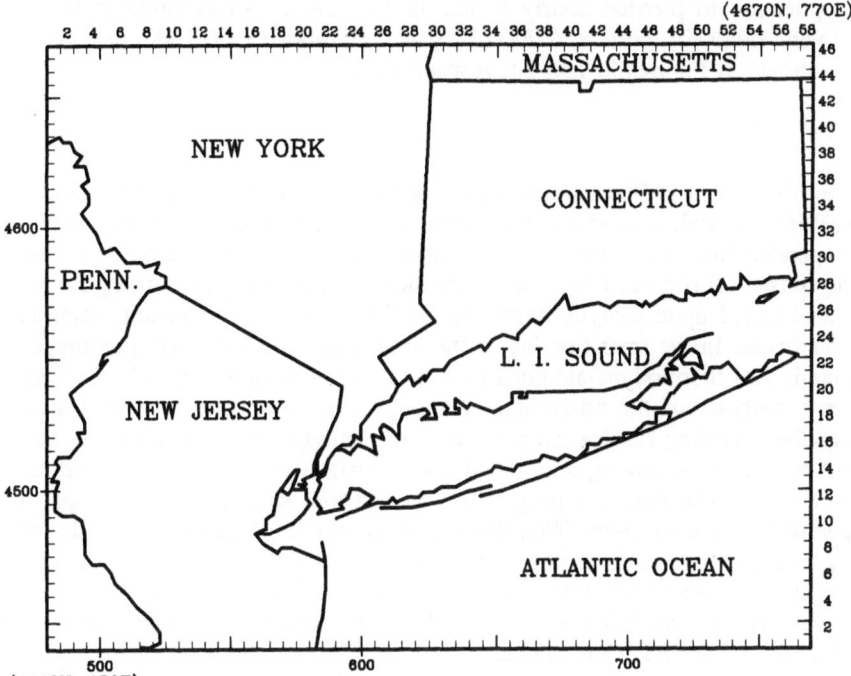

(4670N, 770E)

(4440N, 480E)
500   600   700

Figure 1. Modeling Domain for the New York Airshed on the
Universal Transvers Mercator (UTM) System.

Table 1. Measured Highest and Second Highest Ozone Concentration (ppb) and
Maximum Predicted Ozone Concentration (ppb) in New York Airshed.

Wind Field	Base	M2575[1]	M7525[2]
ROM-UAM	183	141	97
DWM	213	153	123
URBMET/TVM	199	145	108

1   Scenario with 25% NO_x and 75% VOCs reduction in emissions

2   Scenario with 75% NO_x and 25% VOCs reduction in emissions

Table 2. Maximum Predicted Ozone Concentration (ppb) for July 8, 1988 -
Base Case and Emissions Control Scenarios.

Day	Measured Highest	Second Highest	Predicted ROM-UAM	DWM	URBMET/TVM
July 6	178	177	160	188	160
July 7	193	182	192	219	197
July 8	212	200	183	213	199

current application to provide hourly winds for the UAM. An example of the DWM generated wind fields for morning (0700-0800) and afternoon (1500-1600) periods of July 8, 1988 is shown in Figures 2c and 2d, respectively.

URBMET/TVM

The TVM[7] is a 3-dimensional mesoscale vorticity mode numerical model based upon URBMET[8] model. It consists of a sub-surface layer and an atmospheric layer. The latter is seperated into two layers, one a constant flux surface and the other a transition layer. The constant flux surface layer in which the time dependent meteorological profiles are estimated based upon analytic equations that are a function of height, stability and surface roughness. In the transition layer, the hydrostatic and Boussinesq assumption of hydrodynamic and thermodynamic equations are solved numerically. The vertical grid spacing is a function of the horizontal location, with the surface level following the terrain. Spatially varying surface temperature and humidity are calculated using soil heat and moisture fluxes. Assuming a constant water surface temperature, the soil surface temperature is obtained from the prognostic force-restore equation[9]. An example of the resulting winds for the morning (0700-0800) and afternoon (1500-1600) periods for July 8, 1988 is shown in Figures 2e and 2f, respectively. The URBMET/TVM produces winds at 20 vertical levels which are translated to the five-layer vertical structure of the UAM. This has resulted in smoothing away the urban and coastal near-surface flow details produced by the URBMET/TVM model.

COMPARISON AMONG THE WIND FIELDS

An examination of the wind fields generated by the three methods for the morning hour (see Figure 2a, 2c, and 2e) indicates that the DWM and the URBMET/TVM show similar southwesterly winds while the ROM-UAM exhibits westerly winds. Also, wind speed derived from the ROM-UAM are stronger than those of DWM or URBMET/TVM. In the case of the afternoon winds (shown in Figure 2b, 2d, and 2f) for 1500-1600 hours the DWM and the URBMET/TVM wind fields again indicate a substantial southerly wind component than the ROM-UAM wind field.

UAM SIMULATIONS

UAM simulations were performed for the July 5 to 8, 1988 ozone episode with three sets of wind fields, keeping same the other UAM inputs. The observed and predicted maximum ozone concentrations for the simulation period are listed in Table 1. Of the three wind fields, the simulation with the DWM derived winds predicts a higher peak ozone concentration than the other two for all three days. The simulations with URBMET/TVM winds predicts slightly higher ozone concentrations for July 7, 1988, while the other two wind fields predict higher ozone concentrations for July 8, 1988. The agreement between the maximum predicted and measured concentrations, is within the acceptable range of $\pm$ 30%.

Isopleths of maximum ozone concentrations predicted at each grid by the three wind fields along with the maximum measured ozone concentration at several monitoring stations within the domain for July 8, 1988 are presented in Figure 3.. Ozone levels greater than 200 ppb were measured over New Jersey, with concentrations in the 170-180 ppb range extending along the eastern New York and western Connecticut border region with a north-south orientation for the ozone plume. Ozone levels measured along the

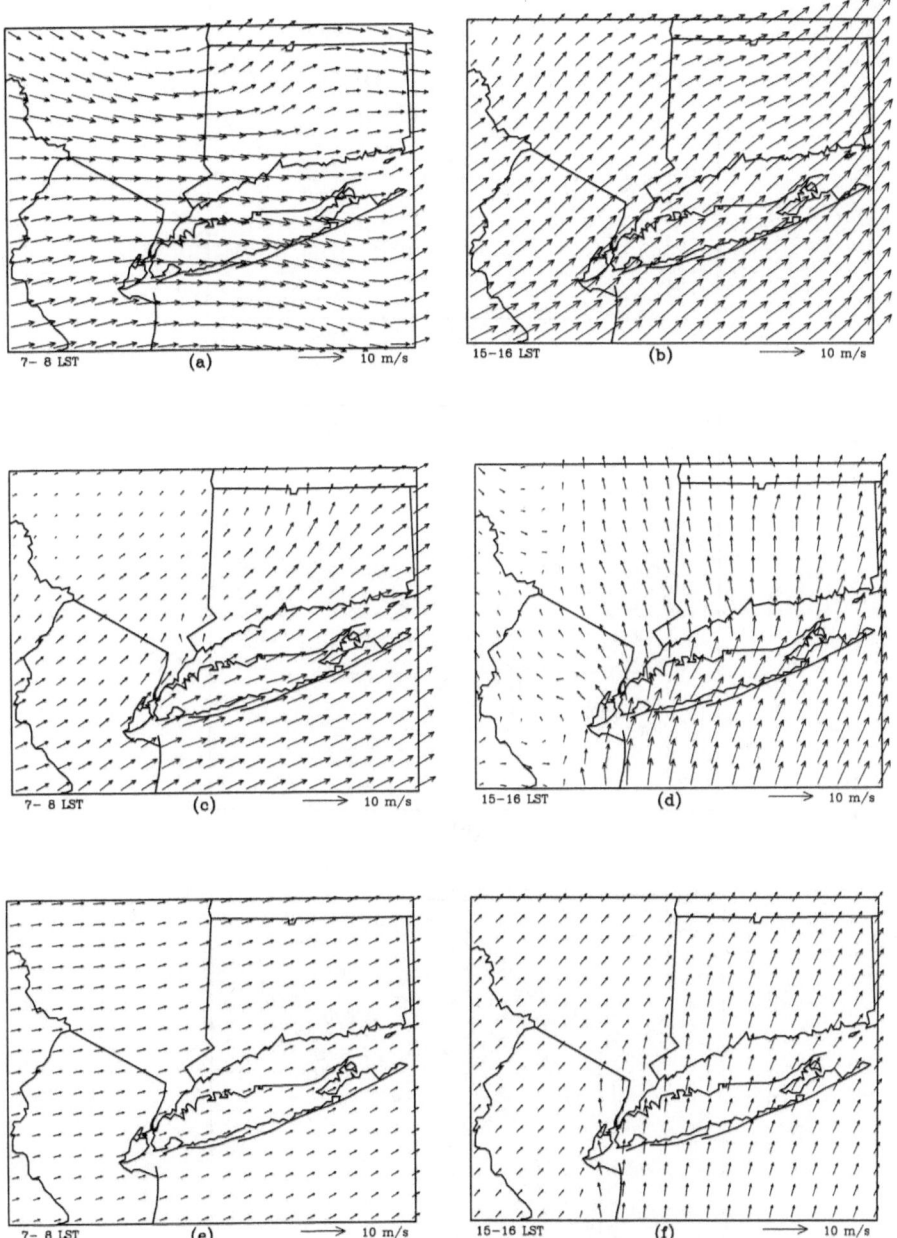

Figure 2. Wind Fields for the New York Airshed Modeling Domain:
(a) and (b) from ROM-UAM Interface,
(c) and (d) from DWM Model, and
(e) and (f) from URBMET/TVM Model

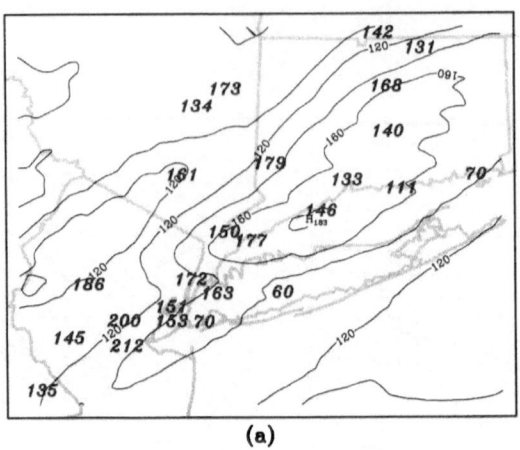

(a)

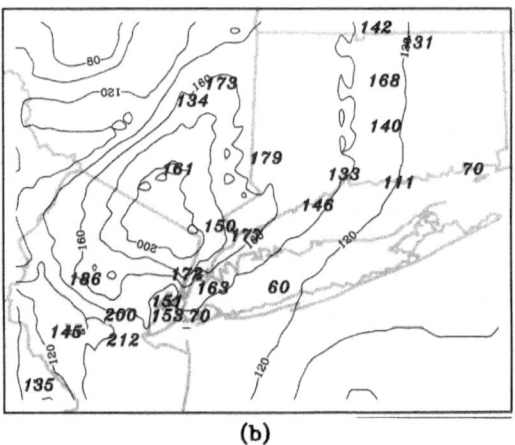

(b)

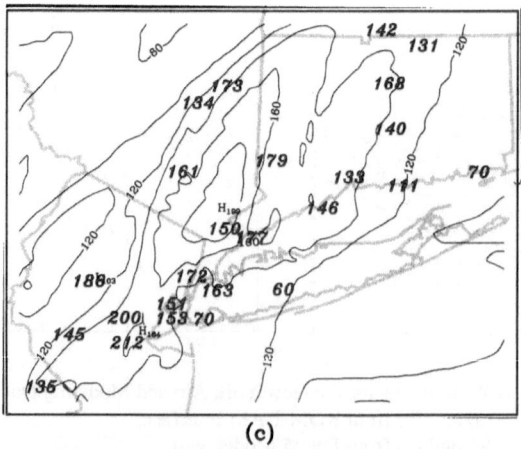

(c)

Figure 3. Measured Maximum Ozone Concentrations (ppb) shown in bold,
and Isopleths of Maximum Ozone Concentration (ppb) for
July 8, 1988 based on Wind Fields from:
(a) ROM-UAM Interface,
(b) DWM Model, and
(c) URBMET/TVM Model

coastal region of New York and Connecticut were at or below the ozone NAAQS. The predicted ozone plume is oriented southwest to northeast for the ROM-UAM simulation with the 160 ppb contour extending from central New Jersey to Hartford, CT and a maximum of 183 ppb over the coastal areas of Bridgeport, CT. In the case of the DWM based simulation, the ozone peak of 213 ppb is predicted over the northern New Jersey-New York region, with the 180 ppb contour level oriented in the north-south direction, a feature which is similar to the measured concentration field. The URBMET/TVM based ozone plume also follows a pattern similar to that of the DWM with the peak predicted concentration of 199 ppb to the north of New York city. While the orientation of the ozone plume simulated with DWM and URBMET/TVM winds are similar, the plume width resulting from the latter is narrow indicating sharp concentration gradients over the domain when compared with the DWM predicted pattern. An examination of the maximum ozone predicted by the three methods for July 8, 1988 indicates that the predicted peaks are displaced spatially with respect to the location of the measured maximum.

Regulatory applications of the UAM are designed to assess the degree to which the various emissions reductions strategies reduce the predicted peak ozone concentrations to the level of the NAAQS. To examine this, several UAM simulations have been performed with the three wind fields for two across-the-board emissions reduction scenarios. The two emissions reduction scenarios considered are: 25% NO_x and 75% VOCs (M2575), and 75% NO_x and 25% VOCs (M7525). Initial/boundary conditions associated with these scenarios were obtained from the corresponding ROM simulations[8]. Table 2 lists the predicted maximum ozone concentration from each of these simulations for July 8, 1988. For the M2575 case, all three wind fields predict ozone levels in excess of the ozone NAAQS where as for the M7525 case, only the DWM method yields ozone peak exceeding the NAAQS level. Although the predicted changes in ozone levels are different, all three wind fields show similar "directions" for controls. The spatial distribution of ozone from these three methods indicates that the location of the occurence of the peak ozone is similar to that of the base case simulation, and indicates that a greater reduction of ozone concentrations in the New York airshed can be achieved with NO_x-focussed emissions controls than with VOC focussed controls. The differences in the response of the peak ozone concentration to the emissions reduction suggest the importance of the wind field in evaluating the effectiveness of an emissions control strategy in reducing ozone concentrations to the level of NAAQS.

SUMMARY AND CONCLUSIONS

In this study, we have investigated the effect of the wind field on the UAM-predicted ozone concentrations for the New York airshed. The predicted ozone patterns based upon the DWM and URBMET/TVM wind fields appears to compare more favorably to the measured distribution than those predicted by the ROM-UAM wind field. The DWM based simulations showed higher ozone concentrations than from the other two wind fields. This may be due to the presence of localized circulations and lower wind speeds associated with the DWM where as smooth and higher wind speeds are associated with the ROM-UAM winds or the prognostic wind fields from URBMET/TVM model. While the predicted ozone distribution pattern has remained the same for all across-the-board reductions in emissions, the ROM-UAM based simulation has provided a greater response to NO_x emissions controls than that provided by the other two wind fields. Future simulations of the UAM will utilizie the full vertical resolution along with the 4-D turbulent kinetic energy values of the URBMET/TVM model.

REFERENCES

1. N. C. Possiel, R. D. Scheffe, S. H. Chu, and R. A. Wayland, *Regional Modeling Protocol -- Ozone SIP Development Support*, EPA-OAQPS, 1992.

2. R. E. Morris and T. C. Myers, *User's Guide for the Urban Airshed Model -- Volume I: User's Manual for UAM(CB-IV)*, EPA-450/4-90-007A, 1990.

3. R. Wayland, Personal communication, USEPA-OAQPS, 1993.

4. R. E. Morris, T. C. Myers, E. L. Carr, M. C. Causley, S. G. Douglas, and J. L. Haney, *User's Guide for the Urban Airshed Model -- Volume II: Preprocessors and Post Processors for the UAM Modeling System*, EPA-450/4-90-007B, 1990.

5. S. D. Douglas, R. C. Kessler, and E. L. Carr, *User's Guide for the Urban Airshed Model -- Volume III: User's Manual for the Diagnostic Wind Model*, EPA-450/90-4-007C, 1990.

6. R.-T. Tang, S. C. Gerry, J. S. Newsom, A. R. VanMeter, R. A. Wayland, J. M. Godowitch, and K. L. Schere, *User's Guide for the Urban Airshed Model -- Volume V: Description and Operation of the ROM-UAM Interface Program System*, EPA-450/4-90-007E, 1990.

7. G. Schayes and P. Thunis, *A Three-Dimensional Mesoscale Model in Vorticity Mode*, Contribution No. 60, Institute of Astronomy and Geophysics, Catholic University of Louvain, Belgium, 1991.

8. R. Bornstein, et al., *Application of Linked Three Dimensional PBL and Dispersion Models to New York City*, in Air Pollution Modeling and its Application V, ed. D. Wispeleare et al., pp543-564, 1986.

9. J. Deardorff, *Efficient Prediction of Ground Surface Temperature and Moisture with Inclusion of a Layer of Vegetation*, J. Geophys. Res., 83, 1198, 1978.

DISCUSSION

G. KALLOS: Have you ever done sensitivity tests (for the VAM model)for different lateral boundary conditions? How your results are modified from changes into the supplied lateral boundary conditions?

R. BORNSTEIN: We have not tested VAM to see how it responds to the lateral B.C.s in TVM.

ASSESSMENT AND VERIFICATION OF DIFFERENT TYPES OF DISPERSION MODELS IN COMPLEX TERRAIN

Volker R. D. Hermberger and Patrick Doria

Institute Paul Scherrer
CH-5232 Villigen PSI, Switzerland

INTRODUCTION

In connection with the evaluation and selection of calculational models, which are well suited for real time dispersion simulation of airborne radioactivity in complex terrain, 10 dispersion models of 4 different types are evaluated, compared and verified by the analysis of 2 different experimental episodes of the tracer experiment SIESTA. The result of the analysis had to be assessed with respect to the performance of the models, which was done qualitatively by visualization and quantitatively by statistical analysis as well as by soft-/hardware characteristics. Specific conclusions had to be drawn with respect to the final selection of a dispersion model operational in complex terrain and under real time conditions. 2 conditions which still present a real challenge for modern scientific computing. More general conclusions are drawn.

SELECTED MODELS

A general review was performed by a questionnaire to 34 modellers, covering 38 dispersion models (Hermberger et al., 1992). The questionnaire concerned general and project specific characteristics, mathematical/physical and technical models, input, output, essential issues and varia in about 90 single items. Five basically different methods were found,
- Analytical or **Gaussian puff** models, where the cloud is simulated by a time series of puffs. Their growth and inner concentration pattern follow a Gaussian distribution..
- Statistical/Monte Carlo simulation of advection/dispersion or **Langrangian particle** models, where the advection and dispersion is simulated by random walk (Monte Carlo). The physical process can be described in any detail. The results contain inherent statistical fluctuations.
- Deterministic advection/diffusion or **Eulerian grid** model, where the process is approximated by an advection/diffusion differential equation, which is integrated numerically by discrete ordinates in space and time (K-theory). The method suffers from deficiencies in the nearby region of point or line sources and for small diffusion times. The results contain inherent numerical errors called "numerical diffusion".
- Combination of statistical and deterministic or **Hybrid** models, where the advection is simulated by Langrangian particle and the dispersion by Eulerian grid or concentration gradient /diffusion method. In addition
- Flow models with integrated dispersion part or **Fluid dynamic processors,** which contain either Lagrangian or Eulerian type dispersion models and have been recently applied to atmospheric dispersion problems. In general they are advanced in turbulence simulation and numerical iteration schemes.
More details of the theoretical base of the different methods can be found in a previous evaluation of flow and dispersion models (Martens et al., 1987).
The modelers' answers to the questionnaire were assessed semi-quantitatively by attributing merit points to the different items weighted by their importance.. Models with the highest number of merit points were considered for every method as the "best referenced" and were selected for the SIESTA-analysis. Evidently

this ranking is rather global so that specific requirements have to be observed seperately as acceptable limitations for spatial or time scale, source code availability for development, useful documentation e.g.
The selected models were
2 Gaussian puff (PUFF_...): (Thykier-Nielsen, 1991), (Wendum, 1988),
3 Eulerian grid (EUL_...): (Gidhagen, 1992), (Perdriel, 1991), (Winkler, 1992)
3 Lagrangian particle (LAGR_...): (Janicke, 1992), (Lamprecht, 1992), (Winkler, 1992)
2 Hybrid (HYBR_...): (Chino, 1992), (Rodriguez 1992).
The top favourites of the Fluid dynamic processors could not be applied to the analysis, because they were neither ready nor fully operational.

SIESTA TRACER EXPERIMENT ANALYSIS

SIESTA was an international meso scale tracer experiment (Gassmann et al., 1987), which has simulated the transport of toxic gases and aerosols in complex terrain around the NPP site Gösgen, Switzerland and was chosen as reference data base for the model assessment and qualification.

Tracer Experiment SIESTA as Reference Data Base

Mean wind and turbulence data were measured in conjunction with atmospheric dispersion of SF6-tracer during weak-wind situations over the Jura ridge and the hilly prealpine region. To limit the effort, the SIESTA analysis was restricted to two episodes different in wind and stratification: Experiment No. 4 on November 24 with a wind from NEE in a neutral situation, and experiment No. 6 on November 30 with a weak wind from WSW to S in a stable situation. Therefore, two different topographies of 47 x 35 and 35 x 35 km were chosen, so that they contain all sampling arcs. The tracer was sampled on 4 and 6 arcs at distances up to 35 and 25 km respectively. The windfields were generated from meteorological measurements of meteo towers, tethered balloons, routine midday soundings and the Swiss topographical data with a resolution of 250 m by the mass-consistent diagnostic flow model CONDOR (Flassak, 1990). Terrain following coordinates with a "horizontal" grid spacing of 1 km were used. The windfield and turbulence data were averaged over the emission period of 6 h. The tracer is released at 6 m and sampled at 1 m above ground level during the last hour of the emission period. Turbulence data and release rates are:

Episode	November 24	November 30
Monin-Obukow length [m]	- 120	+ 150
Friction velocity [m/s]	0.14	0.10
Surface roughness [m]	0.8	0.8
Mixing layer depth [m]	600	200
SF_6-release rate [g/s]	3.15	3.16

Comments to the Reference Data Base

The most severe simplification is the reduced experimental data set with the 6 h averaged windfield and turbulence data. It was chosen for several reasons:
- The time variation of the meteorology was fairly modest.
- A common meteorological data base for all dispersion models was chosen to concentrate on dispersion model capabilities and performance and to eliminate the influence of the individual diagnostic flow models.
- To facilitate the analysis for the modeler by offering a data base of moderate size..
- To simulate a realistic emergency situation, in which a forecast of the environmental impact for the next 6 hours is required on the basis of actual meteo data.
Clearly, this simplification is not thought to be best suited for a SIESTA based model validation study, because the smaller the averaging time interval is the more realistic wind direction and speed fluctuations can be taken into account by the model, which leads to broader plume width and concentration fields. A full analysis of the SIESTA experiment is underway, in which the modeler generates the (non-stationary) windfields and turbulent parameters from the original meteorological measurements by averaging over reasonably small time intervals depending on his flow model and expertise. However, in this work the central question is: **To what extent do the different dispersion models match the SIESTA tracer data with the same simplified (averaged) wind and turbulence information?**

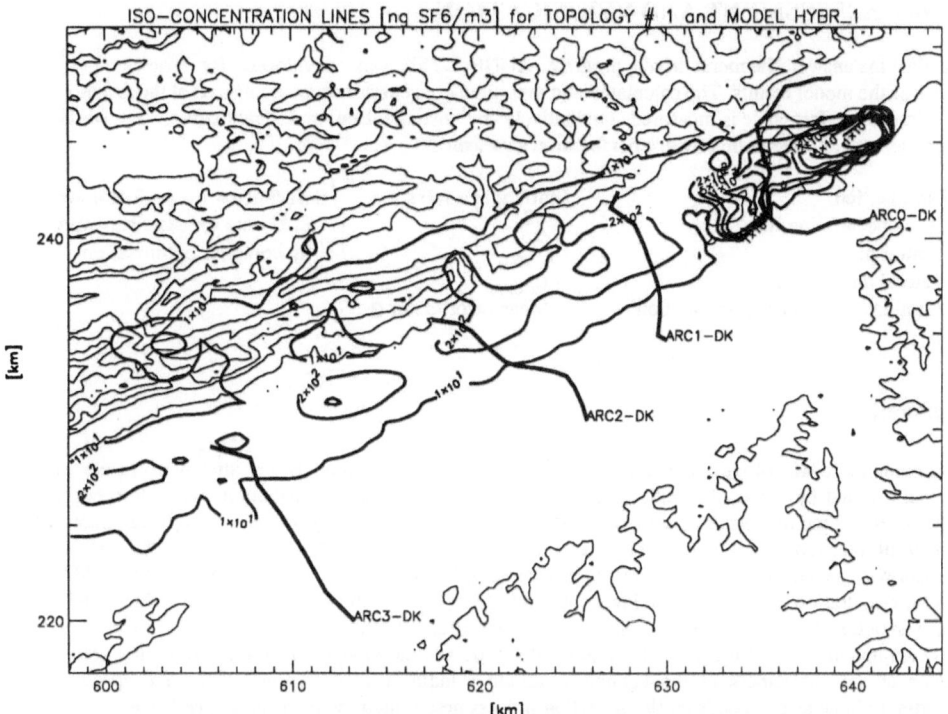

FIG. 1 Iso-concentration lines for model HYBR_1 and SIESTA episode on 24th.

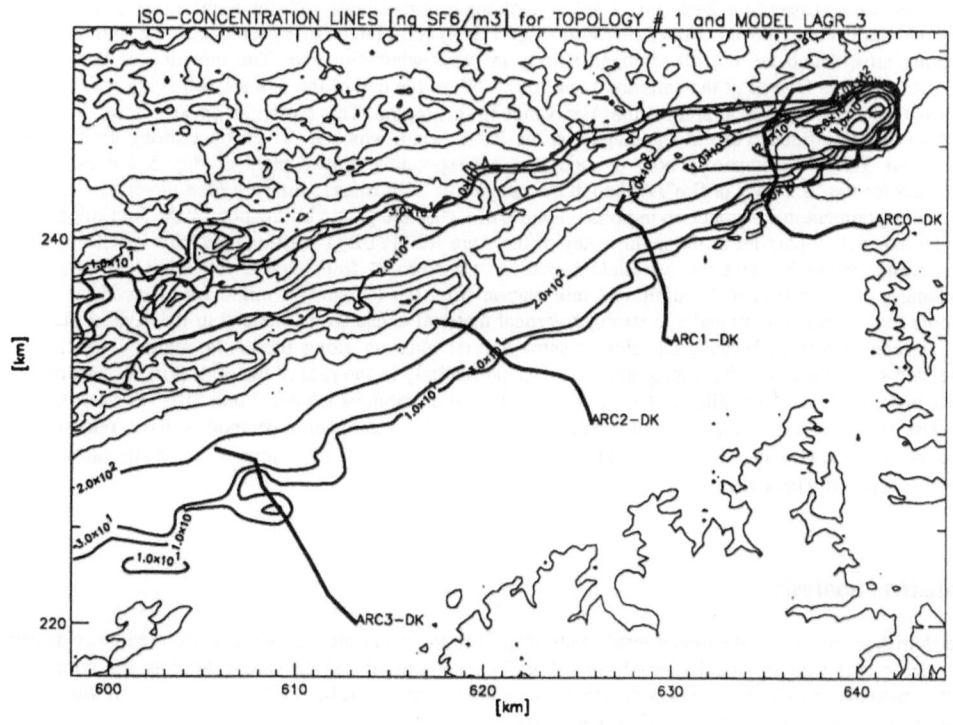

FIG. 2 Iso-concentration lines for model LAGR_3 and SIESTA episode on 24th.

MODEL ASSESSMENT AND VERIFICATION

To facilate the answer the menu driven program ANNE_LYSE was developed for a quick and easy analysis of the model results. The calculated and measured concentrations are compared at the ground level and the model performance is measured qualitatively by **visualization** and quantitatively by **statistical analysis** as well as by **soft-/hardware performance measure**.

Visualization	Statistical analysis	Soft-/hardware performance
- Iso-concentration lines	- χ^2	- CPU time for same hardware
- Scatter plots	- χ^2_{REL}	& compiler optimization
- Lateral distribution	- c_{CORR}	
- Cumulative frequency distribution	- Mean error factor Q	

Visualization

Less convincing results of particular Eulerian grid and Gaussian puff models are already given elsewhere and are illustrated by iso-concentration lines and cumulative frequency distributionswere (Hermberger et al., 1992). Here 2 top models are discussed, which either are systematically and successfully validated with a variety of relevant tracer experiments, as the Hybrid model HYBR_1, or simulate the physics of atmospheric dispersion properly without any severe simplifications, as the Langrangian model LAGR_3. The iso-concentration lignes at ground level with underlying topography of Fig. 1 - 4 give rather conclusive physical insights. All lines have the same relative concentration scale, and their spatial distributions are directly comparable. The position of the plume with respect to the topography and the sampling arcs is an indicator of the correctness of the applied windfield (main wind directions, shear and speed). The longitudinal plume size depends on the advection process determined by main wind speed. The cross wind spread or the lateral plume size depends on the treatment of the horizontal and vertical material exchanges, or on the order of closure and type of parametrization in the turbulence modelling. Both characteristics, maximum concentration and mean plume width or σ value, are compared and analysed per episode, arc and model. In the case of the 24th (Fig. 1 & 2) the plume centre-regions with the highest concentration all lie on the Jura ridges. In the case of the 30th (Fig. 3 & 4) the plume centre-regions all lie within the 6 sampling arcs, which are not all arranged strictly in cross wind direction. The direction of the plume migration changes from E at the emission point to N - NE at the end of the plumes, following roughly the changing direction and topography of the Aare valley in this region. This particular episode is an excellent example for the influence of the topography on the local wind, which is driven by a strong geostrophic wind from SW. In comparing both episodes the corresponding scatter plots of Fig. 5 are excellent indicators for the "accuracy of fire". Although the case of the 24th is less satisfying for a direct comparison with the experiment, it illustrates an important feature of a possible plume bifurcation by the Jura ridge into a small side valley parallel to the main valley of the Aare for HYBR_1 but much less for LAGR_3. The main reason might be, that the windfield had to be transformed from terrain following to cartesian coordinates only for HYBR_1. Additional information from the topography might have forced the flow patterns to bifurcate. Additionally, a stronger vertical material exchange is induced in the Hybrid than in the Lagrangian model, because the plumes show several maxima along the centre-regions, which are correlated sometimes to higher topographical levels, particularly in the case of the 30th (Fig. 3, "where the plume hits the slope of the hills!"). Concerning the lateral distribtions HYBR_1 and SIESTA show larger and less peaked profiles than LAGR_3 (Fig. 6 & 7). All Lagrangian type models have remarkable differences in the σ values for the episode of the 30th. Nevertheless, all 3 refer to the same turbulence parametrization of (Janicke, 1992).

Statistical analysis

The statistical analysis of the model results with respect to the experimental data is performed individually for every arc as well as globally for all arcs (Fig. 7). The same threshold values defining the ranges of higher, moderate and lower model performance were used as proposed by (Päsler, 1986) and are illustrated by differently coloured background of the bar charts.

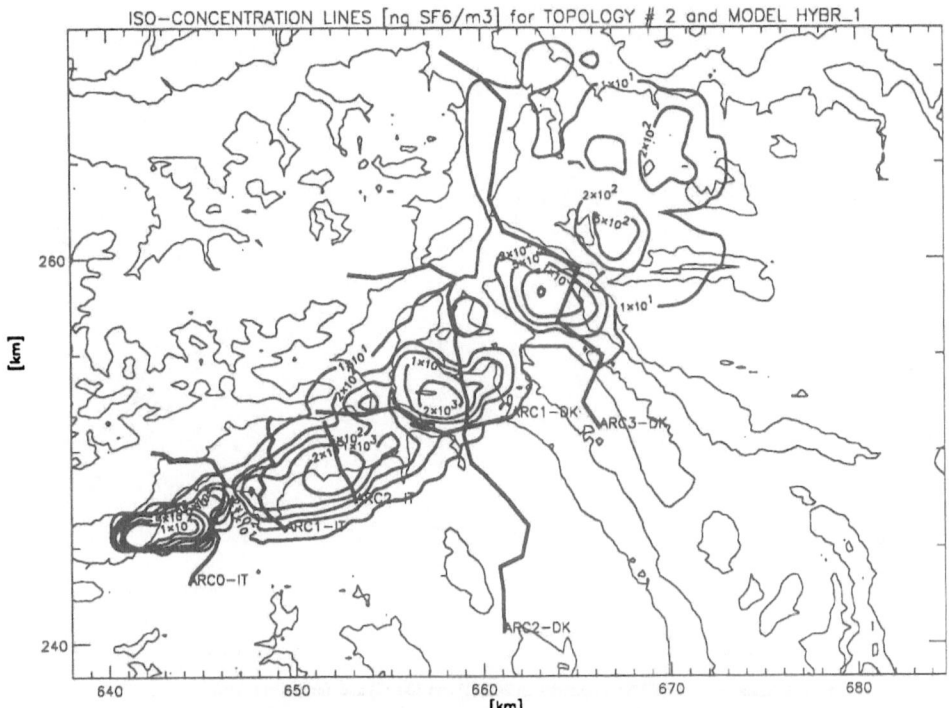

FIG. 3 Iso-concentration lines for model HYBR_1 and SIESTA episode on 30th.

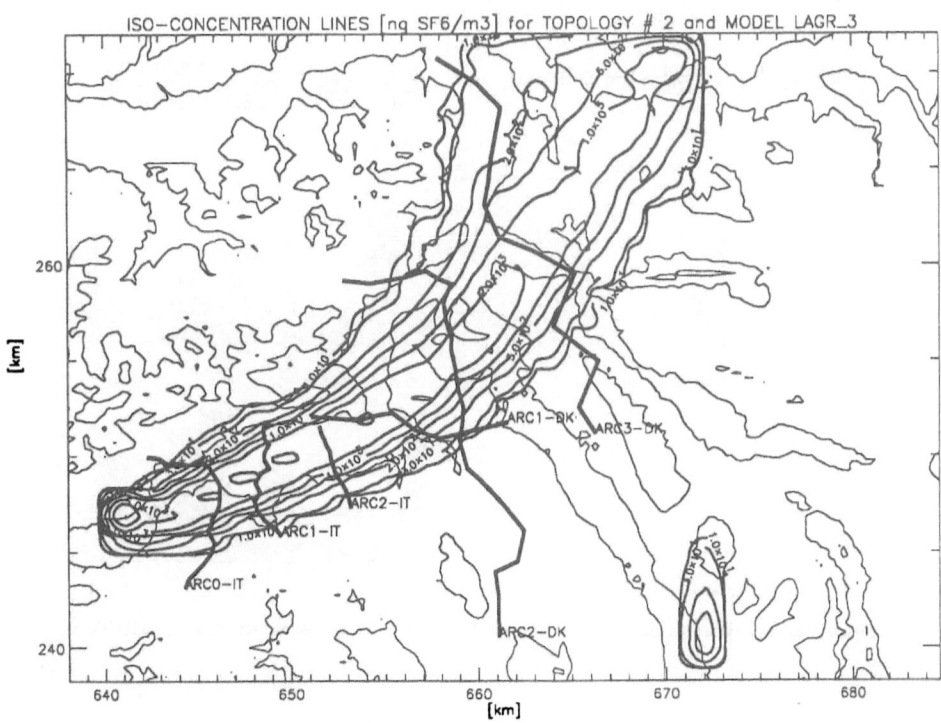

FIG. 4 Iso-concentration lines for model LAGR_3 and SIESTA episode on 30th.

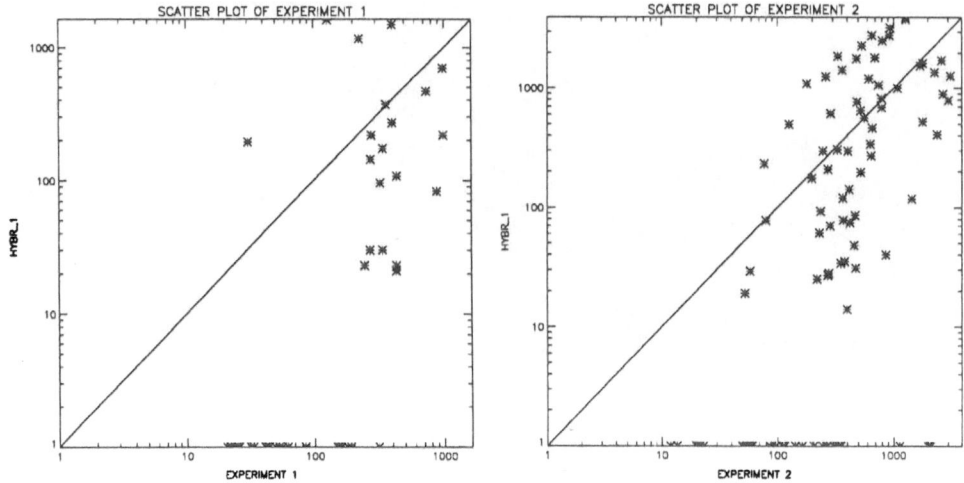

FIG. 5 Scatter plots for SIESTA episodes on 24th (1) and 30th (2) and for model HYBR_1.

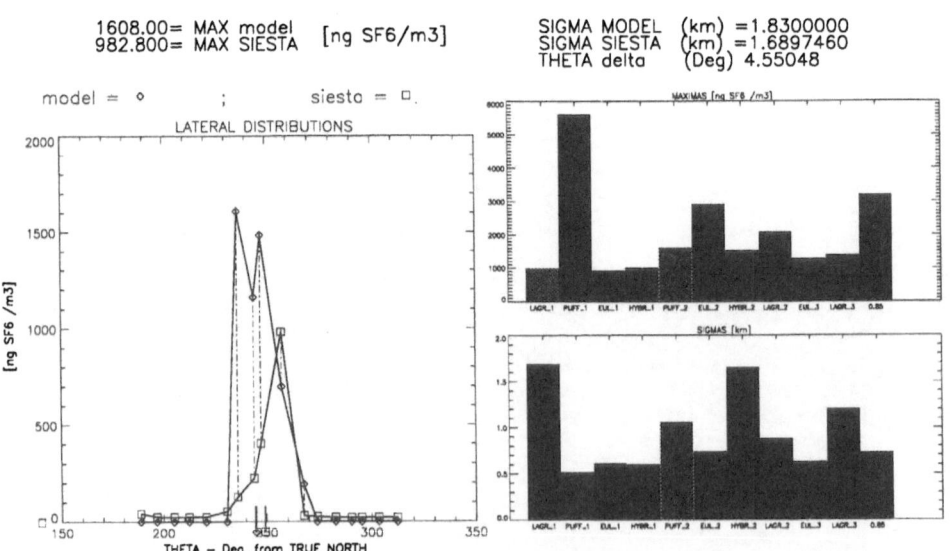

FIG. 6 Lateral concentration distributions on inner arc (0-DK) of SIESTA episode on 24th and of model HYBR_1.

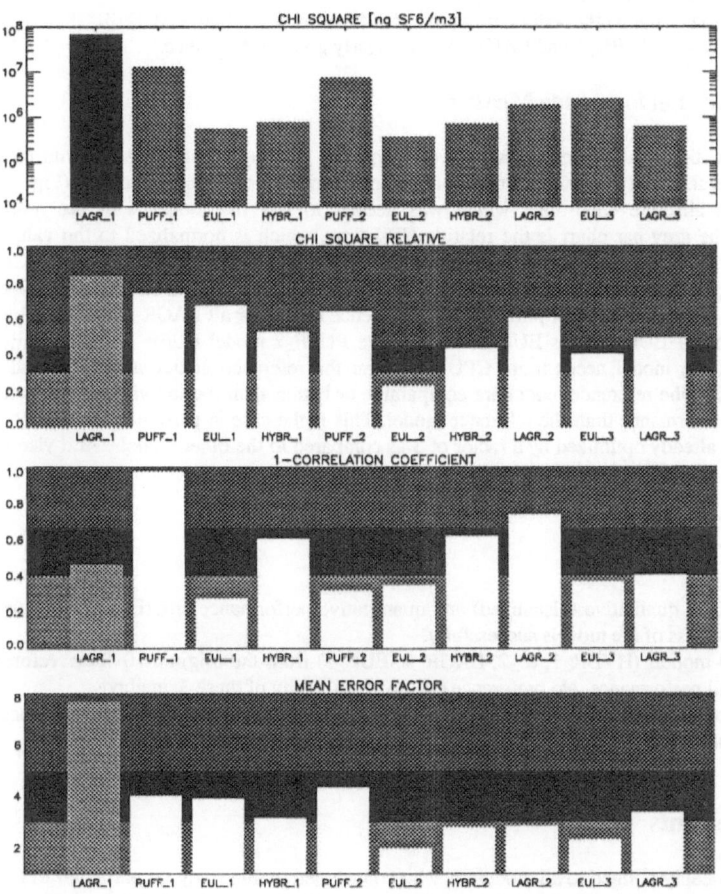

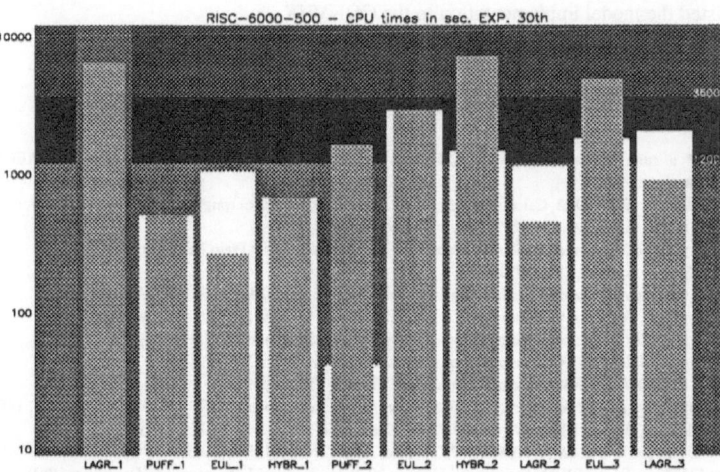

FIG. 7 Statistical results and CPU-times for SIESTA episode on 30th. Different background colours indicate the model performance of bright grey/good, black/modest and dark grey/low quality counted. Absolute and relative CPU-times are indicated as white and grey bar charts respectively.

χ^2: HYBR_1 and LAGR_3 show good performance with respect to the other models.
χ^2_{REL}: Both models have nearly moderate performance.
C_{CORR}: HYBR_1 has a distinctly better correlation with measurements than LAGR_3.
Q: With Q = 3 HYBR_1 and LAGR_3 show nearly good performance.

Soft-/Hardware Performance Measure

For relevant CPU-times the model softawrae were compiled without optimization and run on comparable hardware in general. Two types of CPU-times are presented in form of double bar charts (Fig. 7). The white bar chart is the absolute CPU-time, which was needed to run the model with the modeler's input specifications. The grey bar chart is the relative CPU-time, which is normalized to the values of critical input parameters as specified for one of the models, the so called reference model. Critical input parameters, which essentially determine the CPU-times of the models, depend on the model type as number of particles, grid cells, time steps or puffs e.g. The reference model for all LAGR_ models and for HYBR_2 was HYBR_1 for all EUL_models EUL_3, and for the PUFF_2 model PUFF_1. If the white bar charts exceed the grey, the model needs more CPU-time than the reference model and vice versa. If also the statistical results of the reference model are comparable or better, than the individual models have a lower soft-/hardware performance than the reference model. This is the case in particular for LAGR_3, of which compilation was already optimized by a factor of 3 as compared to the other models. And vice versa for the model HYBR_2 e.g.

CONCLUSIONS

- Different kinds of qualitative (visualized) and quantitative performance criteria are necessary to identify strength and weakness of the models successfully.
- On this basis 4 models (HYBR_1, &_2, LAGR_2, EUL_3) from the original 10 "best referenced" show acceptable overall performance. No preference can be given to any of these 3 methods.
- Transformation of windfields from terrain following to Cartesian coordinates can influence dispersion and the topography can force the windfield to bifurcate.

Acknowledgements

The authors are deeply obliged to all modelers, who contributed to this study. In particular to D. Bürki, D. Buty, M. Chino, T. Flassak, L. Gidhagen, L. Janicke, R. Lamprecht, B. Lawver, K. Massmeyer, J. Moussafir, S. Thykier-Nielsen, Ch. Winkler.
K. Padiyath assisted the model implementation to the CONVEX
The support of the Nuclear Safety Inspectorate, Switzerland, is appreciated.

REFERENCES

Chino, M.: Manual of a suite of computer codes, EXPRESS (EXact PREparedness Supporting System) JAERI - M 92-082, Tokai-mura, Japan, June 1992

Flassak, T., Moussiopoulos, N.: CONDOR3, Calculation of non-divergent flowfields over rough terrain, level 3.0, Version PSI_3.37 from Kaufmann, P. and Gallus, M., CH-5232 Villigen PSI, 1990.

Gassmann, F., Bürki, D.: Experimental Investigation of Atmosph. Disper. over the Swiss Plain - Experiment "SIESTA" Boundary-Layer Meteorology 41, 295-307, 1987

Gidhagen, L.: Indic Airviro: Eulerian advection-diffusion grid model. Indic AB, S-600 45 Norrköping, August 1992

Hermberger V. R. D., Doria, P., Prohaska, G.: Model Evaluation and Selection for Real-time Emergency Applications. Proc. of a Seminar on "Environmental Impact of Nuclear Installations", September 1992, Fribourg, Switzerland

Janicke, L.: Ausbreitungsmodell LASAT Handbuch Version 2.17, D-7770 Überlingen, Juli 1992

Lamprecht, R.: Program PARTRAC, Private communication. PSI, CH-5323 Villigen, 1992

Martens, R. et al.: Bewertung der derzeit genutzten Atmosphärischen Ausbreitungsmodelle. GRS-1-1300, D-5000 Köln, 1987

Perdriel, S.: Notice d'utilisation du code HERMES. EdF, F-78400 Chatou, Réf: HE-33/91.01, Janvier 1991

Rodriguez, D. J. et al.: User's guide to the MATHEW/ADPIC models. UCRL-MA-103581, Livermore, California 94550, April 1992

Thykier-Nielsen, S.: Mikkelsen, T., RIMPUFF users guide. Version 30, Risoe National Laboratory Dk, December 1991

Wendum, D., Biscay, P.: Calculs de transport et de diffusion de polluants par la méthode des bouffées gaussiennes. EdF, F-78400 Chatou, April 1988

Winkler, C.: Dispersion code evaluation by analysis of SIESTA experiment. Step 1: Dispersion model intercomparison and qualification. AG Atmosphärische Ausbreitungsvorgänge, Lehrstuhl & Institut für technische Thermodynamik, Universität, D-7500 Karlsruhe, Interner Bericht, August 1992

DISCUSSION

H. SCHLÜNZEN — Where did you get the windfield for the model comparison?

V. HERRNBERGER — Mean wind and turbulence data were measured in conjunction with atmospheric dispersion of SF6-tracer during weak-wind situations over the Jura ridge and the hilly prealpine region. The windfields were generated from meteorological measurements of meteo towers, tethered balloons, routine midday soundings and the Swiss topographical data with a resolution of 250 m by the mass-consistent diagnostic flow model CONDOR. Terrain following coordinates with a "horizontal" grid spacing of 1 km were used. The windfield and turbulence data were averaged over the emission period of 6 h.

H.SCHLÜNZEN — After decision for one dispersion model: how will you prepare the wind fields in general for the model?

V. HERRNBERGER — Up to 30 km around every site of a major nuclear or chemical plant a network of meteorological ground stations combined with a few vertical soundings as SODARS will be operated with high temporal and spatial resolution over one to two years. As already demonstrated by the project MIS-TRAL around Basel they show all important windfields. A classification of the observed wind fields by cluster analysysis results in a reduced set of typical situations, which are stored in a data bank, and at the same time indicates the most representative stations, which inversely allow to select the typical class from the data bank. Both form the site specific, meteorological data base for the future on-line operation of the selected dispersion model. Such a system helps to estimate the time evolution of inadvertent immissions for emergency planning as well as for emergency management during actual situations without relying on an operational time-dependent hydrodynamical model for the local scale with all its difficulties.

J. LUTZ — The Langrangian modelling results showed rather small plumes. Could this be due to the fact that Lagrangian models have no intrinsic diffusion and the given stationary wind field prohibited a modelling of the meandering of the wind field?

V. HERRNBERGER — This is not the whole truth, because first, Langrangian models show a general tendency of too small plumes and too high concentrations as it was already stated during this conference, and second, horizontal and vertical dispersion was taken into account by measured mean sigma parameters or Monin-Obukow-lengths.

EXAMINATION OF THE EFFICACY OF VOC AND NOx EMISSIONS REDUCTIONS ON OZONE IMPROVEMENT IN THE NEW YORK METROPOLITAN AREA

Kuruvilla John[1], S. T. Rao[1], Gopal Sistla[2], Nianjun Zhou[2], Winston Hao[2], Kenneth Schere[3], Shawn Roselle[3], Norman Possiel[3], and Richard Scheffe[3]

[1]State University of New York at Albany
Albany, New York 12222, U.S.A.

[2]New York State Department of Environmental Conservation
Albany, New York 12233-3259, U.S.A.

[3]U. S. Environmental Protection Agency
Research Triangle Park, North Carolina 27711, U.S.A.

INTRODUCTION

Ozone is not directly emitted into the atmosphere, but is instead a secondary pollutant that is formed from a variety of atmospheric reactants in the presence of sunlight. The magnitude of ozone concentrations in urban areas, for example, along the eastern sea-board of the United States depends upon the transport of ozone and its precursors into the domain, precursors emitted within the domain, the rate at which chemical reactions take place, deposition of pollutants within the domain, and transport and diffusion of pollutants out of the domain.

Two decades after passage of the Federal Clean Air Act (CAA), many urban areas in the United States still violate the ozone standard. Recognizing the severity of the ozone non-attainment problem in the country, the CAA was amended in 1990, requiring the use of grid-based models in complex urban areas in designing emission control strategies. Several states in the Northeast have embarked on photochemical modeling analyses to examine the relationship between ozone and its precursor emissions (US EPA, 1990; Rao and Sistla, 1993). The modeling analysis performed to date indicates that there is no one emissions reduction strategy that will attain the ozone standard throughout the northeastern United States. The challenge is to find attainment strategies which are acceptable to industry and society and leads to ozone attainment as quickly as possible.

The object of this paper is to examine the efficacy of hydrocarbon and nitrogen oxides emissions reduction scenarios in reducing ozone concentrations in the New York metropolitan area using an urban scale photochemical model. To this end, we have applied the Urban Airshed Model (UAM-IV), to a July 1988 high ozone episode in the Northeast using two sets of emissions inventories - EPA's 1985 National Acid Precipitation Assessment Program (NAPAP) emissions inventory adjusted for 1988 meteorological conditions and EPA's interim 1990 emissions inventory. This paper presents the results of a series of simulations with varying amounts of reduction in

Air Pollution Modeling and Its Application X, Edited by S-V. Gryning
and M.M. Millán, Plenum Press, New York, 1994

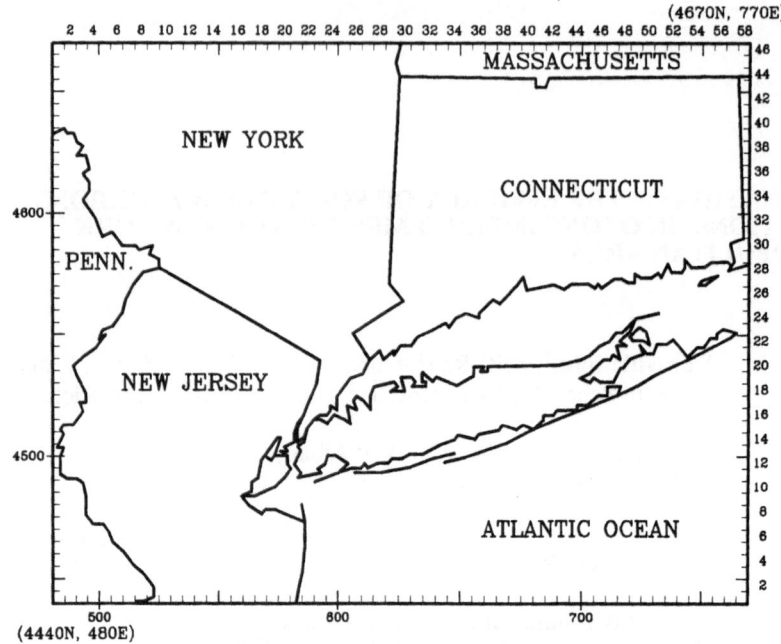

Figure 1. New York Urban Airshed Modeling Domain.

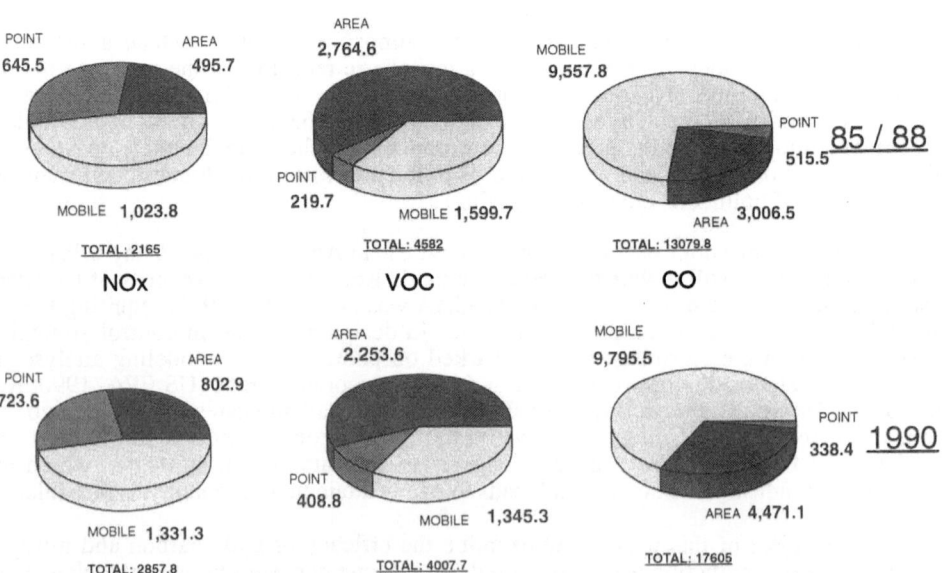

EPA'S ANTHROPOGENIC EMISSIONS INVENTORY FOR NEW YORK AIRSHED (TONS / DAY)

Figure 2. Comparison between EPA's 1985/88 and interim 1990 emissions inventory - anthropogenic emissions for the New York airshed.

hydrocarbon and nitrogen oxides emissions for the episode selected. In all these simulations, meteorological data, and initial/boundary conditions needed for the UAM-IV application to the New York metropolitan area were derived by interfacing the UAM with the EPA's Regional Oxidant Model (ROM). The modeling results reveal that NOx-focused control provided greater benefits than VOC-focused controls in improving ozone air quality within the New York urban airshed for the episode and emissions inventories considered here.

BRIEF DESCRIPTION OF THE MODELS

EPA's Regional Oxidant Model (ROM) is an episodic, Eulerian grid model designed to simulate various physical and chemical processes affecting ozone formation and transport over scales of 1000 km, covering multi-day scenarios. The model employs the Carbon Bond IV (CB-IV) chemical mechanism of Gery et al. (1989) with 83 chemical reactions involving 35 chemical species. The mathematical framework for the ROM can be found in Lamb (1983). The Urban Airshed Model (UAM) is a three dimensional photochemical (Eulerian) grid model designed to calculate the concentrations of both inert and chemically reactive pollutants by simulating the physical and chemical processes in the atmosphere that affect pollutant concentrations(Scheffe and Morris, 1993). The UAM also contains the CB-IV chemical mechanism and is designed to predict hourly ozone concentrations on urban-scale domains. A complete description of the model can be found in Morris et al. (1990).

MODELING ANALYSES

During the summer of 1988, there was a significant increase in ozone exceedances above the ozone standard (120 ppb) in many areas of the country. High ozone levels above 200 ppb were observed in the NY-NJ-CT tri-state region of the Northeast. Modeling analysis was performed using the ROM for the period of July 1-15, 1988. UAM-IV was used for simulating the urban scale ozone episode. The modeling domain chosen for the UAM simulations encompassed the New York - New Jersey - Connecticut tristate region (Figure 1). The base case selected for the UAM simulations covered the high ozone period of July 5 through 8, 1988.

In this study, the ROM-UAM interface (Tang et al., 1990) was used to provide the meteorological inputs along with the initial and boundary conditions for the UAM. In order to assess the sensitivity of the model to changes in emission characteristics, two different emissions inventories, obtained from EPA, were used in this study - i.e. 1985 NAPAP emissions modified for mobile source emissions for 1988 and EPA's 1990 interim emission inventory. This resulted in two separate baseline simulations - i.e. 1985/88 and 1990 utilizing the ROM-UAM meteorological inputs representative of the July 1988 episode. Some of the differences between the emission inventories for the New York airshed are highlighted in Figure 2. Between these two emissions inventories, the total NOx emissions increased by about 32%, while the total VOC emissions decreased by about 12.5%. Such changes in emission loading within an urban airshed can have significant impacts on the ozone air quality as well as on the efficacy of emission reduction strategies.

In order to examine the efficacy of emissions reduction strategies in improving urban-scale ozone air quality, a set of three separate emission reduction scenarios were simulated for both sets of the emissions inventories. These include across-the-board reductions of - (a) 25% NOx and 25% VOC, (b) 25% NOx and 75% VOC, and (c) 75% NOx and 25% VOC. These emission reduction runs provide information regarding the efficacy of control strategies for the New York airshed. The results from the base line simulations as well as the emission reduction cases are analyzed in detail in the following section.

561

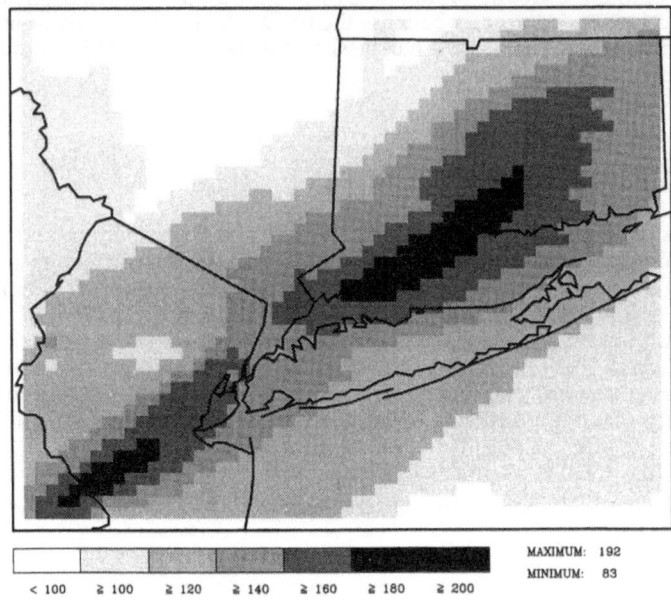

Figure 3. UAM-IV predicted domain maximum ozone concentrations for EPA's 1985/88 emissions inventory with baseline meteorological episode of July 1988 (Base Case).

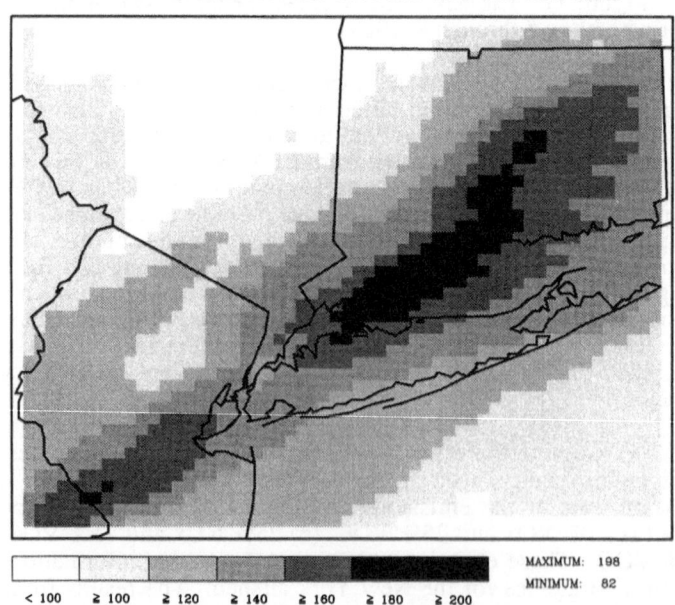

Figure 4. UAM-IV predicted domain maximum ozone concentrations for EPA's interim 1990 emissions inventory with baseline meteorological episode of July 1988.

25% NO$_x$, 75% VOC EMISSIONS REDUCTION SCENARIO – UAM IV
MAXIMUM OZONE CONCENTRATIONS (ppb) FOR JULY 5 TO 8, 1988

WITH RE-GRIDDED 1985/88 ROM EMISSIONS AND ROM–UAM WIND FIELD

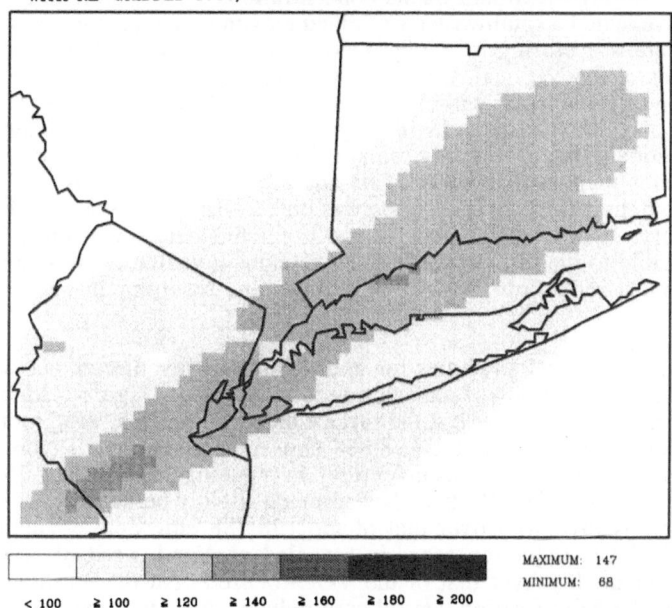

MAXIMUM: 147
MINIMUM: 68

< 100   ≥ 100   ≥ 120   ≥ 140   ≥ 160   ≥ 180   ≥ 200

Figure 5. UAM-IV predicted domain maximum ozone concentrations for 25% NOx and 75% VOC emissions reduction scenario with EPA's 1985/88 emissions inventory.

25% NO$_x$, 75% VOC EMISSIONS REDUCTION SCENARIO – UAM IV
MAXIMUM OZONE CONCENTRATIONS (ppb) FOR JULY 5 TO 8, 1988

WITH EPA'S INTERIM 1990 EMISSIONS AND ROM–UAM WIND FIELD

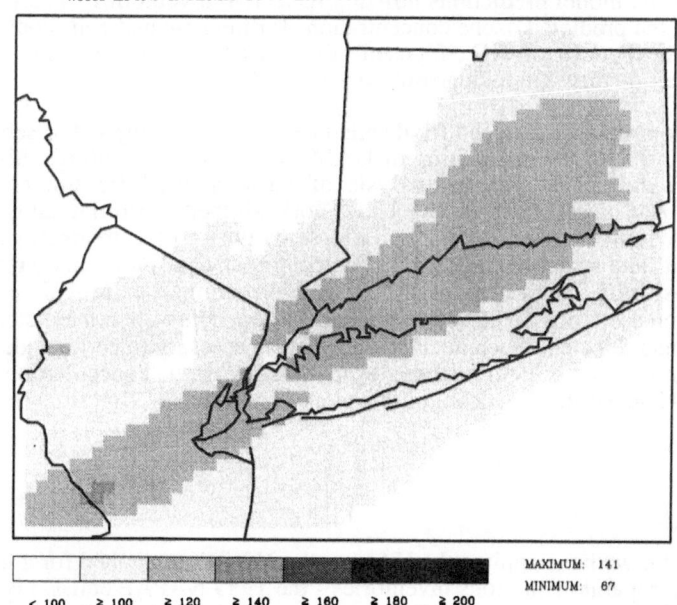

MAXIMUM: 141
MINIMUM: 67

< 100   ≥ 100   ≥ 120   ≥ 140   ≥ 160   ≥ 180   ≥ 200

Figure 6. UAM-IV predicted domain maximum ozone concentrations for 25% NOx and 75% VOC emissions reduction scenario with EPA's interim 1990 emissions inventory.

RESULTS AND DISCUSSION

With the 1985/88 emissions inventory and the wind fields derived from ROM, the UAM prediction of 3-day maximum ozone concentrations for the simulation period of July 5-8, 1988 has revealed a southwesterly oriented maximum ozone concentrations field with the highest values occurring in southern Connecticut. Figure 3 highlights this and reveals exceedances of the National Ambient Air Quality Standards (NAAQS) for ozone (120 ppb) along the urban corridor within New Jersey, New York and Connecticut. When the 1990 interim emissions was applied with July 5-8, 1988 meteorological conditions, the geometry of the ozone plume essentially remained similar. However, there were higher ozone predictions downwind of New York City and in the central regions of Connecticut (Figure 4). The domain maximum of 198 ppb was higher than predicted with the 1985/88 emissions inventory. These differences in the predicted ozone concentrations were essentially attributable to the differences in the emissions inventories; an increase in the NOx emissions and a reduction in the VOCs emissions resulting in ozone increases downwind.

The efficacy of control strategies for both emissions inventories is evaluated by examining the model predictions of emission reduction cases. Figure 5 highlights the domain maximum prediction of ozone at the surface layer for the 75% VOC reduction and 25% NOx reduction scenario using the 1985/88 emissions inventory. Figure 6 reveals similar details using the 1990 emissions inventory. In reducing the VOCs by 75% and the NOx emissions by 25%, the New York airshed revealed slightly better benefits in the case of the 1990 emissions inventory over that of the 1985/88 emissions. The 75% NOx reduction and 25% VOC reduction scenario revealed greater benefit of NOx-focused control for ozone air quality over that of the VOC-focused control. In the case of the 1985/88 emissions, the NOx-focused control case reduced concentrations over the entire domain to well below the level of the ozone NAAQS (Figure 7). However, in the case of the 1990 emissions, there still remained exceedances at few grid cells downwind of New York metropolitan area (Figure 8); the maximum predicted ozone concentration being higher in the 1990 case (i.e. 131 ppb) than the 1985/88 case (103 ppb).

Despite differences in model predictions of emissions reduction scenarios, the NOx-focused control case proved to be more beneficial than the VOC-focused control case for both sets of emissions inventories. Figure 9 (a and b) provides box plot representations of the model predictions utilizing these two inventories. These statistical representations of the predicted ozone concentrations for the base and emissions reduction cases highlight the benefit of NOx-focused controls over VOC-focused controls in improving ozone air quality within the New York airshed.

These results suggest the need for detailed examination of the UAM sensitivity to input parameters prior to the application of UAM in a regulatory setting. Sistla et. al. (1993) provide detailed sensitivity analysis of various wind field/meteorological treatments for UAM application in the New York airshed. Similar analyses were performed by the authors to examine the UAM sensitivity to the treatment of mixing heights, wind fields and spatial/temporal allocation of emissions. Preliminary results from these model sensitivity studies suggest that the "directions" or model response to emissions reduction scenarios could vary significantly depending upon the meteorological and emissions inputs selected. Such detailed analysis is necessary to better understand the efficacy of various emissions control strategies in reducing ozone concentrations in urban airsheds, such as New York, to the level of the ozone NAAQS.

SUMMARY

In this study, we have applied the UAM to the New York airshed for a high ozone episode using two separate emissions inventories - the 1985 NAPAP emissions inventory adjusted for 1988 mobile source emissions and meteorological conditions and EPA's interim 1990 emissions inventory. The interim 1990 inventory had greater amount of

75% NO$_x$, 25% VOC EMISSIONS REDUCTION SCENARIO – UAM IV
MAXIMUM OZONE CONCENTRATIONS (ppb) FOR JULY 5 TO 8, 1988

WITH RE–GRIDDED 1985/88 ROM EMISSIONS AND ROM–UAM WIND FIELD

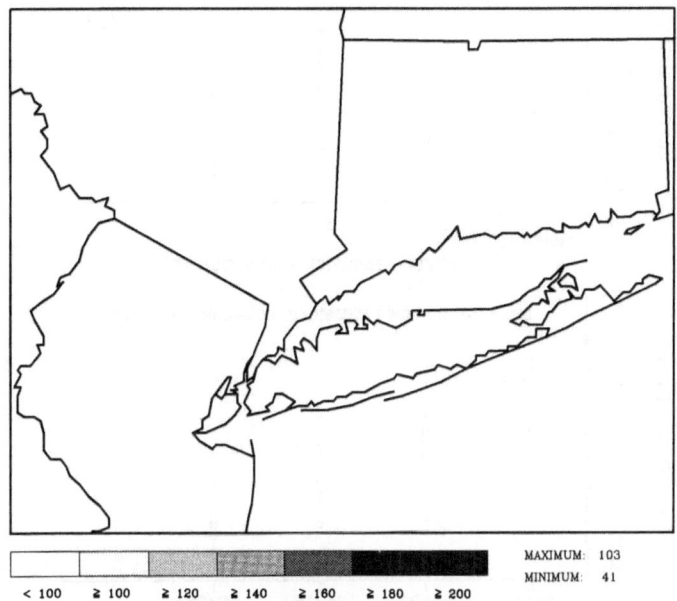

MAXIMUM: 103
MINIMUM: 41

< 100   ≥ 100   ≥ 120   ≥ 140   ≥ 160   ≥ 180   ≥ 200

Figure 7. UAM-IV predicted domain maximum ozone concentrations for 75% NOx and 25% VOC emissions reduction scenario with EPA's 1985/88 emissions inventory.

75% NO$_x$, 25% VOC EMISSIONS REDUCTION SCENARIO – UAM IV
MAXIMUM OZONE CONCENTRATIONS (ppb) FOR JULY 5 TO 8, 1988

WITH EPA'S INTERIM 1990 EMISSIONS AND ROM–UAM WIND FIELD

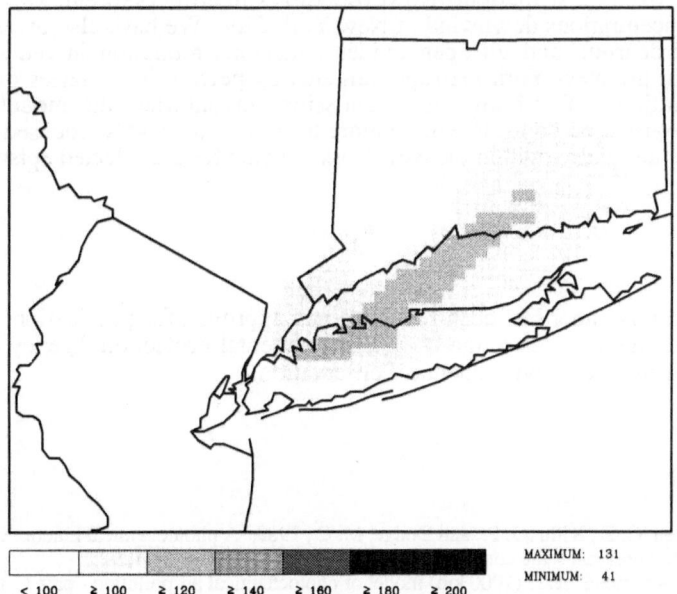

MAXIMUM: 131
MINIMUM: 41

< 100   ≥ 100   ≥ 120   ≥ 140   ≥ 160   ≥ 180   ≥ 200

Figure 8. UAM-IV predicted domain maximum ozone concentrations for 75% NOx and 25% VOC emissions reduction scenario with EPA's interim 1990 emissions inventory.

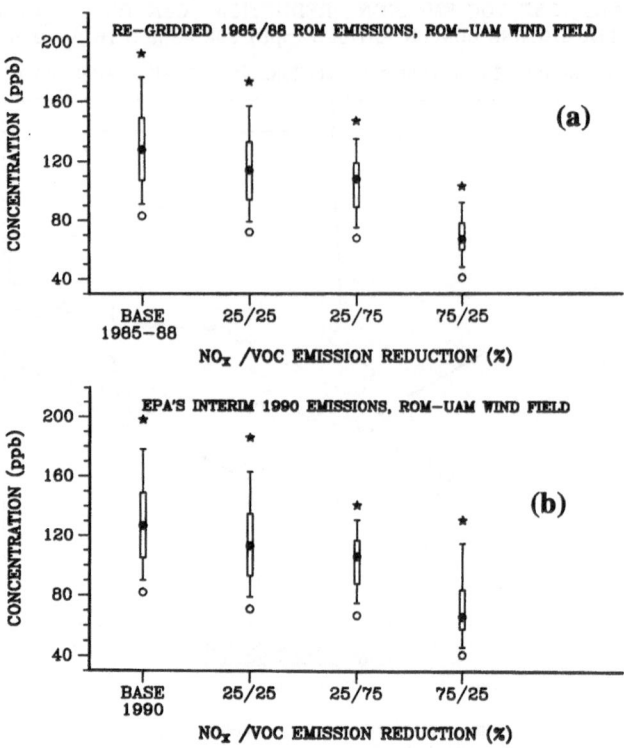

Figure 9. Distribution of UAM-IV predicted episode maximum ozone concentrations (box plots) in the New York airshed - (a) 1985/88 emissions inventory and (b) EPA's interim 1990 emissions inventory.

NOx and slightly lower VOCs when compared with the 1985/88 emissions inventory. The UAM simulations for the 1988 episode using two separate baseline emissions revealed similar ozone distribution. However, EPA's interim 1990 emissions produced higher ozone concentrations downwind of New York City. We have also investigated the efficacy of hydrocarbon and nitrogen oxides emissions reduction in reducing ozone concentrations in the New York metropolitan area by performing a series of emissions reduction simulations. For both sets of emissions inventories, the modeling results suggest that NOx-focused controls will be more beneficial than VOC-focused controls in improving ozone air quality within the New York airshed for the selected episode.

DISCLAIMER

Although this paper has been reviewed and approved for publication, it does not necessarily reflect the policies of the U. S. Environmental Protection Agency or the New York State Department of Environmental Conservation.

REFERENCES

Gery, M. W., Whitten, G. Z., Killus, J. P., and Dodge, M. C., 1989, A photochemical kinetics mechanism for urban and regional scale computer modeling, *J. Geophys. Res.*, 94:12925.
Lamb, R. G., 1983, A regional scale (1000 km) model of photochemical air pollution - part 1. theoretical formulation, EPA-600/3-83-035, U.S. E.P.A., Research Triangle Park.
Morris, R. E., Myers, T. C., Haney, J. L., 1990, User's guide for the urban airshed model, Volume I: User's manual for the UAM (CB-IV), EPA-450/4-90-007A, U.S. E.P.A., Research Triangle Park.

Rao, S. T., and Sistla, G., 1993, Efficacy of nitrogen oxides and hydrocarbons emissions controls in ozone attainment strategies as predicted by the urban airshed model, *Water, Air and Soil Pollution*, 67:95.

Scheffe, R., and Morris, R. E., 1993, A review of the development and application of the urban airshed model, 1993, *Atmos. Envir.*, 27B:23.

Sistla, G., Rao, S. T., Bornstein, R. D., Freedman, F., and Thunis, P., 1993, Sensitivity analysis of the urban airshed model to wind fields derived from the regional oxidant model, diagnostic wind model, and the URBMET/TVM mesoscale model, Proceedings of the Twentieth ITM on Air Pollution Modelling and Its Applications, Valencia, Spain.

Tang, R. T., Gerry, S. C., Newsome, J. S., Van Meter A. R., Godowitch, J. M., and Schere K. L., 1990, User's Guide for the Urban Airshed Model, Volume V: Description and Operation of the ROM-UAM Interface Program System, EPA-450/4-90-007E, U.S. E.P.A., Research Triangle Park.

U.S. Environmental Protection Agency, 1990, Regional Ozone Modeling for Northeast Transport (ROMNET), EPA-450/4-90-002a, U.S. E.P.A., Research Triangle Park.

DISCUSSION

N. MOUSSIOPOULOS: Do you think that the role of biogenic emissions is properly taken into account in the UAM version you are using? What about the natural VOCs "imported" into the model domain?

S.T. RAO: The CB-IV chemical mechanism in the UAM version 6.3 treats isoprene explicitly. In our study, biogenic emissions have been temperature adjusted and allocated spatially. The initial and boundary conditions for our UAM applications have been derived from the Regional Oxidant Model (ROM) and, therefore, reflect the influx of pollutant from both anthropogenic and biogenic sources of emissions into our modeling domain. This is accomplished through a one-way nesting of the urban-scale model (UAM) with regional-scale model (ROM).

R. BORNSTEIN: Do you foresee a renewed emphasis on the meteorological factors required for photochemical regulatory modeling?

S.T. RAO: Yes. It is very important that we characterize meteorological conditions as realistically as possible. Unfortunately, detailed data needed for the UAM simulations are not available from routine measurements. Sensitivity studies have demonstrated the strong influence of meteorological variables such as wind fields, mixing etc on model-predicted ozone concentrations. In the USA, under the guidance of the EPA, a concerted effort is underway on renewing the emphasis on meteorological variables as utilized within the regulatory photochemical modeling activity.

R. SAN JOSE: What is the importance of temporal/spatial resolution in the emission inventory?

S.T. RAO: Both are extremely important. The distribution of sources and their temporal emissions profile determines the chemistry and, thus, the formation/destruction of photochemical oxidants. We have utilized a 5 km horizontal grid spacing, and temporal resolution of 1 hour in our study.

T. IVERSEN: In your rather limited area covered by your model, I would expect that a significant part of ozone in the area would be imported from outside, in particular from the free troposphere. Can you comment on to what degree your results of the NO_x/VOC emission reduction strategies will be of limited value because of this.

S.T. RAO: The initial and boundary conditions used in the emissions reduction strategies were derived from the Regional Oxidant Model's (ROM) simulation for the eastern half of the United States. The ROM assumed background concentrations of pollutants are at the tropospheric level, with ozone set at about 40 ppb. The pollutant concentrations seen by the UAM at the top of the modeling domain are obtained from the ROM, which also accounts for the influx of the free troposphere. Moreover, in our study, the top layer of the modeling domain extends to about 3 km, and is, thus, far removed from the free troposphere. Therefore, the results reflect effects of pollutant burden at both urban and regional levels, and the impact of emission reduction strategies on ozone air quality.

DEPOSITION OF GASES AND PARTICLES IN THE PBL: EVALUATION OF THE INFLUENCE OF A VERTICAL RESOLUTION IN ATMOSPHERIC TRANSPORT MODELS

Ole Hertel,[1] Jesper Christensen,[1] Erik Runge,[1] Ruwim Berkowicz,[1] Willem A.H. Asman,[1] Kit Granby,[1] Mads F. Hovmand,[1] and Øystein Hov[2]

[1]National Environmental Research Institute
Department of Emissions and Air Pollution
Frederiksborgvej 399, 4000 Roskilde, Denmark
[2]University of Bergen, Geophysical Institute
Allegaten 70, 5007 Bergen, Norway

1. INTRODUCTION

The deposition processes determines the lifetime of a variety of gases and particles in the atmosphere. Computed depositions depend highly on the parameterisation and model concept. A vertical resolution of the tested model influences not only directly the description of dry and wet deposition processes, but also indirectly, as the chemical reactions are affected by the vertical distribution of gases and particles in the PBL. This indirect influence is due to the common assumption of full mixing in the PBL in models of no or only coarse resolution in the vertical. The effect of the vertical model resolution on estimations of concentrations and depositions of gases and particles is the subject of this paper.

We present results obtained with the Atmospheric Chemistry and Deposition (ACDEP) model, which is developed under the Danish Sea Research Programme 90 with the purpose of estimating the nitrogen deposition to Danish coastal waters. The ACDEP model is a one-dimensional trajectory model that includes chemical reactions, dry and wet deposition, and vertical transport described by eddy diffusion.

Results for nitrogen and sulphur compounds, H_2O_2 and O_3 are shown for three sites in Denmark: Anholt, an island in the Kattegat; Lille Valby, a rural site on Sealand; and the Copenhagen metropolitan area. For Anholt and Lille Valby the model results are compared with measurements.

The three sites are chosen in order to investigate the dependence of model parameterisation on location of receptor points with respect to main emission areas and surrounding surface conditions.

2. THE MODEL

Only a short description of the ACDEP model is presented here. More details can be found in Asman et al. (1993a).

Two different versions of the ACDEP model are used in the following tests: the basic version (in the following referred to as ACDEP), with a vertical resolution of 10 layers having a logarithmic distribution with a fine mesh close to the ground and coarser as the free troposphere is approached; and a version with only 2 layers for the PBL and the free troposphere, respectively. The 2-layer

Air Pollution Modeling and Its Application X, Edited by S-V. Gryning
and M.M. Millán, Plenum Press, New York, 1994

version is further tested in 2 versions having either constant dry depositions velocities of the gases and aerosols (2GC) or dry deposition depending on wind speed and surface conditions (2GV).

Concentrations of each of the chemical species in the model are calculated from the continuity equation for a one dimensional column of air transported along a trajectory:

$$\frac{Dc}{Dt} = \frac{\partial}{\partial z} K_z \frac{\partial c}{\partial z} + E + P - Lc - (S_{ic} + S_{bc})c \tag{1}$$

where c is the concentration, K_z is the vertical eddy diffusion coefficient for mass transport ($=K_h$ (the eddy diffusion for heat transport)), E is the emission, P and L are the chemical production and loss terms, respectively, and S_{ic} and S_{bc} are the scavenging ratio of in-cloud and below cloud scavenging, respectively. Equation (1) is solved by split step for chemistry, transport, and dry and wet deposition.

The one dimensional column is advected along trajectories to selected receptor points. Backwards, 96-hour trajectories are calculated using 925 hPa wind felds which are assumed to be representative for the boundary layer, disregarding wind turning with height. Trajectories are calculated for arrival times with 6-hour intervals. Positions along a trajectory are computed every 2 hours.

In ACDEP, the vertical diffusion is calculated using an eddy diffusivity model. The boundary conditions are no flux through the upper boundary at 2 km height, while at the lower boundary the vertical flux equals the dry deposition flux,

$$K_z \frac{\partial c}{\partial z} = V_d c \tag{2}$$

where V_d is the dry deposition velocity.
The eddy diffusivity coefficient is given by

$$K_z = \frac{\kappa u_* z}{\phi_h(z/L)}\left(1 - \frac{z}{ZI}\right) \tag{3}$$

where κ is the von Karman constant, u_* is the friction velocity, ϕ_h is the similarity function for heat flux, z is the height, L is the Monin-Obukhov length and ZI is the mixing height. The diffusivity equation is solved by a semi implicit method (theta metod).

In the 2-layer version of the model (2GC and 2GV), vertical transport is simulated by the variation in mixing height only. The lowest layer represents the distance from ground and up to the top of the PBL, and the upper layer a part of the free troposphere from the PBL and up to 2 km height. When the mixing height increases, air from the upper layer is mixed into the lowest layer and vice versa. Full mixing is assumed in both layers.

Meteorological data used for model calculations are from the Norwegian routine numerical weather prediction model. The data are distributed on a 150 x 150 km^2 grid system which is identical to the grid system used in the European Monitoring and Evaluation Programme (EMEP).

2.1 Chemistry

The chemical mechanism used in the ACDEP model is the Carbon Bond Mechanism IV (CBM-IV) presented in Gery et al. (1989a,b). The mechanism is extended to include the chemistry of the ammonium/ammonia system (Hertel et al., 1993a; Asman et al., 1993).

A detailed description on the numerical method used for the solution of the chemistry, Eulerian Backward Iterative method (EBI), is given in Hertel et al. (1993b).

2.2 Emissions

Emission inventories on a 15x15 km^2 grid are used for the computations. These emission data are based on the European 150x150 km^2 grid emission inventory from EMEP for 1990 (Sandnes,

1993). The 150 x 150 km^2 grid squares are subdivided into 10 x 10 grid squares with evenly distributed emissions. A detailed emission inventory is used for Denmark (Asman et al, 1993a; Asman et al, 1993c; Asman, 1992).

In order to simulate the horizontal dispersion, the emissions are averaged over areas which size depend on the distance to the receptor point. The size of the emission area is a tenth of the distance to the receptor point calculated along the trajectory.

2.3 Dry deposition velocities

The dry deposition velocities are computed by the resistance method,

$$V_d = (r_a + r_b + r_s)^{-1} \tag{4}$$

where r_a is the aerodynamic resistance, r_b is the laminar sublayer resistance and r_s is the bulk surface resistance. In ACDEP r_a is computed for each time step for a reference height of 2 m using meteorological data on EMEP grid. A similar procedure is used in 2GV, but with a reference height of 50 m. In 2GC constant dry deposition velocities, corresponding to a wind speed of 5 m s^{-1}, are used. Values for the surface resistance and examples on dry depositions velocities for each species are given in Asman et al. (1993).

2.5 Wet deposition

In the ACDEP model a distinction is made between in-cloud and below-cloud scavenging. In-cloud scavenging is supposed to take place in the layers between 250 m and the upper boundary of the model area (2 km), whereas below-cloud scavenging is assumed to take place below 250 m.

Computation of scavenging coefficients for in-cloud and below-cloud scavenging are described in Asman and Jensen (1993).

Normally a rain event will not cover a whole EMEP grid on 150 x 150 km^2. Rain intensity will also vary during the 6 hour period the rain data are given for. To take this into account, an empirical relation between rain intensity and rain amount, proposed by Iversen et al. (1990), is used.

3. RESULTS

All results shown in this paper are calculated for July 1992. In Figure 1 are shown air concentrations of O_3, H_2O_2, NO_2, SO_2, $SO_4^=$ and NH_4^+ for Anholt. Corresponding calculations for Lille Valby and Copenhagen are shown in Figures 2 and 3, respectively. Results from all three model versions are shown in the figures. For Anholt (Figure 1) are furthermore shown daily averages of measured aerosol concentrations of $SO_4^=$ and NH_4^+, while hourly averages of NO_2 and O_3, and 15 min. averages of H_2O_2 are shown for Lille Valby (Figure 2).

The calculated accumulated dry deposition of SO_2-gas and NH_3-gas are shown in Figure 4. The dry depositions of the aerosols $SO_4^=$ and NH_4^+ are shown in Figure 5. Results from the three model versions are presented.

4. DISCUSSION

The results shown in this paper can only be considered as general illustration of the influence of a vertical resolution and parameterisation method for dry deposition on air concentrations and depositions. This is due to the fact that the vertical resolution of a model influences not only the description of dry deposition directly, but also indirectly as the chemical reactions are affected by the vertical distribution of gases and particles in the PBL. The behaviour is thus expected to vary depending on the chemical species in question. Especially significant difference is expected between primary and secondary pollutants. The influence of the parameterisation method for dry deposition on predicted concentrations and depositions also depends on the specific component.

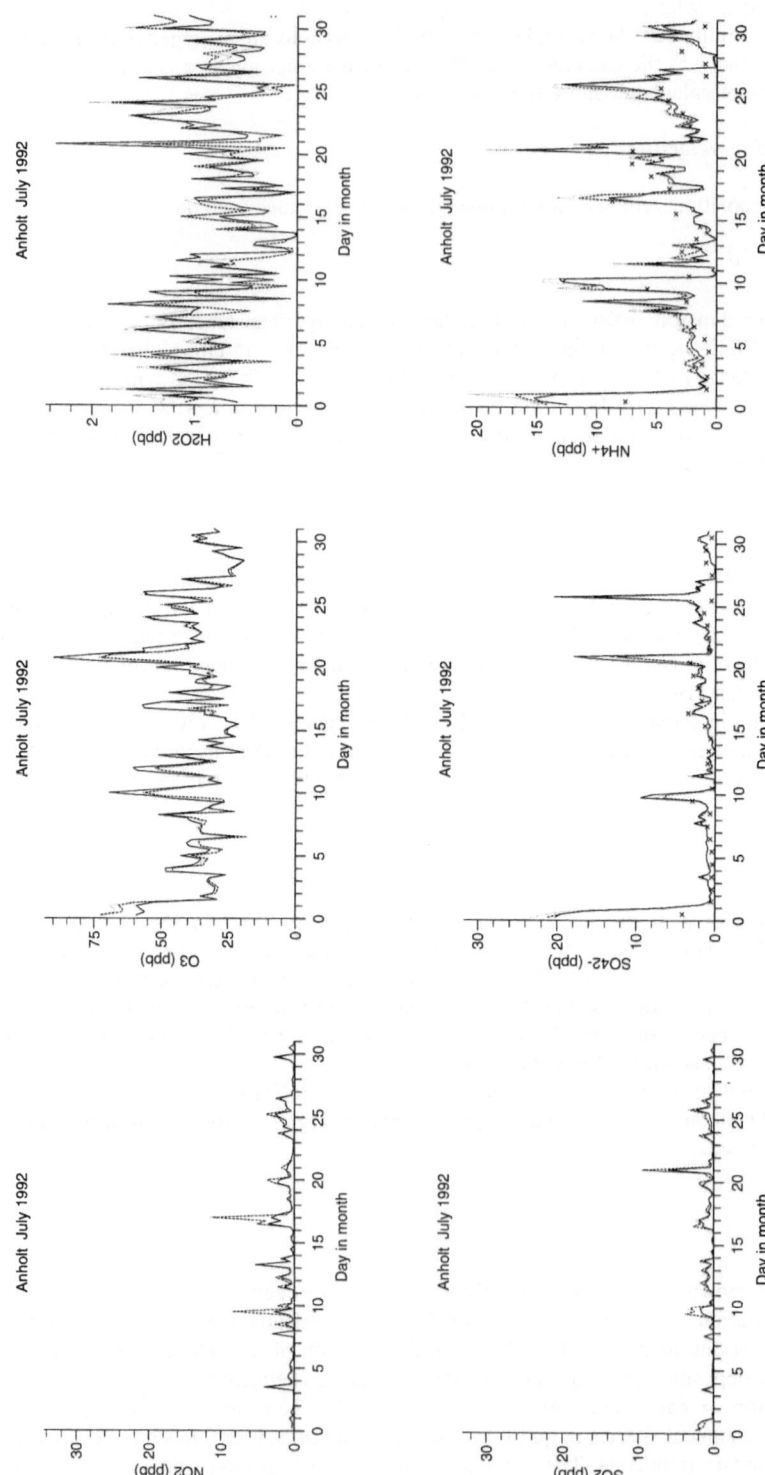

Figure 1. Calculated air concentrations for the island of Anholt. Measured daily averages of $SO_4^=$ and NH_4^+ are also shown. —— ACDEP; --- 2GC; ···· 2GV; × measured.

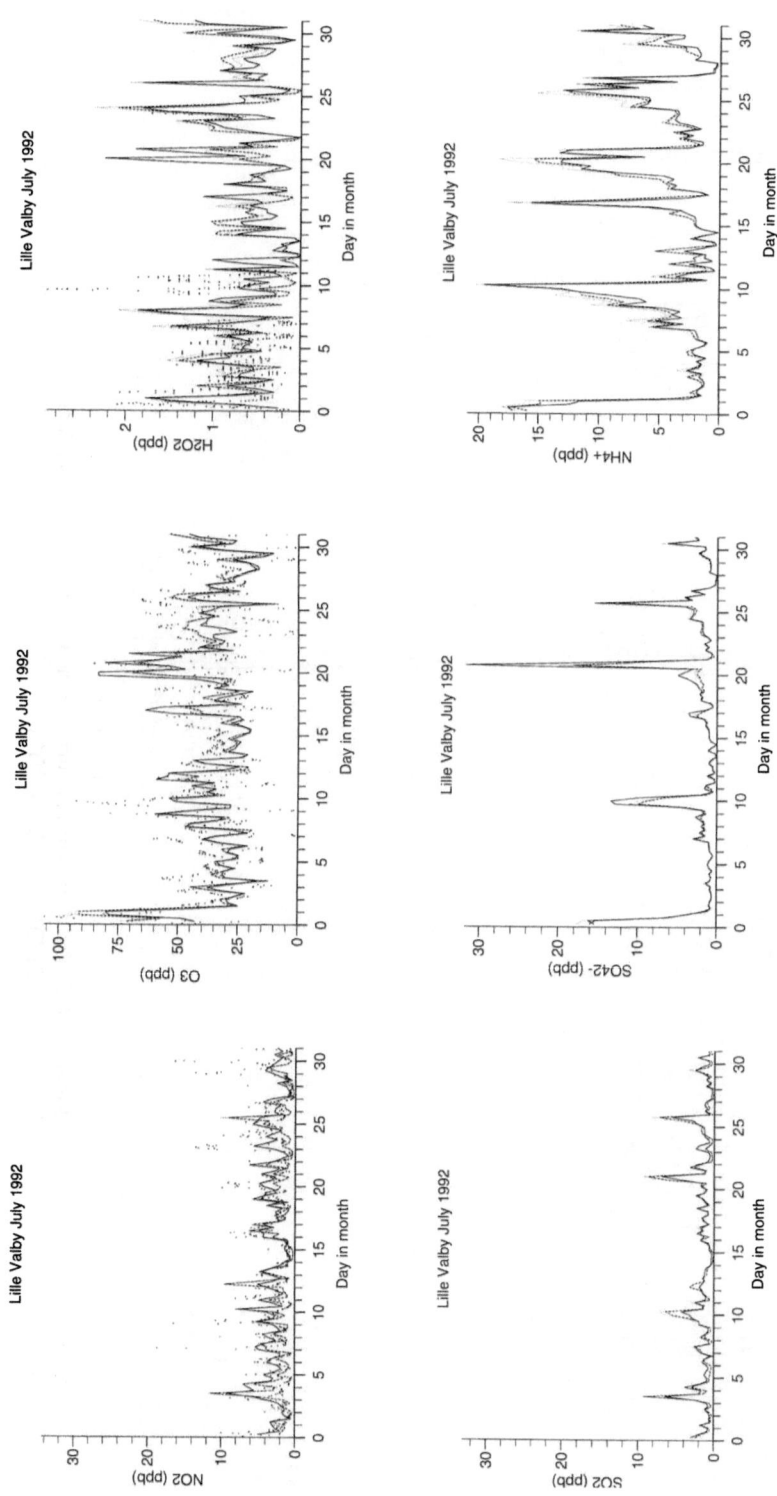

Figure 2. Calculated air concentrations for the rural site Lille Valby. Measured hourly averages of NO_2 and O_3 and 15 min. averages of H_2O_2 are also shown. —— ACDEP; --- 2GC; ···· 2GV; ∎ measured.

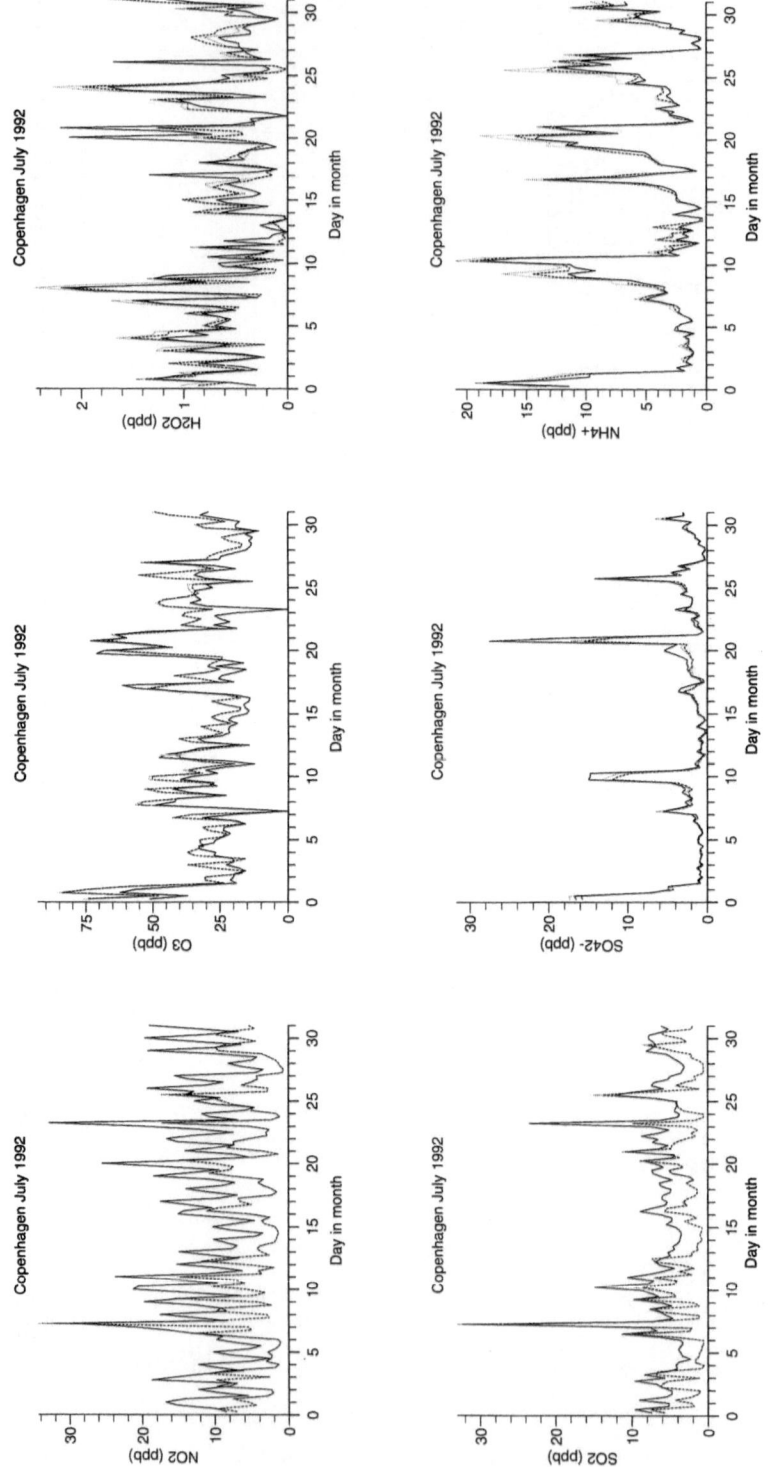

Figure 3. Calculated air concentrations for the Copenhagen metropolitan. —— ACDEP; – – 2GC; ···· 2GV.

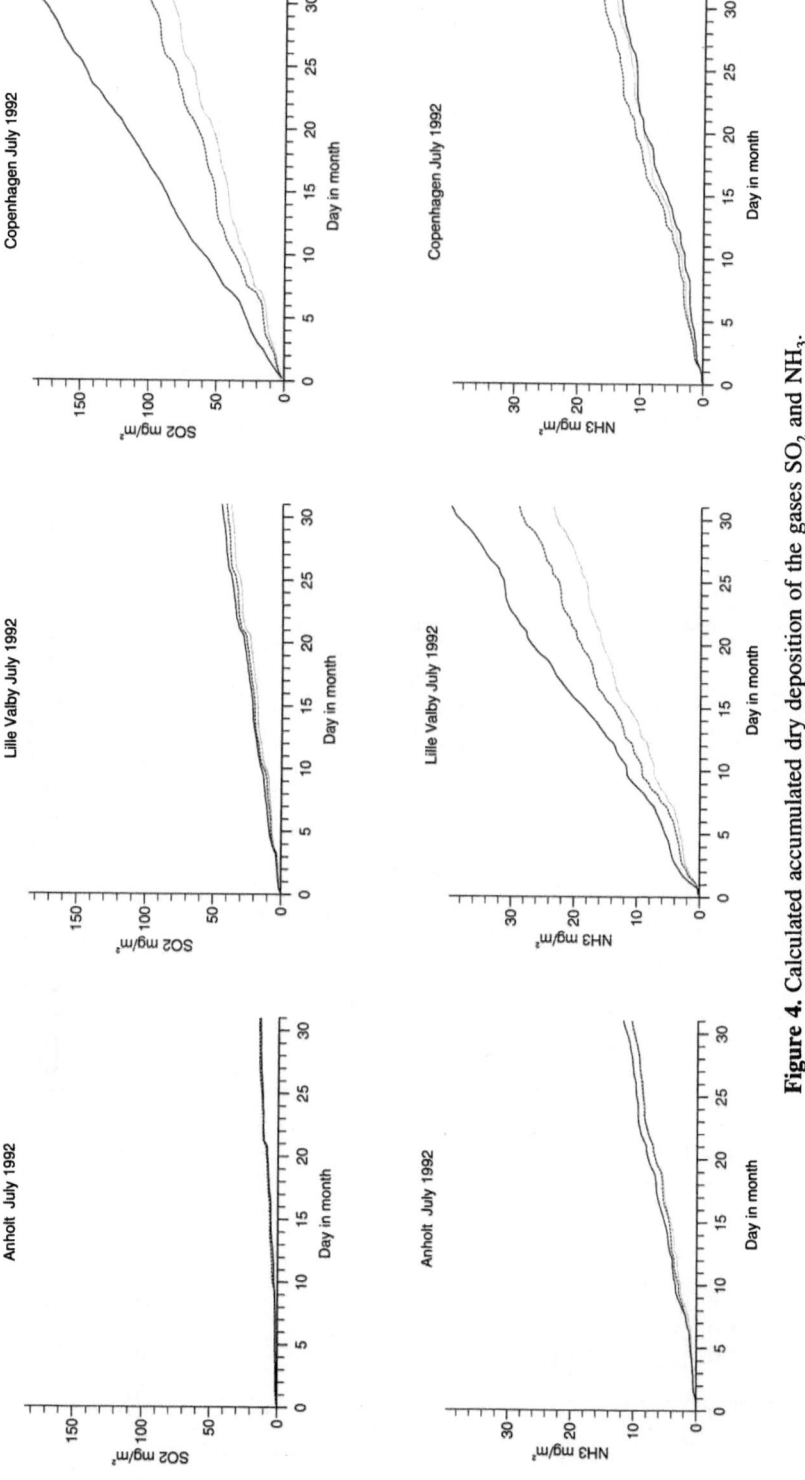

Figure 4. Calculated accumulated dry deposition of the gases SO_2 and NH_3. —— ACDEP; --- 2GC; ···· 2GV.

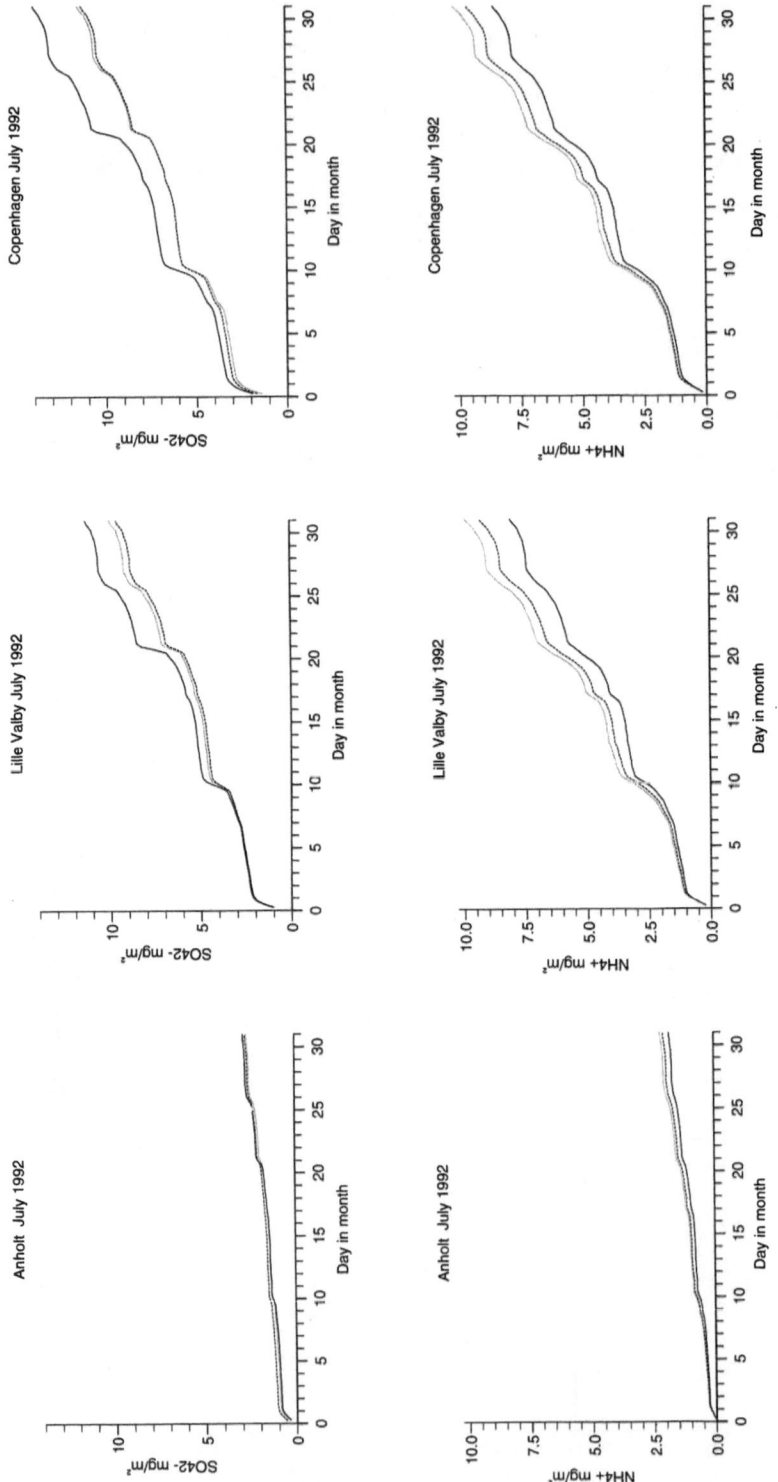

Figure 5. Calculated accumulated dry deposition of the aerosols $SO_4^=$ and NH_4^+:
—— ACDEP; --- 2GC; ···· 2GV.

The differences between calculated air concentrations by the 10-layer version (ACDEP) and the two 2-layer versions (2GV and 2GC) is most pronounced for the gases NO_2, SO_2 and O_3 and especially for the area of Copenhagen where the influence of local emissions is large. Here the concentrations of NO_2 and SO_2 predicted by ACDEP are significantly higher than predicted by 2GC or 2GV. The O_3 concentrations are affected by the chemical loss due to reaction with NO and therefore the concentrations predicted for Copenhagen by the 10-layer version are lower than by the 2-layer versions. In the 2-layer versions the influence of local emissions, due to the assumption of instantaneous mixing in the PBL, is much less than in the 10-layer version. For the rural location Lille Valby and the sea site Anholt the differences are still pronounced, especially for episodes but the tendency is not clear. This is due to that at these locations the air concentrations are more influenced by contributions from remote sources and processes taking place during the transport.

Considering the H_2O_2 concentrations, one can see that the difference between the three model versions is significant for all the locations shown here, but with no clear tendency. This is undoubtedly due to that H_2O_2 is a secondary pollutant influenced by a large number of other chemical compounds.

The concentrations of aerosols, which are predominantly secondary pollutants, are not so significantly influenced by the vertical diffusion processes. Parameterisation of the local deposition of precursors has, however, an indirect influence on the predicted concentrations of aerosols. This influence is most pronounced during episodes of long range transport.

The difference between air concentrations predicted by the two 2-layer versions are not so significant. Largest differences can be observed for H_2O_2, which has very small surface resistance. Due to this, the deposition velocity of H_2O_2 is limited by the aerodynamic resistance which depends on wind speed. Some differences can also be seen for aerosols, for which the deposition velocities are in general very small. Here the difference is caused by effects on the precursors which influence the formation of aerosols.

Comparing model results with the few available measurements presented in the paper, do not permit a detailed evaluation of the model performance in the three different versions. Especially the lack of measurements for Copenhagen, where the differences are largest, is prohibitive.

The differences in accumulated dry deposition are in general small for Anholt both for gases and aerosols. This is due to, that Anholt is located far away from large emission areas, so the differences in parameterisation of mixing processes are not so significant. The most pronounced difference can be seen for SO_2 in Copenhagen and NH_3 in Lille Valby. This can be explained by, that the local emissions of the respective species dominate the air concentrations and depositions at these locations.

In general the difference due to parameterisation of dry deposition velocity is less significant than the difference due to parameterisation of the vertical mixing.

ACKNOWLEDGEMENT

The Danish Research Academy is acknowledged for their financial support of one of the authors (O. Hertel) Ph.D. study, which the present work is a part of. This work is supported by the Danish Environmental Research Programme. Meteorological data for the trajectory calculations were kindly provided by Helge Styve from the EMEP MSC/W at the Norwegian Meteorological Institute in Oslo.

REFERENCES

Asman, W.A.H., 1992, Ammonia emission in Europe: Updated emission and emission variation. Report no. 22841008, National Institute of Public Health and environmental Protection, Bilthoven, The Netherlands.

Asman, W.A.H., Berkowicz, R., Christensen, J., Hertel, O., and Runge, E., 1993a, Deposition of nitrogen and phosphorus compounds to Kattegat (In Danish: Afsætning af kvælstof og fosfosforbindelser til Kattegat). Under the series "Marine Research from the Danish Environmental Protection Agency". In preparation.

Asman, W.A.H., Sørensen, L.L., Berkowicz, R., Granby, K., Nielsen, H., Jensen, B., Runge, E., og Lykkelund, C., 1993b, Dry deposition processes (In Danish: Processer for tørdeposition). Under the series "Marine Research from the Danish Environmental Protection Agency". In press.

Asman, W.A.H., Runge, E., and Kilde, N.A., 1993c. Emission of NH_3, NO_x, SO_2 and NMVOC to the atmosphere in Denmark (In Danish: Emission af NH_3, NO_x, SO_2 og NMVOC til atmosfæren i Danmark). Report Nr. 19 under the series "Marine Research from the Danish Environmental Protection Agency".

Asman, W.A.H. and Jensen, P.K., 1993, Wet deposition processes (In Danish: Processer for våddeposition). Under the series "Marine Research from the Danish Environmental Protection Agency". In press.

Gery, M.W., Whitten, G.Z., and Killus, J.P., 1989a. Development and testing of the CBM-IV for urban and regional computer modelling. **EPA-600/3-88-012. US EPA,** Research Triangle Park, N.C.

Gery, M.W., Whitten, G.Z., Killus, J.P. and Dodge, M.C., 1989b, A Photochemical kinetics mechanism for urban and regional computer modelling. **J. Geophys. Res., 94D,** 12925-12956.

Hertel, O., Berkowicz, R., Christensen, J., and Hov, Ø., 1993, Tests of two Numerical Schemes for use in Atmospheric transport-Chemistry models. Accepted for publication in Atmospheric Environment.

Hertel, O., Berkowicz, R., Asman, W.A.H., Christensen, J., and Sørensen, L.L., 1993, Processes of atmospheric chemistry (In Danish: Beskrivelse af atmosfærekemiske processer). Report Nr 24 under the series "Marine Research from the Danish Environmental Protection Agency". pp 48.

Hovmand, M.F., Grundahl, L., Runge, E., Kemp, K., and Aistrup, W., 1993, Atmospheric deposition of nitrogen and phorphor (In Danish: Atmosfærisk deposition af kvælstof og fosfor). Scientific report nr. xx. National Environmental Research Institute, under the Danish Ministry of the Environment.

Iversen, T., Halvorson, N.E., Saltbones, J., and Sandnes, H., 1990, Calculated budgets for airborne airborne sulphur and nitrogen in Europe.EMEP/MSC-W Report 2/90.

Sandes, H., 1993, Calculated budgets for airborne acidifying components in Europe, 1985, 1987, 1988, 1989, 1990, 1991 and 1992.EMEP/MSC-W Report 1/93.

DISCUSSION

H. SCHLÜNZEN: What is the grid size of the meteorological data used for the model calculations?

O. HERTEL: In the presented computations we have used meteorological data from the EMEP/MSCW on 150x150 km^2 grids. We hope in the future to get access to meteorological data on a higher resolution. Such data are present from the HIRLAM model at the Danish Meteorological Institute.

H. SCHLÜNZEN: Would the influence of the deposition velocity increase if a higher resolution meteorological data set were used?

O. HERTEL: It is not certain that a higher resolution in meteorological data would give larger differences between the two model versions. It would, however, give pronounced differences in the local deposition for both models.

R. SAN JOSE: What kind of deposition model are you using?

O. HERTEL: We model both dry and wet deposition. For the wet deposition we use separate scavenging coefficients for incloud and below cloud scavenging, respectively. For the dry deposition we use the resistance method.

R. SAN JOSE:	What kind of canopy resistance are you using?

O. HERTEL:	For deposition over land we apply constant value of surface resistance, which is, however, different for different species. For sea we have made a more detailed description, where we take into account the roughness of the sea. The sea roughness is again a function of the friction velocity, and a iteration method is implemented to solve this system.

J.I. IBARRA:	Does the model take into account topography?

O. HERTEL:	No, topography is not taken into account in the model. The model is a one- dimensional column advected along a 96 h back-trajectory to a given receptor point. It is assumed that the horizontal transport is independent of height. It is a rough but in mean probably fair assumption.

J.I. IBARRA:	How is thermal stratification modelled?

O. HERTEL:	The thermal stratification is included in the equation for the eddy diffusion coefficient.

J.I. IBARRA:	Is the vertical diffusivity (K_z) a constant value in the mixed layer?

O. HERTEL:	No, the K_z is not a constant value in the mixing layer. The equation used for the eddy diffusion coefficient is, $K_z = \frac{\kappa u_* z}{\phi_h(z/L)} \left(1 - \frac{z}{ZI}\right)$, where κ is von Karman constant ($= 0.35$), u_* is the friction velocity, ϕ_h is the similarity function for heat flux, z is the height, L is the Monin-Obukhov length, and ZI is the mixing height.

VIDEO SESSION

chairman: S.E. Gryning

rapporteur: E. Batchvarova

DESIGNING AN AIR QUALITY NETWORK FOR BRISBANE USING A MESOSCALE MODEL

W. Physick[1], P. Best[2], K. Lunney[3] and G. Johnson[4]

[1]CSIRO Division of Atmospheric Research, Aspendale, Victoria 3195,
  Australia
[2]Katestone Scientific, Paddington, Queensland 4064, Australia
[3]School of Australian Environmental Studies, Griffith University, Nathan,
  Queensland 4111, Australia
[4]CSIRO Division of Coal and Energy Technology, North Ryde, New South
  Wales 2113, Australia

INTRODUCTION

Brisbane is a subtropical city of one million people situated on the east coast of Australia and flanked by mountains to the west and south. The city has very few major industrial sources but is rapidly expanding. Fifteen years of multi-parameter monitoring at three inner suburban sites have shown a considerable inter-annual variability of ozone concentrations at greater than background levels, with maximum readings usually occurring between midday and early afternoon, and from late winter through to late autumn. On all of these days it is likely that areas of greater ozone concentrations exist elsewhere in the region later in the day. It is the aim of this study to identify those areas and to design an expanded meteorological and air quality network. Data from such a network will be invaluable to environmental and planning authorities.

METHOD

An analysis of surface wind, temperature, O_3, NO, and NO_2 data from three inner suburban sites between 1978 and 1992, and similar data over a 12-month period at Deception Bay, 30 km north of central Brisbane, has found that the wind and trace gas behaviour on days with higher ozone concentrations can be classified into four categories. Significant events occur for quite well-defined synoptic conditions and typically persist for 2-3 days. For the warmer months, records at the southwest Brisbane site of Rocklea often reveal an early-morning ozone maximum accompanying the regular afternoon peak associated with the arrival of the sea breeze. However in late autumn and late-winter/early spring, the highest readings are found in onshore flow around midday on the coast well to the north of Brisbane.

We have used a mesoscale model (Physick et al., 1993) to simulate the winds in the Brisbane region on two case-study days. Surface wind traces at six locations and upper

Air Pollution Modeling and Its Application X, Edited by S-V. Gryning
and M.M. Millán, Plenum Press, New York, 1994

winds at 6-hourly intervals at one location were available for model verification. Wind and turbulence predictions from the model were used to disperse inert and neutrally buoyant particles released from locations around Brisbane during the morning and evening peak traffic periods. Trajectories of these tracers were used (a) to predict the late afternoon locations of the emissions (when the highest daily ozone levels are likely to occur), and (b) to help interpret the peaks in the ozone traces at monitor locations. Also useful in understanding the NO_x and ozone data were interpretations performed with the Integrated Empirical Rate (IER) equations of smog formation (Johnson et al., 1990). This approach can determine the photochemical age of air samples and when used in conjunction with trajectories from the mesoscale model constitutes a powerful tool for identifying the origin of smog precursors far from their source.

RESULTS

Observations and model predictions show that the windfield on days of elevated ozone levels is dominated by sea breezes. At night, drainage flows in valleys to the south and southwest of Brisbane spill out onto the coastal plain. As a result, urban emissions are transported to valleys over 50 km from Brisbane by late afternoon, reside there for much of the night and are recirculated to the city early next morning. The model suggests that the early morning ozone peak at Rocklea on the summer case-study day is due to recirculation of the previous evening's traffic emissions, a prediction which is reinforced by the conclusion of the IER analysis that the early morning air contained precursors emitted the previous afternoon. Similarly, the model prediction that the early-afternoon maximum in the ozone trace at Rocklea results from the morning traffic emissions is consistent with the IER finding that precursors in the air are relatively young.

It appears that morning traffic emissions are also responsible for the midday ozone maximum to the north of Brisbane in late winter. Although the sea breeze/drainage flow mechanism is once again dominant, the shorter days and longer nights at this time of year produce stronger and longer-lasting drainage flows, which pick up emissions over Brisbane and transport them offshore. These pollutants then cross the coast in the sea breeze later in the morning.

The recommendations of this study have been accepted and an expanded network, including monitors at six extra locations and measurement of surface and upper-level winds at additional sites, is expected to be in place by the end of 1993.

Acknowledgments

This work was funded by the Queensland Department of Environment and Heritage. The authors are appreciative of the encouragement and close interest shown in the study by Dr. Neville Bofinger of the Brisbane City Council.

REFERENCES

Johnson, G.M., Quigley, S.M., and Smith, J.G., 1990, Management of photochemical smog using the Airtrak approach, in:"Proc. 10th International Clean Air Conference, Clean Air Society of Australia and New Zealand, Auckland", P.Gibson, ed..

Physick, W.L., Noonan, J.A., McGregor, J.L., Hurley, P.J., Abbs, D.J., and Manins, P.C., 1993, LADM: A Lagrangian Atmospheric Dispersion Model, *CSIRO Division of Atmospheric Research Technical Report No. 24*, 137pp.

DISCUSSION

M. ROTACH

Were the particles "released as being ozone" from the beginning or was there any chemistry built into the model to take account of transformations?

W. PHYSICK

There is no chemistry in the results shown in this video. The particles are only tracers. However we have just incorporated photochemistry (via the Integrated Empirical Rate equations of Johnson) into the particle model.

AN OPERATIONAL SYSTEM FOR EMERGENCY RESPONSE TO LARGE SCALE RELEASES OF POLLUTANTS IN THE ATMOSPHERE

Réal D'Amours, Michel Jean, Joseph-Pierre Toviessi and Serge Trudel

Canadian Meteorological Centre
Dorval, Québec
Canada

INTRODUCTION

The Canadian Meteorological Centre (CMC) has been designated by the World Meteorological Organization (WMO) as a Regional Specialized Meteorological Centre (RSMC) for the provision of products and guidance on atmospheric transport-dispersion-deposition of pollutants to member countries during Environmental Emergencies. The Canadian Meteorological Centre is the national meteorological centre for Canada. CMC uses a Global Spectral Model (T119, 21 levels) for medium range forecast, and a variable resolution Regional Finite Element Model (50 km horizontal resolution over North America, 23 levels) for short range forecast. In order to fulfil this international commitment as well as other national responsibilities for environmental emergency response, CMC has implemented in its operations a system for quick response to requests of products and guidance on the large scale movement of pollutants in the atmosphere.

TRANSPORT MODELS

Two models are available for atmospheric transport: a simple 3-D trajectory model, and a complex eulerian 3-D Transport-Dispersion-Deposition model the Canadian Emergency Response Model (CANERM). A detailed description of CANERM can be found in Pudykiewicz (1989). Transport is calculated using the semi-Lagrangian algorithm. Diffusion is modelled according to the gradient theory, and the diffusion coefficients are constant in the free atmosphere, but depend on the boundary layer properties near the surface. Dry deposition is simulated using the concept of a deposition velocity. Wet scavenging is parameterized in terms of a scavenging ratio which is a function of relative humidity.

CANERM operates on a polar stereographic grid and can be executed on the Northern Hemisphere and on the Southern Hemisphere. The horizontal resolution is 150

Air Pollution Modeling and Its Application X, Edited by S-V. Gryning
and M.M. Millán, Plenum Press, New York, 1994

km in the hemispheric configuration, 50 km in the continental configuration, and 25 km in the regional configuration. The model has 11 vertical levels in the σ terrain following coordinates. CANERM can be executed in hindcast mode using a sequence of objective meteorological analyses, or in forecast mode using a sequence of NWP forecasts.

VOLCANIC ASH TRANSPORT

CANERM was used in real time to provide advice to aviation meteorologists after the eruption of Mount Spurr on September 17, 1992; the plume persisted for several days and caused serious perturbation of airline traffic over Eastern North America.

Figure 1 shows the result of a 54 hours hindcast simulation of CANERM, valid on September 19, 1200 UTC. Figure 2 shows a visible image from GOES valid at 1400 UTC. A cloud band can be seen, extending from lake Michigan, through the southern tip of lake Huron, then along the Saint-Lawrence river before being lost in the extensive cloud deck of central Québec. This band has been positively identified as part of the ash plume produced by the eruption of Mount Spurr. The correspondence with the estimated position of the tail of the plume by CANERM is quite good. However it is possible that only part of the actual plume is visible on the satellite picture. The detectability in the visible spectrum appears to be dependant on several variables like the angle of incidence of solar light. A few hours later, the ash plume could not be detected on GOES images, but was still reported by airline pilots.

CONCLUSION

CANERM is used successfully in real time at CMC. In the poster presentation more details on the models capabilities will be presented, along with a description of the interface between the dispersion model and the operator from the input perspective as well as from the output side. A video presentation will also be shown.

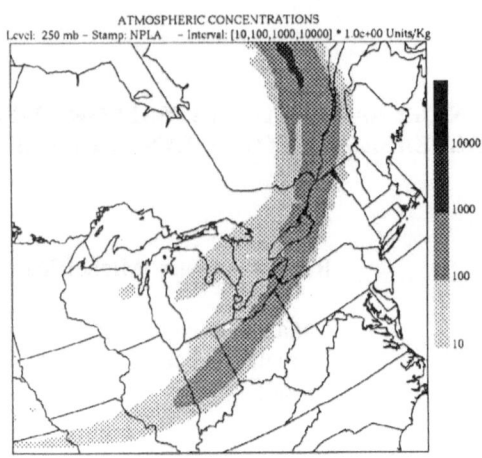

ATMOSPHERIC CONCENTRATIONS
Level: 250 mb - Stamp: NPLA   - Interval: [10,100,1000,10000] * 1.0e+00 Units/Kg

54 hour fcst valid 12Z September 19 1992

Figure 1. CANERM 54 hour simulation

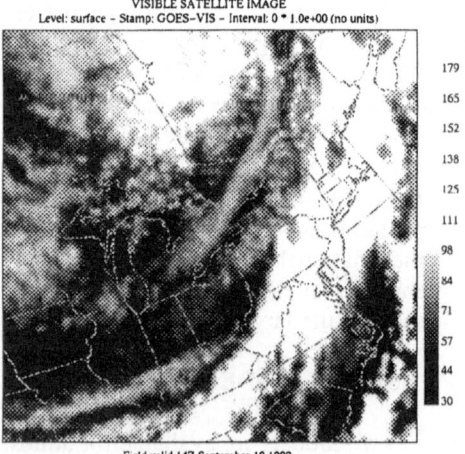

VISIBLE SATELLITE IMAGE
Level: surface - Stamp: GOES-VIS - Interval: 0 * 1.0e+00 (no units)

Field valid 14Z September 19 1992

Figure 2. GOES image

REFERENCE

Pudykiewicz, J., 1989, "Simulation of the Chernobyl dispersion with a 3-D hemispheric tracer model", *Tellus*. 41B:391

REAL-TIME APPLICATION SOFTWARE "TRACER IMAGER PACKAGE" (TRIP)

Wieland Weiß and Eberhard Reimer

Institut für Meteorologie · Freie Universität Berlin
Tropospärische Umweltforschung TrUmF
C-H-B-Weg 6-10
D-12165 Berlin, GERMANY

INTRODUCTION

Operationally running computer programs at the IfM continously collect and archives actual meteorological using the worldwide desseminated WMO-bulletins (GTS), geostationary METEOSAT satellite data (PDUS quality), and orbital NOAA satellite data. After an extended diagnostic cycle the resulting 3D-analyses on isentropic surfaces are used to calculate trajectories by a mixed dynamic/kinematic approach (REIMER, 1991). Starting from selected source positions in forward mode or immission locations in backward mode container files are prepared that contains the synoptic 3D spreading of air particles. Also short term trajectories are calculated at each gridpoint of the analysis area to determine broad scale vertical velocity. These container files then can be interactively displayed on a personal computer using TRIP ("TRacer Imager Package") which is the user interface of TRAMPER (Tropospheric Realtime Applied Meteorological Procedures for Environmental Research). The resulting consecutive images simulate the motion of particles or other meteorological variables that change over time. TRIP so acts as a viewer of accompanied analysis data of the given weather situation too.

SPECIFICATIONS

The package runs in a DOS environment and is programmed since spring '91 utilizing Pascal. This language allows inserting assembler code within time-critical pascal modules - without any loss of source code transparency. The assembler modules are register-level tuned to the graphics controllers of TSENG LABS (ET4000 and ET4000/W32). Its a cost-efficient (<130$) solution of the new generation of SuperVGA and accelerator boards that excelled with fast memory access (1MB DRAM) and full 256 color support in high resolution mode (1024*768 pixels). Further hardware requirements see below in Fig.1.

TRIP consists of an easy-handling menu system as this is an important criteria for the user's acceptability. The design of TRIP's interactive menu structure bases on "SAA" (Standardized Application Architecture) - an international standard which corresponds to the demands of the report "Quality Criteria for Computer Models..." (Ministry of Housing, PPE). TRIP fully supports mouse control within the menu dialogs, and furtheron provides for context sensitive help at every menu level.

The menu system's source code is programmed in object oriented event-driven Pascal (OOP) which separates the menu from the later displayed graphics information. That is advantageous for future supplements and extensions because it eases changing the code.

minimal (necessary) configuration	standard configuration	recommended configuration
IBM-compatible PC (≥80286, ≥16MHz, ≥1MB RAM)	80386-, better 80486-CPU (≥25MHz, CPU-Cache, 8MB RAM)	80486-Tower >33MHz, 16 MB -> 2.500 DM
hard disk ≥ 40MB	hard disk ≥120MB mean access time ≤20ms	210MB hard disk -> 600 DM
graphics card with TSENG LABS ET4000 controller (1MB video RAM)	ET4000 graphics card (1MB RAM) ≥70Hz refresh rate (at 800*600 pixels)	ET4000/W32 local bus card like MegaEva32 -> 300 DM
analog 14" multifrequence color monitor (MultiSync/Multi-/FlexScan)	multifrequence color monitor ≥15" low radiation (MPR I)	EIZO F550i (17") low.rad. MPR II -> 2.400 DM
LaserJet-compatible printer (PCL III, like HP Deskjet 500)	LaserJet-compatible laser printer (PCL III)	HP LaserJet 4L -> 1.400 DM
total costs: ca. 2.000	total costs: 3.000 -	7.200 DM

Fig. 1. Hardware requirements for running TRIP (IX '93)

TRANSFER OF THE DATA TOWARDS OUTSIDE THE IfM

Up to now the main processing of the trajectory calculations and the analysis scheme is managed by our local VMS-based VAX-cluster. Several PCs running TRIP (386, 486) directly are connected to our Ethernet-LAN to have access to the output data. This kind of interface is also an example for a data connection towards outside the LAN by use of FTP (TCP/IP) on INTERNET. In order to handle the high volume of data within a reasonable time the best way to get real-time data from the IfM is ISDN using telecommunication lines. We strongly recommend the digital ISDN solution, because it guarantees a data transfer rate of 8 kByte/s (respectively 16 kBytes/s using two B-channels) as well as allows international data links due to the introduction of EURO-ISDN.

REAL-TIME APPLIANCE

The local Weather Service of the FU Berlin uses TRIP to display processed real-time satellite and radar data via our LAN. This online information represents a valuable assistance to the forecasters.

In course of time some environmental communal departments and research institutes in Germany make use of trajectory data of the IfM to interpret smog relevant weather situations (up to now qualitative and quantitative diagnostics, later on forecasts as well). Twice a day actual data like trajectories and tracer, grid deformation, tropospheric analysis data and processed satellite images can be uploaded from an ISDN-PC-server. The interested external user then uses TRIP to display the actual meteorological information on his PC.

POSTER SESSION

VERTICAL DIFFUSION PARAMETER IN THE ATMOSPHERIC BOUNDARY LAYER

Laura E. Venegas

Dept. of Atmospheric Sciences
Faculty of Sciences. University of Buenos Aires
Ciudad Universitaria. Pab.II
1428 - Buenos Aires. Argentina

This paper presents relations between the mean vertical displacement ($\bar{z}$) of pollutants and their mean horizontal displacement ($\bar{x}$) in the atmospheric boundary layer. A modification of the lagrangian similarity theory applied to atmospheric diffusion extended to the atmospheric boundary layer is developed. According to Pasquill and Smith (1983) the similarity theory of Monin-Obukhov applied to atmospheric diffusion states

$$\frac{d\bar{x}}{d\bar{z}} = \frac{\bar{u}(c\bar{z})\,\bar{z}}{K(\bar{z})} \tag{1}$$

where $\bar{u}$ is the mean wind speed, c depends on stability and K is the vertical material diffusivity. Considering the eddy diffusivity for momentum and the form of the friction velocity introduced by Yokoyama et al.(1979) a form of $\bar{u}(c\bar{z})$ for the atmospheric boundary layer is developed. Substituting $\bar{u}(c\bar{z})$ in Eq.(1), equating K to the heat diffusivity given by Yokoyama et al.(1979) and integrating, the following expressions can be obtained (where z_1 is the boundary layer depth, L is the Monin-Obukhov length, $\theta = \bar{x}/z_1$; $\alpha = \bar{z}/z_1$; $\alpha_0 = z_0/z_1$; k=0.41, β=6.9 and γ=0.92):
- neutral and stable conditions: ($\eta = z_1/L \geq 0$)

$$
\begin{aligned}
\theta = k^{-2} \sum_{i=0}^{\infty} \Bigg\{ & \frac{\alpha^{i+1}}{1+1}\ln\left[\frac{c\,\alpha}{\alpha_0}\right] - \frac{\alpha_0^{i+1}}{1+1}\ln c - \frac{\alpha^{i+1}}{(1+1)^2} + \frac{\alpha_0^{i+1}}{(1+1)^2} - c\,\frac{\alpha^{i+2}}{1+2} + c\,\frac{\alpha_0^{i+2}}{1+2} + \\
& + \alpha_0\,\frac{\alpha^{i+1}}{1+1} - \frac{\alpha_0^{i+2}}{1+1} + \beta\eta\Bigg[c\,\frac{\alpha^{i+2}}{1+2} - c\,\frac{\alpha_0^{i+2}}{1+2} - \alpha_0\,\frac{\alpha^{i+1}}{1+1} + \frac{\alpha_0^{i+2}}{1+1} - \frac{c^2}{2}\,\frac{\alpha^{i+3}}{1+3} + \\
& + \frac{c^2}{2}\,\frac{\alpha_0^{i+3}}{1+3} + \frac{\alpha_0^2}{2}\,\frac{\alpha^{i+1}}{1+1} - \frac{1}{2}\,\frac{\alpha_0^{i+3}}{1+1} + \gamma\eta\Big[c\,\frac{\alpha^{i+3}}{1+3} - c\,\frac{\alpha_0^{i+3}}{1+3} - \alpha_0\,\frac{\alpha^{i+2}}{1+2} + \frac{\alpha_0^{i+3}}{1+2} - \\
& - \frac{c^2}{2}\,\frac{\alpha^{i+4}}{1+4} + \frac{c^2}{2}\,\frac{\alpha_0^{i+4}}{1+4} + \frac{\alpha_0^2}{2}\,\frac{\alpha^{i+2}}{1+2} - \frac{1}{2}\,\frac{\alpha_0^{i+4}}{1+2} \Big] \Bigg] + \gamma\eta\Big[\frac{\alpha^{i+2}}{1+2}\ln\left[\frac{c\,\alpha}{\alpha_0}\right] - \\
& - \frac{\alpha_0^{i+2}}{1+2}\ln c - \frac{\alpha^{i+2}}{(1+2)^2} + \frac{\alpha_0^{i+2}}{(1+2)^2} - c\,\frac{\alpha^{i+3}}{1+3} + c\,\frac{\alpha_0^{i+3}}{1+3} + \alpha_0\,\frac{\alpha^{i+2}}{1+2} - \frac{\alpha_0^{i+3}}{1+2} \Big] \Bigg\}
\end{aligned} \tag{2}
$$

- unstable conditions $\eta = z_1/L < 0$
   a. $0 < \alpha = \bar{z}/z_1 < 0.1$

Air Pollution Modeling and Its Application X, Edited by S-V. Gryning
and M.M. Millán, Plenum Press, New York, 1994

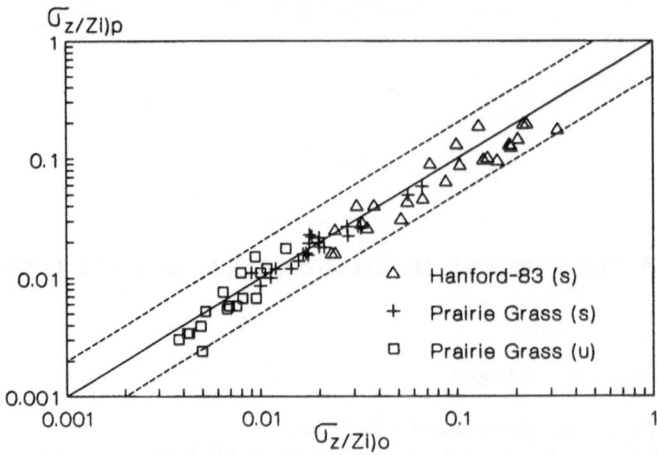

Fig.1 Predicted and observed σ_z/z_i values. (s: stable; u: unstable)

$$\theta - k^{-\frac{2}{3}} \mfancy{m} \Gamma^{\frac{1}{3}} c_1^{-1} \left\{ \frac{3}{2} \alpha^{\frac{2}{3}} \ln \left[\frac{c\alpha}{\alpha_0} \right] - \frac{3}{2} \alpha_0^{\frac{2}{3}} \ln c - \frac{3}{4} \left[\alpha^{\frac{2}{3}} - \alpha_0^{\frac{2}{3}} \right] - [A_0(\alpha - \alpha_0) + \right.$$

$$+ A_1(\alpha\ln\alpha - \alpha - \alpha_0\ln\alpha_0 + \alpha_0) + A_2[\alpha(\ln\alpha)^2 - 2(\alpha\ln\alpha - \alpha) - \alpha_0(\ln\alpha_0)^2 +$$

$$+ 2(\alpha_0\ln\alpha_0 - \alpha_0)] + A_3 \langle \alpha(\ln\alpha)^3 - 3[\alpha(\ln\alpha)^2 - 2(\alpha\ln\alpha - \alpha)] -$$

$$\left. - \alpha_0(\ln\alpha_0)^3 + 3[\alpha_0(\ln\alpha_0)^2 - 2(\alpha_0\ln\alpha_0 - \alpha_0)]] \rangle] \right\} \qquad (3)$$

b. $0.1 \leq \alpha = \bar{z}/z_1 < 1$

$$\theta - \frac{\left[\ln\left[\frac{0.1c}{\alpha_0} \right] + 1 - (1 + 2.2c\,\mfancy{m})^{\frac{1}{4}} \right]}{k^{\frac{2}{3}} c_1 \mfancy{m}^{\frac{1}{3}}} \sum_{i=0}^{n} \left[\frac{\alpha^{1 + \frac{2}{3}} - (0.1)^{1 + \frac{2}{3}}}{1 + \frac{2}{3}} \right] + \bar{\theta} \qquad (4)$$

where $\bar{\theta}$ is the value of θ at $\alpha = 0.1$ evaluated from Eq.(3), $C_1 = 0.75$ and A_1 satisfies: $\alpha^{-1/3}(1 - 22c\,\mfancy{m}\alpha)^{1/4} = A_0 + A_1\ln\alpha + A_2(\ln\alpha)^2 + A_3(\ln\alpha)^3 + A_4(\ln\alpha)^4$.

Thus, given $\bar{z}$ the vertical spread (σ_z) of pollutants emitted from a ground level source can be estimated from:

$$\sigma_z - \left[\frac{\Gamma\left[\frac{3}{s} \right] \Gamma\left[\frac{1}{s} \right]}{\Gamma\left[\frac{2}{s} \right]} \right]^{\frac{1}{2}} \bar{z} \qquad (5)$$

where $\Gamma(\ )$ is the Gamma function and s is a non-dimensional parameter that depends on atmospheric stability and surface roughness (z_0). The comparison between predicted values of $\sigma_z/z_1)_p$ obtained from Eq.(5) combined with Eqs.(2), (3) and (4) with the observational values $\sigma_z/z_1)_o$ from Projects Prairie Grass (23 stable runs and 17 unstable runs) and Hanford-83 (25 stable runs) is shown in the Fig.1. The predicted values are within a factor of 2 of the observed values.

REFERENCES

Pasquill, F. and Smith, F.B. 1983. Atmospheric Diffusion, Wiley, N.Y. 105.
Yokoyama, O.; Gamo, M. and Yamamoto, S. 1979. The vertical profiles of the
    turbulence quantities in the atmospheric boundary layer. J. Met. Soc.
    of Japan, 57, 3, 264.

TRAJECTORY ANALYSIS OF HIGH-ALPINE AIR POLLUTION DATA

P. Seibert[1], H. Kromp-Kolb[1], U. Baltensperger[2], D.T. Jost[2], and
M. Schwikowski[2]

[1]Institute of Meteorology and Geophysics, University of Vienna
  Hohe Warte 38, A-1190 Wien, Austria
[2]Paul Scherrer Institute, CH-5232 Villigen PSI, Switzerland

INTRODUCTION

The EUROTRAC subproject ALPTRAC (High Alpine Aerosol and Snow Chemistry
Study) is devoted to the investigation of air and snow pollution at high Alpine sites. The
aerosol surface concentration is continuously recorded at Jungfraujoch (3450 m a.s.l.,
$7°59'E, 46°32'N$) in the Swiss Alps and Sonnblick (3106 m a.s.l, $12°57'E, 47°03'N$) in the
Austrian Alps with a time resolution of 30 min with an epiphaniometer (Gäggeler et al.,
1989; Baltensperger, et al. 1991). The measurements showed a pronounced seasonal cycle
with mean summer concentrations more than one order of magnitude higher than mean
winter concentrations, and the occurrence of episodes with especially high or low concentrations (Seibert et al., 1993). While the seasonal cycle is mainly to be explained by the atmospheric stability, the short-term variations are caused by synoptic-scale transports. These
have been investigated using isobaric back trajectories at 700 hPa with a length of 72 h,
computed twice daily for a period of three years (July 1990 - June 1993). Due to technical
problems at Sonnblick, only 925 trajectories were available for the analysis; most of the mis-
sing data fall on winter.

METHOD OF TRAJECTORY EVALUATION

Since the concentration data were distributed approximately log-normally, the concentration time series was transformed by taking the common logarithm and then deducting the
arithmetic mean. In order to exclude the variations introduced by the seasonal cycle of stability, the sine and cosine waves with the period 1 year as determined by a least-square fit
were deducted, too. One-hour means at the arrival times or the trajectories were used in the
analysis. The computational domain was projected on a polar stereographic map and divided
into grid cells with a side length of about 150 km (Fig. 1). For each grid cell, the arithmetic
mean of the transformed concentrations observed on arrival of the trajectories passing
through this grid cell was calculated, using the residence time in the cell as a weight factor
because the amount of pollutants emitted into an air parcel is proportional to the time spent
over the source area (and the source strength).

The resulting fields exhibit small-scale variations which are not statistically significant
and disturbing in a contour plot. A special procedure was developed which should remove
the noise while preserving the significant features. This was achieved by the calculation of
the confidence interval (on a 10% error level) and repeated smoothing with a 9-point filter,
imposing the restriction that the values must not be modified beyond the confidence interval;

Air Pollution Modeling and Its Application X, Edited by S-V. Gryning
and M.M. Millán, Plenum Press, New York, 1994

the smoothing process was continued until the modifications by the filter at each grid point were within 0.5% of the data range. This required about 20 smoothing cycles. However, the selection of the error level and the abortion criterion is subjective, and the fact that subsequent trajectories are not independent events is a violation of the assumptions underlying the computation of the confidence interval.

RESULTS

The concentration fields computed according to the method described above are shown in Figure 1. Lower-than-average aerosol concentrations were observed at both stations with trajectories from the Atlantic and from Africa, while higher-than-average concentrations are associated with trajectories from the European continent. At Jungfraujoch, a pronounced east-west gradient was found within Europe. At Sonnblick, Northern Italy, Hungary, and the region of former Yugoslavia seem to be important potential source areas. Unexpected maxima were found for both the stations in Northern Greece and the Iberian Peninsula. On the contrary, the trajectories spending time over the region England-Belgium-Netherlands-Ruhrgebiet, known for having considerable emissions, are not associated with higher mean aerosol concentrations. Both features may be explained by precipitation scavenging, which is probably more frequent in air parcels advected from the northwest than in those coming from the southwest or southeast. Enhanced photochemical reaction rates in the Mediterranean region could be an additional reason.

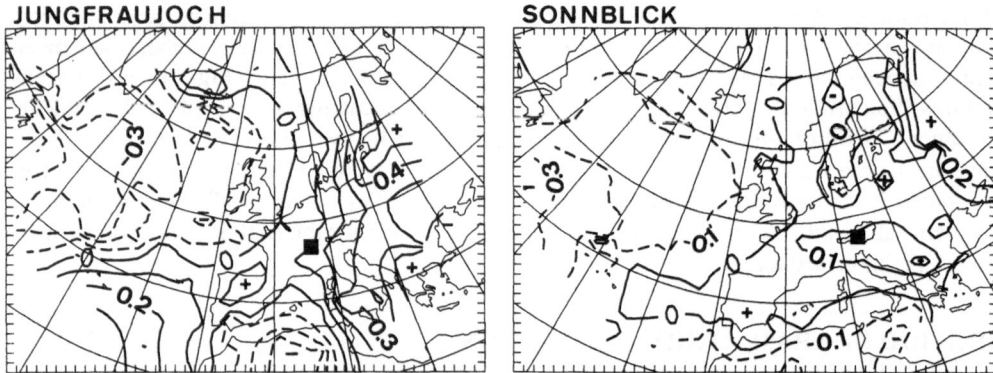

Figure 1. Deviations from the mean aerosol surface concentrations at Jungfraujoch (left) and Sonnblick (right) as a function of the trajectory residence time. Contour spacing is 0.1 logarithmic units, negative deviations are dashed (e.g. -0.2/-0.1/+0.1/+0.2 contours correspond to 63/79/126/158 % of the mean concentration). Contour analysis is restricted to the area where trajectories actually passed. The stations are indicated by a filled square. The grid used is indicated on the margins.

Acknowledgements. The work was supported by the Austrian "Fonds zur Förderung der wissenschaftlichen Forschung" under grant P7809-GEO and the Swiss National Science Foundation.

References

Baltensperger, U., Gäggeler, H.W., Jost, D.T., Emmenegger, M., and Nägeli W., 1991, Continuous background aerosol monitoring with the epiphaniometer, *Atmos. Environ.* 25A:629-634.

Gäggeler, H.W., Baltensperger, U., Emmenegger, M., Jost, D.T., Schmidt-Ott, A., Haller, P., and Hofmann, M., 1989, The epiphaniometer, a new device for continuous aerosol monitoring, *J. Aerosol Sci.* 20:557-564.

Seibert, P., Kromp-Kolb, H., Kovar, A., Puxbaum, H., Winiwarter, W., Jost, D.T., Schwikowski, M., 1993, Meteorological interpretation of an episode of clean air at high-alpine sites, in: "Proceedings EUROTRAC Symposium '92", P.M. Borrell et al., eds., SPB Academic Publishing, The Hague.

AN EDDY DIFFUSIVITY MODEL FROM A THEORETICAL SPECTRAL

MODEL FOR THE STABLE BOUNDARY LAYER

Osvaldo L. Moraes[1] ,Gervásio A. Degrazia[1] , Amauri P. Oliveira[2]

[1]Departamento de Física, UFSM, Santa Maria, Brasil and
[1]Radar Meteorológico, UFPel, Pelotas, Brasil
[2]Departamento de Ciências Atmosféricas, USP, São Paulo , Brasil

INTRODUCTION

Much has been written in recent years concerning modeling and parameterizations of the planetary boundary layer (PBL), particularly in formulations that require turbulent fluxes of momentum, heat and moisture.These turbulent fluxes are primarily responsible for shaping the structure of the PBL and from the mathematical point of view are related with the closure problem. Thus the solution of the PBL equations require some type of closure schemes.

The turbulent Kinetic Energy (TKE) closure model represents an improvement to the simplicity of first-order closure in sense that it takes into account more of the physics of the atmosphere. It is essentially a first-order scheme in sense that closure is obtained by the same relations, but the eddy diffusivities are determined from more realistic relationships.

In this work we adopt the TKE 1-model to derive an expression for the eddy diffusivity. We adopt a turbulence length scale developed by Degrazia et alli(92) and a spectral model formulated by Moraes et alli(92). The novel feature of the Degrazia's model was the provision for multiple length scales, one for each different spacial direction. It was tested comparing with the resulting energy containing eddy size as proposed by Brost and Wyngaard (1978).

The energy spectra model utilized in this study was derived from a dynamical point of view. It was achieved through the use of the energy balance relation obtained by Rotta from Navier-Stokes equations. The resulting spectral equation is an appropriate physical rooting of the Minnesota and Cabauw observed spectra.

Air Pollution Modeling and Its Application X, Edited by S-V. Gryning
and M.M. Millán, Plenum Press, New York, 1994

THE MODEL

Holt and Ramam(1988) separate the differents TKE closure schemes in three basic models based on the prognostic variables considered. The basic of the 1 model is derived from the Prandtl-Kolmogorov hipothesis relating eddy viscosity to turbulent Kinetic energy:

$$K_m = C_e \times l \times \varepsilon^{\frac{1}{2}}$$

(1)

where 1 is the mixing length, C_e is a constant and ε is the turbulent kinetic energy. In a recent paper Degrazia et alli (1992) proposed a method to evaluate characteristic length scales for turbulent flows, based on analysis of the spectral properties of the turbulent flow. An interesting feature of this derivation is the provision for three length scales, one for each different spacial direction.
The vertical length scale based on this method is expressed by the following relation.

$$\frac{1}{l_w} = \frac{1}{0.22z} + \frac{16.8}{\Lambda}$$

(2)

where Λ is the local Obukhov length defined by

$$\Lambda = L \times \left(1 - \frac{z}{h}\right)^{1.5\alpha_1 - \alpha_2}$$

(3)

L being the Obukhov length at the surface and h the heigth of the stable boundary layer (SBL).
Moraes et alli(92) developed a spectral energy function for stable atmospheric boundary layer using the Local Similarity Theory. The starting point of their work is Boussinesq's approximations as employed by Rotta(1951) and it is a dynamical model to be applied in the whole energy spectrum range. They end up with

$$\frac{kE(k)}{U_*^2 \psi_\varepsilon 2/3} = (kz)^{-2/3}_{exp} - A(kz)^{-\frac{4}{3}}$$

(4)

with A being a constant dependent of the Local Similarity Functions.
From equation (4) we obtain $\varepsilon^{1/2}$ to be used in equation (1).

$$\varepsilon = \frac{3}{4}\sqrt{\frac{\pi}{4}} \times U_*^2 \psi_\varepsilon^{2/3}$$

(5)

RESULTS

In order to obtain an expression for the eddy diffusivity following the E-1 model, i.e. equation (1),we use the Degrazia scale (equation (2)) and the expression (5),the resulting model is

$$K_m = c_e \times \left(\frac{0.22z}{1+3.7\frac{z}{\Lambda}} \right) \times \left(\frac{3}{4} \sqrt{\frac{\pi}{A}} U_*^2 \, \psi_\varepsilon^{2/3} \right)^{\frac{1}{2}}$$

(6)

The specification for the dimensionless velocity and dissipation rate is adopted from the LST (Nieuwstadt 84, Sorbjan 85). As a means of testing the expression (6), i.e.direct comparision of data with model prediction we normalize Km by u∗h. Based on data collected at Minnesota, Sorbjan (86) suggested the following eddy diffusivity coefficient for momentum

$$\frac{K_m}{u_* h} = \frac{0.35 \left(1 - \frac{z}{h} \right) \left(\frac{z}{h} \right)}{1 + 4.7 \left(\frac{z}{h} \middle/ \frac{h}{l} \right)}$$

(7)

Assuming $\alpha_1=2$ and $\alpha_2=3$, values indicated at that experiment field when nonstationary process were present, the comparision between (6) and (7) is presented in figure 1.

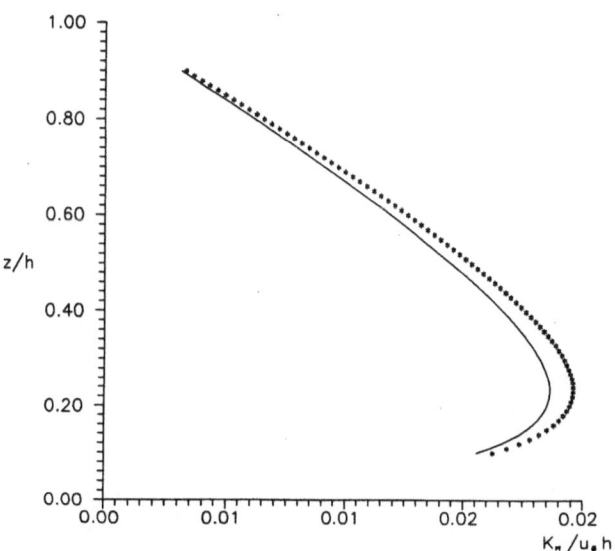

Figure 1. Comparison between theoretical model (solid line)
and Sorbjan's parameterization from the Minnesota
data (dot line).

ACKNOWLEDGEMENTS: This research was supported by Conselho Nacional de Desenvolvimento Científico e Tecnológico (CNPq), Fundação de Amparo a Pesquisa do Estado do Rio Grande do Sul (FAPERGS) and Fundação de Amparo a Pesquisa do Estado de São Paulo (FAPESP).

REFERENCES

Brost, R.A. and Wyngaard, J.C.,1978,A model study of the stably stratified
  planetary boundary layer,J.Atmos.Sci.35:1427.
Degrazia,G.A.and Moraes,O.L.L.,1992,A model for eddy diffusivity in a stable
  boundary layer, Bound.-Layer Meteor.58:205.
Holt T.and Ramam,S.,1988, A review and comparative evaluation of multilevel
  boundary layer parameterization for first-order and turbulent kinetic
  energy closure schemes,Rev. Geophys.26:761.
Moraes,O.L.L., Degrazia,G.A. and Goedert, J.,1992,Energy spectra of the stable
  boundary layer:a theoretical model, Il Nuovo Cim.14D:75.
Nieuwstadt, F.T.M.,1984, The turbulent structure of the stable ,nocturnal
  boundary layer,J.Atmos.Sci.41:2202.
Rotta, J.C.,1951,Z.Phys.129:547.
Sorbjan,Z.,1985, On similarity in the atmospheric boundary layer,Bound.-Layer
  Meteor.34:377.
Sorbjan,Z.,1986,Local similarity of spectral and cospectral characteristics in the
  stable-continuous boundary layer , Bound.-Layer Meteor.35:257.

ON THE RELATIONSHIP BETWEEN SYNOPTIC SCALE PARAMETERS AND PASQUILL STABILITY CLASSES FOR THE PURPOSES OF AIR POLLUTION MODELING

Dimiter Yordanov,[1] Dimiter Syrakov,[2] Maria Kolarova[2]

[1]Geophysical Institute, Bulgarian Academy of Sciences,
Acad.G.Bonchev Str., Block 3, Sofia-1113, Bulgaria

[2]National Institute of Meteorology and Hydrology,
Bulgarian Academy of Sciences, Mladost 1, Sofia-1184

Quite often, for the tasks connected with environmental impact assessment it is necessary to estimate pollutant concentrations on the basis of standard meteorological data such as sunshine, cloud amount, and wind velocity using the Pasquill stability classes. They are usually determined from the atmospheric characteristics estimated near the ground surface using a table classification with 10 m wind velocity and insolation intensity (or cloud cover at night). These meteorological data are not allways available and they are difficult for prognostication. For the purposes of the prognostical estimation of the atmospheric diffusion in case of emergency releases when regulatory type models are used it is useful to find the relation between the synoptic scale parameters obtained from the numerical weather prediction and the Pasquill stability classes.

Using the planetary boundary layer (PBL) models developed on the basis of similarity theory by Yordanov et al.(1983, 1989) and the experimental relation found by Golder (1972) between the Monin-Obukhov length scale L and the roughness parameter zo with the Pasquill-Turner stability classes, a numerical approach is developed by Yordanov et al.(1993) aproximating the Golder curves and finding the relation between the synoptic or "external to the PBL" parameters and the Pasquill stability classes. Instead of estimating L from the measurements, which is not easy, here it is proposed to calculate it using an appropriate PBL parameterization.

Basically L is defined from the vertical heat flux at the surface and the friction velocity (see Yordanov et al., 1990). The turbulent heat and momentum fluxes at the ground are obtained using a generalized similarity theory which includes the following external parameters: the external stratification parameter ($S = \beta(\theta_t - \theta_b)/f\,V_g$), and the surface Rossby number ($Ro = V_g/fz_0$). For the case of convective conditions a different parameterization is developed (Kolarova et al., 1989; Yordanov et al.,1990) taking into account the height of the inversion layer and the roughness parameter.

The resistance and heat exchange laws which give the relation between the external (S, Ro) and the internal (c_g, α, μ) to PBL parameters are numerically solved as a system of nonlinear algebric equations for the case of nonconvective and convective conditions. The solution is approximated with polynomials and the following expressions are obtained:

$$c_g = \frac{u_*}{|V_g|} = \sum_{j=1}^{3} \sum_{i=0}^{2} a_{i+j} \, \tilde{R}_o^i \, \tilde{S}^j \tag{1}$$

$$\mu = \frac{\kappa u_*}{fL} = \sum_{i=0}^{3} (c_i \tilde{R}_o^i)S, \tag{2}$$

c_g is the geostrophic drag coefficient, u_* - friction velocity, V_g - geostrophic wind, μ - internal stratification parameter, κ - Karman constant, f - Coriolis parameter, $\tilde{R}_o = lgR_o$, $R_o = |V_g|/fz_o$ - Rossby number, $\tilde{S} = 0.001S$, $S = \beta \Delta \vartheta / f |V_g|$ - the external stratification parameter, $\beta = g/\vartheta_0$ - buoyancy parameter, g - gravity acceleration, ϑ_0 - mean potential temperature in PBL, $\Delta \vartheta = \vartheta_t - \vartheta_b$ - the difference between potential temperatures at the top of the PBL and at the ground, α - the angle between the geostrofic and the surface wind. The coefficients a_{i+j} and c_i are estimated by Yordanov et al. (1993). As it is shown there the Monin-Obuhkov length scale can be easily expressed using the internal to PBL parameters as:

$$L^{-1} = \frac{f}{\kappa |V_g|} \frac{\mu}{c_g} \tag{3}$$

This way from the synoptic parameters V_g and ϑ_t, which may be considered as wind velocity and potential temperature at 850 mb standard level and having the surface temperature, roughness length and Coriolis parameter, the nondimensional external to PBL parameters R_o and S can be determined. Using (1) and (2) the nondimensional internal PBL parameters can be calculated and finally from equation (3) the Monin - Obuhkov length can be estimated.

This algorithm is realized in the computer code YORDAN for nonconvective conditions and in the computer code YORCON for convective conditions. In addition a two layer PBL K-model consistent with the resistance and heat exchange lows is built in the programs, so to calculate the vertical velocity and eddy diffusivity profiles in PBL. A block for determining the Pasquill stability class is included as well, approximating the relations between the Pasquill classes, the M.- O. length, and the roughness parameter.

When applying the Pasquill-Gifford curves or the revised by Briggs formulas for estimation of the pollution concentrations from a point source, besides the stability class, the wind velocity at the source height is necessary to be known. It can be measured, but in many cases it can be calculated extrapolating the velocity at the standard 10m level to the source height applying EPA power lows for example. Another possibility is to calculate the wind velocity right at the source height using the calculated wind profile with models like YORDAN and YORCON. The second approach has the advantage to take into account the wind rotation with height in PBL due to the friction. The developed algorithm is used in the bulgarian plume dispersion operational models.

REFERENCES

Golder, D., 1972, Boundary Layer Met., 3:47.
Kolarova, M.,Yordanov, D.,Syrakov,D. et al., 1989, Izv.AN USSR,FAO,25:659.
Yordanov, D.,Syrakov, D.,Djolov, G., 1983, Boundary Layer Met.,25:81.
Yordanov, D.,Kolarova,M.,Syrakov,D.,Djolov,G., 1990 ,Proc.9 Symp. on turb. and diffusion, Apr.- May 1990, RISO, Roskilde.
Yordanov, D.,Syrakov, D.,Kolarova, M., 1993, Compt.rend.Acad.bulg.Sci. 7.

A LAGRANGIAN MODEL OF LONG-RANGE
TRANSPORT OF SULPHUR

Zvjezdana Klaić

Hydrometeorological Institute of Croatia
Centre for Meteorological Research
41000 Zagreb, Grič 3, Croatia

A Lagrangian receptor-oriented one-layer model has been developed in order to simulate a synoptic-scale transport of airborne sulphur. Mass-balance equations for sulphur dioxide and particulate sulphate were integrated along 3-day backward trajectories arriving at 0000 and 1200 GMT at selected receptor points. Advective winds used in trajectory calculations were based on the wind profile from radiosonde reports, taking an average over the ground based layer up to an 850 hPa level. These vertically averaged winds were thereafter objectively analysed using the '$1/r^2$ aligned' technique (Kahl and Samson, 1986), which gives the greatest weight to observations upwind and downwind of the interpolation point. Sulphur emissions were taken from EMEP inventory (Sandnes and Styve, 1992). They were assumed to vary linearly over the year, with the maximum and minimum occuring in January (multiplication factor=1.3) and July (multiplication factor=0.7), respectively. Background concentrations proposed by Szepesi and Fekete (1987) were used. Other model parameters had diurnal variation and they were estimated from routine synoptic observations, where $1/r^2$ spatial and linear temporal interpolation techniques were employed. Parameters were calculated at the beginning of each 3-hour time step. Mixing height varied from 500 to 2000 m, depending on stability conditions. Dry deposition velocities of sulphur dioxide (v_d) and particulate sulphate (w_d) over the ground and transformation rate of sulphur dioxide to particulate sulphate were taken as proposed by Renner et al. (1985). Over the sea v_d=0.8 cm/s and w_d=0.1 cm/s were taken. Wet deposition rates for both pollutants depended on   precipitation intensity and mixing height (Eliassen et al., 1988).

The model has so far been applied at 19 receptor points in Central Europe for the 01.11.1991.-30.04.1992. period. Daily (Fig. 1) and monthly mean concentrations for both pollutants and dry and wet depositions of sulphur accumulated during each simulated month (Fig. 2) were calculated. In model application to real data problems arose due to incomplete input of meteorological and measured concentration data.

Modelled monthly mean concentrations for chosen receptor points varied between 2.6-15.5 μg/m3 and 1.0-5.1 μg/m3 for sulphur dioxide and particulate sulphate, recpectively (both expressed as S), while total monthly sulphur depositions were between 49 and 907 mg/m2.

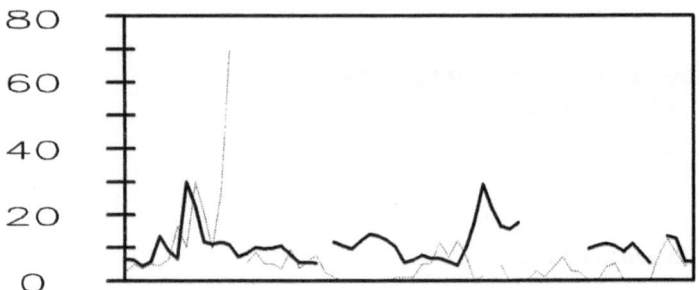

Figure 1. Calculated (solid line) and measured (dot line) daily mean sulphur dioxide concentrations (μg(S)/m3) for Puntijarka (45°54'N 15°58'E) for the period 16.01.1992.-21.03.1992.

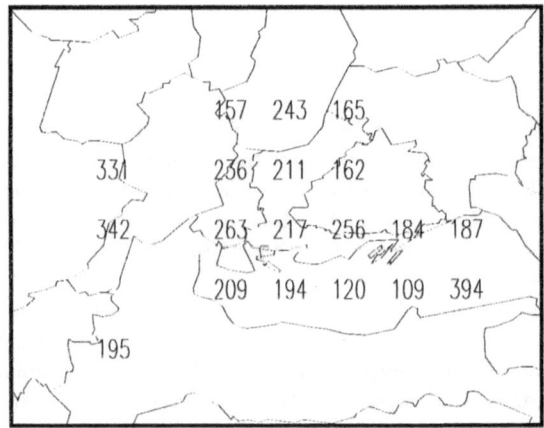

Figure 2. Total (dry+wet) sulphur deposition for April 1992 for selected receptor points (units: mg/m2).

REFERENCES

Eliassen, A., Hov, O., Iversen, T., Saltbones, J. and Simpson, D., 1988, Estimates of airborne transboundary transport of sulphur and nitrogen over Europe, EMEP/MSC-W Report 1/88, DNMI, Oslo, Norway.

Kahl, J. D. and Samson, P. J., 1986, Uncertainty in trajectory calculations due to low resolution meteorological data, J. Climate Appl. Meteor. 1816:1831.

Renner, E., Ratzlaff, U. and Rolle, W., 1985, A Lagrangian multi-level model of transport, transformation and deposition of atmospheric sulfur dioxide and sulfate, Atmos. Environ. 1351:1359.

Sandnes, H. and Styve, H., 1992, Calculated budgets for airborne acidifying components in Europe, 1985, 1987, 1988, 1989, 1990 and 1991, EMEP/MSC-W Report 1/92, DNMI, Oslo, Norway.

Szepesi, D. J. and Fekete, K. E., 1987, Background levels of air and precipitation quality for Europe, Atmos. Environ. 1623:1630.

LOCAL BACKGROUND AIR POLLUTION
IN RESPONSE TO COASTAL CIRCULATION

Edita Lončar,[1] and Nadežda Šinik[2]

[1]Meteorological and Hydrological Service
Zagreb, Croatia
[2]Geophysical Institute University of Zagreb
Zagreb, Croatia

INTRODUCTION

Background pollution is a stationary initial pollution which should be taken into account in every monitoring system and control strategies. Its definition and consequently the methods of its evaluation are not generally adopted and may vary from one country to another.

Our paper defines the local background pollution as a possible minimum concentration which cannot be cleaned away any longer by the wind and turbulent diffusion. Following such a definition a special theoretical approach has been accomplished . It solves the simplified form of the equation of diffusion, applied to a stationary case of the background pollution which also implies constant sources and sinks of suspended material. With proper boundary conditions the equation can be solved analytically. Making use of the surface layer similarity one finally gets to the following form of the solution:

$$ C \ = \ C_o \ \exp\left[\frac{u_* s}{vS} + \left(\frac{z}{kS} \right)^{1/2} \Phi_m \right] \tag{1} $$

where C is the concentration of the given pollutant, C_o its minimum value at the lower boundary, u_* friction velocity, Φ_m universal similarity function, v wind velocity, S length of the area under consideration, s and z horizontal and vertical coordinates and k von Karman's constant. A background pollution C_b is now defined as

$$ C_b = \lim_{\substack{v \to \infty \\ \Phi_m \to 0}} C \tag{2} $$

Based on (2) a practical method has been evaluated to determine the local value of background pollution by means of simultaneous series of C and v with ϕ_m implicit in the procedure.

Air Pollution Modeling and Its Application X, Edited by S-V. Gryning
and M.M. Millán, Plenum Press, New York, 1994

LOCAL BACKGROUND POLLUTION AT THE BAKAR BAY

The method has been applied to determine background value of SO_2 at Bakar bay, northeastern Adriatic coast of Croatia, by means of simultaneous series of half-hourly values of SO_2 concentration and wind velocity.

Due to the orientation of the bay and mountains surrounding it, the air streams from NE and from SW prevail. Northeasterly winds (from land toward sea) are usually strong and have a ventilating effect in this region. On the other hand, southwesterly winds (from sea toward land) are usually light and persist during days when the local circulation dominates the synoptic one.

These facts suggest to estimate a possible background pollution in relation to main wind directions. In accordance with the method the concentration of SO_2 and wind velocity have been translated to $\ln C$ against u_*/v diagram with $u_*=0.2$ for the winds which blow from the land and $u_*=0.5$ for the on shore winds.

For the most probable wind directions NNE-E practical evaluation of local background air pollution has been illustrated on the figure 1. The straight line, drawn just bellow the lowest values of $\ln C$ intersects the ordinate $u_*/v=0$ (which corresponds to $v\rightarrow\infty$) at $\ln C_0=2.0$, what means $C_0=7\mu gm^{-3}$. A use of the straight line just above the highest concentrations includes the influence of turbulent diffusion. It intersects the lowest line at the point which is lower than C_0 point on $u_*/v=0$ axes. A projection of the intersectional point upon $u_*/v=0$ axes gives $C_b=6\mu gm^{-3}$

The background pollution for other not so frequent wind directions has been estimated on the same way. So for:

-NNW-NNE directions $\qquad C_b = 3\mu gm^{-3}$

-ESE-SSE directions $\qquad C_b=15\mu gm^{-3}$

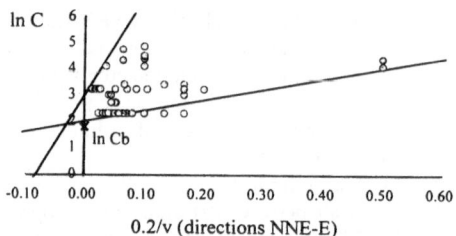

Figure 1.The graphical determination of C_b.

However when applied to the seabreeze (SSW-WSW), which in the Bakar bay appear to be fairly light (velocities less than 3 ms^{-1}), the method fails since such light winds can no ways approximate the condition (2). On the contrary, they describe the effect of pollutant transport from close emission sources to Bakar bay.

LAGRANGIAN STOCHASTIC DISPERSION MODELLING FOR VARYING BOUNDARY LAYER STABILITIES

Mathias Rotach[1], Sven-Erik Gryning and Caterina Tassone
Risø National Laboratory, 4000 Roskilde, Denmark

[1]permanent affiliation: Swiss Federal Institute of Technology
GGIETH, Winterthurerstr. 190, CH-8057 Zürich, Switzerland

MODEL CONSTRUCTION

In a given flow field the particle velocities are modelled as a Markov process, i.e., through the following stochastic differential equation

$$u_i = a_i(\mathbf{x}, \mathbf{u}, t)dt + b_{ij}d\xi_j, \quad d\mathbf{x} = \mathbf{u} \cdot dt. \tag{1}$$

Bold symbols indicate vectors, the subscripts denote the Cartesian components and the functions a_i and b_i are specified by the Fokker-Planck equation for the process (1) to ensure the well-mixed condition (Thomson, 1987). Following the usual procedure (Tassone et al., 1993) the functions a_i can be described as

$$a_i P_{tot} = \frac{\partial}{\partial u_i}(B_{ij}P_{tot}) + \Phi_i, \qquad \frac{\partial \Phi_i}{\partial u_i} = -\frac{\partial}{\partial x_i}(u_i P_{tot}). \tag{2}$$

where $B_{ij} = 1/2\, b_{ik}b_{jk} = \delta_{ij}C_o\varepsilon / 2$. In this expression C_o is a universal constant, set to 5, and ε is the dissipation of TKE. A two-dimensional probability density function (pdf), P_{tot}, for the velocity components $u_1 = u$ and $u_2 = w$ is constructed as follows

$$P_{tot} = \mathcal{F}P_u P_c + (1 - \mathcal{F})P_g = \left[\mathcal{F}P_c + (1 - \mathcal{F})P_w P_{uw}\right]P_u. \tag{3}$$

Here, $P_g = P_u P_w P_{uw}$ denotes a two-dimensional Gaussian pdf and P_c is a skewed one-dimensional (w) Pdf as given by Luhar and Britter (1989). The function $\mathcal{F}$ describes a smooth transition from zero to one to allow for varying shapes of P_{tot}. If $\mathcal{F}$ approaches zero the turbulence is modelled as jointly Gaussian, whereas, in the other limiting case, $\mathcal{F}=1$, the vertical fluctuations are skewed and uncorrelated to the horizontal ones. The function $\mathcal{F}$ is determined by boundary layer stability (z_i/L and z/z_i or z/L) and parameterised using model constraints in the limiting cases, the functions a_i are determined using (2) and (3) (Rotach, et al., in prep.). The parameterisations for the velocity moments are those given in Tassone et al. (1993) which allow for different boundary layer stabilities.

VARYING BOUNDARY LAYER STABILITY

Three cases are simulated in order to establish the dispersion characteristics under different boundary layer stability conditions: Case A: $u_* = 0.05$ ms^{-1}; $w_* = 2$ ms^{-1}; Case B: $u_* = 0.3$ ms^{-1}, $w_* = 2$ ms^{-1}; Case C: $u_* = 0.4$ ms^{-1}; $w_* = 0.6$ ms^{-1}. The boundary layer height is 1 km for all cases and the source height is 115 m.

A comparison of scaled average particle height $\bar{z}$ and plume spread σ_z is given in Figure 1. The convective cases A and B show a slight decrease of the mean plume centre line before it rises to its equilibrium value of 0.5. No 'overshooting' before reaching the equilibrium values of $\bar{z}$ and also σ_z is observed in contrast to the one-dimensional model of Luhar and Britter (1989). This must probably be attributed to the terms in Φ_w that appear through the height dependence of $\overline{u_g^2}$, the variance in P_g(cf. equation (2)). Case C shows a much slower increase of the plume centre line and spread than the two convective cases. This is comparable to the 'R20 case' of Mason (1992), who used LES to calculate the turbulence field but in connection with a much simpler stochastic dispersion model.

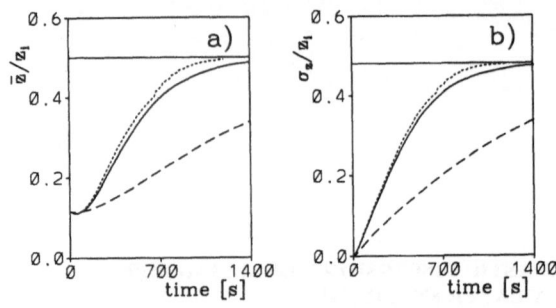

Figure 1. Scaled average plume height $\bar{z}/z_i$ a) and plume spread σ_z/z_i b). Case A: full line; case B: dotted line; case C: dashed line.

THE EFFECT OF THE CORRELATION

Case C was re-run in a different mode, where the fluctuating horizontal wind component as well as the correlation term in P_{uw} were held at zero to investigate the effect of the correlation between u and w introduced in the two-dimensional approach. The resulting 'model' does not completely correspond to a one-dimensional approach due to the presence of vertical derivatives of P_u in Φ_w. However, for the stability in case C the differences are estimated to be small. Figure 2 shows that the 'one-dimensional' approach reproduces the time (or location) maximum of the crosswind integrated surface concentration, $\overline{Q}_y$, quite well. In turn, the fully two-dimensional model exhibits a reduced maximum $\overline{Q}_y$ and a relaxation that is slower.

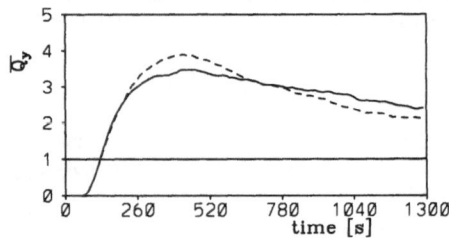

Figure 2. Crosswind integrated surface concentration $\overline{Q}_y$ for case C (dashed line) and the pseudo one-dimensional version (see text) of case C (solid line).

REFERENCES

Luhar, A.L. and Britter, R.E.: 1989, 'A Random Walk Model for Dispersion in Inhomogeneous Turbulence in a Convective Boundary Layer', *Atmos. Environ.*, **23**, 1911-1924.

Mason, P.J.: 1992: 'Large-Eddy Simulation of Dispersion in Convective Boundary Layers with Wind Shear', *Atmos.Env,iron* **26A**, 1561-1571.

Tassone, C; Gryning, S.-E. and Rotach, M.W.: 1993, 'A Random-walk Model for Atmospheric Dispersion in the Daytime Boundary Layer', *Proceedings of the 20th Int. Technical Meeting on Air Pollution Modelling and its Applications*, Nov, 29 -Dec 3, Valencia, Spain.

Thomson, D.J.: 1987: 'Criteria for the Selection of Stochastic Models of Particle Trajectories in Turbulent Flows', *J. Fluid Mech.*, **180**, 529-556.

WAKE FLOWS OVER MOUNTAINOUS AREAS

Gerhard Adrian and Franz Fiedler

Institut für Meteorologie und Klimaforschung
Universität Karlsruhe, Kernforschungszentrum Karlsruhe
D-76128 Karlsruhe, Germany

INTRODUCTION

The wind field near the earth's surface is strongly influenced by the topography which can be observed in all scales of atmospheric motions above all in the mesoscale. Often observed phenomena are secondary circulations which are known to increase the efficiency of transport processes of momentum, heat, humidity and also of airborne material. Therefore it seems to be important to describe circulations induced by the topography with a good accuracy in air pollution modeling as well as in regional climatology. Flow processes influenced by the topography can be thermally or dynamically induced, whereby in nature both mechanisms work often together. Here only dynamically induced phenomena like wake flows with stable stratification are discussed. An overview about the phenomena is given by Etling (1989).

Mesoscale models like KAMM (*K*arlsruhe *A*tmospheric *M*esoscale *M*odel) (Adrian and Fiedler, 1991), which is used here, are able to simulate the phenomena caused by the flow over and around mesoscale mountains for single episodes. But for many problems in the mentioned fields of air pollution modeling or regional climatology it is helpful to have criteria of occurrence of these phenomena based on data which are available from operational stations. Additionally such criteria can be applied to transfer information from larger scales available from operational weather forecasting or global circulation models to the mesoscale. For that purpose a similarity hypotheses of flows over mountainous areas with stable stratification is formulated and tested against model simulations and observations taken in the upper Rhine valley.

DIMENSIONLESS NUMBERS

For the formulation and around mountains dimensionless numbers are are derived from the dimensionless Euler - equations, because the influence of friction and turbulent processes is assumed to be neglectable as found in many experimental and theoretical studies. The obstacles have a typical horizontal scale a and a typical height h. The state

Air Pollution Modeling and Its Application X, Edited by S-V. Gryning
and M.M. Millán, Plenum Press, New York, 1994

of the undisturbed atmosphere is given by a velocity scale U (e. g. geostrophic wind) and the Brunt – Väisälä frequency N. One complete set of dimensionless numbers are a Froude number $Fr_a = U/(aN)$ calculated with the horizontal length scale a, a Froude number $Fr_h = U/(hN)$ calculated with the vertical length scale and a Rossby number $Ro_a = U/(af)$ (Adrian, 1992).

In the linear theory of the flow around mountains the vertical component of the wind vector behaves in inverse proportion to Fr_h and the horizontal components in proportion. Those facts illustrates that a nonlinear behaviour of the flow can be expected at low values of Fr_h which can be proved by numerical nonlinear solutions and experiments in the laboratory (Boyer et al, 1987). The numbers Fr_a and Ro_a describe the form of the solution.

WAKE PHENOMENA

It has been shown (Adrian, 1992) that in the upper Rhine valley two different flow regimes are observed, one is a channeled flow in the valley which has a width of ca. 50 km between the Black Forest mountains and the Vosgues mountains. The other regime is marked by low wind velocities with wind directions varying in space. This behaviour can be explained by large vortices with a scale of the width of the valley shedding behind the Vosgues or Black Forest mountains. This happens at $Fr_h < 0.4$, similar to wake flows behind islands, proved by numerical simulations and observations.

An other phenomenon described by this Froude number is the "Moehlin Jet" in the Basel area between the Black Forest mountains and the Swiss Jura, where often a jet like flow can be observed in the height between the valley and the summits of the mountains with wind velocities which are double as high as the geostrophic wind flow. This jet is important for the regional climate also because of its influence onto the fog distribution. Numerical simulations show that the amplitude of the wind disturbances depend strongly on the Froude number Fr_h and that similar profiles of the wind components can be given which allow to estimate wind profiles from few observations in this area. The results fit well to observations done by Dütsch (1985).

SUMMARY

The two phenomena, the wake flows in the Rhine valley and the Moehlin jet, exemplify that similarity laws can be formulated to describe atmospheric processes in the mesoscale which is a first step for constructing parameterisation schemes of the effects of mesoscale processes on larger scales.

REFERENCES

G. Adrian and F. Fiedler. Simulation of unstationary wind and temperature fields over complex terrain and comparison with observations. *Beitr. Phys. Atmosph.*, 64:27–48, 1991.

G. Adrian. Wake flows in the upper Rhine valley. *Beitr. Phys. Atmosph.*, 65, 1992.

D. L. Boyer, P. A. Davies, W. R. Holland, F. Biolley, and H. Honji. Stratified rotating flow over and around isolated three-dimensional topography. *Phil. Trans. R. Soc. Lond. A*, 322:213–241, 1987.

H. U. Dütsch. Large - scale domination of a regional circulation during winter - time anticyclonic conditions. *Meteor. Rundsch.*, 38:65–75, 1985.

D. Etling. On atmospheric vortex streets in the wake of large islands. *Meteorol. Atmos. Phys.*, 41:157–164, 1989.

FLUID PARTICLE MOTION IN INHOMOGENEOUS AND NON-GAUSSIAN TURBULENCE

Stefan Heinz

IFU
Kreuzeckbahnstr. 19
D-82467 Garmisch-Partenkirchen, F.R.G.

INTRODUCTION

It has been proven that modelling turbulent dispersion in the Lagrangian framework is a successful and flexible approach. The effects of inhomogeneity, instationarity and non-Gaussianity of turbulence have to be taken into account (Thomson, 1987; Sawford, 1991) and there are questions concerning the modelling of buoyancy (van Dop, 1991) and chemical reactions. As shown recently (Heinz and Schaller, 1993), equations for particle motion, particle potential temperature and mass fractions of chemical components can be determined completely consistent with the exact ones for the means and variances, assuming the approximations of Kolmogorov and Rotta. Moreover, the influence of non-linear terms to the systematic particle motion is discussed for locally Gaussian-distributed fluctuations. However, there arises the question in which extent the consistency with the exact transport equations for the first and second moments determines particle motion, which influence arises from non-Gaussian distributed fluctuations for instance for transport coefficients like the diffusion coefficient or the Lagrangian timescale. To get more insight in the contributions of non-Gaussianity and inhomogeneity transport coefficients are calculated here for different non-Gaussian distributed vertical velocity fluctuations coinciding in it means and variances.

DIFFUSION COEFFICIENT

Assuming according to Kolmogorov theory $B^{ij} = 1/2 \; C_0 \; <\varepsilon> \delta_{ij}$, where $<\varepsilon>$ is the ensemble-average rate of dissipation of energy and C_0 a universal constant, and gradients of the mean wind and potential temperature field limited by the scaling of the corresponding fluxes the symmetric component K_s of the 3-dimensional diffusion coefficient matrix is given by (Heinz, 1990) $K_s = <V^2> \; B^{-1}$, V being completely determined by the Eulerian velocity distribution function g_E. For the latter a two-mode model is considered, $g_E = g_G \; g^3/g_G^3$, g_G being a 3-dimensional Gaussian distribution function, g_G^3 the Gaussian distribution of vertical fluctuations and g^3 is given by

Air Pollution Modeling and Its Application X, Edited by S-V. Gryning
and M.M. Millán, Plenum Press, New York, 1994

$$g^3 = \frac{a_-}{(2\pi)^{1/2}\sigma_-} \exp\left\{-\frac{(w+w_-)^2}{2\sigma_-^2}\right\} + \frac{a_+}{(2\pi)^{1/2}\sigma_+} \exp\left\{-\frac{(w-w_+)^2}{2\sigma_+^2}\right\},$$

where it is assumed for simplicity $w_+ = \sigma_+$, $w_- = \sigma_-$, and $\gamma = \sigma_+ / \sigma_-$ as measure of inhomogeneity (w being the vertical velocity). Hence, all parameters are given in dependence on γ, $a_+ = (1+\gamma)^{-1}$, $a_- = \gamma(1+\gamma)^{-1}$, $\sigma_+^2 = <w^2> \gamma /2$, $\sigma_-^2 = <w^2> \gamma^{-1} /2$, ensuring that the distribution function has a mean equals zero and a variance equals $<w^2>$. Then, the diffusion coefficient can be calculated by $K_s = B^{-1} V^2 \Gamma$, where V being the matrix of the second moments of the velocity distribution function and Γ a matrix depending on γ only and representing the influence of non-Gaussianity. Moreover, the skewness $s^3 = <w^3> / <w^2>^{3/2}$ and the curtosis Ku $= <w^4> / <w^2>^2$ are simple related by γ, so that the latter can be calculated from the skewness and the curtosis.

RESULTS

By variation of γ the effect of non-Gaussianity of vertical velocity fluctuations to the diffusion coefficient can be shown. On the other hand, by changing the skewness and the curtosis and calculating the corresponding γ values the influence of third and fourth moments to non-Gaussianity can be explained. Thereby, the relation of the two mode-parts to the statistics of the full vertical velocity fluctuations can be compared with results of concepts and measurements for the convective boundary layer (Hunt, Kaimal, Gaynor, 1988). Moreover, the influence of variations of the vertical profile of the skewness and it derivation is given.

REFERENCES

Heinz, S., 1990, Ph. D. Thesis, Berlin (in German)
Heinz, S., Schaller, E., 1993, (submitted to publication)
Hunt, J.C.R., Kaimal, J.C., Gaynor, J.E., 1988, Q. J. R. Meteorol. Soc., 114, 827
Sawford, B.L., 1991, Bound. Layer Met., 62, 197
Thomson, D.J., 1987, J. Fluid Mech., 180, 529
van Dop, H., 1991, Atm. Env., 26A, 1335

MESOSCALE MODELLING OF THE ATMOSPHERIC INPUT INTO COASTAL WATERS

K. Heinke Schlünzen, Klaus Bigalke, and Ulrike Niemeier

Meteorologisches Institut
Universität Hamburg
Bundesstraße 55
20146 Hamburg, Germany

Coastal waters and the oceans are not only polluted from rivers and by direct input but also from the atmosphere by a considerable amount. In the paper examples are given for the input of SO_2, NO_2, and Pb into the German Bight in a high temporal and spatial resolution. The input figures are calculated from a mesoscale model. The modelled meteorological situations are representative for typical weather conditions in May when the atmospheric input might become an important factor to control the growing of phytoplankton.

For the presented case studies the three-dimensional, non-hydrostatic mesoscale transport and fluid model METRAS is used (Schlünzen, 1990). Wind, temperature, humidity, and up to 89 tracer concentrations are calculated from prognostic equations. Atmospheric exchange processes are parameterized by a first oder closure scheme with a formulation for the exchange coefficient given by Dunst (1982). The dry deposition of gases (SO_2, NO_2) and particles (Pb) is modelled following the resistance model concept (Schlünzen, Pahl, 1992). Chemical transformations and wet deposition processes have been neglected in the studies for simplification. Total SO_2, NO_2, and Pb input (dry deposition only) are calculated for a full diurnal cycle. The integrated daily deposition values are generalized on a statistical basis and result in typical input figures for May.

The model results show that the calculated input does not generally decrease with increasing distance from the coast. The main daily input is close to the coastline but especially at nighttime local maxima occure further offshore over water. They are caused by the diurnal variations in atmospheric stratification and wind field as well as by the emittand distribution.

REFERENCES

Dunst, M. , 1982, On the vertical structure of the eddy diffusion coefficient in the PBL. *Atmosph. Envir.*, 16: 2071-2074

Schlünzen, K.H., 1990, Numerical studies on the inland penetration of sea breeze fronts at a coastline with tidally flooded mudflats. *Beitr. Physik Atm.*, 63:243-256

Schlünzen, K.H. and Pahl, S., 1992, Modification of dry deposition in a developing sea-breeze circulation - a numerical study. *Atmosph. Envir.*, 26A:51-61

REGIONAL SCALE TRANSPORT MODEL OF ATMOSPHERIC ACID COMPOUNDS, AND ITS APPLICATION FOR HUNGARY

Katalin E. Fekete

Hungarian Meteorological Service

H-1675 Budapest, P.O.B. 39, Hungary

A regional scale operational model for the simulation of air and precipitation quality originating from potentially acidifying substances has been worked out (Fekete, 1986; Fekete and Szepesi, 1987; Fekete and Gyenes, 1993). The poster shown at the 20th ITM will present the results of the Hungarian regional scale model for 1987.

REGIONAL TRANSPORT MODEL

The regional transport model estimates annual average pollutant species concentrations originating from sulphur and nitrogen oxides and ammonia. Concentrations of these species in the ambient air originate from domestic area sources, domestic tall stacks, as well as from foreign sources (continental and hemispheric scale background levels are separately considered). The contributions of domestic area sources and the continental pollution effect are simulated by an air trajectory moving-box model, and the pollutants from tall stack are estimated by a Gaussian model. The hemispheric background pollution levels are analyzed from the territorial distributions of measured data in the European region.

The wet deposition of sulphate, nitrate and ammonium is estimated using another method, based on the regression analysis of measured air and precipitation quality data. Combining the contributions from wet and dry deposition and

Air Pollution Modeling and Its Application X, Edited by S-V. Gryning
and M.M. Millán, Plenum Press, New York, 1994

converting them to acid flux, we determine the yearly potential acid deposition.

MODEL RESULTS

Using the regional transport model worked out, calculations were carried out for different years and emission scenarios for Hungary. A run of the model took about 10 min for an IBM AT 486-type personal computer. The computed values correlated well with the measurements.

As a result of the calculations, the annual rate of potential acid deposition is given in Figure 1. It can be seen that values above 0.30 H^+eq m^{-2} $year^{-1}$ occurred along a NE-SW axis.

Figure 1. Territorial distribution of potential acid deposition (10^{-2} H^+eq $m^{-2}year^{-1}$) for Hungary in 1987

REFERENCES

Fekete, K., 1986, Savas eső országos modellje, *Időjárás* 90:103.

Fekete, K. E. and L. Gyenes, 1993, Regional scale transport model
   for ammonia and ammonium. *Atmospheric Environment* 27A:1099.

Fekete, K. E. and D. J. Szepesi, 1987, Simulation of atmospheric acid
   deposition on a regional scale. *Environmental Management* 24:17.

REGIONAL SCALE MODELING CASE STUDIES FOR AEROSOL TRANSPORT OVER HUNGARY

Ernő Mészáros,[1] László Bozó[2] and Ágnes Molnár[1]

[1]Department of Analytical Chemistry
University of Veszprém
H-8201 Veszprém, P.O.Box 158
Hungary
[2]Institute for Atmospheric Physics
H-1675 Budapest P.O.Box 39
Hungary

Aerosol samples were daily taken simultaneously on Nuclepore filters in suburban Budapest and at the regional background air pollution station (K-puszta) 80 km far from Budapest. The samples were analysed among other elements for V, Cr, Co, Ni, Cu, Zn, As and Pb by means of PIXE method. The cases with constant air mass flow between the source and the receptor determined on the basis of radiosonde observation for 925 hPa level carried out in Budapest and Szeged were chosen for modeling consideration. K-puszta station is located between these two cities. A simple source-receptor model of Lagrangian-type was used for the simulations. Budapest was considered as an area source for the elements mentioned above. Aerosol concentration data measured in suburban Budapest were applied as input parameters altogether with the actual mixing heights, precipitation intensities, dry deposition velocities and washout ratios. It was assumed that there are no anthropogenic sources between the two points - except for Pb where road transport has been taken into consideration. Dry and wet cases were separately examined. The aerosol concentrations computed for K-puszta were compared to the data measured there.

The averages computed for three dry cases (Fig. 1.), except for Pb are within a factor of two as compared to the averages of the measured data. For lead, the computed/measured ratio is about 3. For the two wet cases (Fig. 2.) the agreement between the measured and computed data is poorer as compared to dry cases. Model results generally underestimate the measured values for all the elements examined by a factor of 3. The possible reason for the systematical underestimation is that data for precipitation intensity were interpolated only from 3 sites for the computation and wash-out ratios, referring to average conditions can be very different for individual cases. Furthermore, the spatial distribution and heights of emission sources are not exactly mapped for Budapest.

Air Pollution Modeling and Its Application X, Edited by S-V. Gryning
and M.M. Millán, Plenum Press, New York, 1994

617

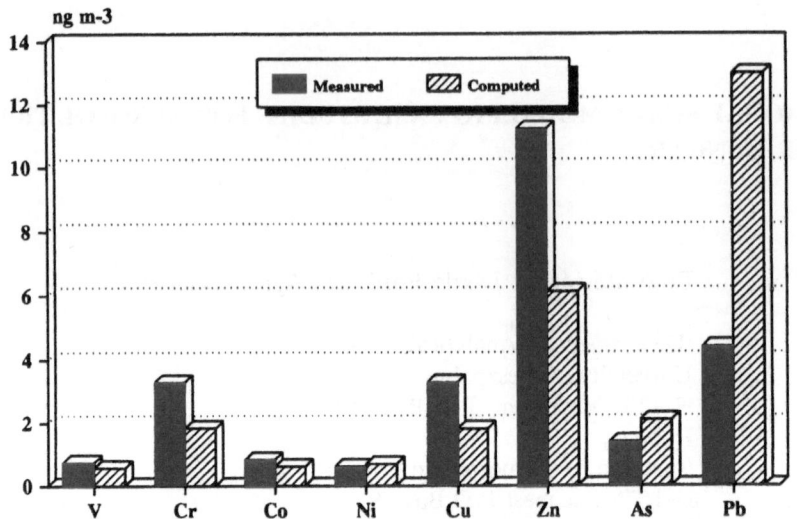

Fig. 1. Comparison of data measured at and computed for K-puszta. Averages of dry cases.

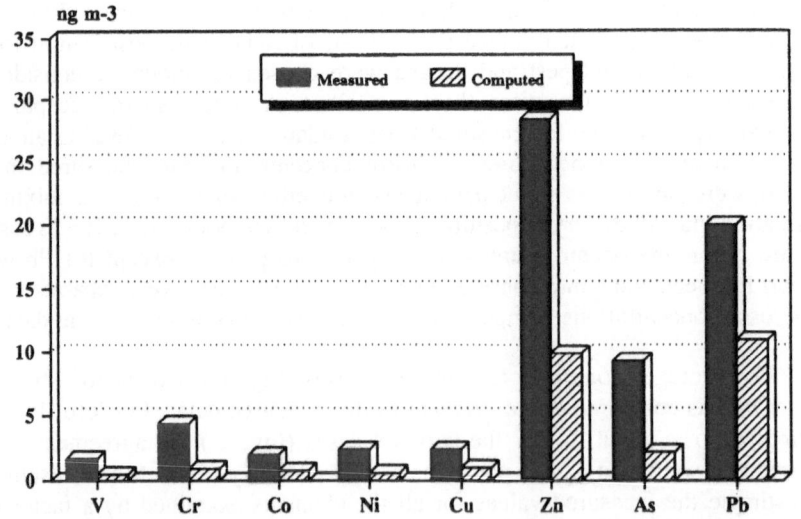

Fig. 2. Comparison of data measured at and computed for K-puszta. Averages of wet cases.

Fluxes over Complex Terrain, Analytical Evaluation

G.A. Dalu[1], M. Baldi[1], R.A. Pielke[2], G. Kallos[3]

1 - CIRA-CSU, Fort Collins, CO 80523; also: IFA-CNR, I - 00144 Rome, Italy

2 - ATM-CSU, Fort Collins, CO 80523; 3 - Applied Physics, Athens, Greece

It is long been recognized that land and sea contrasts produce sea breezes. More recently, it has become evident that landscape variations which result in spatial gradients in surface heat flux can produce mesoscale circulations as strong as sea breezes. However, it remains to determine how small a horizontal variation in surface heating can still produce a significant mesoscale circulation. It order to address this question, in this paper we present results for one Fourier component. While only the linear response can be evaluated using this tool, such models have been shown to be effective at identifying dominant physical processes associated with sea breezes (Rotunno, 1983; Dalu and Pielke, 1989, 93). This approach can also be used to assist in the development of a parameterization of mesoscale effects generated by spatial surface sensible heat variability for use in larger scale models. The primitive equations can be reduced to an equation for the streamfunction:

$$\left[\left(\frac{\partial}{\partial x} + k_p\right)^2 + k_0^2\right]\frac{\partial^2\psi}{\partial z^2} + \left[\left(\frac{\partial}{\partial x} + k_p\right)^2 + l^2\right]\frac{\partial^2\psi}{\partial x^2} = -\frac{1}{U^2}\frac{\partial Q}{\partial x} \tag{1}$$

$$\frac{\partial\psi}{\partial z} = u \quad \text{and} \quad \frac{\partial\psi}{\partial x} = -w; \qquad l^2 = \frac{N_0^2}{U^2} - \frac{U_{zz}}{U}; \qquad p = s + \lambda; \qquad k_p = \frac{p}{U} + \frac{Kk^2}{U} \quad \text{and} \quad k_0 = \frac{f}{U}$$

$$h = h_0\frac{1}{2}[1 + a_1\sin(kx) + a_2\cos(kx)] = h_0\frac{1}{2}[1 + a\sin(kx + b)] \tag{2}$$

$$a_1 = \frac{\lambda(\lambda + Kk^2)}{(\lambda + Kk^2)^2 + k^2U^2}; \qquad a_2 = -\frac{\lambda kU}{(\lambda + Kk^2)^2 + k^2U^2}; \qquad \lim_{k,U\to\infty} a_1, a_2 = 0 \tag{3}$$

$$h_a = \frac{1}{2}h_0(1 + a) \qquad Q = \frac{Q_0 q(s)}{2}[1 + a\sin(kx + b)]\,He(h_a - z) \tag{4}$$

Posing $\psi = \varphi_1(kx)\cos(kx) + \varphi_2(kx)\sin(kx)$ and separating the in-phase from the out-of-phase component, we have to solve the following two equations:

$$\left(\frac{\partial^2}{\partial z^2} - \nu_1^2\right)\varphi_1 = -\frac{\beta_1}{U^2}\frac{1}{2}Q_0\,He(h_a - z)\,q(s) \quad ; \quad \left(\frac{\partial^2}{\partial z^2} - \nu_2^2\right)\varphi_2 = \frac{\beta_2}{U^2}\frac{1}{2}Q_0\,He(h_a - z)\,q(s) \tag{5}$$

Air Pollution Modeling and Its Application X, Edited by S-V. Gryning
and M.M. Millán, Plenum Press, New York, 1994

The temperature perturbation is:

$$\theta = \theta_D - \frac{\partial \theta_D}{\partial x}\Delta_x - \left[\frac{\partial \theta_D}{\partial z} + (1 - r(x,z))\Theta_z\right]\Delta_z \tag{6}$$

$$r(x,z) = \frac{He(h_a - z)}{2}\left[1 + a_1\sin(kx) + a_2\cos(kx)\right] \quad \text{and} \quad \theta_D = (h_a - z)\Theta_z r(x,z) \tag{7}$$

$$\Delta_x(x,z,t_0) = \int_0^{t_0} dt'\, u(x,z,t'); \quad \Delta_z(x,z,t_0) = \int_0^{t_0} dt'\, w(x,z,t')$$

The vertical heat and momentum fluxes, horizontally averaged, are:

$$\Phi_\theta = <w\,\theta>_x; \quad \Phi_M = <w\,u>_x$$

The vertical heat flux, in degree km/day, as function of the horizontal wave number in km^{-1}, for different intensities of the ambient flow, $U = 1$, 2.5, 5, 10 m s^{-1}, is shown in Fig.(1), positive values are fluxes in PBL, negative values in the free atmosphere. The vertical wave number, in km^{-1}, is shown in Fig.(2), note that when the vertical wave number is negative, the perturbation is in form of propagating wave which deeply penetrates in the free atmosphere.

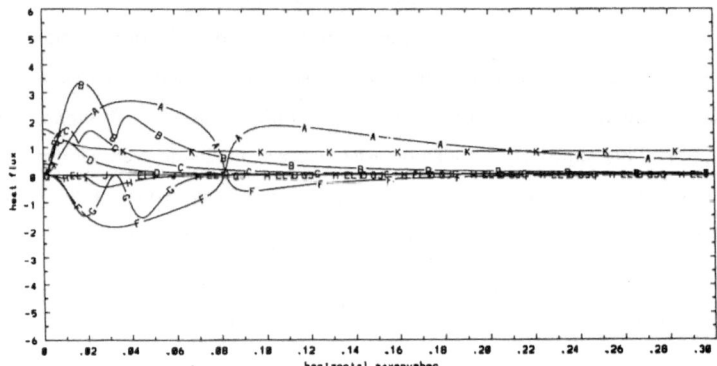

Figure 1. Vertical heat flux

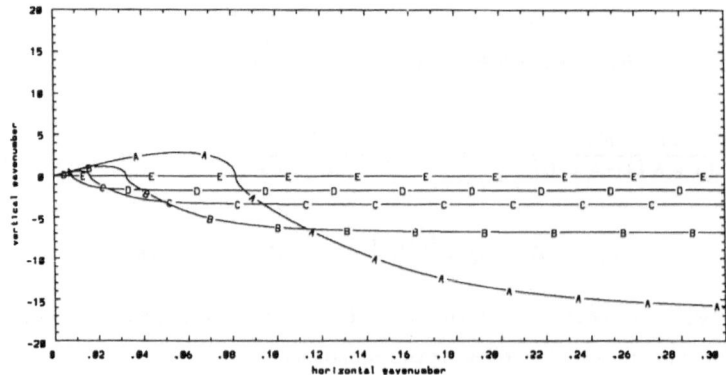

Figure 2. Vertical wave number

Acknowledgments. We acknowledge the support of CIRA-CSU, and of the Italian *CNR-ENEL* and *CNR-Aree Metropolitane* projects.

LAGRANGIAN MODEL SIMULATION OF 3-D CONCENTRATION DISTRIBUTION IN COMPLEX TERRAIN

G. Tinarelli,[1] D. Anfossi,[2] G. Brusasca,[1] E. Ferrero,[3] U. Giostra,[4] M.G. Morselli,[1] F. Tampieri,[5] F. Trombetti[5]

[1] ENEL/CRAM Unita' studi e monitoraggi ambientali, Milano, Italy
[2] Istituto di Cosmogeofisica, C.N.R., Torino, Italy
[3] Istituto di Fisica Generale, Universita' di Alessandria, Italy
[4] ISIATA, C.N.R., Lecce, Italy
[5] Istituto FISBAT, C.N.R., Bologna, Italy

INTRODUCTION

Most of the up to date dispersion simulations are discussed using ground level concentration (g.l.c.) values only. However it is straightforward that the spatial distribution of the concentration should be carefully computed, in order to assess the capability of a given model to correctly simulate the atmospheric dispersion. Therefore, the vertical and horizontal concentration profiles at various downwind distances and at different levels above ground should also be computed and compared to the corresponding observed values if any, even if this is generally thought of as very severe test for any kind of model.

Our team designed a Lagrangian Monte-Carlo particle model named SPRAY based on Thomson's formulation (Thomson, 1984, Tampieri et al. 1992), for the simulation of dispersion over complex terrain. SPRAY was tested against an EPA wind tunnel dispersion experiment both in flat terrain (Anfossi et al, 1992) and in presence of a schematic two-dimensional hill (Tinarelli et al., 1993), verifying its good performances.

The purpose of the present study is to compare our model results regarding the vertical and horizontal concentration profiles with the dispersion data obtained in the same wind tunnel experiment.

SIMULATIONS AND RESULTS

The wind tunnel experiments (Khurshudyan et al. 1981) simulate a boundary layer flow and tracer dispersion from point sources over a simple and schematic 2-D hill. The comparisons refer to cases with a hill having an aspect ratio H/a=8 (where H is the hill height and a its half lenght) and the source located at its downwind base. Two experiments with different source heights were taken into account. In the first one the source height was $H_S=H/4$ while in the second one was $H_S=H$. During the first experiment, 5 vertical and horizontal (crosswind) concentration profiles have been measured in the lee side of the hill, at normalized distances x/a 0.08, 0.25, 0.5, 1, and 2 respectively, referred to the source

Air Pollution Modeling and Its Application X, Edited by S-V. Gryning
and M.M. Millán, Plenum Press, New York, 1994

position. During the second experiment three profiles have been measured at x/a 0.95, 1 and 2 respectively. SPRAY model simulated horizontal and vertical concentration profiles at the same positions. A total number of 477 concentration values (including the horizontal downwind g.l.c. profiles) has been taken into account, 282 for the lower source and 195 for the higher one and a model evaluation has been separately performed. The Fractional Bias FB, the Normalized Mean Square Error NMSE and the Correlation Coefficient CORRE indexes have been computed. Table 1 describes the results.

Table 1. Values of model evaluation indexes

Source	FB	= 0.014	Source	FB	= 0.015
height H/4	NMSE	= 0.208	height H	NMSE	= 0.030
	CORRE	= 0.970		CORRE	= 0.988

In addition, a scatter diagram showing the point-to-point comparison between measured and simulated concentrations is shown in Fig. 1. These results are quite satisfactory showing the overall capability of SPRAY to not only describe the g.l.c. fields but the 3D concentration fields too.

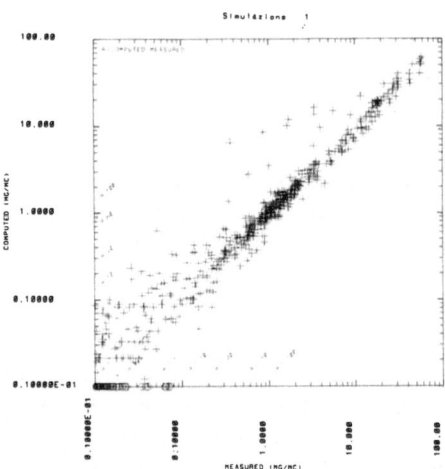

Figure 1. Scatter diagram of normalized concentrations. Measured concentrations are represented on X axis while predicted ones are represented on Y axis.

REFERENCES

Anfossi, D. , Ferrero, E., Brusasca, G., Tinarelli, G., Tampieri, F., Trombetti, F and Giostra, U., 1992,
    Dispersion simulation of a wind tunnel experiment with lagrangian particle models,
    Il Nuovo Cimento C, 15,139:158.
Khurshudyan, L.H., Snyder, W.H. and Nekrasov, I.V., 1981, Flow and dispersion of pollutants
    over two-dimensional hills, EPA rep. 600/4-81-067.
Tinarelli G., Anfossi, D., Brusasca, G., Ferrero, E., Giostra, U., Morselli, M.G., Moussafir, J.,
    Tampieri, F. and Trombetti F., 1993, Lagrangian particle simulation of tracer dispersion in the lee of a
    schematic two-dimensional hill, submitted to *Jou. Appl. Met.*
Tampieri F., Scarani, C., Giostra, Brusasca, G., Tinarelli, G., Anfossi, D. and Ferrero E., 1992,
    On the application of random flight dispersion models in inhomogeneous turbulent flows,
    Ann. Geoph., 10, 749:758.
Thomson D.J., 1984, Random walk modelling of diffusion in inhomogeneous turbulence, *Q.J.R.Met.Soc.*
    110, 1107:1120.

TREATMENT OF TRANSPORT IN MOGUNTIA

Maarten C. Krol

Institute of Marien and Atmospheric research Utrecht (IMAU)
Princetonplein 5
3584 CC Utrecht
The Netherlands

INTRODUCTION

At the IMAU, the program MOGUNTIA has been implemented. MOGUNTIA is a climatological three-dimensional global tropospheric transport model, designed by Zimmermann (1988). The current resolution amounts to 10^0 x 10^0 x 100 hPa. The upper layer is located at 100 hPa. The model is used for extensive tropospheric chemistry calculations in which the effects of heterogeneous chemistry and aerosols can be taken into account. The global character of the model has proved to be valuable for example to calculate budgets of ozone (Dentener and Crutzen, 1993; Crutzen and Zimmermann, 1991).

Currently, the transport is described by monthly averaged wind fields derived from the Oort climatological dataset. Unresolved transport is parametrized by first order K-theory. Furthermore, a deep cumulus convection parametrization scheme is used (Feichter and Crutzen, 1989). Due to a highly efficient numerical scheme, 50 years of transport can easily be modelled.

The use of Oort's dataset for the transport has certain drawbacks. It is not possible to calculate the tracer transport for a specific year. Furthermore, some flexibility is required when the resolution is modified or boundary conditions are changed. Therefore, it would be desirable to obtain a flexible and straightforward method for generating a mass conserving tropospheric wind field from ECMWF data.

CONSTRUCTING A MASS CONSERVING WIND FIELD

In order to obtain a mass-conserving tropospheric wind field from ECMWF analysis data, the following procedure is developed.

Six-hourly ECMWF analysis data are averaged over a month. These data are provided by the Royal Netherlands Meteorological Institute (KNMI) on a 5^0 x 2.5^0 resolution at 14 pressure levels. The ECMWF vertical winds are not reliable and mass conserving. Therefore, only the horizontal winds are interpolated at the boundaries of the desired grid boxes. From these horizontal winds the vertical winds are calculated with the conservation of mass constraint. From the divergence of the horizontal wind at the first layer and the lower boundary condition (no vertical wind at the bottom) the vertical wind at the top of the first layer is calculated. From these vertical winds and the divergence of the second layer, the vertical wind at the top of the second layer is calculated, and so on.

A well-known problem that arises when applying this procedure is the accumulation of errors to the top of the model region. Here, unrealistic high vertical wind speeds are found. O'Brien (1970) suggested to distribute the resulting error at the top layer over all underlying layers. At the same time, a boundary condition at the top of the model region can be introduced (i.e. no vertical wind at the top). The most realistic way to distribute the error over the layers is to scale with respect to the total horizontal wind speed. The horizontal wind components in each model layer are slightly modified in order to fulfil the continuity equation. The procedure described above is repeated several times to obtain absolute mass conservation.

The poor representation of the high latitudes in the ECMWF model may cause problems. This may also be the result of sparse measurements in these regions or orography effects. As a result, an artificial circulation is observed in these regions. This problem is circumvented by replacing the horizontal winds at higher latitudes by the rotational part.

Figure 1 shows the zonal mean circulation that results from the procedure described above. Data are shown for the month October 1987. Notice the two branches of the Hadley circulation and the position of the ITCZ. These features are accurately described by the new wind field approximation.

CONCLUSIONS

A straightforward procedure for the construction of a mass conserving, tropospheric wind field has been developed. The resulting wind field can in principle be calculated on any grid. Moreover, boundary conditions for the wind field can be taken into account easily. Implementation of the new wind field in MOGUNTIA is now taking place and more results will be presented during the conference.

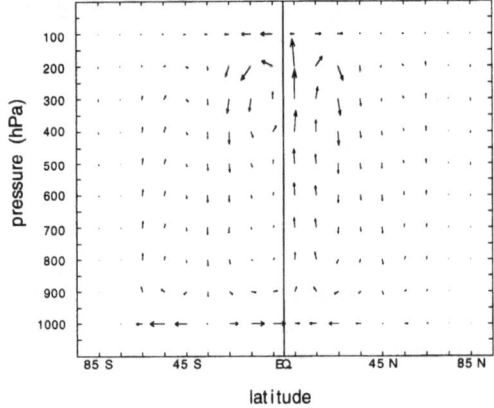

Figure 1. Zonal mean circulation for the month October 1987. For clarity, the vertical wind speed is multiplied by a factor 500. The largest arrow corresponds to ca. 3 m/s

REFERENCES

Crutzen, P.J., Zimmermann, P.H., 1991, The changing photochemistry of the troposphere, *Tellus AB*, 136.
Dentener, F.J. and Crutzen, P.J., 1993, Reaction of N_2O_5 on tropospheric aerosols: impact on the global distributions of NO_x, O_3, and OH, *J. Geophys. Res.*, 98:7149.
Feichter, J. and Crutzen, P.J., 1990, Parametrization of deep cumulus convection in a global tracer transport model and its evaluation with ^{222}Rn, *Tellus B*, 42:100.
O'Brien, J.J., 1970, Alternative solutions to the classical vertical velocity problem, *J. Appl. Meteor.*, 9:197.
Zimmermann, P.H., MOGUNTIA: A handy global tracer model, *in* : "Air Pollution, Modelling and its Application VI,". H. van Dop, ed., Plenum, New York (1988).

METHODOLOGY FOR MAPPING LOCAL SCALE DEPOSITION OF ACIDIFYING COMPONENTS OVER EUROPE

W.A.J. van Pul, J.W. Erisman, J.A. van Jaarsveld and F.A.A.M. de Leeuw

Laboratory for Air Research, LLO
National Institute of Public Health and Environmental Protection, RIVM
P.O. Box 1, 3720 BA Bilthoven, the Netherlands

Current methods and models for estimating deposition of acidifying components differ largely in the used horizontal resolution i.e. varying from 5x5 km (Erisman,1992) up to 150x150 km (EMEP, Iversen et al. 1991). In describing the effects of acidification on the level of ecosystems, acid loads should be available at least on the size of ecosystems. No deposition maps on this resolution are available leaving a serious gap in estimating exceedences of critical loads in Europe.

Here a method is presented based on the combination of concentrations and detailed estimates on the dry deposition process for mapping actual and future deposition fluxes on a European scale or for parts of Europe. The acidifying components studied here are oxidized sulphur and nitrogen and reduced nitrogen compounds.

An overview of the input for and calculation scheme of the method is presented in Figure 1.

In this figure the central map is formed by the concentration data. The concentration fields can originate from measurements or long range transport (LRT) model calculations or a combination of both (However, to obtain a whole coverage of concentration data over Europe, model calculations are necessary). The local deposition velocity of a component is calculated using a resistance model in which the transport to and absorption or uptake by the surface are described (following a scheme proposed in Erisman et al.,1993). From the concentration and the parameterized deposition velocity the dry deposition is inferred (called the inference method Hicks, 1986). In the parameterizations only routinely available data (e.g. land-use maps, synoptical meteorological data) are used. Given detailed information on the surface and environmental conditions a time and space dependent description of the dry deposition process is obtained. In this way it is possible to construct annual average dry deposition maps for the whole of the European continent.

The emphasis in this method lies in modelling local scale dry deposition fluxes. Wet deposition is also estimated to present a complete acidification map. In using calculated concentration maps the relation between emissions and deposition is maintained and so scenario studies can be carried out. A more detailed description of this method is given in van Pul et al., 1993.

An example of a dry deposition field for SO_2 is presented in Figure 2.

Air Pollution Modeling and Its Application X, Edited by S-V. Gryning
and M.M. Millán, Plenum Press, New York, 1994

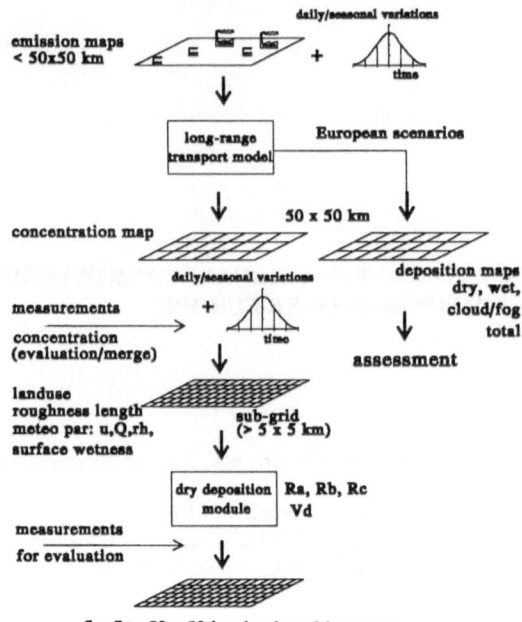

Figure 1 Overview of the input for and calculation scheme for mapping small scale acid deposition fluxes. Emissions (with variations in time) are needed in LRT models to describe European wide concentration fields. Concentration maps are constructed from LRT calculations and measurements. From the land use data and meteorological data (among others: wind velocity,u, global radiation, Q, relative humidity rh, surface wetness) local deposition velocities (constructed from the parameterized resistances: aerodynamic (r_a), boundary layer (r_b) and surface resistance, (r_c)) are modelled. In the last step dry deposition maps are inferred from the constructed concentration and dry deposition velocity fields.

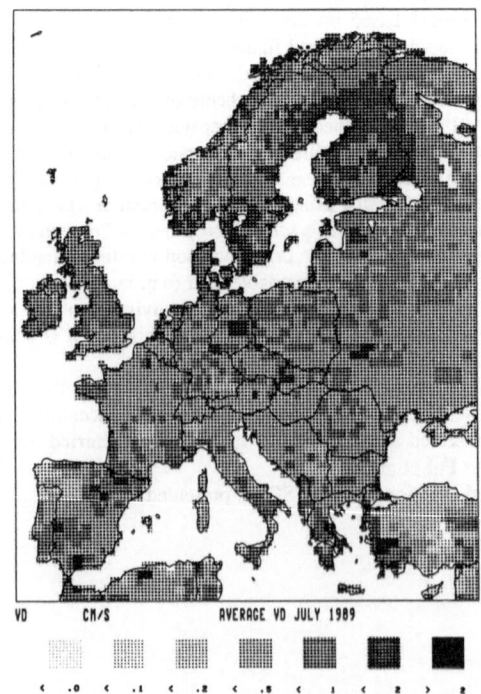

Figure 2 Dry deposition velocity for SO_2 for July 1989 constructed from Olson (1985) land use data base and synops data obtained from ECMWF (Reading, UK). Note that differences in dry

References

Erisman, J.W., 1992, Atmospheric deposition of acidifying compounds in The Netherlands. PhD thesis, University of Utrecht, The Netherlands.

Erisman, J.W., W.A.J. van Pul and G.P.Wyers, 1993, Parameterization of dry deposition mechanisms for the quantification of atmospheric input to ecosystems. in proceedings CEC/Biatex Workshop: General assessment of biogenic emissions and deposition ofnitrogen compounds, sulphur compounds and oxidants in Europe. 4-7 May 1993 Aveiro, Portugal.

Hicks, B.B., 1986, Measuring dry deposition: a re-assessment of the state of the art. Water,air and soil pollution 30:75-90.

Iversen, T, N. Halvorsen, S.Mylona and H.Sandnes, 1991, Calculated budgets for airborne acidifying components in Europe, 1985,1987,1988,1990. Meteorological synthesizing Centre-West, the Norwegian Meteorological Institute, Oslo, Norway.

Pul van, W.A.J., J.W.Erisman, J.A. van Jaarsveld and F.A.A.M. de Leeuw, 1993, Methodology for mapping acidifying components over Europe. in proceedings CEC/Biatex Workshop: General assessment of biogenic emissions and deposition ofnitrogen compounds, sulphur compounds and oxidants in Europe. 4-7 May 1993 Aveiro, Portugal.

GROUND LEVEL CONCENTRATIONS OF OZONE, OXIDANT, PAN AND PRECURSORS IN THE NETHERLANDS DURING THE LAST TWO DECADES AND THE RELATION WITH THE LOTOS-MODEL

P. Esser, M.G.M. Roemer, P.J.H. Builtjes, R.G. Guicherit and Th. Thijsse

TNO Institute of Environmental Sciences
P.O. Box 6011, 2600 JA Delft, The Netherlands

ABSTRACT

PeroxyAcetylNitrate (PAN), ozone (O_3), nitrogen oxides (NO_x) and hydrocarbons (C_2-C_4) are measured at several ground level stations in The Netherlands from 1973 onwards. The hydrocarbon measurements at Delft and Moerdijk show the existence of large industrial sources. Analysis of the annual mean values of the compounds has brought about the following results:

- In the seventies levels of C_2-C_4, NO_x and PAN have generally gone up. For O_3 and O_x no significant trend is observed.
- In the eighties levels of C_2-C_4 and NO_x have slightly gone down. For PAN, O_3 and O_x no significant trend is observed.

The effect of a reduction in precursors on PAN and oxidant levels is estimated with the LOTOS model. The model calculations over April 1985 show, for The Netherlands, that a 30% emission reduction of NO_x and VOC results in a 6% reduction of PAN and no reduction of O_x.

INTRODUCTION

Long-term measurements have established a significant positive trend of 1-2% per year of ozone concentrations at a number of rural sites in Europe over the last 20-30 years. Long-term records of ozonesondes in Europe (Hohenpeissenberg and Payerne), Canada and Japan show also an upward trend in the free troposphere (Bojkov, 1992). Model studies indicate that increasing anthropogenic emissions of CH_4, NO_x, CO and VOCs result in an increase of ozone concentrations. Observations demonstrate increasing concentrations of CH_4 on a global scale and increasing CO concentrations in the Northern Hemisphere (Zander *et al*, 1989). Much less is known of global distributions of VOCs and NO_x, and certainly global scale trends of these species are highly speculative. Not only on a global scale but also in Europe trends of VOCs are very uncertain since monitoring networks are only just began to measure other components than O_3 and NO_x. In this study we present the ground level observations of C_2-C_4, NO_x, PAN, O_3, and O_x (=O_3+NO_2) monitored at several locations in The Netherlands from 1973 onwards.

DESCRIPTION OF DATA AND MEASUREMENT STATIONS

The TNO site in Delft, halfway Rotterdam and The Hague is located in an urban area. Kloosterburen is a rural station at the north coast. Kloosterburen is located very close to the TOR-site Kollumerwaard. Moerdijk is located halfway Rotterdam and Antwerpen and

Air Pollution Modeling and Its Application X, Edited by S-V. Gryning
and M.M. Millán, Plenum Press, New York, 1994

receives relatively clean air with wind from the west. The National Monitoring Network of Air Quality is operated by the National Institute of Public Health and Environmental Protection (RIVM) and includes about 50 NO_x stations in rural as well as in urban areas. Ozone and NO_2 are measured at Delft (1973-1981) and Kloosterburen (1980-1989). The PAN monitoring site is located in an urban area at the TNO site in Delft. Hydrocarbons are measured at Delft and Moerdijk. All measurements are hourly measurements, except for PAN which is measured every 15 minutes.

OZONE AND OXIDANT MEASUREMENTS

Ozone values at Delft are much lower than at Kloosterburen also in the common measurement period 1980 and 1981. This difference can be explained by the different NO_x levels at the stations. From figure 1 it is observed that oxidant levels at Delft (1973-1981) are comparable to oxidant levels at Kloosterburen (1980-1989). Kloosterburen shows a slight negative trend (not significant) of 0.4 ±0.5 ppb per year.

PAN MEASUREMENTS

The highest PAN values, on average, are recorded with easterly winds. The lowest values are measured with westerly and northerly winds. From figure 1 one can notice a significant upward trend (+25% per year) in PAN values from 1973 to 1981. No significant trend is measured from 1981 to 1988.

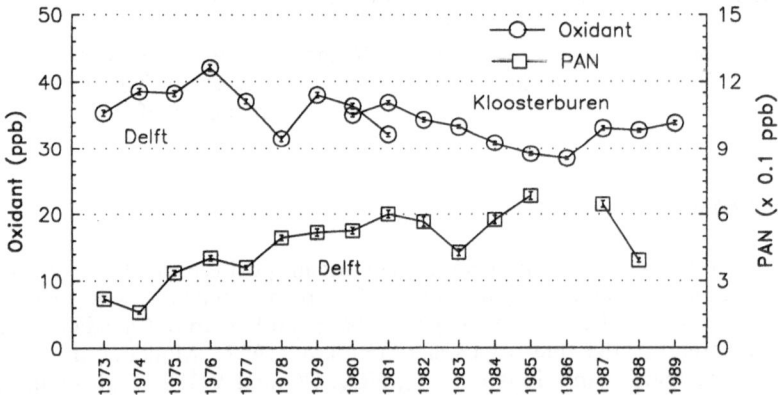

Figure 1   Annual mean values of PeroxyAcetylNitrate (PAN) at Delft, oxidant at Delft (1973-1981) and Kloosterburen (1980-1989).

C$_2$-C$_4$ MEASUREMENTS

The distribution of C_2-C_4 concentrations in Delft very clearly shows the existence of large industrial hydrocarbon emissions in the south sector (Rotterdam and the harbour area). The highest acetylene concentrations (C_2H_2) which is a good indicator of traffic emissions are observed in the east-to-south quadrant. The west and north sector is relatively clean (see figure 2a and 2b).

The Moerdijk data shows elevated concentrations in the Rijnmond sector, low concentrations in the west sector and elevated concentrations in the whole east-to-south quadrant (see figure 2a and 2b). Superimposed peaks are found in the northeast (a local industry complex) and in the south (probably Antwerpen and the harbour area).

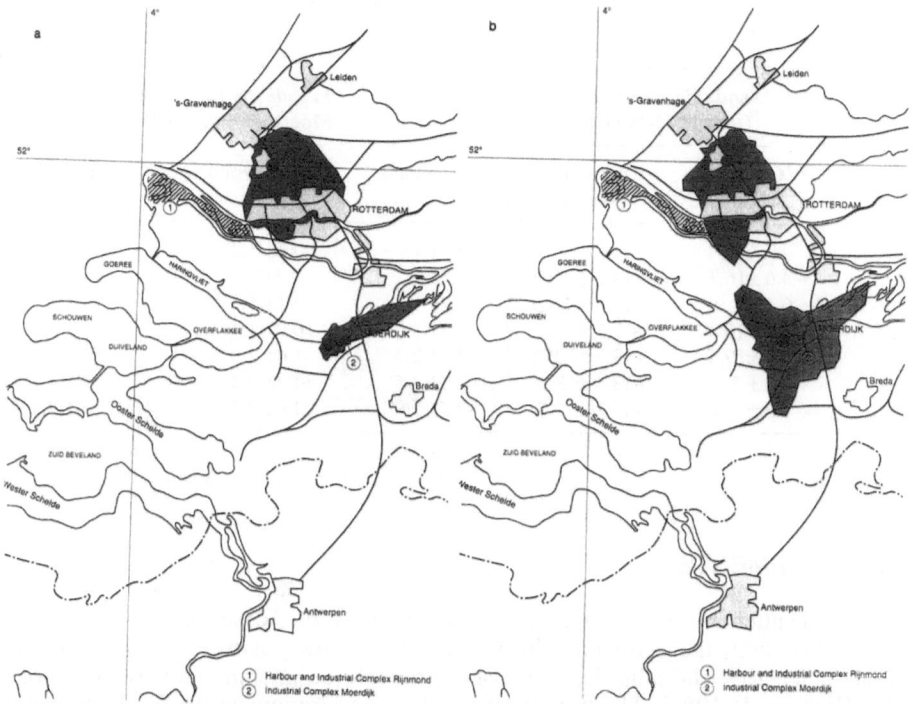

Figure 2a and 2b   Averaged concentrations per sector of 10° of ethene (figure 2a) and iso-butane (figure 2b) at Delft (1982-1984) and Moerdijk (1981-1984).

The trend in the Delft C_2-C_4 data of the clean west sector over 1973-1977 period is positive but not significant at the 2-sigma level, except for i-butane of which the upward trend is significant (table 1). The concentrations in the 1982-1984 period are somewhat lower than in the midseventies. Since we have changed the instrument and lowered the detection limit no conclusions can be drawn from that. In the Moerdijk data over 1981-1991 downward trends of

2-4% per year are observed for ethane, acetylene, propane and n-butane in the clean sector. For the other species the trends are downward but not significant.

The picture that emerges from the hydrocarbon measurements according to table 1 is increasing concentrations in the midseventies, except possibly some industrial contributions which might have stabilized or even have been reduced with respect to the early seventies. During the eighties most of the components and in most of the wind sectors hydrocarbon levels have gone down.

Table 1   Trends of C_2-C_4 for different wind direction sectors at Delft (1973-1977) and Moerdijk (1981-1991).

	Trends (% per year) Delft 1973-1977		Trends (% per year) Moerdijk 1981-1991		
	Clean sector	Continental sector	Rijnmond	Antwerpen	Clean
Ethane			-2.3± 1.7	-0.8± 1.2	-2.0± 1.1
Ethene	+ 4.6 ±16.2	+14.2 ± 10.2	-5.9± 3.2	-1.7± 2.5	-4.0± 4.3
Acetylene	+ 0.4 ± 8.9	+13.5 ± 20.0	-4.2± 1.3	-2.1± 1.5	-3.7± 1.6
Propane	+ 0.7 ±10.8	+ 9.5 ± 9.2	-2.7± 3.0	-0.9± 1.6	-2.3± 2.0
Propene	+ 5.4 ±19.6	+18.2 ± 18.3	-4.3± 2.3	-1.6± 2.6	-3.8± 4.4
i-Butane	+ 7.0 ± 3.2	+13.4 ± 3.8	-2.9± 3.5	+1.3± 2.1	-1.7± 2.2
n-Butane	+ 4.3 ± 4.4	+14.8 ± 1.6	-2.5± 2.0	+1.0± 2.1	-2.5± 2.1

NO$_x$ MEASUREMENTS

In this study the attention to NO$_x$ was restricted to trend studies described in the literature by previous studies. NO$_x$ measurements during the 1970s show positive trends in three of the four urban data series (the other being constant) though the interannual variations are large (Van den Hout et al., 1983). The trend in the NO$_x$ concentrations at the rural site is upward too. From 1979 onwards the annual national average of the 50-percentiles of the NO$_x$ concentrations shows a slight downward trend for the rural as well as the urban sites probably caused by meteorological influences (RIVM, 1992).

LOTOS MODEL CALCULATIONS

The LOTOS (LOng Term Ozone Simulation) model (Builtjes, 1992) was used to estimate, for April 1985, the effect of a 30% reduction of NO$_x$ and VOC emissions on oxidant and PAN levels. A 30% reduction of anthropogenic NO$_x$ and VOC emissions reflects a realistic and achievable scenario by the year 2000 relative to 1985 emissions. In regions with low NO$_x$ values a reduction of oxidant values of more than 5% is calculated in the south and east of Europe and less than 5% reduction is calculated in the north of Europe (figure 3a). No reduction or even an increase in oxidant levels is calculated in Western Europe (high NO$_x$ values). PAN levels are reduced by less than 10% in the north and west part of Europe (figure 3b).

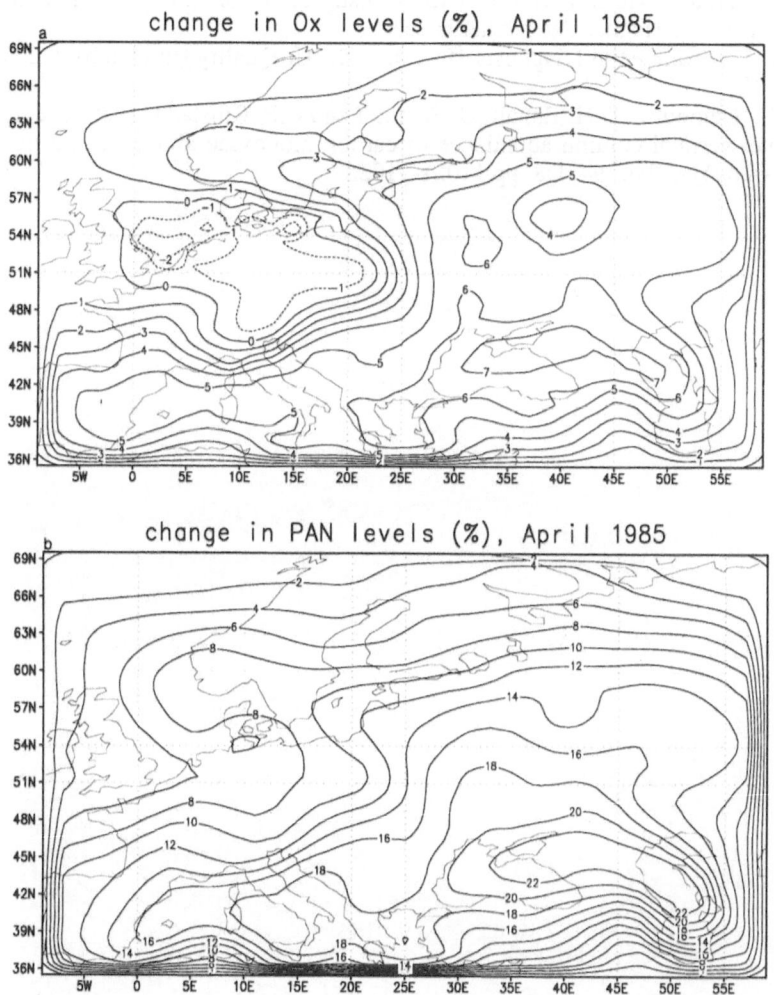

Figure 3a and 3b Effect on oxidant (figure 3a) and PAN (figure 3b) levels due to a reduction of 30% in anthropogenic NO_X and VOC emissions calculated with the LOTOS model over April 1985.

REFERENCES

Bojkov R.D., 1992, Long Term increase of tropospheric ozone, Quadrennial Ozone Symposium Charlottesville, Virginia, Usa, June 1992.

Builtjes P.J.H., 1992, The LOTOS -Long Term Ozone Simulation- project. Summary report, IMW-R 92/240, Delft, The Netherlands.

Hout K.D. van den, C. Huygen, W. den Tonkelaar and N. van Egmond, 1983, Nitrogen oxide in The Netherlands. (in Dutch). Lucht 15, Ministry of Housing, Physical Planning and the Environment, The Hague, The Netherlands.

RIVM, 1992, Environmental Diagnosis 1991, part 2, Air Quality (in Dutch), Bilthoven, The Netherlands.

Zander R., Ph. Demoulin, D.H. Enhalt, U. Schmidt and C.P. Rinsland, 1989, Secular increase of the total vertical column abundance of carbon monoxide above central Europe since 1950, *J. Geophys. Res.*, 94, D8, pp.11021-11028.

METEOROLOGICAL ASPECT OF CHEMICAL COMPOSITION OF PRECIPITATION/DEPOSITION - LONG RANGE POLLUTION TRANSPORT IN A MOUNTAIN REGION

G. Kmieć[1], A. Zwoździak[1], K. Kacperczyk[1], J. Zwoździak[1]

[1] Institute of Environment Protection Engeneering
Technical University of Wrocław
50-370 Wrocław, Wyb.Wyspiańskiego 27, Poland

The acidification of forest ecosystems in the southwestern Poland (Sudety Mountains) has been a topic of intensive research and modeling efforts in recent years. It is generally believed that this problem is a consequence of transport and deposition of acidic atmospheric pollutants, both gases and particles. In the period from 1988 to 1992 sampling was underway at five locations of the mountain region (810 + 1490 m a.s.l.)[1,2]. The main objective of the network was to measure routinely the daily concentration: SO_2, NH_4^+, NO_2, $SO_4^={}_{aer}$, total suspended particles, sulphates in air and the following species: pH value, $SO_4^=$, Cl^-, $N(NO_3^-)$, $N(NH_4^+)$, K^+, Na^+, Ca^{++}, Mg^{++} in precipitation/deposition (rain, snow, rime). Moreover, meteorological parameters were registered: windspeed (V) and direction, humidity (H), temperature (T), the circulation type, the amount of precipitation and trajectory wind roses. Air acidic spacies episodes and precipitation/deposition concentrations of ionic species are found to be strongly dependent upon meteorological circumstances. On a basis of frequency histogram of daily average values (concentration species, meteorological parameters), the range of values were determined for which frequency of occurence was 55 + 80%:

- meteorological parameters

-12.5<T< 6.5 [°C], 2.0<V<14.0 [m/s], 70<H<100 [%], wind direction - SW quadrant (180° + 270°), NW quadrant (270° + 360°), mean precipitation 2.9 + 21.8 [mm/d] (rain, snow) and mean deposition 169 + 770 [g/d] (rime)

- air species concentrations

SO_2: 12.1 + 62.7 [$\mu g/m^3$], NH_4^+: 7.6 + 61.5 [$\mu g/m^3$], NO_2: 4.9 + 26.4 [$\mu g/m^3$], $SO_4^={}_{aer}$: 30 + 110.9 [$\mu g/m^3$], TSP: 5.5 + 130.0 [$\mu g/m^3$], Pb, Cd, Cu, Fe, Zn, Ti, Ca in atmospheric aerosol (as a sum): 0.67 + 8.07 [$\mu g/m^3$]

- atmospheric precipitation/deposition

3.51<pH<5.0, $N(NO_3^-)$: 1.91 + 2.68 [mg/dm^3], $N(NH_4^+)$: 2.33 + 3.87 [mg/dm^3], K^+: 1.14 + 3.77 [mg/dm^3], Na^+: 1.54 + 1.96 [mg/dm^3], Ca^{++}: 1.97 + 2.74 [mg/dm^3], Mg^{++}: 0.26 + 0.46 [mg/dm^3], Cl^-: 3.75 + 5.01 [mg/dm^3], $SO_4^=$: 9.04 + 15.37 [mg/dm^3].

Air Pollution Modeling and Its Application X, Edited by S-V. Gryning
and M.M. Millán, Plenum Press, New York, 1994

The carried out factor analysis of metal concentrations in atmospheric aerosol and the correlation analysis among the concentrations of ionic species in precipitation allows to conclude that considerable effect upon the pollution level in Sudety Mountains have the major pollutant sources located to the west and southwest to the mountain region (products of fossil fuel combustiom and metal processing)[2,3]. Relationship between concentrations of ionic species in precipitations and depositions has been determined by matrix of correlation coefficiens (r), given in Table 1. Determination of acidic species transferred from the atmosphere to the surface (other ecological components) versus "meteorological situations" made possible to enact pollution control strategies. At present, integrated investigations of precipitations, soil, water and plants are carried out in this region.

Table 1. Matrix of correlation coefficiens (r) between ion concentrations in samples of precipitations and depositions in the top of Szrenica (1360 m a.s.l.)[4].

	pH	$N(NO_3^-)$	$SO_4^=$	Cl^-	$N(NH_4^+)$	K^+	Na^+	Ca^{++}	Mg^{++}
pH	-	0	0	0	0	0	0	0	0
$N(NO_3^-)$		-	1	1	1	1	2	0	1
$SO_4^=$			-	0	3	0	1	0	1
Cl^-				-	0	1	3	1	3
$N(NH_4^+)$					-	0	1	2	3
K^+						-	0	0	0
Na^+							-	2	3
Ca^{++}								-	3
Mg^{++}									-

0 for $r < 0.3$; 1 for $0.3 \leq r < 0.4$; 2 for $0.4 \leq r < 0.6$; 3 for $r \geq 0.6$; $\alpha = 0.05$

REFERENCES

1. J.Zwoździak, G.Kmieć, A.Zwoździak, Microscopic and chemical analyses of aerosol and rime samples collected in a mountains area, *Procc. of 12th Int. Conf. on Atm. Aerosols and Nucleation 22 - 27 Aug. 1988*, Berlin (1988).
2. J.Zwoździak, A.Zwoździak, Chemical composition and potential sources of atmospheric aerosols during air pollution episodes in Karkonosze range, *Procc. of the 8th World Clean Air Congress 11 - 15 Sept. 1989*, Hague (1989).
3. G.Kmieć, Agressive and corrosive effects of the atmospheric aerosol in a mountain region - long range transport of pollutants, *Procc. of the 8th World Clean Air Congress 11 - 15 Sept. 1989*, Hague (1989).
4. J.Zwoździak, A.Lisowski, G.Kmieć, A.Zwoździak, Z.Matyniak, R.Jagiełło, Określenie rodzaju i stopnia skażenia chemicznego poszczególnych komponentów środowiska. Identyfikacja zanieczyszczeń napływowych. *Rep. Inst. Environ Protect. Eng. Tech. Univ. of Wrocław, SPR 27/89, SPR 63/90, SPR 50/91*, (unpubl.).

SEA BREEZE IN SUMMER, ALONG THE WEST COAST OF PORTUGAL

Renato A. C. Carvalho and Victor Prior

Instituto de Meteorologia
Rua C, Aeroporto
Lisboa, Portugal   .

INTRODUCTION

Along the west coast of Portugal, in a narrow strip of 20 to 35 Km, more than 70% of the portuguese population is concentrated, and more than 90% of air pollutants in Portugal are released in this region. The atmospheric circulations detailed study as well as the thermal structure in the lower troposphere are very important because they are determinant factors in the air pollutants transport and dispersion conditions. During the dry period, since middle May to middle October, it is frequent the formation of a thermal low pressure over the central region of the Iberian peninsula; in association with this low pressure, a persistent ridge over the north of the peninsula is connected with the Azores anticiclone. Under these conditions, along the west coast of Portugal, especially during the months since June to September, the sea breeze regime is a direct consequence of the differential heating of the surface air, which reaches more than 20 ºC between the land and the sea, in the horizontal scale along the coastal strip.

COASTAL   DATA BANK

In order to become available for studies on sea breezes at surface in the coastal regions of Portugal, a meteorological data bank was organized including hourly values of the surface wind at 11 meteorological stations and air temperature, and relative humidity at surface in 5 of those meteorological stations located in the coastal regions of Portugal, for the period June to September, 1988-91. Relating the same period, the results of the Lisboa radiosoundings below 250 hPa, at 00 and 12 UTC, were archived in a dedicated magnetic file. In order to complete the climatological data bank, we intend to join, as soon as available, the SO2 and O3 hourly values data measured in various areas in the coastal regions of Portugal.

OBSERVATIONAL FIELD CAMPAIGNS

For the study of the lower troposphere structure in the west coast of Portugal (37º to 42ºN), particulary related with the sea breeze characteristics in summer, and the influence of the thermal low pressure located during the summer over the southwest of the Iberian peninsula, four observational campaigns were carried out in the period 1989-1992. During these field campaigns, more than 400 surface observations, 15000 tower observations (5 levels in a 12 m hight tower), 190 air temperature soundings and 200 pillot-balloon observations, 55 tethered-balloon observations and 1250 "doppler" acoustic hourly soundings were made.

Air Pollution Modeling and Its Application X, Edited by S-V. Gryning
and M.M. Millán, Plenum Press, New York, 1994

SEA BREEZE OBSERVATIONAL RESULTS

Most important characteristic of the sea breeze, along the west coast of Portugal in summer, are presented in the Figures 1, 2 and 3. The sea breeze devellops from midmorning and reaches the maximum intensity by the end of the afternoon, when the vertical extent is about 1000 m. The most frequent maximum wind speed observed in the sea breeze layer is 5-7 m/s at the height of 400-600 m and reaching some times 8 m/s. The main feature of the lower troposphere thermal structure is a strong subsidence inversion below 1.5 Km, frequently bellow 800 m, with thickness of 100 to 400 m. Another typical feature is the ocurrence of different layers (3 or more) with air temperature inversions bellow 2 Km, resulting from recirculation phenomena taking place along the coast in sucessive days. During daytime, over land the temperature lapse rate in the surface layer below the first subsidence inversion is nearly dry adiabatic or even superadiabatic in the first 100 m to 200 m during midday as a consequence of the intense heating of the land surface, in contrast to the sea. Horizontal temperature gradients at surface, along the central west coast in summer are very weak in the morning, less than 0.5 ºC/Km, and reaches 5 ºC/10 Km or more in the afternoon.

Aknowledgements. This research was supported by the CEC under the contract nº EV5V-CT91-0050 (Proj. SECAP).

TYPICAL LOWER TROPOSPHERE STRUCTURE OF A SEA BREEZE DAY IN THE WEST COAST OF PORTUGAL, IN SUMMER LEIROSA (WEST COAST), 31st JULY 1989

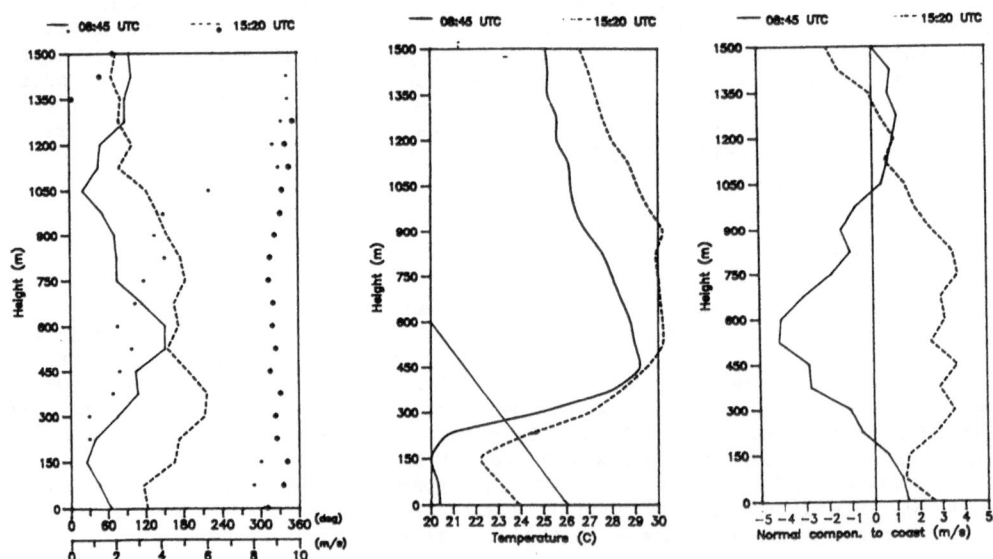

Fig.1-Vertical Wind profile

Fig.2-Vertical temperature profile

Fig.3-Vertical profile of the normal component to the coast

--- Wind speed
: Wind direction

RESEARCHES ON DISPERSION OF THE POLLUTANTS EMITTED INTO THE ATMOSPHERE FROM A NUCLEAR POWER PLANT UNDER RUGGED CONDITIONS IN VIEW OF VALIDATING VARIED MATHEMATICAL MODELS

Traian Pop and Livia-Mihaela Pop

Research and Engineering Institute for Environment
Bucharest 78, Romania

INTRODUCTION

The air concentrations field determination of the exhausted noxes through sources of low emissions height, from the energy systems (NPP,TEPP) or industrial plants situated on a complex roughness soil (hills, valleys, high buildings, and so on) are of major concearn for selecting a mathematical model. In this case both the influence of the buildings in the emission area on the plume rise (Δh) and of the complex soil roughness on the air currents circulations are manifested. These influences are reflected through the local changes of the air dispersion parameters as transport support.

EXPERIMENTING VERSIONS AND RESULTS

The experiments were performed on wind tunnels for the two and three-dimensional study of the flow, equipped with special devices generating a variable intensity wind. The wind direction changing in the tunnel, for the three-dimensional study is carried out by using a rotating modelling platform.The equippement for one Nuclear Unit is : cooling tower, reactor building, additional building, ventilator chimney .

Two-dimensional Flow Study (complex roughness of the soil)

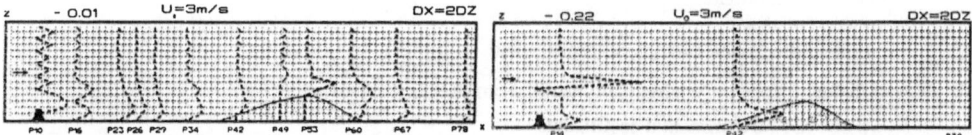

Fig.1 Relative turbulence intensity $(\overline{u'^2})^{1/2}/u$ Fig.2 Coefficients of turbulent dispersion $K_x(x,z)/K_{xo}$ - vertical profiles

Air Pollution Modeling and Its Application X, Edited by S-V. Gryning
and M.M. Millán, Plenum Press, New York, 1994

Three-dimensional Flow Study (flat ground)

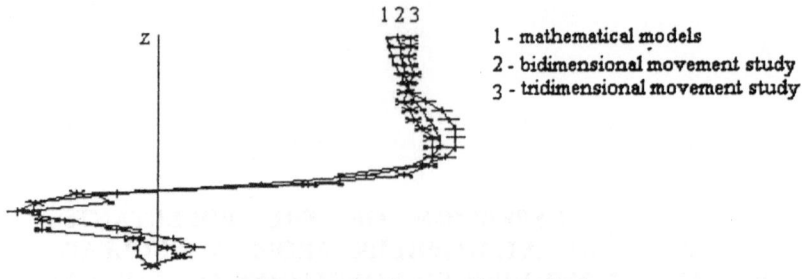

Fig.3 Comparative diagram - longitudinal
component u/u$_o$ of the speed in P10

CONCLUSIONS

The obtained experimental results partially presented in this pape characterise the air currents flow structure in the site area and in that neighbouring a NPP with a single unit. Taking into account that the atmospheric air has an important role in the atmospheric dispersion, as transport support of the pollutants, it results the necessity of knowing the currents flow structure, especially in the areas where the soil roughness produces significant changes on this one.

The obtained results also represent useful information for a correct selection of the mathematical models of the dispersion in the atmosphere of the gas pollutants. For instance in the analysed version (see the two-dimensional study) it is remarked that in the vertical profile P78 the turbulence decreases very much and the dispersion coefficient K_x (x,z) is almost invariable. It results that from this section in the flow direction, the dispersion calculation can be solved with a Gaussian model for long distances from the emission source (evacuation chimney).

In this case the concentrations field up to this profile can be determined with a numerical model that should take into account the local dispersion parameters - the concentrations determined in P78 becoming input data for the Gaussian model. The virtual source coordinates can be determined in this case based on the data of P78.

NOMENCLATURE

u	longitudinal component of the current speed, time averaged
u$_0$	average main current speed in vertical profile upstream the site
u'	speed pulsation
u	average speed in vertical profile in the main current
x	longitude coordinate
z	vertical coordinate
K$_{xo}$	turbulent dispersion coefficient, near the soil

K_x(x,z) turbulent dispersion coefficient in vertical profile : $K_x(x,z)=k((\overline{u'^2})^{1/2}/u)^f$

where : k, f - experimental coefficients.

LONG—TERM AVERAGE AIR POLLUTION OVER CITIES:
OPERATIVE CALCULATION TECHNIQUE FOR ELEVATED SOURCES

Irina A. Krotova, Larissa Melikhova

State Hydrometeorological Institute
of Russian Federation
St.-Petersburg
Russia

Long-term average field of pollutant concentrations in the atmospheric surface layer is one of the criteria of air quality in different urban districts as well as a basis for accepting adequate city layout decisions. An applied technique for calculating such a field formed by urban multiple elevated point source emissions is proposed. Superposition principle for single source average fields is used. The technique takes into account a seasonal joint frequency for wind velocity, wind direction, thermal stratification and precipitation gradations, seasonal changes in emission rates, plume rise depending on seasonal emission characteristics and meteorological gradations, dry deposition of different spesies and washout. An original method of the stratification parameter calculation based on routine meteorological data is proposed. The whole algorithm is implemented as a set of FORTRAN routines.

The operativeness of calculations for a single source is reached by using a simple 4-parameter approximation formula for the concentration distribution over distance from the source. Parameters of the formula are determined by preliminary joint solution of the advection-diffusion equation and atmospheric boundary layer model equations over surface of high roughness. The model allows wind and turbulence vertical variability and provides automatic calculation of mixing height. Peculiarity of the dispersion in calm conditions is taken into account. Arrays of approximation coefficients for different combinations of influencing parameters are contained in the software.

The air and ground pollution fields formed by principal ingredients emitted from 104 stacks of the city Yakutsk was calculated by the proposed technique. Real 8-term routine data over the 15-year period of the urban meteorological station as well as mean seasonal pollutant and heat   emission rates for each source were used as input data. The qualitative correspondence with the snow cover pollution field in the city during 1982 winter period was obtained.

INFLUENCE OF THE TRAFFIC CONDITIONS ON THE AIR QUALITY OF BARCELONA DURING THE OLYMPIC GAMES '92

J.M. Baldasano[1], M. Costa[1], L. Cremades[1], Th. Flassak[2] and M. Wortmann-Vierthaler[2]

[1]Instituto de Tecnología y Modelización Ambiental (ITEMA), Universidad Politécnica de Cataluña, 08220 Terrassa (Barcelona, Spain)
[2]Institut für Technische Thermodynamik, Universität Karlsruhe, 7500 Karlsruhe (Germany)

OVERVIEW

During the last Olympic Games '92 in Barcelona (July 25th - August 9th), 1) the city council of Barcelona adopted special traffic restriction measures in some areas (several streets were closed and there were parking limitations) and 2) some traffic jams were expected between 11[00] and 13[00] LST, and between 19[00] and 21[00] LST in several highways going into the city. The aim of this contribution is to present the results of a numerical simulation to forecast the influence that these traffic conditions could have on the air quality over the city with respect to the normal situation existing before the Olympic Games.

The simulations have been performed for a typical summer non-workable day (24 hours) within a 39x39x6-km^3 region with a horizontal grid spacing of 1 km x 1 km.

Concentration fields for NO_x and O_3 have been calculated for selected scenarios on the Olympic Games '92 in Barcelona by applying the Eulerian dispersion model for reactive species MARS.[1] For the numerical modeling of photochemical oxidant formation the chemical transformation of pollutants should be considered together with their transport in the atmospheric boundary layer. 3D wind fields necessary to run the model were calculated by the non-hydrostatic mesoscale model MEMO[2] and the emission inventory calculated was done by the model EMITEMA-EIM.[3]

This work has been supported by the Servei de Control del Medi de l'Area Metropolitana de Barcelona, IBM España, S.A., and Comisión Interministerial de Ciencia y Tecnología (CICYT) grant no. NAT91-0987.

The following discussion refers to differences in the pollutant concentration levels between the situation forecast during the Olympic Games '92 and the normal situation existing before the Olympics.

Because of the traffic restrictions, surface concentration patterns obtained by simulation show lower ozone concentrations. From 9^{00} to 14^{00} LST, a spread area of lower ozone concentrations appears (minus 5-20 ppbv O_3), which is originated in Barcelona Downtown Area (BDA) and driven towards the sea. Due to the sea breeze the afternoon ozone concentration patterns also show lower ozone concentrations (minus 5-10 ppbv O_3) in the North-western part of the modelling domain and in the Llobregat valley. It can be said that the decrease of emissions due to traffic restrictions leads to decreased ozone concentrations in widely spread areas, whereas significant decrease of NO and NO_2 concentrations (minus 5-20 ppbv), are located almost only near BDA, except in the morning hours (until 10^{00} LST), when differences of NO and NO_2 concentrations are also observed above the sea. During the whole day lower NO_2 concentration levels in BDA are promoted by decreased NO_2 formation which is caused by less NO and O_3 reactions.

As to the influence of traffic jams, it can be said that at the beginning of the traffic jams higher ozone and lower NO concentrations can be seen only in a small area surrounding the concerned highway sections. Noon concentration patterns show lower NO concentrations (minus 5-20 ppbv) in the surroundings of concerned highway sections and higher ozone concentrations (plus 3-10 ppbv) in more widely spread regions. At 13^{00} and 14^{00} LST the sea breeze is stronger and thus transports air with higher ozone concentrations (plus 3-10 ppbv) along the Llobregat valley. Ozone concentrations in Barcelona and above the sea are slightly higher in the 13^{00} LST concentration pattern. These ozone concentration levels are almost fully dilluted by fresh air in the afternoon. The influence of the evening traffic jams can be seen in the concentration patterns between 19^{00} and 21^{00} LST. Again higher ozone concentrations (plus 3-20 ppbv) are observed near the concerned highway sections, but also in the valley of Llobregat and above the sea.

As conclusion, the influences of both lower emissions due to traffic restrictions in BDA and higher VOC emissions caused by traffic jams on some highway sections, are almost spatially separated, except for a small region in BDA where decrease of ozone due to traffic restrictions is almost counterbalanced by an increase of ozone formation due to traffic jams. In the evening lower ozone concentrations in the Llobregat valley because of traffic restrictions are overcompensated by higher ozone concentrations due to the traffic jams.

These preliminary results are now being validated.

REFERENCES

1. N. Moussiopoulos. "Mathematische Modellierung Mesoskaliger Ausbreitung in der Atmosphäre". Fortschrittberichte VDI, Reihe 15, Nr. 64, VDI-Verlag (1989).

2. Th. Flassak. "Ein nicht-Hydrostatisches Mesoskaliges Modell zur Beschreibung der Dynamik der Planetaren Grenzschicht". Fortschrittberichte VDI, Reihe 15, Nr. 74, VDI Verlag (1990).

3. J.M. Baldasano, M. Costa, L. Cremades, Th. Flassak, L. Pardina and M. Wortmann. Inventory of gaseous emissions in the Barcelona geographical area during the Olympic Games, in: "Air Pollution", Zannetti P. et al., ed., Comp. Mech. Pub. and Elsevier (1993).

EVALUATION OF THE THREE-DIMENSIONAL DISTRIBUTION OF DENSE GAS CONCENTRATION ESTIMATED BY NUMERICAL MODELS

Fernando Martín, Inmaculada Palomino and Begoña Aceña

Unidad de Medio Ambiente Convencional, CIEMAT
Avda. Complutense, 22. 28040 Madrid, Spain

INTRODUCTION

Generally, the evaluations of heavy gas models have consisted in checking the performance in predicting the plume centerline concentration at ground level and the width of the plume (see, for example, Hanna et al., 1991), but they have not considered the vertical distribution of dense gas. Experimental studies have shown that vertical distributions of gas inside a dense-than-air cloud is an important factor to describe the dispersion of the cloud impacting an obstacle. In this paper, the three-dimensional distributions of dense gas concentration have been computed from the outputs of the SLAB and HEGADAS models which were run for some field experiments of dense gas release in the atmosphere. The resultant distributions have been statistically compared.

RUNNING THE MODELS

The Burro experiments consisted of releases of LNG onto a pool of water (Koopman et al., 1982) and have been considered as suitable for evaluating models for dispersion of steady-state plume of dense gas. The Burro 3, 5 and 9 tests were selected to run SLAB (Ermak, 1990) and HEGADAS (Colebrander and Puttock, 1988) models. Gas concentration were measured at heights of 1, 3 and 8 m at about 20 gas-sampling stations sited at several distances downwind the emission point. The Burro 3, 5 and 9 data has been averaged over a period of "steady" release conditions. The SLAB and HEGADAS models were run taking into account these averaging periods. The outputs of these models were post-processed by using the similarity profiles assumed by the models in order to obtain estimations of the horizontal distribution of averaged concentrations for three levels (1, 3 and 8 m AGL).

Air Pollution Modeling and Its Application X, Edited by S-V. Gryning
and M.M. Millán, Plenum Press, New York, 1994

STATISTICAL EVALUATION

The statistical evaluation have consisted in the comparison of observed and modeled concentration data paired in space. Although identical pairing in space may be unnecessary in some comparisons, it is necessary to achieve a detailed evaluation of the performance of the models in simulating the three-dimensional distribution of the gas concentration. Five statistical measures has been used: fractional bias (FB), normalized mean square error (NMSE), geometric mean bias (MG) of the ratios between observed and predicted values (R), fraction of data for which R is between 0.5 and 2.0 (FAC2) and the correlation coefficient (CORR) (Hanna et al., 1991).

Including all the data of the three experiments, both models overpredict the concentration but the overprediction is clearly stronger for SLAB (MG equal to 0.43). The differences between observed and modeled data were lesser for HEGADAS. The predictions of HEGADAS were very well correlated with the observations. However, the FAC2 were clearly better for SLAB predictions than for those of HEGADAS. Nevertheless, these results are affected by some pairs of data where the observed concentration is relatively low. We have also evaluated the performance of the models in predicting high concentrations. Both models slightly underpredict the concentrations higher than 0.1% in volume (specially HEGADAS with MG equal to 2.4). However, it was observed these models overpredict the concentration maxima. A clearly different performance of the models was also observed when analyzing the predictions against the distance from the emission point (underprediction for distances shorter than 200 m and overprediction for regions far away the source).

Concerning the horizontal distribution of gas concentration for 1, 3 and 8 m AGL, it was concluded that SLAB and HEGADAS cannot predict very well the concentrations for the upper level. Actually, the upper level is very close to the top of the cloud and the concentration of gas varies very suddenly with the height in this layer. However, the results were clearly better for the heights of 1 and 3 m. HEGADAS have a better performance for the lower level, whereas the SLAB predictions are pretty well correlated with the observations (CORR equal to 0.78) for the height of 3 m.

Although this results may be still inconclusive and it is desirable to extent this evaluation using more field experiments, it seems to be clear that there are some uncertainties in the vertical distribution of gas concentration predicted by the SLAB and HEGADAS models.

REFERENCES

Colenbrander G.W. and Puttock J.S. (1988). Dispersion of Releases of Dense Gas: Development of the HEGADAS model. Shell Research Ltd. Thornton Research Centre, UK.

Ermak D.L. (1990). User's Manual for SLAB: an Atmospheric Dispersion Model for Denser-Than-Air Releases. Lawrence Livermore National Laboratory. CA, USA.

Hanna S.R. et al. (1991). Hazard Response Modeling Uncertainty. Volume II. Sigma Research Corporation. Final Report. MA, USA.

Koopman R.P. et al. (1982). Burro Series Data Report LLNL/NWC 1980 LNG Spills Tests". UCID 19075. Lawrence Livermore National Laboratory. CA, USA.

INFLUENCE OF THE TOPOGRAPHY ON THE LONG-TERM AVERAGE CONCENTRATION COMPUTED BY DISPERSION MODELS

Fernando Martín[1], Inmaculada Palomino[1] and Rosa Salvador[2]

[1]Unidad de Medio Ambiente Convencional, CIEMAT
Avda. Complutense 22, 28040 Madrid, Spain
[2]CEAM, Plaza del Carmen,4, Palacio de Pineda
46003 Valencia, Spain

INTRODUCTION

Although the straight-line Gaussian plume models are still recommended for regulatory use for routine release of nuclear installations, it is well known that they could be inaccurate under certain complex meteorological or/and topographical conditions. The development of computer capabilities, measurement instruments and modeling approaches made possible the use of more realistic models.

THE MODELS

Two climatological dispersion models have been used in order to check how the results of the models are depending on the way in which topography is assumed. One is a straight-line Gaussian model, XOQDOQ (Sangerdorf et al.1982), and the other is a Lagrangian trajectory dispersion model, MESOILT2 (Ramsdell and Burk 1991). A narrow valley in the South of Spain where a radioactive waste treatment facility (El Cabril) is sited was selected for running the models. A complete year of data from August 1988 to July 1989 obtained from seven meteorological towers was used to achieve this study. Whereas homogeneus, uniform wind fields unaffected by the topography are assumed in XOQDOQ, MESOILT2 computes the wind field in a three-dimensional domain in order to estimate the movement of the puffs. MESOILT2 uses the topography in two ways: in the diffusion calculations, and in the adjusment of the wind field to account for mechanical effects. The wind field is estimated by objective analysis techniques. The computations to modify the wind field are based on two adjustment factors for winds at each node of the grid. We have developed an objective method for computing them. Using a finer resolution topography, slopes were computed at every node. It was assumed that the values of the adjustment factor (AD) are an inverse function of the slopes:

Air Pollution Modeling and Its Application X, Edited by S-V. Gryning
and M.M. Millán, Plenum Press, New York, 1994

$$AD_{i,j} = \frac{a}{(\nabla H)^b_{i,j}} \qquad (1)$$

where H is the elevation of the terrain. Gradients of elevations were computed applying a finite difference scheme to a finer grid centered in every (i,j) node. The best results were obtained assuming $a = 0.1$ and $b = 1.0$. The values of AD were then corrected to take into account the channeling effects in the bottom of the valleys. The following criteria were used:

$$\begin{aligned}
AD_{i,j} &= 0.5AD_{i,j} & &\text{if } \max(DE_{l,m}) \geq 150m. \\
AD_{i,j} &= 0.0 & &\text{if } \max(DE_{l,m}) \geq 250m. \qquad (2) \\
DE_{l,m} &= H_{i+l,j+m} - H_{i,j} & &l = -L,...,-1,0,1,...,L \quad ; \quad m = -M,...,-1,0,1,...,M
\end{aligned}$$

where $(L+1)\times(M+1)$ is the number of nodes of the finer topographical grid included in one cell of the 16x16 MESOILT2 grid. A computer code called TERRAN was made to compute the adjustment factors based on the above mentioned criteria. We must point out that the criteria used for computing adjustment factors only consider dynamically distorted flows, i.e., thermal effects have not been taken into account.

RESULTS AND DISCUSSION

Monthly averaged relative concentrations were computed by running XOQDOQ and MESOILT2 programs for August 1988 to July 1989. Two main features are observed from the comparisons between the XOQDOQ and MESOILT2 results. Firstly, the XOQDOQ estimates are one order of magnitude higher than those of MESOILT2. This is more clearly observed close to the source location. This may be due to either the different methods of computing the transport or the MESOILT2 assumption of taking a puff series for continuous releases. Secondly, the distributions computed by MESOILT2 generally follow the topography. The area of higher relative concentrations are along the main valley while the lowest concentrations are behind the ridges and the highest hills. This feature is not observed in XOQDOQ distributions. Contour lines for XOQDOQ are clearly smoother than those for MESOILT2 and the shape of the maximum concentration area does not follow the topographical obstacles.

A more detailed comparison has been made for August 1988 and March 1989. We have analyzed the dependence of the XOQDOQ-MESOILT2 log concentration differences on the distance from the source and on the direction. The differences range from 0.6 to 1.1 for both months. Minima of the differences are between 4 and 6 Km from the source. Non-significant changes were detected when the patterns of log concentration differences were studied for along-and cross-valley directions.

REFERENCES

Ramsdell Jr. J.V and Burk K.W.(1991).MESOILT2, A Lagrangian Trajectory Climatological Dispersion Model. Handford Environmental Dose Reconstruction Project. Pacific Northwest Laboratory, Richland, Washington 99352.

Sagendorf J.F., Goll J.T. and Sandusky W.F.(1982).XOQDOQ: Computer Program for the Meteorological Evaluation of Routine Effluent Release at Nuclear Power Station. NUREG/CR-2919.PNL-4380.

COUPLING THE PHOTOCHEMICAL, EULERIAN TRANSPORT AND 'BIG-LEAF' DEPOSITION MODELLING IN A THREE DIMENSIONAL MESOSCALE CONTEXT

Roberto San José, Luis Rodríguez, Magdalena Palacios and
Javier Moreno

Group of Environmental Software and Modelling
Computer Science School - Technical University of Madrid
Boadilla del Monte - 28660 (Madrid, Spain)

In this paper we show the different modules of an air dispersion modelling study for the Madrid Area. We show the general background of a non-hydrostatic meteorological mesoscale model as a key module for obtaining the wind, temperature and humidity fields by using a k-Theory turbulence model. Because we solve the pressure in the vertical axe, we have to solve explicitly the Helmohltz equation because the complexity of the terrain (in a general sense) request to use the so-called "terrain following coordinates". This approach introduces non linear terms in the first and second order spatial coordinates which complicates the solution of the Helmohltz equation, Pielke (1984). The solar radiation module is also essential for knowing the radiation -in all the wave lengths- at every cell of the three dimensional domain. This radiation -global, diffuse, scattered and net- allows to parameterize the complex photochemical reactions which are undergoing in the atmosphere engined by the solar radiation. This photochemical package works in parallel with the prognostic and diagnostic meteorological model by using different time steps.

The terrain has been treated by applying a land use classification based on fifteen different types and hand made prepared. After this classification a matrix transformation allowed to prepare a seven land use types according with some European mesoscale models (Moussiopoulos et. al. 1991). In addition, altimetry at every surface resolution cell has been implemented.

The most accurate transport model for this type of application is the so-called Eulerian transport model. This approach is based on fixed coordinates and implicit or explicit direct solution of the transport advection-diffusion equation. Sources are establish by using a dynamic emission data base and the deposition or removal terms are implemented as a lower boundary condition of the transport equation.

The deposition or removal term is implemented by using the resistance or inferential method which is based on the Similarity Theory. The canopy resistance takes into account up to seven different effects of absorbing and receiving the pollutant over a

Air Pollution Modeling and Its Application X, Edited by S-V. Gryning
and M.M. Millán, Plenum Press, New York, 1994

canopy cover, Wesely (1989). The parameterization of these effects are based on a correct evaluation of the net and global solar radiation, surface temperature and chemical parameters as the reactivity of the pollutant and the Henry's law constant. Some data is shown from a Field Experiment in the Valladolid (Spain) Area in September, 1991.

ACKNOWLEDGEMENTS

The authors are grateful to Professor N. Nicolas Moussiopoulos for providing most valuable information on the non-hydrostatic mesoscale model MEMO and photochemical model MARS and the code of MEMO which has been used to make initial runs for the Madrid Area.

REFERENCES

Moussiopoulos N., Flassak T and Kessler C. 1991. Modelling of photosmog forma
    tion in Athens, in: 'Air Pollution Modelling and it Application VOL IX'. H. van
    Dop and G. Kallos, ed, Plenum Publishing Corp., New York.
Pielke R.A. 1984. Mesoscale meteorological modelling. Academic Press Inc. (Lon
    don).
Wesely M.L. 1989. Parameterization of surface resistances to gaseous dry deposition
    in regional-scale numerical models. Atmospheric Environment 23, 6, 1293-1304.

TEST OF MESOSCALE NUMERICAL MODEL ON SWISS MIDLANDS

D. Schneiter [1], C. Thurre [2]

[1] Swiss Meteorological Institute, CH-1530 Payerne
[2] Currently at the University of Quebec, Montreal, Canada

Introduction

Mountain and valley systems exert a particularly strong influence on air flow, through channelling effects of the topography, and through the generation of mountain and valley breezes. The purpose of this limited study was to see how far a mesoscale numerical model can simulate regional effects due to the complex topography of the Swiss Midlands, and to compare these results with measurements carried out in the summers of 1990 and 1991 during the two first campaigns of the POLLUMET (POLLution and METeorology) project.

Numerical model DREAMS

The chosen model, called DREAMS (Differential-equation REgional Atmospheric Modelling System) developed at the Swiss Federal Institute of Technology, Lausanne (Beniston 1987, 1991, Beniston et al. 1990) is a three-dimensional atmospheric model which makes use of the finite difference technique to solve a series of time dependent equations for processes acting typically on a regional scale. Dynamic processes are assumed to be hydrostatic. It may be criticized that this approximation may no longer be valid over complex terrain. It has been found that model solutions for relatively stable stratification and low wind velocities, as those encountered in many environmental studies, are in good accord with reality. The vertical velocity field is diagnosed from the mass continuity equation. The anelastic approximation has been retained in order to account for vertical density variations in the atmosphere. The studied region covers a surface of 80 km x 66 km, with a grid interval of 1.4 km. 20 levels are unevenly distributed between ground and 5000 m ASL. The vertical coordinate system is not terrain-following, but remains Cartesian in nature. This allows the study of regions characterized by steep slopes, such as in the Alps. The initialization procedure makes use of interpolated wind profiles, combining aerological soundings and Doppler sodar measurements, and horizontal homogeneous potential temperature and specific humidity fields. The topography altitude is given for each grid point.

Air Pollution Modeling and Its Application X, Edited by S-V. Gryning
and M.M. Millán, Plenum Press, New York, 1994

Results

To compare the numerical results of DREAMS with meteorological measurements obtained during the first two POLLUMET campaigns, undertaken on the Swiss Midlands in the summers of 1990 and 1991, mainly for ozone building study, three different sets of data have been used: surface meteorological measurements, trajectories of constant level balloon (CLB) and atmospheric profiles, obtained by aeroplane over the investigated region.

As expected, comparison with surface stations can give some discrepancies due to the fact that altitude of the model can be different to the real altitude of the surface measurements, especially for mountain stations, which are also under the influence of very local effects, disturbing the air flow. In established windflow, the model simulates the channelling effect of the Jura Chain and gives trajectories corresponding well to the ones measured by CLB. Comparison in the free atmosphere with profiles taken by aeroplane flying up and down through the investigated domain, show that only a part of the complicated stratification of wind direction and wind speed appears in calculated fields.

Conclusions

This first approach of the model shows us that such a tool can be very useful for describing mesoscale channelling effects due to the topography. With the version of DREAMS used for this study, it was difficult to give the model the necessary pressure boundary conditions in order to obtain, after stabilisation, the right profile of wind and thermodynamic values corresponding to the aerological sounding given in the middle of the investigated region. The small amount of computed cases cannot allow a complete validation of the model, but these first results are very promising and encourage us to use such numerical tools for environmental studies, to get a detailed three-dimensional description of wind fields in complex terrain.

Acknowledgements

The authors wish to thank M. Beniston for placing his model DREAMS at our disposal and gratefully acknowledge financial support from the Swiss National Science Foundation, grant No. 21-27828.89.

References

Beniston, M. 1987: A Numerical study of atmospheric pollution over complex terrain in Switzerland. In: "Energy Transformations and Interactions with Atmospheric Processes", M. Beniston and R.A. Pielke, Eds., D.Reidel Publishing Company, pp. 75 - 96

Beniston, M., J.P. Wolf, M. Beniston-Rebetez, H.J. Kölsch, P. Rairoux, and L. Wöste, 1990: Use of Lidar measurements and numerical models in air pollution research. *J. Geophys.Res.*, **95**,D7, 9879-9894

Beniston, M., 1991: DREAMS. A numerical atmospheric modelling system for regional climate and air quality studies. User's guide to theory and applications on supercomputers. Swiss National Climate Program (ProClim). P.O. Box 7613, CH 3000 Bern, Switzerland.

Schneiter, D., C. Thurre, P. Jeannet, 1992: Projet VMTP. Campagnes de mesure et vérification d'un modèle numérique pour le transport de pollution atmosphérique. Institut suisse de météorologie, Les Invuardes, CH 1530 Payerne, Switzerland

APPLICATION OF THE ABATEMENT STRATEGIES ASSESSMENT

MODEL, ASAM, TO ABATEMENT OF SO2 EMISSIONS IN EUROPE

Helen M. ApSimon and Rachel F. Warren

Imperial College Centre for Environmental Technology
48 Princes Gardens, London SW7 2PE, United Kingdom

In preparing a new international protocol on sulphur dioxide emissions under the auspices of the UN Economic Commission for Europe, the Task Force on Integrated Assessment Modelling has been investigating scenarios for future abatement. The ASAM model (see below) has been used together with 2 other models (RAINS and CASM) to derive emission reductions required in different countries to reach given objectives. The calculations are based on estimates of atmospheric transport between different European sources and receptors derived with the EMEP model of the Norwegian Meteorological Institute.

The work takes account of critical loads as the levels of annual deposition of sulphur sustainable without causing harm. However it is not feasible to reduce current deposition to these levels with currently available technology. Hence intermediate target loads have been proposed- in particular the "60% gap closure" aiming to reduce the current exceedance by 60%. The corresponding emission reductions required across Europe to achieve this as estimated by ASAM are shown in figure 1. The resulting deposition still exceeds the critical loads over significant areas of Europe.

The ASAM model

The Abatement Strategies Assessment Model, ASAM, combines projected emissions across Europe ,atmospheric transport and deposition (based on the EMEP model), critical loads or target load, and control measures and costs (results here based on costs by IIASA). The ASAM model generally achieves the target loads with lower emission reductions than CASM or RAINS because it makes use of selective reductions geographically within countries. This is especially significant for larger countries with localised sensitive areas requiring lower deposition.

In contrast to the other models the ASAM model takes a step-wise approach, scanning emissions in each grid-square in each step to select which can be reduced with the greatest ratio of environmental benefit to cost. A sequence of abatement measures is derived in order of priority, converging towards the desired goal. Since an excess gram of sulphur will cause more damage on a sensitive area, ASAM can incorporate weighting functions to reflect this. It is hoped to extend this in future to reflect the potential damage avoided by abatement more specifically and to address nitrogen species too.

The results from ASAM illustrate how the environmental improvement decreases once the most effective steps have been taken to reduce emissions- see figure 2 where ASAM has been applied to an alternative strategy aiming straight towards the 5%ile critical loads. This strategy achieves a larger area brought under protection for the same cost than the 60% gap closure scenario described above.We can also deduce the increased investment per hectare of land brought under protection at different stages of overall expenditure.

Air Pollution Modeling and Its Application X, Edited by S-V. Gryning
and M.M. Millán, Plenum Press, New York, 1994

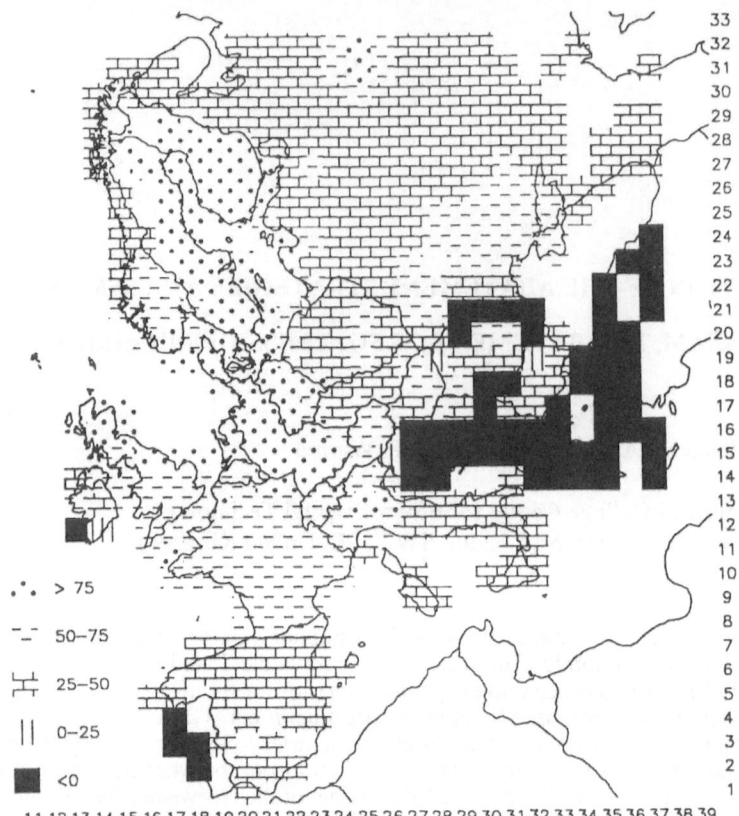

Figure 1. SO2 emission reductions required to achieve the 60% gap
closure target load (as % of emissions in 1980)

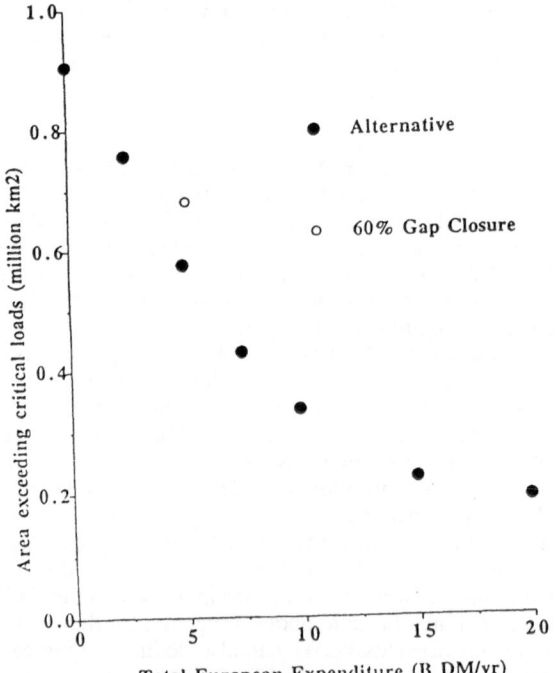

Figure 2.

REVIEW AND EVALUATION OF THE RATCHET MODEL USED FOR THE HANFORD DOSE RECONSTRUCTION PROJECT

C. J. Nappo, W. R. Pendergrass, and R. M. Eckman

Atmospheric Turbulence and Diffusion Division
Air Resources Laboratory
National Oceanic and Atmospheric Administration
Oak Ridge, Tennessee 47830

The Regional Atmospheric Transport Code for Hanford Emission Tracking (RATCHET) long-term, trajectory puff air dispersion computer model was developed by the Pacific Northwest Laboratory as part of the Hanford Environmental Dose Reconstruction (HEDR) project. The purpose of HEDR is to estimate monthly-averaged radiation doses that individuals could have received from operations at Hanford since 1945. The U.S. Centers for Disease Control (CDC) has responsibility to insure that all tools used in environmental radiation dose reconstruction performed under its direction be technically sound and defensible. In order to address concerns about the validity of atmospheric transport and deposition models, the CDC coordinated a through technical review of the RATCHET model by an expert panel of air-pollution scientists. It has been suggested that a statistical (i.e., a sector-averaged Gaussian plume) model might be better suited for calculating monthly-averaged air concentrations than the RATCHET trajectory-puff model. In order to check this suggestion, we implemented a version of the widely used AIRDOSE-EPA model, and compared this model with RATCHET for the month of January 1945. Hourly meteorological input data for AIRDOS for January 1945 were taken from the RATCHET input meteorological file. We used the winds from the 200 ft level of the Hanford tower because these correspond closely to the winds at the estimated plume height and at the source location. Figure 1 shows the contours of monthly-averaged concentration (Ci m^{-3}) calculated by the RATCHET model, and Figure 2 shows the results from the AIRDOSE model. The highest concentration contours extend to the southeast for both models; however, beyond about 100 km from the source, RATCHET predicts that the isopleths turn to the northeast, but no similar swing is predicted by the AIRDOSE model. This comparison helps to highlight the critical nature of dose reconstruction over long averaging times. Over the complex terrain of eastern Washington, a single meteorological station cannot represent the effects of varying topography and changing meteorological conditions. It is concluded that sector-averaged models such as AIRDOSE are not suitable for dose reconstruction calculations over regions of complex terrain.

Air Pollution Modeling and Its Application X, Edited by S-V. Gryning
and M.M. Millán, Plenum Press, New York, 1994

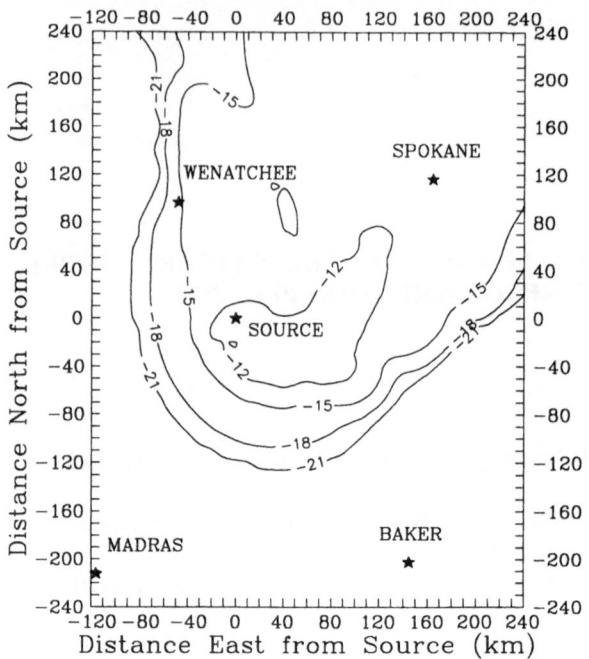

Figure 1. RATCHET concentration field for January 1945.

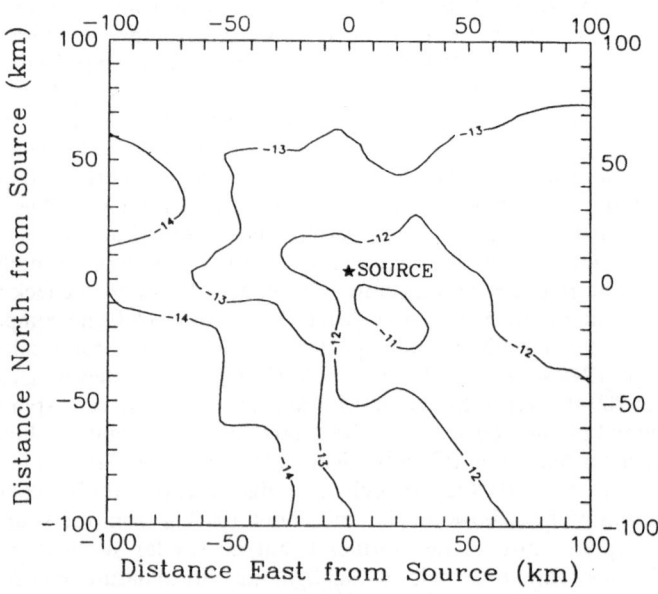

Figure 2. AIRDOSE concentration field for January 1945.

REFERENCES

Ramsdell J.V., and K.W. Burk, 1992: Regional Transport Code for Hanford Emission Tracking (RATCHET). PNL-8003 HEDR, Pacific Northwest Laboratory, Richland, WA 99352.

LINEAR ADVECTION SCHEME FOR AIR POLLUTION TRANSPORT

MODELLING FROM INDIVIDUAL SOURCES

Margarita Pekar

Meteorological Sinthesizing Center-East of EMEP
Moscow

Presented new technique of advection numerical modelling may be attributed to methods simulating continuous flow by the motion of ensemble of discrete elements. To these methods belong "particles-in-cell" techniques,[1] Egan-Mahoney's,[2] "large particles" by Belotserckovsky[3]. The scheme considered differs from the indicated methods in the following:

1. Pollution is imitated by discrete extended mass elements, which widths are 2 times higher than grid interval. In the course of the motion elements are imposed on each other.

2. The ensemble of these elements unambiguously corresponds to continuous distribution of grid elements two times narrower which result from conditions of first three moments conservation (for one - dimensional variant) of the elements and defines current mass distribution with the grid.

3. "Summing" of mobile elements which mass centers hit one cell is a transformation of these elements distribution again with conservation of the first three moments. In case when elements number is not growing (instantaneous source), when computer resources provide operations with full array of elements the "summing" procedure is omitted and in this case the algorithm is particularly simple, effective and its precision is increased.

Consider on one-dimensional grid with interval Δx the motion of an element of mass M_i with width $2\Delta x$ and initial co-ordinate of the mass centre 0 (index i means the allocation of mass centre to the i-th grid cell). During time interval τ element M_i is shifted at the distance $u_i \tau$ (u_i - advection velocity) and now its center has relative co-ordinate $X_i = u_i \tau / \Delta x$. Following distribution M_i with co-ordinate X_i with grid cells providing conservation of mass, center co-ordinates, dispersion is possible:

$$\mu_i^{i-1} = 0.5 - X_i + 0.5 X_i^2,$$
$$\mu_i^i = 0.5 + X_i - X_i^2,$$
$$\mu_i^{i+1} = 0.5 X_i^2,$$

(1)

where weights $\mu_i^k = \Delta m_k / M_i$, $k = i-1, i, i+1$, Δm_k - contribution of M_i in k-cell.

The distribution of all elements M_i according to weights (1) and summing in cells determine complete distribution of the mass at a given step:

$$m_k = \sum_i M_i \mu_i^k, \quad k = i-1, i, i+1$$

(2)

The scheme is readily extended for two dimensional variant:

$$m_{kl} = M_{ij} \mu_i^k v_j^l; \quad k = i-1, i, i+1; \quad l = j-1, j, j+1$$

(3)

where: $v_j^l = v_j^l (Y_j)$ - similar to μ_i^k weights for Y - co-ordinate.

Total mass distribution is formed by summing the contributions from all M_{ij} elements. It's one part of the algorithm. The following part is summing of mobile elements on a cell (if required).

Air Pollution Modeling and Its Application X, Edited by S-V. Gryning
and M.M. Millán, Plenum Press, New York, 1994

At each time step elements which mass centers hit a given cell form the distribution characterized by 6 moments (assume that n elements hit the k-th cell):

$$M_k = \sum_n M_n, \qquad\qquad M_k \overline{X_k^2} = \sum_n M_n X_n^2,$$

$$M_k \overline{X_k} = \sum_n M_n X_n, \qquad M_k \overline{Y_k^2} = \sum_n M_n Y_n^2, \qquad (4)$$

$$M_k \overline{Y_k} = \sum_n M_n Y_n, \qquad M_k (\overline{XY})_k = \sum_n M_n X_n Y_n,$$

which can be substituted with the conservation of all moments (4) by the distribution of two elements with equal mass $M_{k1} = M_{k2} = 0.5 M_k$ with co-ordinates:

$$X_{k1} = \overline{X}_k + \Delta X_k, \qquad\qquad X_{k2} = \overline{X}_k - \Delta X_k, \qquad (5)$$
$$Y_{k1} = \overline{Y}_k + \rho \Delta Y_k, \qquad\qquad Y_{k2} = \overline{Y}_k - \rho \Delta Y_k,$$

where: $\Delta X_k = (\overline{X_k^2} - \overline{X}_k^2)^{\frac{1}{2}},$ $\Delta Y_k = (\overline{Y_k^2} - \overline{Y}_k^2)^{\frac{1}{2}},$ $\rho = sign((\overline{XY})_k - \overline{X}_k \overline{Y}_k).$ (6)

To demonstrate the scheme operation figures 1-2 give some testing results.

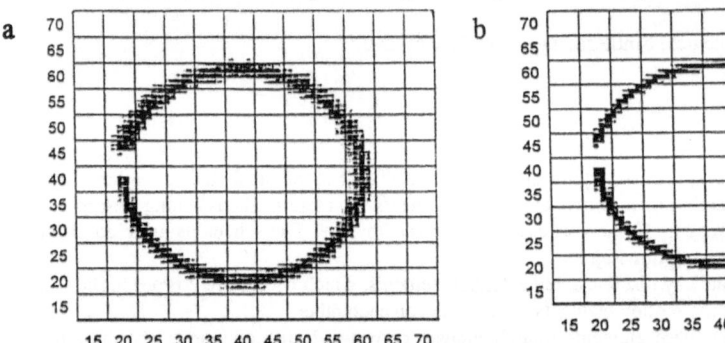

Figure 1. Test with continuous source in the field of rotation velocities. Full revolution is
equal to 400 steps. Position on the 385-th step:
a) without allowance for correlation moment,
b) the same, but with allowance for correlation moment.

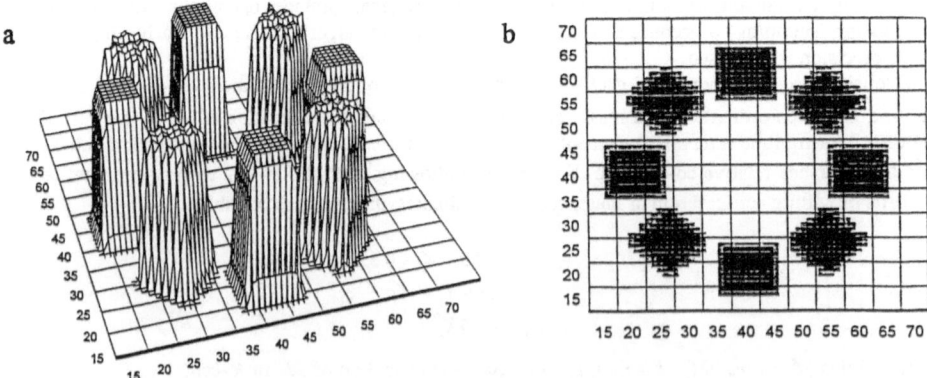

Figure 2. Rotation of square distribution around axis (40,40) with full revolution
800 steps. Initial position of square center is in cell (20,40).
Positions are given with time interval of 100 steps:
a) three - dimensional picture,   b) the same, in plane.

REFERENCES

1. R. Lange, ADPIC - A three dimensional particle-in-cell model for the dispersal of atmospheric pollutants and its comparison to regional tracer studies, *J. of Appl. Meteor.* 17:230(1978).
2. B.A. Egan, J.R. Mahoney, Numerical modelling of advection and diffusion of urban area source pollutants, *J. of Appl. Meteor.* 11:312(1972).
3. O.M. Belotserckovsky, Yu.N. Davidov, Non-stationary techniques of "large-particles" for gasedynamic calculations, *J. of comp. math. and math. physic.* 11:37(1980).

NEURAL NETWORKS PREDICT POLLUTION

Primož Mlakar,[1] Marija Božnar,[1] Martin Lesjak[2]

[1]"Jožef Stefan" Institute
[2]AMES d.o.o.
Jamova 39
Ljubljana, Slovenia

INTRODUCTION

Air pollution is a very big problem in Slovenia. The greatest pollutants are two big thermal power plants (TPP-s) which are placed in the valleys near the coal mines. The coal has a very big percentage of the sulphur (up to 2%) and the TPP-s do not have wet desulphurisation. The results of this are the episodes of very high air pollution in the vicinity of the TPP-s. The local government intends to omit this episodes by forcing the TPP-s to reduce the power significantly (from 30% up to 70%). Such quick reduction of the TPP power is certainly an economical and technological problem.

These are the reasons, why the TPP-s want to have a reliable short-term (prediction of half an hour average values up to 2 hours in advance) air pollution prediction models.

Since the classical statistical prediction models perform much better for long-term predictions than for short-term ones, we tried to find a better solution.

NEURAL NETWORKS-BASED METHOD OF PREDICTION

Neural networks[1] are mathematical structures that mock the behaviour of the human brain. The basic element of the neural network is a neuron that receives information from the input links, proceeds it and send the result to the output links. Usually the neurons are organised in several layers. The input layer neurons take the information from the outside world and the output layer neurons give us the results. Different topologies (number of neurons and their interconnections) are known in the literature. Basically the topology determines the capabilities of the network.

Neural networks are used in many fields of science: speech recognition, computer vision, process control, financial forecasting e.t.c..

We think that neural networks are also suitable for air pollution prediction[2] in particular because of their capability of learning the implicit rules of non-linear multidimensional systems from training set of patterns (pattern is a couple composed of input vector and its corresponding output vector from a system).

We should solve a lot of problems while implementing neural networks to the air pollution prediction. First we must choose the proper topology of the neural network. We decided that the multilayer perceptron is the most suitable one. Then we should determine the environmental parameters that describe the particular pollution situation (input and output features). It is very important that the feature can be any type of information (for instance: pressure, temperature, time, whether forecasts ...). Choosing the proper learning method together with selecting large enough set of learning patterns is also very important.

APPLICATION OF THE NEURAL NETWORK-BASED METHOD OF PREDICTION TO THE PROBLEM OF THE SO$_2$ POLLUTION AROUND THE SOSTANJ TPP

The Sostanj TPP (700 MW) is the biggest thermal power plant in Slovenia. It has a modern automatic environmental informational system[3] (EIS) (6 remote stations in the surroundings and an emission station). The TPP is placed in a very complex terrain.

The data base obtained by the EIS consist of all basic meteorological parameters, gas pollution measurements and emission data (half hour average values). The data have been collected since January 1990. This huge data base is the basis for our research work.

We started with half hour advance predictions of SO$_2$ concentrations on all six remote stations. We get the best results at the station which has pollution during the thermal inversion situations and the worst results at the station where the pollution is a result of direct wind from the direction of the TPP stacks (it is very difficult to predict changeable wind direction).   We concentrate our research work on the first type of pollution.

We have put a lot of effort in finding the proper selection of the input features from the available data base. We usually use beetwen 20 to 30 input features to the three layer perceptron neural network and one output feature (SO$_2$ concentration after half an hour at particular station). The training and the testing set of patterns usually consists of data collected over few weeks.

The results are very encouraging, because the neural network was capable of predicting the shape and the peak of the pollution event, although it was not able to predict the exact value of the concentrations (usually the error was within the factor 4).

We have also done the preliminary comparison with a statistical model (Cyclo Stationary Auto Regressive Predictor). This results are also encouraging.

CONCLUSIONS

The proposed method seems to be worth further research. It is especially useful for the situations where we do not have the explicit knowledge about the pollution phenomena. This include the modelling in very complex terrain and pollution in the urban areas where it is very difficult to define the emissions. We tested the method on the SO$_2$ pollution, but it can also be used for other gas pollutants.

REFERENCES

1. J. Lavrence. "Introduction to Neural Networks," California Scientific Software, Grass Valley (1991)
2. M. Božnar, M. Lesjak, P. Mlakar. Neural network-based method for short-term predictions of ambient SO$_2$ concentrations in highly polluted industrial areas of complex terrain, *Atmospheric Environment*, 27B:221(1993).
3. M. Lesjak, B. Diallo, P. Mlakar, Z. Rupnik, J. Snajder and B. Paradiz. Computerized ecological monitoring system for thermal power plant Sostanj, *Man and his ecosistem, Proc. 8th Word Clean Air Congress, Hague, The Netherlands*, 3:31(1989).

SPECIAL SESSION:

ATHENIAN PHOTOCHEMICAL SMOG INTERCOMPARISON OF
SIMULATIONS (APSIS)

RESULTS OF NESTED WIND FLOW SIMULATIONS FOR THE ATHENS BASIN USING THE NON-HYDROSTATIC MODEL MEMO

Rainer Kunz[1] and Nicolas Moussiopoulos[2]

[1]Institut für Technische Thermodynamik
Universität Karlsruhe, Kaiserstraße 12
D-76128 Karlsruhe, Germany

[2]Laboratory of Heat Transfer and Environmental Engineering
Aristotle University Thessaloniki, Box 483
GR-54006 Thessaloniki, Greece

INTRODUCTION

In the frame of the APSIS project a nested version of the nonhydrostatic mesoscale model MEMO was used to simulate the wind flow in Athens. The nested model version was applied as former applications of the non-nested version of MEMO for the Greater Athens Area proved lateral boundary conditions to falsify the wind field close to the boundaries [Moussiopoulos et al. 1993].

In the non-nested simulations with MEMO only a single vertical sounding each 12 hours was available to describe meteorological conditions at the lateral boundaries. In order to numerically provide boundary values at a higher temporal and spatial resolution, an one-way nesting technique was applied: Model results of an additional simulation with MEMO using an expanded model domain ('coarse grid') were taken to generate boundary conditions for the simulation on the APSIS domain ('fine grid').

CASE SPECIFICATION

The coarse grid simulation was performed on a 360 x 360 km² domain covering most of Southern Greece at a resolution of 5 x 5 km². The fine grid domain, which essentially coincides with the Attica Peninsula, streched over 72 x 72 km² at a horizontal resolution of 2 x 2 km².

In the present study the simulation was performed for June 10, 1987. On this day weak pressure gradients were prevailing over Southern Greece thus leading to a typical land/sea breeze situation in the Athens Basin. For this day detailed meteorological measurements are available [Asimakopoulos et al. 1992].

Air Pollution Modeling and Its Application X, Edited by S-V. Gryning
and M.M. Millán, Plenum Press, New York, 1994

RESULTS

In the remainder of this article only a brief discussion of the fine grid results is given. More details about coarse and fine grid results can be found elsewhere [Kunz and Moussiopoulos 1993].

The model results reveal the typical diurnal cycle of a land/sea breeze circulation in the Greater Athens Area. The nocturnal flow pattern shows a land breeze which is partly enhanced by katabatic winds from the mountain slopes. In the morning the land breeze calms down and is replaced by the onset of the sea breeze, the latter being fully established about noon. Intense turbulent mixing leads to a rather homogeneous wind field in the afternoon. In the evening the sea breeze starts to deform and is finally replaced by the land breeze over the Petalic Gulf, wheras over the Saronic Gulf still an onshore air motion occurs due to the synoptic forcing.

Predictions were compared with ground based wind measurements at 6 locations within the fine grid domain. As an example, Figure 1 shows the predicted and observed wind velocities at the Refinery to the West of the Athens Basin. For all locations the agreement between measurements and model results can be classified as very good.

As the most important feature of the nested simulations with MEMO, the lateral boundaries have no falsifying influence on the resulting wind fields [Kunz and Moussiopoulos 1993].

In summary the applied one-way nesting technique appears to describe successfully the wind field in the Greater Athens Area. The predicted wind fields can therefore be regarded as a good basis for air pollution studies.

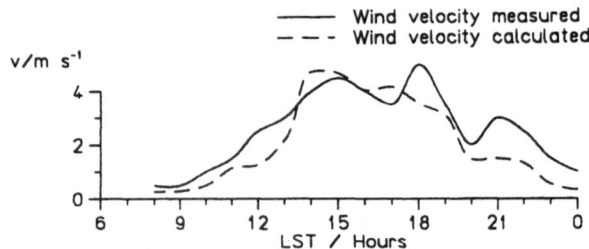

Figure 1. Predicted and observed wind velocities at the Refinery to the West of Athens on June 10, 1987

REFERENCES

Asimakopoulos, D. N., Deligiorgi, D. G., Drakopoulos, C., Helmis, C. G., Kokkori, K., Lalas, D. P., Sikiolis, D.,Varotsos, C., 1992, An Experimental Study of Nighttime Air-Pollutant Transport over Complex Terrain in Athens, *Atmosph. Environ.* 26B, 59:71.

Kunz, R., Moussiopoulos, N., 1993, Simulations of the Wind Field in Athens Using Refined Boundary Conditions, submitted to *Atmospheric Environment*.

Moussiopoulos, N., Flassak, Th., Berlowitz, D., Sahm, P., 1993, Simulations of the Wind Field in Athens With the Nonhydrostatic Mesoscale Model MEMO, *Environmental Software* 8, 29:42.

INFLUENCE OF THE SEA BREEZE ON THE AIR POLLUTION OVER THE ATTICA PENINSULA

K. Nester

Institut für Meteorologie und Klimaforschung
Kernforschungszentrum Karlsruhe/Universität Karlsruhe

Introduction

In the frame of the APSIS project simulations of the dispersion of air pollutants over the Attica Peninsula are carried out for May 25, 1990. This day is characterized by clear sky conditions and a general flow from Northwest. Such conditions support the development of a sea breeze around the Peninsula, which is especially pronounced in the basin of Athens and the neighbouring coastal zone. The flow and turbulence fields are simulated with the the non hydrostatic model KAMM (Adrian and Fiedler, 1991). Based on these fields the dispersion of the air pollutants are calculated with the DRAIS model (Nester and Fiedler, 1992). The simpler chemical mechanism RADM1 was selected because only measurements of NO, NO_2, SO_2, CO and O_3 are available, which are well predicted by the RADM1 mechanism.

Behaviour of NO, NO_2, SO_2, and CO

The general behaviour of the simulated NO, NO_2, SO_2 and CO concentrations are quite similar. During the night the week land breeze transports these species from the city of Athens to the sea. In the morning the wind changes from land to sea breeze and the pollutants are partly transported back into the city of Athens. Before 11 o'clock the concentrations in the city usually reach their first maxima. After that time the concentrations decrease strongly. Increasing vertical mixing and higher wind speeds in the fully developed sea breeze cause this dilution. During daytime the concentrations don't vary very much. In the evening an increase of the concentrations is simulated, because of lower wind speed and lower turbulence in the boundary layer. The second maximum concentrations are reached about midnight. With the beginning of the land breeze the air pollution decreases again. The simulated diurnal cycle is confirmed by the observations. But in the morning the peak concentrations are usually underestimated by the model, whereas the opposite behaviour is found in the night.

Behaviour of Ozone

The simulated behaviour of ozone is completely different to that of the other species. In the industrialized area at the bay of Eleusis a puff of high ozone concentration is formed in the morning hours. This puff moves in the developing sea breeze over the peninsula. During this transport the puff becomes larger and the peak ozone concentration increases. About noon this puff combines with the ozone plume of the city of Athens (see fig.1), which is formed north-east of Athens by the city traffic emissions. In the afternoon the ozone concentration in the plume decreases and moves to the sea on the opposite side of the peninsula. The comparison between the simulated and measured maximum concentrations in the city of Athens show a satisfying agreement. During night time the model calculates no ozon in the city, whereas the measurements show low but non zero values.

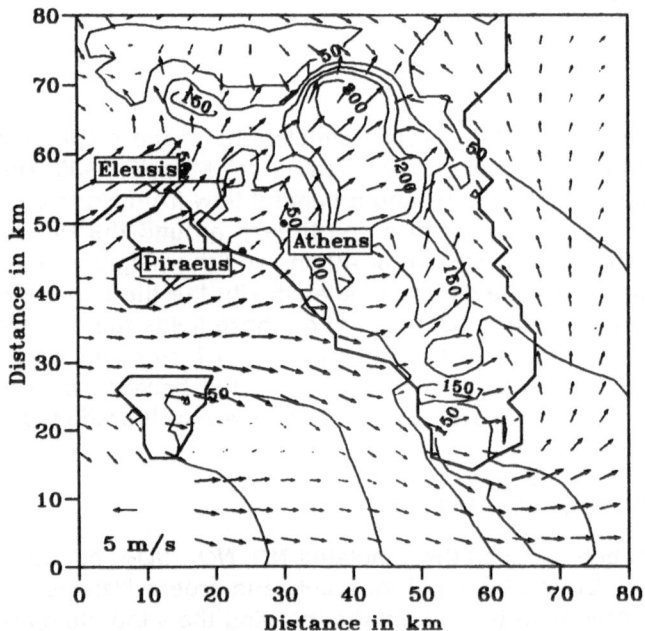

Figure 1. Simulated ground level ozone concentration($\mu g/m^3$) at 12 UTC

References

Adrian,A. and Fiedler,F.,1991,Simulation of unstationary wind and temperature fields over complex terrain and comparison with observations, Beitr. Phys.Atmosph. 27:48
Nester,K. and Fiedler, F.,1992, Modeling of the diurnal variation of air pollutants in a mesoscale area, Proceedings of the 9th Clean Air Congress, Montreal, August 30-September 4, 1992

NUMERICAL SIMULATION OF THE FLOW REGIME IN ATHENS AREA

D. Melas, I. Ziomas, and C. Zerefos

Laboratory of Atmospheric Physics, Physics Department
University of Thessaloniki, 54006 Thessaloniki, Greece

INTRODUCTION

The flow regime in Greater Athens Area (GAA) is numerically investigated by a mesoscale, higher-order turbulence closure model. The model is three-dimensional, hydrostatic, with a terrain following coordinate system and it solves the mean quantities as well as the turbulent energy equation prognostically. This enables the explicit treatment of horizontal inhomogeneity and unsteadiness, which are key features in the area under study. It has been developed at the Department of Meteorology in Uppsala (MIUU) (Enger, 1986).

The numerical investigation was performed within the frame of the Athenian Photochemical Smog Intercomparison of Simulations (APSIS) and the day chosen for simulation is June 10, 1987. The values of potential temperature at the lower boundary were prescribed according to the observations (Asimakopoulos et al., 1992). The diurnal change of the potential temperature was represented by a sinusoidal wave. The temperature of the sea surface was set to 20 °C.

A detailed description of the Athens area is found in Melas and Enger (1993).

RESULTS

During the simulation day of the 10th of June, 1987, Greece was under the influence of a high pressure system which covered the east part of the Mediterranean. The pressure gradients were rather weak and the associated synoptic wind was also week (~ 5.5 ms^{-1}) from the north (Asimakopoulos et al., 1992).

The simulated wind fields are quite variable in time as well in space. In the early morning hours, the flow is NW over the sea, while the winds over Athens basin are light with variable directions (mostly within the northern sector). This is in agreement with the observations reported by Asimakopoulos et al. (1992). During daytime, the situation changes drastically. Thermally induced local circulations start to develop and the flow regime in GAA is more complicated. At 14.00 local time (Fig. 1), the flow over the sea is generally from the W with speeds ~ 6 ms^{-1} but there is a splitting over Saronikos Gulf and part of the air moves over the Athens basin. This is due to the development of the sea

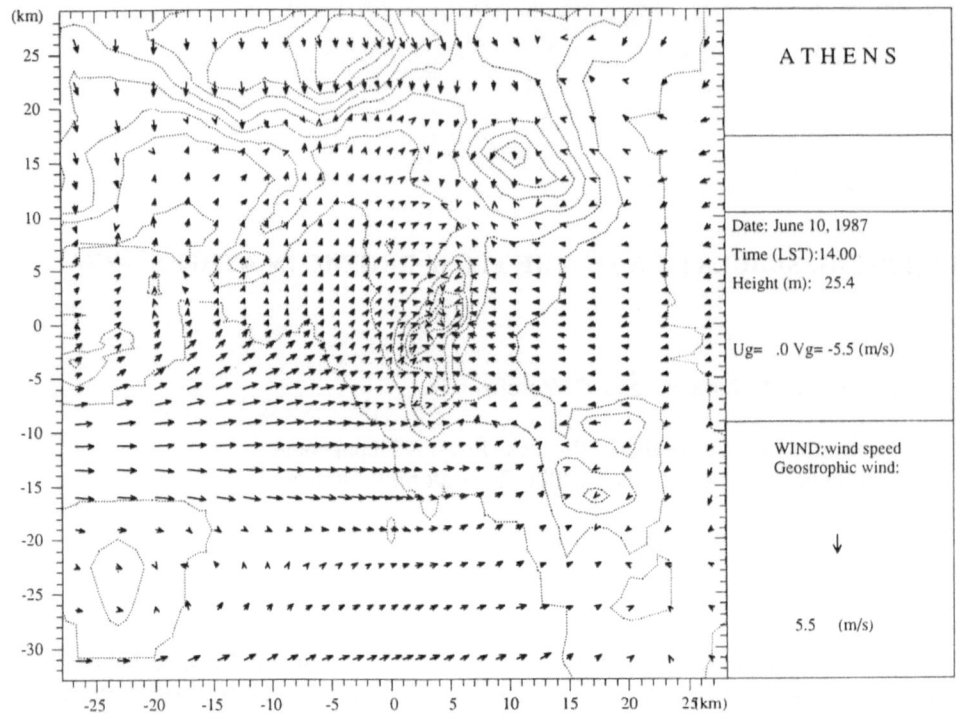

Figure 1. Simulated wind field at 25 m AGL at 14.00 LST.

breeze circulation which diverts part of the flow towards Athens basin. The winds over the urban area of Athens at 25 m AGL are coming from SW and they are rather light, 2-3 ms^{-1}. This notable wind deceleration downwind the shoreline is resulting from the high roughness of the city and has large consequences on pollutant dispersion in the area.

There are two more sea-breeze systems which can be readily identified from the simulation results. One is formed in the east part of Attica peninsula, at Mesogia plain, while the other is generated in the west side of mountain Aigaleo, at Thriassion plain.

Another interesting feature of the flow field is the blocking of low level winds by the mountains. A large portion of the air is forced to move around the mountains, through the gaps between them, while another portion flows over the mountains. Air masses from the industrialized Thriassion plain are entering the Athens area mainly through the gap between Aigaleo and Parnitha.

References

Asimakopoulos, D., Deligiorgi, D., Drakopoulos, C., Helmis, C. Kokkori, K., Lalas, D., Sikiotis, D., and Varotsos, C.: 1992, An experimental study of nighttime air-pollutant transport over complex terrain in Athens, *Atmos. Environ.*, **26B**, 59:71.

Enger, L.: 1986, A higher order closure model applied to dispersion in a convective PBL, *Atmos. Environ.*, **20**, 879:894.

Melas, D. and Enger, L., 1993, A numerical study of flow in Athens area using the MIUU model, *Environmental Software*, **8**, 55:63.

PREDICTION OF WIND FLOW AND OZONE FORMATION IN ATHENS
FOR THE APSIS B₂ EXERCISE USING THE EUMAC ZOOMING MODEL

Nicolas Moussiopoulos and Peter Sahm

Laboratory of Heat Transfer and Environmental Engineering
Aristotle University, 54006 Thessaloniki, Greece

INTRODUCTION

The APSIS activity, which was initiated in October 1991 in the frame of the EURO-TRAC subproject EUMAC, aims to intercompare model simulations of photosmog formation in Athens. In addition to more experience with regard to model evaluation, obvious benefits from such an intercomparison will be the excessive testing of mesoscale models and the determination of advantages and weaknesses of individual model concepts.

Among the exercises defined within APSIS, exercise B_2 refers to pollutant transport and transformation simulations. As a suitable time period to be simulated, the photochemical smog episode of May 25, 1990 was selected. A discussion of the synoptic conditions prevailing on that day can be found elsewhere (Moussiopoulos, 1993a).

MODEL APPLICATION AND RESULTS

Ozone formation in Athens on May 25, 1990 was analysed with the EUMAC Zooming Model (EZM), i.e. the photochemical dispersion model MARS using results of nested wind flow simulations with the nonhydrostatic mesoscale model MEMO. Descriptions of both models are given by Moussiopoulos (1993b).

Emission rates were obtained for all relevant species on the basis of available emission data. Chemical transformations were modelled on the basis of KOREM, a modified version of the reaction mechanism of Bottenheim and Strausz (1982). The numerical simulations were performed on a 72×72 km^2 grid at a resolution of 2 km over a period of three days. The results for the third day were proved to be independent of the assumed initial concentrations. In the vertical direction nineteen non-equidistantly distributed layers were used with a minimum spacing of 20 m.

Detailed simulation results as well as an extensive comparison of the predictions with available observations are given by Moussiopoulos and Sahm (1993). As an example, Fig. 1 illustrates predicted and observed diurnal variations of the NO, NO_2 and ozone concentrations at the Athinas measuring station in the centre of Athens. Apparently, the EZM proves capable reproducing the major mechanisms leading to the dispersion and chemical transformation of air pollutants in Athens. Yet, in spite of the satisfactory general agreement, several deviations between prediction and observation can be detected, the reason for which remains to be analysed. An example is the delayed decrease of the NO and NO_2 concentrations during the day, which might be caused by an underestimation of the daytime turbulent diffusive transport. It is probable that the ongoing model evaluation effort in the APSIS activity will elucidate the major sources for discrepancies between observation and calculation with all models involved in the activity.

Air Pollution Modeling and Its Application X, Edited by S-V. Gryning
and M.M. Millán, Plenum Press, New York, 1994

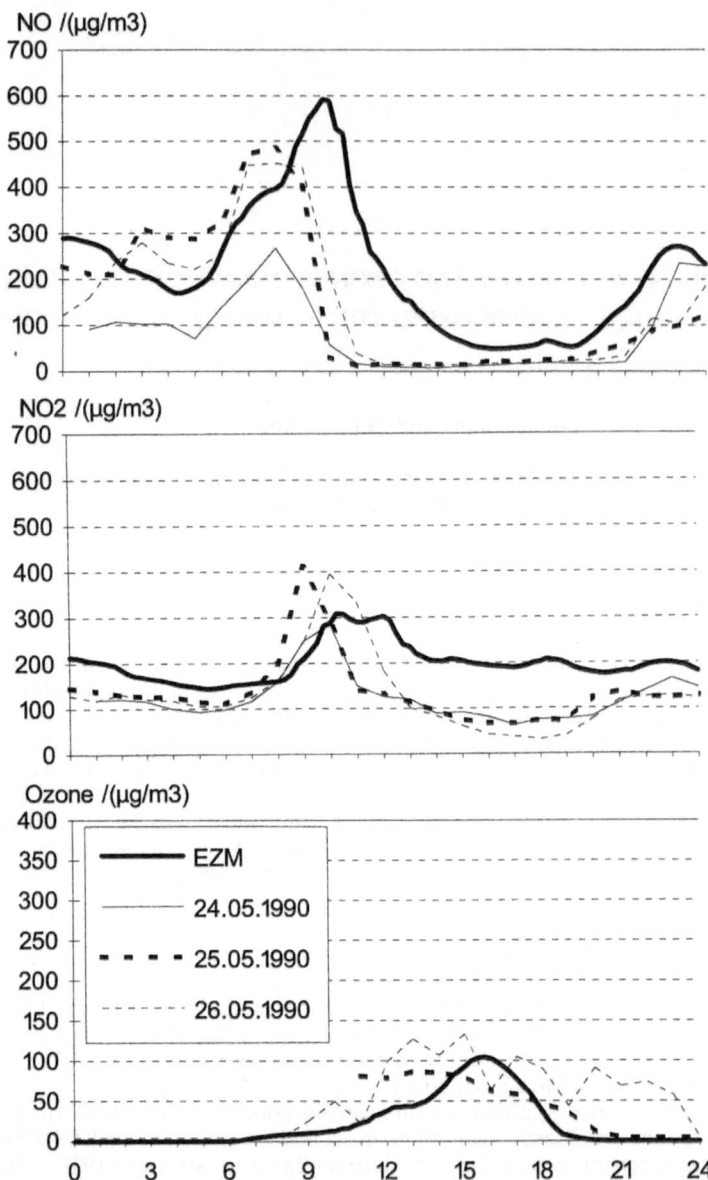

Figure 1. Predicted and observed diurnal variations of the ground level NO, NO_2 and ozone concentration at the Athinas station (close to the centre of Athens) on May 25, 1990. Observed concentrations for May 24 and May 26 are also shown for convenience.

REFERENCES

Bottenheim, J.W. and Strausz, O.P., 1982, Modelling study of a chemically reactive power plant plume, Atmos. Environ. 16:85.

Moussiopoulos, N., 1993a, Athenian photochemical smog: intercomparison of simulations (APSIS), background and objectives, Environmental Software 8:3.

Moussiopoulos, N., 1993b, The Athens Experience, Tutorial Notes, Air Pollution '93 Conference, Monterrey, Mexico.

Moussiopoulos, N. and Sahm, P., 1993, Application of the EUMAC Zooming Model (EZM) to simulate the ozone formation in Athens, submitted to the Atmos. Environ.

INTERCOMPARISON ON THE FLOW FIELD OVER THE ATTIC PENINSULA WITH TWO MODELS

G. Schayes (*), H. Gallée (*), G. Graziani (+), P. Thunis (+)

(+) Environment Institute, JRC Ispra (Italy)
(*) Institut d'Astronomie et de Géophysique G. Lemaitre
    Univ. catholique de Louvain-la-Neuve (IAG-UCL) (Belgium)

1 INTRODUCTION

The circulation of air masses for the control of air pollution is particularly difficult to estimate in areas of complex terrain, due to the presence of mountains and hills or to the effect of large water basins. In the Athens area, the local circulation is mainly driven by the sea-land breezes, which are strongly influenced by a mountain ridge culminating at 1460m (Parnitha mountain fairly close to the sea-shore). Actual terrain slopes can be locally very large, which renders the simulation particularly challenging for mesoscale models. In this paper two different models participating in the APSIS intercomparison exercice are presented and initialized in order to describe the typical summer sea-breeze in the region of Athens.

2 MODELS PRESENTATION AND INITIALISATION

The models used in this study are TVM (Topographic Vorticity Mesoscale) model developed at IAG-UCL (G. Schayes, P. Thunis, 1990) from the URBMET code, and MAR (Modèle Atmosphérique Régional) also developed at IAG-UCL (H. Gallée, G. Schayes, 1991). TVM is hydrostatic and Boussinesq. It solves the vorticity form of the momentum equations thereby eliminating the pressure and the density from the equations at the expense of a few additional computations. MAR is hydrostatic as well; it keeps however the full form of the continuity equation. The two codes are written in a terrain influenced coordinate system (relative to pressure for MAR) to resolve complex terrain. In both models, soil temperature is calculated via a two layers force-restore method and soil latent heat flux is treated by a Penmann-Monteith formulation. Table 1 hereafter synthetizes the main characteristics of the two mesoscale models.

Table 1. Main Characteristics of the models

	T. V. M.	M.A.R
Main characteristics	Hydrostatic Boussinesq Incompressible Vorticity formulation	Hydrostatic Compressible Primitive equations
Vertical coord. system	Sigma-z	Sigma-p
Vertical diffusion	TKE closure Businger surf. layer	Duynkerke E-ε closure Duynkerke surf. layer
Horizontal diffusion	Constant	Variable (Smagorinsky)
IR flux divergence	Sasamori	Morcrette

Air Pollution Modeling and Its Application X, Edited by S-V. Gryning
and M.M. Millán, Plenum Press, New York, 1994

Meteorological Measurements and Initialisation

The meteorological measurements available for the region consist of wind, humidity and temperature profiles at the Hellenicon radiosounding station every 12 hours; wind and temperature profiles at the Refinery station every 90 min (from tethered balloon); and hourly wind speeds and directions (at 10 m) from 9 stations.

Unfortunately, these latter stations (as in the previous APSIS study, Thunis et al 1993) are only representative of a small part of the modelled domain.centered on the Athens Basin.

The modelling domain consists of a 2 km resolution 36x36 horizontal grid. As far as possible, the same conditions and characteristics have been imposed on both models: e.g. identical vertical grid spacing ranging from 10 m to 15000 m. The original topography field included slopes as large as 35%, but was smoothed to remove $2\Delta x$ waves. The resulting topography has maximum slopes of 10% while keeping peak altitudes above 1100 m.

June 10 was chosen and is characterized by quasi-constant geostrophic forcing. Table 2 shows the initial set-up for the two models.

Table 2

Geostrophic wind speed and direction	6.0 m/sec, 360 deg
Sea surface temperature	293 K
Potential temperature lapse rate	4.3 K/km below 2500 m
	3.5 K/km above 2500 m
Latitude, longitude	37° N, 23° E
Upper Absorbing layer	From 5000 to 15000 m
Starting time	21:00 LST on June 9

3 RESULTS AND CONCLUSIONS

The models were run for 36 h. The simulations show reasonable agreement with the available data for both models. The main differences occur on the sea (near the borders) and in the mountains where unfortunately no observatioanl data can be used for comparison (the measurements were all taken in a relatively flat region near Athens). It was already found in Thunis et al (1993) that, even for such a complex geographical situation, the hydrostatic simulation gave a rather reasonable description of the wind and temperature fields. Also despite their large conceptual differences, the two models predicted a similar sea-breeze intensity and front pattern in the Athens region. However, possibly because of the different turbulence closures, the soil temperature wave in MAR has a slightly larger amplitude than in TVM, producing a more marked night land breeze. MAR also generates stronger katabatic winds in the mointains.

A report containing a complete description of this APSIS simulations and the models intercomparison will soon be available.

5 REFERENCES

Gallée, H. and Schayes, G.: 1992. Dynamical Aspects of the katabatic winds evolution in the Antartic coastal zone. *Boundary-Layer Meteorol.*, **59**, 141-161.

Schayes, G. and Thunis, P.: 1990. A three-dimensional mesoscale model in vorticity model. *Contribution nb. 60*, Institut d'Astronomie et de Géophysique, UCL, Louvain-la-Neuve, 42p.

Thunis, P., Grossi, P., Graziani, G., Gallée, H., Moyaux, B. and Schayes, G.: 1993. Preliminary simulations of the flow field over the Attic Peninsula. *Environmental Software*, **8**, 43-54.

STUDIES WITH THE THREE-DIMENSIONAL EULERIAN PHOTOCHEMICAL DISPERSION MODEL MARS FOR THE APSIS B CASE: VIABILITY OF THE INCLUSION OF HORIZONTAL DIFFUSION INTO THE IMPLICIT SOLVER OF THE MODEL

D. Berlowitz[*] and N. Moussiopoulos[+]

[*]Paul Scherrer Institute
CH-5232 Villigen PSI, Switzerland
[+]Laboratory of Heat Transfer and Environmental Engineering
Aristotle University Thessaloniki
GR-54006 Thessaloniki, Greece

INTRODUCTION

The Eulerian photochemical dispersion model MARS (Graf and Moussiopoulos, 1991) solves the three-dimensional dispersion equation for chemically reactive species by coupling the vertical diffusion to the chemical reaction mechanism, using a predictor-corrector integration scheme similar to the one proposed by Gear, while the horizontal diffusion and the advection terms are computed explicitly.

The aim of this study is to investigate methods to connect the horizontal diffusion to the chemistry of the model, i.e. linking the horizontal to the vertical diffusion and incorporating the whole process in the implicit solver of the model. First results indicate, that operator splitting coupled with an ADI method should be applied in order to achieve this objective, taking present computational limits into consideration. The APSIS B case study is used as a test case for the modified program versions.

INCLUSION OF THE HORIZONTAL DIFFUSION INTO THE IMPLICIT SOLVER OF MARS

The general equation for the dispersion of chemically reactive species is given by

$$\frac{\partial \underline{c}}{\partial t} + \underbrace{\nabla \left(\underline{v} \underline{c} \right)}_{advection} = \nabla \left(\underline{\underline{K_c}} \nabla \underline{c} \right) + \underline{R} + \underline{S}, \tag{1}$$

where $\underline{c}$ denotes the concentration of the considered species, $\underline{v}$ the wind velocity, $\underline{\underline{K_c}}$ the matrix of the turbulent diffusion coefficients and where the source and sink terms are denoted by $\underline{R}$ and $\underline{S}$. If the advection part of equation (1) is set aside and solved explicitly, the remaining three-dimensional equation to be integrated in the implicit solver, transformed into the applied terrain following coordinate system (Flassak, 1990 and Moussiopoulos et al., 1993), reads

$$\frac{\partial \underline{c}}{\partial t} = \frac{\partial}{\partial x} \left[K_H \left(\frac{\partial}{\partial x} \left(\underline{c} \right) + \frac{\partial}{\partial y} \left(\underline{c} \right) + \frac{\partial}{\partial z} \left(G^{31} \underline{c} \right) + \frac{\partial}{\partial z} \left(G^{32} \underline{c} \right) + \frac{\partial}{\partial z} \left(G^{33} \underline{c} \right) \right) \right]$$
$$+ \frac{\partial}{\partial y} \left[K_H \left(\frac{\partial}{\partial x} \left(\underline{c} \right) + \frac{\partial}{\partial y} \left(\underline{c} \right) + \frac{\partial}{\partial z} \left(G^{31} \underline{c} \right) + \frac{\partial}{\partial z} \left(G^{32} \underline{c} \right) + \frac{\partial}{\partial z} \left(G^{33} \underline{c} \right) \right) \right]$$

$$+ \quad \frac{\partial}{\partial z} \left[G^{31} K_Z \left(\frac{\partial}{\partial x} (\underline{c}) + \frac{\partial}{\partial y} (\underline{c}) + \frac{\partial}{\partial z} (G^{31}\underline{c}) + \frac{\partial}{\partial z} (G^{32}\underline{c}) + \frac{\partial}{\partial z} (G^{33}\underline{c}) \right) \right. \tag{2}$$

$$+ \quad G^{32} K_Z \left(\frac{\partial}{\partial x} (\underline{c}) + \frac{\partial}{\partial y} (\underline{c}) + \frac{\partial}{\partial z} (G^{31}\underline{c}) + \frac{\partial}{\partial z} (G^{32}\underline{c}) + \frac{\partial}{\partial z} (G^{33}\underline{c}) \right)$$

$$+ \quad \left. G^{33} K_Z \left(\frac{\partial}{\partial x} (\underline{c}) + \frac{\partial}{\partial y} (\underline{c}) + \frac{\partial}{\partial z} (G^{31}\underline{c}) + \frac{\partial}{\partial z} (G^{32}\underline{c}) + \frac{\partial}{\partial z} (G^{33}\underline{c}) \right) \right] + \underline{R} + \underline{S},$$

where G^{31}, G^{32} and G^{33} denote the metric coefficients as defined in Flassak (1990) and Moussiopoulos et al. (1993). K_H and K_Z are the horizontal and vertical turbulent diffusion coefficients, respectively.

Discretization of the above equation leads to a 15 (or 25, depending on the discretization method) point operator (for grid levels k-1, k and k+1, or grid levels k-2, k-1, k, k+1 and k+2, respectively) which, when included directly into the existing implicit solver, requires the solution of a (IM × JM × KM × N)-equation system where IM, JM and KM denote the dimensions of the computational domain and N represents the number of species considered in the applied chemical reaction mechanism. The expression *n point operator* denotes the discrete operator at the position (i,j,k), which depends on the concentration value at the position itself as well as on *(n-1)* surrounding concentration values; 5 values at the following positions need to be considered for each k-level of the grid: (i-1,j), (i+1,j), (i,j-1), (i,j+1) and (i,j). Obviously the computational requirements for a direct solution of such an equation system are considerably higher, than what is currently provided by modern supercomputers.

Several approaches are thus attempted to apply an appropriate splitting mechanism to reduce the dimension of the operator in question.

One such approach splits the operator in eq. (2) in such a way that terms concerning the x-z and y-z planes can be integrated separately, using an *Alternating Direction Implicit* (ADI) method for solving eq. (2). Each plane then consists of a 9 point operator (only three concentration values per k-level are required, i.e. for i-1, i and i+1 for the x-z planes and j-1, j, and j+1 for the y-z planes, respectively). The x-z plane represents the first "direction", the y-z plane the second and the remaining terms along the vertical axis denote the third "direction" of the solution process. In this case the applied implicit solver would have to be adjusted to accomodate an additional dimension (x- or y-axis) and thus looses some of its slickness. Computational demands would be multiplied by a factor of IM^3 or JM^3, respectively.

Alternate solutions include using a finer parametrization of the horizontal diffusion coefficients, which, combined with an extensive scale analysis, could help eliminate some of the "superfluous" terms in eq. (2). This scale analysis is also necessary to estimate the relevance of a finer treatment of the horizontal diffusion in a specific case study, i.e. it enables to decide for a given meteorological and topographical situation, whether a more extensive calculation of the turbulent diffusion is of any relevance for the overall dispersion simulation.

Furthermore the ADI process can be accelerated, by further splitting the x-z (or y-z) diffusion operator: the second derivatives with respect to x and z (or y and z) in eq. (2) are separated from the rest of the respective x-z or y-z operator and integrated by a fast solver for elliptic partial differential equations in two dimensions. The remaining terms are dealt with by a Block-Iteration Method as described in Moussiopoulos and Flassak (1986). Those remaining terms include the discretizised mixed derivations as well as the derivation with respect to x or y as denoted in eq. (2), respectively.

A third ADI process can be envisaged by splitting the diffusion operator into three main terms, each pertaining to one of the spacial axes, while the remaining mixed terms are dealt with as perturbation terms. Depending on their relevance according to the abovementioned scale analysis they are calculated explicitly, while the main terms in the three axes are connected to the applied chemical reaction mechanism using the existing implicit solver of the model for a third of the normal time step.

REFERENCES

Flassak Th., 1990, Ein nicht-hydrostatisches mesoskaliges Modell zur Beschreibung der Dynamik der planetaren Grenzschicht, Reihe 15, Nr. 74, VDI Verlag, Düsseldorf.

Graf J. and Moussiopoulos N., 1991, Intercomparison of two models for the dispersion of chemically reacting pollutants, Beitr. Phys. Atmosph. 64:13-25.

Moussiopoulos N. and Flassak Th., 1986, Two vectorized algorithms for the effective calculation of mass-consistent flow fields, J. Clim. Appl. Met, 25:847.

Moussiopoulos N., Flassak Th., Sahm P. and Berlowitz D., 1993, Simmulation of the wind field in Athens with the nonhydrostatic mesoscale model MEMO, Env. Software 8:29-42

SPECIAL SESSION:

AIR POLLUTION TRANSPORT AND DIFFUSION OVER COASTAL URBAN AREAS (ATHENS)

A NUMERICAL STUDY OF AIR FLOW IN THESSALONIKI AREA

D. Melas, I. Ziomas, and C. Zerefos

Laboratory of Atmospheric Physics, Physics Department
University of Thessaloniki, 54006 Thessaloniki, Greece

INTRODUCTION

A three-dimensional, higher order turbulence closure, mesoscale model for complex terrain is applied to simulate the flow dynamics in Thessaloniki area which is located in the north-east coast of Mediterranean. The present version of the model is hydrostatic and has a terrain following coordinate system. A more detailed description can be found in other literature references (Enger, 1986, Melas and Enger, 1993).

As suitable simulation day, we have selected the 18th September 1987. During this day, the synoptic wind was rather weak ($=1.0$ ms^{-1}) from SW, allowing the development of a sea breeze circulation. The simulation results were compared with measurements performed with a tether balloon in the urban area of Thessaloniki.

RESULTS

The sea breeze in Thessaloniki area blows throughout the year, but during the warm period of the year the average frequency of occurrence is seventeen days per month (Sahsamanoglou, 1976). The results of the numerical simulations reveal some interesting features of the flow regime in the area and specially the development of the sea breeze circulations. During night and early morning hours, the winds are very light in the whole model domain with variable directions. Since the temperature of the land surface was slightly higher than the sea surface temperature, a land breeze circulation did not occur. During daytime the situation changes drastically. Differential heating of land and sea surfaces generates a horizontally inhomogeneous temperature field with the subsequent development of sea breeze circulations. Due to the complex topographical conditions and the coastline orientation, the wind field is complicated with more than one sea breeze cells. The prevailing winds in the model domain, at 9:00 LST, are generally light with variable directions. A small scale sea breeze cell is established in Thessaloniki Bay. During the coarse of the day, a larger scale sea breeze cell develops in the central and southern part of Thermaikos gulf which prevails upon the smaller scale sea breeze cell (Fig. 1). This implies that the low level winds in the urban area of Thessaloniki are veering from WSW to SW and their speed at 25 m AGL increase to approximately 3 ms^{-1}.

Air Pollution Modeling and Its Application X, Edited by S-V. Gryning
and M.M. Millán, Plenum Press, New York, 1994

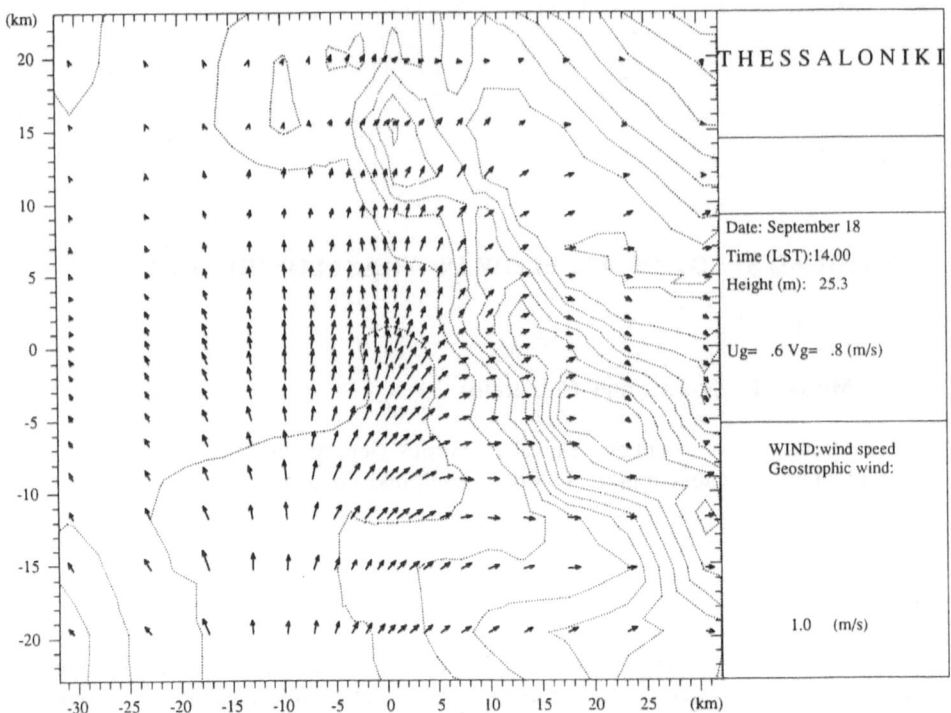

Figure 1. Vector plot showing the wind field at 25 m height at 14.00 LST.

Figure 2 shows the vertical profiles of the wind speed and direction at a grid point corresponding approximately the position of the University of Thessaloniki which is located near the centrum of the city approximately 1 km from the shoreline. From this figure it is recognized that the depth of the sea breeze circulation is limited to a few hundrend meters.

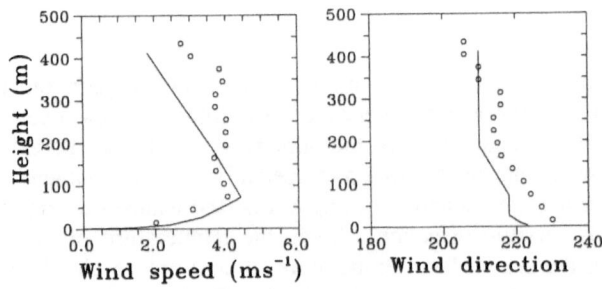

Figure 2. Vertical profiles of wind speed and direction at 14:00 LST. Solid line model simulated and circles measured data.

References

Enger, L., 1986, A higher order closure model applied to dispersion in a convective PBL, *Atmos. Environ.*, **20**, 879-894.

Melas, D. and Enger, L., 1993, A numerical study of flow in Athens area using the MIUU model, *Environmental Software*, **8**, 55-63.

Sahsamanoglou, C.S., 1976, Sea breeze in Thessaloniki, *Bulletin of the hellenic Meteorological Society* **1**, 19-33.

PARTICIPANTS

The 20th International Technical Meeting on Air Pollution Modelling and its Application.
Valencia, Spain, 29 November - 3 December 1993

ARGENTINA

Mazzeo N.A.

Dept. of Atmospheric Sciences.
Faculty of Sciences
University of Buenos Aires,
Ciudad Universitaria.
1428 Buenos Aires

Venegas L.E.

Dept. of Atmospheric Sciences.
Faculty of Sciences
University of Buenos Aires,
Ciudad Universitaria.
1428 Buenos Aires

AUSTRALIA

Noonan J.A.

CSIRO
Division of Atmospheric Research
Mordialloc, Victoria 3195

Physick W.L.

CSIRO
Division of Atmospheric Research
Mordialloc, Victoria 3195

AUSTRIA

Seibert P.

Institute of Meteorology and Geophysics
University of Vienna
Hohe Warte 38
1190 Wien

Stohl A.

Institute of Meteorology and Geophysics
University of Vienna
Hohe Warte 38
1190 Wien

Piringer M.

Institute of Meteorology and Geophysics
University of Vienna
Hohe Warte 38
1190 Wien

BELGIUM

Kretzschmar J.G.

Energi Division
VITO
Boeretang 200
2400 Mol

Loosveldt P.

Faculte Des Sciences Agronomiques
Uert De Biologie Vegetale
2a Avenue de La Faculté
5030 Gembloux

Schayes G.

UCL
Institut d'Astronomie et de Geophysique
2 Chemin du Cyclotron
1348 Louvain-la-Neuve

BRAZIL

Oliveira A.P.

Instituto Astronomico E Geofisico
Departamento De Ciencias Atmosphericas
Universidade De Sao Paulo
CP 9638, 01.065.970 Sao Paulo, SP

BULGARIA

Batchvarova E.

National Institute of Meteorology and Hydrology
Bulgarian Academy of Sciences
66 Tzarigradsko Chaussee
1784 Sofia

Kolarova M.P.

National Institute of Meteorology and Hydrology
Bulgarian Academy of Sciences
66 Tzarigradsko Chaussee
1784 Sofia

CANADA

D'amours R.

Canadian Meteorological Center
2121 North Service Road, 2nd Floor
Trans-Canada Highway
Dorval, Quebec H9P 1J3

Olson M.   Atmospheric Environment Service
                                  4905 Dufferin Street
                                  North York, Ontario, M3H 5T4

Tangirala R.S.   Atmospheric Environment Service
                                  4905 Dufferin Street
                                  North York, Ontario M3H 5T4

Walmsley J.L.   Atmospheric Environment Service
                                  4905 Dufferin Street
                                  North York, Ontario, M3H 5T4

CROATIA

Klaic Z.   Center Meteorological Research
                                  Hydrometeorogical Institute of Croatia
                                  Kranjceviceva, 37-5
                                  41000 Zagreb

Loncar E.   Meteorological and Hydrological Service
                                  Centrer Meteorological Research
                                  Gric, 3
                                  41000 Zagreb

Sinik N.   Geophysical Institute, University of Zagreb
                                  Horvatovac 66, P.O. 224
                                  41000 Zagreb

CZECH REPUBLIC

Pretel J.   Belcicka 2826
                                  14100 Prague 4

DENMARK

Christensen J.   National Environmental Research Institute
                                  Frederiksborgvej 399
                                  4000 Roskilde

Gryning S.   Risø National Laboratory
                                  Department of Meteorology and Wind Energy
                                  Postboks 49
                                  4000 Roskilde

Hertel O.   National Environmental Research Institute
                                   Frederiksborgvej 399
                                   4000 Roskilde

Jegede O.   National Environmental Research Institute
                                   Frederiksborgvej 399
                                   4000 Roskilde

Jensen A.B.   DK-Teknik
                                   Gladsaxe Møllevej 15
                                   2860 Søborg

Kristensen L.   Risø National Laboratory
                                   Department of Meteorology and Wind Energy
                                   Postbox 49
                                   4000 Roskilde

Mikkelsen T.   Risø National Laboratory
                                   Department of Meteorology and Wind Energy
                                   Postboks 49
                                   4000 Roskilde

Olesen H.R.   National Environmental Research Institute
                                   Frederiksborgvej 399
                                   4000 Roskilde

Svensmark H.   DMI
                                   Lyngbyvej 100
                                   2100 Copenhagen

Sørensen, J.H.   DMI
                                   Lyngbyvej 100
                                   2100 Copenhagen

Tassone C.   Risø National Laboratory
                                   Department of Meteorology and Wind Energy
                                   Postboks 49
                                   4000 Roskilde

Thykier-Nielsen S.   Risø National Laboratory
                                   Department of Meteorology and Wind Energy
                                   Postboks 49
                                   4000 Roskilde

FINLAND

Kukkonen J.

Air Quality Department
Finnish Meteorological Institute
Vuorikatu 24
00101 Helsinki

Valkonen, E.P.

Air Quality Department
Finnish Metorological Institute
Vuorikatu 24
00101 Helsinki

FRANCE

Borrel L.

Meteo-France
SCEM/OSAS
42, Avenue Gustave Coriolis
31057 Toulouse Cedex

Bottenheim J.

CNRS.
Laboratoire de Glaciologie et Geophysique de
L'environnement
BP 96
38402 Saint Martin d'heres Cedex

Carissimo B.C.

Electricite de France
Quai Watier, 6 Edif
78401 Chatou

Riboud P.M.

Electricite de France
Quai Watier, 6 Edif
78401 Chatou

Rosset R.

Laboratoire d'Aerologie
118 Route de Narbonne
31062 Toulouse

Suhre K.

Laboratoire d'Aerologie
118 Route de Narbonne
31062 Toulouse

GERMANY

Adrian G.E.

Institut für Meteorologie und Klimaforschung
Universitat Karlsruhe
Kaiserstrasse 12
76128 Karlsruhe

Fay B.   Deutscher Wetterdienst
                          Frankfurter Strasse 135
                          3067 Offenbach Am Main

Graf J.   Institut für Physik der Atmosphäre
                          82230 Oberpfaffenhofen Wessling

Heinz S.   IFU
                          Kreuzeckbahnstrasse 19
                          Garmisch-Partenkirchen

Janicke L.E.   Ing.-Buro
                          Primelweg 8
                          7770 Überlingen

Knoth O.   Inst. für Troposphärenforschung
                          Permosestrasse 15
                          04303 Leipzig

Kunz R.   Lehrstuhl und Institut für Technische
                          Thermodynamik
                          Kaiserstrasse 12
                          76128 Karlsruhe

Martens R.   Gesellschaft für Anlagen und Reaktorsicherheit
                          Schwertnergasse 1
                          50667 köln

Massmeyer K.   Gesellschaft für Anlagen und Reaktorsicherheit
                          Schwertnergasse 1
                          50667 köln

Nester K.   Institut für Meteorologie und Klimaforschung
                          KFK,
                          Universität Karlsruhe
                          76128 Karlsruhe

Petersen G.   GKSS-Research Center Geestacht
                          Institute of Physics
                          Max-Planck Strasse 1
                          2054 Geestacht

Rall A.M.   TUV Bayern Sachsen
                          80674 Münich

Scherer B.	Free University Berlin Inst. F. Met. Carl-Heinrich Becker-Weg 6-10 12165 Berlin
Schlünzen K.H.	Meteorologisches Inst. University of Hamburg Bundesstr. 55 20146 Hamburg
Wichmann F.M.	Landesanstalt für Immissionsschultz Wallneyerstrasse 6 4300-Essen
Wolke R.	Institut für Troposphärenforschung Permoserstrasse 15 04303 Leipzig
Zilitinkevich S.	Alfred Wegener Institute for Polar and Marine Research Am Handelshafen,12 27570 Bremerhaven

GREECE

Amanatidis G.	NCSR "Demokritos" Institut of Nuclear Technology P.O. Box 60228 15310 Aghia Paraskevi - Attiki
Kallos G.	University of Athens Department of Applied Physics Ippocratous 33 10680 Athens
Kambezidis H.	Institute of Meteorology and Physics of the Atmospheric Environment National Observatory of Athens P.O. Box 20048 11810 Athens
Kassomenos P.	School of Physics Meteorology Laboratory Ippocratous 33 10680 Athens

Melas D.

Laboratory of Atmospheric Physics
Aristotle University of Thessaloniki
Box 483
54006 Thessaloniki

Moussiopoulos N.

Laboratory of Heat Transfer
Aristotle University of Thessaloniki
Box 483
54006 Thessaloniki

Psyloglou V.

Institute of Meteorology and Physics of the
Atmospheric Environment
National Observatory of Athens
P.O. Box 20048
11810 Athens

Ziomas I.

Laboratory of Atmospheric Physics
Aristotle University of Thessaloniki
Box 483
54006 Thessaloniki

HUNGARY

Bozo L.

Institut Atmospheric Physics
P.O. Box 39
H-1675 Budapest

Deme S.

KFKI Atomic Energy Research Institute
Health Physics Department
P.0. Box 49
H-1525 Budapest

Fekete K.E.

Hungarian Meteorological Service
P.O.Box 39
H-1675 Budapest

Meszaros E.

Department for Analytical Chemistry
University of Veszprem
P.O.Box 158
H-8201 Veszprem

Molnar A.

Department for Analytical Chemistry
University of Veszprem
P.O.Box 158
H-8201 Veszprem

ISRAEL

Tokar Y.

Department of Geophysics and Planetary
Sciences
Tel-Aviv University
Ramat-Aviv 69978

Goldstein J.

Department of Geophysics and Planetary
Sciences
Tel-Aviv University
Ramat-Aviv 69978

Mahrer Y.

Department of Soil and Water Science
The Hebrew University Jerusalem
Faculty of Agriculture
76100 Rehovot

ITALY

Anfossi D.

Istituto Di Cosmogeofisica
CNR
Corso Fliume, 4
10133 Torino

Bellasio R.

Via Dolomiti, 8
Cap 20017
RHO (Milano)

Carboni G.

ENEL DCO/ULC
Via N. Bixio 39
29100 Piacenza

Giovannini I.

Kamprogetti
Via Toniolo, 1
61032 Fano, Pesaro

Graziani G.

Joint Research Centre
Environmental Institute
21020 Ispra (Varese)

Gualdi R.

Department Salud
Papa Giovanni XXIII, 43
20091 Bresso

Lanzani G.	Via Confalonieri 15 CAP-20030 Seveso (Milano)
Nanni A.	Tecnoinformatica Via Agnesi, 145 20033 Desio (Milan)
Nodop K.	CEC Joint Research Centre Environmental Institute 21020 Ispra (Varese)
Puglisi F.	Deputacion Provinciale Suzzani 286 20162 Milano
Sempreviva A.	Instituto di Fisica Dell'atmosfera CNR PL Luigi Sturzo 31 00144 Roma
Tamponi M.	PMIP/USSL 75/III 4a Unit'a Operativa Sezione di Fisica e Tutela Ambiente Via F. Juvara 22 20129 Milan
Tinarelli G.	ENEL Cram Via Rubattino, 54 20134 Milan
Zanini G.	ENEA Viale Ercolan 18 40138 Bologna

LITHUANIA

Perkauskas D.	Institute of Physics Savanoriu 231 2053 Vilnius

THE NETHERLANDS

van Dop H.

Utrecht Universitat
IMAU
Department of Physics and Astromy
P.O. Box 80005
3508 TA Utrech

Esser P.

IMW-TNO
P.O. BOX 6011
2600 JA Delft

Evers C.W.A.

Ministry of Environment
P.O. Box 30945
2500 GX The Hague

van Jaarsveld J.A..

RIVM
P.O. Box 1
3720 BA Bilthoven

Krol M.

IMAU
Utrecht University
Princetonplein 5
3584 CC Utrecht

Matthijsen J.

TNO
P.O. Box 6011
2600 JA Delft

van der Most P.F.J.

IMET-TNO
Postbus 342
7300 AN Apeldoorn

van Pul W.A.J.

Laboratory for Air Research
National Institute for Public Health
and Environmental Protection, RIVM
P.O. Box 1
3720 BA Bilthoven

Versluis A.H.

Ministry of Transports, Publics Works and
Water Management, Directorate General for
Public Works and Water Management
Road and Hydraulic Engineering Division
P.O. Box 5044
2600 GA Delft

Vila-Guerau de Arrellano J.   IMAU
Utrecht University
Princetonplein 5
3584 CC Utrecht

van Weele M.   IMAU
Utrecht University
P.O. Box 80005
3508 TA Utrecht

Wolff C.J.M.   Shell Internationale Petroleum Maatschappijj
P.O. Box 162
2501 AN The Hague

NORWAY

Grønskei K.E.   Norwegian Institute for Air Research
P.O. Box 64
2001 Lillestrøm

Iversen T.   Institute of Geophysics
University of Oslo
P.O. Box 1022
0315 Oslo

POLAND

Juda-Rezler K.   Inst. Enviroment Engineering Systems
Warsaw University of Technology
Nowowiejska 20
00-653 Warsaw

Kmiec G.   Technnical University of Wroclaw
Wyb. Wyspianskiego 27
50-370 Wroclaw

PORTUGAL

Borrego C.   Departamento de Ambiente E Ordenamento
Universidade de Aveiro
3800 Aveiro

Carvalho R.A.C.   Instituto de Meteorologia
Rua C do Aeroporto de Lisboa
1700 Lisboa

Conceico A.M.   Departamento de Ambiente E Ordenamento
                               Universidade de Aveiro
                               3800 Aveiro

Coutinho M.   Departamento de Ambiente E Ordenamento
                               Universidade de Aveiro
                               3800 Aveiro

Cruz Azevedo Barros N.A.   Departamento de Ambiente E Ordenamento
                               Universidade de Aveiro
                               3800 Aveiro

Fernandes Thomaz S.   Departamento de Ambiente E Ordenamento
                               Universidade de Aveiro
                               3800 Aveiro

Martins J.M.   Departamento de Ambiente E Ordenamento
                               Universidade de Aveiro
                               3800 Aveiro

Prior V.   Instituto de Meterologia
                               Rua C do Aeroporto de Lisboa
                               1700 Lisboa

Silva Miranda A.I.   Departamento de Ambiente E Ordenamento
                               Universidade de Aveiro
                               3800 Aveiro

ROMANIA

Pop L.   Ministry of Environment
                               INIM Spl. Independentei nr 294
                               Cod 77703 Sector 6
                               Bucharest

Pop T.   Ministry of Environment
                               INIM Spl. Independentei nr 294
                               Cod 77703 Sector 6
                               Bucharest

RUSSIA

Pekar M.

Meteorological Synthesizing Centre-East (MSC-E)
Str. Kedrova, 8 117874 12-47
117874 Moscow

Zayar Lina L.

Meteorological Synthesizing Centre-East (MSC-E)
Str. Kedrova 8
117874 Moscow

SLOVENIA

Boznar M.

Jozef Stefan Institute
Jamova 39
61000 Ljubliana

Mlakar P.

Jozef Stefan Institute
Jamova 39
61000 Ljubliana

SPAIN

Acena Moreno M.

CIEMAT
Avenida Complutense 22
28040 Madrid

Albizuri Churruca A.

Environment & Systems
Luis Brinas, 9-1 Izda.
48013 Bilbao

Aparisi Garcia F.J.

Universidad Politecnica Valencia
Camio de Vera S/N
46010 Valencia

Artiano Rodriguez de
Torres B.

CIEMAT
Avenida Complutense 22
28040 Madrid

Baldasano Recio J.

ITEMA
Universidad Politecnica de Cataluna
Apartado de Correos 508
08220 Tarrasa (Barcelona)

Calbo Angrill J.

ITEMA
Universidad Politecnica de Cataluna
Apartado de Correos 508
08220 Tarrasa (Barcelona)

Corregidor Sanz D.

UNESA
Francisco Gerejs, 3
28020 Madrid

Costa Francitorra M.

ITEMA-UPC
Apartado de Correos 508
608220 Tarrasa (Barcelona)

Cremades O. L.

ITEMA
Universidad Politecnica de Cataluna
Apartado de Correos 508
08220 Tarrasa (Barcelona)

Diaz G. G.

ETS II
Alameda de Urquijo S/N
48013 Bilbao

Gangoiti B. G.

ETS II
Alameda de Urquijo S/N
48013 Bilbao

Garcia F. J.A.

ETS II
Alameda de Urquijo S/N
48013 Bilbao

Gasulla F. N.

Gerencia de Proteccion Civil
Generalitat de Catalunya
San Ermengildo, 10 ENTL
08006 Barcelona

Hernandez B. J.F.

ITEMA-UPC
Apartado Correos 508
08220 Tarrasa (Barcelona)

Ibarra M.J.I.

Iberdrola
Area de Generacion
Goya, 4-3
28001 Madrid

Ilardia G. J.L.

ETS II
Universidad Del Pais Vasco
Alameda de Urquijo S/N
48013 Bilbao

Lucas J.

Servicio Vasco de Meteorologia
Paseo Arriola, 17-2-D
20009 San Sebastian

Martin Florente F.

CIEMAT
Unidad De Medio Ambiente
Avenida Complutense, 22
28040 Madrid

Millan M.M.

CEAM
Plaza Del Carmen, 4 (Palacio Pineda)
46003 Valencia

Moreno G.F.J.

Universidad Politecnica de Madrid
Avenida Buenos Aires, 61-2-c
28058 Madrid

Moreno S.J.

ITEMA-UPC
Apartado Correos 508
08220 Tarrasa (Barcelona)

Palacios G.M.

EST II
Universidad Politecnica de Madrid
St. M. Magdalena, 24-4-A
28050 Madrid

Rodriguez B.L.M.

Avda. Seneca, 16
28040 Madrid

Salvador R.

CEAM
Plaza del Carmen, 4 (Palacio Pineda)
46003 Valencia

San Jose Garcia R.

Facultad de Informatica
Universidad Politecnica de Madrid
28660 Boadilla Del Norte (Madrid)

Soriano O.C.   Universidad Politecnica de Cataluna
                                ITEMA
                                Apartado Correos 508
                                08220 Tarrasa (Barcelona)

SWEDEN

Enger L.   Department Meteorology
                                Uppsala University
                                P.O. Box 516
                                75120 Uppsala

Johansson P.E.   FOA 4
                                901 82 Umeå

Langner J.   SMHI
                                60176 Norrköping

Näslund E.   FOA 4
                                901 82 Umeå

Persson C.   SMHI
                                60176 Norrköping

Svensson G.   University of Uppsala
                                Department of Meteorology
                                Box 516
                                75120 Uppsala

SWITZERLAND

Berlowitz D.   Paul Scherrer Institute
                                Würenlingen und Villigen
                                5232 Villigen

Herrenberger V.R.D.   Paul Scherrer Institute
                                Würenlingen und Willigen
                                5232 Villigen

Rotach M.   ETH Swiss Federal Institute of Technology
                                Winterthurerstr. 190
                                8057 Zürich

Schneiter D.   Swiss Meteorological Institute
                                Station Aerologico
                                1530 Payerne

TURKEY

Yenigun O.

Inst. Enviro. Sci.
Bogazici Univer.
80630 Bebek/Estabum

UNITED KINGDOM

ApSimon H.M.

Air Pollution Group
Imperial College
Centre for Environmental Technology
48 Princes Gardens
London SW7 2PE

Carruthers D.J.

CERC Ltd
3D King's Parade
Cambridge CB2 1SJ

Fisher B.

National Power
Windmill Hill Business Park
SN5 6P8 Swidon (Wiltshire)

Gimson N.R.

University of Reading
Department of Meteorology
2 Early Gate - P.O. Box 239
Whiteknights

Hawkings C.

Environment Resources Management
106 Gloucester Place
6W1H 3OB London

Smith F.B.

Imperial College (ICCET)
48 Princes Gardens
London SW7 2PE

U.S.A.

Bornstein R.D.

Dept. of Meteorology
San Jose State University
San Jose
California 95192 - 0104

DiCristofaro S.C.

Sigma Research Corporation
196 Baker Avenue
Concord, Massachusetts 01742

Easter R.C.

Battlle NW Laboratories
P.O. Box 999
Richland, WA 99352

Koracin D.

Department Research Institute
Box 60220 Reno
Nevada 89506

McCaslin P.

CIRA
Broadway, 325 R.E.
80303 Boulder (Colorado)

Nappo C.J.

National Oceanic and Atmospheric
Administration
Atmospheric Turbulence
and Diffusion Division
Oak Ridge
Tennessee 37830

Rao S.T.

New York State Department of Environmental
Conservation
Albany, New York 12233-3259

Schiermeier F.A.

U.S. EPA/NOAA
Atmospheric Sciences Modelling Division
Research Triangle Park
NC 27711

Weil J.C.

Cooperative Institute for Research and
Environmental Sciences
University of Colorado
Boulder, CO 80309

AUTHOR INDEX

Aceña, B., 645
Adrian, G., 609
Alcamo, J., 129
Alpert, P., 45
Amanatidis, G.T., 305
Anfossi, D., 329, 621
ApSimon, H., 63, 653
Asman, W.A.H., 569

Baldasano, J.M., 643
Baldi, M., 619
Balmor, Y., 45
Baltensperger, U., 595
Barker, B., 63
Bartnicki, J., 129
Bartochowska, M., 27
Bartzis, J.G., 305
Baskett, R.L., 447
Batchvarova, E., 253
Bellasio, R., 439
Benjelloun, F., 357
Berge, E., 157
Berkowicz, R., 569
Berlowitz, D., 413, 673
Best, P., 583
Bigalke, K., 613
Bornstein, R.D., 101, 541
Borrego, C., 53
Bourouag, T., 357
Božnar, M., 659
Bozó, L., 129, 617
Braverman, T.N., 503
Brusasca, G., 329, 621
Builtjes, P.J.H., 195, 629

Carmichael, G., 213
Carras, J., 533
Carruthers, D.J., 491
Carvalho, R.A., 637
Cautenet, S., 273
Chapman, E.G., 185
Chaumerliac, N., 273
Christensen, J., 119, 569
Costa, M., 643

Cotton, W.R., 19
Coutinho, M., 53
Cox, W.M., 503
Cremades, L., 643
Crist, K.C., 213

Daggupaty, S.M., 367
Dalu, G.A., 619
D'Amours, R., 587
Davies, B., 491
Degrazia, G.A., 325, 597
Delvosalle, C., 357
DiCristofaro, D.C., 503
Dop van, H., 177
Doria, P., 549
Duynkerke, P.G., 203, 295
Dzisiak, J.P., 357

Easter, R.C., 185
Eckman, R.M., 655
Erisman, J.W., 625
Esser, P., 629
Everberg, E., 357

Fay, B., 395
Fekete, K.E., 615
Ferrero, E., 329, 621
Fiedler, F., 609
Flassak, Th., 643
Freedman, F., 541
Freis, R.P., 447

Gallée, H., 671
Giostra, U., 621
Glaab, H., 395
Glaser, E., 45
Goldstein, J., 45
Graf, J., 3
Gram, F., 91
Granby, K., 569
Graziani, G., 671
Gryning, S.E., 243, 253, 607
Grønskei, K.E., 91
Guicherit, R.G., 629

Hagen, L.O., 91
Hao, W., 559
Heinz, S., 611
Herrnberger, V., 549
Hertel, O., 569
Hov, Ø., 569
Hovmand, M.F., 569
Huret, N., 273
Hurley, P., 235

Ibarra, J.I., 19, 523
Iverfeldt, Å., 167
Iversen, T., 157

Jaarsveld van, J.A., 143, 625
Jacobsen, I., 395
Janicke, L., 405
Jean, M., 587
John, K., 213, 559
Johnson, G., 583
Jost, D.T., 595

Kacperczyk, K., 635
Kallos, G., 35, 513, 619
Karpik, S.R., 263
Kassomenos, P., 35, 513
Kayin, S., 63
Klaic, Z., 603
Kmieć, G., 635
Knoth, O., 287
Kolarova, M., 137, 601
Krautstrunk, M., 3
Kristensen, L., 341
Krol, M.C., 623
Kromp-Kolb, H., 595
Krotova, I.A., 641
Kukkonen, J., 431
Kulmala, M., 431
Kunz, R., 663

Lamprecht, R., 413
Langner, J., 9
Lanzani, G., 81
Larssen, S., 91
Laursen, L., 373

Leeuw de, F.A.A.M., 143, 625
Lesjak, M., 659
Levert, J.M., 357
Lončar, E., 605
Luhar, A.K., 315
Lunney, K., 583
Lyons, W.A., 19, 423

Madany, A., 27
Martín, F., 645, 647
Matthijsen, J., 195
McHugh, C.A., 491
Melas, D., 667, 677
Melikhova, L.G., 641
Mészáros, E., 617
Mikelinskene, A., 137
Mikkelsen, T., 383
Mlakar, P., 659
Molnár, A., 617
Montgomery, M., 491
Moraes, O.L.L., 325, 597
Moreno, J., 73, 649
Morselli, M.G., 621
Moussiopoulos, N., 109, 663, 669, 673
Moyaux, B., 357
Munthe, J., 167

Nappo, C.J., 177, 655
Nasstrom, J.S., 447
Nester, K., 665
Niemeier, U., 613
Nikmo, J., 431
Noonan, J., 533

Olesen, H.R., 481
Oliveira, A.P. de, 325, 597

Palacios, M., 73, 649
Palomino, I., 645, 647
Papadopoulos, A., 513
Parfiniewicz, J., 27
Pekar, M., 657
Pendergrass, W.R., 655
Perantonis, S.J., 305
Perkauskas, D., 137

Persson, C., 9
Petersen, G., 167
Physick, W., 235, 533, 583
Pielke, R.A., 19, 619
Piwkowski, H., 27
Pop, L., 639
Pop, T., 639
Possiel, N., 559
Prior, V., 637
Proyou, A., 109
Pul van, W.A.J., 143, 625

Rao, K.S., 315, 541, 559
Rasmussen, A., 373
Reimer, E., 589
Robertson, L., 9
Robins, A.G., 491
Rocha, A., 53
Rodríguez, L., 73, 649
Roemer, M.G.M., 195, 629
Ronday, F., 357
Roselle, S., 559
Rosset, R., 279
Rotach, M., 243, 607
Rozkrut, M., 27
Runge, E., 569

Sahm, P., 109, 669
Sahota, H., 367
Salvador, R., 647
San José, R., 73, 649
Santabàrbara, J.M., 383
Saylor, R.D., 185
Schayes, G., 101, 357, 671
Scheffe, R., 559
Schere, K., 559
Schlager, H., 3
Schlünzen, K.H., 613
Schneiter, D., 651
Schrodin, R., 395
Schwikowski, M., 595
Seibert, P., 595
Senuta, K., 137
Šinik, N., 605
Sistla, G., 541, 559
Strimaitis, D.G., 503

Suhre, K., 279
Syrakov, D., 137, 601
Sørensen, J.H., 373

Tampieri, F., 329, 621
Tamponi, M., 81, 439
Tangirala, R.S., 367
Tassone, C., 243, 607
Taylor, P.A., 263
Thijsse, Th., 629
Thomson, D.J., 491
Thorson, W.P., 423
Thunis, P., 101, 541, 671
Thurre, C., 651
Thykier-Nielsen, S., 383
Tinarelli, G., 329, 621
Tokar, Y., 45
Toviessi, J.P., 587
Tremback, C.J., 19, 423
Trombetti, F., 329, 621
Trudel, S., 587

Uliasz, M., 19, 27

Varinou, M., 513
Varoufakis, S., 305
Vassilas, N., 305
Venegas, L., 593
Vesala, T., 431
Vila-Guerau de Arellano, J., 203, 295

Walker, S.E., 91
Walko, R.L., 19, 423
Walmsley, J.L., 263
Warren, R., 653
Webber, D.M., 431
Weele van, M., 203
Weil, J.C., 457
Weiß, W., 589
Weng, W., 263
Williams, D., 533
Wolke, R., 287
Wortmann-Vierthaler, M., 643
Wren, T., 431

Xu, D., 263

Yordanov, D., 601

Zeller, K.F., 295
Zerefos, C., 667, 677

Zhou, N., 559
Zilitinkevich, S., 223
Ziomas, I., 667, 677
Zwoździak, A., 635
Zwoździak, J. 635

SUBJECT INDEX

Abatement strategy, 653
Accidental releases, 395
ACDEP, 569
Acid precipitation, 137, 615, 635
Advection scheme, 657
Aerosol transport, 617
Aerosol vaporisation, 431
Air quality network, 583
Alpine pollution, 595
Alpine-region, 3
APSIS, 663, 665, 667, 669, 671, 673
Aqueous phase, 195
ARAC, 447
Athens, 513, 663, 667
Average concentration, 647
Bakar Bay, 605
Big-leaf model, 73, 649
Black triangle, 27
Blashavel Hill, 263
Boundary layer parametrization, 177
Brisbane, 583
Carbon Bond IV, 541
Carbon monoxide, 185, 503
Chemical kinetic solvers, 287
Chemistry model, 185
Cities, 641
Climate change, 213
Cloud water, 273
Clouds, 195, 203
Coastal area, 45, 533, 613, 637
Coastal circulation, 605
Complex terrain, 329, 357, 367, 549, 621, 647
Composite performance measure, 503
Concentration fluctuations, 341
Confidence intervals, 503
Convective boundary layer, 223, 457
Cooling towers, 405
Dense gas dispersion, 431, 439, 645
Deposition acidifying compounds, 625
Deposition, 73, 569, 635
DMS, 119, 279
ECMWF, 119, 373
Emergency response, 423, 587
Emission reduction strategy, 559

Entrainment zone, 253
Episode prediction SO_2, 305
ERDAS, 423
Eulerian model, 9
EUMAC, 109, 669
EUTREND, 143
Evaluation of models, 481
Exceedance statistics, 341
Explosive sources, 447
Extreme concentrations, 341
Farming, 9
Flow field, 667, 671
Flux-gradient relationships, 295
Fluxes complex terrain, 619
Fokker Plank equation, 243
Footprint, 315
Gaussian plume model, 457
GEM, 81
Global model, 177, 185, 623
Guardo field study, 523
Guardo valley, 19
HEDAGAS, 645
Hemispheric dispersion, 119
HIRLAM, 373
Hydrogenperoxide, 569
Iberian thermal low, 53
Industrial hazards, 9, 357
Inhomogeneous turbulence, 611
Integrated measurements, 395
KAMM, 609
Kincaid field study, 491
LADM, 533
Lagrangian particle model, 27, 81, 235, 243, 315, 329, 373, 395, 405, 533, 607, 621
Lake breeze, 19
LASAT, 405
Length scales, 325
Lillestrøm field study, 491
Limited area model, 35, 367, 373
LINCOM, 395
Liquid drops, 431
Lisbon region, 53
Local scale, 481
Local weather forecasting, 423

Long-range transport, 119, 137, 143
Marine areas, 143
Marine boundary layer, 279
MARS, 673
MATCH, 9
MDGP, 439
MEMO, 413
Mercury modelling, 167
Mesoscale circulation, 53
METRAS, 613
Mid-troposphere, 185
Mixed layer height, 253
Model performance, 413
MOGUNTIA, 623
MOHAVE, 27
Monitoring system, 405
Monthly averaged radiation dose, 655
Mountains, 609, 635
Nested models, 35
Neural network, 305, 659
Nitrogen compounds, 569
Nitrogen oxides,91, 559, 629
NM4-model, 45
Non-Gaussian turbulence, 611
Olympic games '92, 643
Organic pollutants, 143
Ozone, 3, 19, 195, 541, 559, 569, 629, 669
PAN, 629
Pasquill classes, 601
Photochemical model, 19, 73, 91, 109, 213, 287, 541, 649
Photochemistry, 673
Photodissociation, 203
Precipitation, 119
Precursors, 629
Primary pollutants, 81
Radiative flux, 213
Radical formation, 203
Rain water, 273
RAMS, 19, 513
RATCHET, 655

Real time model, 589
Reduced nitrogen, 63
Regulatory modelling, 457, 481
RIMPUFF, 395
Routine calculations, 157
Scalar fluxes, 315
Scaling, 223
Scavenging, 273
Sea breeze, 143, 513, 637, 665
Semi-Lagrangian approach, 533
SEVEX, 357
Shipping, 9
Similarity theory, 223
Skewed turbulence, 235, 243
SLAB, 645
Source geometry, 447
Spectral model, 597
Stable boundary layer, 457, 597
Sulphate chemistry, 157, 569, 603, 653
Swiss Midland, 651
Synoptic scale, 601
TERN, 63
Thessaloniki, 109, 667
Trace metals, 129
Traffic, 9, 583, 643
Trajectory model, 395
TRANSALP, 413
Tropospheric wind-field, 623
Turbulence, 325
TVM/URBMET, 101
Two-phase releases, 431
UK-ADMS, 491
Urban Airshed model, 541
Urban canyons, 81
Urban dispersion model, 91
Urban topography, 101
Vertical diffusion, 593
VOC, 559
Wake-flow, 609, 639
Wet deposition, 273
Wind tunnel, 639